HEAVY FLAVOR PHYSICS

Other Related Titles from the AIP Conference Proceedings Subseries on High Energy Physics

602 QCD@Work: International Workshop on Quantum Chromodynamics: Theory and Experiment
Edited by Pietro Colangelo and Giuseppe Nardulli, December 2001, 0-7354-0046-6

601 Theoretical High Energy Physics: MRST 2001: A Tribute to Roger Migneron
Edited by V. Elias, D. G. C. McKeon, and V. A. Miransky, December 2001, 0-7354-0045-8

578 Physics and Experiments with Future Linear e^+e^- Colliders: LCWS 2000
Edited by Adam Para and H. Eugene Fisk, July 2001, 0-7354-17-2

562 Particles and Fields: Ninth Mexican School
Edited by Gerardo Herrera Corral and Lukas Nellen, April 2001, 1-56396-998-X

549 Intersections of Particle and Nuclear Physics: 7th Conference, CIPANP2000
Edited by Zohreh Parsa and William J. Marciano, December 2000, 1-56396-978-5

540 Particle Physics and Cosmology: Second Tropical Workshop
Edited by José F. Nieves, October 2000, 1-56396-965-3

531 Particles and Fields: Seventh Mexican Workshop
Edited by Alejandro Ayala, Guillermo Contreras, and Gerardo Herrera, July 2000, 1-56396-954-8

482 RHIC Physics and Beyond: Kay Kay Gee Day
Edited by Berndt Müller and Robert Pisarski, July 1999, 1-56396-878-9

424 Twenty Beautiful Years of Bottom Physics: Proceedings of the *b*20 Symposium
Edited by R. A. Burnstein, D. M. Kaplan, and H. A. Rubin, March 1998, 1-56396-745-6

To learn more about these titles, or the AIP Conference Proceedings Series, please visit the webpage **http://proceedings.aip.org**

HEAVY FLAVOR PHYSICS

Ninth International Symposium on Heavy Flavor Physics

Pasadena, California 10–13 September 2001

EDITORS

Anders Ryd

Frank C. Porter

California Institute of Technology
Pasadena, California

AMERICAN INSTITUTE OF PHYSICS

Melville, New York, 2002

AIP CONFERENCE PROCEEDINGS ■ VOLUME 618

Editors:

Anders Ryd
Frank Clifford Porter

California Institute of Technology
Mail Stop 356-48
Pasadena, CA 91125-4800
USA

E-mail: ryd@hep.caltech.edu
fcp@hep.caltech.edu

L.C. Catalog Card No. 2002104591
ISBN 0-7354-0064-4
ISSN 0094-243X
Printed in the United States of America

CONTENTS

PREFACE

Caltech hosted the Ninth International Symposium on Heavy Flavor Physics in Pasadena from September 10-13, 2001. The papers from the symposium are collected in these proceedings.

Looking back over the program, it is striking how broad a range of technique is being brought to bear on the questions being addressed in heavy flavor physics. A significant portion of the program involved the measurement and understanding of the CKM matrix elements, under attack from a variety of experimental and theoretical angles. This, of course, is part of the broader program of understanding the Standard Model, and of trying to find the chinks in it that we expect to be there, somewhere. Heavy flavor physics seems a promising path to such discovery. At this conference, we had clearly finally entered the "factory era", and in the course of the presentation of already exciting results there was much anticipation of rapid advances, as well as discussions on what might be the next steps.

The success of the conference relied in no small measure on the many hours devoted by Caltech's high energy physics administrative staff: Debra Kingston, Virginia Licon, Mark Rincon, and Meiske van der Eb, with the direction of Betty Smith, and by our computing system manager, Juan Barayoga. The conference was chaired by David Hitlin, who led the organization with characteristic enthusiasm. Gregory Dubois-Felsmann, Matt Weaver, Alan Weinstein, and Songhoon Yang all dedicated considerable effort to make sure that we had the myriad aspects of the conference covered. We are grateful for the help and advice of our International Organizing Committee, which was important in putting together a comprehensive and balanced physics program in such a broad and active field. This symposium was supported in part with funding from Caltech.

As we arose from our slumbers for the second day of sessions, we heard the shocking news from the eastern United States that has put September 11 into the history books. This caused considerable disruption, both as the organizers were faced with a change of venue for that day, and as the participants struggled with enormous travel uncertainties and delays. Through it all, however, we somehow maintained attention on the interesting physics discussions. We are extremely grateful for the patience and understanding shown by the participants.

Anders and Frank

INTERNATIONAL ORGANIZING COMMITTEE

H. Aihara	Tokyo
J. Alexander	Cornell
G. Altarelli	CERN
J. Appel	Fermilab
G. Bonneaud	Ecole Polytechnique
A. Buras	Munich
P. Dauncey	Imperial College
J. Dorfan	SLAC
F. Gilman	Carnegie-Mellon
M. Giorgi	INFN Pisa
D. Hitlin	Caltech
D. MacFarlane	UC San Diego
M. Neubert	Cornell
R. Peccei	UCLA
K. Pitts	Illinois
H. Quinn	SLAC
J. Richman	UC Santa Barbara
A. I. Sanda	Nagoya
K. Schubert	Dresden
A. Schwarz	DESY
S. Stone	Syracuse
H. Sugawara	KEK

LOCAL ORGANIZING COMMITTEE

C. Campagnari	UC Santa Barbara
G. Dubois-Felsmann	Caltech
B. Filippone	Caltech
B. Grinstein	UC San Diego
D. Hitlin	Caltech
D. MacFarlane	UC San Diego
R. Peccei	UCLA
F. Porter	Caltech
A. Ryd	Caltech
V. Sharma	UC San Diego
P. Vogel	Caltech
M. Weaver	Caltech
A. Weinstein	Caltech
M. Wise	Caltech
S. Yang	Caltech

Dedication to Nathan Isgur

Mark B. Wise

California Institute of Technology, 452-48 Caltech, Pasadena CA 91125, USA

Nathan passed away in July after a lengthy illness. I am sure most of you are familiar with his many contributions to heavy quark physics and it is certainly fitting that we take a few minutes to honor him at the beginning of this meeting. Actually Nathan's main physics interest was the strong interactions rather than heavy quark physics per se. He was already very well known for work he did with Gabriel Karl and others on the nonrelativistic quark model before the work that he did on heavy quark symmetry and its applications. However, Nathan understood the limitations of the nonrelativistic quark model, and was thrilled that the methods he helped develop allowed one to derive systematically from the theory of the strong interactions many properties of hadrons that contain a heavy quark.

Since many of you may not be familiar with Nathan's work outside of the area of heavy quark physics, I give below a list of some of his most important non heavy quark physics papers. They are not in any particular order and might not even be the most cited. But I think they are at least some of the ones he was particularly proud of:

- "Why the pseudoscalar-meson mixing angle is $-10°$", *Phys. Rev.* **D12**, 3770 (1975).
  In this paper Nathan gave a simple explanation based on quark model mixing for the observed $\eta - \eta'$ mixing angle. It is a beautiful piece of simple physics that still forms the basis of our understanding for the value of this angle.

- "P-wave baryons in the quark model", (with G. Karl) *Phys. Rev.* **D18**, 4187 (1978); "Positive-parity excited baryons in a quark model with hyperfine interactions", (with G. Karl) *Phys. Rev.* **D19**, 2653 (1979).
  These two papers form the basis of the Isgur-Karl quark model for baryons. It is shocking how many properties of the baryons this nonrelativistic quark model approach is able to explain.

- "$K\bar{K}$ molecules", (with J. Weinstein) *Phys. Rev.* **D41**, 2236 (1990).
  This paper gave some very simple but powerful arguments that what used to be called the $\delta(980)$ should be interpreted not as the usual type of quark model state but rather as a weakly bound kaon-anti-kaon state.

- "Perturbative QCD in Exclusive Processes", (with C. Lewellyn Smith) *Phys. Lett.* **B217**, 535 (1989).
  Here Nathan argues that the perturbative predictions for exclusive processes made by Brodsky and Lepage should not be a good approximation until very large Q^2.

CP618, *Heavy Flavor Physics: Ninth International Symposium*, edited by A. Ryd and F. C. Porter

This was a very important paper because at the time it was written much of the community thought that these predictions are relevant at quite low Q^2.

- "Flux-tube model for hadrons in QCD", (with J. Paton) *Phys. Rev.* **D31**, 2910 (1985).
  There are many properties of low energy QCD that cannot be described with the non relativistic quark model, for example, the properties of exotic mesons. Also one would like to have a deeper understanding of why the nonrelativistic quark model works at all. Nathan made a start on this problem in this paper.

When I called my wife Jackie from Rome to tell her that Nathan passed away she was saddened that as she put it "such a good and gentle man" had left us. I never heard Nathan raise his voice or get angry during any discussion, physics or otherwise. It's not that he could not be critical, but he was always composed and never felt the need to make someone feel small if they had not appreciated the proper reasoning in a particular argument. But still I have a hard time viewing Nathan as "gentle". I think this is because most of my interactions with him centered on physics. He was fiercely driven to understand the physics issues that he was puzzling over, and he could perform even the most bruising calculations if they would help him gain the intuition he was after. In physics being a great calculator is a little like being a great athlete, it takes discipline and perseverance as well as natural talent. Just as it is hard to picture professional athletes as gentle souls when they are off the field it is sometimes hard for me to see Nathan in that light.

I first met Nathan when I was a sophomore at the University of Toronto. He had recently come to Toronto as a graduate student and was teaching an undergraduate seminar class on statistical mechanics. He chose as the book Kittel's thermal physics, which is one of great undergraduate texts. It was a marvelous course. Nathan was a gifted teacher both in the seminar style of that course and in the more conventional lecture style as well.

Canadians are not as daring as Americans. Like most of my friends I lived at home during my undergraduate years and was more than a little nervous about leaving home to do graduate work in the United States. By the time I was looking into graduate schools Nathan was a professor at the University of Toronto. He told me that he would be happy to have me as his graduate student but insisted the best thing for me was to go to the US and get my PhD there.

The first time I tried, I didn't get into any of the US graduate schools that I wanted to attend, so I stayed at University of Toronto for a masters. It ended up being one of the best moves of my career. I took a graduate course on the quark model from Nathan and he also supervised my reading of Bjorken and Drell Volume 1. I have often thought what Nathan taught me that year gave me an advantage over many of my contemporaries who had more experience with tools from field theory but lacked some of the intuition that came from the type of quark model calculations that Nathan had exposed me to.

After that time I would work with Nathan whenever I came home to visit my family during the summer. Usually it was a two week visit and I would go into University of Toronto to talk with Nathan almost every weekday. That's how our collaboration on heavy quark physics started. Actually, it might have started off with a phone call before a visit home, but certainly much of the serious work was done during my visits home. I remember vividly that the idea for the paper where we treated semileptonic B to D and $D^{(*)}$ decays away from zero recoil came on the last day of one of my visits to Toronto. I only had a few hours before my flight left. The idea came to us while we were in front of the blackboard. Time was so short I didn't have time to use the restroom before heading out into traffic to get to the airport. It was a rather uncomfortable drive but I did make the flight and satisfy my other needs at the airport.

I have very fond memories of the work we did together on heavy quark physics. Intellectually it was the most exciting period of my career in physics. Even after the early 1990's and well into the period when Nathan was not well we continued to talk about particle physics. He never lost his passion for the subject and continued to publish important work. Few particle theorists have been as influential as Nathan and so universally well liked. I miss him and think about him often, as I am sure many of you also do.

Results on CP Violation from Belle

T.E. Browder representing the Belle Collaboration

Department of Physics, University of Hawaii, Honolulu, Hawaii

Abstract. I describe the recent measurement of the CP violating parameter $\sin 2\phi_1 = (0.99 \pm 0.14 \pm 0.05)$ from the Belle experiment at KEK.

INTRODUCTION

In 1973, Kobayashi and Maskawa (KM) first proposed a model where *CP* violation is incorporated as an irreducible complex phase in the weak-interaction quark mixing matrix [1]. The idea was remarkable and daring because it required the existence of six quarks at a time when only the u, d and s quarks were known to exist, The subsequent discoveries of the c, b and t quarks, and the compatibility of the model with the *CP* violation observed in the neutral kaon system led to the incorporation of the KM mechanism into the Standard Model, even though it had not been conclusively tested experimentally.

In 1981, Sanda, Bigi and Carter pointed out that a consequence of the KM model was that large *CP* violating asymmetries could occur in certain decay modes of the B mesons.[2] These asymmetries may occur when a neutral B meson decays to a *CP* eigenstate. In this case the amplitude for the direct decay interferes with that for the process where the B meson first mixes to a $\overline{B}$ meson that then decays to the *CP* eigenstate.

The unitarity of the CKM matrix implies that the existence of three measurable phases. In the "Nihongo" convention, these are denoted

$$\phi_1 \equiv arg\left(-\frac{V_{cd}V_{cb}^*}{V_{td}V_{tb}^*}\right),\ \phi_2 \equiv arg\left(-\frac{V_{ud}V_{ub}^*}{V_{td}V_{tb}^*}\right),\ \phi_3 \equiv arg\left(-\frac{V_{cd}V_{cb}^*}{V_{ud}V_{ub}^*}\right). \quad (1)$$

while at SLAC these angles are usually referred to as β, α and γ, respectively.

A non-zero value of ϕ_{CP} results in the time dependent asymmetry

$$A_f = \frac{R(B^0 \to f_{CP}) - R(\bar{B}^0 \to f_{CP})}{R(B^0 \to f_{CP}) + R(\bar{B}^0 \to f_{CP})} = \xi_f \sin 2\phi_{CP} \cdot \sin(\Delta m \cdot (t_2 \pm t_1)), \quad (2)$$

where ξ_f is the CP eigenvalue (± 1), Δm denotes the mass difference between the two B^0 mass eigenstates and t_1 and t_2 are the proper time for the tagged-B and *CP* eigenstate decays, respectively. The $+$ sign corresponds to the case where the B^0 and $\overline{B^0}$ are in an even L orbital angular momentum state, the $-$ sign obtains for odd L states such as the $\Upsilon(4S)$. A determination of A_f thus provides a measurement of $\sin 2\phi_{CP}$.

We note that due to the restrictions of quantum mechanics, time integrated asymmetries at the $\Upsilon(4S)$ resonance (which corresponds to the $-$ sign in the above equation)

CP618, *Heavy Flavor Physics: Ninth International Symposium*, edited by A. Ryd and F. C. Porter

are identically zero. Therefore, one must make time dependent measurements. Since the pairs of B mesons are produced nearly at rest in the usual arrangement at threshold, the $\Upsilon(4S)$ center of mass frame must be boosted. This is accomplished by the use of beams with asymmetric energies. For example at KEK-B, $\beta\gamma \sim 0.43$, and as a result the typical B meson decay length is dilated from 20μm to about 200μm, which is measurable with double-sided silicon strip vertex detectors close to the interaction point.

The measurement therefore requires:

- a large sample of reconstructed $B \to (c\bar{c})K^0$ eigenstate decays;
- a determination of the flavor of the accompanying B ("tagging");
- a measurement of Δz, the vertex separation between the CP eigenstate and flavor tag decays; and
- a fit to the flavor-tagged vertex distribution to extract $\sin 2\phi_1$.

The KEKB high luminosity double storage ring facility was commissioned with remarkable speed starting in late 1998. This accelerator facility allows us to satisfy the first requirement. By the summer of 2001, 29.1 fb^{-1} was integrated on the $\Upsilon(4S)$. This data sample was used for the $\sin 2\phi_1$ measurement and corresponds to 31.3 million $B\bar{B}$ pairs. KEKB uses a ± 11 mrad crossing angle to separate the incoming and outgoing beams and minimize parasitic collisions. So far no special limitations associated with the crossing angle have been observed (e.g. synchrobetatron oscillations). KEKB now routinely achieves peak instantaneous luminosities above $5 \times 10^{33}/\mathrm{cm}^2/\mathrm{s}$ with acceptable experimental backgrounds and trigger rates in the Belle detector. The beam currents are still far below the design values and therefore there is room for further improvements in luminosity. Much larger data samples are expected in the near future.

The Belle detector has good lepton identification and high efficiency for both charged and neutral particles. It allows the CP eigenstate decays to be efficiently reconstructed. Belle is a large-solid-angle magnetic spectrometer that consists of a three-layer silicon vertex detector (SVD), a 50-layer central drift chamber (CDC), a mosaic of aerogel threshold Čerenkov counters (ACC), time-of-flight scintillation counters (TOF), and an array of CsI(Tl) crystals (ECL) located inside a superconducting solenoid coil that provides a 1.5 T magnetic field. An iron flux-return located outside of the coil is instrumented to identify muons and K_L's (KLM). The detector is described in detail elsewhere [3]. Examples of its performance are given in the following section.

The Belle experiment starting physics data taking in 1999. In the summer of 2001, Belle (together with BABAR) announced the observation of the first statistically significant signals for CP violation outside of the kaon system. In this report I will describe some of the details of the measurement.

SIN $2\phi_1$ MEASUREMENT

We reconstruct B^0 decays to the following CP eigenstates [5]: $J/\psi K_S$, $\psi(2S)K_S$, $\chi_{c1}K_S$, $\eta_c K_S$ for $\xi_f = -1$ and $J/\psi K_L$ for $\xi_f = +1$. We also use $B^0 \to J/\psi K^{*0}$ decays where $K^{*0} \to K_S\pi^0$. Here the final state is a mixture of even and odd CP, depending on the relative orbital angular momentum of the J/ψ and K^{*0}. The CP content is determined

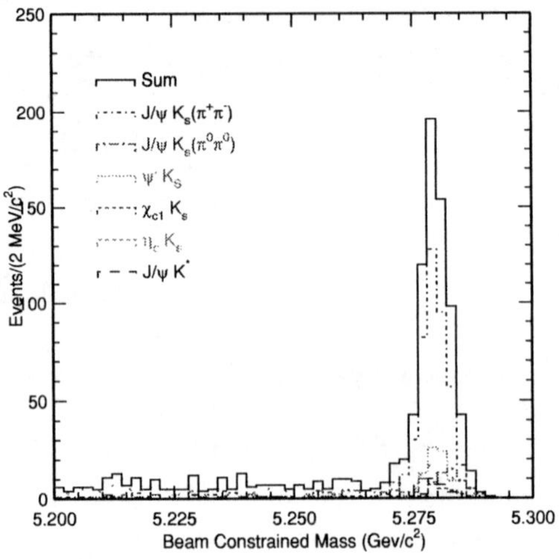

FIGURE 1. The beam-energy constrained mass distribution for all decay modes combined other than $J/\psi K_L$. The shaded area is the estimated background. The signal region is the range $5.27 - 5.29$ GeV/c ^{2}.

from a fit to the full angular distribution of all $J/\psi K^*$ decay modes other than $K^{*0} \to K_S\pi^0$. We find that the final state is primarily $\xi_f = +1$; the $\xi_f = -1$ fraction is $0.19 \pm 0.04(\text{stat}) \pm 0.04(\text{syst})$ [9].

J/ψ and $\psi(2S)$ mesons are reconstructed via their decays to $\ell^+\ell^-$ $(\ell = \mu, e)$. The $\psi(2S)$ is also reconstructed via $J/\psi\pi^+\pi^-$, and the χ_{c1} via $J/\psi\gamma$. The η_c is detected in the $K^+K^-\pi^0$ and $K_SK^-\pi^+$ modes. For the $J/\psi K_S$ mode, we use $K_S \to \pi^+\pi^-$ and $\pi^0\pi^0$ decays; for other modes we only use $K_S \to \pi^+\pi^-$.

The J/ψ and $\psi(2S) \to \mu^+\mu^-$ candidates are reconstructed from oppositely charged track pairs where at least one track is positively identified as a muon by the KLM system and the other is either positively identified as a muon or has an ECL energy deposit consistent with that of a minimum ionizing particle. For e^+e^- decays, we use oppositely charged track pairs where at least one track is a well identified electron and the other track satisfies minimal dE/dx and E/p requirements. For dielectrons, we correct for final state radiation or bremsstrahlung and recover additional ψ candidates by including the four-momentum of every photon detected within 0.05 radians of the original e^+ or e^- direction in the e^+e^- invariant mass calculation. Candidate $K_S \to \pi^+\pi^-$ decays are oppositely charged track pairs that have an invariant mass between 482 and 514 MeV$/c^2$, which corresponds to $\pm 3\sigma$ around the K_S mass peak.

To reconstruct $\chi_{c1}K_S$ decays, we select $\chi_{c1} \to J/\psi\gamma$ decays, rejecting γ's that are consistent with $\pi^0 \to \gamma\gamma$ decays, and impose the requirement $385 < M_{\gamma\ell\ell} - M_{\ell\ell} < 430.5$ MeV$/c^2$. For η_c decays, we distinguish kaons from pions using a combination of CDC dE/dx measurements and information from the TOF and ACC systems. Candidate $\eta_c \to K^+K^-\pi^0$ $(K_SK^-\pi^+)$ decays are selected with a $KK\pi$ mass requirement that

TABLE 1. The numbers of observed events ($N_{\rm ev}$) and the estimated background ($N_{\rm bkgd}$) in the signal region for each f_{CP} mode.

Mode	$N_{\rm ev}$	$N_{\rm bkgd}$
$J/\psi(\ell^+\ell^-)K_S(\pi^+\pi^-)$	457	11.9
$J/\psi(\ell^+\ell^-)K_S(\pi^0\pi^0)$	76	9.4
$\psi(2S)(\ell^+\ell^-)K_S(\pi^+\pi^-)$	39	1.2
$\psi(2S)(J/\psi\pi^+\pi^-)K_S(\pi^+\pi^-)$	46	2.1
$\chi_{c1}(J/\psi\gamma)K_S(\pi^+\pi^-)$	24	2.4
$\eta_c(K^+K^-\pi^0)K_S(\pi^+\pi^-)$	23	11.3
$\eta_c(K_SK^-\pi^+)K_S(\pi^+\pi^-)$	41	13.6
$J/\psi K^{*0}(K_S\pi^0)$	41	6.7
Sub-total	747	58.6
$J/\psi(\ell^+\ell^-)K_L$	569	223

takes into account the natural width of the η_c. For $J/\psi K^{*0}(K_S\pi^0)$ decays, we use $K_S\pi^0$ combinations that have an invariant mass within 75 MeV/c^2 of the nominal K^* mass. We reduce background from low-momentum π^0's by requiring $\cos\theta_{K^*} < 0.8$, where θ_{K^*} is the angle between the K_S momentum vector and the K^{*0} flight direction calculated in the K^{*0} rest frame.

Reconstructed B meson decays are identified using the beam-constrained mass $M_{beam} \equiv \sqrt{E_{beam}^2 - p_B^2}$ and the energy difference $\Delta E \equiv E_B - E_{beam}$, where E_{beam} is the cms beam energy, and p_B and E_B are the B candidate three-momentum and energy calculated in the cms.

Figure 1 shows the combined $M_{\rm bc}$ distribution for all channels other than $J/\psi K_L$ after a mode-dependent requirement on ΔE. The B meson signal region is defined as $5.270 < M_{\rm bc} < 5.290$ GeV/c^2. Table 1 lists the numbers of observed candidates ($N_{\rm ev}$) and the background ($N_{\rm bkgd}$) determined by extrapolating the rate from the ΔE *vs.* $M_{\rm bc}$ sideband region into the signal region. About 65% of the exclusive CP signal events (44% of the full CP sample) are reconstructd in the $B^0 \to \psi K_S$, $K_S \to \pi^+\pi^-$ mode.

Candidate $B^0 \to J/\psi K_L$ decays are selected by requiring ECL and/or KLM hit patterns that are consistent with the presence of a shower induced by a neutral hadron. K_L candidates with ECL information only are treated separately from the K_L candidates with KLM hits. The centroid of the shower is required to be within a 45° cone centered on the K_L direction that is inferred from two-body decay kinematics and the measured four-momentum of the J/ψ. To reduce the background we cut on a likelihood ratio that depends on the J/ψ cms momentum, the angle between the K_L and its nearest-neighbor charged track, the charged track multiplicity of the event, the extent to which the event is consistent with a $B^+ \to J/\psi K^{*+}(K_L\pi^+)$ hypothesis, and the polar angle with respect to the z direction of the reconstructed B^0 meson in the cms. In addition, events that were reconstructed as $B^0 \to J/\psi K_S$, $J/\psi K^{*0}(K^+\pi^-, K_S\pi^0)$, $B^+ \to J/\psi K^+$, or $J/\psi K^{*+}(K^+\pi^0, K_S\pi^+)$ decays are removed. Finally, K_L clusters with positions that match photons from reconstructed π^0's are also rejected.

Figure 2 shows the $p_B^{\rm cms}$ distribution, calculated with the $B^0 \to J/\psi K_L$ two-body decay hypothesis. The histograms are the results of a fit to the signal and background distributions. The shapes are derived from the Belle GEANT based Monte Carlo (MC) simulation. However, the normalization and peak position of the signal are allowed to vary. There are 397 entries in the $0.2 \leq p_B^{\rm cms} \leq 0.45$ GeV/c signal region with KLM clusters. There are 172 entries in the range $0.2 \leq p_B^{\rm cms} \leq 0.40$ GeV/c with clusters in the ECL only. The fit finds a total of 346 ± 29 $J/\psi K_L$ signal events, and a signal purity of 61%. Thus, about 33% of the signal in the full CP eigenstate sample is reconstructed in the $B^0 \to \psi K_L$ mode.

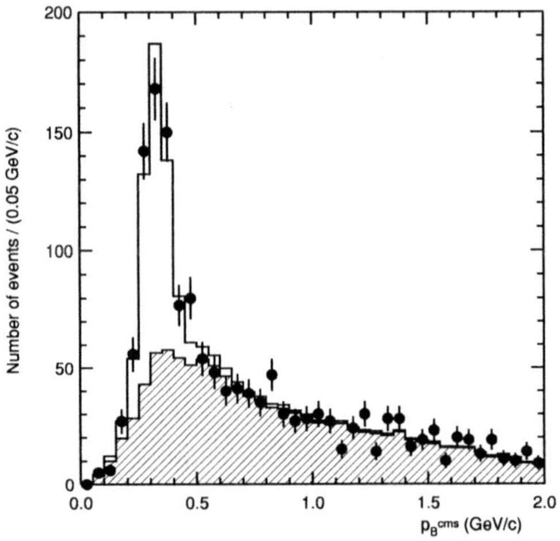

FIGURE 2. The $p_B^{\rm cms}$ distribution for $B^0 \to J/\psi K_L$ candidates with the results of the fit. The solid line is the signal plus background; the shaded area is background only. The signal region for KLM (ECL-only) clusters is $0.2 \leq p_B^{\rm cms} \leq 0.45(0.40)$ GeV/c.

To identify the flavor of the accompanying B meson, leptons, kaons, Λ's, charged slow pions from $D^* \to D^0\pi^+$ decays, and energetic pions from two-body B decay (e.g. $\bar{B}^0 \to D^{*+}\pi^+$) are used. A likelihood based method, described in detail below, is used to combine information from the different categories and to take into account their correlations. The figure of merit for flavor tagging performance is the effective efficiency, ε_{eff}, which is $\varepsilon(1-2w^2)$ summed over all tagging categories. This method gives $\varepsilon_{eff} = 0.270 \pm 0.008^{+0.006}_{-0.009}$.

A Monte Carlo simulation is used to determine a table for a category-dependent variable that indicates whether a particle originates from a B^0 or $\overline{B}^0$. The values of this variable range from -1 for a reliably identified $\overline{B}^0$ to $+1$ for a reliably identified B^0 and depend on the tagging particle's charge, cm momentum, polar angle, particle-identification probability, as well as other kinematic and event shape quantities. For lepton tags, the missing momentum and recoil momentum are included in the likelihood determination. For slow pion tags, the angle between the pion and the thrust axis of the non-f_{CP} tracks is used. Charged kaon tags accompanied by K_S mesons, which have additional strange quark content and a lower tagging value, are treated as a separate

TABLE 2. The event fractions (f_l) and incorrect flavor assignment probabilities (w_l) for each r interval. The errors include both statistical and systematic uncertainties.

l	r	f_l	w_l
1	0.000 – 0.250	0.405	$0.465^{+0.010}_{-0.009}$
2	0.250 – 0.500	0.149	$0.352^{+0.015}_{-0.014}$
3	0.500 – 0.625	0.081	$0.243^{+0.021}_{-0.030}$
4	0.625 – 0.750	0.099	$0.176^{+0.022}_{-0.017}$
5	0.750 – 0.875	0.123	$0.110^{+0.022}_{-0.014}$
6	0.875 – 1.000	0.140	$0.041^{+0.011}_{-0.010}$

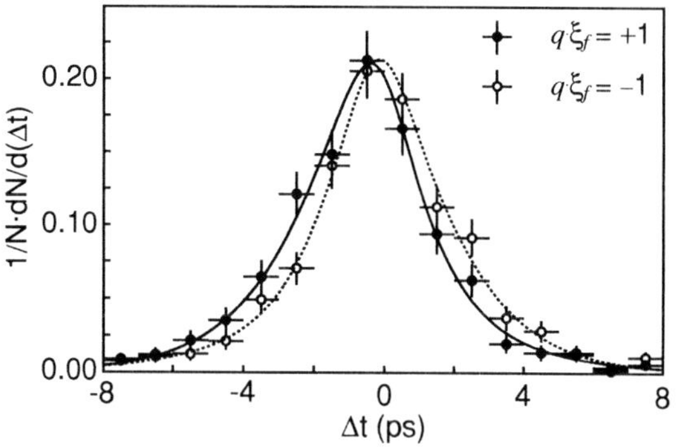

FIGURE 3. Δt distributions for the events with $q\xi_f = +1$ (solid points) and $q\xi_f = -1$ (open points). The results of the global fit (with $\sin 2\phi_1 = 0.99$) are shown as solid and dashed curves, respectively.

tagging category.

The results from the separate, particle-level categories are then combined in a second stage that takes correlations for the case of multiple particle-level tags into account. This second stage determines two event-level parameters, q and r. The first, q, has the discrete values $q = +1$ when the tagged B meson is more likely to be a B^0 and -1 when it is more likely to be a $\overline{B}^0$. The parameter r is an event-by-event flavor-tagging dilution factor that ranges from $r = 0$ for no flavor discrimination to $r = 1$ for unambiguous flavor assignment. The value of r is used only to sort data into six intervals of flavor purity; the wrong-tag probabilities that are used in the final CP fit are determined from data.

To avoid dependence on the MC, the probabilities of an incorrect flavor assignment, w_l ($l = 1,6$), are determined directly from the data for six r intervals using exclusively reconstructed, self-tagged $B^0 \to D^{*-}\ell^+\nu, D^{(*)-}\pi^+, D^{*-}\rho^+$ and $J/\psi K^{*0}(K^+\pi^-)$ decays. The b-flavor of the accompanying B meson is assigned according to the flavor-tagging algorithm described above. The exclusive decay and tag vertices are reconstructed using the same vertexing algorithm that is used in the CP fit. The values of w_l are obtained from the amplitudes of the time-dependent $B^0\overline{B}^0$ mixing oscillations: $(N_{\rm OF} - N_{\rm SF})/(N_{\rm OF} +$

TABLE 3. The values of $\sin 2\phi_1$ for various subsamples (statistical errors only).

Sample	$\sin 2\phi_1$
$f_{\rm tag} = B^0$ $(q = +1)$	0.84 ± 0.21
$f_{\rm tag} = \overline{B}^0$ $(q = -1)$	1.11 ± 0.17
$J/\psi K_S(\pi^+\pi^-)$	0.81 ± 0.20
$(c\bar{c})K_S$ except $J/\psi K_S(\pi^+\pi^-)$	1.00 ± 0.40
$J/\psi K_L$	1.31 ± 0.23
$J/\psi K^{*0}(K_S\pi^0)$	0.85 ± 1.45
All	0.99 ± 0.14

$N_{\rm SF}) = (1-2w_l)\cos(\Delta m_d \Delta t)$. Here $N_{\rm OF}$ and $N_{\rm SF}$ are the numbers of opposite flavor and same flavor events. The value of Δm_d is fixed at the world average [6]. Table 2 lists the resulting w_l values together with the fraction of the events (f_l) in each r interval.

The f_{CP} vertex is determined using lepton tracks from J/ψ or $\psi(2S)$ decays, or prompt tracks from η_c decays. The $f_{\rm tag}$ vertex is determined from well reconstructed tracks not assigned to f_{CP}. Tracks that form a K_S are not used. The tracks used for vertexing must have at least one three dimensional point with consistent hits in $r-\phi$ plus at least one additional z hit. Each vertex position is required to be consistent with the interaction point profile smeared in the r-ϕ plane by the B meson decay length. We use an iterative procedure: if the quality of the vertex fit is poor, the track that is the largest contributor to the χ^2 is removed and the fit is repeated. The typical vertex-finding efficiency and vertex resolution (rms) for z_{CP} ($z_{\rm tag}$) are 92 (91)% and 75 (140) μm, respectively. Note that the resolution on the tag side includes a large additional contribution from charm decay. Once the tag and CP eigenstate vertices are reconstructed, the proper time is calculated from $\Delta z/\gamma\beta$.

After vertexing there are 560 events with $q = +1$ flavor tags and 577 events with $q = -1$. Figure 3 shows the observed Δt distributions for the $q\xi_f = +1$ (solid points) and $q\xi_f = -1$ (open points) event samples. In the raw data there is a clear asymmetry between the two distributions; this demonstrates visually that CP symmetry is violated.

To extract the measured value of the CP violating parameter, we perform an unbinned maximum likelihood fit to the time distributions of the tagged and vertexed events. The fit takes into account the effects of background, vertex resolutions and incorrect tagging.

For modes other than $J/\psi K^{*0}$ the pdf expected for the signal is

$$\mathcal{P}_{\rm sig}(\Delta t, q, w_l, \xi_f) = \frac{e^{-|\Delta t|/\tau_{B0}}}{2\tau_{B0}}\{1 - \xi_f q(1-2w_l)\sin 2\phi_1 \sin(\Delta m_d \Delta t)\},$$

where we fix τ_{B0} and Δm_d at their world average values [6]. For the $B^0 \to \psi K^{*0}$ mode a more complex fit to Δt and the transversity angle[8] is used to take into account the two CP eigenstates which contribute[9].

The pdf used for the background distribution is $\mathcal{P}_{\rm bkg}(\Delta t) = f_\tau e^{-|\Delta t|/\tau_{\rm bkg}}/2\tau_{\rm bkg} + (1 - f_\tau)\delta(\Delta t)$, where f_τ is the fraction of the background component with an effective lifetime $\tau_{\rm bkg}$ and δ is the Dirac delta function. For all f_{CP} modes other than $J/\psi K_L$, a study using

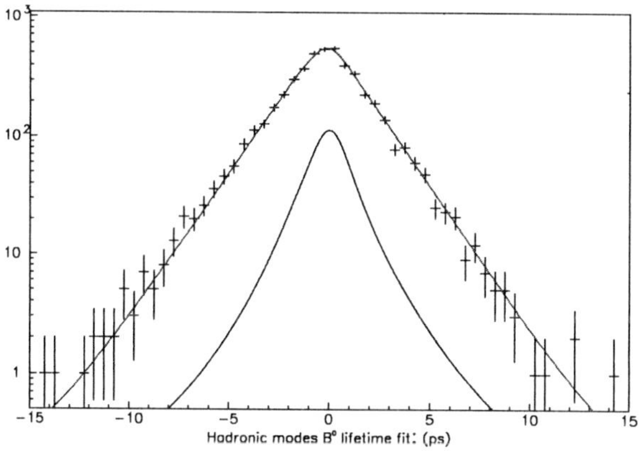

FIGURE 4. The proper decay time distributions for $B^0 \to D^{(*)-}\pi^+$ decays. The upper curve is the result of the fit for the B^0 lifetime; the lower curve represents the background contribution.

events in the ΔE *vs.* M_{bc} sideband regions shows that the f_τ component is negligible. For these modes we use $\mathcal{P}_{bkg}(\Delta t) = \delta(\Delta t)$.

In the case of $B^0 \to \psi K_L$, the background is dominated by $B \to \psi X$ decays where some final states are CP eigenstates. We estimate the fractions of the background components with and without a true K_L cluster by fitting the p_B^{cm} distribution to the expected shapes determined from MC. We also use the MC to determine the fraction of events with definite CP content within each component. The result is a background that is 71% from non-CP modes with $\tau_{bkg} = \tau_B$. For the CP-mode backgrounds, we use the signal pdf given above with the appropriate ξ_f values. For $\psi K^*(K_L\pi^0)$, which is 13% of the background, we use the $\xi_f = -1$ content determined from the full ψK^* sample. The remaining backgrounds are $\xi_f = -1$ states (10%) including ψK_S, and $\xi_f = +1$ states (5%) including $\psi(2S)K_L$, $\chi_{c1}K_L$ and $\psi\pi^0$.

The last ingredient needed for the CP fit is the proper-time interval resolution, $R(\Delta t)$. This resolution function is parameterized by convolving a sum of two Gaussians (a *main* component due to the SVD vertex resolution and charmed meson lifetimes, plus a *tail* component caused by poorly reconstructed tracks) with a function that takes into account the cm motion of the B mesons. The relative fraction of the main Gaussian is determined to be 0.97 ± 0.02 from a study of $B^0 \to D^{*-}\pi^+$, $D^{*-}\rho^+$, $D^-\pi^+$, ψK^{*0}, ψK_S and $B^+ \to \overline{D}^0\pi^+$, ψK^+ events. The means (μ_{main}, μ_{tail}) and widths (σ_{main}, σ_{tail}) of the Gaussians are calculated event-by-event from the f_{CP} and f_{tag} vertex-fit error matrices and the χ^2 values of the fit; typical values are $\mu_{main} = -0.24$ ps, $\mu_{tail} = 0.18$ ps and $\sigma_{main} = 1.49$ ps, $\sigma_{tail} = 3.85$ ps. An example of a fit for hadronic non-CP eigenstate using the resolution function modes is shown in Fig. 4. The fit agrees well with data out to 10 lifetimes on the logarithmic scale of the figure. As a consistency check, we obtain lifetimes for the neutral and charged B mesons using the same vertexing procedure that is used for the CP fit. The results of the fit agree well with the world average B^0 lifetime value.[7]

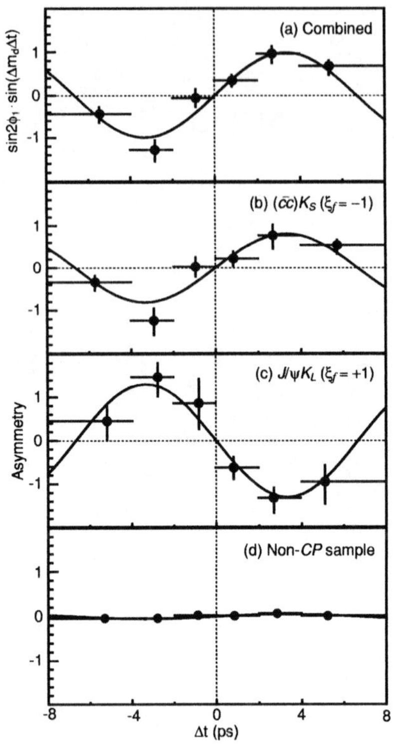

FIGURE 5. (a) The asymmetry obtained from separate fits to each Δt bin for the full data sample; the curve is the result of the global fit. The corresponding plots for the (b) $(c\bar{c})K_S$ ($\xi_f = -1$), (c) $J/\psi K_L$ ($\xi_f = +1$), and (d) B^0 control samples are also shown. The curves are the results of the fit applied separately to the individual data samples.

The pdfs for signal and background are convolved with $R(\Delta t)$ to determine the likelihood value for each event as a function of $\sin 2\phi_1$:

$$\begin{aligned}\mathcal{L}_i &= \int \{f_{\rm sig}\mathcal{P}_{\rm sig}(\Delta t', q, w_l, \xi_f)R_{sig}(\Delta t - \Delta t') \\ &\quad +(1-f_{\rm sig})\mathcal{P}_{\rm bkg}(\Delta t')R_{bkg}(\Delta t - \Delta t')\}d\Delta t',\end{aligned}$$

where $f_{\rm sig}$ is the probability that the event is signal. The most probable $\sin 2\phi_1$ is the value that maximizes the likelihood function $L = \prod_i \mathcal{L}_i$, where the product is over all events. (Note that the signal and background resolution functions are different.)

The result of the fit is

$$\sin 2\phi_1 = 0.99 \pm 0.14(\text{stat}) \pm 0.06(\text{syst}).$$

In Fig. 5(a) we show the asymmetries for the combined data sample that are obtained by applying the fit to the events in each Δt bin separately. The smooth curve is the result of the global unbinned fit. Figures 5(b) and (c) show the corresponding asymmetry displays for the $(c\bar{c})K_S$ ($\xi_f = -1$) and the $J/\psi K_L$ ($\xi_f = +1$) modes separately. The observed asymmetries for the different CP states are indeed opposite, as expected. The curves are the results of unbinned fits applied separately to the two samples; the resultant $\sin 2\phi_1$ values are 0.84 ± 0.17(stat) and 1.31 ± 0.23(stat), respectively.

The systematic error is dominated by uncertainties in the tails of the vertex distributions, which contribute 0.04. Other significant contributions come from uncertainties (a) in w_l (0.03); (b) in the parameters of the resolution function (0.02); and (c) in the $J/\psi K_L$ background fraction (0.02). The errors introduced by uncertainties in Δm_d and τ_{B^0} are negligible.

A number of checks on the measurement were performed. Table 3 lists the results obtained by applying the same analysis to various subsamples. All values are statistically consistent with each other. The result is unchanged if we use the w_l's determined separately for $f_{\rm tag} = B^0$ and $\overline{B}^0$. Fitting to the non-CP eigenstate self-tagged modes $B^0 \to D^{(*)-}\pi^+, D^{*-}\rho^+, J/\psi K^{*0}(K^+\pi^-)$ and $D^{*-}\ell^+\nu$, where no asymmetry is expected, yields 0.05 ± 0.04. The asymmetry distribution for this control sample is shown in Fig. 5(d).

A single consistent reconstruction and analysis procedure is used for all data samples. We verify that, as expected, the $\sin 2\phi_1$ values for the different run periods are consistent. With toy Monte Carlo studies we also verify that the analysis procedure and the errors in the analysis are well behaved when $\sin 2\phi_1$ is close to the physical boundary. As a further check, we used three independent CP fitting programs and two different algorithms for the $f_{\rm tag}$ vertexing and found no discrepancy.

Finally, we comment on the possibility of direct CP violation. The signal pdf for a neutral B meson decaying into a CP eigenstate can be expressed in the more general form

$$\mathcal{P}_{sig}(\Delta t) = \frac{e^{-|\Delta t|/\tau_{B^0}}}{2\tau_{B^0}(1+|\lambda|^2)}\{\frac{1+|\lambda|^2}{2} + q(1-2w_l)[\xi_f Im\lambda \sin(\Delta m_d \Delta t) - \frac{1-|\lambda|^2}{2}\cos(\Delta m_d \Delta t)]\}$$

where λ is a complex parameter that depends on both $B^0\overline{B}^0$ mixing and on the amplitudes for B^0 and $\overline{B}^0$ decay to a CP eigenstate. The presence of a cosine term ($|\lambda| \neq 1$) indicates direct CP violation. For the primary analysis, we assumed $|\lambda| = 1$ which is the SM expectation. In order to test this assumption, we also performed a fit using the above expression with $Im\lambda$ (= "$\sin 2\phi_1$") and $|\lambda|$ as free parameters. We obtain $|\lambda| = 1.03 \pm 0.09$ and $Im\lambda = 0.99 \pm 0.14$ for all CP modes combined, where the errors are statistical only. This result confirms the assumption used in our analysis.

CONCLUSIONS

In the summer of 2001, Belle (along with BABAR) presented its first significant measurement of the CP violating parameter $\sin 2\phi_1$. BELLE found

$$\sin 2\phi_1 = 0.99 \pm 0.14 \pm 0.06$$

with a statistical significance of greater than six standard deviations[10]. This can be compared to the BABAR result of $\sin 2\phi_1 = 0.59 \pm 0.14 \pm 0.05$[11]. The two results are based on data samples of comparable size (31 million and 32 million $B\bar{B}$ pairs, respectively). The efficiencies and resolutions of the two experiments are also quite similar. However, although the weighted average of the two results agrees well with indirect determinations that assume the Standard Model, the two measurements themselves are only marginally consistent. Larger data samples and additional more precise measurements will be required to fully reconcile these two results for $\sin 2\phi_1$. In parallel, a program for the measurement of the other angles, ϕ_2 and ϕ_3, has started.

ACKNOWLEDGMENTS

I wish to acknowledge the extraordinary achievement of my colleagues on Belle and the heroic effort of the KEKB accelerator team. I would also like to thank the conference organizers at Caltech for their hospitality and for a well-organized and interesting meeting held under difficult circumstances.

REFERENCES

1. M. Kobayashi and T. Maskawa, *Prog. Theor. Phys.* **49**, 652 (1973).
2. A.B. Carter and A.I. Sanda, *Phys. Rev.* **D23**, 1567 (1981); I.I. Bigi and A.I. Sanda, *Nucl. Phys.* **B193**, 85 (1981).
3. K. Abe *et al.* (Belle Collab.), *The Belle Detector*, KEK Report 2000-4, to be published in *Nucl. Instrum. Methods.* and *KEKB B-Factory Design Report*, KEK Report 95-7.
4. KEKB B Factory Design Report, KEK Report 95-1, 1995, unpublished.
5. Throughout this report, whenever a mode is quoted the inclusion of the charge conjugate mode is implied.
6. D.E. Groom *et al.* (Particle Data Group), *Eur. Phys. J.* **C15**, 1 (2000).
7. The measured B-lifetimes are: $\tau_{B^0} = 1.547 \pm 0.021$ ps and $\tau_{B^+} = 1.641 \pm 0.033$ ps (statistical errors only). These results are slightly different from recent Belle lifetime results due to the use of an improved resolution function.
8. θ_{tr} is defined as the angle between the ℓ^+ direction in the J/ψ rest frame and the z-axis, where the x-axis is defined as the direction of motion of the J/ψ in the $\Upsilon(4S)$ rest frame. The x-y plane is defined by the K^* decay products in the J/ψ rest frame.
9. K. Abe *et al.* (Belle Collab.), *Measurements of Polarization and CP Asymmetry in $B \to J/\psi + K^*$ decays*, paper submitted to LP01, Rome, July 2001; BELLE-CONF-0105.
10. K. Abe *et al.* (Belle Collaboration), *Phys. Rev. Lett.* 87, 091802 (2001).
11. B. Aubert *et al.* (BABAR Collaboration), *Phys. Rev. Lett.* 87, 091801 (2001).

Measurements of *CP* Violation, Mixing and Lifetimes of *B* Mesons with the *BABAR* Detector

Sören Prell representing the *BABAR* Collaboration

University of California at San Diego
Department of Physics
9500 Gilman Drive, La Jolla, CA 92093

Abstract. We report the observation of *CP* violation in the B^0 meson system. Using a novel technique for time-dependent measurements, we measure a non-zero value for the *CP*-violating amplitude $\sin 2\beta$ at the 4.1 σ level. We also report on precision measurements of the B^+ and B^0 lifetimes and the $B^0\bar{B}^0$ mixing frequency Δm_d obtained with the same technique, and on a first measurement of the time-dependent *CP*-violating amplitude in $B^0 \to \pi^+\pi^-$ decays.

INTRODUCTION

CP violation has been a central concern of particle physics since its discovery in 1964 in the decays of K_L^0 mesons [1]. An elegant explanation of the origin of *CP* violation was proposed by Kobayashi and Maskawa, as a complex phase in the three-generation CKM quark-mixing matrix [2]. In this picture, measurements of *CP*-violating asymmetries in the time distributions of B^0 decays to charmonium final states are expected to be large and provide a direct test of the Standard Model of electroweak interactions [3].

We present measurements of time-dependent *CP*-asymmetries in samples of fully reconstructed *B* decays to charmonium-containing *CP* eigenstates ($b \to c\bar{c}s$) and to the $\pi^+\pi^-$ final state. The data for these studies were recorded at the $\Upsilon(4S)$ resonance by the *BABAR* detector at the PEP-II asymmetric-energy e^+e^- collider at the Stanford Linear Accelerator Center.

When the $\Upsilon(4S)$ decays, the *P*-wave $B^0\bar{B}^0$ state evolves coherently until one of the mesons decays. In one of four time-order and flavor configurations, if the tagging meson $B_{\rm tag}$ decays first, and as a B^0, the other meson must be a $\bar{B}^0$ at that same time $t_{\rm tag}$. It then evolves independently, and can decay into a *CP* eigenstate B_{CP} at a later time t_{CP}. The time between the two decays $\Delta t = t_{CP} - t_{\rm tag}$ is a signed quantity made measurable by producing the $\Upsilon(4S)$ with a boost $\beta\gamma = 0.56$ along the collision (z) axis, with nominal energies of 9.0 and 3.1 GeV for the electron and positron beams. The measured distance $\Delta z \approx \beta\gamma c\Delta t$ between the two decay vertices provides a good estimate of the corresponding time interval Δt; the average value of $|\Delta z|$ is $\beta\gamma c\tau_{B^0} \approx 250\,\mu$m.

We examine each of the events in the B_{CP} sample for evidence that the other neutral *B* meson decayed as a B^0 or a $\bar{B}^0$ (flavor tag). The distribution $f_+(f_-)$ of the decay rate

CP618, *Heavy Flavor Physics: Ninth International Symposium*, edited by A. Ryd and F. C. Porter

when the tagging meson is a $B^0(\bar{B}^0)$ is given by

$$f_{\pm}(\Delta t) = \frac{e^{-|\Delta t|/\tau}}{4\tau}[1 \pm S\sin(\Delta m_d \Delta t) \mp C\cos(\Delta m_d \Delta t)], \quad (1)$$

where Δt is the time between the two B decays, τ is the B^0 lifetime [4], Δm_d is the $B^0\bar{B}^0$ mixing frequency [4], and the lifetime difference between neutral B mass eigenstates is assumed to be negligible. The sine term in Eq. 1 is due to interference between direct decay and decay after mixing, and the cosine term is due to direct CP violation. The CP-violating parameters S and C are defined in terms of a complex parameter λ that depends on both $B^0\bar{B}^0$ mixing and on the amplitudes describing $\bar{B}^0$ and B^0 decay to a common final state f [5]:

$$S = \frac{2\,Im\lambda}{1+|\lambda|^2} \quad \text{and} \quad C = \frac{1-|\lambda|^2}{1+|\lambda|^2}. \quad (2)$$

A difference between the B^0 and $\bar{B}^0$ Δt distributions or a Δt asymmetry for either flavor tag is evidence for CP violation.

In the Standard Model $\lambda = \eta_f e^{-2i\beta}$ for charmonium-containing $b \to c\bar{c}s$ decays, η_f is the CP eigenvalue of the state f and $\beta = \arg\left[-V_{cd}V_{cb}^*/V_{td}V_{tb}^*\right]$ is an angle of the Unitarity Triangle of the three-generation CKM matrix [2]. Thus, the time-dependent CP-violating asymmetry is

$$A_{CP}(\Delta t) \equiv \frac{f_+(\Delta t) - f_-(\Delta t)}{f_+(\Delta t) + f_-(\Delta t)} = -\eta_f \sin 2\beta \sin(\Delta m_{B^0} \Delta t). \quad (3)$$

The analogous time-dependent CP-violating asymmetry in the decay $B^0 \to \pi^+\pi^-$ arises from interference between mixing and decay amplitudes, and interference between the $b \to uW^-$ (tree) and $b \to dg$ (penguin) decay amplitudes. If the decay proceeds purely through the tree process, the complex parameter λ is directly related to CKM matrix elements and $|\lambda| = 1$ and $Im\lambda = \sin 2\alpha$, where $\alpha = \arg\left[-V_{td}V_{tb}^*/V_{ud}V_{ub}^*\right]$.

However, recent theoretical estimates suggest that the contribution from the gluonic penguin amplitude can be significant [6] leading to $|\lambda| \neq 1$ and $Im\lambda = |\lambda|\sin 2\alpha_{\text{eff}}$, where α_{eff} depends on the magnitudes and strong phases of the tree and penguin amplitudes.

We also present precise measurements of the B^0-$\bar{B}^0$ mixing frequency Δm_d and the neutral and charged B lifetimes. These measurements use the same vertexing algorithm and Δt calculation as the measurements of the CP-asymmetries. In addition, for the Δm_d measurement, the same flavor tagging algorithm as for the CP analyses is used. These measurements are amongst the most precise available and provide a good validation of the novel technique to study time-dependent B decays.

In all analyses the values of the parameters under study (B lifetimes, Δm_d, $\sin 2\beta$, $S_{\pi\pi}$ and $C_{\pi\pi}$) were hidden to eliminate possible experimenter's bias until event selection, Δt reconstruction method, and fitting procedures were finalized and systematic errors were determined.

THE *BABAR* DETECTOR

A detailed description of the *BABAR* detector can be found in Ref. [7]. Charged particles are detected and their momenta measured by a combination of a silicon vertex tracker (SVT) consisting of five double-sided layers and a central drift chamber (DCH), in a 1.5-T solenoidal field. The average vertex resolution in the z direction is 70μm for a fully reconstructed B meson. We identify leptons and hadrons with measurements from all detector systems, including the energy loss (dE/dx) in the DCH and SVT. Electrons and photons are identified by a CsI electromagnetic calorimeter (EMC). Muons are identified in the instrumented flux return (IFR). A Cherenkov ring imaging detector (DIRC) covering the central region, together with the dE/dx information, provides K-π separation of at least three standard deviations for B decay products with momentum greater than $250\,\mathrm{MeV}/c$ in the laboratory.

MEASUREMENT OF B LIFETIMES AND ΔM_D

The measurements of the charged and neutral B lifetimes and the B^0-$\bar{B}^0$ mixing frequency Δm_d are based on a sample of approximately 23 million $B\bar{B}$ pairs.

Exclusive B Reconstruction

Samples of B^0 and B^+ mesons $B_{\rm rec}$ are reconstructed in the modes $B^0 \to D^{(*)-}\pi^+$, $D^{(*)-}\rho^+$, $D^{(*)-}a_1^+$, $J/\psi K^{*0}$ and $B^+ \to \bar{D}^{(*)0}\pi^+$, $J/\psi K^+$, $\psi(2S)K^+$. Charged and neutral $\bar{D}^*$ candidates are formed by combining a $\bar{D}^0$ with a π^- or π^0. $\bar{D}^0$ candidates are reconstructed in the decay channels $K^+\pi^-$, $K^+\pi^-\pi^0$, $K^+\pi^+\pi^-\pi^-$ and $K_S^0\pi^+\pi^-$ and D^- candidates in the decay channels $K^+\pi^-\pi^-$ and $K_S^0\pi^-$. We reconstruct J/ψ and $\psi(2S)$ in the decays to e^+e^- and $\mu^+\mu^-$ and the $\psi(2S)$ decay to $J/\psi\ \pi^+\ \pi^-$.

Continuum $e^+e^- \to q\bar{q}$ background is suppressed by requirements on the normalized second Fox-Wolfram moment [8] for the event and on the angle between the thrust axes of $B_{\rm rec}$ and of the other $B = B_{\rm opp}$ in the event. B candidates are identified by the difference ΔE between the reconstructed B energy and the beam energy $\sqrt{s}/2$ in the $\Upsilon(4S)$ frame, and the beam-energy substituted mass $m_{\rm ES}$ calculated from $\sqrt{s}/2$ and the reconstructed B momentum. We require $m_{\rm ES} > 5.2\,\mathrm{GeV}/c^2$ and $|\Delta \mathrm{E}| < 3\sigma_{\Delta \mathrm{E}}$. The distributions of $m_{\rm ES}$ for selected B candidates in 30 fb^{-1} is shown in Fig. 1.

Δt Measurement

The decay time difference, Δt, between B decays is determined from the measured separation $\Delta z = z_{rec} - z_{opp}$ along the z axis between the $B_{\rm rec}$ and the $B_{\rm opp}$ vertices. This Δz is converted into Δt using the $\Upsilon(4S)$ boost and correcting on an event-by-event basis for the direction of the B mesons. The resolution of the Δt measurement is dominated by the z resolution of the $B_{\rm opp}$ decay vertex. This vertex uses all tracks in the event

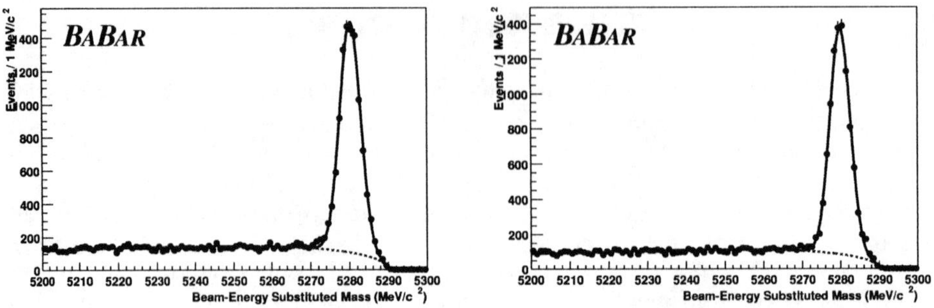

FIGURE 1. Beam energy substituted mass distribution for selected B^0 (left) and B^+ (right) candidates. In 30 fb^{-1}, we reconstruct 9400 B^0 and 8500 B^+ signal events. Average signal purities for $m_{\mathrm{ES}} > 5.27\,\mathrm{GeV}/c^2$, are 83 % and 85 % for B^0 and B^+, respectively.

except those incorporated in B_{rec}. An additional constraint is provided by a calculated B_{opp} production point and three-momentum, determined from the three-momentum of the B_{rec} candidate, its decay vertex, and the average position of the interaction point and the $\Upsilon(4S)$ boost. Reconstructed K_S^0 or Λ candidates are used as input to the fit in place of their daughters in order to reduce bias due to long-lived particles. Tracks with a large contribution to the χ^2 are iteratively removed from the fit, until all remaining tracks have a reasonable fit probability. Candidates with $|\Delta z| < 3.0\,\mathrm{mm}$ and $\sigma_{\Delta z} < 400\,\mu\mathrm{m}$ are retained. For the measurement of the B lifetimes, we require that at least two tracks are included in the B_{opp} vertex .

Two different parameterizations are used to model the decay-time difference resolution functions. In the measurements of Δm_d, $\sin 2\beta$, and $\sin 2\alpha$, the resolution function is approximated by a sum of three Gaussian distributions with different means δ_k and widths σ_k,

$$\mathcal{R}(\delta_t, \sigma_{\Delta t}|\hat{a}) = \sum_{k=1}^{3} \frac{f_k}{\sigma_k\sqrt{2\pi}} e^{-(\delta_t - \delta_k)^2/2\sigma_k{}^2}, \tag{4}$$

where δ_t is the difference between the measured and true Δt values. For the core and tail Gaussians, the widths $\sigma_{1,2} = S_{1,2} \times \sigma_{\Delta t}$ are the scaled event-by-event measurement error, $\sigma_{\Delta t}$, derived from the vertex fits. The third Gaussian, with a fixed width of $\sigma_3 = 8\,\mathrm{ps}$, accounts for less than 1% of *outlier* events with incorrectly reconstructed vertices. The three Gaussian resolution function is not suited for the measurement of the B lifetimes because strong correlations lead to increased statistical errors. Studies with Monte Carlo simulation and data show that the sum of a zero-mean Gaussian distribution and its convolution with an exponential provides a good trade-off between statistical and systematic uncertainties:

$$\mathcal{R}(\delta_t, \sigma_{\Delta t}|\hat{a} = \{h, s, \kappa\}) = h \frac{1}{\sqrt{2\pi} s \sigma_{\Delta t}} \exp\left(-\frac{\delta_t^2}{2s^2\sigma_{\Delta t}^2}\right) \tag{5}$$
$$+ \int_{-\infty}^{0} \frac{1-h}{\kappa\sigma_{\Delta t}} \exp\left(\frac{\delta_t'}{\kappa\sigma_{\Delta t}}\right) \frac{1}{\sqrt{2\pi} s \sigma_{\Delta t}} \exp\left(-\frac{(\delta_t - \delta_t')^2}{2s^2\sigma_{\Delta t}^2}\right) d(\delta_t') \,.$$

The parameters $\hat{a}$ are the fraction h in the core Gaussian component, a scale factor s for the per-event errors $\sigma_{\Delta t}$, and the factor κ in the effective time constant $\kappa\sigma_{\Delta t}$ of the exponential which accounts for charm decays. Δt *outlier* events are modeled the same way as in the three Gaussian resolution function. The resolution function parameters are assumed to be the same for all B^0 and B^+ decay modes. This assumption is confirmed by Monte Carlo simulation studies. The resolution functions differ only slightly between B^0 and B^+ mesons due to different mixtures of D^- and $\bar{D}^0$ mesons in the $B_{\rm opp}$ decays and we use a single set of resolution function parameters for both B^0 and B^+ in the lifetime fits.

B Lifetime Results

We extract the B^+ and B^0 lifetimes from an unbinned maximum likelihood fit to the Δt distributions of the selected B candidates. The probability for an event to be signal is estimated from $m_{\rm ES}$ fits (Fig. 1) and the $m_{\rm ES}$ value of the $B_{\rm rec}$ candidate. In the likelihood, the probability density for the signal events is given by

$$\mathcal{G}(\Delta t, \sigma_{\Delta t}|\tau, \hat{a}) = \int_{-\infty}^{+\infty} e^{-|\Delta t|/\tau}/(2\tau)\,\mathcal{R}(\Delta t - \Delta t', \sigma_{\Delta t}|\hat{a})\,\mathrm{d}(\Delta t'), \tag{6}$$

and the background Δt distribution for each B species is empirically modeled by the sum of a prompt component and a lifetime component convolved with the same resolution function, but with a separate set of parameters. The likelihood fit involves 17 free parameters in addition to the B^0 and the B^+ lifetimes: 12 to describe the background Δt distributions and 5 for the signal resolution function. The charged B lifetime τ_{B^+} is replaced with $\tau_{B^+} = r \cdot \tau_{B^0}$ to estimate the statistical error on the ratio $r = \tau_{B^+}/\tau_{B^0}$.

We determine the B^0 and B^+ meson lifetimes and their ratio to be:

$$\begin{aligned}
\tau_{B^0} &= 1.546 \pm 0.032\,(\text{stat}) \pm 0.022\,(\text{syst})\ \text{ps},\\
\tau_{B^+} &= 1.673 \pm 0.032\,(\text{stat}) \pm 0.023\,(\text{syst})\ \text{ps, and}\\
\tau_{B^+}/\tau_{B^0} &= 1.082 \pm 0.026\,(\text{stat}) \pm 0.012\,(\text{syst}).
\end{aligned}$$

These are the most precise measurements to date [9] and are consistent with the world averages [4]. The resolution function parameters are consistent with those found in a Monte Carlo simulation that includes detector alignment effects. Figure 2 shows the results of the likelihood fit superimposed on the Δt distributions for B^0 and B^+ events. With the current data sample these measurements are still statistically limited. The dominant systematic errors arise from uncertainties in the description of the combinatorial background and of events with large Δt values, the use of a common time resolution function for B^0 and B^+ and from limited Monte Carlo statistics.

B Flavor Tagging

The measurements of Δm_d, sin2β and sin2α require knowledge of the $B_{\rm opp} = B_{\rm tag}$ flavor. We use the same tagging algorithm in the three analyses to determine the $B_{\rm tag}$

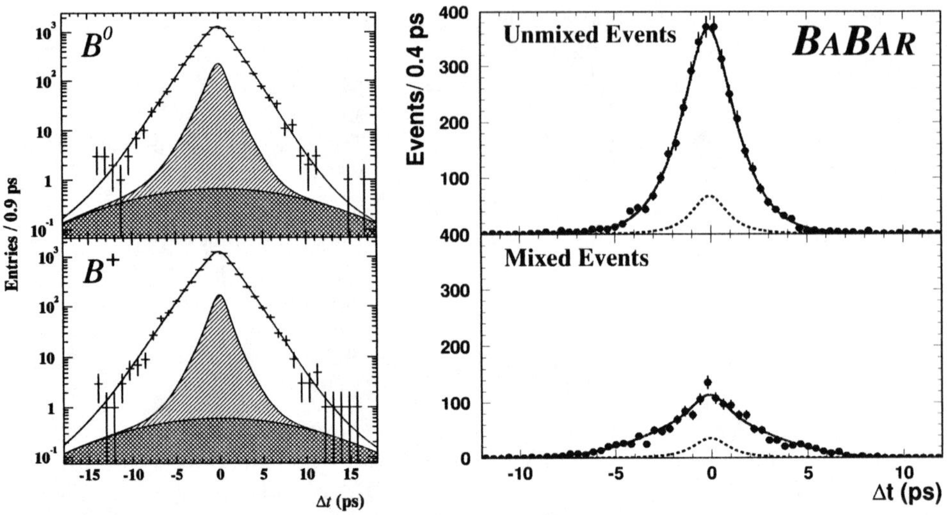

FIGURE 2. Left: Δt distribution for the B^0 (top) and B^+ (bottom) events within 2σ of the B mass with superimposed fit results. The single-hatched areas are the background contributions and the cross-hatched areas represent the Δt outliers. The probability of obtaining a lower likelihood is 7.3%. Right: Δt distributions in data for the mixed and unmixed events ($m_{\rm ES}(B_{rec}) > 5.27\,{\rm GeV}/c^2$), with overlaid the projection of the likelihood fit (solid) and the background contributions (dashed).

flavor from the charges of its decay products.

The charge of energetic electrons and muons from semileptonic B decays, kaons, soft pions from D^* decays, and high momentum charged particles is correlated with the flavor of the decaying b quark. Each event is assigned to one of four hierarchical, mutually exclusive tagging categories or has no flavor tag. A lepton tag requires an electron (muon) candidate with a center-of-mass momentum $p_{\rm cm} > 1.0\ (1.1)\,{\rm GeV}/c$. This efficiently selects primary leptons and reduces contamination due to oppositely-charged leptons from charm decays. Events meeting these criteria are assigned to the `Lepton` category unless the lepton charge and the net charge of all kaon candidates indicate opposite tags. Events without a lepton tag but with a non-zero net kaon charge are assigned to the `Kaon` category. All remaining events are passed to a neural network algorithm whose main inputs are the momentum and charge of the track with the highest center-of-mass momentum, and the outputs of secondary networks, trained with Monte Carlo samples to identify primary leptons, kaons, and soft pions. Based on the output of the neural network algorithm, events are tagged as B^0 or $\bar{B}^0$ and assigned to the `NT1` (more certain tags) or `NT2` (less certain tags) category, or not tagged at all. The tagging power of the `NT1` and `NT2` categories arises primarily from soft pions and from recovering unidentified isolated primary electrons and muons. The yields, efficiencies, purities and mistag rates w for each tagging category are listed in Table. 1.

TABLE 1. Event yields for the different tagging categories obtained from fits to the m_{ES} distributions. The purity is quoted for $m_{ES} > 5.270\,\text{MeV}/c^2$ and average mistag fractions w_i extracted for each tagging category i from the likelihood fit to the time distribution for the fully-reconstructed flavor eigenstate sample.

Category	Tagged	Efficiency (%)	Purity (%)	w
Lepton	754 ± 28	11.3 ± 0.4	97.1 ± 0.6	0.085 ± 0.018
Kaon	2317 ± 54	34.8 ± 0.6	85.2 ± 0.8	0.167 ± 0.014
NT1	556 ± 26	8.3 ± 0.3	88.7 ± 1.5	0.195 ± 0.026
NT2	910 ± 36	13.7 ± 0.4	83.0 ± 1.3	0.326 ± 0.024
Total	4538 ± 75	68.1 ± 0.9	86.7 ± 0.5	

$B^0\overline{B}^0$ Mixing Result

The value of Δm_d is extracted from the tagged flavor-eigenstate B^0 sample B_{flav} with a simultaneous unbinned likelihood fit to the Δt distributions of both unmixed ($B^0\overline{B}^0$) and mixed (B^0B^0 and $\overline{B}^0\overline{B}^0$) events. The PDFs for the unmixed $(+)$ and mixed $(-)$ signal events for the ith tagging category are given by

$$\mathcal{H}_{\pm}(\Delta t|\Delta m_d, w_i, \hat{a}_i) = \frac{e^{-|\Delta t|/\tau}}{4\tau}\left[1 \pm (1-2w_i)\cos\Delta m_d \Delta t\right] \otimes \mathcal{R}(\delta_t|\hat{a}_i), \tag{7}$$

Some resolution function parameters are allowed to differ for each tagging category to account for shifts due to inclusion of charm decay products in the tag vertex. The PDFs are extended to include background terms, different for each tagging category. The probability that a B^0 candidate is a signal event is determined from a fit to the observed m_{ES} distribution for its tagging category. The Δt distributions of the combinatorial background are described with a zero lifetime component and a non-oscillatory component with non-zero lifetime. Separate resolution function parameters are used for signal and background to minimize correlations.

The likelihood fit involves a total of 34 parameters, including Δm_d (1), the mistag rates w_i and mistag differences $\Delta w_i = w_i(B^0) - w_i(\overline{B}^0)$ (8), Δt resolution function parameters (9) and background parameters (16). We display the result of the likelihood fit by using the mixing asymmetry,

$$\mathcal{A}_{mix}(\Delta t) = \frac{N_{unmixed}(\Delta t) - N_{mixed}(\Delta t)}{N_{unmixed}(\Delta t) + N_{mixed}(\Delta t)}. \tag{8}$$

If flavor tagging and Δt determination were perfect, the asymmetry as a function of Δt would be a cosine with unit amplitude. The amplitude is diluted by mistag probabilities and the Δt resolution. The Δt distributions of mixed and unmixed events, and their asymmetry, $\mathcal{A}_{mix}$, are shown in Fig. 2 and 3 along with projections of the fit result. The probability to obtain a smaller likelihood is 28 %.

Systematic uncertainties in the Δm_d measurement arise from various sources. The conversion of Δz to Δt introduces an uncertainty ($\pm 0.007\,\text{ps}^{-1}$) due to the limited knowledge of the PEP-II boost, the z length scale of *BABAR* and the B_{rec} momen-

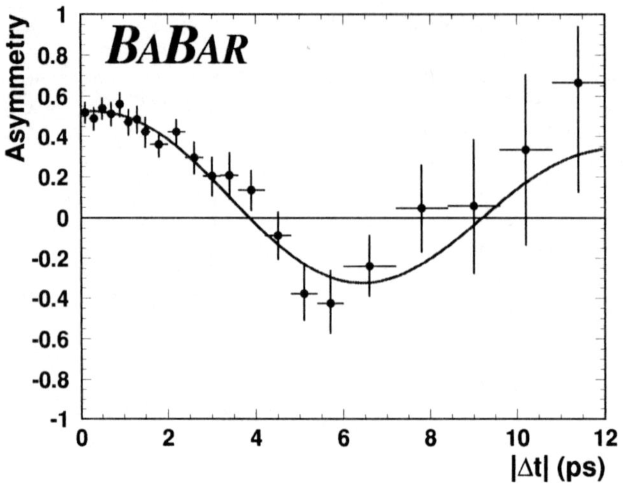

FIGURE 3. The asymmetry $\mathcal{A}_{mix}(\Delta t)$ between unmixed and mixed events as a function of $|\Delta t|$ overlaid with the result from the likelihood fit.

tum vector in the $\Upsilon(4S)$ frame. Systematic uncertainties related to the resolution function ($\pm 0.005\,\text{ps}^{-1}$), are attributed to the choice of the parameterization, the description of outliers, and the capability of the resolution model to deal with various plausible misalignment scenarios applied to the Monte Carlo simulation. The parameters of the background Δt distribution are left free in the likelihood fit, but systematic errors ($\pm 0.005\,\text{ps}^{-1}$), are introduced by the uncertainty in signal probabilities, parameterization of the background Δt distributions and resolution function, and the small amount of correlated B^+ background. Finally, statistical limitations of Monte Carlo validation tests ($\pm 0.004\,\text{ps}^{-1}$), the full size of a (negative) correction obtained from Monte Carlo ($\pm 0.009\,\text{ps}^{-1}$), and the variation of the B^0 lifetime [4] ($\pm 0.006\,\text{ps}^{-1}$) contribute. These contributions added in quadrature yield a total systematical error of $0.016\,\text{ps}^{-1}$.

In conclusion, the B^0-$\bar{B}^0$ mixing frequency Δm_d is determined with a new time-dependent technique from tagged fully-reconstructed B^0 decays to be

$$\Delta m_d = 0.519 \pm 0.020(stat) \pm 0.016(syst)\,\text{ps}^{-1}.$$

This preliminary result is one of the single most precise measurements available, and is consistent with the current world average [4] and a recent *BABAR* measurement with a dilepton sample [10]. The error on Δm_d is still dominated by statistics, leaving substantial room for further improvement as more data are accumulated.

MEASUREMENT OF SIN2β

In 32 million $B\bar{B}$, we extract sin2β from a sample of fully reconstructed B^0 decays (B_{CP}) to final states with $\eta_f = -1$ ($J/\psi K^0_S$, $\psi(2S)K^0_S$, $\chi_{c1}K^0_S$), $\eta_f = +1$ ($J/\psi K^0_L$) and

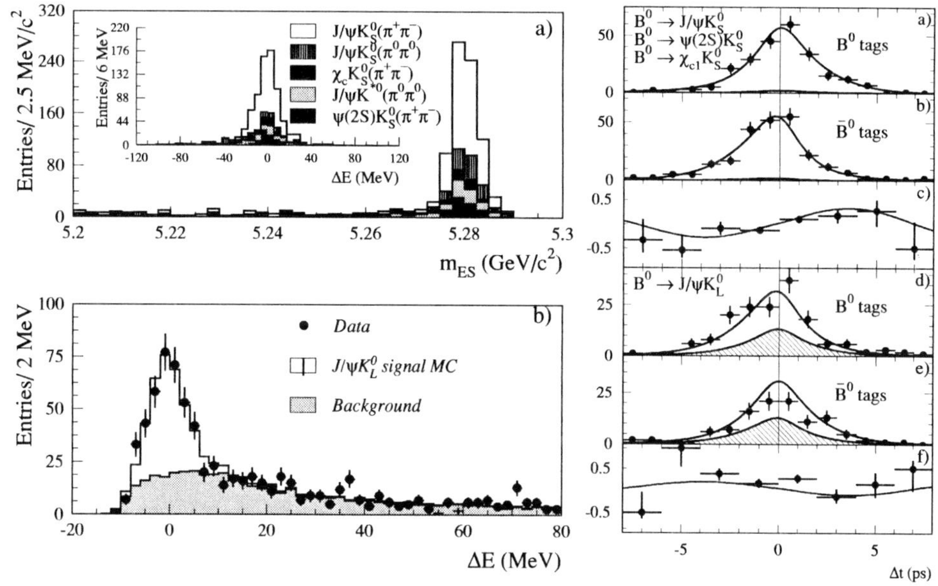

FIGURE 4. Left: a) distribution of m_{ES} for B_{CP} candidates having a K_S^0 in the final state; b) distribution of ΔE for $J/\psi K_L^0$ candidates. Right: Number of $\eta_f = -1$ candidates ($J/\psi K_S^0$, $\psi(2S)K_S^0$, and $\chi_{c1}K_S^0$) in the signal region a) with a B^0 tag N_{B^0} and b) with a $\bar{B}^0$ tag $N_{\bar{B}^0}$, and c) the asymmetry $(N_{B^0} - N_{\bar{B}^0})/(N_{B^0} + N_{\bar{B}^0})$, as functions of Δt. The solid curves represent the result of the combined fit to all selected *CP* events; the shaded regions represent the background contributions. Figures d)–f) contain the corresponding information for the $\eta_f = +1$ mode ($J/\psi K_L^0$).

$\eta_f(\text{effective}) = 0.65 \pm 0.07$ ($J/\psi K^{*0}$ with $K^{*0} \to K_S^0\pi^0$) [11].

The sin2β measurement is made with a simultaneous unbinned likelihood fit to the Δt distributions of the tagged B_{CP} and B_{flav} samples. The Δt distribution of the former is given by Eq. 1, with $|\lambda| = 1$. The B_{flav} sample evolves according to the PDF for B^0 flavor oscillations as described in the previous section. The amplitudes for the B_{CP} *CP*-asymmetries and for the B_{flav} flavor oscillations are reduced by the same factor $(1-2w)$ due to wrong tags. Both distributions are convolved with a common Δt resolution function and backgrounds are accounted for by adding terms to the likelihood, incorporated with different assumptions about their Δt evolution and convolved with a separate resolution function. Events are assigned signal and background probabilities based on the m_{ES} (all modes except $J/\psi K_L^0$) or ΔE ($J/\psi K_L^0$) distributions shown in Fig. 4. Separate Δt resolution functions parameters have been used for the data collected in 1999-2000 and 2001, due to the significant improvement in the SVT alignment.

A total of 45 parameters are varied in the likelihood fit, including sin2β (1), mistag fractions w and differences Δw (8), parameters for the signal Δt resolution (16), and parameters for background time dependence (9), Δt resolution (3) and mistag fractions (8). The determination of the mistag fractions and signal Δt resolution function is dominated by the large B_{flav} sample. The largest correlation between sin2β and any linear combination of the other free parameters is only 0.13. We fix τ_{B^0} and Δm_d [4].

Figure 4 shows the Δt distributions and A_{CP} as a function of Δt overlaid with the likelihood fit result for the $\eta_f = -1$ and $\eta_f = +1$ samples. The probability of obtaining a lower likelihood is 27%. The simultaneous fit to all *CP* decay modes and flavor decay modes yields

$$\sin 2\beta = 0.59 \pm 0.14 \text{ (stat)} \pm 0.05 \text{ (syst)}.$$

The dominant sources of systematic error are the parameterization of the Δt resolution function (0.03), due in part to residual uncertainties in SVT alignment, possible differences in the mistag fractions between the B_{CP} and B_{flav} samples (0.03), and uncertainties in the level, composition, and *CP* asymmetry of the background in the selected *CP* events (0.02). The systematic errors from uncertainties in Δm_{B^0} and τ_{B^0} and from the parameterization of the background in the B_{flav} sample are small; an increase of 0.020 ps^{-1} in the value for Δm_{B^0} decreases $\sin 2\beta$ by 0.015.

TABLE 2. Number of tagged events, signal purity and observed *CP* asymmetries in the *CP* samples and control samples. Errors are statistical only.

Sample	N_{tag}	Purity (%)	$\sin 2\beta$
$J/\psi K^0_S, \psi(2S)K^0_S, \chi_{c1}K^0_S$	480	96	0.56±0.15
$J/\psi K^0_L$	273	51	0.70±0.34
$J/\psi K^{*0}, K^{*0} \to K^0_S \pi^0$	50	74	0.82±1.00
Full *CP* sample	803	80	0.59±0.14
B_{flav} non-*CP* sample	7591	86	0.02±0.04
Charged *B* non-*CP* sample	6814	86	0.03±0.04

The large sample of reconstructed events allows a number of consistency checks, including separation of the data by decay mode, tagging category and B_{tag} flavor. The results of fits to some subsamples and to the samples of non-*CP* decay modes are shown in Table 2. For the latter, no statistically significant asymmetry is found.

If $|\lambda|$ is allowed to float in the fit to the $\eta_f = -1$ sample, which has high purity and requires minimal assumptions on backgrounds, we obtain $|\lambda| = 0.93 \pm 0.09$ (stat) $\pm$ 0.03 (syst). The sources of systematic error are the same as in the $\sin 2\beta$ analysis.

The measurement of $\sin 2\beta = 0.59 \pm 0.14$ (stat) ± 0.05 (syst) establishes *CP* violation in the B^0 meson system at the 4.1σ level. This significance is computed from the sum in quadrature of the statistical and additive systematic errors. The probability of obtaining this value or higher in the absence of *CP* violation is less than 3×10^{-5}. The corresponding probability for the $\eta_f = -1$ modes alone is 2×10^{-4}.

MEASUREMENT OF SIN2α

We reconstruct neutral B mesons decaying to $h^+h'^-$, where h and h' refer to π or K in a sample of 33 million $B\bar{B}$. The data set includes 30.4 fb^{-1} collected on the $\Upsilon(4S)$ resonance and 3.3 fb^{-1} collected below the $B\bar{B}$ threshold used for continuum background studies.

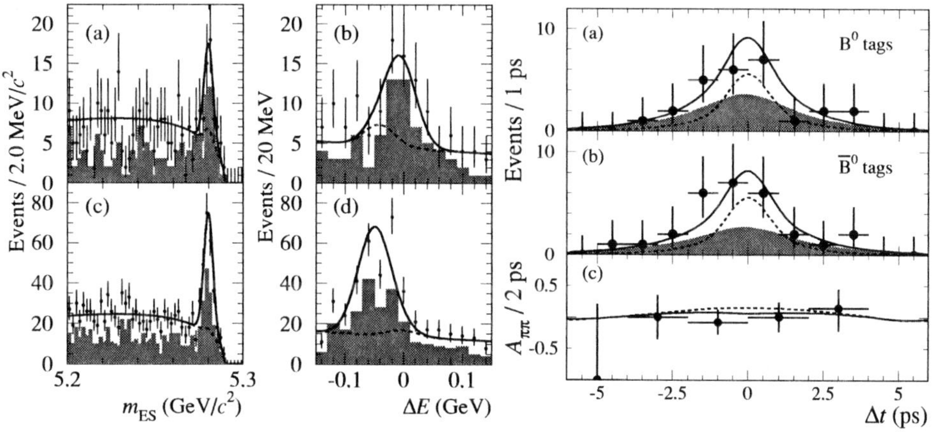

FIGURE 5. Left: Distributions of $m_{\rm ES}$ and ΔE (unshaded histograms) for events enhanced in signal (a), (b) $\pi\pi$ and (c), (d) $K\pi$ decays. Solid curves represent projections of the maximum likelihood fit result, while dashed curves represent $q\bar{q}$ and $\pi\pi \leftrightarrow K\pi$ cross-feed background. Shaded histograms show the subset of events that are tagged. Right: Distributions of Δt for events enhanced in signal $\pi\pi$ decays. Figures (a) and (b) show events (points with errors) with $B_{\rm tag} = B^0$ or $\bar{B}^0$. Solid curves represent projections of the likelihood fit, dashed curves represent the sum of $q\bar{q}$ and $K\pi$ background, and the shaded region represents the contribution from signal $\pi\pi$. Figure (c) shows $\mathcal{A}_{\pi\pi}(\Delta t)$ for data (points with errors), as well as fit projections for signal and background (solid curve), and signal only (dashed curve).

We select $B \to h^+h'^-$ candiates in the region $5.2 < m_{\rm ES} < 5.3\,{\rm GeV}/c^2$ and $|\Delta E| < 0.15\,{\rm GeV}$ and apply requirements on track multiplicity and event topology. The total number of events satisfying these criteria is 9741. This sample contains 97% background, mostly due to random combinations of tracks produced in $e^+e^- \to q\bar{q}$ events. We extract signal and background yields for $\pi^+\pi^-$, $K^+\pi^-$, and K^+K^- decays, and the amplitudes of the $\pi\pi$ sine ($S_{\pi\pi}$) and cosine ($C_{\pi\pi}$) oscillation terms simultaneously from an unbinned likelihood fit. We parameterize the $K\pi$ component in terms of the total yield and the CP-violating charge asymmetry $\mathcal{A}_{K\pi} \equiv (N_{K^-\pi^+} - N_{K^+\pi^-})(N_{K^-\pi^+} + N_{K^+\pi^-})$. Background parameters are determined from $m_{\rm ES}$ and ΔE sideband regions.

Discrimination between signal and background is based on $m_{\rm ES}$, ΔE, and a Fisher discriminant $\mathcal{F}$ [12] constructed from the scalar sum of the CM momenta of tracks and photons (excluding tracks from the $B_{\rm rec}$ candidate) and between pions and kaon tracks on the Cherenkov angle measurement from the DIRC. The inclusion of events with no flavor tag improves the signal yield estimates and provides a larger sample for determining background shape parameters in the likelihood fit.

There are 18 free parameters in the fit. In addition to the CP-violating parameters $S_{\pi\pi}$, $C_{\pi\pi}$, and $\mathcal{A}_{K\pi}$ (3), the fit determines signal and background yields (6), the background $K\pi$ charge asymmetry (1), and parameters describing the background shapes in $m_{\rm ES}$, ΔE, and $\mathcal{F}$ (8). We fix τ and Δm_d [4]. The Δt PDF for signal $\pi^+\pi^-$ decays is given by Eq. 1, modified to include mistags and convolved with the signal resolution function. The Δt PDF for signal $K\pi$ events takes into account B^0–$\bar{B}^0$ mixing and $B^0 \to K^+K^-$ decays are parameterized as an exponential convolved with the resolution function.

Figure 5 shows distributions of $m_{\rm ES}$ and ΔE for events enhanced in signal decays based on likelihood ratios and the Δt distributions and *CP* asymmetry $\mathcal{A}_{\pi\pi}(\Delta t) = (N_{B^0}(\Delta t) - N_{\bar{B}^0}(\Delta t))/(N_{B^0}(\Delta t) + N_{\bar{B}^0}(\Delta t))$ for tagged events enhanced in signal $\pi\pi$ decays (approximately 24 $\pi\pi$, 22 $q\bar{q}$, and 5 $K\pi$ events satisfy this selection).

In conclusion, in a sample of 33 million $B\bar{B}$, we find 65^{+12}_{-11} $\pi\pi$, 217 ± 18 $K\pi$, and $4.3^{+6.3}_{-4.3}$ KK events. These yields are consistent with the branching fractions reported in Ref. [12]. We measure the following *CP* parameters:

$$\begin{aligned} S_{\pi\pi} &= 0.03^{+0.53}_{-0.56}\,(\text{stat}) \pm 0.11\,(\text{syst}), \quad \mathcal{A}_{K\pi} = -0.07 \pm 0.08\,(\text{stat}) \pm 0.02\,(\text{syst}), \\ C_{\pi\pi} &= -0.25^{+0.45}_{-0.47}\,(\text{stat}) \pm 0.14\,(\text{syst}). \end{aligned}$$

The correlation between $S_{\pi\pi}$ and $C_{\pi\pi}$ is -21%, while $\mathcal{A}_{K\pi}$ is uncorrelated with $S_{\pi\pi}$ and $C_{\pi\pi}$. Systematic errors on $S_{\pi\pi}$, $C_{\pi\pi}$, and $\mathcal{A}_{K\pi}$ arise primarily from uncertainties in PDF shapes, tagging efficiencies and dilutions, τ, and Δm_d.

SUMMARY AND OUTLOOK

We have observed *CP* violation in the neutral *B* system at the 4.1 σ level using a sample of fully reconstructed B^0 decays to *CP* eigenstates. With the same novel technique of time-dependent measurements, we have determined the B^+ and B^0 lifetimes and the $B^0\bar{B}^0$ mixing frequency Δm_d with high precision. In addition, we have presented the first measurement of the time-dependent asymmetry in $B^0 \to \pi^+\pi^-$ decays. All results are limited by the data sample size and we expect improved measurements from the rapidly growing *BABAR* data sample in the near future especially for the *CP* violating asymmetries.

REFERENCES

1. J.H. Christenson *et al.*, *Phys. Rev. Lett.* **13**, 138 (1964).
2. N. Cabibbo, *Phys. Rev. Lett.* **10**, 531 (1963);
   M. Kobayashi and T. Maskawa, *Prog. Th. Phys.* **49**, 652 (1973).
3. A.B. Carter and A.I. Sanda, *Phys. Rev.* **D23**, 1567 (1981);
   I.I. Bigi and A.I. Sanda, *Nucl. Phys.* **B193**, 85 (1981).
4. Particle Data Group, D.E. Groom *et al.*, *Eur. Phys. Jour.* **C 15**, 1 (2000).
5. See, for example, L. Wolfenstein, *Eur. Phys. Jour.* **C 15**, 115 (2000).
6. M. Beneke, G. Buchalla, M. Neubert, and C.T. Sachrajda, *Nucl. Phys.* **B 606**, 245 (2001);
   Y.Y. Keum, H-n. Li, and A.I. Sanda, *Phys. Rev.* **D 63**, 054008 (2001);
   M. Ciuchini *et al.*, *Phys. Lett.* **B 515**, 33 (2001).
7. BaBar Collaboration, B. Aubert *et al.*, BaBar-PUB-01/08, to appear in *Nucl. Instrum. Methods.*
8. G.C. Fox and S. Wolfram, *Phys. Rev. Lett.* **41**, 1581 (1978).
9. BaBar Collaboration, B. Aubert *et al. Phys. Rev. Lett.* **87**, 201803 (2001).
10. T. Brandt, contribution to this conference.
11. BaBar Collaboration, B. Aubert *et al. Phys. Rev. Lett.* **86**, 2515 (2001);
    BaBar Collaboration, B. Aubert *et al. Phys. Rev. Lett.* **87**, 091801 (2001).
12. BaBar Collaboration, B. Aubert *et al.*, *Phys. Rev. Lett.* **87**, 151802 (2001).

CKM Matrix: Status and New Developments

A. Höcker*, H. Lacker*, S. Laplace* and F. Le Diberder*

*Laboratoire de l'Accélérateur Linéaire, IN2P3-CNRS et Université de Paris-Sud, BP 34, F-91405 Orsay Cedex, France

Abstract. An analysis of the CKM matrix parameters within the *R*fit approach is presented using updated input values with special emphasis on the recent $\sin 2\beta$ measurements from BABAR and Belle. The QCD Factorisation Approach describing $B \to \pi\pi, K\pi$ decays has been implemented in the software package CKMfitter. Fits using branching ratios and CP asymmetries are discussed.

STATISTICAL FRAMEWORK AND INPUTS

In the Standard Model (SM) with three families, CP violation is generated by a single phase in the CKM matrix [1]. This picture can be probed quantitatively by means of a global fit to all quantities sensitive to CKM elements in the SM. The analysis presented here is performed within the *R*fit statistical approach [2], which is implemented in the software package CKMfitter [3].
The quantity $\chi^2 = -2\ln\mathcal{L}(y_{\rm mod})$ is minimized in the fit, where the likelihood function is defined by $\mathcal{L}(y_{\rm mod}) = \mathcal{L}_{\rm exp}(x_{\rm exp} - x_{\rm theo}(y_{\rm mod})) \cdot \mathcal{L}_{\rm theo}(y_{\rm QCD})$. The experimental part, $\mathcal{L}_{\rm exp}$, depends on measurements, $x_{\rm exp}$, and theoretical predictions, $x_{\rm theo}$, which are functions of model parameters, $y_{\rm mod}$. The theoretical part, $\mathcal{L}_{\rm theo}$, describes the knowledge on the QCD parameters, $y_{\rm QCD} \in \{y_{\rm mod}\}$, where the theoretical uncertainties are considered as allowed ranges. The agreement between data and the SM is gauged by the global minimum $\chi^2_{{\rm min};y_{\rm mod}}$, determined by varying all model parameters $y_{\rm mod}$. For $\chi^2_{{\rm min};y_{\rm mod}}$, a confidence level (CL), expressing the goodness-of-fit, is computed by means of a Monte Carlo simulation. If the hypothesis "the CKM picture of the SM is correct" is accepted, CLs in parameter subspaces a, *e.g.* $a = (\bar\rho, \bar\eta)$ [4], are evaluated. For fixed a, one calculates $\Delta\chi^2(a) = \chi^2_{{\rm min};\mu}(a) - \chi^2_{{\rm min};y_{\rm mod}}$, where μ stands for all model parameters (including $y_{\rm QCD}$) with the exception of a. The corresponding CL is obtained from ${\rm CL}(a) = {\rm Prob}(\Delta\chi^2(a), N_{\rm dof})$, where $N_{\rm dof}$ is the number of degrees of freedom, in general the dimension of of the subspace a. Since the CL depends on the choice of the ranges for the $y_{\rm QCD}$, the results obtained in the fit have to be interpreted with care.
The input values used in this analysis are listed in Tab. 1. For $|V_{ub}|$, inclusive measurements from LEP and exclusive measurements from CLEO have been used. The preliminary CLEO lepton endpoint analysis [5] using moments obtained from $B \to X_s\gamma$ is not yet included. For $|V_{cb}|$, inclusive measurements from LEP, the measurements of $B \to D^*\ell\nu$ at zero-recoil and the moments analysis from CLEO [5], using inclusive $B \to X_c\ell\nu$ and $B \to X_s\gamma$ decays, have been combined. The uncertainty on Δm_d has been significantly reduced due to the measurements from the B-factories [6]. However, the constraint on

CP618, *Heavy Flavor Physics: Ninth International Symposium*, edited by A. Ryd and F. C. Porter

TABLE 1. Left: input values for the global fit. Right: fit results quoted for CL > 5 %.

Input Parameter	Value	Fit Output	Range (CL > 5 %)
$\|V_{ud}\|$	0.97394 ± 0.00089	λ	0.2221 ± 0.0041
$\|V_{us}\|$	0.2200 ± 0.0025	A	$0.763 - 0.905$
$\|V_{cd}\|$	0.224 ± 0.014	$\bar{\rho}$	$0.07 - 0.37$
$\|V_{cs}\|$	0.969 ± 0.058	$\bar{\eta}$	$0.26 - 0.49$
$\|V_{ub}\|$	$(3.49 \pm 0.24 \pm 0.55) \cdot 10^{-3}$	$J(10^{-3})$	$2.2 - 3.7$
$\|V_{cb}\|$	$(40.4 \pm 1.3 \pm 0.9) \cdot 10^{-3}$	$\sin 2\alpha$	$-0.90 - 0.51$
Δm_d	$(0.489 \pm 0.008)\ \mathrm{ps}^{-1}$	$\sin 2\beta$	$0.59 - 0.88$
Δm_s	Amplitude Spectrum	γ	$37° - 80°$
$\|\varepsilon_K\|$	$(2.271 \pm 0.017) \cdot 10^{-3}$	$\Delta m_s\ (\mathrm{ps}^{-1})$	$14.6 - 32.0$
$\sin 2\beta$	0.793 ± 0.102	$f_{B_d}\sqrt{B_d}$ (MeV)	$192 - 284$
$m_t(\overline{\mathrm{MS}})$	(166 ± 5) GeV	B_K	$0.52 - 1.68$
m_c (GeV)	1.30 ± 0.10	m_t (GeV)	$95 - 405$
B_K	$0.87 \pm 0.06 \pm 0.13$	$\mathcal{B}(K_L \to \pi^0 \nu\bar{\nu}) \cdot (10^{-11})$	$1.6 - 4.4$
η_{cc}	1.38 ± 0.53	$\mathcal{B}(K^+ \to \pi^+ \nu\bar{\nu}) \cdot (10^{-11})$	$4.8 - 9.4$
$f_{B_d}\sqrt{B_d}$ (GeV)	$(0.230 \pm 0.028 \pm 0.028)$ GeV	$\mathcal{B}(B^+ \to \tau^+ \nu_\tau) \cdot (10^{-7})$	$5.6 - 23.8$
$\eta_B(\overline{\mathrm{MS}})$	0.55 ± 0.01	$\mathcal{B}(B^+ \to \mu^+ \nu_\mu) \cdot (10^{-5})$	$2.2 - 9.4$
ξ	$1.16 \pm 0.03 \pm 0.05$		

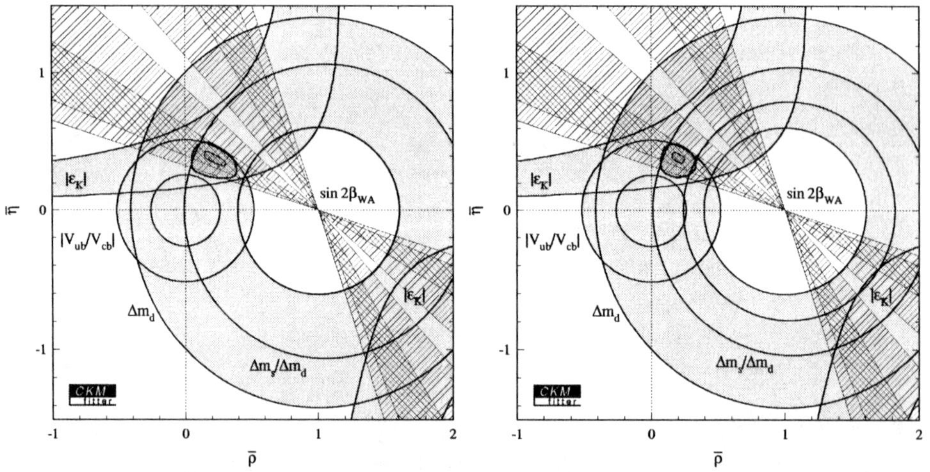

FIGURE 1. Left: constraints in the $(\bar{\rho}, \bar{\eta})$ plane where $\bar{\rho}$ and $\bar{\eta}$ are defined by $\bar{\rho} = \rho(1 - \lambda^2/2)$, $\bar{\eta} = \eta(1 - \lambda^2/2)$. Shown are the 5 % CLs for the individual constraints as shaded areas and the 1σ- and 2σ- contours from $\sin 2\beta$. In addition, the 5 % and 90 % CL contours for the combined fit are drawn. Right: constraints in the $(\bar{\rho}, \bar{\eta})$ plane if the likelihood ratio is used for the Δm_s constraint.

$(\bar{\rho}, \bar{\eta})$ is not improved since it is dominated by the theoretical uncertainty on $f_{B_d}\sqrt{B_d}$. For Δm_s, the most recent combined amplitude spectrum from [7] is included in the fit using a modified version of the standard amplitude method [2]. If the amplitude spectrum is translated into a likelihood ratio [8, 9], a stronger constraint is obtained. However, to

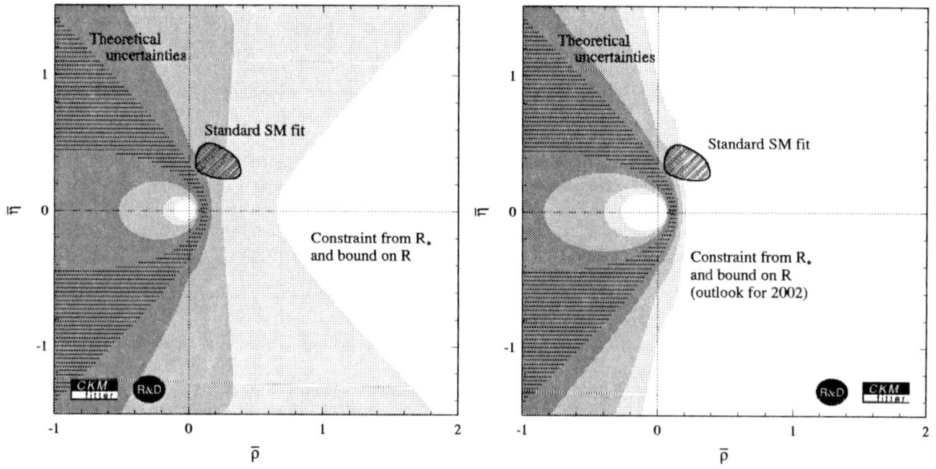

FIGURE 2. Left: Constraints in the $(\bar{\rho},\bar{\eta})$ plane using R_* (see text) and the bound on R. The CL is indicated by the shaded areas; dark-grey: > 90 %, gray: 90 % $- 32$ %, light-gray: 32 % $- 5$ %. Right: Hypothetical constraints for summer 2002.

our knowledge, it has not been demonstrated so far that the likelihood ratio can be interpreted as a probability density function. Hence, we use the more conservative method of Ref. [2] for the numerical analysis presented here. For sin2β, the world average is used. It should be noted that the most precise measurements from BABAR and Belle [6] differ presently by about two standard deviations.

FIT RESULTS

The global minimum of the CKM fit is found to be $\chi^2_{\mathrm{min};y_{\mathrm{mod}}} = 2.3$, resulting in a goodness-of-fit of 71%. It quantifies the excellent agreement between experimental data and the CKM picture of the SM. Fig. 1 shows the $(\bar{\rho},\bar{\eta})$ plane. Drawn are 5% CL contours from the single constraints using Δm_d, Δm_s, $|V_{ub}/V_{cb}|$ and $|\varepsilon_K|$, respectively, and the 1σ- and 2σ-contours for the four-fold ambiguity on β from $\sin 2\beta$. Shown in addition are the contours for the combined fit including sin2β. The statistical precision of the sin2β measurement already competes with the indirect, theoretically limited constraints. Fig. 1 (right) illustrates the improved constraint when using the likelihood ratio for Δm_s. Selected numerical results are summarized in Tab. 1.

CHARMLESS TWO BODY DECAYS

Constraints on the angle γ can be obtained from $B \to \pi\pi, K\pi$ decays. Based on color transparency arguments, theoretical calculations such as the QCD Factorisation Approach (FA) [10] and the QCD hard scattering approach [11] have been developed. Re-

cently, the FA has been implemented in CKMfitter. At present, it is premature to infer reliable constraints on the basis of these calculations due to open theoretical questions [10, 11, 12]. Data from the B-factories are not yet precise enough to probe the calculations in detail. Hence, all fit results are marked by an appropriate "R&D" logo. For this review, a global fit to $B \to \pi\pi, K\pi$ branching fractions and direct CP asymmetries measured in self-tagging $B \to K\pi$ decays has been performed within the framework of the FA, where most recent experimental results from BABAR [13], Belle [14] and CLEO [15] have been used. The numerous theoretical parameters are let free to vary within the ranges given in Ref. [10]. We find $\chi^2_{\mathrm{min};y_{\mathrm{mod}}} = 2.0$ and conclude that data are consistently described within the FA. The best FA fits are found at $\gamma \approx 80^{\circ}$ and are in agreement with the constraints from the standard fit.
Using less theoretical assumptions, ratios of branching fractions can be formed to derive constraints in the $(\bar{\rho}, \bar{\eta})$ plane. As an example, the CP-averaged ratio

$$R = \frac{\tau_{B^{\pm}}}{\tau_{B^0}} \frac{\mathcal{B}(B^0 \to K^{\pm}\pi^{\mp})}{\mathcal{B}(B^{\pm} \to K^0\pi^{\pm})}, \tag{1}$$

provides the bound $R > \sin^2\gamma$ which is independent of the strong phases [16]. Unfortunately, the present world average $R = 1.07^{+0.19}_{-0.15}$ leads to weak constraints only owing to the tails of the experimental errors. The ratio

$$R_* = \frac{\mathcal{B}(B^{\pm} \to K^0\pi^{\pm})}{2 \cdot \mathcal{B}(B^{\pm} \to K^{\pm}\pi^0)}, \tag{2}$$

measured to be $R_* = 0.70^{+0.16}_{-0.13}$, can be used to derive bounds in the $(\bar{\rho}, \bar{\eta})$ plane [17]. An important input for the theoretical prediction of R_* is the tree-to-penguin ratio (P/T) which can be determined experimentally using the relation

$$\bar{\varepsilon}_{3/2} = R_{\mathrm{th}} \cdot \tan\theta_C \cdot \frac{f_K}{f_\pi} \sqrt{\frac{2 \cdot \mathcal{B}(B^{\pm} \to \pi^{\pm}\pi^0)}{\mathcal{B}(B^{\pm} \to K^0\pi^{\pm})}}, \tag{3}$$

where R_{th} stands for SU(3) breaking corrections estimated in the FA to be [10] $R_{\mathrm{th}} = 0.98 \pm 0.05$. The bound on R_* can be translated into a prediction if additional information on the strong phases is inserted [10]. Adopting the values for theoretical ranges quoted in Ref. [10], one obtains the constraints shown in Fig. 2. At present, the constraints remain rather weak due to the limited experimental precision. The slight deformation of the shape pattern around $\gamma \approx 90^{\circ}$ is due to the bound on R. For summer 2002, an integrated luminosity of 100 fb^{-1} is expected to be collected by each experiment, BABAR and Belle. The experimental precision will then start to provide interesting constraints, as can be seen from Fig. 2 obtained assuming the present central experimental values and appropriately rescaling their errors. However, the constraints would still rely on the validity of some theoretical assumptions not yet fully explored.
Within the FA, the P/T ratio for $B \to \pi^+\pi^-$ is predicted. Compared to the present experimental error on the time-dependent asymmetry $S_{\pi\pi} = 0.03^{+0.53}_{-0.56} \pm 0.11$ from BABAR, the quoted theoretical uncertainty is much smaller [10]. In Fig. 3 (left), the constraints in $(\bar{\rho}, \bar{\eta})$ from $S_{\pi\pi}$ are shown using P/T from FA where theoretical uncertainties have been neglected. The right plot shows the constraints when also using $\sin 2\beta$.

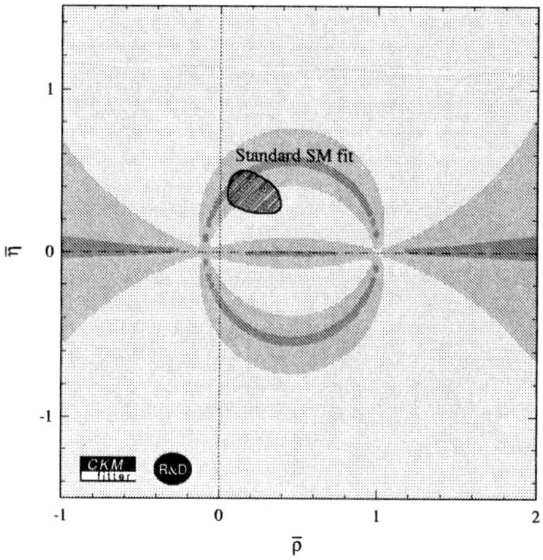

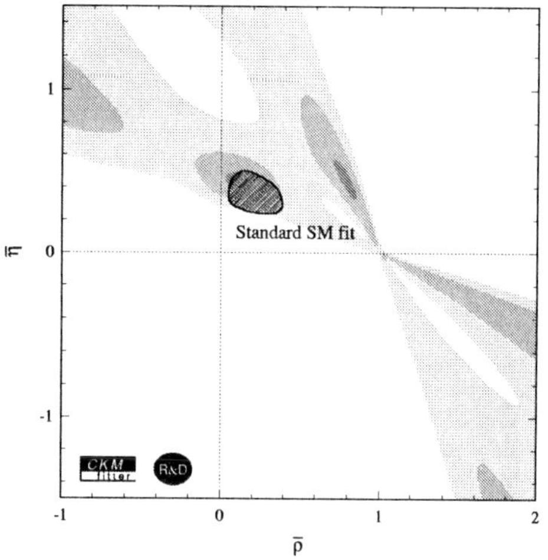

FIGURE 3. Left: Constraints on $(\bar{\rho}, \bar{\eta})$ from $S_{\pi\pi}$ using the penguin-over-tree ratio from FA. Right: Constraints when using also $\sin 2\beta$.

ACKNOWLEDGMENTS

It was a pleasure to attend this conference which was very successful despite the tragic events of the 11th September. We are indebted to Martin Beneke and Matthias Neubert for their help implementing the QCD Factorisation Approach in CKMfitter. HL was supported by the Fifth Framework Programme of the European Community Research under the grant No. HPMF-CT-1999-00032.

REFERENCES

1. N. Cabibbo, *Phys. Rev. Lett.* **10**, 531 (1963); M. Kobayashi and T. Maskawa, *Prog. Th. Phys.* **49**, 652 (1973).
2. A. Höcker, H. Lacker, S. Laplace and F. Le Diberder, *Eur. Phys. Jour.* **C21/2**, 225 (2001).
3. "CKMfitter: code, numerical results and plots", http://ckmfitter.in2p3.fr/.
4. L. Wolfenstein, *Phys. Rev. Lett.* **51**, 1945 (1983).
5. R.A. Briere, these proceedings.
6. T. Browder and S. Prell, these proceedings.
7. LEP *B* Oscillation WG, http://lepbosc.web.cern.ch/LEPBOSC/combined_results/budapest_2001/.
8. H.G. Moser and A. Roussarie, *Nucl. Instr. Meth.* **A384**, 491 (1997).
9. M. Ciuchini *et al.*, *Jour. High Ener. Phys.* **0107**, 013 (2001).
10. M. Beneke, G. Buchalla, M. Neubert and C.T. Sachrajda, hep-ph/0104110, (2001).
11. Y.-Y. Keum, H.-n. Li, and A.I. Sanda, hep-ph/0004173, (2000).
12. M. Ciuchini *et al.*, hep-ph/0104126, (2001).
13. BABAR coll., hep-ex/0105061, (2001); hep-ex/0107074, (2001).
14. Belle coll., hep-ex/0104030 (2001); hep-ex/0106095, (2001).
15. CLEO coll., *Phys. Rev. Lett.* **85**, 515 (2000); *Phys. Rev. Lett.* **85**; 525 (2000).
16. R. Fleischer and T. Mannel, *Phys. Rev.* **D57**, 2752 (1998).
17. M. Neubert and J.R. Rosner, *Phys. Lett.* **B441**, 403 (1998).

Review of LEP Results

F. Parodi representing the LEP collaborations

Dipartimento di Fisica, Via Dodecaneso 33, 16146 Genova, Italy

Abstract. I present a review of the results obtained during 10 years of activity in b-physics at LEP. Special emphasis will be put on measurements that attained precisions not even envisaged at the beginning of the LEP programme (V_{ub} and Δm_s). Finally the impact of these measurements on the CKM parameters determination will be presented.

INTRODUCTION

The b-physics program at LEP have covered different domains. In this paper I will concentrate on the measurements related to determination of the parameters of the CKM matrix, neglecting, because of lack of space, electroweak and spectroscopy results.
The leading role of LEP in the last decade of b-physics is evident not only in the experimental results but also in the averaging activities. LEP Working Groups have pushed LEP, SLD and Tevatron communities to compare results and to invent new methods for combining measurements in order to exploit at best the available data.
This work should be continued by new experiments.
Most of the results presented here comes, except where explicitly stated, from the summary paper produced by the LEP Heavy Flavour Working Groups for Summer Conferences 2001 [1].

B-LIFETIME MEASUREMENTS

The measurements of average lifetime of weakly decaying B-hadrons are an important test of the B-decay dynamics. In the naive spectator model all the B lifetimes should be equal the contribution of other processes (exchange, annihilation) contributing to the total decay rate modify this simple picture.
The Operator Product Expansion, within the factorization approximation, give, for the lifetime ratio, the following predictions [2]:

$$\frac{\tau(B^+)}{\tau(B_d^0)} = 1 + 0.05 \times \frac{f_B^2}{(200\,MeV)^2}, \frac{\tau(B_s^0)}{\tau(B_d^0)} = 1 \pm O(1\%), \frac{\tau(\Lambda_b^0)}{\tau(B_d^0)} = 0.90 - 0.95 \quad (1)$$

The study of the B-lifetimes has been also stimulating new methods for selecting pure samples of a definite B specie. At the beginning of LEP the first signal of B_s^0 and Λ_b^0 have been selected looking for right sign correlations, in the same hemisphere of a selected event, between the sign of the lepton issued by a semi-leptonic decay and the accom-

CP618, *Heavy Flavor Physics: Ninth International Symposium*, edited by A. Ryd and F. C. Porter

panying charm hadron. Afterwards semi-inclusive reconstructions of the charm hadron have been used to increase the available statistics and finally, in the last years, charged and neutral B hadrons have been separated efficiently using neural network methods. The average lifetimes from LEP/SLD/Tevatron computed for the 2001 Summer Conference[1] [3] show remarkable precisions:

$$
\begin{aligned}
\tau(B_d^0) &= 1.545 \pm 0.020\ ps \qquad (1.3\%)\\
\tau(B^+) &= 1.642 \pm 0.017\ ps \qquad (1.0\%)\\
\tau(\mathrm{B_s^0}) &= 1.464 \pm 0.057\ ps \qquad (3.9\%)\\
\tau(\Lambda_b) &= 1.208 \pm 0.051\ ps \qquad (4.2\%)
\end{aligned}
$$

Comparing these results with the predictions of Equation 1 one can conclude that:

- the ratio $\tau(B^+)/\tau(B_d^0)$ show a 3σ deviation from unity in agreement with theory;
- the ratio $\tau(B_s^0)/\tau(B_d^0)$ is in agreement with theory;
- the measurements of the Λ_b^0 lifetime are already precise enough to spot an inconsistency with the prediction and push for a better understanding of the theory.

The experimental results show that the hierarchy has been correctly predicted by theory. More stringent tests will be performed by B-factories and Tevatron.

LIFETIME DIFFERENCE: $\Delta\Gamma_S$

The lifetime difference between the weak eigenstates in the B_s^0-$\bar{B}_s^0$ system is expected to be small, to the first approximation $\Delta\Gamma_s/\Delta m_s \simeq 3/2\pi(m_b/m_t)^2$ [4]. Theoretical calculations [5] of the ratio $\Delta\Gamma_s/\Gamma_s$ at next-to-leading order give:

$$\frac{\Delta\Gamma_s}{\Gamma_s} = \left(\frac{f_{B_s}}{230\ MeV}\right)^2 0.007B_b(m_b) + (0.132^{0.011}_{0.027})\bar{B}_s(m_b) - (0.078 \pm 0.018)] \qquad (2)$$

where f_{B_s} is the B_s^0 decay constant, $B_b(m_b)$ and $\bar{B}_b(m_b)$ are bag parameters.
The study of $\Delta\Gamma_s$ benefit from the work done on lifetimes. Experimental informations on $\Delta\Gamma_s$ have been extracted by studying the proper time distribution of data samples enriched in B_s mesons or by measuring the branching fraction $B_s \to D_s^{(*)+}D_s^{(*)-}$.
In order to obtain an improved limit on $\Delta\Gamma_s$, the results based on fits to the proper time distributions are used to constrain $1/\Gamma_s$ to the world average lifetime $\tau(B_d^0)$. This is well motivated theoretically, as the total widths of the B_s^0 and B_d^0 mesons are expected to be equal within less than one percent and $\Delta\Gamma_d$ is expected to be small.
The combined results from LEP/Tevatron on $\Delta\Gamma_s$ are:

Assuming $\tau(B^0) = \tau(\mathrm{B_s^0})$	No assumptions
$\Delta\Gamma_s/\Gamma_s = 0.16^{+0.08}_{-0.09}$	$\Delta\Gamma_s/\Gamma_s = 0.24^{+0.16}_{-0.12}$
$\Delta\Gamma_s/\Gamma_s < 0.31$ 95% C.L.	$\Delta\Gamma_s/\Gamma_s < 0.53$ 95% C.L.

[1] This average includes also BaBar measurements

The possibility of using these combined limits for the indirect determination of Δm_s is still limited by the theoretical uncertainties in the evaluation of $\Delta\Gamma_s/\Delta m_s$ [6].

N_C vs. $BR(B \to \ell\nu)$

The measured value of the b-hadron semileptonic decay branching fraction is, since several years, on the low side of the theoretical expectations [7]. One way to reconcile theory with experiments consists in assuming that the c-quark effective mass is lower than used in these evaluations; this implies that decays of the type $b \to c\bar{c}s(d)$ correspond to larger decay rate. The average number of c and $\bar{c}$ quarks contributing in b-hadron decays is thus negatively correlated with the expected value of $BR(b \to \ell X)$: the simultaneous measurement of these two quantities may help to clarify the theoretical picture.
Experimentally, the number of c and $\bar{c}$ quarks contributing to b-hadron decays can be obtained by measuring the production fractions of charmed hadrons and charmonium states. Measurements originate from four sources:

- open-charm counting using exclusively reconstructed charmed hadrons,
- charmonium production,
- inclusive measurements of the distribution of charged track allowing the determination of $b \to D\bar{D}X$ and $b \to 0D$,
- $b \to D_i\bar{D}_jX$ branching fraction measurements in which $D_{i(j)}$ are completely reconstructed.

Measurements done at $\Upsilon(4S)$ and at the Z have been combined separately and they are showing in Figure 1. These results, summarizing 10 years of experimental activity, are

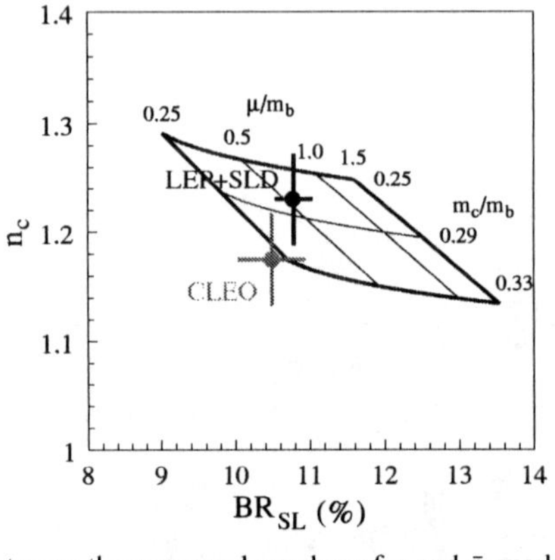

FIGURE 1. Comparison between the measured number of c and $\bar{c}$ quarks in b-hadron decays and of inclusive semileptonic fraction, with theoretical predictions.

compatible with the theoretical predictions and favour a rather standard value for the charm quark mass and a low scale, μ, at which QCD corrections have to be evaluated.

$|V_{CB}|$

$|V_{cb}|$ can be determined with two methods: the inclusive method, which uses semileptonic decay width of b-decays and the OPE, and the exclusive method, where $|V_{cb}|$ is extracted by studying the exclusive $\bar{B}_d^0 \to D^{*+}\ell^-\bar{\nu}_\ell$ decay process using HQET (measuring the product $\mathcal{F}(1)|V_{cb}|$).
Measurements done at LEP give, using the inclusive method:

$$|V_{cb}|^{incl.} = (40.7 \pm 0.5(exp.) \pm 2.4(theo.)) \times 10^{-3} \tag{3}$$

and using the exclusive method with $\mathcal{F}(1) = 0.88 \pm 0.05$ [8]:

$$|V_{cb}|^{incl.} = (40.5 \pm 1.9(exp.) \pm 2.3(theo.)) \times 10^{-3} \tag{4}$$

The LEP $|V_{cb}|$ Working Group computed a combined average taking into account the correlations between the two methods. The combined value is:

$$|V_{cb}| = (40.6 \pm 1.9) \times 10^{-3} \tag{5}$$

where 1.0×10^{-3} comes from correlated sources.

$|V_{UB}|$

The LEP Collaborations have measured, using different techniques, the inclusive yield of $b \to u$ transitions in semi-leptonic B decays. ALEPH and OPAL used a neural network discriminant based in kinematical variables, DELPHI preferred a classification based on the reconstructed mass M_X, decay topology and presence of secondary kaons, while L3 adopted a sequential cut analysis based on the kinematics of the two leading hadrons produced in the same hemisphere as a tagged lepton.
Each experiment optimize the performance of its analysis by choosing different working points of efficiency and signal-to-background ratio (S/B). Starting from a natural S/B of $\simeq 0.02$, ALEPH obtained $S/B = 0.07$ with an efficiency $\varepsilon = 11\%$, DELPHI had $S/B = 0.10$ with $\varepsilon = 6.5\%$ and L3 $S/B = 0.16$ with $\varepsilon = 1.5\%$.
Figure 2 show the evidence of inclusive charmless decays in a recent analysis by OPAL [9].

The LEP average give [10]:

$$BR(b \to \ell^-\nu X_u) = (1.67 \pm 0.31(stat. + exp.) \pm 0.37(b \to c) \pm 0.2(b \to u)) \times 10^{-3} \tag{6}$$

The magnitude of the matrix element V_{ub} has been extracted using the following relationship derived in the context of OPE [11]:

$$|V_{ub}| = 0.00445\sqrt{\frac{BR(b \to X_u\ell^-\bar{\nu}_\ell)1.55ps}{0.002\tau_b}} \times (1 \pm 0.010(pert) \pm 0.030(1/m_b^5) \pm 0.035(m_b)) \tag{7}$$

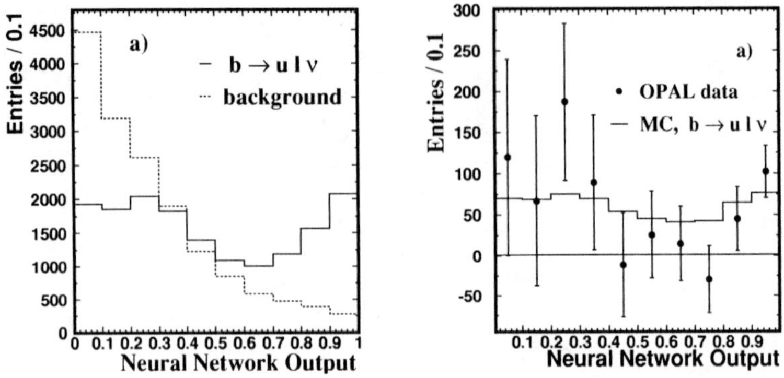

FIGURE 2. OPAL Collaboration: evidence of charmless decays as a function of the neural network output.

by assuming $m_b = (4.58 \pm 0.06) GeV/c^2$; this give

$$|V_{ub}| = (40.5 \pm 6.2(exp.) \pm 3.1(theo.)) \times 10^{-4} \tag{8}$$

This result is compatible and competitive with the determination based on the decays $B \to \rho \ell \nu$:

$$|V_{ub}| = (32.5 \pm 2.9(exp.) \pm 5.5(theo.)) \times 10^{-4} \tag{9}$$

The two determinations are also complementary since the theoretically uncertainties are totally uncorrelated.

B^0-$\bar{B}^0$ OSCILLATIONS

The oscillations between particle and anti-particle in neutral B mesons systems is a process at the second order in the electroweak perturbation theory. The frequency of this oscillation is given by the mass difference between the two weak eigenstates:

$$\Delta m_q \propto m_W^2 f_{B_q}^2 B_{B_q} |V_{tq}|^2 |V_{tb}|^2 \tag{10}$$

where $f_{B_q}^2 B_{B_q}$ is a non-perturbative QCD parameter and V_{ij} are parameters of the CKM matrix. The hierarchy of CKM parameters makes $\Delta m_s \approx 20 \Delta m_d$ and consequently the B_s^0 oscillations difficult to observe.
From the theoretical side it is important to measure precisely both Δm_d and Δm_s because in the ratio $\Delta m_d / \Delta m_s$ the non perturbative QCD part is believed to be better under control than in the absolute values.

B_d^0-$\bar{B}_d^0$ system

The time depend study of B_d^0 oscillations has been pioneered at LEP. The precision of world average of Δm_d receives, at present, similar contributions by B-factories and LEP/SLD/Tevatron:

$$\Delta m_d(LEP/SLD/CDF) = (0.492 \pm 0.013)\ ps^{-1}$$
$$\Delta m_d(B - Factories) = (0.485 \pm 0.010)\ ps^{-1}$$
$$\Delta m_d(World\ avg.) = (0.489 \pm 0.008)\ ps^{-1}$$

B_s^0-$\bar{B}_s^0$ system

The study of the B_s^0 oscillations requires, because of the expected high value of Δm_s, good proper time resolution and clean samples of B_s^0 decays.
At LEP/SLD/Tevatron in the last few years several analyses have been tried ranging from the most inclusive (several 10000 events) with low B_s purity ($\simeq$10%) to the most exclusive (few 100 events) with high B_s purity and high proper time resolution. Moreover, in most of the analyses, discrimant methods have been developed in order to separated the bulk of precisely measured candidates from the rest of the sample.
No experiment has yet directly observed B_s^0 oscillations but the sensitivity of the combined world average analysis has improved over the years.
The combination is performed, by the LEP Oscillations Working Group, in the framework of the amplitude method [12]. At each value of Δm_s an amplitude is measured in each analysis, where the expected value of the amplitude is unity at the true frequency. An overall limit on Δm_s is then inferred from the combined amplitude spectrum by excluding regions where the amplitude is incompatible with unity. The sensitivity of the analisys is defined as the limit one would get if all the amplitude values were put at 0.
The present world average give the lower limit

$$\Delta m_s > 14.8\ ps^{-1}\ @\ 95\%\ \ C.L.\quad \text{with a sensitivity of}\quad \Delta m_s = 18\ ps^{-1}$$

How to use Δm_s in CKM fits ?

The amplitude spectrum contains more information than the $95\% C.L.$ limit. In the present case, for instance, there is a "hint" of a signal at 17 ps^{-1} with 2.6σ significance. The matter is not to decide if this is a signal or not[2] but to use the information coming from data without introducing any bias.

---

[2] The 3σ or 5σ criteria are only useful conventions.

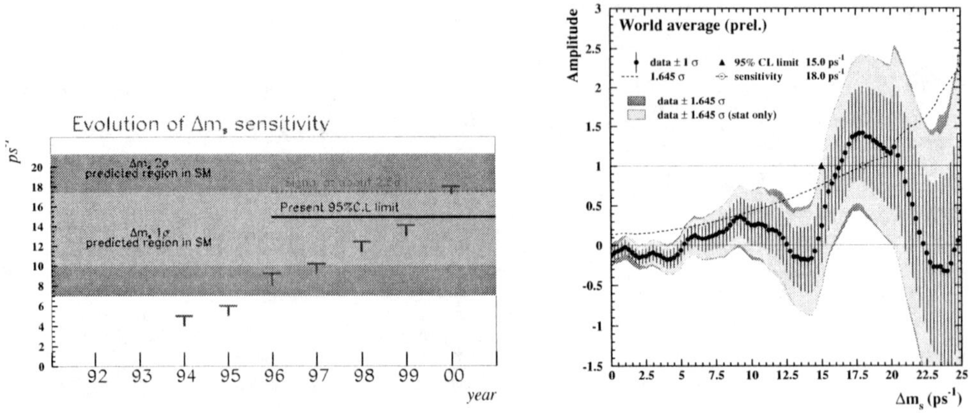

FIGURE 3. a) Evolution of the Δm_s sensitivity. b) World average amplitude spectrum

Likelihood ratio R method

Recently it has been proposed to use the log-likelihood function $\Delta \log \mathcal{L}^{\infty}(\Delta m_s)$ referenced to its value obtained for $\Delta m_s = \infty$ [13]. Similar considerations, developed in a different context, have been detailed in [14]. The log-likelihood values can be easily deduced from $\mathcal{A}$ and $\sigma_{\mathcal{A}}$ using the expressions given in [12]; the likelihood ratio R is then defined as:

$$\mathrm{R}(\Delta m_s) = e^{-\Delta \log \mathcal{L}^{\infty}(\Delta m_s)} = \frac{\mathcal{L}(\Delta m_s)}{\mathcal{L}(\Delta m_s = \infty)} \,. \tag{11}$$

The function R corresponds to the ratio of the probability density for a given Δm_s value over the probability density for $\Delta m_s = \infty$.

Modified χ^2 method

In the first CKM fits χ^2 with respect to 1 was used:

$$\chi^2 = \left(\frac{1-\mathcal{A}}{\sigma_{\mathcal{A}}}\right)^2 \tag{12}$$

This method has two main drawbacks:

- the sign of the deviation of the amplitude with respect to the value $\mathcal{A} = 1$ was not used, whereas it is expected that an evidence for a signal would manifest itself by giving an amplitude value which is simultaneously compatible with $\mathcal{A} = 1$ and incompatible with $\mathcal{A} = 0$;
- values of $\mathcal{A} > 1$ are disfavoured w.r.t. $\mathcal{A} = 0$, while it is expected that, because of statistical fluctuations, the amplitude value corresponding to the "true" Δm_s value

could be higher than 1. This problem was solved in the early days of the use of Δm_s in CKM fits putting $\mathcal{A} = 1$ whenever higher.

In [15] the χ^2 has been modified ad hoc to solve the second problem:

$$\chi^2 = 2 \cdot \left[\text{Erfc}^{-1}\left(\frac{1}{2}\text{Erfc}\left(\frac{1-A}{\sqrt{2}\sigma_A} \right) \right) \right]^2 \tag{13}$$

Comparison of the two methods

The amplitude as a function of Δm_s and the corresponding $\Delta \log \mathcal{L}(\Delta m_s)$ are shown in Figure 4-a),b). The corresponding Likelihoods obtained using the Likelihood ratio and the Modified χ^2 methods are shown in Figure 4-c). It is clear that the two methods give

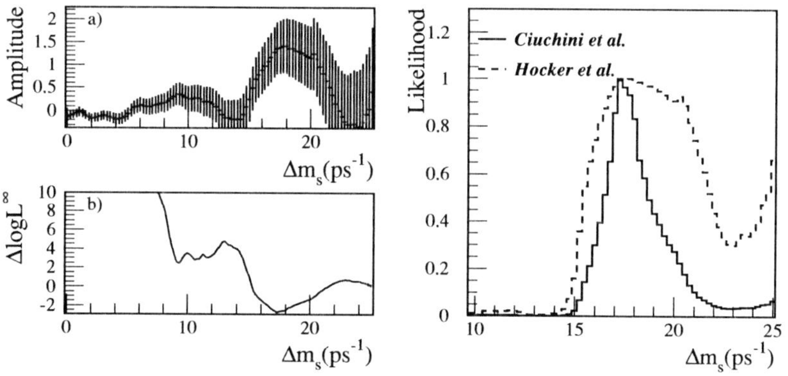

FIGURE 4. World average amplitude analysis: a) amplitude spectrum, b) $\Delta \log \mathcal{L}^{\infty}(\Delta m_s)$, c) comparison between R $\equiv$ Likelihood ratio method and the Modified χ^2 method.

very different Likelihood functions. In particular the Modified χ^2 method give a less tight constraint.

However no conclusion can be deduced from a single experiment.

The minimal requirement for a "good" method is to give the correct probability density function in case of an oscillation signal. To test this case a toy Montecarlo has been generated with $\Delta m_s = 17\ ps^{-1}$ with an average significance at that value of 6σ. The results are shown in Figure 5. It is clear that the Likelihood ratio method is able to see the signal at the correct Δm_s value, whereas the Modified χ^2 method failed. The same exercise was repeated at different generated values of Δm_s giving similar results.

Only the Likelihood ratio method will be then used in the following.

UNITARITY TRIANGLE FIT

The unitarity triangle fit presented here is based on the Bayesian approach described in [16]. The constraint are $|V_{ub}|/|V_{cb}|$, ε_K, Δm_d and Δm_s (used as described in the previous

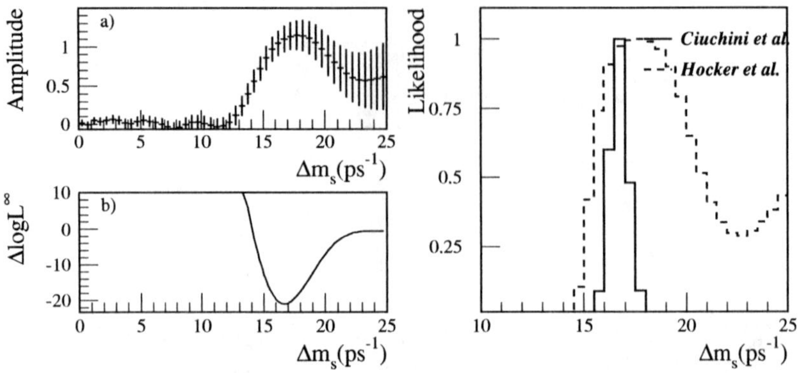

FIGURE 5. Toy-MC analyses with the same $\sigma(\mathcal{A})$ versus Δm_s behaviour as the world average analysis : a) amplitude spectrum, b) $\Delta \log \mathcal{L}^{\infty}(\Delta m_s)$, c) comparison between R $\equiv$ Likelihood ratio method and the Modified χ^2 method.

section) and the list of parameter of [16] has been updated with the most recent values, presented in this review, of $|V_{ub}|$, $|V_{cb}|$ and Δm_s.
The allowed region in the ρ-η plane in the Standard Model framework is shown in Figure 6. The fitted values of $\bar{\rho}$, of $\bar{\eta}$ and of the angles of unitarity triangle are accurately

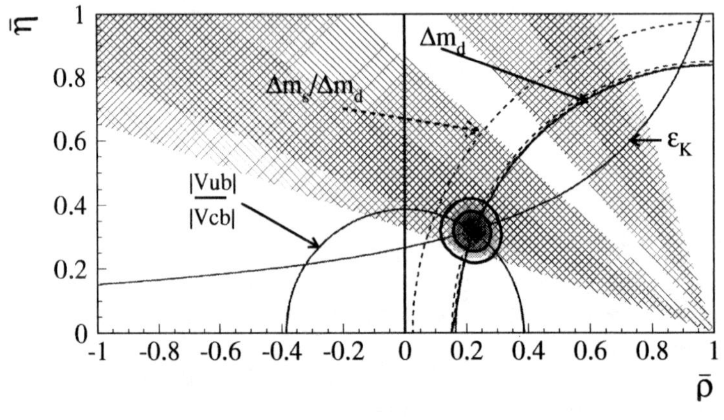

FIGURE 6. Allowed region in the $\bar{\rho}$-$\bar{\eta}$ plane selected by the $|V_{ub}|/|V_{cb}|$, ε_K, Δm_d and Δm_s constraints. The $sin2\beta$ constraint, not used in the fit, is superimposed.

determined:

$$\bar{\rho} = 0.218 \pm 0.038$$
$$\bar{\eta} = 0.316 \pm 0.040$$
$$sin2\beta = 0.696 \pm 0.067$$
$$sin2\alpha = -0.42 \pm 0.23$$
$$\gamma = (55.5 \pm 6.2)^o$$

The indirect determination of $sin2\beta$ is compatible with the world average of $sin2\beta$ (dominated by BaBar and Belle measurements): $sin2\beta = 0.79 \pm 0.10$.
The result of the fit show that the measurements performed at LEP/SLD/Tevatron constrained efficiently the sides of the unitarity triangle allowing the determine, in indirect way, $sin2\beta$ and γ at 10%. The comparison between the sides and the angles of unitarity triangle will become more stringent as soon as the $\sin 2\beta$ measurements will improve and new experimental constraints (other angles, constraints from Kaon physics, ...) will be added.

ACKNOWLEDGMENTS

The results presented in the section "Unitarity triangle fit" are based on work done in collaboration with M. Ciuchini, G. D' Agostini, E. Franco, V. Lubicz, G. Martinelli, P. Roudeau, L. Silvestrini and A. Stocchi.

REFERENCES

1. ALEPH, CDF, DELPHI, L3, OPAL, SLD Collaborations, *CERN-EP*, **2001-050**, 1 (2001).
2. G. Bellini, I.I. Bigi and P.J. Dornan, *Phys. Rep.*, **289**, 1 (1997).
3. LEP Lifetime Working Group, Averages for Summer Conferences 2001 (2001), URL `http://claires.home.cern.ch/claires/lepblife.html`, and references therein.
4. A. J. Buras, W. Slominski and H. Steger, *Nucl. Phys.*, **B284**, 369 (1984).
5. M. Beneke *et al.*, *Phys. Lett.*, **B459**, 631 (1999).
6. D. Becirevic, D. Meloni, A. Retico, V. Gimenez, V. Lubicz and G. Martinelli, *Phys. J.*, **C18**, 157 (2000).
7. M. Neubert and C.T. Sachrajda, *Nucl. Phys.*, **B483**, 339 (1997).
8. I.I. Bigi, M. Shifman and N. Uraltsev, *Annu. Rev. Nucl. Part. Sci.*, **47**, 591 (1997).
9. The OPAL Collaboration, G. Abbiendi *et al.*, *Eur. Phys. J.*, **C21**, 399 (2001).
10. The LEP V_{ub} Working Group, *LEPVUB*, **01/01** (2001).
11. N. Uraltsev *et al.*, *Eur. Phys. J.*, **C4**, 453 (1998).
12. G. D'Agostini and G. Degrassi, *Eur. Phys. J.*, **C10**, 633 (1999).
13. P. Checchia, E. Piotto, F. Simonetto, *hep-ph*, **9907300**, 1 (1999).
14. G. D'Agostini, *hep-ph*, **0002055**, 1 (2000).
15. A. Hocker et al., *Eur. Phys. J.*, **C21/2**, 225 (2001).
16. M. Ciuchini *et al.*, *JHEP*, **0107**, 013 (2001).

SLD Results on B Physics

H. Neal representing the SLD collaboration

Yale University, P.O. Box 208120, New Haven, CT 06520

Abstract. Results on B-lifetimes, B_d^o and B_s^o mixing, and B decay charm counting studies from the SLD experiment at the Stanford Linear Accelerator Center are presented. These results which exploit the small stable beam spots and high precision detector and the polarized electron beam are among the most precise measurements to date.

INTRODUCTION

The SLD experiment collected a sample of about 550K hadronic Z^0 decays from data taken between 1993 and 1998. These events were the results of e^+e^- collisions at the Stanford Linear Collider(SLC) using an average electron beam polarization of 73%. The small and stable beam spot at the interaction point allowed the SLD detector to make full use of its high precision CCD vertex detector to reconstruct secondary and tertiary vertices in heavy flavor decays. The polarization is used as a powerful tool in discriminating between initial state B's and $\bar{B}$'s. The combination of the assets of the accelerator and detector resulted in some of the most precise heavy flavor measurements. Several of the important contributions of the SLD experiment to heavy flavor physics are discussed here. These include B lifetimes, B_d^o and B_s^o mixing, and B decay charm counting studies.

DETECTOR

For the final 400K Z^o decays (from the 1996→1998 runs), the tracking systems, consisting of a 3 layer 300 million pixel CCD Detector (VXD3)[1], and a 80 layer 5120 wire Central Drift Chamber (CDC)[2] yield impact parameter resolutions of 8 μm ($r\phi$ projection) and 10 μm (rz projection) for high momentum particles and multiple scattering contributions of 33 $\mu\text{m}/(psin^{3/2}\theta)$ in both projections where p is in GeV/c. The earlier 150K Z^o sample used a 120 million pixel CCD detector (VXD2) which when combined with the same CDC yielded impact parameter resolutions of 11 μm ($r\phi$ projection) and 38 μm (rz projection) for high momentum particles and multiple scattering contributions of 70 $\mu\text{m}/(psin^{3/2}\theta)$ in both projections. The excellent resolutions in both projections allow efficient 3-D vertexing. The micron-sized SLC Interaction Point (IP) contributes only $4 \pm 2\mu$m to the overall uncertainty in the $r\phi$ plane. A Cherenkov Ring Imaging Detector (CRID) is used for K/π separation. Electron identification and event shape measurements use a Liquid Argon Calorimeter (LAC) with an energy resolution

CP618, *Heavy Flavor Physics: Ninth International Symposium*, edited by A. Ryd and F. C. Porter

for electromagnetic showers of $\sigma/E = 15\%/\sqrt{E(GeV)}$. Muon identification is provided by a Warm Iron Calorimeter (WIC).

B-LIFETIME MEASUREMENTS

The comparison of lifetimes of charged and neutral B-mesons is important to verify predictions of Heavy Quark Expansion, which predicts that the lifetime of different b hadrons differ by no more than 10%. The analysis uses an inclusive topological vertexing technique to exploit the 3-D vertexing capabilities of the SLD. Secondary vertices are found in 65% of b hemispheres but in only 20% of c hemispheres and in less than 1% of uds hemispheres. After the vertices have been identified, a cut on the mass of the vertex corrected for the transverse component of the total momentum of tracks relative to the vertex axis is used to eliminate charm and light flavor contamination. Requiring $M > 2$ GeV/c^2 yields a b-hadron sample with 98% b purity and 50% efficiency (for normalized decay length $> 5\sigma$). The lifetime is determined from the decay length of the reconstructed vertex and the charge is determined from the total charge of tracks associated with the vertex. The B-lifetimes are obtained from the decay length distributions using a binned χ^2 fit to reweighted Monte Carlo distributions. The SLD results for the B^0 and the B^+ lifetimes and their ratio are:

$$\tau_{B^0} = 1.585 \pm 0.048 \text{ ps}, \; \tau_{B^+} = 1.623 \pm 0.039 \text{ ps}$$
$$\tau_{B^+}/\tau_{B^0} = 1.037 \pm 0.035.$$

Only the new result from BaBar [3] surpasses the precision of the SLD result.

B-MIXING

B mixing proceeds via second order weak interactions of the type shown in Figure 1. The oscillation rate between the B^0 and $\bar{B}^0$ depends on the mass difference Δm_d in the B^0_d system or Δm_s in the B^0_s system. Measurements of the $B^0 - \bar{B}^0$ mixing rate allow one to constrain the location of the apex of the unitarity triangle (a.k.a. "Bjorken triangle" - shown in Figure 1), given by (ρ,η), which determines CP violation in the Standard Model. A measurement of Δm_d constrains $|V_{td}|^2 \propto (1-\rho)^2 + \eta^2$ and Δm_s constrains $|V_{ts}|^2$ but the ratio $\Delta m_s/\Delta m_d$ is more powerful because the sources of uncertainty from hadronic matrix elements cancel out. The ratio yields $\Delta m_s/\Delta m_d = (1.15 \pm .05)^2 \, |V_{ts}/V_{td}|^2$ which if one uses the near equality $|V_{ts}| = |V_{cb}|$ allows a measurement of the ratio $|V_{td}|/|V_{cb}|$, which is the least known side of the triangle.

To extract B-mixing from the data, the production and decay flavor (B or $\bar{B}$) are tagged and the proper decay time is determined from the decay length and boost.

The initial state tag uses all of the strengths of the detector and accelerator. It combines a tag exploiting the large forward-backward asymmetry of polarized $Z^0 \to b\bar{b}$ decays with a neural net applied to the hemisphere opposite that being analyzed. The neural net uses jet charge, vertex charge, the lepton charge from $(b \to \ell^-)$ decays, the kaon charge in $(b \to c \to s)$ decay and the dipole charge between the D and B vertices. An initial state tag is possible on every event thanks to the polarization. The initial state b-quark probability is shown in Figure 2. The average mistag rate is 22-25%, for the different analyses.

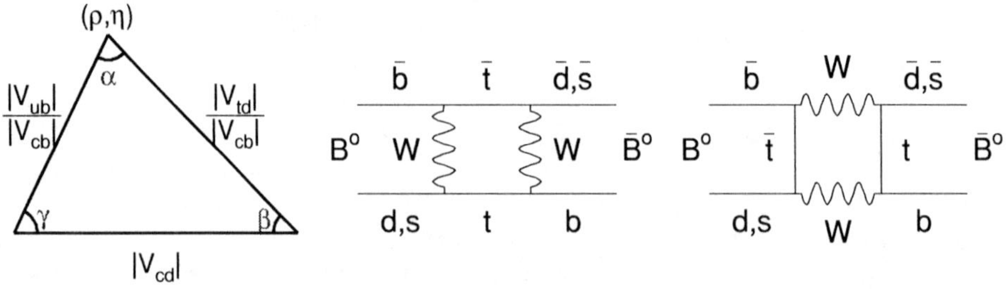

FIGURE 1. (left) The Bjorken triangle. (middle and right) Diagrams of $B^0 - \bar{B}^0$ mixing.

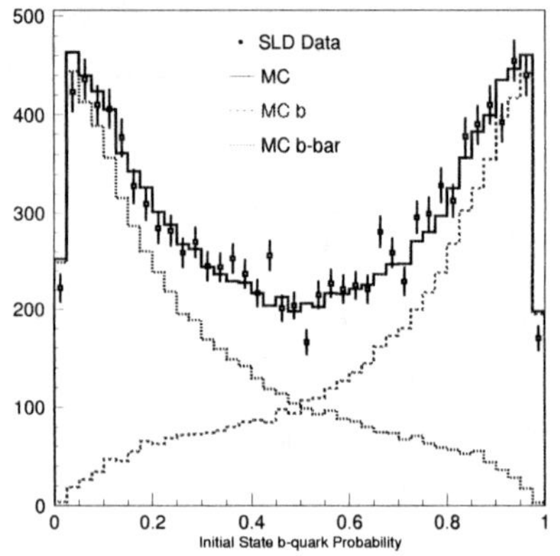

FIGURE 2. Distribution of the computed initial state b-quark probability for data (points) and Monte Carlo (histograms) showing the b and $\bar{b}$ components for the events selected in the Charge Dipole analysis.

The significance (signal to noise) of the mixing signal is expressed by the equation: $S = \sqrt{\frac{N}{2} f_{B_s}(1-2w)e^{-\frac{1}{2}(\Delta m_s \sigma_t)^2}}$ where N is the number of events, w is the mistag rate, f_{B_s} is the purity and ${\sigma_t}^2 = (\frac{\sigma_L}{\gamma\beta c})^2 + (\frac{\sigma_P}{p}t)^2$. This shows the importance of having excellent decay length resolution (σ_L). The decay length resolution is very important for measurements of Δm_s where the oscillation rate is expected to be about a factor of 20 ($\approx \Delta m_s/\Delta m_d$) higher than in the B_d^o system. Figure 3 shows a comparison of the fraction of decays observed to have mixed versus time is shown for $\Delta m_s = 20\text{ps}^{-1}$ for the typical decay length resolution obtained from LEP analyses (200 μm) and SLD analyses (60 μm).

B_d *Mixing*

The analysis of ${B_d}^0 - \bar{B_d}^0$ mixing uses kaons to identify the final state in the decays $B_d \to \bar{D^-}/\bar{D}^0 \to K^+$ and $\bar{B}_d \to D^+/D^0 \to K^-$. Here, the sign of the kaon differentiates

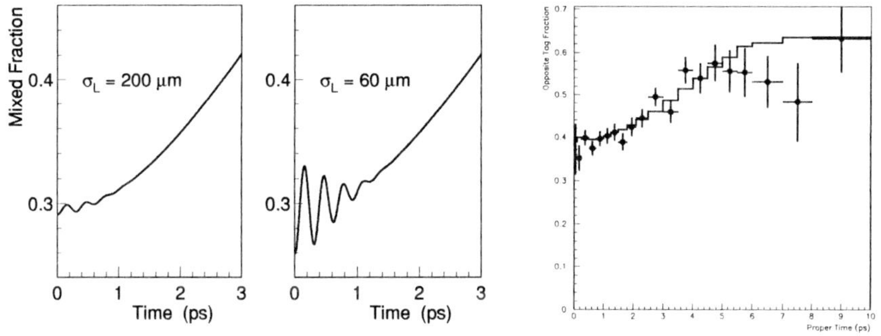

FIGURE 3. (left) The fraction of b-hadron decays tagged as mixed versus proper time for decay lengths resolutions of 200 μm and 60 μm for Δm_s=20 ps^{-1}, f_{B_s}=18%, w=0.25 and σ_p/p=10%. (right) The B_d mixed fraction plot versus proper time showing the data as points and the curve of the likelihood fit overlayed.

between the B_d and $\bar{B}_d$ final states. The value of Δm_d is extracted by performing a likelihood fit to the mixed fraction distribution shown in Figure 3. The preliminary result for the ${B_d}^0 - \bar{B_d}^0$ mass difference is

$$\Delta m_d = 0.503 \pm 0.028\ (\text{stat.}) \pm 0.020\ (\text{syst.})\ ps^{-1}.$$

The largest contribution to the systematic uncertainty is the B_u right sign fraction which contributes 0.012 to the uncertainty on Δm_d. Each of the other sources contribute less than 0.007.

B_s *Mixing*

The B_s mixing study combines three analyses exploiting different decay topologies and final states. The three analyses ("D_s+tracks", "lepton+D" and "Charge Dipole") make use of the inclusive topological vertexing technique [4] developed for B lifetime [5] and R_b [6] analyses to tag and reconstruct b-hadron decays. This inclusive vertexing technique has been adapted for semileptonic decays to reconstruct the D decay topology.

The "D_s+tracks" analysis [7] does a full reconstruction of the D_s in the decays $D_s \to \phi\pi^-, K^{*o}K^-$ using the CRID to discriminate between π's and K's. A total of 280 $D_s \to \phi\pi^-$ candidates and 81 $D_s \to K^{*o}K^-$ candidates pass the selection. As shown in Table 1, this analysis has excellent decay length resolution, a high B_s fraction (compared to the B_s^0 production fraction in the $Z^0 \to b\bar{b}$ which is 10.0%), and a clean final state tag.

In the lepton+"D" analysis, semileptonic decays are selected and the B decay point is reconstructed by intersecting a lepton track with the trajectory of a topologically reconstructed D meson. A neural-network is used to clean up the D vertex candidates and reduce the contamination from cascade ($b \to c \to l$) charm semileptonic decays. Only multiprong D decays were used at this time. The final state B^0 or $\overline{B^0}$ flavor is tagged by

TABLE 1. Decay length resolution, momentum resolution, and purities of the three B_s analyses.

Method	D_s+tracks	lepton+D	Vertex Charge Dipole
σ_L core (60%)	48 μm	54 μm	81 μm
σ_L tail (40%)	152 μm	213 μm	297 μm
σ_p/p core (60%)	0.08	0.07	0.07
σ_p/p tail (40%)	0.19	0.17	0.21
B_s fraction	38%	16%	16%
final state mistag	10%	4%	22%

the sign of the lepton charge. To enhance the fraction of B_s^0 decays, the sum of lepton + D vertex track charges is required to be $Q = 0$. A sample of 2087 decays is obtained in the 1997-98 data. An opposite sign lepton-kaon tag is used to enhance the B_s^0 fraction. The overall performance is shown in Table 1. The analysis also has good decay length resolution and exceptionally good final state tagging.

The polarization-dependent forward-backward asymmetry is shown in Fig. 4. A clear asymmetry is observed, in reasonable agreement with the Monte Carlo, indicating that the final state tag purity is adequately modeled in the simulation.

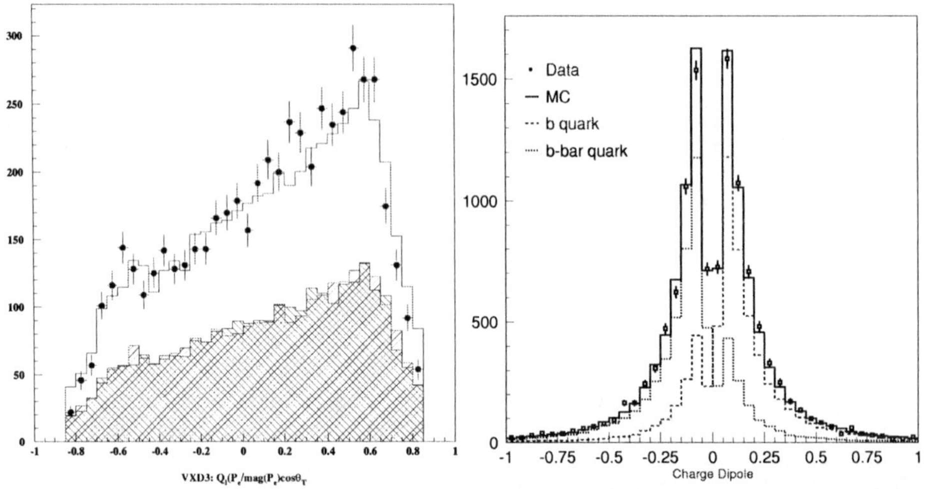

FIGURE 4. (left) Distribution of $\cos\theta$ for the thrust axis direction signed by the product $(Q_{lept} \times P_e)$ for data (points) and Monte Carlo (histograms) from the lepton+D analysis. (right) Distribution of the charge dipole for data (points) and Monte Carlo (solid histogram). Also shown are the contributions from b hadrons containing a b quark (dotted histogram) or a $\bar{b}$ quark (dashed histogram).

The Charge Dipole analysis reconstructs the B and D vertex topologies in inclusive decays and tags the B^0 or $\overline{B^0}$ decay flavor based on the charge difference between the B and D vertices. This analysis technique is unique to SLD and relies extensively on the excellent resolution of the vertex detector. A "Charge Dipole" is defined as $\delta Q \equiv D_{BD} \times SIGN(Q_D - Q_B)$, where D_{BD} is the distance between the two vertices and Q_B (Q_D) is the charge of the B (D) vertex. Positive (negative) values of δQ tag $\overline{B^0}$ (B^0)

decays. The total track charge Q (from both secondary and tertiary vertices) is required to be 0 to enhance the fraction of B_s^0 decays in the sample and to increase the quality of the charge difference reconstruction for neutral B decays. The distribution of δQ for the data and Monte Carlo is shown in Figure 4. The performance of the method is shown in Table 1. While the resolutions and purities are comparable to the other two analyses, this method greatly benefits from a an event sample that is more than three times that of the lepton+D analysis and twenty times that of the D_s+tracks analysis.

The D_s+tracks, lepton+D, and Charge Dipole analyses are combined taking into account correlated systematic errors. Events shared by two or more analyses are assigned to the analysis with the best sensitivity such as to produce statistically independent analyses. Figure 5 shows the measured amplitude as a function of Δm_s for the combination. As noted earlier, the measured values are consistent with $A = 0$ for the whole range of Δm_s up to 25 ps^{-1} and no evidence is found for a preferred value of the mixing frequency. Using the condition $A + 1.645\,\sigma_A < 1$, the range $\Delta m_s < 13.2$ ps^{-1} is exclude at 95% C.L.. The combined sensitivity to set a 95% C.L. lower limit is found to be at a Δm_s value of 13.2 ps^{-1}. These results are preliminary. The overall sensitivity is expected to continue improving as further analysis refinements are implemented.

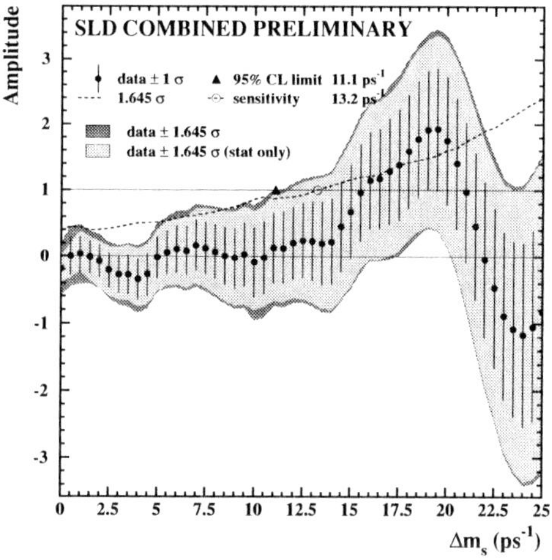

FIGURE 5. Measured amplitude as a function of Δm_s for the D_s+tracks, lepton+D, and Charge Dipole analyses combined.

Charm counting in B-decays

The ability of the SLD detector to identify tertiary decays from charm in B-decays has been exploited [8] to help answer why the world averaged b-semileptonic branching ratio is lower than the prediction and why the current value of the average number of charm in B-decays (N_c) is low. An increased width from non-semileptonic decay modes could resolve the problem. The analysis measures the rates for having zero, one or two open charm mesons by fitting the nearest neighbor vertex seperations for the number of secondary vertices (N_{vtx}) being $N_{vtx} = 0, 1, 2$ and ≥ 3. In the case of two charm vertices the average distances with respect to the B vertex are 0.5 mm and 1.0 mm.

Event hemispheres opposite B-tagged hemispheres are used. The same topological vertexing algroithm as for the B-lifetime and B-mixing analyses is used to reconstruct vertices in the sample hemispheres. Samples containing zero, one or two tertiary reconstructed vertices are simultaneously fitted to linear combinations of Monte Carlo shapes for zero, one and two open charm mesons to determine the rates. The number of found secondary vertices, the vertex seperations for two and three vertices are shown in Figure 6 for the data and Monte Carlo. The prelimnary results of the fit are:

$$BR(b \to 0D) = (5.6 \pm 1.1 \text{ (stat.) } \pm 2.0)\% \text{ (syst.)}$$
$$BR(b \to 2D) = (24.6 \pm 1.4 \text{ (stat.) } \pm 4.0)\% \text{ (syst.)}$$
$$N_c = 1.238 \pm 0.027 \text{ (stat.) } \pm 0.048 \text{ (syst.) } \pm 0.006 \text{ (charmonia)}$$

The value of N_c pulls the world average towards the region preferred by the Standard Model.

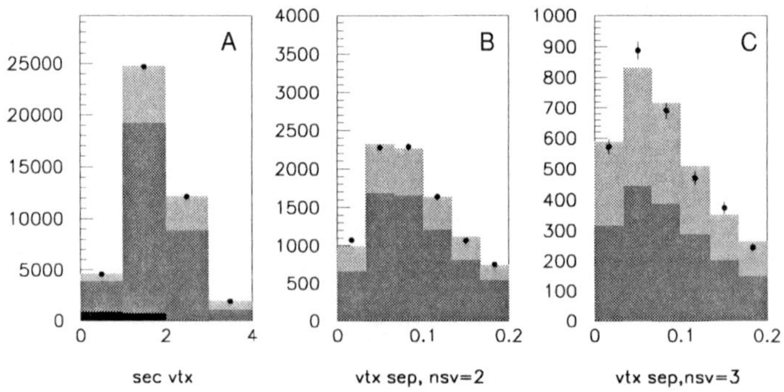

FIGURE 6. The number of found secondary vertices is shown in plot A. The measured $b - D$ vertex seperation for two and three vertices is shown in plots B and C. The contributions to each histogram from bottom to top are *udsc*, zero, one and two open charm decays.

REFERENCES

1. K. Abe *et al.*, *Nucl. Inst. and Meth.* **A400**, 287 (1997).
2. K. Abe *et al.*, *Phys. Rev.* **D53**, 1023 (1996).
3. B. Aubert *et al.* [BABAR Collaboration], *Phys. Rev. Lett.* **87**, 201803 (2001).
4. D. J. Jackson, *Nucl. Inst. and Meth.* **A388**, 247 (1997).
5. K. Abe *et al.* [SLD Collaboration], arXiv:hep-ex/0012043.
6. K. Abe *et al.*, *Phys. Rev. Lett.* **80**, 660 (1998).
7. K. Abe *et al.* [SLD Collaboration], arXiv:hep-ex/0011041.
8. A. Chou *et al.*, *b Decay Charm Counting Via Topological Vertexing*, *SLAC-PUB-8686*, January 2001.

CDF B Physics: Run I Results and Run II Status

Greg Feild representing the CDF collaboration

Physics Department, Yale University, New Haven CT 06520

Abstract. This paper reviews the CDF b physics program from Tevatron Run I and presents prospects and early results from Run II.

INTRODUCTION

In Run I (1992-1996), the Fermilab Tevatron provided $p\bar{p}$ collisions at $\sqrt{s}$ = 1.8 TeV with a bunch crossing rate of 3.5 μs and an average instantaneous luminosity of 10^{30} $\text{cm}^{-2}\text{s}^{-1}$. During this time 110 pb^{-1} of data were collected by the CDF detector. This data sample enabled us to make many interesting and unique b physics measurements although some analyses were limited or prohibited by low statistics and/or triggering constraints. During Run I the triggers employed to collect b physics data samples were high p_T leptons (e, μ) and low p_T dimuons (e.g., $J/\psi \to \mu^+\mu^-$).

In Run II, the Tevatron will provide collisions at $\sqrt{s}$ = 2 TeV every 132 ns (the current rate is every 396 ns) and is expected to achieve an instantaneous luminosity of 10^{32} $\text{cm}^{-2}\text{s}^{-1}$ delivering an integrated luminosity of 2 fb^{-1} over the next several years. The CDF detector has been upgraded extensively to function in this new environment and provide greater acceptance and precision for b physics measurements. The trigger system has been completely rebuilt to function efficiently at the higher crossing rate and to include the capability of triggering on purely hadronic decay modes (e.g., $B^0 \to \pi\pi$, $B_s \to D_s\pi$) [1].

CDF RUN I

Hadronic collisions at the Tevatron allow for the study of strong and weak interactions in the production and decay of b quarks and a broad spectrum of b hadrons as well as the study of mixing and CP violation in the B^0 and B_s systems. Studies involving the heavier hadrons (e.g., B_s, B_c, Λ_b masses and lifetimes) are best suited to or unique to the Tevatron due to the higher energies and large production cross sections.

Measurements from CDF Run I include [2]:

- Individual b hadron masses (B^+, B^0, B_s, B_c, Λ_b)
- Individual b hadron lifetimes (B^+, B^0, B_s, B_c, Λ_b)
- Polarization in $B^0 \to J/\psi K^{*0}$ and $B_s \to J/\psi\phi$
- b quark and B meson cross sections and $b\bar{b}$ production correlations

CP618, *Heavy Flavor Physics: Ninth International Symposium*, edited by A. Ryd and F. C. Porter

- Quarkonium studies including J/ψ and $\psi(2S)$ cross sections and polarization, $\Upsilon(1S)$ cross section and polarization and $\Upsilon(2S)$, $\Upsilon(3S)$ cross sections
- Search for rare decays ($B^0, B_s \rightarrow \mu^+\mu^-$; $B^\pm \rightarrow \mu\mu K^{*0}$; $B^0, B_s \rightarrow \mu e$)

CDF measurements of b quark, b meson and quarkonium production cross sections are consistently larger than QCD predictions and this discrepancy between experiment and theory has yet to be satisfactorily or uniquely resolved.

In addition to the above studies, CDF has made several independent measurements of the mass difference between the light and heavy B^0 mesons Δm_d, observed CP violation (measurement of sin2β) in the $B^0/\bar{B}^0$ system and set limits on the the B_s mixing parameter x_s $(= \Delta m_s/\Gamma_s)$.

Observation of mixing or flavor oscillations in the $B^0/\bar{B}^0$ system requires a determination of the flavor of the B^0 meson both at the time of production and decay. Identification of the B mesons in these studies is done by searching for semileptonic decays ($B \rightarrow e$, μX). The position of the secondary vertex gives the time of decay and the charge of the lepton gives the flavor at the time of decay. The flavor at the time of production has been determined by three separate flavor tagging methods:

- Soft Lepton tagging (SLT) uses the sign of the opposite side muon or electron to determine the flavor of the B^0
- Jet charge tagging (JETQ) uses the average charge of the jet found in the opposite hemisphere to the B^0
- Same side tagging (SST) uses charge correlations between the B^0 and nearby hadrons to tag the flavor of the B^0

Each method has an effective tagging efficiency εD^2 of approximately 2 % although they have different efficiencies (ε) and dilutions (D). The values of the mixing parameter Δm_d from the various CDF measurements are displayed in Figure 1 (left). Also shown in Figure 1 (right) are the time dependent mixing results from the SST measurement.

The observation of CP violation and the determination of sin2β entails measuring the mixing asymmetry of $B^0/\bar{B}^0$ decays to a specific CP eigenstate, in this case the decay mode $B^0 \rightarrow J/\psi K_s$. The mixing asymmetry is given by $A_{cp}(t) = \sin 2\beta \sin(\Delta m_d t)$. The flavor of the B^0 at the time of production is determined by combining all three tagging methods described above for a combined εD^2 of $6.3 \pm 1.7\%$ The signal sample for this analysis is shown in Figure 2 (left). The extraction of sin2β is carried out using a unbinned maximum likelihood fit to the normalized mass, the decay time and the tag pattern. Parameters in the fit include the B lifetime τ_B, the mixing frequency Δm_d, scale factors for mass and decay time errors, efficiencies, and dilutions. The result yields $\sin 2\beta = 0.79^{+0.41}_{-0.44}$ and is displayed in Figure 2 (right). The statistical error is 0.39 and the systematic error, dominated by the error in dilution, is 0.16.

Studies of the B_s meson in Run I have yielded values for the mass as well as limits on the lifetime difference between the light and heavy eigenstates and the mixing parameter x_s. A fit to the pseudo-$c\tau$ distributions in the decays $B_s \rightarrow D_s l\nu$ returns a lifetime of $\tau(B_s) = 1.36 \pm 0.09^{+0.06}_{-0.05}$ ps. A fit assuming contributions to the lifetime from light and heavy states returns a limit on the lifetime difference between these states of $\Delta\Gamma_s/\Gamma_s < 0.83$ at 95% CL. A search for B_s mixing was performed using the decay mode

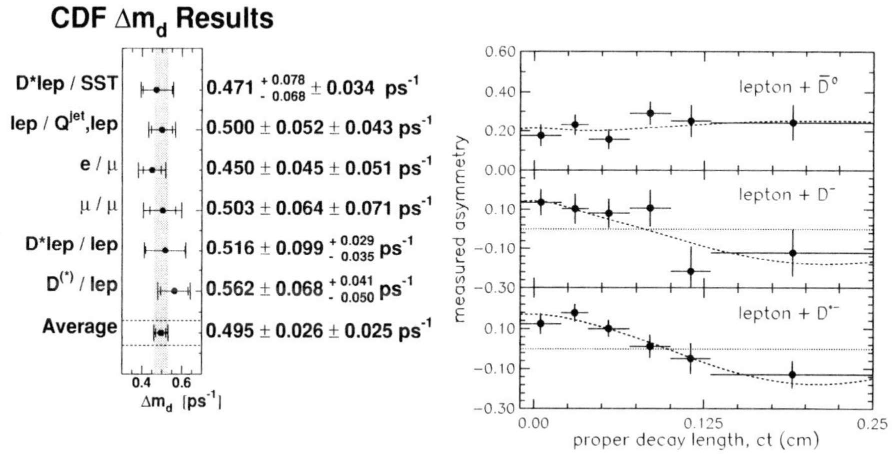

FIGURE 1. Left: A compilation of Δm_d values from the various CDF measurements. Right: Mixing asymmetry vs. proper time in the Same Side Tagging measurement. Lepton $+\bar{D}^0$ (top) is a signature for B^+ and the asymmetry is seen to be flat with time. Lepton $+D^-$ (middle) and Lepton $+D^{*-}$ (bottom) are signatures of B^0. The asymmetry is cosine like and the amplitude gives the dilution of the tag.

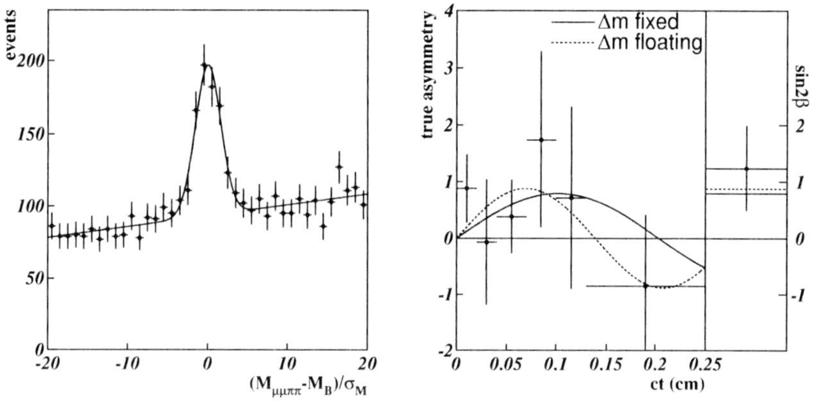

FIGURE 2. Left: The normalized mass for the $B^0 \to J/\psi K_s$ signal (395 ± 31 events) used in the sin2β measurement. Right: The fit to $A_{cp}(t)$. The data points represent the time dependent asymmetry for the SVX sample and the one inclusive bin (on the right) is the non SVX sample. The solid line is a projection of the fit with Δm_d constrained to the world average. The dotted line represents the fit result with Δm_d floating.

$B_s \to \nu l \phi X$; $\phi \to KK$. Flavor tagging was performed as in the B^0 mixing analysis. The amplitude fit method was employed where the data is fit assuming various values for Δm_s. When the correct value of Δm_s is inserted, the resulting amplitude is 1. The results of such a scan return a limit of $\Delta m_s > 5.8\text{ps}^{-1}$ and are displayed in Figure 3.

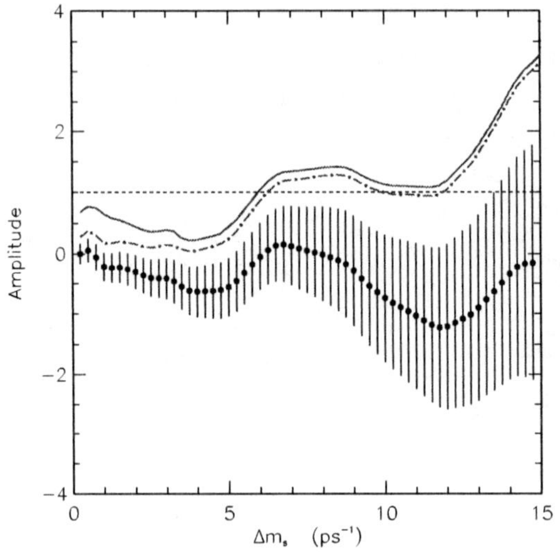

FIGURE 3. The amplitude fit result $A(\Delta m_s)$ vs. Δm_s. The points represent the fitted amplitudes and their errors. The dot-dashed line corresponds to $A + 1.645\sigma_A$, the statistical uncertainty while the solid line represents the inclusion of the systematic uncertainties.

CDF RUN II

For the past seven years the CDF collaboration has been preparing for Tevatron Run II replacing the three level trigger system and much of the detector as well as refining measurement techniques to collect and precisely analyze as much data as possible.

Detector upgrades

For Run II we have replaced the two-barrel (51 cm long), 4 layer, $r - \phi$ only silicon vertex detector (SVX) with the 3 barrel (96 cm long), eight layer, 3D SVXII. The new SVX detector covers an angular region of $|\eta| < 2$ and will provide precision z and $c\tau$ information.

A completely new detector for Run II is the Time of Flight (TOF) located between the SVXII and the new Central Outer Tracker (COT). The TOF has a timing resolution of

100 ps and will provided $K\pi$ separation of 2σ for particles of $p_T < 1.6$ GeV/c. Particle identification at higher p_T will be achieved using dE/dx information from the COT.

Figure 4 shows J/ψ and $B^{\pm} \to J/\psi K^{\pm}$ signals reconstructed from COT data taken during the recent Run II commissioning run.

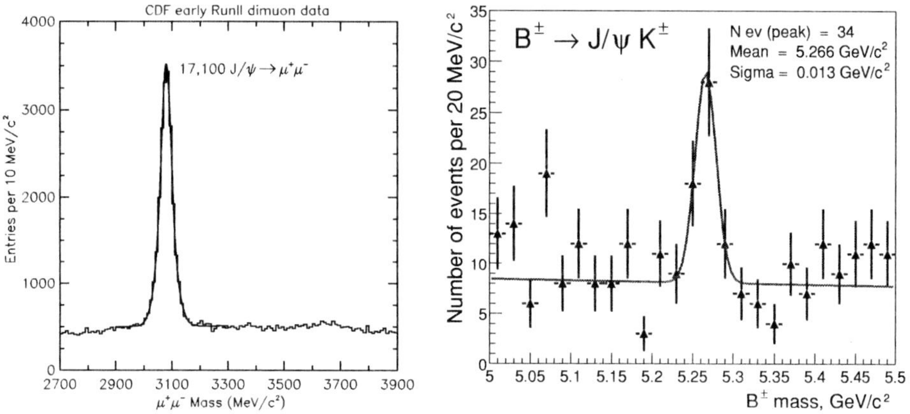

FIGURE 4. J/ψ and $B^{\pm} \to J/\psi K^{\pm}$ signals reconstructed from Run II commissioning data.

Trigger upgrade

The CDF trigger for Run II is synchronous, pipelined and "deadtimeless". The Level 1 trigger processes every event reducing the rate to 40 kHz. Data is then stored in 1 of 4 buffers during Level 2 processing which reduces the rate to 300 Hz. Finally, Level 3 is able to write 50 Hz to tape.

A new feature in Run II is the addition of track information from the COT at Level 1 and silicon information from the SVX at Level 2. At Level 1 the Extremely Fast Tracker (XFT) reconstructs tracks of $p_T > 1.5$ GeV/c with a resolution of $\delta p_T / p_T^2 = 1.8\%$. The trigger efficiency of the XFT from early Run II commissioning data is shown in Figure 5. The addition of track matching to muon stubs at Level 1 allows us to reduce the p_T threshold to 1.5 GeV/c, increasing the acceptance of the muon trigger by approximately 50%. The XFT also provide two-track triggers which feed the Silicon Vertex Trigger (SVT) at Level 2 where cuts are made on impact parameter to trigger on hadronic decay modes such as $B \to \pi\pi$. The SVT has an impact parameter resolution of $\delta d = 35\mu$m as measured from recent commissioning data.

Physics goals and projections

In addition to pursuing and refining many of the measurements made in Run I, the major goals of the CDF b physics program in Run II are:

- The measurement of $\sin 2\beta$ in $B^0 \to J/\psi K_s$ decays with an error of ± 0.09.

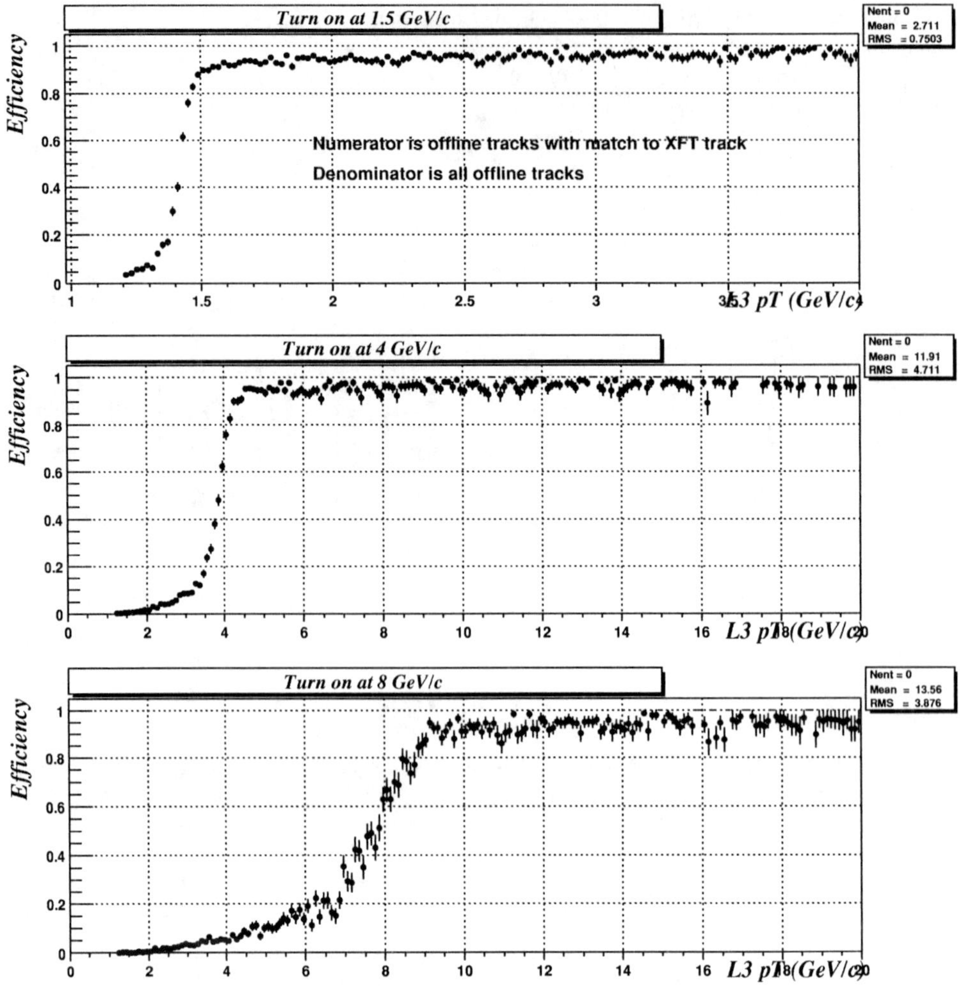

FIGURE 5. XFT trigger efficiency versus momentum for three different momentum thresholds: 1.5 GeV/c, 4.0 GeV/c, and 8.0 GeV/c.

- Observation of CP violation in $B^0 \to \pi^+\pi^-$ decays and the measurement of sin2α to ± 0.15.
- Observation of B_s mixing and a determination of the ratio of the CKM matrix elements $|V_{td}/V_{ts}|$ with a precision of 20%
- Search for CP violation in $B_s \to J/\psi\phi$ decays
- Observation of the rare decays $B^0 \to \mu\mu K^{*0}$ and $B^\pm \to \mu\mu K^\pm$

The measurement of sin2β will be performed much as in the Run I analysis. Accounting for the increased luminosity, increased SVX acceptance and lower muon trigger thresholds we expect to collect $\sim 10,000\, B^0 \rightarrow J/\psi K_s$ events fully contained within the SVX. With this increase in statistics we can make a conservative estimate of the error on sin2β of 0.09. The addition of opposite side kaon tagging with the Time of Flight detector and other improvements could reduce the error even more.

The measurement of sin2α will be performed by looking for time dependent *CP* asymmetry in the decay $B^0 \rightarrow \pi^+\pi-$. The two major challenges in this analysis will be triggering on the signal and separating the signal from the background. With the two track trigger at Level 1 we will trigger on events with two tracks each of transverse momentum $p_T > 1.5$ GeV/c at a rate of 16 KHz. These events then feed the SVT at Level 2 where a cut on impact parameter of $d > 100\,\mu$m will reduce the rate to 20 Hz. With these trigger requirements and a branching ratio of $BR(B^0 \rightarrow \pi\pi) \approx 5\text{x}10^{-6}$ we expect to collect 5000 events in 2 fb^{-1} of running. Backgrounds to the $B \rightarrow \pi\pi$ signal will arise both from combinatorics and decays due to similar physics processes. The physics backgrounds have been estimated from a Monte Carlo simulation and are shown in Figure 6. The $B \rightarrow \pi\pi$ signal will be extracted utilizing both the mass resolution of the COT as well as its dE/dx particle ID capabilities. With a conservative estimate of $S/N = 1/4$ and assuming an effective tagging efficiency εD^2 of 7.8 % we expect an overall uncertainty on sin2α of 0.15.

The search for B_s oscillations will be made in the decay channels $B_s \rightarrow D_s$ nπ; $D_s \rightarrow \phi\pi$, $K^{*0}K$. Since Δm_s is expected to be large, its measurement will require fully reconstructed B_s decays in order to achieve the necessary momentum and time resolution. The data will be collected using a trigger path similar to that for $B \rightarrow \pi\pi$. If one includes data from the TOF detector, the estimated effective flavor tagging efficiency for the B_s mesons at production is $\varepsilon D^2 = 11.3$ %. Given the expected proper time resolution of the Run II silicon detectors of $\sigma_t = 0.0045$ ps, the ultimate limit on the x_s reach will be given by the number of events collected. Monte Carlo simulations predict $\approx$ 20,000 events in 2 fb^{-1} and a projected upper limit on the mixing measurement of $x_s \leq 60$. This dependence on integrated luminosity is graphically illustrated in Figure 7.

REFERENCES

1. *The CDF Collaboration*, **1** (1996), URL `http://www-cdf.fnal.gov/upgrades/tdr/tdr.html`.
2. *The CDF Collaboration*, **2** (2001), URL `http://www-cdf.fnal.gov/physics/new/bottom/bottom.html`.

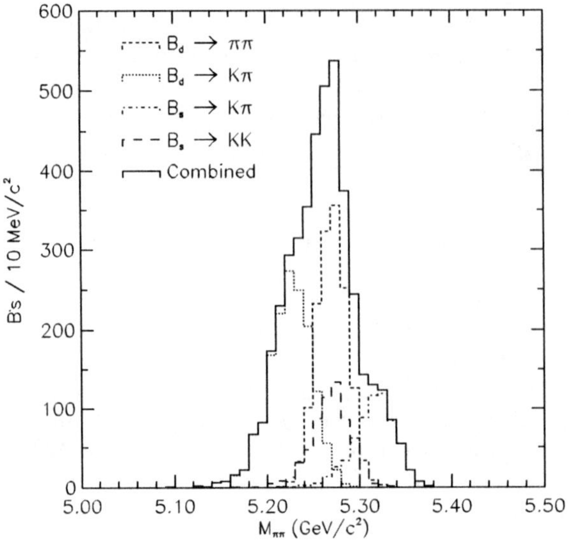

FIGURE 6. Mass distribution for the combination $B^0 \to \pi\pi$, $B^0 \to K\pi$, $B_s \to K\pi$, and $B_s \to KK$ assuming all charged kaons to be pions. The assumed mass resolution is $\sim 20\ \mathrm{MeV}/c^2$.

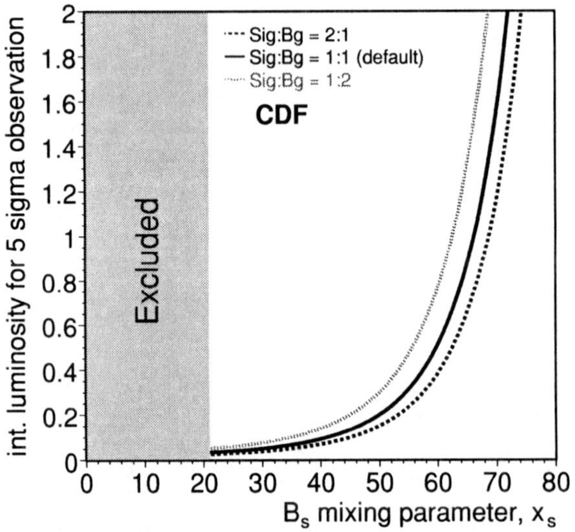

FIGURE 7. Required luminosity for a 5 σ observation of B_s mixing as a function of the value of the mixing parameter x_s.

DØ Results and Run II Status

Brad Abbott representing the DØ collaboration

University of Oklahoma, 440 West Brooks St., Norman, OK 73019

Abstract. Run II at the Tevatron has begun. The DØ detector is expected to collect nearly 2 fb^{-1} of data at $\sqrt{s}$=2.0 TeV for Run II in approximately 2 years with a design instantaneous luminosity of 2×10^{32} cm^{-2} s^{-1}. This large data set will allow precision measurements in a number of important B physics areas and new measurements in B decays. This paper describes the prospects for several important B physics measurements, including sin(2β) using $B \to J/\psi\ K_s^0$ decays, Δm_s using B_s mixing, the Λ_b lifetime and rare B decays.

INTRODUCTION

In this era of asymmetric e^+e^- B factories, the Tevatron provides an important and complimentary tool for studying b physics. Although the environment at the B factories is much cleaner and the backgrounds at the Tevatron are much larger, the Tevatron will play an important role in many areas of b physics. The Tevatron has a number of advantages over e^+e^- machines with sit at the $\Upsilon(4S)$. The b cross section in $p\bar{p} \to b\bar{b}$ at $\sqrt{s}$ = 2.0 TeV is orders of magnitude larger (150 μb) than at the B factories (1 nb). Additionally, with the larger center of mass energy available at the Tevatron, all flavors of B hadrons are produced including B_u, B_d, B_c, B_s, and Λ_b. This provides DØ a large sample for many different B physics measurements.

The DØ detector has recently undergone a number of important upgrades [1] to allow much improved performance in this high luminosity environment. The detector has enhanced its strengths of excellent calorimetry and extended muon coverage by installing a much improved tracking system. A silicon vertex detector consisting of both barrels and disks resides closest to the beam pipe. The silicon detector provides excellent vertexing capabilities with a primary vertex resolution of σ=15-30 μm in r-ϕ and σ=75-100 μm in z. A secondary vertex resolution of σ=40 μm in r-ϕ will allow DØ to accurately reconstruct and measure displaced vertices. A new tracker consisting of 74,000 scintillating fibers mounted on 8 carbon fiber cylinders provides tracking at larger radii. Both of these trackers reside in a 2 Tesla magnetic field. The combination of the silicon and fiber tracker provide excellent tracking capabilities. Tracks with $|\eta| < 3$ can be reconstructed with a momentum resolution of σ_{p_T}/p_T^2=0.002.

Other important upgrades include an improved muon system, a new preshower detector and an improved trigger system. These upgrades will allow DØ to trigger on muons with P_T of 1.5 GeV for $|\eta| < 2$ and to trigger on electrons with P_T of 1 GeV for $|\eta| <$ 2.5.

CP618, *Heavy Flavor Physics: Ninth International Symposium*, edited by A. Ryd and F. C. Porter

CP VIOLATION

Sin2β

One of the primary measurements expected from DØ is an accurate measurement of sin(2β). We expect to use the "Golden" decay mode $B \to J/\psi\, K_s^0$ with the $J/\psi \to \mu^+\mu^-$ and the $K_s^0 \to \pi^+\pi^-$. Figure 1 shows the invariant mass of the B for Monte Carlo $B \to J/\psi\, K_s^0$ decays after a constrained fit was performed. DØ expects to reconstruct nearly 40,000 events in this decay mode with an additional 30,000 events in the mode $J/\psi \to e^+e^-$. The time dependent asymmetry

$$A(t) = \frac{\Gamma(\bar{B}^0 \to J/\psi K_s^0) - \Gamma(B^0 \to J/\psi K_s^0)}{\Gamma(\bar{B}^0 \to J/\psi K_s^0) + \Gamma(B^0 \to J/\psi K_s^0)}$$

can be related to the CKM parameter β by

$$A(t) = \sin(2\beta)\sin(\Delta m_d t)$$

In order to measure this asymmetry, the flavor of the B at production must be known. Many different flavor tagging techniques will be used in order to maximize the tagging performance. The flavor of the B will be determined using the charge of the lepton from semileptonic decays, the charge of the "away side" jet, and the charge of the soft pion nearest to the $B^0 \to J/\psi K_s^0$ decay. With these tagging techniques, studies have shown that an effective tagging efficiency εD^2 of 10% can be achieved, where ε is the tagging efficiency and D is the dilution factor D=2P-1, where P is the probability to determine a correct tag. We can estimate the expected measurement accuracy of sin(2β) using:

$$\sigma(sin2\beta) \approx e^{x_d^2\Gamma^2\sigma_t^2}\sqrt{\frac{1+4x_d^2}{2x_d}}\frac{1}{\sqrt{\varepsilon D^2 N}}\sqrt{1+\frac{B}{S}}$$

where Γ and $x_d = \Delta m_d/\Gamma$ are the decay width and mixing parameter of the B, N is the number of events, B/S is the background to signal ratio and σ_t, the proper time resolution, is approximately 100 fs. Using a S/B ratio of 0.75, an uncertainty on sin(2β) of 0.04 is expected after collecting 2 fb^{-1} using $J/\psi \to \mu^+\mu^-$. A similar accuracy is expected for $J/\psi \to e^+e^-$ giving a combined accuracy on sin 2β of 0.03 after only 2 years of data.

DØ will also study CP violation in $B_s \to J/\psi\,\phi$ decays which have similar trigger and reconstruction efficiencies as $B \to J/\psi\, K_s^0$. The asymmetry in this channel is expected to be small so any measured asymmetry could be a signal for new physics.

α and γ

By measuring an asymmetry in the decays $B_d \to \pi^+\pi^-$ and $B_s \to K^+K^-$, a measurement of α and γ may be possible [2]. One difficulty which arises in this measurement is the background due to $B_d \to K^+\pi^-$ and $B_s \to \pi^+K^-$ decays. DØ has limited particle ID

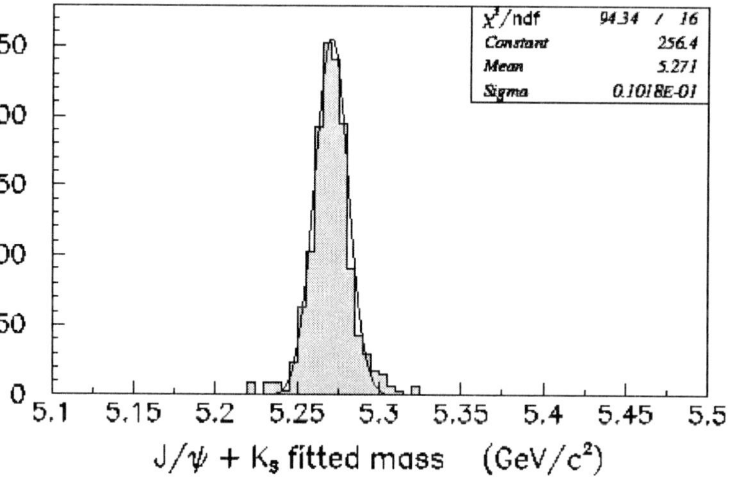

FIGURE 1. Reconstructed B mass using a constrained fit to $B \to J/\psi K_s^0$.

capabilities so separating pions from kaons cannot be done. Therefore we expect large backgrounds in this measurement. Also triggering on these hadronic decay modes of the B will prove challenging. Due to limitations in our trigger bandwidth, DØ cannot trigger on two tracks with low p_T. The additional presence of a lepton in the event must be required to reduce the trigger rate. Since the flavor of the B must be known in order to measure an asymmetry and a lepton provides a very good flavor tag, DØ will require the opposite side B to decay semileptonically. Although this constraint seems harsh, it allows for a much easier method of triggering and flavor tagging. Simulations have shown a detection efficiency of $\sim 0.5\%$ which includes both trigger and reconstruction efficiency. With 2 fb^{-1}, DØ expects to reconstruct approximately 500 $B_d \to \pi^+\pi^-$ events and 1000 $B_s \to K^+K^-$. Figure 2 shows the reconstructed B mass for four different decays. By combining the reconstructed mass and lifetime information, CP asymmetries can be extracted. Another difficulty which arises in determining α and γ are theoretical uncertainties due to possible large penguin contributions. A measured asymmetry may not directly translate into a measurement of α or γ. Due to these difficulties a determination of α and γ may be possible, but the level of uncertainty on these measurements is not yet known.

B_S MIXING

One of the measurements not possible at the e^+e^- B factories is B_s mixing. Measuring Δm_s is one of the highest priorities of DØ's B physics program. The oscillation frequency for B_s is expected to be much higher than for B_d ($\Delta m_d/\Delta m_s \approx 1/20$) making this

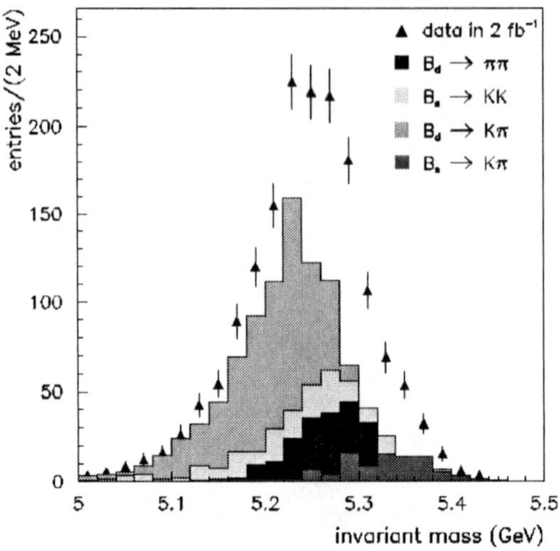

FIGURE 2. The reconstructed B mass for four different B decays, $B_d \to \pi\pi$, $B_d \to K\pi$, $B_s \to KK$, and $B_s \to K\pi$. Since DØ cannot separate kaons from pions, in all cases the pion mass has been assigned to the two tracks coming from the B vertex.

measurement challenging. Due to the high frequency of oscillation, one needs both very good decay length resolution and momentum resolution in order to measure the proper time with sufficient accuracy. There are many different decays which can be used to measure B_s mixing. The mode providing the easiest triggers are semileptonic decays of the B_s such as $B_s \to \phi l^+ X \nu$. By requiring the other B in the event to decay semileptonically, one can trigger on the two leptons in the event. Tagging of the initial flavor and final flavor of the B is achieved by using the sign of the leptons. Large event samples are expected even using more exclusive decays of the B_s such as $B_s \to D_s^- l^+ \nu X$ where $D_s \to \phi X$ and $\phi \to K^+ K^-$. In 2 fb^{-1} of data, we expect to reconstruct $\sim$ 40,000 events. However, due to the presence of the neutrino the B_s momentum is not accurately measured, limiting the Δm_s reach. It is expected that DØ can measure $x_s < 30$ using semileptonic decays of the B_s.

To more accurately measure the B_s momentum, hadronic decays of the B such as $B_s \to D_s \pi$ and $B_s \to D_s \pi\pi\pi$ must be used. Because the trigger rates will be too high without requiring a lepton in the event, the other B in the event must decay semileptonically. The initial flavor of the B is tagged by the charge of the lepton and the final flavor of the B by the charge of the D_s. Due to the strict restrictions required by the trigger, DØ expects to fully reconstruct $\sim$ 1000 hadronic B_s decays. Although this event sample is much smaller than in the semileptonic decay of the B_s, the superior momentum resolution allows a much more accurate measure of the proper time. A reach of $x_s < 30$ for the hadronic decays of the B_s is expected with 2 fb^{-1} of data.

RARE B DECAYS

With 2 fb^{-1} of data, DØ will have enough data to search for rare and exotic B decays. Flavor Changing Neutral Currents processes such as the decay $b \to dl^+l^-$ are forbidden at tree-level in the Standard Model. These processes require loop diagrams and are, therefore, sensitive to new physics making them very interesting.

The decay $B_d \to K^*\mu^+\mu^-$ with $K^* \to \pi K$ has a Standard Model branching ratio of $(1\text{-}1.5) \times 10^{-6}$. Non Standard Model physics will change the branching ratio and decay asymmetry. Since there are two muons in the event, these events can be triggered by requiring a two muon trigger. Requiring a secondary vertex well displaced from the primary and removing events with the dimuon mass around known resonances, will allow DØ to reconstruct approximately 700 events. Additionally $B_d \to K^*e^+e^-$ and $B^+ \to K^+\mu^+\mu^-$ will be studied.

The rare decay $b \to s\mu^+\mu^-$ is theoretically very interesting but experimentally very difficult. Large backgrounds are difficult to separate from signal. To reduce background only a small dimuon mass window can be used. Assuming a Standard Model branching ratio of 6×10^{-6}, DØ expects to reconstruct about 1000 events in this channel. However, there will remain a large signal to background ratio of nearly 1:100. With this large background and theoretical uncertainties this measurement will be challenging.

Very rare decays such as $B_s \to \mu^+\mu^-$ will also be studied. Assuming the ratio between B_s and B_d is 40%, and assuming the Standard Model branching ratio of 4×10^{-9}, less than 10 events are expected to be found. However since the Standard Model branching ratio is so small for this decay, any signal cannot be Standard Model physics and may be a hint for physics beyond the Standard Model.

B_c decays will also be studied in the channel $B_c \to J/\psi l\nu$. By triggering on the leptons from the J/ψ, we expect to reconstruct approximately 600 events which will allow much improved measurements on the mass and lifetime of the B_c .

Λ_B^0 LIFETIME

Current measurements of the Λ_b^0 lifetime show $\tau(\Lambda_b^0)/\tau(B^0)$=0.79±0.05 [3] while naive spectator model predicts this ratio to be unity. In the standard model, non spectator processes such as final state quark interference and W-boson exchange are too small to account for this deviation. Previous measurements often used semileptonic decays of the Λ_b^0 in order to achieve a large event sample. These measurements have the disadvantage that the momentum of the Λ_b^0 cannot be directly measured due to the presence of the neutrino. With the large data available in Run II, it is expected that DØ will be able to fully reconstruct nearly 15,000 events in the decay mode $\Lambda_b^0 \to J/\psi\Lambda^0$. Using fully reconstructed events a lifetime resolution for the Λ_b^0 of 0.11 ps is expected (see Figure 3). It will be very interesting to see if the discrepancy between theory and data is present in this measurement.

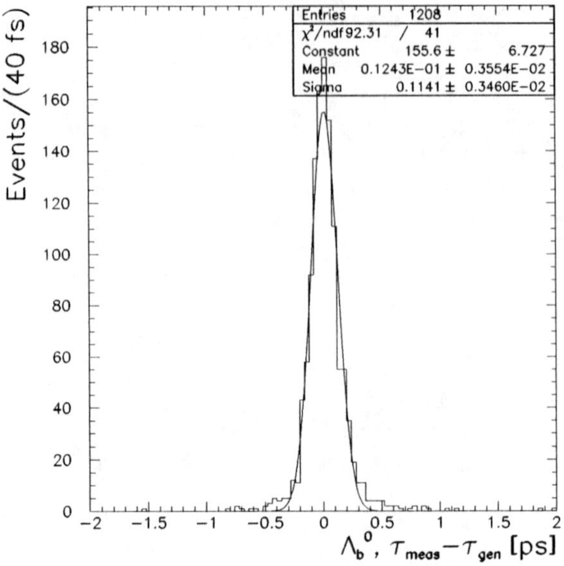

FIGURE 3. Distribution of measured minus generated lifetime for $J/\psi\Lambda^0$ candidates.

CONCLUSIONS

The newly upgraded DØ detector is ready to collect high quality physics data. Many exciting *B* physics measurements are planned. In the next 2 years, we expect to measure Sin(2β) to an accuracy of better than 0.04. We can measure B_s mixing if x_s is less than 30. Rare decays will be studied with increased precision. Mass and lifetimes of *b* baryons will be accurately measured. These new results will significantly constrain the Standard Model or perhaps indicate previously unknown physics.

REFERENCES

1. The DØ upgrade, http://www-d0.fnal.gov/hardware/upgrade/upgrade.html.
2. R.Fleischer, *Phys. Lett B* **459** 306 (1999).
3. LEP B Lifetime Working Group, http://home.cern.ch/~claires/lepblife.html.

New Measurement of Direct CP Violation by NA48 at CERN

Flavio Marchetto representing the NA48 collaboration

Istituto di Fisica dell' Universita', via Pietro Giuria, 1 - 10125 Torino - Italy

Abstract. The measurement of the direct CP violation in the neutral kaon system has been performed by the NA48 collaboration at the CERN SPS. The result of $\mathrm{Re}(\varepsilon'/\varepsilon)$ from the 98 and 99 data is $(15.0 \pm 2.7) \times 10^{-4}$, which confirms the existence of a direct CP violation.

INTRODUCTION

The importance of the $\mathrm{Re}(\varepsilon'/\varepsilon)$ measurement has been addressed by theoretical talks during this Symposium, to which the reader should refer for a comprehensive approach to the subject. This paper will only discuss the experimental aspects of the NA48 measurement.

Direct CP violation has been investigated in the last decades by several experiments with non conclusive results: NA31 [1] measured $\mathrm{Re}(\varepsilon'/\varepsilon) = (23.0 \pm 6.5) \times 10^{-4}$ while E731 [2] found $\mathrm{Re}(\varepsilon'/\varepsilon) = (7.4 \pm 5.9) \times 10^{-4}$.

The experimental challenge remains on the required precision and I'll first summarize the main requirements.

The value of $\mathrm{Re}(\varepsilon'/\varepsilon)$ comes directly from the measurement of $\mathcal{R}$, the double ratio, via the relation: $\mathrm{Re}(\varepsilon'/\varepsilon) \cong (1 - \mathcal{R})/6$, where $\mathcal{R} = \frac{N(K_L \to \pi^0\pi^0)}{N(K_S \to \pi^0\pi^0)} / \frac{N(K_L \to \pi^+\pi^-)}{N(K_S \to \pi^+\pi^-)}$. One is basically dealing with a counting experiment. The statistical error is largely dominated by the number of $K_L \to \pi^0\pi^0$ due to the combination of the larger lifetime and smaller branching fraction. 3×10^6 $K_L \to \pi^0\pi^0$ induces a statistical error $\Delta\mathcal{R} \cong 7.8 \times 10^{-4}$.

An ideal experiment would necessitate:

- simultaneous K_S and K_L beams;
- identical K_S and K_L energy spectra;
- identical detector illumination for charged and neutral decays;
- identical trigger and acquisition biases;
- overlapping decay volumes.

A good experiment design tries to be as close as possible to the ideal design to minimize the corrections on $\mathcal{R}$ by exploiting the cancellations for at least a pair (numerator/denominator) of the four decay modes.

In the next section I'll present the solutions identified by NA48 followed by a short description of the detector. Corrections to $\mathcal{R}$ are then briefly discussed and summarized. In the final section the results are reported.

CP618, *Heavy Flavor Physics: Ninth International Symposium*, edited by A. Ryd and F. C. Porter

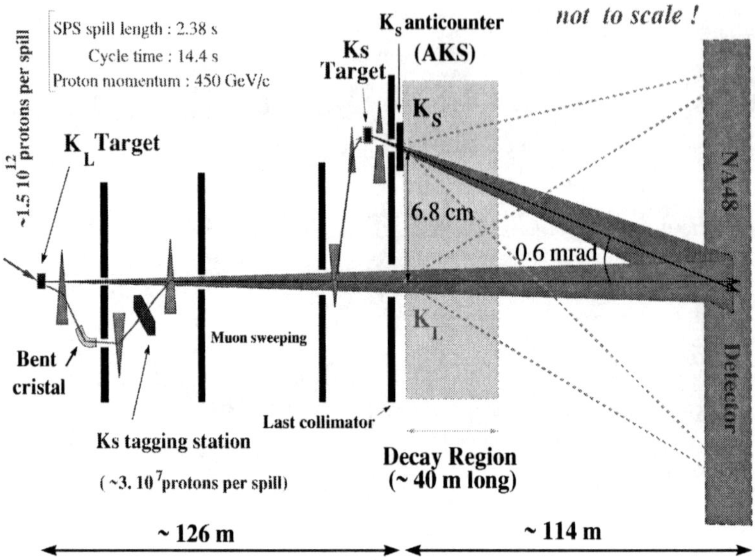

FIGURE 1. Sketch of the $K_S - K_L$ beam system

THE NA48 METHOD

A primary proton beam interacting on a farther upstream target produces the K_L beam. The K_S beam is obtained steering on a downstream target the protons which do not interact on the K_L target [3]. The system provides a relative K_S to K_L intensity which is almost constant. Concerning the geometry of the beams, we have that the K_L beamline is centered with respect to the detector. On the other hand, the K_S beam is $\cong$ 6.8 cm off center at the beginning of the fiducial volume and it is pointing to the electromagnetic calorimeter center. The divergence between the two beams is $\cong$ 0.6 mrad. The sketch of the $K_L - K_S$ beam system is shown in Figure 1.

Due to the large $K_L - K_S$ lifetime difference, the longitudinal (z) decay distributions differ in a substantial way. This would lead to a large acceptance correction. To minimize such a correction, K_L decays are weighted as a function of z to make the K_L vertex distribution similar to the K_S one:

$$\text{weight} \cong \exp(-z/(\beta\gamma c))(1/\tau_{K_S} - 1/\tau_{K_L}).$$

The drawback is that the statistical error increases by $\simeq 35\%$.

Furthermore, the differences of the $K_L - K_S$ energy spectra can induce a bias on $\mathcal{R}$. To limit the effect the accepted range of the kaon energy (70 - 170 GeV) has been splitted into 20 energy bins, each 5 GeV wide. $\mathcal{R}$ was fully and independently computed for each bin and averaged to obtain the final value.

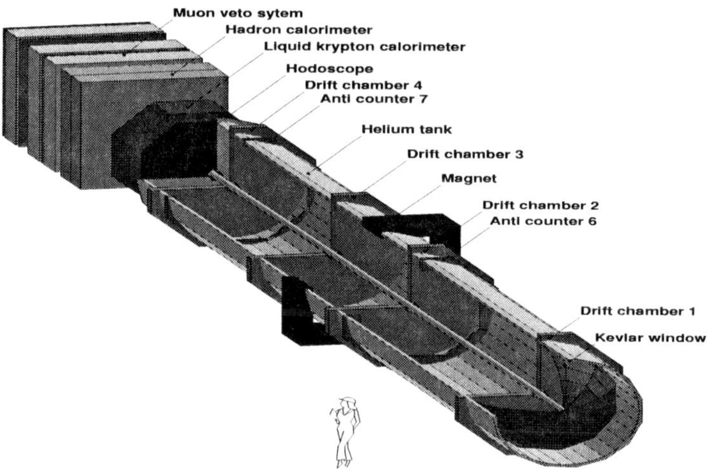

FIGURE 2. The NA48 detector

The trigger biases were minimized by applying offline the same dead time conditions to all the decays.

The detector

The detector is shown in Figure 2. It has been designed to achieve a good mass resolution combined to a precise event time measurement.

The main components of the detector are:

- the magnetic spectrometer with two drift (5 cm pitch) chambers before and two after a 2 Tesla magnet. Each chamber is made of four planes giving a spatial resolution of $\simeq 100\mu$m
- the electromagnetic calorimeter is a quasi-homogeneous detector segmented into 2×2 cm^2 cells with liquid Krypton (LKR) as active medium.

The performances of the magnetic spectrometer are shown by the distribution of the reconstructed $\pi^+\pi^-$ (Figure 3 (left)) which has a kaon mass resolution of $\cong 2.5$ MeV/c^2.
The results of the neutral decay pairing and reconstruction is represented in Figure 3 (right). The resulting π^0 mass resolution is $\cong 1$ MeV/c^2. Both resolutions, charged and neutral, allow a good signal to background separation and the rejection of a large fraction of the background. The event time resolution is $\simeq 200$ ps.

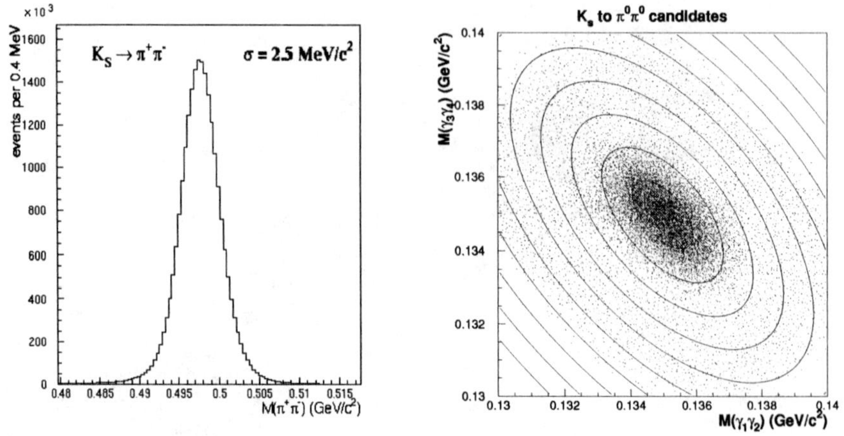

FIGURE 3. $\pi^+\pi^-$ mass distribution (left). Mass of $\gamma_1\gamma_2$ vs. mass of $\gamma_3\gamma_4$ for $K_S \to \pi^0\pi^0$ candidates (right)

A crucial feature of the detector is the $K_L - K_S$ separation, which is done by measuring the timing of every proton directed to the K_S target [4] and comparing it to the event time as measured by the detector. If any of the time differences is smaller than 2 ns the decay is assigned to K_S, otherwise it will be interpreted as a K_L.

Whenever the time measurement of one of the components of the tagging system fails, a K_S can be misinterpreted as a K_L (α_{SL}). Occasionally due to the high rate of protons crossing the tagger, in time coincidence with a genuine K_L one can find a proton, which induces the K_L to be misidentified as a K_S (α_{LS}).

CORRECTIONS TO $\mathcal{R}$

In 1998 and 1999, NA48 collected $\cong 1+2$ million $K_L \to \pi^0\pi^0$ decays. The corrections to $\mathcal{R}$ are briefly discussed in the next subsections.

Trigger and DAQ efficiency

- $\pi^0\pi^0$: selection is based on the number of photons, energy and decay vertex as measured by the LKR. The output rate is $\simeq 2$ kHz, with no dead time. The trigger efficiency has been measured to be ($99.920 \pm 0.009\%$) leading to a negligible $\Delta\mathcal{R}$.
- $\pi^+\pi^-$: the *level* 1 trigger was implemented with a fast logic as: (Scintillator Hodoscope with a hit in opposite quadrants) $\times$ (Calorimeters total energy above a threshold set at 35 GeV) $\times$ (≥ 3 of hits in drift chambers). The output rate was at about 100 kHz, with 0.5% dead time. The efficiencies were measured to be: $(99.535 \pm 0.011)\%$ for K_S and $(99.542 \pm 0.018)\%$ for K_L, leading to a correction

on $\mathcal{R}$ of $(0.9 \pm 2.2) \times 10^{-4}$. The *level* 2 performs track reconstruction and compute the vertex position to reject decays outside the fiducial volume ($0. < \tau < 3.5 \times \tau_S$). The level 2 output rate was $\simeq 2$ kHz with a 1.1 % dead time. The measured efficiencies were $(98.353 \pm 0.022)\%$ for K_S and $(98.319 \pm 0.038)\%$ for K_L, leading to a correction $\Delta\mathcal{R} = (-4.5 \pm 4.7) \times 10^{-4}$
The dead time conditions on the $\pi^+\pi^-$ trigger and readout ($\approx 20\%$) are applied offline to the $\pi^0\pi^0$ candidates to symmetrize eventual biases.

Charged background

The background to $K_S \rightarrow \pi^+\pi^-$ is represented mainly by the decay $\Lambda \rightarrow p\pi^-$, which is completely rejected by the invariant mass cut ($\pm 3\sigma$) around the kaon mass.
On the other hand the background to $K_L \rightarrow \pi^+\pi^-$ is mainly due to $K_{\mu 3}$ and K_{e3}, while $K_L \rightarrow \pi^+\pi^-\pi^0$ decays are rejected by the invariant mass cut. The large bulk of $K_{\mu 3}$ background is tagged with the Muon Veto, while the K_{e3} decays, that mimic a $K_L \rightarrow \pi^+\pi^-$, are recognized from the ratio $E/P > 0.8$ [1] for one of the tracks.
To estimate the remaining background we studied the K_L , K_S , $K_{\mu 3}$, and K_{e3} distributions in the $M_{\pi\pi} - P_t'^2$ plane. [2]
By modelling the $K_{\mu 3}$ and K_{e3} distributions in control regions and then projecting under the signal region, we estimated a correction $\Delta\mathcal{R} = (16.9 \pm 3.0) \times 10^{-4}$.

Neutral background

While $K_S \rightarrow \pi^0\pi^0$ is background free, the background to $K_L \rightarrow \pi^0\pi^0$ is due to $3\pi^0$ decays. A large fraction of the background is rejected by strictly requiring 4 γ's on time and a χ^2 for the $\pi^0\pi^0$ hypothesys < 13.5 (9 ndf).
The remaining background is calculated comparing the χ^2 distribution of $K_L \rightarrow \pi^0\pi^0$ to $K_S \rightarrow \pi^0\pi^0$. From the χ^2 distribution of ($K_L \rightarrow \pi^0\pi^0 - K_S \rightarrow \pi^0\pi^0$) extrapolated from the control region (large χ^2) to the signal region ($\chi^2 < 13.5$), we evaluated a correction $\Delta\mathcal{R} = (-5.9 \pm 2.0) \times 10^{-4}$.

Collimator scattering

Secondary interactions with the final collimators produce neutral kaons which affect the value of $\mathcal{R}$. One can identify two categories of kaon production:

- secondary interactions on the K_S beam collimator: these are characterized by a

[1] E is the energy associated to a track as reconstructed in the LKR and p is the momentum as measured by the magnetic spectrometer.
[2] P_t' is the transverse momentum rescaled taking into account the K_S and K_L target positions.

direction which is pointing away from the detector center [3] while P_t' is well within the accepted values.

- secondary interactions on the two final K_L beam collimators. The situation is completely reversed: c.o.g. small and anomalous P_t' values.

Neutral decays with large c.o.g. value were rejected while no request on P_t' was applied. On the contrary, charged decays were required to fulfill both requirements. The asymmetry of the selection has been properly evaluated and taken into account. The correction to $\mathcal{R}$ is $(-9.6 \pm 2.0) \times 10^{-4}$.

Tagging inefficiency

K_S is identified by a coincidence (± 2 ns) between a proton directed to the K_S target and the kaon event time. A mismeasurement of either one causes the misidentification of the K_S : $\alpha_{SL} \cong (1.63 \pm 0.03) \times 10^{-4}$. The effect tends to cancel out between $\pi^+\pi^-$ and $\pi^0\pi^0$ and only the difference requires a correction on $\mathcal{R}$. In fact we measured $\Delta\alpha_{SL} = (0 \pm 0.5) \times 10^{-4}$, which leads to $\Delta\mathcal{R} = (0 \pm 3.0) \times 10^{-4}$.

A K_L can be misidentified as a K_S whenever a hit in the K_S tagger occurs to be in coincidence with the event time. The probability of the K_L to K_S leakage, α_{LS}, is related to the beam intensity and the measurement gave $\alpha_{LS} = (10.649 \pm 0.008)\%$. Again, it is the neutral - charged difference which is relevant and it was found to be: $\Delta\alpha_{LS} = (4.3 \pm 1.8) \times 10^{-4}$. giving a correction $\Delta\mathcal{R} = (8.3 \pm 3.4) \times 10^{-4}$

Acceptance correction

The weighting method removes most of the asymmetries between K_S and K_L decays. Only second order effects are left and have to be corrected for. As an example one can mention the remaining divergence between the two beams ($\cong 0.6$ mrad). Furthermore the K_S beam is pointing to the center of the e.m. calorimeter, thus at the magnetic spectrometer the charged decays are off-centered.

The correction on $\mathcal{R}$ as a function of the kaon energy is shown in Figure 4 and they are compared to the expected ones if the weighting procedure is not applied.

Accidental activity

Losses and gains of events are due to beam related activity and noise. First order corrections are cancelled out by the constancy of the beams relative intensity, by the comparable K_S and K_L detector illumination and by dead time symmetrization.

[3] The deviation is measured by the center of gravity (c.o.g.) of the decay products.

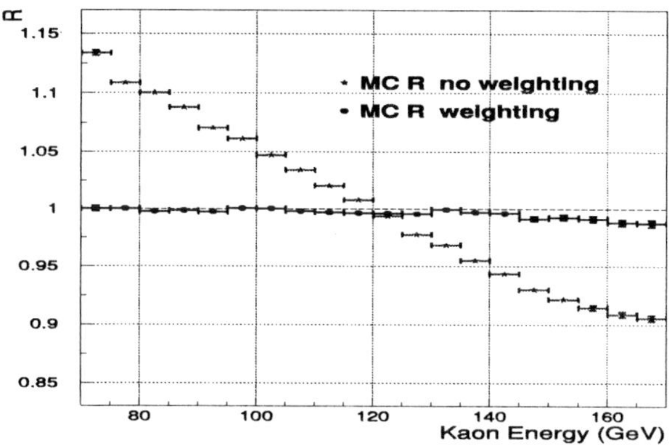

FIGURE 4. Acceptance correction vs. kaon energy

TABLE 1. Reconstructed decays statistics (in million)

	1998	1999	total
$K_L \to \pi^0\pi^0$	1.047	2.243	3.290
$K_S \to \pi^0\pi^0$	1.638	3.571	5.209
$K_L \to \pi^+\pi^-$	4.541	9.912	14.453
$K_S \to \pi^+\pi^-$	6.910	15.311	22.221
statistical error on $\mathcal{R}$ 10^{-4} units	18.0	12.2	10.1

The remaining corrections to $\mathcal{R}$ can be written as the sum of two terms : $\Delta\mathcal{R} = (\Delta\mathcal{R})_{\text{intensity}} + (\Delta\mathcal{R})_{\text{geometry}}$ where $(\Delta\mathcal{R})_{\text{intensity}}$ depends on the residual decorrelation between the two beams, while $(\Delta\mathcal{R})_{\text{geometry}}$ is affected by the differences of the detector illumination.

Both terms have been extensively studied: $(\Delta\mathcal{R})_{\text{intensity}}$ by estimating the possible difference of the beam activities and $(\Delta\mathcal{R})_{\text{geometry}}$ with by overlaying random triggers to reconstructed decays. This last study has been done both with data and MonteCarlo simulated events. We quote $\Delta\mathcal{R} = (0 \pm 3) \times 10^{-4}$ due to relative beam intensity variations and an equivalent correction due to the geometry. The resulting correction for accidental activity is $\Delta\mathcal{R} = (0 \pm 4.2) \times 10^{-4}$.

RESULTS

The $\text{Re}(\varepsilon'/\varepsilon)$ result is based on the statistics reported in Table 1.

The corrections on $\mathcal{R}$ are then summarized in Table 2 in 10^{-4} units.

After applying all the above corrections, we quote: $\mathcal{R} = (0.99098 \pm 0.00101_{stat} \pm 0.00126_{syst})$ from which is derived: $\text{Re}(\varepsilon'/\varepsilon) = (15.0 \pm 2.7) \times 10^{-4}$.

TABLE 2. Summary of the corrections on $\mathcal{R}$ (10^{-4} units)

	correction	uncertainty
$\pi^+\pi^-$ trigger inefficiency	-3.6	±5.2
AKS inefficiency	+1.1	±0.4
Reconstruction of $\pi^0\pi^0$	-	±5.8
Reconstruction of $\pi^+\pi^-$	+2.0	±2.8
Background to $\pi^0\pi^0$	-5.9	±2.0
Background to $\pi^+\pi^-$	+16.9	±3.0
Beam scattering	-9.6	±2.0
Accidental tagging	+8.3	±3.4
Tagging inefficiency	-	±3.0
Acceptance	+26.7	±4.1
Acceptance systematics		±4.0
Accidental activity	-	±4.4
Long term variations of K_S/K_L	-	±0.6
Total	+35.9	±12.6

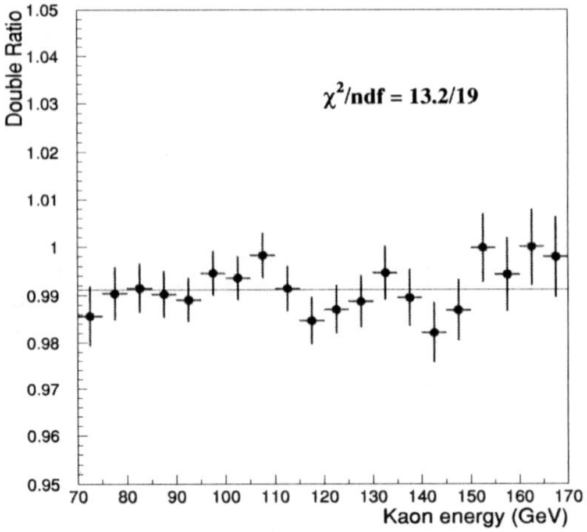

FIGURE 5. $\mathcal{R}$ vs. kaon energy

As mentioned $\mathcal{R}$ has been computed in 20 energy bins 5 GeV wide. Results are plotted in Figure 5.

The stability of the value of $\mathcal{R}$ has been computed against a large number of checks. The checks can be grouped under several types of variation of cuts, namely: acceptance, accidental, tagging, energy scale, neutral and charged background, and beam halo. Results are summarized in Figure 6. For numerous cuts it is shown the change on $\mathcal{R}$, while the bands represent the quoted systematic errors.

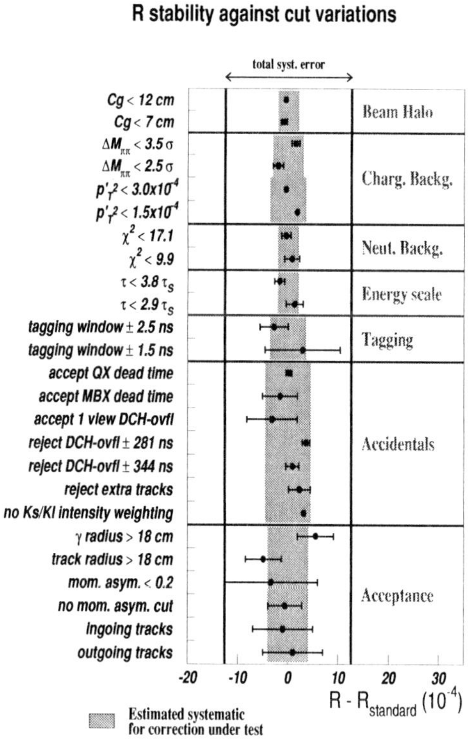

FIGURE 6. Checks of $\mathcal{R}$ stability

By combining the present to the 97 data taking result [5], NA48 quotes: $\mathrm{Re}(\varepsilon'/\varepsilon) = (15.3 \pm 2.6) \times 10^{-4}$.

REFERENCES

1. G. Barr *et al.*, *Phys. Lett. B*, **206**, 233 (1993).
2. L.K. Gibbons *et al.*, *Phys. Rev. Lett.*, **70**, 1203 (1993).
3. N. Doble *et al.*, *Nucl. Instr. and Method*, **B119**, 181 (1996).
4. P. Graftström *et al.*, *Nucl. Instr. and Method*, **A344**, 487 (1994).
5. V. Fanti *et al.*, *Phys. Lett.*, **B465**, 335 (1999).

Recent ε'/ε Results from KTeV

Yee B. Hsiung representing the KTeV collaboration

MS122, P.O.Box 500, Fermilab, Batavia, IL 60510

Abstract. An update of the current status of direct-*CP* violation measurement, ε'/ε, from Fermilab KTeV experiment is reviewed here. Such long-sought effect in the two-pion system of neutral kaon decays has been observed and confirmed in the KTeV data. A recent preliminary result from the combined KTeV 1996-1997 run data finds $Re(\varepsilon'/\varepsilon) = (20.7 \pm 2.7) \times 10^{-4}$. The new world average on this important measurement is then $Re(\varepsilon'/\varepsilon) = (17.2 \pm 1.8) \times 10^{-4}$ with a confidence level of 13% which clearly establishing the existence of "direct" *CP*-violation.

INTRODUCTION

Studying symmetry or the lack of symmetry in nature is a powerful tool in modern physics to understand many of its underlying fundamental laws of nature. The *CP* violation is clearly falling into such category to probe the new physics. Ever since the unexpected discovery [1] of *CP* violation in $K_L \to \pi\pi$ decays in 1964, there has been great interest in understanding the origin of this asymmetry in kaon decays and now also in B decays.

An open question is "Whether the *CP* violation only appears in the asymmetry of the $K^0 - \bar{K^0}$ mixing, the indirect *CP* violation;[1] or it also occurs in the $K \to \pi\pi$ decay process itself, referred to as "direct" *CP* Violation [2]?" The "direct" *CP* violation, parametrized by ε', contributes differently to the rates of $K_L \to \pi^+\pi^-$ versus $K_L \to \pi^0\pi^0$ decays, and would be observed as a nonzero value in the ratio of $\mathrm{Re}(\varepsilon'/\varepsilon)$.

Experimentally we measure the double ratio R from 4 decay modes,

$$R = \frac{\Gamma(K_L \to \pi^+\pi^-)/\Gamma(K_S \to \pi^+\pi^-)}{\Gamma(K_L \to \pi^0\pi^0)/\Gamma(K_S \to \pi^0\pi^0)} \approx 1 + 6\mathrm{Re}(\varepsilon'/\varepsilon). \quad (1)$$

The standard Cabbibo-Kobayashi-Maskawa (CKM) model [3] accomodates *CP* violation with a complex phase in the quark mixing matrix, but the calculations of $\mathrm{Re}(\varepsilon'/\varepsilon)$ depend on several input parameters and on the method used to estimate the hadronic matrix elements. Recent theoretical estimates [4, 5, 6, 7] generally fall in the range from 10^{-4} to 10^{-3} but with large uncertainty.[2]

Alternatively, a "superweak" interaction [8] could produce the observed *CP*-violating mixing effect (ε) but would give $\mathrm{Re}(\varepsilon'/\varepsilon) = 0$. Therefore, a non-zero value of $\mathrm{Re}(\varepsilon'/\varepsilon)$

---

[1] A small mixture of wrong *CP* states in the K_S and K_L which is parametrized by $\varepsilon \approx 0.0023$.

[2] See S. Bertolini's talk in this Proceeding.

CP618, *Heavy Flavor Physics: Ninth International Symposium,* edited by A. Ryd and F. C. Porter

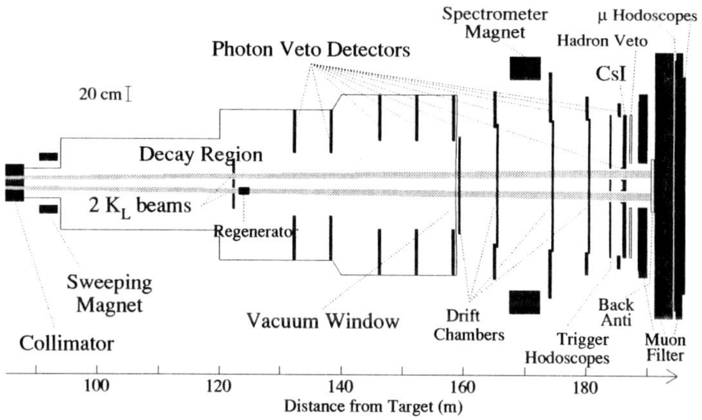

FIGURE 1. The KTeV detector configured for measuring Re(ε'/ε).

rules out the possibility that a superweak interaction is the sole source of *CP* violation, and would establish the "direct" *CP* violation from the decay process itself.

The past experimental measurements in 80s and early 90s of Re(ε'/ε) from Fermilab-E731 [9] and CERN-NA31 [10] were not conclusive. New experiments have been conducted in late 90s at Fermilab-KTeV, CERN-NA48 and Frascati-KLOE to measure Re(ε'/ε) with a precision of $1 \sim 2 \times 10^{-4}$ for the search of "direct" *CP* violation and determining its magnitude.

KTEV DETECTOR AND DOUBLE BEAM METHOD

The KTeV beamline and detector at Fermilab have been described elsewhere in the literature [11]. The KTeV experiment was designed to improve on the previous experiments and ultimately to have the sensitivity to establish direct *CP* violation if Re(ε'/ε) is on the order of 10^{-3}. The experimental technique is similar to E731.

The KTeV detector is shown in Fig. 1. Two parallel neutral K_L beams enter a long vacuum tank, which defines the fiducial volume for accepted decays. One of the beams strikes an active absorber (regenerator), which serves to tag the coherent regeneration of K_S. The regenerator position was moved to the other beam in between Tevatron spill cycles. Behind the vacuum tank, the charged decay products are analyzed by 4 planar drift chambers and an analysis magnet that imparted a 411 MeV/c horizontal transverse kick to the particles. A high precision 3100-element pure Cesium Iodide calorimeter (CsI) is used to measure the energy of $e^{\pm}$ and photons. Photon veto detectors surrounding the vacuum decay volume, drift chambers, and CsI serve to reject events with particles escaping the calorimeter. The regenerator is made of scintillator and is fully active to reduce the scattered background to the coherently regenerated K_S.

To measure the double ratio of decay rates for Re(ε'/ε), we need understand the difference between the acceptances for K_S versus K_L decays to each $\pi\pi$ final state.

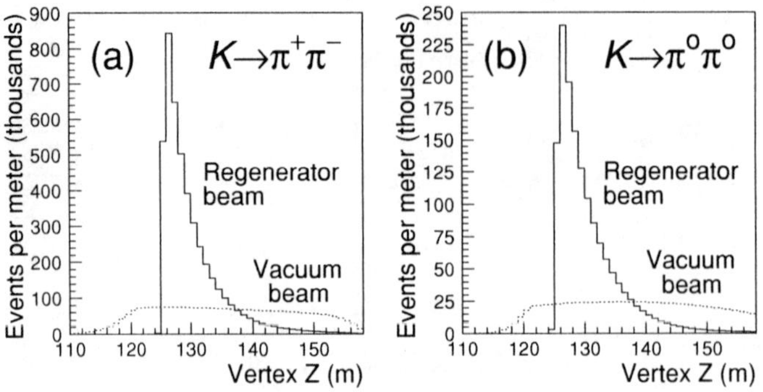

FIGURE 2. Decay vertex distributions for the (a) $K \to \pi^+\pi^-$ and (b) $K \to \pi^0\pi^0$ decay modes, showing the difference between the "regenerator" (K_S) and "vacuum" (K_L) beams.

Event reconstruction and selection are done with identical criteria for decays in either beam, so the principle difference between the K_S and K_L data samples is in the decay vertex distributions, shown in Fig. 2 as a function of Z, the distance from the target. Therefore, the crucial requirement of measuring $\mathrm{Re}(\varepsilon'/\varepsilon)$ with this technique is a precise understanding of the Z-dependence of the detector acceptance.

In the vacuum beam, the acceptance for decays upstream of $Z = 122\ m$ is limited by a lead-scintillator counter, "mask-anti", with two square holes 50% larger than the beams. In the regenerator beam, the beginning of the decay region is sharpely defined by a thin lead-scintillator module at the downstream end of the 1.7 *m* long regenerator.

The typical track position resolution is 110 μm for the drift chamber and the momentum resolution is 0.4% at the mean pion momentum of 36 GeV/c. The CsI calorimeter consists of 3100 pure CsI blocks in a square array 1.9 *m* on a side and 50 *cm* deep. Two 15 *cm* square holes allow the passage of the neutral beams through the calorimeter. The calibration was done by using 424 million $K_L \to \pi e \nu$ decays collected during the normal running. This gives an average energy resolution for photons from $\pi^0\pi^0$ is 0.7% with a mean photon energy of 19 GeV.

The charged mode (for $\pi^+\pi^-$, $Ke3$) trigger is based on a scintillator hodoscope located just upstream of the CsI calorimeter and requirements on the number and pattern of hits in the drift chambers. The neutral mode trigger is based on a fast energy sum from the calorimeter and a hardware cluster finding processor (e.g. for $\pi^0\pi^0$ it must find 4 or 5 clusters of energy in the CsI). Additional fast veto signals from the regenerator, photon vetoes and a downstream muon veto hodoscope are also used in the trigger to keep the event rate at a manageable level. After readout, a CPU-based "Level 3 filter" reconstructs events and applies loose kinematic cuts to select $\pi^+\pi^-$ and $\pi^0\pi^0$ candidates. Large samples of $K_L \to \pi e \nu$ and $K_L \to 3\pi^0$ decays are recorded for detector calibration and acceptance studies.

TABLE 1. Background sources and fraction of background level for KTeV

Background source	**Vacuum Beam** $\times 10^{-3}$	**Regenerator Beam** $\times 10^{-3}$
Charged $K \to \pi l \nu$	0.90	0.03
Charged scattering in collimator	0.10	0.10
Charged scattering in regenerator	-	0.73
TOTAL for Charged Mode	1.00	0.87
$K_L \to 3\pi^0$	1.1	0.3
Neutral scattering in collimator	1.2	0.9
Neutral scattering in regenerator	2.5	11.3
Regenerator hadronic interaction	-	0.1
TOTAL for Neutral Mode	4.8	12.4

THE ε'/ε ANALYSIS

KTeV collected data in the 1996-1997 and 1999 runs. The combined statistics collected in 1996-1997 run reduced the statistical error on $\mathrm{Re}(\varepsilon'/\varepsilon)$ to 1.5×10^{-4}, in which the final sample we have $3.4M$ $K_L \to \pi^0\pi^0$, $11.2M$ $K_L \to \pi^+\pi^-$, $5.6M$ $K_S \to \pi^0\pi^0$ and $19.4M$ $K_S \to \pi^+\pi^-$. Similar statistics was also collected for 1999 run which can further reduce the error, but the analysis of that data has not been finished.

The analysis of entire KTeV 1997 data sample was done with many improvements in calibration and systematic studies. The backgrounds were quite small and were determined well by matching data with the detailed monte carlo simulation in the side bands. The background sources and fractions in each of the 4 decay modes are shown in Table 1. The backgrounds were dominated mainly by the remaining scattered K_S from the regenerator. The systematic uncertainty of background determination for $\mathrm{Re}(\varepsilon'/\varepsilon)$ is 1.1×10^{-4}.

The acceptance corrections was done by a detailed monte carlo simulation of both detector geometry and responses. Although the correction to $\mathrm{Re}(\varepsilon'/\varepsilon)$ is large ($\sim 80 \times 10^{-4}$), most ($\sim 90\%$) is due to the geometry of the detector and can be seen in the slope and ratio of data versus monte carlo. Detailed simulation of detector resolution, inefficiencies and accidental activity were all included, such as individual CsI channel responses, drift chamber inefficiencies and delayed hits in the beam region, etc. High statistics modes of K_{e3} and $K_L \to 3\pi^0$ were used extensively to cross check the performance of monte carlo. The monte carlo also predicted slight acceptance differences between beams due to the accidental activity. Measurement of other kaon parameters (τ_S, Δm) in the regenerated beam also gave us confidence that the systematics of acceptance correction is under control.

Fitting to extract $\mathrm{Re}(\varepsilon'/\varepsilon)$ and other kaon parameters were done with the final data sample binned in kaon energy and decay z vertex bins after the background subtraction, acceptance correction. The result is shown in Figure 3 for $\mathrm{Re}(\varepsilon'/\varepsilon)$ versus kaon energy.

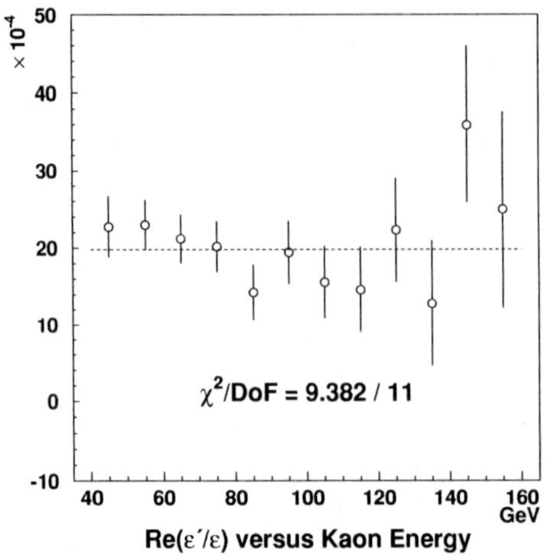

FIGURE 3. ε'/ε versus kaon energy of KTeV data.

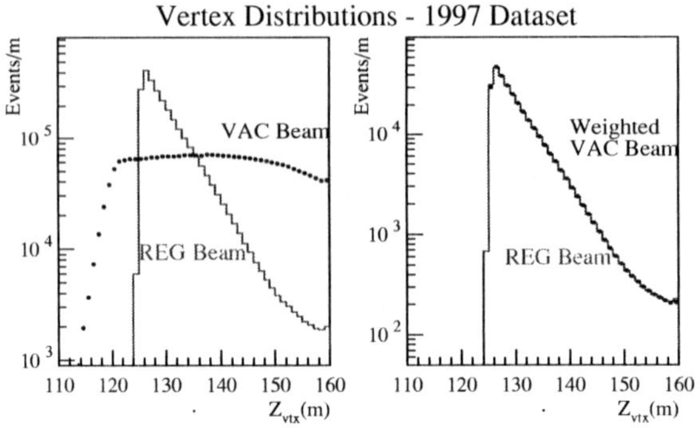

FIGURE 4. Decay z vertex distributions comparison before and after the reweighting method.

Many systematics uncertainties have been studied in detail as shown in Table 2, which included the trigger cut, analysis cuts, detector resolution, inefficiencies, nonlinearity, energy scale, acceptance as well as the fitted kaon parameters.

An alternative re-weighting analysis similar to the method used in CERN-NA48 experiment was also employed to cross check the systematics and results. Figure 4 shows the decay z vertex comparison between the VAC K_L beam and REG K_S beam before and after the re-weighting method. The agreement is good and the difference in $\Delta\mathrm{Re}(\varepsilon'/\varepsilon) = 1.5 \pm 2.1(stat) \pm 3(syst) \times 10^{-4}$.

TABLE 2. Systematic uncertainties on $\mathrm{Re}(\varepsilon'/\varepsilon)$ for KTeV 1996-1997 run analysis.

	Uncertainty ($\times 10^{-4}$)	
Source of Uncertainty	$\pi^+\pi^-$	$\pi^0\pi^0$
Trigger and Level 3 filter	0.56	0.18
Energy/Resolution scale	0.16	1.27
Calorimeter nonlinearity	-	0.66
Detector calibration, alignment	0.18	0.38
Analysis cut	0.32	0.37
Backgrounds	0.20	1.07
Detector aperture	0.30	0.48
Detector resolution	0.15	0.08
Drift chamber inefficiencies	0.21	-
Accidental activity	0.30	-
Z dependence of acceptance	0.89	0.32
Kaon flux and physics parameters	0.44	
TOTAL	2.36	

RESULTS AND CONCLUSION

Our preliminary combined result for 1996-1997 KTeV data is $\mathrm{Re}(\varepsilon'/\varepsilon) = (20.7 \pm 1.5(stat) \pm 2.3(syst) \pm 0.6(MC)) \times 10^{-4}$. This result supersedes the previously published KTeV ε'/ε result [11] which was based on only 23% partial data sample of 1996-1997 run.[3] The results of the other kaon parameters from the fit also give improved precision for Δm and τ_S: $\Delta m = (0.5262 \pm 0.0015) \times 10^{-10}\,\hbar\mathrm{s}^{-1}$ and for $\tau_S = (0.8964 \pm 0.0005) \times 10^{-10}$ s.

This non-zero result of $\mathrm{Re}(\varepsilon'/\varepsilon)$ clearly establishes the existence of direct CP violation in the decay process in the K meson system. We can compare it with the recent NA48 result,[4]

$$\mathrm{Re}(\varepsilon'/\varepsilon) = (20.7 \pm 2.7) \times 10^{-4}, \qquad (KTeV\ \ 96-97) \tag{2}$$

$$\mathrm{Re}(\varepsilon'/\varepsilon) = (15.3 \pm 2.6) \times 10^{-4}, \qquad (NA48\ \ 97-99). \tag{3}$$

[3] The old result [11], $\mathrm{Re}(\varepsilon'/\varepsilon) = (28.0 \pm 4.1) \times 10^{-4}$, has been updated here with an improved re-analysis to be $(23.2 \pm 4.4) \times 10^{-4}$, where a better background determination in neutral mode changes the result by -1.7×10^{-4}; and the remaining changes consistent with statistical variations come from many small improvements in analysis, new monte carlo statistics as well as updated Δm and τ_S.

[4] A. Lai *et al.* to be published in *The European Physical Journal* which supersedes the earlier measurement of partial data sample from 1998 [12], see also F. Marchetto's talk in this Proceeding.

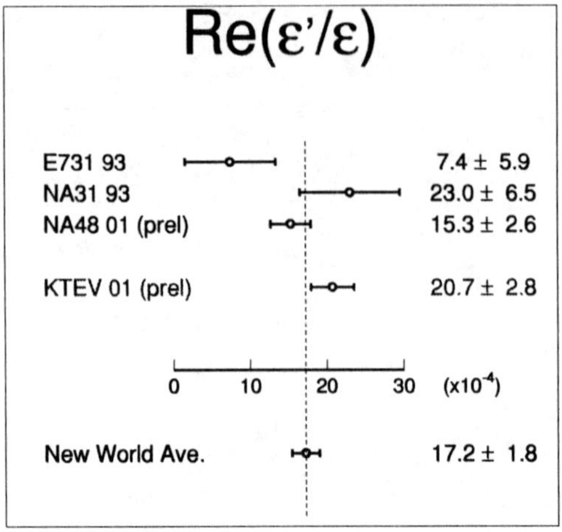

FIGURE 5. Comparison of recent measurements for ε'/ε from E731, NA31, KTeV and NA48.

Figure 5 shows the comparison for the recent measurements of ε'/ε from E731, NA31, KTeV and NA48, as well as the new world average, $Re(\varepsilon'/\varepsilon) = (17.2 \pm 1.8) \times 10^{-4}$ with a confidence level of 13%. Those results are now in much better agreement with the world average and is clearly non-zero. The 1999 KTeV data has about the same statistics of 1996-1997 data sample. Therefore, it will further improve the accuracy of the final measurement.

REFERENCES

1. J.H. Christenson, J.W. Cronin, V.L. Fitch, and R. Turlay, *Phys. Rev. Lett.*, **13**, 138 (1964).
2. B. Winstein and L. Wolfenstein, *Rev. Mod. Phys.*, **65**, 1113 (1993).
3. M. Kobayashi and T. Maskawa, *Prog. Theor. Phys.*, **49**, 652 (1973).
4. A.J. Buras, P. Gambino, M. Gorbahn, S. Jager, and L. Silvestrini , *Nucl. Phys.*, **B592**, 55 (2000).
5. S. Bertolini, M. Fabbrichesi, and J.O. Eeg, *Rev. Mod. Phys.*, **72**, 65 (2000).
6. S. Narison, *Nucl. Phys.*, **B593**, 3 (2001).
7. Y.L. Wu, *Phys. Rev.*, **D64**, 016001 (2001).
8. L. Wolfenstein, *Phys. Rev. Lett.*, **13**, 569 (1964).
9. L. K. Gibbons *et al.*, *Phys. Rev. Lett.*, **70**, 1203 (1993).
10. G. D. Barr *et al.*, *Phys. Lett.*, **B317**, 233 (1993).
11. A. Alavi-Harati *et al.*, *Phys. Rev. Lett.*, **83**, 22 (1999).
12. V. Fanti *et al.*, *Phys. Lett.*, **B465**, 335 (1999).

Calculation of ε'/ε

Stefano Bertolini

INFN and SISSA, Via Beirut 4, I-34013 Trieste, Italy

Abstract. I shortly review the present status of the theoretical calculations of ε'/ε and the comparison with the present experimental results. I discuss the role of higher order chiral corrections and in general of non-factorizable contributions for the explanation of the $\Delta I = 1/2$ selection rule and direct CP violation in kaon decays. Still lacking reliable lattice calculations, analytic methods and phenomenological approaches are helpful in understanding correlations among theoretical effects and the experimental data. Substantial progress from lattice QCD is expected in the coming years.

INTRODUCTION

The results obtained in the last two years by the NA48 [1] and the KTeV [2] collaborations have marked a great experimental achievement, establishing some 35 years after the discovery of CP violation in the neutral kaon system [3] the existence of a much smaller violation acting directly in the decays:

$$\mathrm{Re}(\varepsilon'/\varepsilon) = \begin{cases} (15.3 \pm 2.6) \times 10^{-4} & (\mathrm{NA48}) \\ (20.7 \pm 2.8) \times 10^{-4} & (\mathrm{KTeV}). \end{cases} \tag{1}$$

The average of these results with the previous measurements by the NA31 collaboration at CERN and by the E731 experiment at Fermilab gives

$$\mathrm{Re}(\varepsilon'/\varepsilon) = (17.2 \pm 1.8) \times 10^{-4}. \tag{2}$$

While the Standard Model (SM) of strong and electroweak interactions provides an economical and elegant understanding of indirect (ε) and direct (ε') CP violation in term of a single phase, the detailed calculation of the size of these effects implies mastering strong interactions at a scale where perturbative methods break down. In addition, direct CP violation in $K \to \pi\pi$ decays arises from a detailed balance of two competing sets of contributions, which may hopelessly inflate the uncertainties related to the relevant hadronic matrix elements in the final outcome. All that makes predicting ε'/ε a complex and challenging task [4].

Just from the onset of the calculation the presence in the definition of ε'/ε, written as

$$\frac{\varepsilon'}{\varepsilon} = \frac{1}{\sqrt{2}} \left\{ \frac{\langle (\pi\pi)_{I=2} | \mathcal{H}_W | K_L \rangle}{\langle (\pi\pi)_{I=0} | \mathcal{H}_W | K_L \rangle} - \frac{\langle (\pi\pi)_{I=2} | \mathcal{H}_W | K_S \rangle}{\langle (\pi\pi)_{I=0} | \mathcal{H}_W | K_S \rangle} \right\}, \tag{3}$$

of given ratios of isospin amplitudes warns us of a longstanding and still unsolved theoretical "problem": the explanation of the $\Delta I = 1/2$ selection rule.

CP618, *Heavy Flavor Physics: Ninth International Symposium*, edited by A. Ryd and F. C. Porter

The $\Delta I = 1/2$ selection rule in $K \to \pi\pi$ decays is known since 45 years [5] and it states the experimental evidence that kaons are 400 times more likely to decay in the $I = 0$ two-pion state than in the $I = 2$ component ($\omega \equiv A_2/A_0 \simeq 1/22$). This rule is not justified by any symmetry argument and, although it is common understanding that its explanation must be rooted in the dynamics of strong interactions, there is no up to date derivation of this effect from first principle QCD.

Given the possibility that common systematic uncertainties may a-priori affect the calculation of ε'/ε and the $\Delta I = 1/2$ rule (see for instance the present difficulties in calculating on the lattice the "penguin contractions" for CP violating as well as for CP conserving amplitudes [6]) a convincing calculation of ε'/ε must involve at the same time a reliable explanation of the $\Delta I = 1/2$ selection rule. Both observables indicate the need of large corrections to factorization in the evaluation of the four-quark hadronic transitions. Among these corrections Final State Interactions (FSI) play a substantial role. However, FSI alone are *not* enough to account for the large ratio of the $I = 0$ over $I = 2$ amplitudes. Other sources of large non-factorizable corrections are therefore needed for the CP conserving amplitudes, which might affect the determination of ε'/ε as well. As a consequence, a self-contained calculation of ε'/ε should also address the determination of the CP conserving $K \to \pi\pi$ amplitudes.

OPE: AN "EFFECTIVE" APPROACH

The Operator Product Expansion (OPE) provides us with a very effective way to address the calculation of hadronic transitions in gauge theories. The integration of the "heavy" gauge and matter fields allows us to write the relevant amplitudes in terms of the hadronic matrix elements of effective quark operators and of the corresponding Wilson coefficients (at a scale μ), which encode the information about those dynamical degrees of freedom which are heavier than the chosen renormalization scale. With the SM flavor structure $\Delta S = 1$ transitions are effectively described by

$$\mathcal{H}_{\Delta S=1} = \frac{G_F}{\sqrt{2}} V_{ud} V_{us}^* \sum_i \Big[z_i(\mu) + \tau\, y_i(\mu) \Big] Q_i(\mu) \ . \tag{4}$$

The entries V_{ij} of the 3×3 Cabibbo-Kobayashi-Maskawa (CKM) matrix describe the flavour mixing in the SM and $\tau = -V_{td}V_{ts}^*/V_{ud}V_{us}^*$. For $\mu < m_c$ ($q = u,d,s$), the relevant quark operators are:

$$\begin{array}{lcll}
Q_1 & = & (\bar{s}_\alpha u_\beta)_{V-A}\,(\bar{u}_\beta d_\alpha)_{V-A} & \\
Q_2 & = & (\bar{s}u)_{V-A}\,(\bar{u}d)_{V-A} & \text{Current-Current} \\
Q_{3,5} & = & (\bar{s}d)_{V-A} \sum_q (\bar{q}q)_{V\mp A} & \\
Q_{4,6} & = & (\bar{s}_\alpha d_\beta)_{V-A} \sum_q (\bar{q}_\beta q_\alpha)_{V\mp A} & \text{Gluon ``penguins''} \\
Q_{7,9} & = & \frac{3}{2}(\bar{s}d)_{V-A} \sum_q \hat{e}_q (\bar{q}q)_{V\pm A} & \\
Q_{8,10} & = & \frac{3}{2}(\bar{s}_\alpha d_\beta)_{V-A} \sum_q \hat{e}_q (\bar{q}_\beta q_\alpha)_{V\pm A} & \text{Electroweak ``penguins''}
\end{array} \tag{5}$$

Current-current operators are induced by tree-level W-exchange whereas the so-called penguin (and "box") diagrams are generated via an electroweak loop. Only the latter "feel" all three quark families via the virtual quark exchange and are therefore sensitive to the weak CP phase. Current-current operators control instead the CP conserving transitions. This fact suggests already that the connection between ε'/ε and the $\Delta I = 1/2$ rule is by no means a straightforward one.

Using the effective $\Delta S = 1$ quark Hamiltonian we can write ε'/ε as

$$\frac{\varepsilon'}{\varepsilon} = e^{i\phi} \frac{G_F \omega}{2|\varepsilon| \operatorname{Re} A_0} \operatorname{Im} \lambda_t \left[\Pi_0 - \frac{1}{\omega} \Pi_2 \right] \tag{6}$$

where

$$\begin{aligned} \Pi_0 &= \frac{1}{\cos\delta_0} \sum_i y_i \operatorname{Re}\langle Q_i \rangle_0 \, (1 - \Omega_{\rm IB}) \\ \Pi_2 &= \frac{1}{\cos\delta_2} \sum_i y_i \operatorname{Re}\langle Q_i \rangle_2 \quad , \end{aligned} \tag{7}$$

and $\langle Q_i \rangle \equiv \langle \pi\pi | Q_i | K \rangle$. The rescattering phases $\delta_{0,2}$ can be extracted from elastic $\pi - \pi$ scattering data [7] and are such that $\cos\delta_0 \simeq 0.8$ and $\cos\delta_2 \simeq 1$. Given that the phase of ε (θ_ε) is approximately $\pi/4$, as well as $\delta_0 - \delta_2$, $\phi = \frac{\pi}{2} + \delta_2 - \delta_0 - \theta_\varepsilon$ turns out to be consistent with zero. While G_F, ω, $|\varepsilon|$ and $\operatorname{Re} A_0$ are precisely determined by experimental data, the first source of uncertainty that we encounter in eq. 6 is the value of $\operatorname{Im}\lambda_t \equiv \operatorname{Im}(V_{ts}^* V_{td})$, the combination of CKM elements which measures CP violation in $\Delta S = 1$ transitions. The determination of $\operatorname{Im}\lambda_t$ depends on B-physics constraints and on ε [8]. In turn, the fit of ε depends on the theoretical determination of B_K, the $\bar{K}^0 - K^0$ hadronic parameter, which should be self-consistently determined within every analysis. The theoretical uncertainty on B_K was in the past the main component of the final uncertainty on $\operatorname{Im}\lambda_t$. The improved determination of the unitarity triangle coming from B-factories and hadronic colliders [9] has weakened and will eventually lift the dependence of $\operatorname{Im}\lambda_t$ on B_K, allowing for an experimental measurement of the latter from ε . Within kaon physics, the decay $K_L \to \pi^0 \nu\bar{\nu}$ gives the cleanest "theoretical" determination of $\operatorname{Im}\lambda_t$, albeit representing a great experimental challenge. At present, a typical range of values for $\operatorname{Im}\lambda_t$ is $(0.94 - 1.60) \times 10^{-4}$ [10].

We come now to the quantities in the square brackets. While the calculation of the Wilson coefficients is well under control, thanks primarily to the work done in the early nineties by the Munich [11] and Rome [12] groups, the evaluation of the "long-distance" factors in eq. 7 is the crucial issue for the ongoing calculations. The isospin breaking (IB) parameter $\Omega_{\rm IB}$, gives at the leading-order (LO) in the chiral expansion a *positive* correction to the A_2 amplitude (proportional to A_0 via the $\pi^0 - \eta$ mixing) of about 0.13 [13]. At the next-to-leading order (NLO) the full inclusion of the $\pi^0 - \eta - \eta'$ mixing lift the value of $\Omega_{\rm IB}$ to 0.16 ± 0.03 [14]. On the other hand, the complete NLO calculation of IB effects beyond the $\pi^0 - \eta - \eta'$ mixing (of strong and electromagnetic origin, among which the presence of $\Delta I = 5/2$ transitions) involves a number of unknown NLO chiral couplings and is presently quite uncertain. Dimensional estimates show that IB effects may be large and affect ε'/ε sizeably in both directions [15]. Although a partial cancellation of the indirect ($\Delta\omega$) and direct $\Delta\Omega_{\rm IB}$ NLO isospin breaking corrections in eq. 6 may reduce their final numerical impact on ε'/ε , we must await for further analyses in order to confidently assess their relevance. At present one may use $\Omega_{\rm IB} = 0.10 \pm 0.20$ [16, 17, 18] as a conservative estimate of the IB effects.

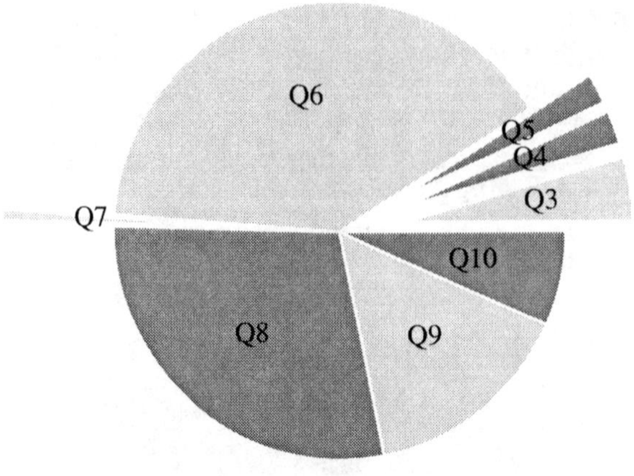

FIGURE 1. Anatomy of ε'/ε in the Vacuum Saturation Approximation. In light (dark) gray the positive (negative) contributions of the effective four-quark operators are shown with proportional weight.

The final basic ingredient for the calculation of ε'/ε is the evaluation of the $K \to \pi\pi$ hadronic matrix elements of the quark operators in eq. 5. A simple albeit naive approach to the problem is the Vacuum Saturation Approximation (VSA), which is based on two drastic assumptions: the factorization of the four quark operators in products of currents and densities and the saturation of the intermediate states by the vacuum state. As an example:

$$\begin{aligned}\langle\pi^+\pi^-|Q_6|K^0\rangle &= 2\,\langle\pi^-|\bar{u}\gamma_5 d|0\rangle\langle\pi^+|\bar{s}u|K^0\rangle - 2\,\langle\pi^+\pi^-|\bar{d}d|0\rangle\langle 0|\bar{s}\gamma_5 d|K^0\rangle \\ &\quad + 2\left[\langle 0|\bar{s}s|0\rangle - \langle 0|\bar{d}d|0\rangle\right]\langle\pi^+\pi^-|\bar{s}\gamma_5 d|K^0\rangle \qquad (8)\end{aligned}$$

The VSA does not exhibit a consistent matching of the renormalization scale and scheme dependences of the Wilson coefficients and it carries potentially large systematic uncertainties [4]. On the other hand it provides useful insights on the main features of the problem. A pictorial summary of the relative weights of the contributions of the various operators to ε'/ε, as obtained in the VSA, is shown in Fig. 1.

As we have already mentioned, CP violation involves loop-induced operators ($Q_3 - Q_{10}$). From Fig. 1 one clearly notices the potentially large cancellation among the strong and electroweak sectors and the leading role played by the gluonic penguin operator Q_6 and the electroweak operator Q_8. Tipical range of values for ε'/ε, obtained using the VSA, are shown in Fig. 2 together with the three most updated predictions available before 1999 (when the first KTeV and NA48 results became known) [19, 20, 26]. The fact that the cancellation among the strong and electroweak sectors turns out to be quite effective (at least in the VSA) warns us about the possibility that the uncertainties in the determination of the relevant hadronic matrix elements may be largely amplified in the calculation of ε'/ε. It is therefore important to asses carefully the approximations related to the various parts of the calculations. In particular, the analysis of the problem suggests that factorization may be highly unreliable.

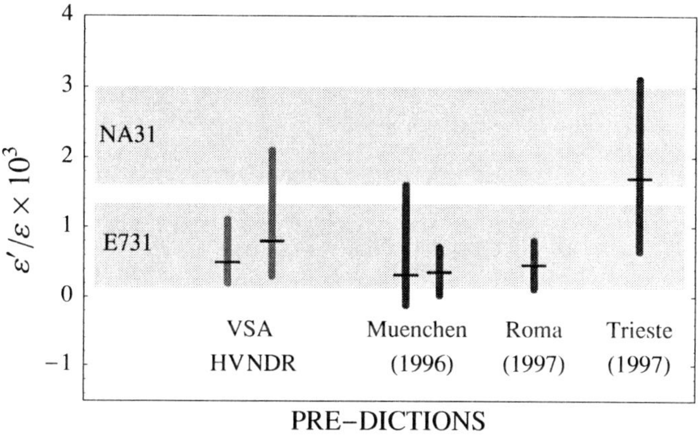

FIGURE 2. The 1-σ results of the NA31 and E731 Collaborations (early 90's) are shown by the gray horizontal bands. The old München, Roma and Trieste theoretical predictions for ε'/ε are depicted by the vertical bars with their central values. For comparison, the VSA estimate is shown using two renormalization schemes.

BEYOND FACTORIZATION

The dark gray bars in Fig. 2 represent the results of three calculations of ε'/ε which are representative of approaches that in principle allow us to go beyond naive factorization. They are based from left to right on the large N_c expansion [19, 21], on lattice regularization [20, 22], and on phenomenological modelling of low-energy QCD (the chiral quark model) [23, 24, 25, 26].

The experimental and theoretical scenarios have changed substantially after the first KTeV data and the subsequent NA48 results. Fig. 3 shows the present experimental world average for ε'/ε compared with the revised or new theoretical calculations that appeared during the last year. Without entering into the details of the results (for a short summary see [27]) they all represent attempts to incorporate non-perturbative information into the calculation of the hadronic matrix elemnts, whether it is based on the large N_c expansion (München [28], Dortmund [29], Beijing [30], Taipei [31], Valencia [32]), phenomenological modelling of low-energy QCD (Dubna [33], Trieste [34, 35], Lund [36]), QCD Sum Rules (Montpellier [37]) or, finally, lattice regularization (Roma [6]).

Overall most of the theoretical calculations are consistent with a non-vanishing positive effect in the SM (with the exception of some very recent lattice results on which I will comment shortly).

At a closer look however, if we focus our attention on the central values, many of the predictions prefer the 10^{-4} regime, whereas only a few of them stand above 10^{-3}. Is this just "noise" in the theoretical calculations? Actually, without entering the many details on which the estimates are based, most of the aforementioned difference is explained in terms of a single effect: the different size of the hadronic matrix element of the gluonic penguin Q_6 obtained in the various approaches. In turn this can be understood in terms

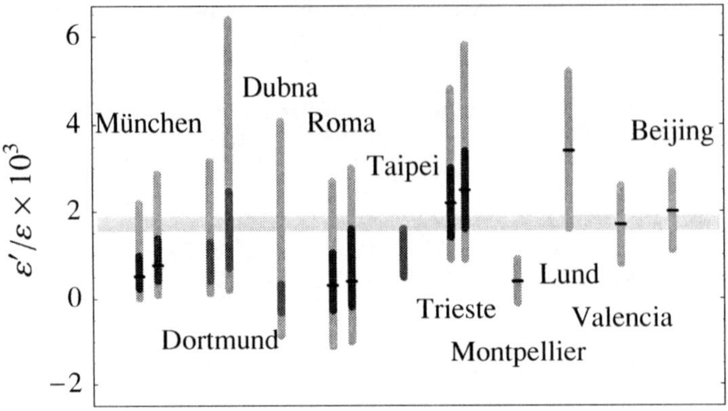

FIGURE 3. Recent theoretical calculations of ε'/ε are compared with the combined 1-σ average of the NA31, E731, KTeV and NA48 results ($\varepsilon'/\varepsilon = 17.2 \pm 1.8 \times 10^{-4}$), depicted by the gray horizontal band.

of sizeable higher order chiral contributions (NLO in the $1/N_c$ expansion) to the $I = 0$ amplitudes.

This effect was stigmatized before the latest experimental round by the work of the Trieste group [25, 26], and appears clearly in the comparison of the leading $1/N_c$ and lattice results with the chiral quark model analysis in Fig. 2. The chiral quark model approach, together with the fit of the CP conserving amplitudes to phenomenologically normalize matching and model parameters, allows one to carry a complete calculation of the hadronic matrix elements beyond the leading order in the chiral expansion (including the needed local counterterms). Non-factorizable chiral contributions (not included in leading $1/N_c$ or lattice calculations) were shown to produce a substantial enhancement of the $I = 0$ transitions lifting the expectation of ε'/ε at the 10^{-3} level.

Since then a number of groups have attempted to improve the calculation of $K \to \pi\pi$ matrix elements in a model independent way. The present status of the art is summarized in Table 1. Due to the leading role played by Q_6 and Q_8 we may write a simplified version of eq. 6, namely

$$\frac{\varepsilon'}{\varepsilon} \approx 13 \left(\frac{\Lambda^{(4)}_{\overline{MS}}}{340\ \text{MeV}} \right) \text{Im}\,\lambda_t \left[\frac{110\ \text{MeV}}{m_s\ (2\ \text{GeV})} \right]^2 \left[B_6 (1 - \Omega_{\text{IB}}) - 0.4 B_8^{(2)} \right] , \qquad (9)$$

which in spite of the fact that it should "not be used for any serious analysis" [28] gives an effective and practical way to test and compare different calculations.

The B-factors $B_i \equiv \langle Q_i \rangle / \langle Q_i \rangle_{\text{VSA}}$ represent a convenient parametrization of the hadronic matrix elements, albeit tricky, in that their values are in general scale and renormalization-scheme dependent, and a spurious dependence on the quark masses is introduced in the result whenever quark densities are involved. The latter is the case for the Q_6 and Q_8 penguins. As a consequence the VSA normalization may vary from author to author thus introducing systematic ambiguities. By defining the VSA matrix

TABLE 1. Comparison of various calculations of penguin matrix elements

Method	B_6 (NDR)	$B_8^{(2)}$ (NDR)	
Lattice [CP-PACS]	< 0.3	~ 0.9	[38]
Lattice [RBC]	~ 0.3	~ 1	[39]
Lattice [APE]	–	0.47 ± 0.05	[40]
Lattice [SPQcdR]	–	0.45 ± 0.06	[41]
Sum rules+LMD (χ-limit)	–	3.2 ± 1.0	[42]
Dispersive+data (χ-limit)	–	1.9 ± 0.7	[43]
	–	1.4 ± 0.6	[44]
Large N_c + data	1.0 ± 0.3	0.8 ± 0.2	[28]
NLO $1/N_c$ CHPT	$1.5 \sim 1.7$	$0.4 \sim 0.7$	[29]
NLO $1/N_c$ ENJL (χ-limit)	2.9 ± 0.5	1.5 ± 0.2	[36]
NLO χ-Quark Model	1.5 ± 0.4	0.84 ± 0.04	[26, 34]
FSI + large N_c	1.55 ± 0.10	0.92 ± 0.03	[32]

elements at the scale $\mu = 2$ GeV we obtain [4]

$$\begin{aligned} \langle(\pi\pi)_{I=2}|Q_8|K^0\rangle_{\rm VSA} &= \sqrt{6}\, f\, m_K^4\, (m_s+m_d)^{-2} \simeq 1.1\ \text{GeV}^3\,, \\ \langle\pi\pi|Q_6|K^0\rangle_{\rm VSA} / \langle(\pi\pi)_{I=2}|Q_8|K^0\rangle_{\rm VSA} &= 2\sqrt{2}\,(f_K - f_\pi)/f_\pi \simeq 0.63\,, \end{aligned} \tag{10}$$

where I have used $(m_s + m_d)(2\ \text{GeV}) = 110$ MeV and the chiral value $f = 86$ MeV for the octet decay constant.

It is known that B_6 and $B_8^{(2)}$ are perturbatively very weakly dependent on the renormalization scale [4]. Therefore it makes sense to compare the B's obtained in different approaches, where the matrix elements $\langle Q_i\rangle$ are computed at different scales. In Table 1 I collected some recent results for the relevant penguin matrix elements coming from various approaches, having care of normalizing the results in a homogeneous way (as far as detailed information on definitions and renormalization schemes was available).

As a guiding information, taking $\text{Im}\,\lambda_t = 1.3 \times 10^{-4}$, the present experimental central value of ε'/ε is reproduced by $B_6\,(1-\Omega_{\rm IB}) - 0.4\, B_8^{(2)} \approx 1$.

The most important fact is the first evidence of a signal in lattice calculations of $\langle\pi\pi|Q_6|K\rangle$, obtained by the CP-PACS [38] and RBC [39] collaborations. Both groups use the Domain Wall Fermion approach which allows to control the chiral symmetry on the lattice as a volume effect in a fifth dimension. This approach softens in principle the problem of large power subtractions which affects the lattice extraction of $I = 0$ amplitudes (penguin contractions). Still the $\langle\pi|Q_i|K\rangle$ transition is directly computed on the lattice and chiral perturbation is needed to extrapolate to the physical amplitude. Both groups obtain comparable values of Q_6 and Q_8 matrix elements thus leading to negative ε'/ε . On the other hand the calculations are at an early stage and do not fully include higher order chiral contributions which may be responsible for the dynamical enhancement of the gluonic penguin, as large N_c approaches beyond LO and the Chiral Quark Model suggest.

Important results have been obtained using data on spectral functions in connection with QCD sum rules and dispersive relations in the attempt to obtain model independent information on the relevant matrix elements. These approaches have produced as of today calculations of Q_8 (in the chiral limit) which are subtantially larger than the factorization (and lattice) results. In addition there is still substantial disagreement among different methods, in particular between the result of ref. [42], which is based on large N_c and Lowest Meson Dominance (LMD), and that of ref. [44] obtained via dispersive relations and data.

Calculations which sofar have gone beyond LO in the large N_c expansion, based on chiral perturbation theory and/or models of low-energy QCD, have shown the crucial role of higher order non-factorizable corrections in the enhancement of the $I = 0$ matrix elements [26, 29, 32, 36]. Chiral loop corrections drive the final value of ε'/ε in the ballpark of the present data. However the calculation of higher order chiral corrections cannot be fully accomplished in a model independent way due to the many unknown NLO local couplings. In the chiral quark model approach all needed local coupling are computed in terms of quark masses, meson decay constants and a few non-perturbative parameters as quark and gluon condensates. The latter are determined self-consistently in a phenomenological way via the fit of the CP conserving $K \to \pi\pi$ amplitudes, thus encoding the $\Delta I = 1/2$ rule in the calculation [25, 34]. The analysis shows that the role of local counterterms is subleading to the chiral logs when using the Modified Minimal Subtraction as opposed to the commonly used Gasser-Leutwyler prescription.

Among non-factorizable chiral corrections FSI play a leading role. As a matter of fact, one should in general expect an enhancement of ε'/ε with respect to the naive VSA due to FSI. As Fermi first argued [45], in potential scattering the isospin $I = 0$ two-body states feel an attractive interaction, of a sign opposite to that of the $I = 2$ components thus affecting the size of the corresponding amplitudes. This feature is at the root of the enhancement of the $I = 0$ amplitude over the $I = 2$ one and of the corresponding enhancement of ε'/ε beyond factorization. An attempt to resum these effects in a model independent way has been worked out by the authors of ref. [32], using a dispersive approach a la Omnès-Mushkelishvili [46]. Their analysis shows that resummation does not substantially modify the one-loop perturbative result and, as it appears from Table 1, a 50% enhancement of the gluonic penguin matrix element is found over the factorized result. However the analysis suffers from a sistematic uncertainty due to the value of the factorized amplitude at the chosen subtraction point which is identified with the large N_c result [47]. Even when the authors in the most recent work match the dispersive resummation with the perturbative one-loop calculation, again a systematic uncertainty remains hidden in the unknown polinomial parts of the local chiral counterterms. In this respect it is important to stress that gauging the scheme dependence of the calculation by varying the renormalization scale (in the chiral logs) as the authors do, cannot account for the arbitrariness related to the subtraction (as a consequence of an incomplete calculation).

In addition, it has been recently emphasized [48] that cut-off based approaches should pay attention to higher-dimension operators which become relevant for matching scales below 2 GeV and may represent one of the largest sources of uncertainty in present calculations. The results of ref. [43] include these effects. The calculations based on dimensional regularization may be safe if phenomenological input is used in order to encode in the relevant hadronic matrix elements the physics at all scales (this is done in the Trieste approach).

In summary, while some model dependent calculations suggest no conflict between theory and experiment for ε'/ε , a precise and "pristine" prediction of the observable is still quite ahead of us.

OUTLOOK AND CONCLUSIONS

Higher-order chiral corrections are taking the stage of $K \to \pi\pi$ physics. They are needed in order to asses the size of crucial parameters (as Ω_{IB}) and the effect of non-factorizable contributions in the penguin matrix elements.

Lattice, as a regularization of QCD, is *the* first-principle approach to the problem. However, lattice calculations still heavily depend on chiral perturbation theory [49]. Presently, very promising developments are being undertaken to circumvemt the technical and conceptual shortcomings related to the calculation of weak matrix elements (for a recent survey see ref. [50]). Among those are the Domain Wall Fermion approach [51] which allows us to decouple the chiral symmetry from the continuum limit, and the very interesting observation that the Maiani-Testa theorem [52] can be overcomed using the fact that lattice calculations are performed in finite volume [53], thus allowing for the direct calculation of the physical $K \to \pi\pi$ amplitude on the lattice. All these developments need a tremendous effort in machine power and in devising faster algorithms. Preliminary results for lattice calculations of both ε'/ε and the $\Delta I = 1/2$ selection rule are already available and others are currently under way [41].

In the meantime analytical and semi-phenomenological approaches have been crucially helpful in driving the attention of the community on some systematic shortcomings of "first-principle" calculations. The amount of theoretical work triggered by the NA48 and KTeV data promises rewarding and perhaps exciting results within the next few years.

ACKNOWLEDGMENTS

I wish to thank the organizers of *Heavy Flavours 9* for the excellent organization of the meeting at a time when world events played the lead in everyone's mind.

REFERENCES

1. V. Fanti *et al.* , *Phys. Lett. B* **465**, 335 (1999); A. Lai *et al.* , *Eur. Phys. J. C* **22**, 231 (2001); F. Marchetto, these Proceedings.
2. A. Alavi-Harati *et al.* , *Phys. Rev. Lett.* **83**, 22 (1999); Y. Hsiung, these Proceedings.
3. J.H. Christenson *et al.* , *Phys. Rev. Lett.* **13**, 138 (1964).
4. S. Bertolini, J.O. Eeg and M. Fabbrichesi, *Rev. Mod. Phys.* **72**, 65 (2000); A.J. Buras, *Erice School Lectures 2000*, hep-ph/0101336.
5. M. Gell-Mann and A. Pais, *Proc. Glasgow Conf.* , 342 (1955).
6. M. Ciuchini and G. Martinelli, hep-ph/0006056.
7. J. Gasser and U.G. Meissner, *Phys. Lett. B* **258**, 219 (1991); E. Chell and M.G. Olsson, *Phys. Rev. D* **48**, 4076 (1993).
8. M. Ciuchini *et al.* , *JHEP* **0107**, 013 (2001).

9. H. Lacker, these Proceeedings.
10. A.J. Buras, Proceedings of *Kaon 2001*, Pisa, Italy, hep-ph/0109197.
11. A.J. Buras *et al.* , *Nucl. Phys. B* **370**, 69 (1992); *Nucl. Phys. B* **400**, 37 (1993); *Nucl. Phys. B* **400**, 75 (1993); *Nucl. Phys. B* **408**, 209 (1993).
12. M. Ciuchini *et al.* , *Phys. Lett. B* **301**, 263 (1993); *Nucl. Phys. B* **415**, 403 (1994).
13. J. Gasser and H. Leutwyler, *Nucl. Phys. B* **465**, 1985 (.)
14. G. Ecker *et al.* , *Phys. Lett. B* **467**, 88 (2000).
15. S. Gardner and G. Valencia, *Phys. Lett. B* **466**, 355 (1999).
16. S. Gardner and G. Valencia, *Phys. Rev. D* **62**, 094024 (2000).
17. K. Maltman and C. Wolfe, *Phys. Lett. B* **482**, 77 (2000); *Phys. Rev. D* **63**, 014008 (2001).
18. V. Cirigliano, J.F. Donoghue and E. Golowich, *Phys. Lett. B* **450**, 241 (1999); *Phys. Rev. D* **61**, 093001 (2000); *Phys. Rev. D* **61**, 093002 (2000); *Eur. Phys. J. C* **18**, 83 (2000).
19. A.J. Buras, M. Jamin and E. Lautenbacher, *Phys. Lett. B* **389**, 749 (1996).
20. Ciuchini *et al.* , *Z. Phys. C* **68**, 239 (1995); *Nucl. Phys. Proc. Suppl.* **59**, 149 (1997).
21. W.A. Bardeen, A.J. Buras and J.M. Gerard, *Nucl. Phys. B* **293**, 787 (1987).
22. G. Martinelli, Proceedings of the *Kaon 99* Conference, Chicago, USA, hep-ph/9910237.
23. S. Weinberg, *Physica* **A**, 96 (327)1979; A. Manhoar and H. Georgi, *Nucl. Phys. B* **234**, 189 (1984).
24. A. Pich and E. de Rafael, *Nucl. Phys. B* **358**, 311 (1991).
25. S. Bertolini *et al.* , *Nucl. Phys. B* **514**, 63 (1998).
26. S. Bertolini *et al.* , *Nucl. Phys. B* **514**, 93 (1998).
27. S. Bertolini, Proceedings of *RADCOR 2000*, Carmel, USA, eConf C000911, hep-ph/0101212.
28. A.J. Buras, Proceedings of the *Kaon 99* Conference, Chicago, USA, hep-ph/9908395.
29. T. Hambye *et al.* , *Phys. Rev. D* **58**, 014017 (1998); *Nucl. Phys. B* **564**, 391 (2000).
30. Yue-Liang Wu, *Phys. Rev. D* **64**, 016001 (2001).
31. H.Y. Cheng, *Chin. J. Phys.* **38**, 1044 (2000).
32. E. Pallante and A. Pich, *Phys. Rev. Lett.* **84**, 2568 (2000); *Nucl. Phys. B* **592**, 294 (2000); E. Pallante, A. Pich and I. Scimemi, *Nucl. Phys. B* **617**, 441 (2001).
33. A.A. Bel'kov *et al.* , hep-ph/9907335;
34. S. Bertolini, J.O. Eeg and M. Fabbrichesi, *Phys. Rev. D* **63**, 056009 (2001).
35. M. Fabbrichesi, *Phys. Rev. D* **62**, 097902 (2000).
36. J. Bijnens and J. Prades, *JHEP* **0006**, 035 (2000).
37. S. Narison, *Nucl. Phys. B* **593**, 3 (2001).
38. J. Noaki *et al.* , hep-lat/0108013.
39. T. Blum *et al.* , hep-lat/0110075.
40. A. Donini, V. Gimenez, L. Giusti and G. Martinelli, *Phys. Lett. B* **470**, 233 (1999).
41. G. Martinelli, *Nucl. Phys. Proc. Suppl.* **106**, 98 (2002); Also in *Berlin 2001, Lattice Field Theory* 98-110, hep-lat/0112011.
42. M. Knecht, S. Peris and E. de Rafael, *Phys. Lett. B* **508**, 117 (2001).
43. V. Cirigliano, J.F. Donoghue, E. Golowich and K. Maltman, *Phys. Lett. B* **522**, 245 (2001).
44. J. Bijnens, E. Gamiz and J. Prades, *JHEP* **0110**, 009 (2001).
45. E. Fermi, *Suppl. Nuovo Cim.* **2**, 17 (1955).
46. N.I. Mushkelishvili, *Singular Integral Equations* (Noordhoff, Gronigen, 1953), p. 204; R. Omnès, *Nuovo Cim.* **8**, 316 (1958).
47. A. J. Buras *et al.* , *Phys. Lett. B* **480**, 80 (2000); M. Buechler *et al.* , *Phys. Lett. B* **521**, 29 (2001).
48. V. Cirigliano, J.F. Donoghue and E. Golowich, *JHEP* **0010**, 048 (2000).
49. M. Golterman and E. Pallante, *Nucl. Phys. Proc. Suppl.* **106**, 335 (2002); Also in *Berlin 2001, Lattice field theory* 335-337, hep-lat/0110183.
50. C.T. Sachrajda, Proceedings of *Lepton-Photon 2001*, Roma, Italy, hep-ph/0110304.
51. D. Kaplan, *Phys. Lett. B* **288**, 342 (1992).
52. L. Maiani and M. Testa, *Phys. Lett. B* **245**, 585 (1990).
53. L. Lellouch and M. Lüscher, *Comm. Math. Phys.* **219**, 31 (2001).

Rare K Decays: Results and Prospects

Laurence Littenberg

Brookhaven National Laboratory, Upton, NY 11973

Abstract. Recent results on rare kaon decays are reviewed and prospects for future experiments are discussed.

INTRODUCTION

In recent years the study of the rare decays of kaons has had three primary motivations. The first is the search for physics beyond the Standard Model (BSM). Virtually all attempts to redress the theoretical shortcomings of the Standard Model (SM) predict some degree of lepton flavor violation (LFV). Decays such as $K_L \to \mu^{\pm}e^{\mp}$ have very good experimental signatures and can consequently be pursued to remarkable sensitivities. These sensitivities correspond to extremely high energy scales in models where the only suppression is that of the mass of the exchanged field. There are also theories that predict new particles created in kaon decay or the violation of symmetries other than lepton flavor.

The second is the potential of decays that are allowed but that are extremely suppressed in the SM. In several of these, the leading component is a G.I.M.-suppressed[1] one-loop process that is quite sensitive to fundamental SM parameters such as V_{td}. These decays are also potentially very sensitive to BSM physics.

Finally there are a number of long-distance-dominated decays which can test theoretical techniques such as chiral Lagrangians that purport to explain the low-energy behavior of QCD. Knowledge of some of these decays is also needed to extract more fundamental information from certain of the one-loop processes.

This field is quite active as indicated by Table 1 that lists the decays for which results have been forthcoming in the last couple of years as well as those that are under analysis. Thus in a short review such as this, one must be quite selective.

BEYOND THE STANDARD MODEL

There were several K decay experiments dedicated to lepton flavor violation at the Brookhaven AGS during the 1990's. These advanced the sensitivity to such processes by many orders of magnitude. In addition, several "by-product" results on LFV and other BSM topics have emerged from the other kaon decay experiments of this period. Table 2 summarizes the status of work on BSM probes in kaon decay.

CP618, *Heavy Flavor Physics: Ninth International Symposium*, edited by A. Ryd and F. C. Porter

TABLE 1. *Rare K decay modes under recent or on-going study.*

$K^+ \to \pi^+\nu\bar{\nu}$	$K_L \to \pi^0\nu\bar{\nu}$	$K_L \to \pi^0\mu^+\mu^-$	$K_L \to \pi^0e^+e^-$
$K^+ \to \pi^+\mu^+\mu^-$	$K^+ \to \pi^+e^+e^-$	$K_L \to \mu^+\mu^-$	$K_L \to e^+e^-$
$K^+ \to \pi^+\pi^0\nu\bar{\nu}$	$K^+ \to \pi^+e^+e^-\gamma$	$K^+ \to \pi^+\gamma\gamma$	$K_L \to \pi^0\gamma\gamma$
$K_L \to \pi^0e^+e^-\gamma$	$K^+ \to \pi^+\pi^0\gamma$	$K_L \to \pi^+\pi^-\gamma$	$K^+ \to \pi^+\pi^0e^+e^-$
$K_L \to \pi^+\pi^-e^+e^-$	$K^+ \to \mu^+\nu\gamma$	$K^+ \to \mu^+\nu e^+e^-$	$K^+ \to e^+\nu e^+e^-$
$K^+ \to e^+\nu\mu^+\mu^-$	$K_L \to e^+e^-\gamma$	$K_L \to \mu^+\mu^-\gamma$	$K_L \to e^+e^-e^+e^-$
$K_L \to e^\pm e^\mp\mu^\pm\mu^\mp$	$K_L \to e^+e^-\gamma\gamma$	$K_L \to \mu^+\mu^-\gamma\gamma$	$K^+ \to \pi^0\mu^+\nu\gamma$
$K^+ \to \pi^+\mu^+e^-$	$K_L \to \pi^0\mu^\pm e^\mp$	$K_L \to \mu^\pm e^\mp$	$K^+ \to \pi^-\mu^+e^+$
$K^+ \to \pi^-e^+e^+$	$K^+ \to \pi^-\mu^+\mu^+$	$K^+ \to \pi^+X^0$	$K_L \to e^\pm e^\pm\mu^\mp\mu^\mp$
$K^+ \to \pi^+\gamma$			

TABLE 2. *Current 90% CL limits on K decay modes violating the SM. The violation codes are "LF" for lepton flavor, "LN" for lepton number, "G" for generation number, [9], "H" for helicity, "N" requires new particle*

Process	Violates	Limit	Experiment	Reference
$K_L \to \mu e$	LF	4.7×10^{-12}	AGS-871	[2]
$K^+ \to \pi^+\mu^+e^-$	LF	2.8×10^{-11}	AGS-865	[3]
$K^+ \to \pi^+\mu^-e^+$	LF, G	5.2×10^{-10}	AGS-865	[4]
$K_L \to \pi^0\mu e$	LF	4.4×10^{-10}	KTeV	[5]
$K^+ \to \pi^-e^+e^+$	LN, G	6.4×10^{-10}	AGS-865	[4]
$K^+ \to \pi^-\mu^+\mu^+$	LN, G	3.0×10^{-9}	AGS-865	[4]
$K^+ \to \pi^-\mu^+e^+$	LF, LN, G	5.0×10^{-10}	AGS-865	[4]
$K_L \to \mu^\pm\mu^\pm e^\mp e^\mp$	LF, LN, G	1.36×10^{-10}	KTeV	[6]
$K^+ \to \pi^+f^0$	N	5.9×10^{-11}	AGS-787	[7]
$K^+ \to \pi^+\gamma$	H	3.6×10^{-7}	AGS-787	[8]

It is clear from this table that any deviation from the SM must be highly suppressed. The LFV probes in particular have become the victims of their own success. The specific theories they were designed to test have been killed or at least forced to retreat to the point where meaningful tests in the kaon system would be very difficult. Both kaon flux and rejection of background are becoming problematical. Analysis of data already collected is continuing but no new kaon experiments focussed on LFV are being planned. Interest in probing LFV has migrated to the muon sector.

ONE LOOP DECAYS

In the kaon sector experimental effort has shifted from LFV to "one-loop" decays. These are GIM-suppressed decays in which loops containing weak bosons and heavy quarks dominate or at least contribute measurably to the rate. These processes include $K_L \to \pi^0\nu\bar{\nu}$, $K^+ \to \pi^+\nu\bar{\nu}$, $K_L \to \mu^+\mu^-$, $K_L \to \pi^0e^+e^-$ and $K_L \to \pi^0\mu^+\mu^-$. In some cases the one-loop contributions violate CP. In one, $K_L \to \pi^0\nu\bar{\nu}$, this contribution completely dominates the decay[10]. Since the GIM-mechanism tends to enhance the contribution of heavy quarks in the loops, in the SM these decays are sensitive to the product of cou-

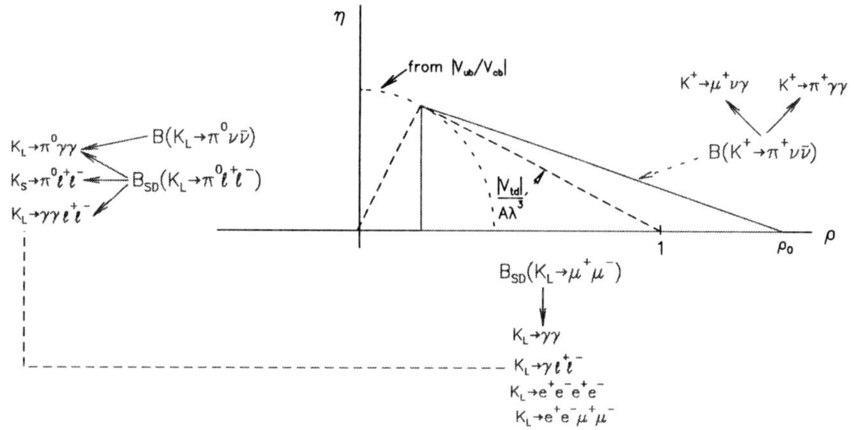

FIGURE 1. *K* decays and the unitarity plane. The usual unitarity triangle is dashed. The triangle that can be constructed from rare *K* decays is solid. See text for further details.

plings $V_{ts}^* V_{td}$, often abbreviated as λ_t. Although one can readily analyze these decays in terms of the real and imaginary parts of λ_t, for comparison with results in the *B* system, it is conventional to parameterize them in terms of the Wolfenstein variables, *A*, ρ, and η. Fig. 1 shows the relationship of rare kaon decays to the unitarity triangle construction. The dashed triangle is the usual one derived from $V_{ub}^* V_{ud} + V_{cb}^* V_{cd} + V_{tb}^* V_{td} = 0$, whereas the solid triangle illustrates the information available from rare kaon decays. Note that the "unitarity point" at the apex, (ρ, η), can be determined from either triangle, and disagreement between the *K* and *B* determinations implies physics beyond the SM. In Fig. 1 the branching ratio closest to each side of the triangle determines the length of that side. The arrows leading outward from those branching ratios point to processes that need to be studied either because they potentially constitute backgrounds, or because knowledge of them is required to relate the innermost branching ratios to fundamental parameters. $K_L \to \mu^+\mu^-$, which can determine the bottom of the triangle (ρ), is the process for which the experimental data is the best, but for which the theory is most problematical. $K_L \to \pi^0 \nu\bar{\nu}$, which determines the height of the triangle is theoretically the cleanest, but experiment is many orders of magnitude short of the SM-predicted level. $K^+ \to \pi^+ \nu\bar{\nu}$, which determines the hypotenuse, is nearly as clean as $K_L \to \pi^0 \nu\bar{\nu}$ and has been observed. Prospects for $K^+ \to \pi^+ \nu\bar{\nu}$ are probably the best of the three since it is already clear it can be exploited.

$K_L \to \mu^+\mu^-$

The short distance component of this decay can be quite reliably calculated in the SM[11]. The most recent measurement of its branching ratio[12] based on some 6200 events gave $B(K_L \to \mu^+\mu^-) = (7.18 \pm 0.17) \times 10^{-9}$. However $K_L \to \mu^+\mu^-$ is dominated by long distance effects, the largest of which, the absorptive contribution mediated by

from several exps. last in 1976!

calculated — *last measured by NA31 in 1987* — $\frac{B(K_L \to \pi^0 \pi 0)}{B(K_L \to \pi^+ \pi^-)}$

$$\frac{B^{abs}_{\gamma\gamma}(K_L \to \mu\mu)}{B(K_L \to \pi^+\pi^-)} = \frac{B^{abs}_{\gamma\gamma}(K_L \to \mu\mu)}{B(K_L \to \gamma\gamma)} \frac{B(K_L \to \gamma\gamma)}{B(K_L \to \pi^0\pi^0)} \frac{B(K_S \to \pi^0\pi^0)}{B(K_S \to \pi^+\pi^-)} \left(1 - 6Re\frac{\epsilon'}{\epsilon}\right)$$

$1.2 \cdot 10^{-5}$

0.632 ± 0.009

$(2.185 \pm 0.027)^{-1}$

$1 - 6\,(17.2 \pm 1.8) \cdot 10^{-4}$

FIGURE 2. Components of the calculation of $\frac{B^{abs}_{\gamma\gamma}(K_L \to \mu^+\mu^-)}{B(K_L \to \pi^+\pi^-)}$.

$K_L \to \gamma\gamma$, accounts for $(7.07 \pm 0.18) \times 10^{-9}$. Subtracting the two, yields a 90% CL upper limit on the total dispersive part of $B(K_L \to \mu^+\mu^-)$ of 0.37×10^{-9}. One can do a little better than this in the following way. The actual quantity measured in Ref [12] was $\frac{B(K_L \to \mu^+\mu^-)}{B(K_L \to \pi^+\pi^-)} = (3.48 \pm 0.05) \times 10^{-6}$ One wants to subtract from this measured quantity the ratio $\frac{B^{abs}_{\gamma\gamma}(K_L \to \mu^+\mu^-)}{B(K_L \to \pi^+\pi^-)}$. Fig 2 shows the components of this latter ratio, obtained from Ref. [13], whose product is $(3.435 \pm 0.065) \times 10^{-6}$.

The subtraction yields $\frac{B^{disp}(K_L \to \mu^+\mu^-)}{B(K_L \to \pi^+\pi^-)} = (0.045 \pm 0.082) \times 10^{-6}$ (where B^{disp} refers to the dispersive part of $B(K_L \to \mu^+\mu^-)$). $\frac{B^{disp}(K_L \to \mu^+\mu^-)}{B(K_L \to \pi^+\pi^-)}$ can then be multiplied by $B(K_L \to \pi^+\pi^-) = (2.056 \pm 0.033) \times 10^{-3}$ to obtain $B^{disp}(K_L \to \mu^+\mu^-) = (0.093 \pm 0.169) \times 10^{-9}$, or $B^{disp}(K_L \to \mu^+\mu^-) < 0.31 \times 10^{-9}$ at 90% CL. Note that some of the components represent quite old measurements. Since $B(K_L \to \mu^+\mu^-)$ and $B^{abs}_{\gamma\gamma}(K_L \to \mu^+\mu^-)$ are so close, small shifts in the component values could have relatively large consequences for $B^{disp}(K_L \to \mu^+\mu^-)$. Several of the components could be remeasured by experiments presently in progress[1]. Now if one inserts the result of even very conservative recent CKM fits into the formula for the short distance part of $B(K_L \to \mu^+\mu^-)$, one gets rather poor agreement with the limit of $B^{disp}(K_L \to \mu^+\mu^-)$ derived above. For example the 95% CL fit of Hocker et al.[15][16], $\bar{\rho} = 0.07 - 0.37$, gives $B^{SD}(K_L \to \mu^+\mu^-) = (0.4 - 1.3) \times 10^{-9}$. So why haven't we been hearing about this apparent violation of the SM?

The answer is that unfortunately $K_L \to \gamma^*\gamma^*$ also gives rise to a dispersive contribution, which is much less tractable than the absorptive part, and which can interfere with the short-distance weak contribution that one is trying to extract. The problem in calculating this contribution is the necessity of including intermediate states with virtual photons of

---

[1] There is a new preliminary result from the KLOE experiment of $B(K_S \to \pi^+\pi^-)/B(K_S \to \pi^0\pi^0) = 2.192 \pm 0.003_{stat} \pm 0.016_{syst}$[14]. Fortuitously, inserting this result in place of the PDG value makes no difference to the final limit.

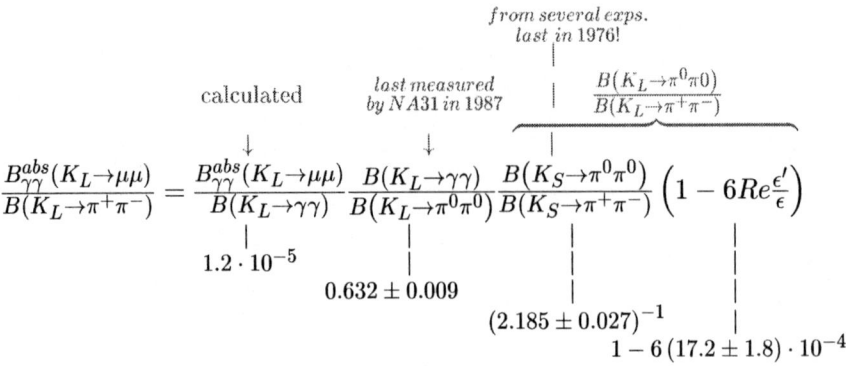

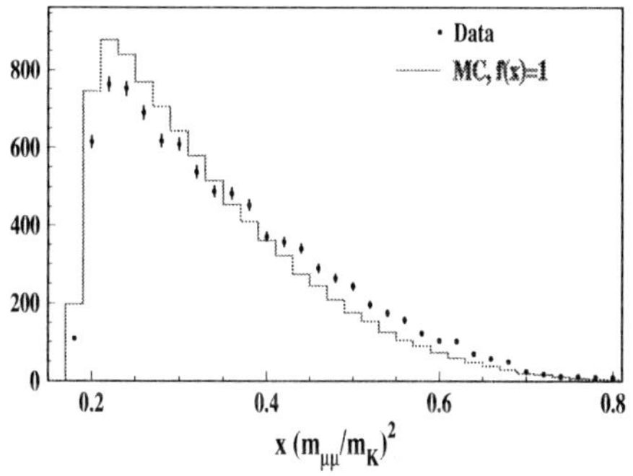

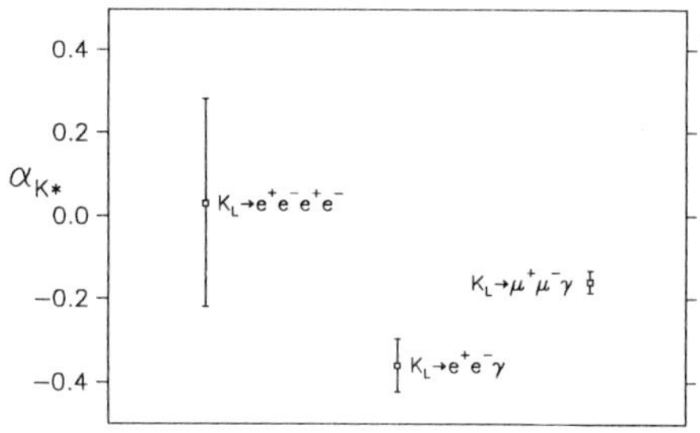

FIGURE 3. Left: Spectrum of $x = (m_{\mu\mu}/m_K)^2$ in $K_L \to \mu^+\mu^-\gamma$ from Ref.[17]. Right: Determinations of the BMS parameter α_{K^*} from three K_L decays involving virtual photons.

all effective masses. Thus such calculations can only partially be validated by studies of processes containing virtual photons in the final state. Recently there have been publications on $K_L \to \gamma\mu^+\mu^-$[17] (9327 events), $K_L \to e^+e^-e^+e^-$[18] (441 events), and $K_L \to \mu^+\mu^-e^+e^-$[6] (38 events) and there exist slightly older high statistics data on $K_L \to \gamma e^+e^-$[19] (6864 events). Figure 3-left shows the spectrum of $x = (m_{\mu\mu}/m_K)^2$ from Ref.[17]. The disagreement between the data (filled circles with error bars) and the prediction of pointlike behavior (histogram) clearly indicates the presence of a form factor. A long-standing candidate for this is provided by the BMS model[20] which depends on a single parameter, α_{K^*}.

Fig. 3-right shows three determinations of this parameter. The level of agreement of these results leaves something to be desired. Fitting to a more recent parameterization of these decays[21] also results in quite marginal agreement. This may improve when radiative corrections are properly taken into account. Thus additional effort, both experimental and theoretical, is required before the quite precise data on $B(K_L \to \mu^+\mu^-)$ can be fully exploited.

$K^+ \to \pi^+\nu\bar{\nu}$

Theoretically $K^+ \to \pi^+\nu\bar{\nu}$ is remarkably clean, suffering from none of the long distance complications of $K_L \to \mu^+\mu^-$. The hadronic matrix element, so often a problem in other processes, can be calculated to a $\sim 2\%$ via an isospin transformation from that of K_{e3}[22]. Interest in $K^+ \to \pi^+\nu\bar{\nu}$ is driven in large part by its sensitivity to V_{td} (it is actually directly sensitive to the quantity $|V_{ts}^* V_{td}|$). Its amplitude is proportional to the dark slanted line at the right in Fig. 1. This is equal to the vector sum of the line proportional to $|V_{td}|/A\lambda^3$ (where $\lambda \equiv sin\theta_{Cabibbo}$) and that from $(1,0)$ to the point marked ρ_0. The length $\rho_0 - 1$ along the real axis is proportional to the amplitude for the charm contribution to $K^+ \to \pi^+\nu\bar{\nu}$. The QCD corrections to this amplitude, which are responsible for the largest uncertainty in $B(K^+ \to \pi^+\nu\bar{\nu})$, have been calculated to NLLA[11]. The residual uncertainty in the charm amplitude is estimated to be $\sim 15\%$ which leads to only a $\sim 6\%$ uncertainty[23] in extracting $|V_{td}|$

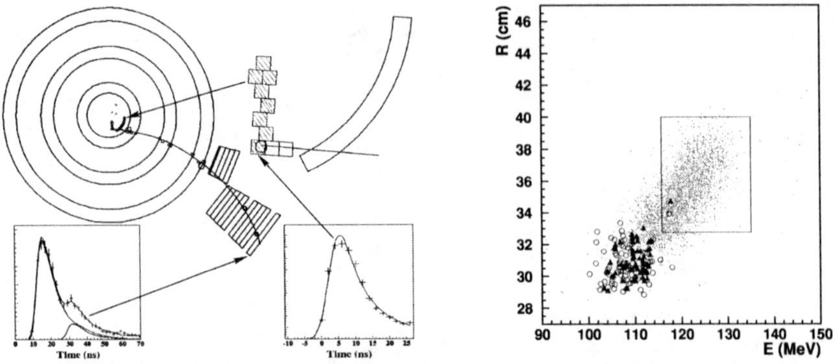

FIGURE 4. Left: new $K^+ \to \pi^+\nu\bar{\nu}$ event. Right: Range vs energy of π^+ in the final sample. The circles are 1998 data and the triangles 1995-7 data. The events around $E = 108$ MeV are $K^+ \to \pi^+\pi^0$ background. The simulated distribution of expected signal events is indicated by dots.

from $B(K^+ \to \pi^+\nu\bar{\nu})$. Recently AGS E787 has seen evidence for a second event of $K^+ \to \pi^+\nu\bar{\nu}$ [7] (see Fig. 4) which, combined with previous data [24], yields a branching ratio $B(K^+ \to \pi^+\nu\bar{\nu}) = (1.57^{+1.75}_{-0.82}) \times 10^{-10}$. By comparison, a fit to the CKM phenomenology yields the expectation $(0.72 \pm 0.21) \times 10^{-10}$[25]. It is notable that E787 has established methods to reduce the residual background to $\sim 10\%$ of the the signal branching ratio predicted by the SM.

A new experiment, AGS E949[26], based on an upgrade of the E787 detector, is about to begin its first physics run. Using the entire flux of the AGS for 6000 hours, it is designed to reach a sensitivity of $\sim 10^{-11}$/event. In June 2001, Fermilab gave Stage 1 approval to an experiment (CKM[27]) to extend the study of $K^+ \to \pi^+\nu\bar{\nu}$ by yet an another order of magnitude in sensitivity. This experiment, unlike all previous ones on this process, uses an in-flight rather than a stopping K^+ technique. This experiment is expected to start collecting data in 2007 or 2008.

Fig. 5 shows the history and expectations of progress in studying $K^+ \to \pi^+\nu\bar{\nu}$.

$K_L \to \pi^0\nu\bar{\nu}$

$K_L \to \pi^0\nu\bar{\nu}$ is the most attractive target in the kaon system, since it is direct CP-violating to a very good approximation[10, 28] ($B(K_L \to \pi^0\nu\bar{\nu}) \propto \eta^2$). Like $K^+ \to \pi^+\nu\bar{\nu}$ it has a hadronic matrix element that can be obtained from K_{e3}, but, it has no significant contribution from charm. As a result, the intrinsic theoretical uncertainty connecting $B(K_L \to \pi^0\nu\bar{\nu})$ to the fundamental SM parameters is only about 2%. Note also that $B(K_L \to \pi^0\nu\bar{\nu})$ is directly proportional to the square of $Im\lambda_t$ and that $Im\lambda_t = -\mathcal{J}/[\lambda(1-\frac{\lambda^2}{2})]$ where $\mathcal{J}$ is the Jarlskog invariant[29]. Thus a measurement of $B(K_L \to \pi^0\nu\bar{\nu})$ determines the area of the unitarity triangles with a precision twice as good as that on $B(K_L \to \pi^0\nu\bar{\nu})$ itself.

$B(K_L \to \pi^0\nu\bar{\nu})$ can be bounded indirectly by measurements of $B(K^+ \to \pi^+\nu\bar{\nu})$

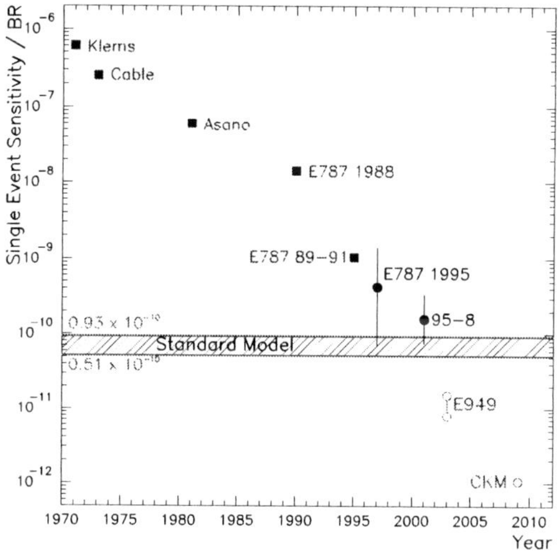

FIGURE 5. History and prospects for the study of $K^+ \to \pi^+\nu\bar{\nu}$. Points without error bars are single event sensitivities, those with error bars are measured branching ratio.

through a nearly model-independent relationship pointed out by Grossman and Nir[30]. The application of this to the new E787 result yields $B(K_L \to \pi^0\nu\bar{\nu}) < 1.7 \times 10^{-9}$ at 90% CL. This is far tighter than the current direct experimental limit, 5.9×10^{-7}, obtained by KTeV[31]. To actually measure $B(K_L \to \pi^0\nu\bar{\nu})$ at the SM level ($\sim 3 \times 10^{-11}$), one will need to improve on this by some five orders of magnitude. The KEK E391a experiment[32] proposes to achieve a sensitivity of $\sim 3 \times 10^{-10}$/event which would better the indirect limit by a factor five, but would not quite bridge this gap. It will serve as a test for a future much more sensitive experiment to be performed at the Japanese Hadron Facility. E391a features a carefully designed "pencil" beam, and a very high performance photon veto. The active photon detector is a CsI-pure crystal calorimeter. The entire rather compact apparatus will operate in vacuum. Beamline construction and tuning started in March 2000 and physics running is expected to begin in Fall, 2003.

The KOPIO experiment[33] at BNL (E926) takes a completely different approach, exploiting the intensity and flexibility of the AGS to make a high-flux, low-energy, microbunched K_L beam. The proposed experiment is shown in Fig. 6. The neutral beam will be extracted at $\sim 45°$ to soften the K_L spectrum sufficiently to permit time-of-flight determination of the K_L velocity. The large production angle also softens the neutron spectrum so that they (and the K_L) are by and large below threshold for the hadro-production of π^0's. The beam region will be evacuated to 10^{-7} Torr to further minimize such production. With a 10m beam channel and this low energy beam, the contribution of hyperons to the background will be negligible. K_L decays from a $\sim$ 3m fiducial region will be accepted. Signal photons impinge on a 2 X_0 thick preradiator capable of measuring their direction. An alternating drift chamber/scintillator plane structure will allow energy measurement as well. A high-precision shashlyk calorimeter downstream of the preradiator will complete the energy measurement. The photon directional information will allow the decay vertex position to be determined. Combined with the target position and time of flight information, this provides a measurement of the K_L 3-momentum so that kinematic constraints as well as photon vetoing are available to suppress backgrounds. The leading expected background is $K_L \to \pi^0\pi^0$, which is initially some eight orders of magnitude larger than the predicted signal. However since π^0's from this background have a unique energy in the K_L center of mass, a very effective kinematic cut can be applied. This reduces the load on the photon veto system surrounding the decay region to the point where the hermetic veto techniques proven in E787 are sufficient. In fact most of the techniques necessary for KOPIO have been proven in previous experiments or in prototype tests. KOPIO aims to collect about 50 $K_L \to \pi^0\nu\bar{\nu}$ events with a signal to background ratio of 2:1. This will permit η to be determined to $\sim$ 10%, given expected progress in measuring m_t and V_{cb}. KOPIO will run during the ~20 hours/day the AGS is not needed for injection into RHIC.

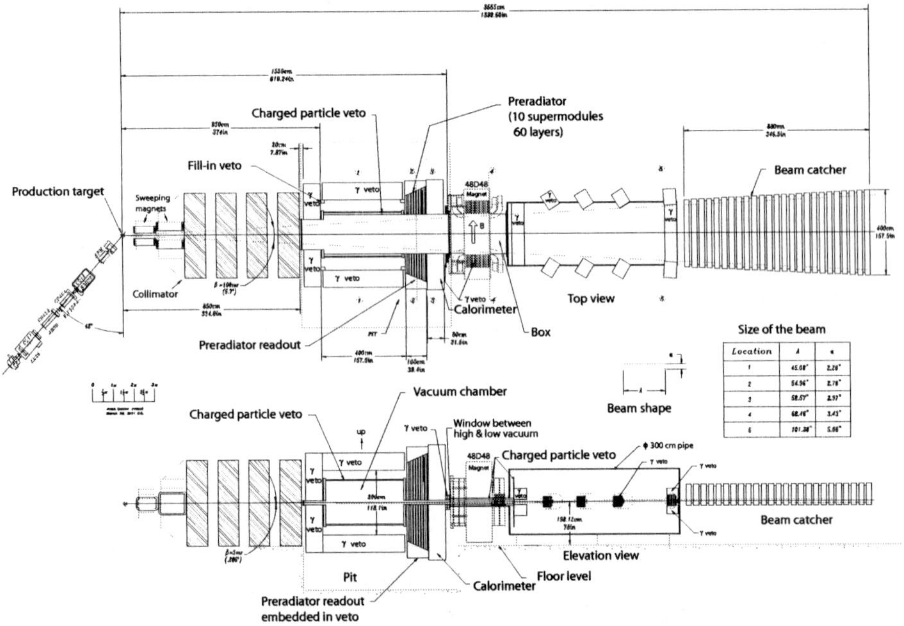

FIGURE 6. Layout of the KOPIO detector.

$K_L \to \pi^0\ell^+\ell^-$

These are reactions initially thought experimentally more tractable than $K_L \to \pi^0\nu\bar{\nu}$. Like $K_L \to \pi^0\nu\bar{\nu}$, in the SM they are sensitive to $Im\lambda_t$, but in general they have different sensitivity to BSM effects[34]. Although their signatures are intrinsically superior to that of $K_L \to \pi^0\nu\bar{\nu}$, they are subject to a serious background that has no analogue in the case of the latter: $K_L \to \gamma\gamma\ell^+\ell^-$. This process, a radiative correction to $K_L \to \gamma\ell^+\ell^-$, occurs roughly 10^5 times more frequently than $K_L \to \pi^0\ell^+\ell^-$. Kinematic cuts are quite effective, but it is very difficult to improve the signal:background beyond about 1 : 5[35]. Both varieties of $K_L \to \gamma\gamma\ell^+\ell^-$ have been observed, $B(K_L \to \gamma\gamma e^+e^-)_{k_\gamma > 5MeV} = (5.84 \pm 0.15(stat) \pm 0.32(syst)) \times 10^{-7}$[36] and $B(K_L \to \gamma\gamma\mu^+\mu^-)_{m_{\gamma\gamma} > 1MeV/c^2} = (10.4^{+7.5}_{-5.9}(stat) \pm 0.7(syst)) \times 10^{-9}$[37]; both agree with theoretical prediction. By comparison, in the SM $B^{direct}(K_L \to \pi^0 e^+e^-)$ is predicted to be [38] $(4.3 \pm 2.1) \times 10^{-12}$ and $B^{direct}(K_L \to \pi^0\mu^+\mu^-)$ about fives times smaller.

In addition to this background, there are two other issues that make the extraction of short-distance information from $K_L \to \pi^0\ell^+\ell^-$ problematical. First, there is an indirect CP-violating contribution from the K_1 component of K_L given by $|\varepsilon|^2 \frac{\tau_{K_L}}{\tau_{K_S}} B(K_S \to \pi^0 e^+e^-)$ which is of the same order of magnitude as the direct CP-violating piece[2]. The exact size of this contribution will be predictable if and when $B(K_S \to \pi^0 e^+e^-)$ is measured, hopefully by the upcoming NA48/1 experiment[39]. Second is yet another contribution of similar size mediated by $K_L \to \pi^0\gamma\gamma$ which is *CP-conserving*. To some extent this contribution can be predicted from measurements of the branching ratio and kinematic distributions of $K_L \to \pi^0\gamma\gamma$, and thousands of these events have been observed. However as indicated in Table 3, a new result from NA48[40] disagrees by nearly 3σ from the previous result from KTeV[41]. The change in the vector meson exchange contribution, characterized by the parameter a_V, reduces the predicted size of $B^{CP-cons}(K_L \to \pi^0 e^+e^-)$ considerably[42] which is good news for the prospects of measuring $B^{direct}(K_L \to \pi^0 e^+e^-)$. However the validity of the current technique for predicting $B^{CP-cons}(K_L \to \pi^0 e^+e^-)$ from $K_L \to \pi^0\gamma\gamma$ has recently been reexamined [43], and questions raised about the functions used to fit the spectrum and about the treatment of the dispersive contribution. Thus both the theoretical and experimental situations are quite unsettled at the moment. Depending on whose data and whose theory one uses, values from 0.25×10^{-12} to 7.3×10^{-12} are predicted for $B^{CP-cons}(K_L \to \pi^0 e^+e^-)$.

TABLE 3. *Results on $K_L \to \pi^0\gamma\gamma$.*

Exp.	$B(K_L \to \pi^0\gamma\gamma) \times 10^6$	a_V	Ref.
KTeV	$1.68 \pm 0.07_{stat} \pm 0.08_{syst}$	$-0.72 \pm 0.05 \pm 0.06$	[41]
NA48	$1.36 \pm 0.03_{stat} \pm 0.03_{syst} \pm 0.03_{norm}$	$-0.46 \pm 0.03 \pm 0.03 \pm 0.02_{theor}$	[40]

The current experimental status of $K_L \to \pi^0\ell^+\ell^-$ is summarized in Table 4. A factor ~ 2.5 more data is expected from the KTeV 1999 run, but as can be seen from the table, background is already starting to be observed at a sensitivity roughly 100 times short of the expected signal level.

To make a useful measurement under these conditions will require markedly increased statistics on the signal and both theoretical and experimental advances in the ancillary modes $K_L \to \pi^0\gamma\gamma$ and $K_S \to \pi^0 e^+e^-$. Various approaches for mitigating these problems have been suggested over the years including studies of the Dalitz Plot [46], the

[2] There is also an interference term between the indirect and direct CP-violating amplitudes.

TABLE 4. *Results on* $K_L \to \pi^0 \ell^+ \ell^-$.

Mode	90% CL upper limit	Est. bkgnd.	Obs. evts.	Ref.
$K_L \to \pi^0 e^+ e^-$	5.1×10^{-10}	1.06 ± 0.41	2	[44]
$K_L \to \pi^0 \mu^+ \mu^-$	3.8×10^{-10}	0.87 ± 0.15	2	[45]

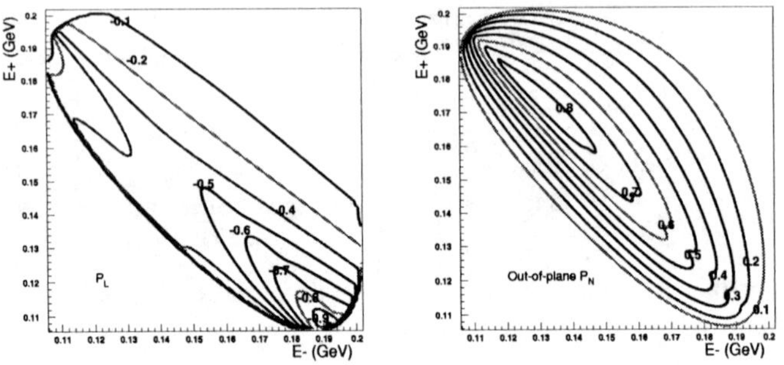

FIGURE 7. μ^+ polarizations in $K_L \to \pi^0 \mu^+ \mu^-$, plotted against the muon cm energies. Left: longitudinal polarization. Right: out-of-plane polarization.

time development [47], or both [48]. However an innovative approach has recently been suggested [49] in which muon polarization in $K_L \to \pi^0 \mu^+ \mu^-$ as well as kinematic distributions are exploited. The μ^+ longitudinal polarization is proportional to the direct CP-violating amplitude, whereas the energy asymmetry and the out-of-plane polarization depend on both indirect and direct CP violating amplitudes. As shown in Fig. 7, the polarizations involved turn out to be extremely large so that enormous numbers of events may not be required.

An alternative parameterization

Although it is customary to write the branching ratios and other observables of the one-loop processes in terms of the Wolfenstein parameterization of the CKM matrix, this parameterization is not really natural to the kaon system, and puts results from this system at a certain disadvantage in comparisons with those from the B system. To extract information on ρ and η, for example, it is necessary to divide the physical measurements by $\lambda^8 A^4$, thereby introducing "external" contributions to the uncertainty of $8\sigma_\lambda$ and $4\sigma_A$. One can avoid this by resorting to expressions for the branching ratios in terms of the quantity λ_t. Since as noted above, the imaginary part of this quantity determines the area of all unitarity triangles, it is no less fundamental than ρ and η.

The formulae for the branching ratios of three of the decays discussed above are:

$$B(K^+ \to \pi^+ \nu\bar{\nu}) = \xi[(\lambda_c \bar{X} + Re(\lambda_t)X_t)^2 + (Im(\lambda_t)X_t)^2] \quad (1)$$

$$B^{SD}(K_L \to \mu^+\mu^-) = \xi'[Re(\lambda_c)Y_{NL} + Re(\lambda_t)Y(x_t)]^2 \tag{2}$$

$$B(K_L \to \pi^0\nu\bar{\nu}) = \xi''(Im(\lambda_t)X_t)^2 \tag{3}$$

where

$$\xi \equiv \frac{3r_{K^+}\alpha^2 B_{K^+e3}}{V_{us}^2 2\pi^2 sin^4\theta_W} = 1.55 \times 10^{-4} \tag{4}$$

$$\xi' \equiv \frac{\tau_{K_L}\alpha^2 B_{K^+\mu\nu}}{\tau_{K^+}V_{us}^2\pi^2 sin^4\theta_W} = 6.32 \times 10^{-3} \tag{5}$$

$$\xi'' \equiv \frac{3r_{K_L}\tau_{K_L}\alpha^2 B_{K^+e3}}{\tau_{K^+}V_{us}^2 2\pi^2 sin^4\theta_W} = 6.77 \times 10^{-4} \tag{6}$$

and to a good approximation the Inami-Lim [50] functions $X_t = 1.56(m_t/170\text{GeV})^{1.15}$, $Y_t = 1.02(m_t/170\text{GeV})^{1.56}$. The quantities $r_{K^+} = 0.901$ and $r_{K_L} = 0.944$ are isospin correction factors that relate the hadronic matrix elements of the $K \to \pi\nu\bar{\nu}$ processes to that of $K^+ \to \pi^0 e^+\nu$ [22]. The terms $\bar{X}$ and Y_{NL} are the Inami-Lim functions for the charm contributions which, after correction to NLLA, are known to about 15%.

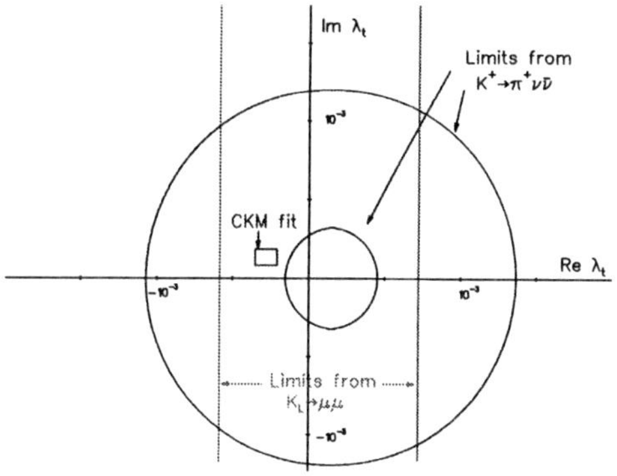

FIGURE 8. Comparison of 90% CL constraints from current data on rare kaon decays with 95% CL constraints from a typical unitarity fit (based on Ref. [16]). The allowed region from kaon decays lies between the two circles and within the outer two vertical lines.

Fig. 8 shows the 90% CL constraints currently available from $K_L \to \mu^+\mu^-$ and $K^+ \to \pi^+\nu\bar{\nu}$. To extract a limit from the former we adopt the value for the maximum long distance dispersive contribution from Ref. [21]. Also shown is the region in the λ_t plane bounded by the CKM fit mentioned above [16]. The two kinds of information are clearly consistent at the moment. Note, however, that if the current central value of $B(K^+ \to \pi^+\nu\bar{\nu})$ should hold through E949, and expected progress is made in the B sector, this agreement could prove short-lived, as shown on the left of Fig. 9. When, eventually, we have 10-15% measurements of $B(K^+ \to \pi^+\nu\bar{\nu})$ and $B(K_L \to \pi^0\nu\bar{\nu})$, comparison of K and B results will become a critical test of the SM. Fig. 9 (right) illustrates a scenario in which such a failure is evident.

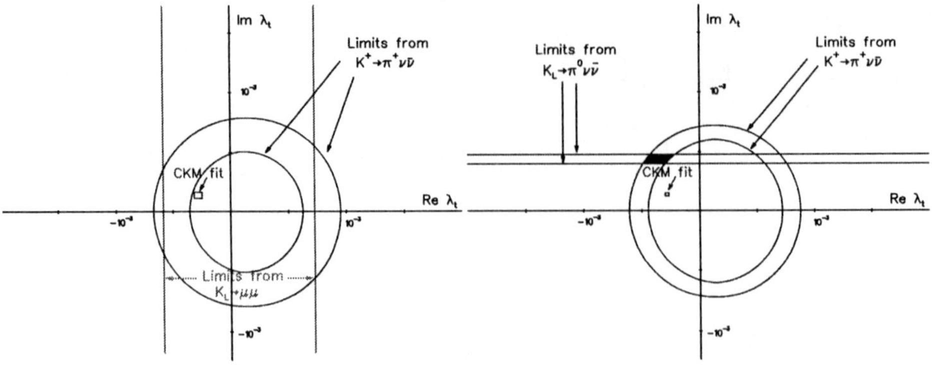

FIGURE 9. Left: Similar plot to Fig. 8 after 10^{-11}/event $K^+ \to \pi^+\nu\bar{\nu}$ experiment. Assumes central value of $B(K^+ \to \pi^+\nu\bar{\nu})$ stays the same and also that CKM fit contours and precision on m_t are improved by a factor 2. Right: Similar plot for possible scenario after 10% measurements of $|\lambda_t|$ and $Im(\lambda_t)$. Further improvements in CKM parameters and m_t assumed.

CONCLUSIONS

The success of lepton flavor violation experiments in reaching sensitivities corresponding to mass scales of well over 100 TeV has helped kill most models predicting accessible LFV in kaon decay. The most popular varieties of SUSY predict LFV at levels far beyond the current experimental state of the art [51]. Thus new dedicated experiments in this area are unlikely in the near future.

The existing precision measurement of $K_L \to \mu^+\mu^-$ will be very useful if theorists can make enough progress on calculating the dispersive long-distance amplitude, perhaps helped by experimental progress in $K_L \to \gamma\ell^+\ell^-$, $K_L \to 4\ leptons$, etc. The exploitation of $K_L \to \mu^+\mu^-$ would also be aided by higher precision measurements of some of the normalizing reactions, such as $K_L \to \gamma\gamma$.

$K^+ \to \pi^+\nu\bar{\nu}$ will clearly be further exploited. Two coordinated initiatives are devoted to this: a 10^{-11}/event experiment (E949) just underway at the BNL AGS and a 10^{-12}/event experiment (CKM) recently approved for the FNAL Main Injector. The first dedicated experiment to seek $K_L \to \pi^0\nu\bar{\nu}$ (E391a) is proceeding and an experiment (KOPIO) at the AGS with the goal of making a $\sim$ 10% measurement of $Im(\lambda_t)$ is approved and in R&D.

Measurements of $K^+ \to \pi^+\nu\bar{\nu}$ and $K_L \to \pi^0\nu\bar{\nu}$ can determine an alternative unitarity triangle that will offer a critical comparison with results from the B system. If new physics is in play in the flavor sector, the two triangles will almost certainly disagree.

ACKNOWLEDGMENTS

I thank D. Bryman, M. Diwan, F. Gabbiani, D. Jaffe, S. Kettell, G. Valencia, and M. Zeller for useful discussions, access to results, and other materials. This work was supported by the U.S. Department of Energy under Contract No. DE-AC02-98CH10886.

REFERENCES

1. S.L. Glashow, J. Iliopoulos and L. Maiani, *Phys. Rev. D*, **2**, 1285 (1970).
2. Ambrose, D., *et al.*, *Phys. Rev. Lett.*, **81**, 5734–5737 (1998).
3. Appel, R., *et al.*, *Phys. Rev. Lett.*, **85**, 2450–2453 (2000).
4. Appel, R., *et al.*, *Phys. Rev. Lett.*, **85**, 2877–2880 (2000).
5. Ledovskoy, A., *Recent Results from KTeV Rare K_L Decays* (2001), talk given at the KAON2001 International Conference on CP Violation.
6. Alavi-Harati, A., *et al.*, *Phys. Rev. Lett.*, **87**, 111802 (2001).
7. Adler, S. C., *et al.*, *Further evidence for the decay $K^+ \to \pi^+\nu\bar{\nu}$* (2001), hep-ex/0111091.
8. Adler, S. C., *et al.*, *Search for the rare decay $K^+ \to \pi^+\gamma$* (2001), hep-ex/0108006.
9. Cahn, R. N., and Harari, H., *Nucl. Phys.*, **B176**, 135 (1980).
10. Littenberg, L. S., *Phys. Rev.*, **D39**, 3322–3324 (1989).
11. Buchalla, G., and Buras, A. J., *Nucl. Phys.*, **B412**, 106–142 (1994).
12. Ambrose, D., *et al.*, *Phys. Rev. Lett.*, **84**, 1389–1392 (2000).
13. Groom, D. E., *et al.*, *Eur. Phys. J.*, **C15**, 1–878 (2000).
14. Di Domenico, A., *Recent Results from KLOE at DAΦNE* (2001), talk given at the fifth KEK topical conference Frontiers in Flavor Physics, KEK, Tsukuba, Japan, Nov 20-22, 2001.
15. Hocker, A., Lacker, H., Laplace, S., and Le Diberder, F., *Eur. Phys. J.*, **C21**, 225–259 (2001).
16. Hocker, A., Lacker, H., Laplace, S., and Diberder, F. L., *CKM Matrix: Status and New Developments* (2001), hep-ph/0112295.
17. Alavi-Harati, A., *et al.*, *Phys. Rev. Lett.*, **87**, 071801 (2001).
18. Alavi-Harati, A., *et al.*, *Phys. Rev. Lett.*, **86**, 5425–5429 (2001).
19. Fanti, V., *et al.*, *Phys. Lett.*, **B458**, 553–563 (1999).
20. Bergstrom, L., Masso, E., and Singer, P., *Phys. Lett.*, **B131**, 229 (1983).
21. D'Ambrosio, G., Isidori, G., and Portoles, J., *Phys. Lett.*, **B423**, 385–394 (1998).
22. Marciano, W. J., and Parsa, Z., *Phys. Rev.*, **D53**, 1–5 (1996).
23. Buchalla, G., and Buras, A. J., *Nucl. Phys.*, **B548**, 309–327 (1999).
24. Adler, S. C., *et al.*, *Phys. Rev. Lett.*, **84**, 3768–3770 (2000).
25. D'Ambrosio, G., and Isidori, G., *$K^+ \to \pi^+\nu\bar{\nu}$: a rising star on the stage of flavour physics* (2001), hep-ph/0112135.
26. Bassalleck, B., *et al.*, *E949: An experiment to measure the branching ratio $B(K^+ \to \pi^+\nu\bar{\nu})$* (1999).
27. Cooper, P. S., *Nucl. Phys. Proc. Suppl.*, **99N3**, 121–126 (2001).
28. Buchalla, G., and Isidori, G., *Phys. Lett.*, **B440**, 170–178 (1998).
29. Jarlskog, C., *Phys. Rev. Lett.*, **55**, 1039 (1985).
30. Grossman, Y., and Nir, Y., *Phys. Lett.*, **B398**, 163–168 (1997).
31. Alavi-Harati, A., *et al.*, *Phys. Rev.*, **D61**, 072006 (2000).
32. Inagaki, T., *A plan for the experimental study on the $K_L \to \pi^0\nu\bar{\nu}$ decay at KEK* (1996), talk given at 3rd International Workshop on Particle Physics Phenomenology, Taipei, Taiwan, 14-17 Nov 1996.
33. Konaka, A., *$K_L \to \pi^0\nu\bar{\nu}$ at the AGS* (1998), hep-ex/9903016.
34. Buras, A. J., Colangelo, G., Isidori, G., Romanino, A., and Silvestrini, L., *Nucl. Phys.*, **B566**, 3–32 (2000).
35. Greenlee, H. B., *Phys. Rev.*, **D42**, 3724–3731 (1990).
36. Alavi-Harati, A., *et al.*, *Phys. Rev.*, **D64**, 012003 (2001).
37. Alavi-Harati, A., *et al.*, *Phys. Rev.*, **D62**, 112001 (2000).
38. Buras, A. J., Flavor dynamics: CP violation and rare decays (2001), hep-ph/0101336.
39. Batley, B., *et al.*, A high sensitivity investigation of ks and neutral hyperon decays, Tech. Rep. SPSC/P253, CERN/SPSC2000-002 (1999).
40. Zinchenko, A., *New results on rare decays and on future NA48* (2001), talk given at HADRON2001.
41. Alavi-Harati, A., *et al.*, *Phys. Rev. Lett.*, **83**, 917–921 (1999).
42. D'Ambrosio, G., and Portoles, J., *Nucl. Phys.*, **B492**, 417–454 (1997).
43. Gabbiani, F., and Valencia, G., *Phys. Rev.*, **D64**, 094008 (2001).
44. Alavi-Harati, A., *et al.*, *Phys. Rev. Lett.*, **86**, 397–401 (2001).
45. Alavi-Harati, A., *et al.*, *Phys. Rev. Lett.*, **84**, 5279–5282 (2000).
46. Donoghue, J. F., Holstein, B. R., and Valencia, G., *Phys. Rev.*, **D35**, 2769 (1987).
47. Kohler, G. O., and Paschos, E. A., *Phys. Rev.*, **D52**, 175–180 (1995).

48. Littenberg, L. S., *CP violation in $K_L \to \pi^0 e^+ e^-$* (1988), in *Vancouver 1988, Proceedings, CP violation at KAON Factory* 19-28.
49. Diwan, M. V., Ma, H., and Trueman, T. L., *Muon decay asymmetries from $K_L^0 \to \pi^0 \mu^+ \mu^-$ decays* (2001), hep-ex/0112350.
50. Inami, T., and Lim, C. S., *Prog. Theor. Phys.*, **65**, 297 (1981).
51. Belyaev, A., *et al.*, *Charged-lepton flavour violation in kaon decays in supersymmetric theories* (2000), hep-ph/0008276.

Semileptonic B Decays at BABAR

Thorsten Brandt
representing the *BABAR* collaboration

Institut für Kern- und Teilchenphysik, Technische Universität Dresden, D-01062 Dresden, Germany

Abstract. We present preliminary results of two inclusive measurements of $\mathcal{B}(B \to Xe\nu)$ performed with the BABAR detector at the PEP-II asymmetric B Factory. In both analyses, electrons from semileptonic B decays are separated from secondary charm semileptonic decays by comparing their charge with the flavor of the other B meson. A measurement based on events with a fully reconstructed B meson decay yields $\mathcal{B}(B \to Xe\nu) = (10.4 \pm 0.5_{(stat)} \pm 0.5_{(syst)})\%$ and $\tau_{B^+}/\tau_{B^0} = 0.99 \pm 0.1_{(stat)} \pm 0.04_{(syst)}$. Using high momentum electrons to tag the B flavor leads to $\mathcal{B}(B \to Xe\nu) = (10.82 \pm 0.21_{(stat)} \pm 0.38_{(syst)})\%$. Further, we present the preliminary results of two analyses of $B^0\overline{B}^0$ mixing using di-lepton events. The mass difference Δm_d between the two mass eigenstates of the B meson is measured to be $\Delta m_d = (0.499 \pm 0.010_{(stat)} \pm 0.012_{(syst)})\hbar\,\mathrm{ps}^{-1}$. A measurement of the parameter ε_B quantifying CP violation in $B^0\overline{B}^0$ mixing yields $Re(\varepsilon_B)/(1+|\varepsilon_B|^2) = (1.2 \pm 2.9_{(stat)} \pm 3.6_{(syst)}) \times 10^{-3}$.

INTRODUCTION

Semileptonic decays, due to their simplicity, provide an excellent laboratory for the study of electro-weak and strong interactions. In particular, inclusive measurements, *i.e.* measurements that do not differentiate between final state hadrons accompanying the charged lepton and neutrino, provide a straight forward, yet model dependent way to measure the coupling to the charged weak current in terms of the CKM matrix elements V_{cb} and V_{ub}. In the standard model, semileptonic decays arise from spectator diagrams only and thus studies of such decays can illucidate the role of these diagrams in weak decays of heavy quarks. From the experimental point of view, such studies rely mainly on the identification of leptons. Since the *BABAR* detector [1] is capable of separating leptons, especially electrons, from hadrons with a very low misidentification rate while retaining a high selection efficiency, inclusive studies of semileptonic decays can be performed with rather small statistical and systematic uncertainties.

Another important feature of semileptonic decays of B mesons is the fact that they are the dominant source of high energetic leptons in $\Upsilon(4S)$ decays. Therefore, the identification of two such leptons can be used to tag the flavor of both B mesons. This allows for investigations of the $B^0\overline{B}^0$ oscillation frequency and CP violation in $B^0\overline{B}^0$ mixing.

CP618, *Heavy Flavor Physics: Ninth International Symposium*, edited by A. Ryd and F. C. Porter

INCLUSIVE MEASUREMENTS OF THE SEMILEPTONIC BRANCHING FRACTION

General strategy

$B\bar{B}$ events are selected and the flavor of one of the two B mesons is tagged either by a high energy lepton from a semileptonic decay or by full reconstruction of a hadronic decay. Among the decays of the second B meson in the event, semileptonic decays are selected by identifying an electron. The flavor of the tagged B meson and the charge of this electron can be used to separate prompt from secondary semileptonic decays:

$$\bar{B}^0/B^- \text{ as tag: } \begin{cases} B^{0,+} \to X_{\bar{c},\bar{u}}\, e^+\, \nu_e & \text{prompt electron, \textit{right-sign}} \\ B^{0,+} \to X_{\bar{c}}\, Y,\; X_{\bar{c}} \to Y'\, e^-\, \bar{\nu}_e & \text{secondary electron, \textit{wrong-sign}} \end{cases}$$

In absence of other backgrounds, which are discussed in the next section, the number of prompt electrons is identical to the number of *right-sign* candidates for charged B mesons. For neutral B mesons, prompt electrons also occur in the *wrong-sign* sample due to $B^0\bar{B}^0$ mixing. Their fraction is given by the mixing parameter χ_0, which is precisely measured. The semileptonic branching ratio $\mathcal{B}(B \to X\, e\, \nu)$ is derived from the number of prompt electrons (N_{prompt}), the number of tagged B mesons (N_{tag}) and the relative efficiency for detecting a tagged event with and without a signal electron (ε_{evt}) in the following way :

$$\mathcal{B}(B \to X e \nu) = \frac{1}{\varepsilon_{evt}} N_{prompt}/N_{tag}\ . \tag{1}$$

Common backgrounds

Electrons originating from processes producing $e^+\, e^-$ pairs contribute to both *right-* and *wrong-sign* samples. Such processes are photon conversions, π^0 Dalitz decays and J/ψ decays. They are suppressed using vertex and invariant mass cuts. We correct the observed number of such background electrons for the efficiencies of these cuts, which are determined using a full detector simulation.

Misidentified hadrons form another source of background tracks. The hadron fake rates are determined from pure samples of positive and negative pions, kaons, and protons which are selected kinematically from the decays $K_s^0 \to \pi^+\pi^-$, $D^* \to D\pi$, $D \to K\pi$ and $\Lambda \to p\pi$. Together with the relative fraction of each species in tagged events, which we determine from Monte Carlo, we estimate the momentum spectrum of this background.

We also have to consider cascade decays from the signal B which produce *right-sign* electrons in unmixed and *wrong-sign* electrons in mixed events:

- Semileptonic decay of a D_s meson from $c\bar{c}s$ production:
  $B \to D_s^{(*)}\bar{D}\ ,\ D_s \to X'e^+\nu_e$
- Semileptonic decay of a D meson from $c\bar{c}s$ production:
  $B \to D^{(*)}\bar{D}^{(*)}K\ ,\ D \to X'e^+\nu_e$

- Semileptonic decays with a τ lepton:
  $B \to \tau^+\nu_\tau, \tau^+ \to e^+\nu_e\bar{\nu}_\tau$ or $D_s \to \tau^+\nu_\tau, \tau^+ \to e^+\nu_e\bar{\nu}_\tau$

We determine these background spectra from Monte Carlo, where the involved branching ratios are taken from the latest measurements, with the systematic errors derived from their uncertainties.

Finally, we have to correct for electrons found in mistagged events. In the analysis using fully reconstructed B tags, these originate from combinatorial background and $B^0 \leftrightarrow B^+$ cross feed, while for the analysis based on lepton tags, these are mainly due to high energetic electrons from secondary charm decays.

Analysis with fully reconstructed B tags

The measurement of $\mathcal{B}(B \to Xe\nu)$ using fully reconstructed B tags is based on 20.6 fb^{-1} of *BABAR* data recorded in 1999 and 2000. We reconstruct the following hadronic B decays: $B \to D^{(*)}\pi$, $B \to D^{(*)}\rho$, $B \to D^{(*)}a_1$, $B \to J/\psi K^{(*)}$, and $B \to \psi(2S)K^{(*)}$. The kinematic selection of these B decays checks the validity of the reconstructed decay candidates using the energy difference $\Delta E = E_B^* - E_{Beam}^*$ and the energy substituted mass $m_{ES} = (E_{beam}^{*2} - p_B^{*2})^{1/2}$. We additionally require the second Fox-Wolfram moment R_2 to be smaller than 0.5 in order to reduce non-resonant background, and at least one more track besides those assigned to the B meson. The yield is 6533 ± 112 neutral and 7684 ± 120 charged B decays. The purities of the B^0 and B^+ samples are $(84.4 \pm 0.4)\%$ and $(81.6 \pm 0.4)\%$, respectively.

Electrons are identified by subsequent cuts on measurements of the Electromagnetic Calorimeter (EMC), the Drift Chamber (DCH) and the Cerenkov Detector (DIRC). In order to retain a high selection efficiency and a reasonable signal-to-background ratio, only tracks within the acceptance of the EMC and a laboratory momentum above 0.5 GeV/c are considered. Fig. 1 shows the momentum spectra of *right-sign* and *wrong-sign* electrons for both B^0 and B^+ tagged events.

For charged B mesons, the number of right-sign leptons $N_{right-sign}$ represents the number N_{prompt} of prompt decays, while for decays of neutral B mesons we have:

$$N_{right-sign} = N_{prompt}(1-\chi_0) + N_{secondary}\,\chi_0 \,,$$

$$N_{wrong-sign} = N_{prompt}\,\chi_0 + N_{secondary}(1-\chi_0)\,.$$

Therefore, we can determine the prompt and secondary electron momentum spectra (Fig. 2) from the spectra of *right-sign* and *wrong-sign* electrons. Extrapolation to p=0 and applying acceptance corrections leads to the following preliminary results:

$$\mathcal{B}(B^+ \to Xe\nu) = (10.3 \pm 0.6_{(stat)} \pm 0.5_{(syst)})\%,$$

$$\mathcal{B}(B^0 \to Xe\nu) = (10.4 \pm 0.8_{(stat)} \pm 0.5_{(syst)})\%,$$

$$\mathcal{B}(B \to Xe\nu) = (10.4 \pm 0.5_{(stat)} \pm 0.5_{(syst)})\%,$$

$$\tau_{B^+}/\tau_{B^0} = \mathcal{B}(B^+ \to Xe\nu)/\mathcal{B}(B^0 \to Xe\nu) = 0.99 \pm 0.1_{(stat)} \pm 0.04_{(syst)}.$$

Table 1 lists the systematic errors for this measurement.

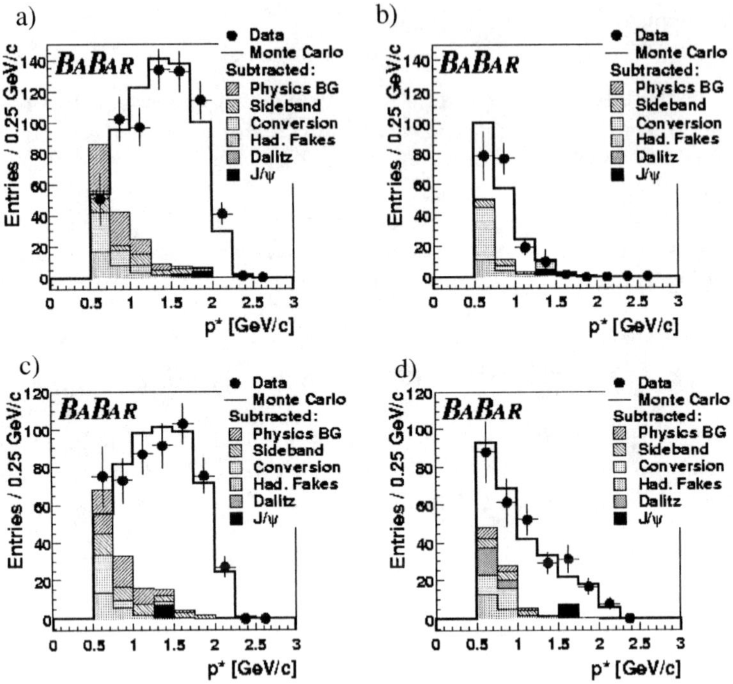

FIGURE 1. Momentum spectra for *right-sign* (a,c) and *wrong-sign* (b,d) electrons in events tagged with fully reconstructed B^0 (a,b) and B^+ (c,d) decays.

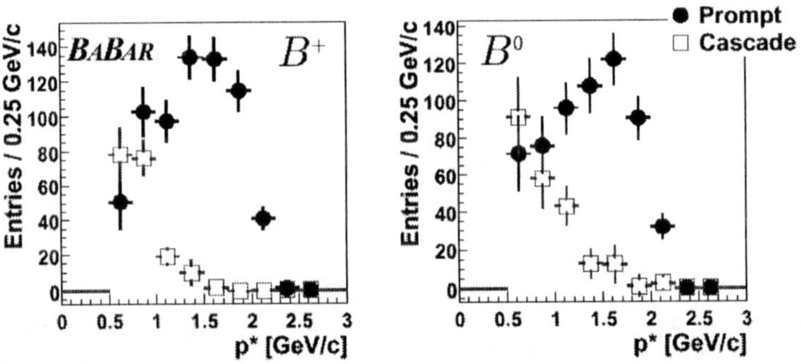

FIGURE 2. Momentum spectra for prompt and secondary electrons in events tagged with fully reconstructed B decays.

Analysis with electron tags

The determination of $\mathcal{B}(B \to X\, e\, \nu)$ by using a high energetic electron to tag the B-flavor is based on 4.1 fb^{-1} and 0.97 fb^{-1} of data recorded on and 40 MeV below the

TABLE 1. Systematic errors for analysis of $\mathcal{B}(B \to Xe\nu)$ with fully reconstructed B tags.

Source	$\Delta\mathcal{B}(B \to Xe\nu)[\%]$	$\Delta\frac{\mathcal{B}(B^+ \to Xe\nu)}{\mathcal{B}(B^0 \to Xe\nu)}$
Analysis efficiency	0.19	0.027
B^0/B^+ cross feed	0.2	0.027
Mixing parameter χ_0	0.07	0.007
Tracking efficiency	0.15	cancels
Extrapolation to p=0	0.18	cancels
Background from $c\bar{c}s$ production and τ decays	0.16	cancels
Electron efficiency	0.18	cancels
Faked electrons	0.16	cancels
Pair background	0.018	cancels
Total	0.5	0.04

$\Upsilon(4S)$ resonance. To suppress non-$B\bar{B}$ events, we require $R_2 < 0.6$ and $N_{charged} \geq 4$ for the charged multiplicity. For events with four charged tracks at least two energy deposits in the calorimeter that are not associated with a charged track are required. Electrons are selected using a likelihood based electron identification algorithm, which combines the measurements of the EMC, the DCH and the DIRC. The momentum dependent electron efficiency and hadron fake rates are shown in Fig. 3.

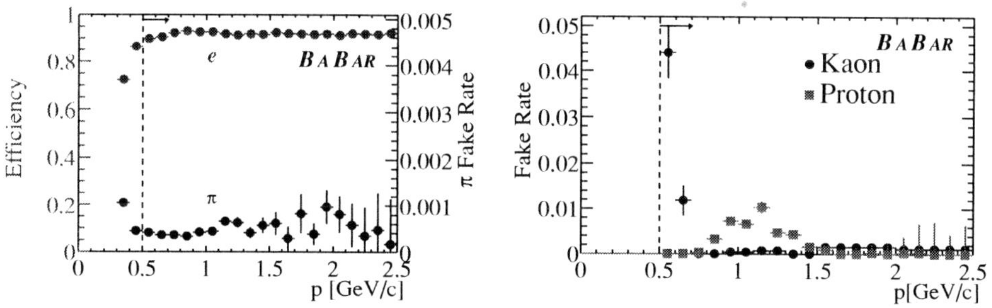

FIGURE 3. Efficiency and fake rates of likelihood based electron identification for tracks within the EMC acceptance, averaged over positively and negatively charged particles.

We consider tracks within the acceptance of the EMC only, and due to an increasing number of background electrons from other physics processes (especially photon conversions) at low momenta, we require a center-of-mass momentum p^* of at least $0.5\,\mathrm{GeV}/c$. Electrons which have not been identified as part of a photon conversion, Dalitz or J/ψ decay are considered as tag if $p^* > 1.4\,\mathrm{GeV}/c$.

In contrast to the analysis based on fully reconstructed B decays as tags, we observe a background to the *right-sign* sample that originates from a semileptonic charm particle decay of the form

$$\bar{B} \to X_c\, e^-_{tag}\, \bar{\nu}_e \quad , \quad X_c \to Y\, e^+_{cand}\, \nu_e \, .$$

With increasing momentum, candidate electrons produced in these cascades show an increasing back-to-back correlation to the tag-lepton (Fig. 4b). On the other hand, since the two B mesons are produced nearly at rest in the $\Upsilon(4S)$ rest frame, *right-sign* prompt electrons produced via

$$\bar{B} \to X_c\, e^-_{tag}\, \bar{\nu}_e \quad , \quad B \to X_{\bar{c}}\, e^+_{cand}\, \nu_e$$

show no angular correlation to the tag (Fig. 4a). Therefore, we suppress these same-side cascades by a momentum dependent cut as depicted in Fig. 4. To obtain an estimate of the number of same-side cascade electrons passing the cut on this opening angle α, the distribution of $\cos\alpha$ is analyzed. Assuming that the only non-flat contribution arises from same-side cascades, we fit a function of the form $f(\cos\alpha) = c_0 + c_1 f_{same}(\cos\alpha)$ with c_0 and c_1 as parameters to this distribution, where f_{same} is extracted from Monte Carlo. From the fit result, we can derive an estimate of the number of same-side cascades passing the cut on $\cos\alpha$ which does not rely on the knowledge of any branching fraction. Fig. 5 summarizes all backgrounds for both *right-sign* and *wrong-sign* electrons.

Since lepton tags cannot distinguish between charged and neutral B mesons, we have to use the measured fraction $f_0 = 0.5 \pm 0.02$ for $B^0\bar{B}^0$ pair production in $\Upsilon(4S)$ decays as additional input to infer prompt and secondary electron spectra from the *right-* and *wrong-sign* spectra. After all background corrections, we have

$$\frac{dN_{right-sign}}{dp^*} = \varepsilon_\alpha \varepsilon_e \left[\frac{dN_{prompt}}{dp^*}(1 - f_0\chi_0) + \frac{dN_{secondary}}{dp^*} f_0\chi_0 \right] ,$$

$$\frac{dN_{wrong-sign}}{dp^*} = \varepsilon_e \left[\frac{dN_{prompt}}{dp^*} f_0\chi_0 + \frac{dN_{secondary}}{dp^*}(1 - f_0\chi_0) \right] .$$

ε_e is the electron identification efficiency and ε_α is the efficiency of the same-side cascade suppression cut. Solving these two equations after dN_{prompt}/dp^* and $dN_{secondary}/dp^*$, we extract primary and secondary momentum spectra (Fig. 6a). The prompt electron spectrum is corrected for external bremsstrahlung using a full detector simulation. Fig. 6b shows the final spectrum, together with the MC prediction, which is based on a parameterization of HQET-derived form factors [2] for the decay mode $B \to D^* l \nu$, the Goity-Roberts model [3] for non-resonant decays, and the ISGW2 model [4] for all other modes.

After all background and efficiency corrections, including geometrical acceptance and extrapolation to p^*=0, we arrive at 32238 prompt and 304051 tag electrons. With Eq. 1 and $\varepsilon_{evt} = 0.98$, this results in

$$\mathcal{B}(B \to X\, e\, \nu) = (10.82 \pm 0.21_{(stat)} \pm 0.38_{(syst)})\% \; .$$

Table 4 shows the individual contributions to the systematic error. In Ref. [5], I. Bigi *et al.* present a relation between $\mathcal{B}(B \to X_c\, e\, \nu)$ and $|V_{cb}|$:

$$|V_{cb}| = 0.0411 \left(\frac{\mathcal{B}(B \to X_c\, e\, \nu)}{0.105} \frac{1.55\,\mathrm{ps}}{\tau_B} \right)^{1/2} (1.0 \pm 0.015_{pert} \pm 0.01_{m_b} \pm 0.012_{1/m_Q^3}) \; .$$

Using the current PDG-values of (1.597 ± 0.024) ps for τ_B and $(1.7 \pm 0.5) \times 10^{-3}$ for $\mathcal{B}(B \to X_u\, e\, \nu)$, and adding the theoretical errors linearly, we determine $|V_{cb}|$ to be

$$|V_{cb}| = 0.0408 \pm 0.0009_{exp} \pm 0.0015_{theory} \; .$$

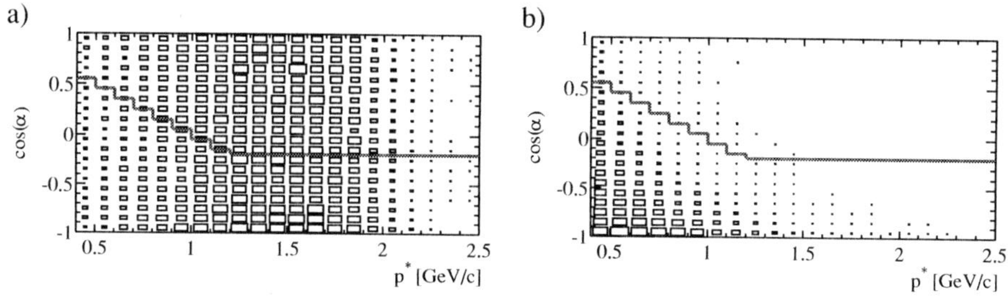

FIGURE 4. MC simulations: Distribution of angle between tag- and candidate electron for prompt electrons (left) and secondary electrons from same-side cascades (right). The line indicates the cut we use to suppress electrons of the latter category.

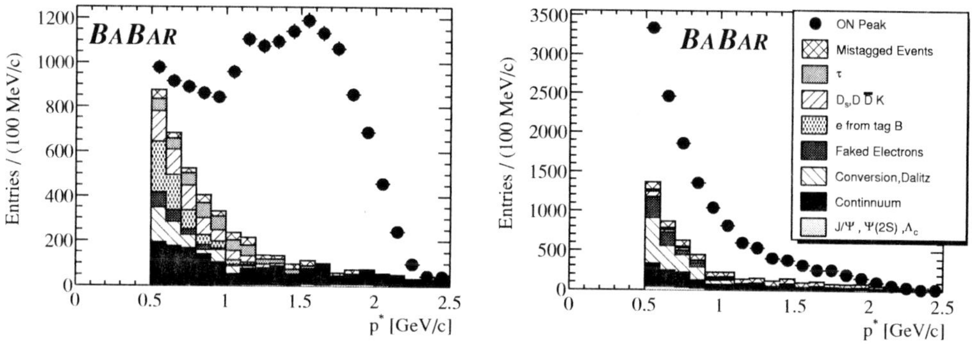

FIGURE 5. Momentum spectra of *right-sign* (left) and *wrong-sign* (right) electrons and backgrounds for lepton tag analysis. The solid points show the electron spectra without any background correction applied.

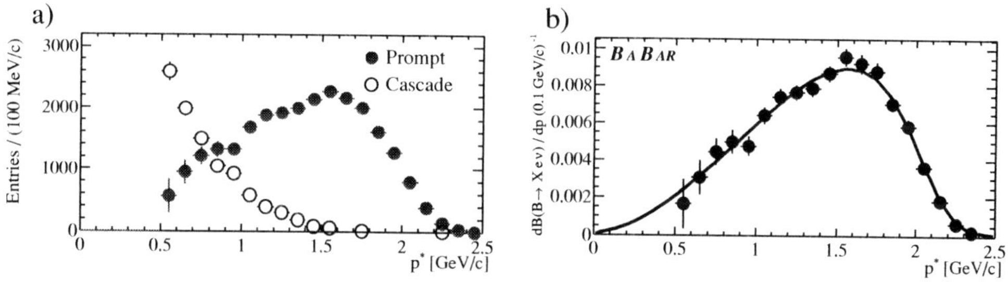

FIGURE 6. a) Momentum spectra of prompt and secondary electrons after correction for electron identification efficiency. b) Spectrum of prompt electrons after correction for external bremsstrahlung, acceptance and normalization to the number of tags.

TABLE 2. Systematic errors of $\mathcal{B}(B \to Xe\nu)$ measured with with electron tags.

Source	$\Delta\mathcal{B}(B \to Xe\nu)$ [%]
Pair background	0.074
Faked electrons	0.045
Cascades from tag B	0.096
Mistagged events	0.068
e from τ decays	0.058
e from cascade charm	0.164 ($D^{(*)}$)
decays of signal B	0.175 ($D_s^{(*)}$)
$f_0\chi_0$	0.045
Electron efficiency	0.182
Tracking efficiency	0.11
Extrapolation to p=0	0.11
Number of tags	0.07
Total	0.38

$B^0\overline{B}^0$ MIXING WITH DILEPTON EVENTS

The following two analyses are based on data collected between October 1999 and October 2000 with integrated luminosities of 20.7 fb^{-1} on and 2.6 fb^{-1} 40 MeV below the $\Upsilon(4S)$ resonance. Identifying both electrons and muons, we select events with two high energetic leptons. Such leptons predominantly originate from semileptonic B decays and therefore can be used to determine the flavors of the B mesons at the time of their decays. Since at *BABAR* the $\Upsilon(4S)$ rest frame is boosted with $\beta\gamma = 0.56$ in the laboratory frame, the time between these two decays can be measured. In this inclusive approach, we use the two lepton tracks to estimate the z-coordinates of the two B meson decays. First, we estimate the position of the interaction vertex in the xy projection from the two lepton trajectories and the beam spot. Second, we determine the difference Δz between the z-coordinates of the points of closest approach of the trajectories to the xy interaction point. With $\langle\beta\gamma\rangle$ being the average boost along the z-axis, we have $\Delta t = \Delta z\,/\,(\langle\beta\gamma\rangle c)$.

Since the two B mesons are not entirely at rest in the $\Upsilon(4S)$ rest frame, this procedure leads to slightly wrong measurements of Δt. While the influence on the boost $\langle\beta\gamma\rangle$ is small, the resolution in Δz deteriorates. This is taken into account by folding the expected distributions of Δz with a resolution function before comparing them with the observed ones. This resolution function is taken from a Monte Carlo simulation and has been validated on data by J/ψ decays. It can be approximated by the sum of three Gaussian functions, where the two dominant contributions have widths of 85μm and 175μm.

Measurement of Δm_d

The asymmetry A in the number of events with equal and oppositely charged lepton pairs is measured as function of Δt. This asymmetry depends on the mass difference

Δm_d between the two B meson mass eigenstates:

$$A(\Delta t) = \frac{N(l^\pm l^\mp)(\Delta t) - N(l^\pm l^\pm)(\Delta t)}{N(l^\pm l^\mp)(\Delta t) + N(l^\pm l^\pm)(\Delta t)} = \frac{e^{-\Gamma_0|\Delta t|}\cos(\Delta m_d \Delta t) + R\, e^{-\Gamma_+|\Delta t|}}{e^{-\Gamma_0|\Delta t|} + Re^{-\Gamma_+|\Delta t|}}, \quad (2)$$

where Γ_0, Γ_+ are the decay widths of neutral and charged B mesons, and R describes the fraction of $B^\pm$ mesons in the sample, which can be reduced by requiring a partially reconstructed charged D^* meson. Dividing the dilepton events into two subsets with and without partially reconstructed $D^{\pm *}$, we determine Δm_d from a simultaneous binned likelihood fit of the expected distributions of $N(l^\pm l^\mp)$ and $N(l^\pm l^\pm)$ to the observed ones. The asymmetries and the result of the fit are shown in Fig. 7. The preliminary result is

$$\Delta m_d = (0.499 \pm 0.010_{(stat)} \pm 0.012_{(syst)})\ \hbar\ \mathrm{ps}^{-1},$$

where the dominating contributions to the systematic error arise from background modeling, lepton misidentification rates and the resolution function. This result is also sensitive to the B lifetimes and the contribution of non-$B\bar{B}$ background.

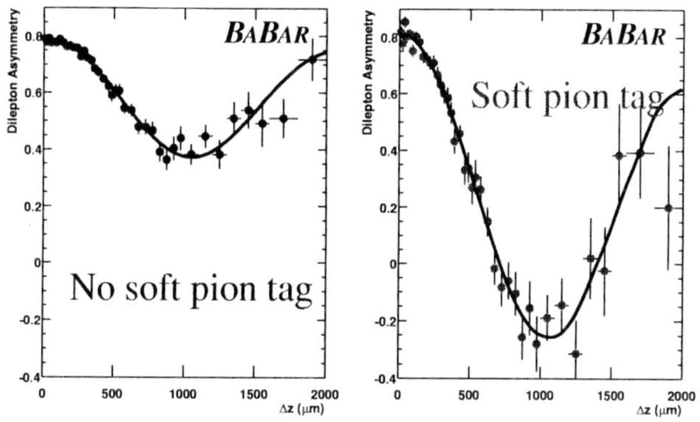

FIGURE 7. Dilepton asymmetry A as defined in Eq. 2 as function of Δz . Left: sample without partially reconstructed $D^{*\pm}$ meson; right: sample with partially reconstructed $D^{*\pm}$ meson.

CP violation in $B^0\bar{B}^0$ mixing

The parameter ε_B connects the mass eigenstates $|B^0_{L,H}\rangle$ and flavor eigenstates $|B^0\rangle, |\bar{B}^0\rangle$ of B mesons in the following way:

$$|B^0_{H,L}\rangle = (2 + 2|\varepsilon_B|^2)^{-1/2}\left[(1+\varepsilon_B)|B^0\rangle \pm (1-\varepsilon_B)|\bar{B}^0\rangle\right] .$$

$\varepsilon_B \neq 0$ results in different probabilities for the transitions $B^0 \to \bar{B}^0$ and $\bar{B}^0 \to B^0$ (*CP* violation in mixing). In this case, an asymmetry between the numbers of positively and negatively charged lepton pairs could be observed. In first order, this time independent asymmetry is related to ε_B by

$$A_{CP}(\Delta t) = \frac{N(l^+l^+)(\Delta t) - N(l^-l^-)(\Delta t)}{N(l^+l^+)(\Delta t) + N(l^-l^-)(\Delta t)} \approx \frac{4Re(\varepsilon_B)}{1+|\varepsilon_B|^2} . \quad (3)$$

Fig. 8 shows A_{CP} vs. Δt, which is consistent with the expected time independence of A_{CP}. Since for $\Delta z < 200\,\mu$m, cascade leptons from B^+ and unmixed B^0 decays contribute significantly to the measurements, we fit a constant function $f(\Delta t) = A_t$ to $A_{CP}(\Delta t)$ for $\Delta z > 200\,\mu$m. The preliminary result is

$$\frac{Re(\varepsilon_B)}{1+|\varepsilon_B|^2} = (1.2 \pm 2.9_{(stat)} \pm 3.6_{(syst)}) \times 10^{-3} .$$

The dominating systematic errors are the charge asymmetry in lepton detection ($\sigma(A_t) = 0.8\%$), and the charge asymmetries in backgrounds from $B\bar{B}$ events ($\sigma(A_t) = 0.9\%$) and non-$B\bar{B}$ events ($\sigma(A_t) = 0.7\%$).

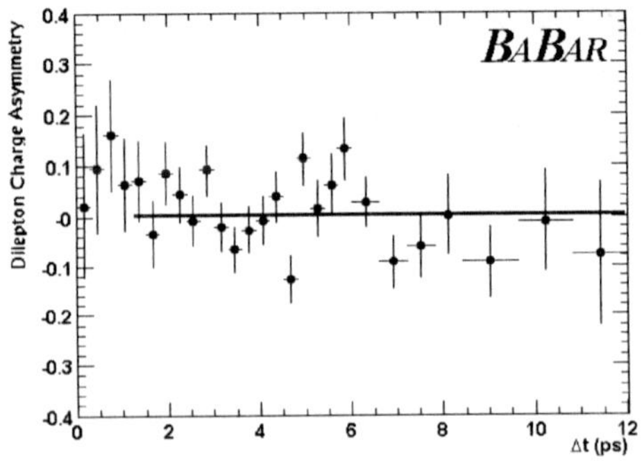

FIGURE 8. Dilepton asymmetry $A_{CP}(\Delta t)$ as defined in Eq. 3.

REFERENCES

1. B. Aubert *et al.*, [BABAR Collaboration], SLAC-PUB-8569, hep-ex/0105044, to appear in *Nucl. Instrum. and Methods* (2001).
2. J.E. Dubosq *et al.*, Measurement of form factors for $\bar{B}^0 \to D^{*+} l^- \bar{\nu}$, *Phys. Rev. Lett.* **76**, 3898 (1996).
3. J.L. Goity and W. Roberts, Soft pion emission in semileptonic B meson decays, *Phys. Rev.* **D51** 3459-3477 (1995).
4. D. Scora and N. Isgur, Semileptonic meson decays in the quark model: An update, *Phys. Rev.* **D52**, 2783 (1995).
5. I. Bigi, M. Shifman and N. Uraltsev in *Ann. Rev. Nucl. Part. Science* **47**, 591 (1997).

B Semileptonic Decays at Belle

Heejong Kim representing the Belle collaboration

Department of Physics, Yonsei University, Seoul 120-749, KOREA

Abstract. We present preliminary results on branching fraction measurements for semileptonic B decays of $\bar{B}^0 \to D^{(*)+}\ell^-\bar{\nu}$, inclusive semiletonic decay of B with lepton tags, and the charmless semileptonic B decay $\bar{B}^0 \to \pi^+\ell^-\bar{\nu}$ with data set of 10.2, 5.1 and 21.2 fb^{-1}, respectively. The CabbiboKobayashi-Maskawa matrix element $|V_{ub}|$ is obtained from our results.

INTRODUCTION

One of the goals of B physics is to measure Cabibbo-Kobayashi-Maskawa(CKM) [1] matrix elements precisely. In the Standard Model of electroweak interactions, the elements of the CKM quark mixing matrix are constrained by unitarity. Semileptonic B decays, due to their simplicity, provide the cleanest way to measure $|V_{cb}|$ and $|V_{ub}|$. In such decays, the effects of strong interactions which enter from the hadronic current can be isolated and parameterized in terms of a small number of form factors. To date $|V_{cb}|$ is measured relatively well. However more accurate measurements of $|V_{cb}|$ are necessary to constrain the Standard Model. For instance, the extraction of the constraint in the $\rho - \eta$ plane from the CP violation parameter $|\varepsilon_K|$ depends on A^4 (where ρ, η and A are the Wolfenstein parameters) [2]. The recent development of Heavy Quark Effective Theory (HQET) [3] makes $|V_{cb}|$ measurements with exclusive semileptonic decays more reliable. HQET yields an expression for the $\bar{B}^0 \to D^{(*)+}\ell^-\bar{\nu}$ decay rate with a single unknown form factor and provides controllable theorectical uncertainties at the zero recoil point. $|V_{cb}|$ can also be obtained by inclusive semileptonic branching fraction measurement. On the other hand, $|V_{ub}|$ is experimentally the least-understood CKM element. The inclusive method using the lepton-endpoint region has a large theoretical uncertainty because it uses only a small part of phase space and relies on an extrapolation to the full lepton momentum range [4]. The exclusive approach to measuring $|V_{ub}|$ uses branching fraction measurements of $b \to u$ decays. $\bar{B}^0 \to \pi^+\ell^-\bar{\nu}$ is the cleanest channel experimentally among various $b \to u$ decays. With the large data samples accumulated at B factories, we can measure the braching fraction of $\bar{B}^0 \to \pi^+\ell^-\bar{\nu}$ more precisely and consequently extract $|V_{ub}|$. q^2 measurements which can differentiate various form factor models will also be possible with large data samples. We present results of branching fraction measurements of B semileptonic decays; $\bar{B}^0 \to D^{(*)+}\ell^-\bar{\nu}$, inclusive semileptonic B decays and $\bar{B}^0 \to \pi^+\ell^-\bar{\nu}$ decays with data samples of 10.2, 5.1 and 21.2 fb^{-1}, respectively , recorded at Belle detector. The $|V_{cb}|$ is obtained as a result and shown in this review. Throught this paper, the inclusion of charge conjugate states is implied.

CP618, *Heavy Flavor Physics: Ninth International Symposium*, edited by A. Ryd and F. C. Porter

MEASUREMENT OF $\bar{B}^0 \to D^+\ell^-\bar{\nu}$

The semileptonic decay $\bar{B}^0 \to D^+\ell^-\bar{\nu}$ is among the cleanest modes that can be used to measure the CKM matrix element, $|V_{cb}|$, in the framework of HQET. The differential decay rate for $\bar{B}^0 \to D^+\ell^-\bar{\nu}$ can be expressed by

$$\frac{d\Gamma}{dy} = \frac{G_F^2|V_{cb}|^2}{48\pi^3}(m_{\bar{B}^0} + m_{D^+})^2 m_{D^+}^3 (y^2-1)^{3/2} F_D^2(y), \quad (1)$$

where G_F is the weak coupling constant and y is the inner product of the B and D^+ meson 4-velocities ($y \equiv \upsilon_{\bar{B}^0} \cdot \upsilon_{D^+} = \frac{P_{\bar{B}^0}}{m_{\bar{B}^0}} \cdot \frac{P_{D^+}}{m_{D^+}}$), which is related to q^2, the mass squared of the lepton-neutrino system. By extraploation to the point of zero recoil of the D^+ meson, $|V_{cb}|$ can be determined using a theoretical prediction for $F_D(1)$.

This analysis is based on the neutrino reconstruction method [5], which exploits the hermiticity of the detector and near zero value of the neutrino mass. We extract information on the neutrino from the missing momentum ($\vec{p}_{miss} = -\Sigma\vec{p}_i$) and missing energy ($E_{miss} = 2E_{beam} - \Sigma E_i$) in each event, where the sum is over all reconstructed particles i in the event. If the only undetected particle in an event is the neutrino, the missing momentum and energy can be attributed to the neutrino, in which case the missing mass($M_{miss}^2 = E_{miss}^2 - |\vec{p}_{miss}|^2$) should be consistent with zero, we require -2.0 $< M_{miss}^2 < 3.0$ GeV^2/c^4. Since the initial state is charge-neutral, we should have a zero charge sum if we detect all the particles. Hence, we require that the net charge, $|\Sigma Q| \leq 1$ to reject events with other missing charged particles. To avoid events where missing particles are last along the beam pipe and mimic neutrinos, we require $|\cos\theta_{\vec{p}_{miss}}| < 0.95$. Neutrino are usually produced together with charged leptons(e or μ). Two or more charged leptons in an event implies that there might be more than one neutrino, in which case we cannot assume that the missing mass is close to zero. Therefore we use events which have only one charged lepton. Leptons are required to have $p_{lab} > 0.8$ GeV/c. This requirement helps to reduce the backgrounds from hadrons misidentified as leptons and random combinations of leptons and D^+ mesons.

D^+ candidates are reconstructed from the $D^+ \to K^-\pi^+\pi^+$ decay mode. Kaons are required to be positively identified by the hadron identification devices. Three charged tracks are then kinematically fit to a D^+ decay vertex, and we require that the impact parameter of each track with respect to the vertex point to be consistent with zero. We further require $P_{D^+} < 2.5$ GeV/c in order to suppress continuum background. We select $K\pi\pi$ combinations where the invariant mass is within 20 MeV/c^2 of the nominal D^+ mass. D^+ mesons that are consistent with being produced in the decay $D^{*+} \to D^+\pi^0$ are vetoed.

A variable $\cos\theta_{B-D\ell} = (2E_B E_{D\ell} - m_B^2 - m_{D\ell}^2)/2|\vec{p}_B||\vec{p}_{D\ell}|$ is defined as the cosine of the angle between $\vec{p}_B$ and $\vec{p}_{D-\ell}(= \vec{p}_D + \vec{p}_\ell)$. The signal events are distributed mostly within the physically allowed region $|\cos\theta_{B-D\ell}| < 1$, while the background events extend to a much wider range. We require that candidates have $|\cos\theta_{B-D\ell}| < 1$.

Since the resolution of the momentum measurement is better than that of energy, we take $(E_\nu, \vec{p}_\nu) = (|\vec{p}_{miss}|, \vec{p}_{miss})$ as the 4-momentum of the neutrino. Combining the energy-momentum 4-vectors for the reconstructed D^+ meson, the signal lepton and

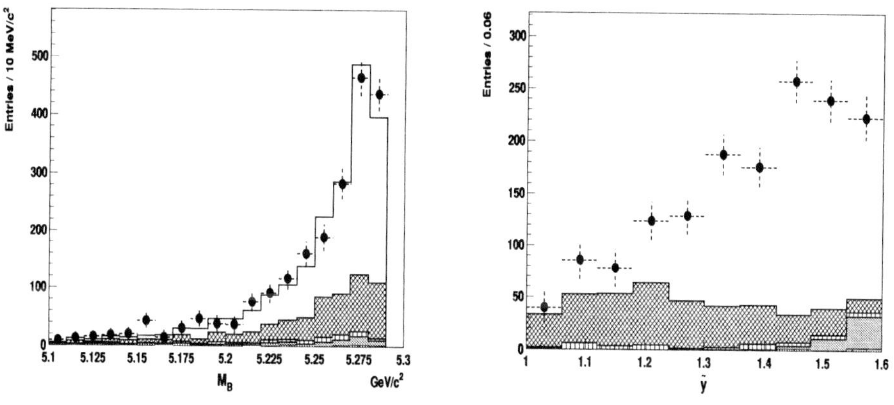

(a) (b)

FIGURE 1. (a) The M_B distribution for $\bar{B}^0 \to D^+\ell^-\bar{\nu}$ after D^+sideband subtractraction. (b) The $\tilde{y}$ distribution after the $M_B > 5.24$ GeV/c^2 cut and D^+sideband subtractraction. The points with error bars show data. The continuum, fake lepton, uncorrelated and correlated backgrounds are shown as the solid, shaded, vertically hatched and cross-hatched histogram respectively from the bottom.

the neutrino, and using the constaint of energy-momentum conservation, we obtain the variables for full B decay reconstruction, *i.e* the beam constrained mass M_B and the energy difference ΔE defined as

$$\Delta E = E_{beam} - (E_{D^+} + E_{\ell^-} + E_{\bar{\nu}}), \tag{2}$$

$$M_B = \sqrt{E_{beam}^2 - |\vec{p}_{D^+} + \vec{p}_{\ell^-} + \alpha\vec{p}_{\bar{\nu}}|^2} \tag{3}$$

We select events with ΔE close to 0, by requiring -0.2 $< \Delta E <$ 1 GeV and impose the condition $\Delta E = 0$ with $\alpha = 1 + \Delta E/E_{\bar{\nu}}$ in the calculation of M_B.

The background sources fall into five categories: combinatoric, correlated, uncorrelated, continuum and misidentified lepton. The dominant background is the combinatoric background in D^+ reconstruction. The magnitude of the combinatoric backgrounds is estimated using the sideband in $M(K^-\pi^+\pi^+)$.

Fig. 1-(a) shows the M_B distribution after all event selection criteria. The M_B signal region is defined as $M_B > 5.24$ GeV/c^2. The signal efficiency is 2.69%. Fig. 1-(b) shows the $\tilde{y}$ distribution for data and backgrounds, where $\tilde{y}$ is the reconstructed value of y. We use ten $\tilde{y}$ bins over the range $1.00 \leq \tilde{y} \leq 1.59$. After all backgrounds are subtracted, we perform a χ^2 fit to the $\tilde{y}$ distribution to obtain $|V_{cb}|$ and form factor. We use the following parameterization,

$$F_D(y) = F_D(1)(1 - \hat{\rho}_D^2(y-1) + \hat{c}_D(y-1)^2) \tag{4}$$

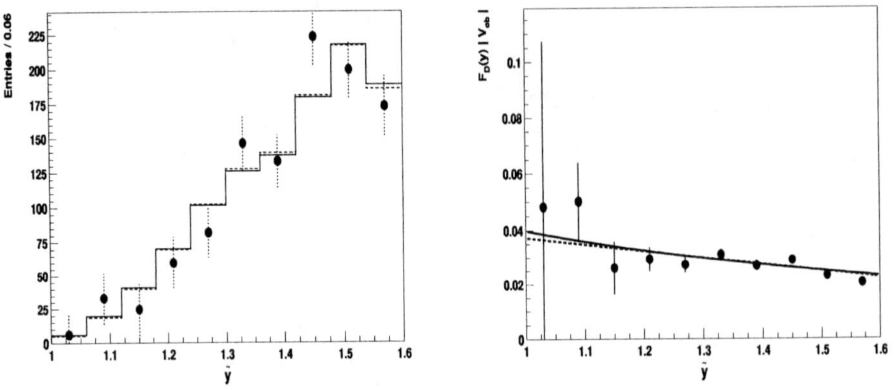

(a) (b)

FIGURE 2. (a) The $\tilde{y}$ distribution and fit. (b) The $F_D(y)|V_{cb}|$ distribution after correction for the smearing of y. The points with error bars are the data and the curves are fit results for linear from-factor(dashed) and the Caprini *et al*. from-factor(solid).

With Caprini *et al.*'s form-factor parameterization [6], we find $|V_{cb}|F_D(1) = (4.11 \pm 0.44) \times 10^{-2}$, $\hat{\rho}_D^2 = 1.12 \pm 0.22$ and $\hat{c}_D = 1.03 \pm 0.23$. The fit results are shown in Fig. 2. By integrating the differential decay rate, $d\Gamma/dy$, with the Caprini *et al.* form-factor parameter determined by the fit, we obtain the total decay rate $\Gamma(\bar{B}^0 \to D^+\ell^-\bar{\nu}) = (13.79 \pm 0.76)\text{ns}^{-1}$ and branching fraction of $B(\bar{B}^0 \to D^+\ell^-\bar{\nu}) = (2.13 \pm 0.12)\%$. Using $F_D(1) = 0.98 \pm 0.07$ [6], we obtain $|V_{cb}| = (4.19 \pm 0.45) \times 10^{-2}$, where the error is statistical only. The dominant uncertainty is the imperfection of the detector simlulation for the neutrino simulation and we assign 10.6% in $|V_{cb}|F_D(1)$ from neutrino simluation uncentainty.

MEASUREMENT OF $\bar{B}^0 \to D^{*+}\ell^-\bar{\nu}$

We reconstruct $\bar{B}^0 \to D^{*+}\ell^-\bar{\nu}$ decays using the decay chain $D^{*+} \to D^0\pi^+, D^0 \to K^-\pi^+$ and require ℓ to be an electron. The partial reconstruction method, which does not reconstruct neutrino explicitly with missing momentum, is used for B reconstruction. In order to suppress $e^+e^- \to q\bar{q}$ continuum background, the ratio of the second to zeroth Fox-Wolfram moments (R_2) [7] is required to be less than 0.4. Each event is required to contain at least one electon candidate with momentum in the range of 1.0 GeV/c to 2.45 GeV/c in the center of mass frame. After a $K^-\pi^+$ combination is selected, its vertex is computed in space and the momentum vector of each particle is recalculated at this position. D^0 candidates are required to satisfy track quality cuts based on their impact parameters relative the the $K^-\pi^+$ vertex: $\Delta r < 0.2$ cm and $\Delta z < 0.5$ cm. The $D^0 \to K^-\pi^+$ candidates must have an invariant mass with 3 σ of the nominal D^0 mass.

We combine D^0 candidates with pion candidates to fully reconstruct D^{*+} in the mode $D^{*+} \to D^0\pi^+$. These pion candidates are called slow pions (π_s^+) due to their low momentum in the laboraratory frame. The $\Delta M(M_{K^-\pi^+\pi_s^+} - M_{K^-\pi^+})$ signal region for D^{*+} is 6 MeV/c^2 wide and is centered on the measured ΔM mean of 145.5 MeV/c^2. Furthemore the momentum of D^{*+} candidates must satisfy $|\vec{p}_{D^{*+}}|/\sqrt{E_B^2 - m_{D^{*+}}^2} < 0.5$ to be consistent with a B decay hypothesis. A D^0-lepton vertex is fit in space and the lepton momentum vector is recalculated by imposing the requirement that it originated from this new vertex.

For a signal event, we expect $P_{miss}^2 = P_\nu^2 = 0$, where

$$P_{miss}^2 = (P_B - P_{D^*\ell})^2 \equiv m_B^2 + m_{D^*\ell}^2 - 2E_B E_{D^*\ell} + 2|\vec{p}_B||\vec{p}_{D^*\ell}|\cos\theta_{B,D^*\ell} \qquad (5)$$

Although the angle $\theta_{B,D^*\ell}$ between the B and $(D^{*+} + \ell^-)$ momentum direction is unkown, one can neglect the last term since the B meson is produced nearly at rest in the $\Upsilon(4S)$ frame. Thus we define $M_{miss}^2 = m_B^2 + m_{D^*\ell}^2 - 2E_B E_{D^*\ell}$ and $C = 2|\vec{p}_B||\vec{p}_{D^*\ell}|$. A cut on $M_{miss}^2 < 1$ GeV^2/c^4 will suppress the bacground from $\bar{B} \to D^{**}\ell\bar{\nu}$ because an additional unreconstructed pion will shift the peak in M_{miss}^2. Using M_{miss}^2 and C, we can extract $\cos\theta_{B,D^*\ell} = -M_{miss}^2/C$. $\cos\theta_{B,D^*\ell}$ which is highly correlated with the estimated value of M_{miss}^2 but allows us to impose a kinematic consistency condition through the requirement $\cos\theta_{B,D^*\ell} < 1$.

The dominant background is from fake D^*s. These fakes come from combining a true D^0 with a random or fake π, from a fake D^0 and the correct slow π, or from combinations where both are fakes. The size of background is determined using the ΔM sideband which is defined as $0.155 < \Delta M < 0.165$ (GeV/c^2). This region is sufficiently far away from the signal region not to be contaminated with signal as shown in Fig. 3. Correlated background occurs when a $\bar{B}^0$ meson decays to a final state with a D^{*+} and

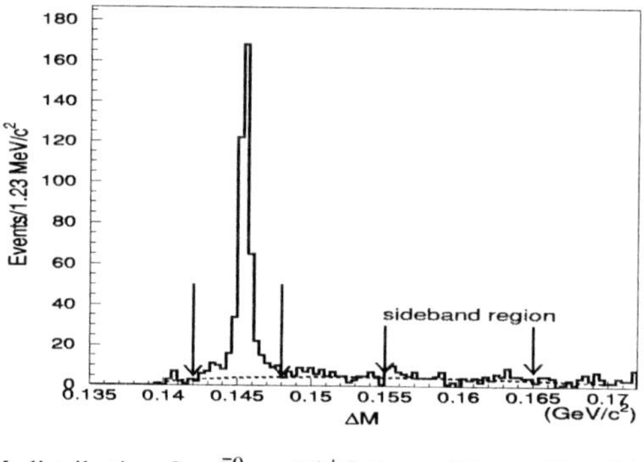

FIGURE 3. The ΔM distribution for $\bar{B}^0 \to D^{*+}\ell^-\bar{\nu}$ candidates. The sideband used to subtract the combinatoric background is shown by short arrows and long arrows indicates the signal region. The data(histogram) are superimposed with the combinatoric background distribution(dashed line).

an e^- through channels other than $\bar{B}^0 \to D^{*+}e^-\bar{\nu}$. Uncorrelated backgrounds are events

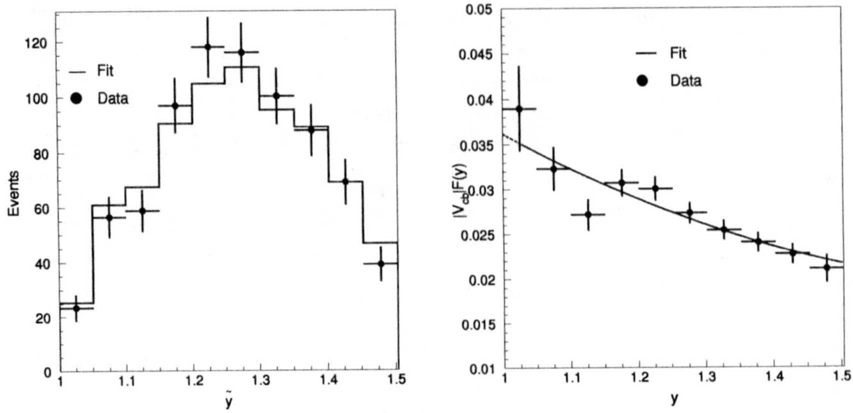

(a) (b)

FIGURE 4. (a) $\bar{B}^0 \to D^{*+}\ell^-\bar{\nu}$ yield as a function of $\tilde{y}$. The histogram represents the results of the fit and points with errors are the data. (b) $|V_{cb}|F_{D^*}(y)$. The data point are derived from the yields after correcting for efficiency and smearing effect. The curve shows the result of the fit.

with a D^{*+} from the decay of the $\bar{B}^0$ and a lepton from the B^0. It is also possible to have leptons from the decay or misidentification of light hadrons and these are classified as fake lepton background.

The partial width for $\bar{B}^0 \to D^{*+}\ell^-\bar{\nu}$ decays is geven by

$$\frac{d\Gamma(\bar{B}^0 \to D^{*+}\ell^-\bar{\nu})}{dy} = \frac{G_F^2}{48\pi^3} m_{D^{*+}}^3 (m_{\bar{B}^0} - m_{D^{*+}})^2 g(y) |V_{cb}|^2 F_{D^*}(y)^2 \tag{6}$$

where $g(y)$ can be found in [3]. We fit the final yield after subtraction of background as a function of y for $|V_{cb}|F_D*(1)$ and ρ^2 , fixing $R_1(1) = 1.3$ and $R_2(1) = 0.8$ which are obtained using HQET and QCD sum rules [3]. The fit result is shown in Fig. 4. We find $|V_{cb}|F_{D^*}(1) = (3.54 \pm 0.19) \times 10^{-2}$ and $\rho^2 = 1.35 \pm 0.17$. We obtain the branching fraction of $B(\bar{B}^0 \to D^{*+}\ell^-\bar{\nu}) = (4.59 \pm 0.23) \times 10^{-2}$ by integrating the fitted function $d\Gamma/dy$ and $|V_{cb}| = (3.88 \pm 0.21) \times 10^{-2}$ using $F_D*(1) = 0.913 \pm 0.042$ [8], where all errors are statistical only. The largest systematic uncertainty comes from the reconstruction efficiency of slow pions. We assign 2.6% in $|V_{cb}|F_{D*}(1)$ as the systematic error due to the low momentum track reconstruction. This error is estimated using track embedding study.

INCLUSIVE *B* SEMILEPTONIC DECAY WITH LEPTON TAGS

We measure the semileptonic branching fraction $\mathbf{B}(B \to Xe\nu)$ with the dilepton method [9]. The dilepton method uses another high momentum lepton which tags B

flavor in addition to the electron used for the measurement. We require a high momentum with $1.4 < p* < 2.2$ GeV/c as tagging lepton, which can be muon or electron. The efficiencies of electron(muon) identification in this momentum range is 91.7(85.7)%. We use the charge and kinematical correlations between the tag lepton and the electron to distinguish primary electrons from secondary leptons. The primary electron and the tag lepton have opposite charges unless the B mesons decay are of the same flavor due to $B^0\bar{B}^0$ mixing. The opposite-sign dileptons are required to satisfy, $p^*_\ell + \cos\theta^*_{le} > 1.2$ (p^*_e in GeV/c) or $\cos^*_{le} > 0.3$, where θ^*_{le} is the opening angle between the tag lepton and the electron in the CM frame. We also require $-0.8 < \cos\theta^*_{le} < 0.998$ for both opposite-sign and same-sign leptons to reduce the electron background from continuum events and fake tracks. A simultaneous binned maximum likelihood fit is performed on the E/p distributions of the electron candidates in each momentum(p_{lab}) and theta(θ_{lab}) bin to minimize the contribution from fake electrons. Events containing leptons from J/ψ decays are rejected if a lepton candidate, when combined with an oppositely-charged track in the event, has an invariant mass consistent with that of the J/ψ. On the other hand, leptons from τ decays($B \to X\tau\nu, \tau \to \ell\nu\nu$) or D_s decays ($B \to X_c D_s, D_s \to Y\ell\nu$) are included as tags.

Taking into account $B^0\bar{B}^0$ mixing, the two electron spectra, opposite-sign(N_{+-}) and same-sign($N_{\pm\pm}$) can be written as

$$\frac{dN_{+-}}{dp} = N_{tag}\eta(p)\varepsilon_{k1}(p)[\frac{dB(b \to x\ell\nu)}{dp}(1-\chi) + \frac{dB(b \to c \to y\ell\nu)}{dp}\chi] \tag{7}$$

$$\frac{dN_{\pm\pm}}{dp} = N_{tag}\eta(p)\varepsilon_{k2}(p)[\frac{dB(b \to x\ell\nu)}{dp}\chi + \frac{dB(b \to c \to y\ell\nu)}{dp}(1-\chi)] \tag{8}$$

where N_{tag} is the number of tag leptons, $\eta(p)$ is the identification efficiency as a function of the momentum, including the acceptance and reconstruction efficiencies, $\varepsilon_k(p)$ is the efficiency of kinematical cuts and χ is the mixing parameter. We obtain the momentum spectra for both the primary and secondary electrons, which are shown in Fig. 5, by solving the equations (7) and (8). We use $\chi = 0.0843 \pm 0.0060$ for the mixing parameter ($\chi \equiv \chi_0 f_{00}$). This value is calculated by using the parameters, $\chi_0 = 0.174 \pm 0.01$, $f_{+-}/f_{00} = 1.04 \pm 0.08$ and $f_{+-} + f_{00} = 1.0$, where $f_{00}(f_{+-})$ is the branching fractioin for $\Upsilon(4S)$ decay into $B^0\bar{B}^0(B^+B^-)$. We fit the primary electron spectrum with the ISGW2 [11] model prediction to extract the semileptonic branching fraction and obtain the inclusive semileptonic branching fraction. Fig. 5 shows the fitted primary lepton momentum spectrum.

$$\mathbf{B}(B \to Xe\nu) = (10.86 \pm 0.14 \pm 0.47)\% \tag{9}$$

The main contributions to the systematic error are a 3% uncertainty in the electron detection efficiency, which includes the tracking efficiency and a 2% uncertainty in the kinematic cut efficiency.

$|V_{cb}|$ is extracted from this measurement using the relation, $\mathbf{B}(B \to X\ell\nu)/\tau_B = \gamma_c|V_{cb}|^2 + \gamma_u|V_{ub}|^2$ neglecting the small value of $|V_{ub}|$, where the factors γ_c and γ_u are taken from theory. Table 1 summarizes the results of V_{ub} for four different models.

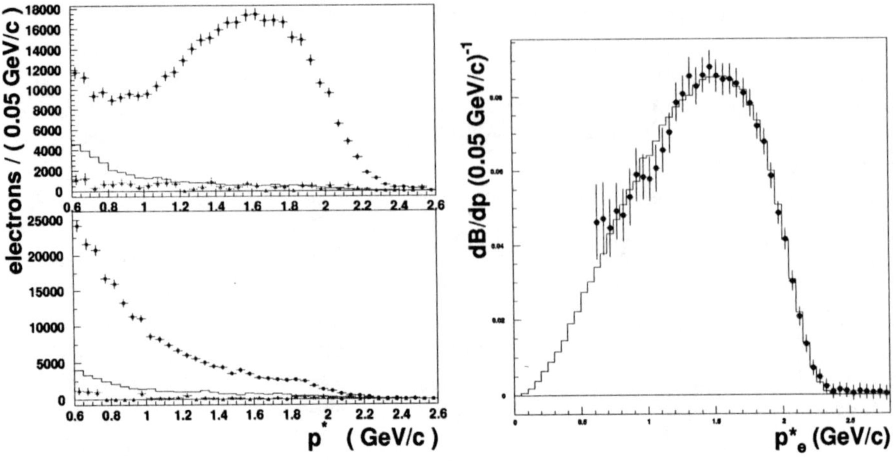

(a) (b)

FIGURE 5. (a) Electron momentum spectra. The top and bottom figures show the spectra of opposite sign and same sign electrons, respectively. The points with error bars are the on-resonance data. The triangles and the histogram are the scaled off-resonance data and background in $B\bar{B}$ events, respectively. (b)The primary electron spectrum of semileptonic deay. The solid line is a fit to the ISGW2 model.

TABLE 1. Determinations of $|V_{cb}|$

Model	γ_c	$\lvert V_{cb}\rvert \times 10^{-2}$
ACCMM [12]	40 ± 8	$4.10 \pm 0.10 \pm 0.40$
ISGW [13]	42 ± 8	$4.00 \pm 0.10 \pm 0.40$
M.Shifman *et al.* [14]	41.3 ± 4.0	$4.04 \pm 0.10 \pm 0.20$
P.Ball *et al.* [15]	43.2 ± 4.2	$3.95 \pm 0.09 \pm 0.19$

MEASUREMENT OF $\bar{B}^0 \to \pi^+ \ell^- \bar{\nu}$

The branching fraction for $\bar{B}^0 \to \pi^+ \ell^- \bar{\nu}$ is measured with the neutrino reconstruction method. The four-momentum of the undected neutrino is inferred from the missing energy E_{miss} and the missing momentum P_{miss}. The missing mass-squared of reconstructed neutrino is required to satisfy $|M^2_{miss}| < 2$ GeV^2/c^4 for the neutrino consistency. The missing momentum P_{miss} misrepresents the neutrino momentum P_{miss}, when the event has more than two netrinos or when some other particle is not detected or is incorrectly reconstructed. In order to suppress such events, we require each event to have only one lepton and also to have total charge ΣQ_i consistent with zero. We allow $\Sigma Q_i = \pm 1$, as well as 0, to maintain a high signal efficiency. Missing particles along the beam axis can simulate a neutrino. We reject events with $|\cos\theta_{miss}| > 0.8$, where θ_{miss} is the polar angle of the missing momentum P_{miss} with respect to the beam axis.

Backgrounds come from $e^+e^- \to q\bar{q}$ continuum events, $b \to c\ell\nu$ decays and other $b \to u\ell\nu$ decays. The continuum background is suppressed using the event shape variable R_2, the ratio of the second to zeroth Fox-Wolfram moments. We require $R_2 < 0.25$ to reduce continuum backgrounds. The largest source of background are $b \to c\ell\nu$ decays, which have branching fractions three orders of magnitude higher than that of the signal decay. Since $b \to u\ell\nu$ decay modes have harder lepton spectra than $b \to c\ell\nu$ decay modes, the $b \to c\ell\nu$ background is effectively reduced with the lepton momentum requirement $P_\ell > 1.2$ GeV/c. We further require $|P_\ell| + |P_\pi| > 3.1$ GeV/c, utilizing the fact that pions from $\bar{B}^0 \to \pi^+\ell^-\bar{\nu}$ decays have a harder spectrum than those from $b \to c\ell\nu$ decays and other $b \to u\ell\nu$ decay modes. With the precisely known constraints $E_B = E_{beam}$ and $p_\nu^2 = (p_B - p_\pi - p_\ell)^2 = 0$, we compute the angle between the B momentum direction and that of the combined $Y = \pi + \ell$ system, $\cos\theta_{B-Y} = (2E_BE_Y - M_B^2 - M_Y^2)/2|P_B||P_Y|$. Signal $\pi - \ell$ combinations should have $\cos\theta_{B-Y}$ between -1 and 1, whereas backgrounds may have unphysical values. We suppress the background by requiring $|\cos\theta_{B-Y}| < 1$.

The $\bar{B}^0 \to \pi^+\ell^-\bar{\nu}$ signal yields are extracted from the distribution of the beam energy constrained mass M_B and the energy difference ΔE. If more than one B candidate remains in the $M_B > 5.1$ GeV/c^2 region, we choose the candidate which has ΔE closest to zero. The signal region is chosen as $5.26 < M_B < 5.29$ GeV/c^2 and $|\Delta E| < 0.3$ GeV based on a MC study. The signal yield is extracted by fitting the M_B distribution in the ΔE signal region($|\Delta E| < 0.3$ GeV) with the predicted signal and background shapes. The signal shape is simulated using the ISGW2 [11] and WSB [16] models. The average of the two distributions is used with the normalization allowed to float in the fitting. The continuum and $b \to c\ell\nu$ background are simulated using detailed descriptions of the $q\bar{q}$ hadronization processes and generic $b \to c$ and subsequent charm decays, respectively. The $b \to u\ell\nu$ background is generated using the ISGW2 model to predict the feed-down into the signal region. Its magnitude is fixed using the branching fraction of the $\bar{B}^0 \to \rho^+\ell^-\nu$ decay mode, $\mathbf{B}(\bar{B}^0 \to \rho^+\ell^-\nu) = (2.6 \pm 0.3) \times 10^{-4}$ [17]. Figs. 6-(a) and (b) show the M_B and ΔE distributions for the signal regions; $|\Delta E| < 0.3$ GeV and $M_B > 5.26$ GeV/c^2. The extracted signal yield is 107 ± 16, corresponding to a branching fraction of $\mathbf{B}(\bar{B}^0 \to \pi^+\ell^-\bar{\nu}) = (1.28 \pm 0.20) \times 10^{-4}$, where error is statistical only. The largest uncertainties of systematic errors are related to the imperfections of the detector simulation of the neutrino reconstruction. We assign a 15.9% errors to the ν simulation.

SUMMARY

We measured branching fractions and $|V_{cb}|$ using semileptonic B decay modes. From $\bar{B}^0 \to D^+\ell^-\bar{\nu}$ mode, we obtained $\mathbf{B}(\bar{B}^0 \to D^+\ell^-\bar{\nu}) = (2.13 \pm 0.12 \pm 0.39)\%$ and $|V_{cb}| = (4.19 \pm 0.45 \pm 0.53 \pm 0.30) \times 10^{-2}$ using the neutrino reconstruction method, where errors are statistical, systematical and theoretical in order [18]. And we measured $\mathbf{B}(\bar{B}^0 \to D^{*+}\ell^-\bar{\nu}) = (4.59 \pm 0.23 \pm 0.40)\%$ and $|V_{cb}| = (3.88 \pm 0.21 \pm 0.20 \pm 0.19) \times 10^{-2}$ from $\bar{B}^0 \to D^{*+}\ell^-\bar{\nu}$ [19]. Semileptonic B decay branching fraction also was measured as $\mathbf{B}(B \to Xe\nu) = (10.86 \pm 0.14 \pm 0.47)\%$. Results on $|V_{cb}|$ from these measurements are consistent with each other. For $b \to u$ decay process, we measured the preliminary result of $\mathbf{B}(\bar{B}^0 \to \pi^+\ell^-\bar{\nu}) = (1.28 \pm 0.20 \pm 0.26) \times 10^{-4}$. We expect much more precise determinations of $|V_{cb}|$ and $|V_{ub}|$ with on-going understanding of systematics and larger data samples which will be accumulated at Belle.

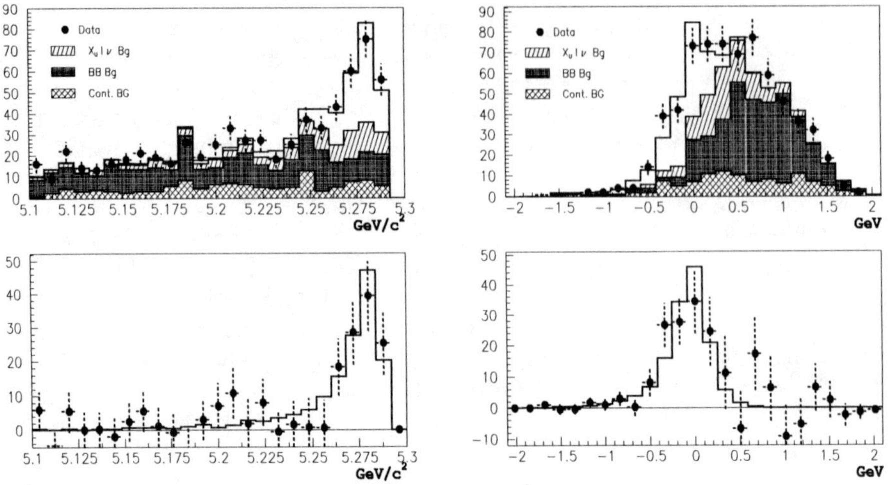

(a) (b)

FIGURE 6. Distribution of (a) M_B in the ΔE signal region ($|\Delta E| < 0.3$ GeV), (b) ΔE in the M_B signal region ($M_B > 5.26$ GeV/c^2). In each figure, closed circles show the real data. The cross-hatched, dark shaded, slant-hatched and unshaded histograms show continuum, $b \to c\ell\nu$, $b \to u\ell\nu$ backgrounds and the signal components, respectively. Bottom plots in each case compare the background subtracted data to the signal MC simulation.

REFERENCES

1. N. Cabibbo, *Phys. Rev. Lett.* **10**, 531 (1963);
   M. Kobayashi and T. Maskawa,*Prog. Theor. Phys.* **49**, 652 (1973)
2. L. Wolfenstein, *Phys. Rev. Lett.* **51**, 1945 (1983)
3. M. Neubert, *Phys. Reports* **245**, 259 (1994)
4. J. Bartelt *et al.*(CLEO Collab.), *Phys. Rev. Lett.* **71**, 4111 (1993).
5. J. P. Alexander *et al.*(CLEO Collab.), *Phys. Rev. Lett* **77**, 5000 (1996).
6. I. Caprini, L. Lellouch and M. Neubert, *Nuclear Physics* **B530**, 153 (1998).
7. G. Fox and S. Wolfram, *Phys. Rev. Lett.* **41**, 1581 (1978).
8. M. Neubert in BaBar Physics Book, P.F. Harrison and H. R. Quinn(editor), SLAC-R-504 (1998).
9. H. Albrecht *et al.* (ARGUS Collab.), *Phys. Lett. B* **318**, 397 (1993).
10. Particle Data Group, D. E. Groom *et al.*, *Eur. Phys. J. C* **15**, 1 (2000).
11. N. Isgur and D. Scora, *Phys. Rev.* **D52**, 2783 (1995).
12. G. Altarelli, N. Cabibbo, G. Corbo, L. Maiani and G. Martinelli, *Nuclear Phys.* **B208**, 365 (1982)
13. N. Isgur, D. Scora, B. Grinstein and M. B. Wise, *Phys. Rev.* **D39** 799(1989)
14. M. Shifman, N. G. Uraltsev, and A. Vainshtein, *Phys. Rev.* **D51**, 2217 (1995).
15. P. Ball, M. Beneke, and V. M. Braun, *Phys. Rev.* **D52** 3932 (1995).
16. M. Wirbel, B. Stech and M. Bauer, *Z.Phys.* **C29**, 637 (1985).
17. B. H. Behrens *et al.* (CLEO Collab.), *Phys. Rev.* **D61**, 052001 (2000).
18. K. Abe *et al.* (Belle Collab.), KEK preprint 2001-150[hep-ex/0111082], submitted to *Phys. Lett. B*.
19. K. Abe *et al.* (Belle Collab.), KEK preprint 2001-149[hep-ex/0111060], submitted to *Phys. Lett. B*.

Present and Future in Semileptonic B Decays

Christian W. Bauer

Physics Department, University of California at San Diego, La Jolla, CA 92093

Abstract. In this talk I review the status of our ability to extract the CKM matrix elements $|V_{ub}|$ and $|V_{cb}|$ from semileptonic decays. I will review both exclusive and inclusive methods and put a strong emphasis on how to ensure keeping the extractions model independent.

INTRODUCTION

One of the main goals in particle physics over the next several years is to test precisely the flavor sector of the standard model (SM). In the SM, all weak interactions are determined by 12 parameters, namely the Fermi coupling constant, the weak mixing angle θ_W, the six quark masses and four parameters in the CKM matrix. B physics plays an integral role in overconstraining the four parameters in the CKM matrix, which leads to a stringent test of the SM. In this talk I give an overview of the progress that has been made on using semileptonic B decays to determine two of these parameters, the magnitude of V_{ub} and V_{cb}. Semileptonic decays provide an ideal way to measure the magnitude of CKM matrix elements, since the strong interaction effects are greatly simplified by the presence of the two leptons in the final state.

I will review how to measure V_{ub} and V_{cb} from both inclusive and exclusive decays. I will present the current status of extracting these CKM parameters from semileptonic decays, and point out possible improvements in the coming years.

DETERMINATION OF $V_{\rm cb}$

Exclusive Decays

One of the main applications of heavy quark effective theory (HQET) [1] is that $V_{\rm cb}$ can be extracted in a model independent way using exclusive $b \to c$ semileptonic decays. Heavy quark spin and flavor symmetry predicts that in the infinite mass limit all heavy to heavy form factors are determined by a single nonperturbative function, the Isgur-Wise function. Moreover, this function is normalized to unity at zero recoil, which allows for a model independent prediction of semileptonic exclusive decays at that kinematic point.

To be more precise, consider the decay $B \to D^* \ell \bar{\nu}$, and define the four-velocities of the B and the D^* mesons

$$v_b^\mu = \frac{p_B^\mu}{m_B}, \qquad v_c^\mu = \frac{p_{D^*}^\mu}{m_{D^*}}. \tag{1}$$

CP618, *Heavy Flavor Physics: Ninth International Symposium*, edited by A. Ryd and F. C. Porter

The decay rate as a function of the recoil $w = v_b \cdot v_c$ is given by

$$\frac{d\Gamma_{B\to D^*}}{dw} = \sqrt{w^2-1} f(m_B, m_D, w) |V_{cb}|^2 F_{D^*}^2(w), \tag{2}$$

where $\sqrt{w^2-1}$ is a phase space suppression factor, $F_{D^*}^2(w)$ is a form factor and

$$f(m_B, m_D, w) = \frac{G_F^2 m_B^5}{48\pi^3} r^3 (1-r)^2 (1+w^2) \left(1 + \frac{4w}{1+w} \frac{1-2wr+r^2}{(1-r)^2}\right), \tag{3}$$

where $r = m_{D^*}/m_B$. The form factor is related to the Isgur Wise function by

$$F_{D^*}(w) = \eta_A \xi(w), \tag{4}$$

where η_A is a correction factor including both perturbative corrections, given by an expansion in $\alpha_s(m_b)$, and nonperturbative corrections, determined by an expansion in $\Lambda/m_{b,c}$. At zero recoil the Isgur-Wise function is normalized, $\xi(1) = 1$, and all $O(\Lambda/m)$ corrections in η_A vanish at that kinematic point [2]. The perturbative contributions to η_A have been calculated to two loop order [3] and the $\Lambda^2/m^2_{(b,c)}$ corrections can be estimated using phenomenological models [4]. This leads to [5]

$$F_{D^*}(1) = \eta_A = 0.913 \pm 0.042, \tag{5}$$

where the uncertainty is mostly due to the unknown Λ^2/m^2 and higher corrections. Recently, this form factor has also been calculated on the lattice, and the result has identical central value and comparable uncertainties[1] [6]

$$F_{D^*}^{\mathrm{Lat}}(1) = 0.913^{+24+17}_{-17-30}. \tag{6}$$

Several experiments have measured this differential decay rate and for illustration we show in Fig. 1 a plot of the BELLE collaboration together with a plot comparing the different measurements. Since the rate vanishes at zero recoil due to the phase space suppression factor, the differential rate has to be extrapolated to zero recoil. This extrapolation is guided by a model independent relationship between the curvature and the slope of $F_{D^*}(1)$ [7]. Combining the experimental measurement with the theoretical calculation of $F_{D^*}(1)$ leads to the values presented at this conference [9, 10, 11]

$$\begin{aligned} V_{cb}^{\mathrm{BELLE}} &= [39.7 \pm 1.6(stat) \pm 2.0(sys) \pm 1.9(theor)] \times 10^{-3} \\ V_{cb}^{\mathrm{CLEO}} &= [46.2 \pm 1.4(stat) \pm 2.0(sys) \pm 2.1(theor)] \times 10^{-3} \\ V_{cb}^{\mathrm{LEP}} &= [40.5 \pm 0.5 \pm 2.0] \times 10^{-3} \end{aligned} \tag{7}$$

I would like to make a few comments on how the accuracy in this measurement might be improved. As can be seen from Fig. 1, there is a large correlation between the slope of the physical form factor and its value at zero recoil. Thus, information about this slope

[1] This result does not include the QED corrections of +0.007, which is included in (5).

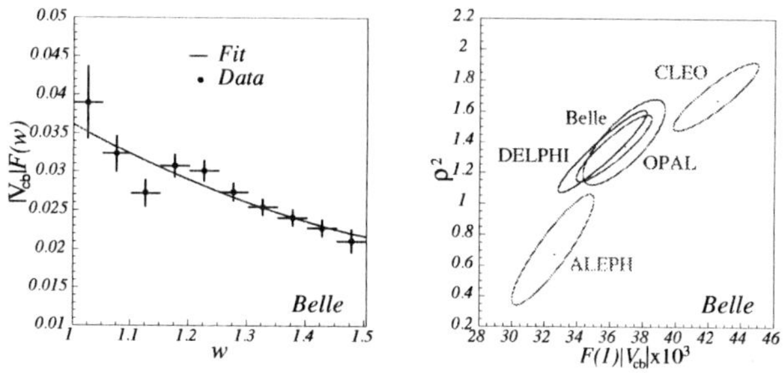

FIGURE 1. Left: Extrapolation of $|V_{cb}|F_{D^*}(w)$ to the zero recoil point by Belle. Right: comparison of results from various collaborations.

can reduce the errors on $|V_{cb}|$ considerably. There have recently been suggestions that QCD sumrules give constraints on the slope of the Isgur-Wise function [12], although one has to be careful in relating these to the slopes of the physical form factor. Since the slopes of the $B \to D$ and the $B \to D^*$ form factor are related to one another in a calculable way, one should use the information from both of these decays at the same time (note that the experimental uncertainty in the slopes in these two decays is roughly the same [9]). In summary, performing a new complete high statistics analysis, including both the D and D^*, as well as the form factor relations R_1 and R_2, should lower the uncertainties below the current level of 5%.

Inclusive Decays

Inclusive semileptonic B decays can be calculated using an operator product expansion (OPE). This leads to a simultaneous expansion in powers of the strong coupling constant $\alpha_s(m_b)$ and inverse powers of the heavy b quark mass. At leading order in this expansion this reproduces the parton model result

$$\Gamma_0 = \frac{G_F^2 |V_{cb}|^2 m_b^5}{192\pi^3} \left(1 - 8\rho + 8\rho^3 - \rho^4 - 12\rho^2 \log\rho\right) , \tag{8}$$

where $\rho = m_c^2/m_b^2$, and nonperturbative corrections are suppressed by at least two powers of m_b. The resulting expression for the total rate of the semileptonic $B \to X_c \ell \bar{\nu}$ can be written as

$$\Gamma^{b\to c} = \Gamma_0 \left[1 + A\left[\frac{\alpha_s}{\pi}\right] + B\left[\left(\frac{\alpha_s}{\pi}\right)^2 \beta_0\right] + 0\left[\frac{\Lambda}{m_b}\right] + C\left[\frac{\Lambda^2}{m_b^2}\right] + O\left(\alpha_s^2, \frac{\Lambda^3}{m_b^3}, \frac{\alpha_s}{m_b^2}\right)\right] , \tag{9}$$

where the three coefficients A, B, C depend on the quark masses $m_{(c,b)}$. The perturbative corrections are known up to order $\alpha_s^2\beta_0$. There are no nonperturbative corrections at order Λ/m_b, and at order Λ^2/m_b^2 they are given in terms of the two matrix elements λ_1 and λ_2, with the dependence on these matrix elements contained in the coefficient

$C \equiv C(\lambda_1, \lambda_2)$. Since the value of λ_2 can be obtained from the $B - B^*$ mass splitting, and the charm quark mass can be related to known meson masses, the b quark mass and λ_1, the only unknowns in extracting $|V_{cb}|$ from the inclusive rate are the precise value of the b quark mass m_b and the matrix element λ_1.

In my opinion the best way to extract m_b and λ_1 is to use the semileptonic decay data itself. Various differential decay spectra can be measured, not only the total decay rate. Several observables have already been constructed out of these spectra which are sensitive to m_b and λ_1 [13]. The goal should be to measure and calculate as many inclusive observables as possible. All these observables are given in terms of m_b and λ_1, so a fit to these observables will determine these parameters with theoretical uncertainties given by Λ^3/m_b^3 terms in the OPE [14]. Recently, the CLEO collaboration performed such an analysis in which they used the first moment of the hadronic invariant mass spectrum in $B \to X_c \ell \bar{\nu}$ and the first moment of the photon energy in the rare decay $B \to X_s \gamma$ to extract m_b and λ_1 and used these values to determine V_{cb} [10]. Defining $\bar{\Lambda} = m_B - m_b + \ldots$ they find

$$\bar{\Lambda} = 0.35 \pm 0.07 \pm 0.10\,\mathrm{GeV}, \qquad \lambda_1 = -0.238 \pm 0.071 \pm 0.078\,\mathrm{GeV}^2, \tag{10}$$

and with that

$$|V_{cb}| = [4.04 \pm 0.09 \pm 0.05 \pm 0.08] \times 10^{-3}. \tag{11}$$

The errors are (in order) from the measurement of the total decay rate, from the uncertainties in $\bar{\Lambda}$, λ_1, and from the higher order terms in both the perturbative and nonperturbative expansion.

Using $\bar{\Lambda}$ as the mass parameter in the semileptonic decay rate has the disadvantage that the parameter $\bar{\Lambda}$ is not an infrared safe quantity. This means that the value of this parameter is ambiguous up to terms of order Λ_{QCD} [15]. This ambiguity cancels in any physical observable by a similar ambiguity in the perturbative expansion [16]. For this cancellation to occur, however, it is essential that one works consistently to a particular order in perturbation theory. Since this is an artifact of the pole mass, it would be convenient to repeat the analysis with an infrared safe definition of the b quark mass [17]. The numerical value of such a short distance mass can then be used safely in other processes as well.

Finally, I want to comment on an uncertainty not included in (11), namely the uncertainty from duality violations, which are present for any inclusive observable in heavy quark decays. While there have been several estimates of these uncertainties [18], I again would like to advocate using the semileptonic decay data itself to test for duality violations. Since many semileptonic observables can be calculated to the same order as (9) and all depend on the same two parameters m_b and λ_1, one can check for duality violations by comparing the theoretical predictions with experimental results. This is an important analysis which still needs to be performed.

DETERMINATION OF V_{ub}

Exclusive Decays

Measuring $|V_{ub}|$ from exclusive decays of B mesons is considerably more difficult than the determination of $|V_{cb}|$ discussed earlier. This is because the heavy quark spin-flavor symmetry, which gives rise to the simplifications of heavy to heavy form factors, does not apply to the heavy to light decays mediated by a $b \to u$ transition. Thus, to extract $|V_{ub}|$ from exclusive decays one has to find other ways to calculate these form factors.

As an example, consider the exclusive decay $B \to \pi\ell\bar{\nu}$. The required matrix element can be written in terms of two form factors

$$\langle\pi|\bar{u}\gamma^{\mu}P_L|B\rangle = f_+(q^2)\left[(p_B+p_\pi)^\mu - \frac{m_B^2}{q^2}q^\mu\right] + f_0(q^2)\frac{m_B^2}{q^2}q^\mu\,, \tag{12}$$

and a similar relation is true for $B \to \rho\ell\bar{\nu}$. The form factor $f_0(q^2)$ vanishes for zero lepton mass, but the form factor $f_+(q^2)$ is required to describe the decay rate. There are various suggestions in the literature how to extract these form factors. For $B \to \rho\ell\bar{\nu}$ the relevant form factors can be related to those in $D \to K^*\ell\nu$ using heavy quark and chiral symmetries[19]. Corrections are first order in both heavy quark and chiral symmetry breaking individually and are at the 30% level. This uncertainty could be reduced considerably if information from $D \to \rho\ell\nu$ and $B \to K^*\ell^+\ell^-$ became available, since then a double ratio can be constructed which has corrections which simultaneously violate heavy quark and chiral symmetry [20]. The resulting uncertainties should be below the 10% level.

Ultimately, lattice calculations will determine the required heavy to light form factors in a model independent way. In the last year several quenched calculation of the form factor $f_+(q^2)$ have been performed. Since lattice calculations can only be performed for slow pions, the lattice results are only available for small pion energies $E_\pi \lesssim 1\,\mathrm{GeV}$. A result from a recent lattice calculation of this form factor [21] is shown in Fig. 2. Experimentally, the situation is reversed with the efficiency to measure slow pions being small. In order to use lattice calculations to determine $|V_{ub}|$ from exclusive decays, however, it is crucial to increase the kinematic overlap of the lattice calculation of the form factors with the experimental measurements of the differential decay rates.

Inclusive Decays

The inclusive decay rate $B \to X_u\ell\bar{\nu}$ is directly proportional to $|V_{ub}|^2$ and can be calculated reliably and with small uncertainties using the operator product expansion. Unfortunately, the $\sim$100 times background from $B \to X_c\ell\bar{\nu}$ makes the measurement of the totally inclusive rate an almost impossible task. Several cuts have been proposed in order to reject the $b \to c$ background, however care has to be taken to ensure that the decay rate in the restricted region of phase space can still be predicted reliably theoretically. The cut which is easiest to implement experimentally is on the energy of the charged lepton, requiring $E_\ell > (m_B^2 - m_D^2)/2m_B$. Unfortunately, this cut restricts the

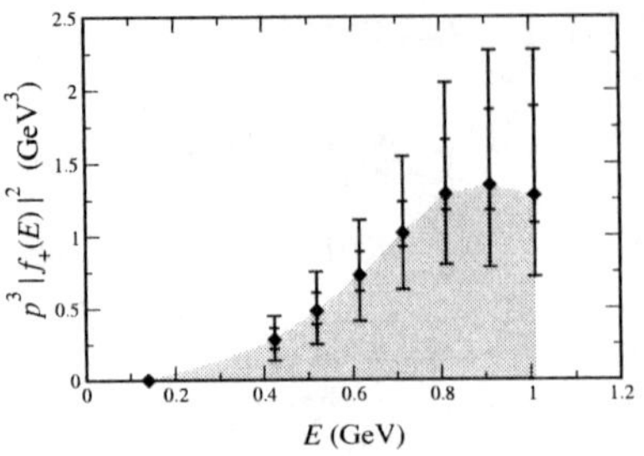

FIGURE 2. Lattice calculation of the form factor f_+ as a function of the pion energy.

remaining region of phase space too much for the OPE to still be valid. Instead, a twist expansion has to be performed [22], and at leading order the decay rate is determined by the light cone distribution function of the B meson, with subleading twist corrections suppressed by powers of Λ/m_b [23]. This distribution function is the matrix element of a nonlocal operator and can not be calculated perturbatively. However, since it is a property of the B meson itself, and thus independent of the decay process, it can be extracted from a different process and then used in the inclusive $b \to u\ell\bar{\nu}$ decay. This is similar to parton distribution functions, which can be measured for example in ep collisions and then be used in other processes. The best way to measure this distribution function is in the decay $B \to X_s\gamma$. At leading order in both α_s and Λ/m_b, the shape of the photon energy spectrum is entirely due to this light cone distribution function, and the perturbative corrections are known to order $\alpha_s^2\beta_0$ [24]. The photon spectrum has recently been measured to good accuracy by the CLEO collaboration [10], which constituted the first determination of this structure function.

Unfortunately, a cut on the lepton energy has the disadvantage that only $\sim 10\%$ of the events pass this cut. Thus, the decay is far from being fully inclusive and one has to worry about duality violations becoming large. It is therefore advantageous to find ways to suppress the background which allows more $b \to u$ events to survive. The optimal such cut is on the hadronic invariant mass $m_X < m_D$ [25]. This cut is optimal in the sense that any other cut is a subset of it, and it has been estimated that $\sim 80\%$ of the $b \to u$ events survive this cut. The same structure function which describes the lepton energy endpoint is needed to describe the region of low invariant mass. In [26] it was shown how to eliminate the structure function entirely and measure $|V_{ub}|$ model independently by using semileptonic data in combination with the photon energy spectrum in $B \to X_s\gamma$. The main uncertainties to this method of measuring $|V_{ub}|$ are given by unknown subleading structure functions, which are parametrically suppressed by $\Lambda_{\rm QCD}/m_b$ [23]. It is hard to quantify the exact size of these uncertainties, and I would conservatively estimate them to be 10-20%.

It was shown in [27] that using a cut on the leptonic invariant mass $q^2 = (p_\ell + p_\nu)^2$ allows to measure $|V_{ub}|^2$ without requiring knowledge of the structure function of the B meson. The number of events surviving such a cut on q^2 can be calculated using the usual local OPE and depending on the exact value of the cut chosen, the fraction of events surviving the cut is 10-20%, with uncertainties on $|V_{ub}|$ ranging from 15% for

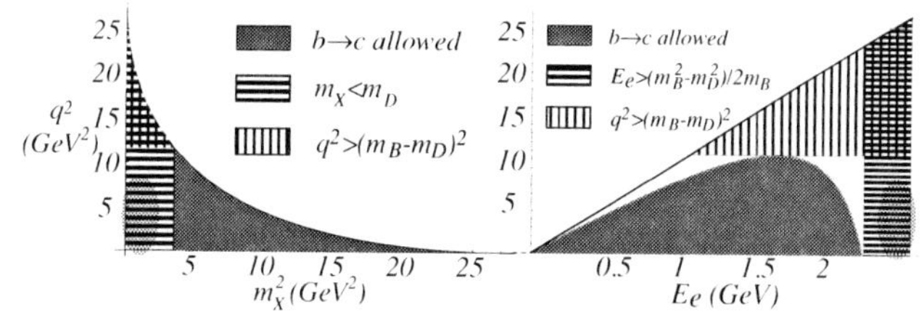

FIGURE 3. The dalitz plot in the q^2/s_H and q^2/E_ℓ plane. The grey blob in the lower right (left) hand corner of the phase space indicates where the local OPE breaks down.

$q^2_{\rm cut} = m_B^2 - m_D^2 = 11.6\,{\rm GeV}^2$ to 25% for $q^2_{\rm cut} = 14\,{\rm GeV}^2$. The advantage of not having to use information from $B \to X_s\gamma$ and of the absence of power corrections at order $\Lambda_{\rm QCD}/m_b$ is offset by the fact that this cut again eliminates a huge fraction of the events and the worries about duality violations reappear.

To improve the situation, one has to understand why the q^2 cut allows to calculate using the local OPE, while the other cuts require the twist expansion. The twist expansion is relevant if the remaining phase space is dominated by hadronic states with energy much larger than their invariant mass. This is the case for both a cut on E_ℓ and a cut on m_X. The cut on q^2, however, restricts the hadron energy to be $E_X < m_B - \sqrt{q_c^2} < m_D$ and this cut therefore eliminates this dangerous region of phase space. The situation is illustrated in Fig. 3.

In [28] a new strategy to measure $|V_{ub}|$ was proposed which combines the advantages of the m_X cut with those of the q^2 cut. The idea is to use a cut on the hadronic invariant mass to reject the $b \to c$ background, and simultaneously use a lower cut on q^2 to avoid the twist region. The pure m_X and q^2 cuts are contained in this approach as the limits $q^2_{\rm cut} = 0$ or $q^2_{\rm cut} = (m_B - m_X^{\rm cut})^2$, respectively. Thus, varying the $q^2_{\rm cut}$ in the presence of a cut on m_X allows to smoothly interpolate between these two cases. Using these results the strategy to extract $|V_{ub}|$ in a model independent way is:

- make the cut on m_X as large as possible, keeping the background from $b \to c$ small.
- for a given cut on m_X, reduce the q^2 such as to minimize the overall uncertainty.

To illustrate this procedure I will present graphically the results of [28]. In Fig. 4 the fraction of events surviving such combined cuts are shown for three values of $m_X^{\rm cut}$, in combination with the effect of the structure function. As advertised, lowering the cut on q^2 increases the fraction of events, while the effect of the structure function grows. Depending on how well the structure function will be known in the future determines how much the cut on q^2 can be lowered. Due to the fact that the structure function can only be determined up to corrections scaling as $\Lambda_{\rm QCD}/m_b$, it will always have an uncertainty at the 10-20% level, even with a perfect measurement of the $B \to X_s\gamma$ photon spectrum. Thus, for a determination of $|V_{ub}|$ with uncertainties below 10% one should ensure that the difference between the prediction with and without the structure function is less than $\sim 30\%$.

As in any inclusive observable, the remaining uncertainties are due to three sources:

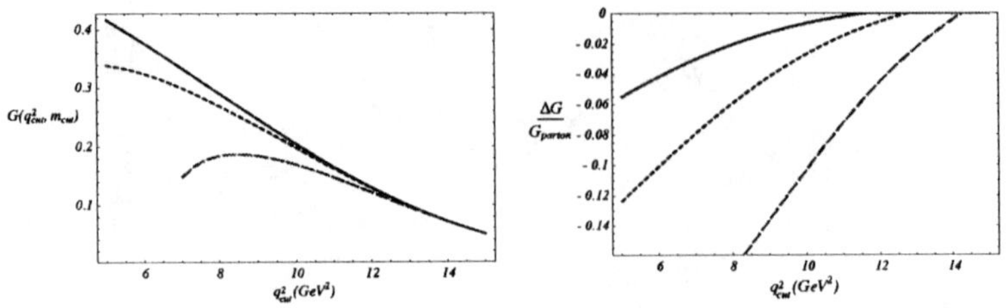

FIGURE 4. Left: $G(q^2_{\text{cut}}, m_{\text{cut}})$, as defined in [28], as a function of q^2_{cut}, for $m_{\text{cut}} = 1.86\,\text{GeV}$ (solid line), $1.7\,\text{GeV}$ (short dashed line) and $1.5\,\text{GeV}$ (long dashed line). The fraction of events is given by $1.21 \times G(q^2_{\text{cut}}, m_{\text{cut}})$. Right: The effect of the structure function.

unknown perturbative corrections, uncertainties due to the parameters m_b and λ_1 and uncertainties due to unknown matrix elements of local operators at $O(\Lambda^3/m_b^3)$ in the OPE. To compare various methods to determine $|V_{ub}|$ it is crucial to have a consistent scheme to determine the theoretical uncertainties. In [28] such an error analysis was performed for the combined q^2, m_X cut method, and in Table. 1 I present the result for various different combinations of cuts. From this table it is obvious that even without any knowl-

TABLE 1. Fraction of events for several different choices of $(q^2_{\text{cut}}, m_{\text{cut}})$, along with the uncertainties on $|V_{ub}|$. The two last lines correspond to pure q^2 cuts, $m_{\text{cut}} = m_B - \sqrt{q^2}$, and are included for comparison. Δ_{struct} gives the fractional effect of the structure function $f(k_+)$ in a simple model; we do not include an uncertainty on this in our error estimate. The overall uncertainty $\Delta|V_{ub}|$ is obtained by combining the other uncertainties in quadrature. The two values correspond to $\Delta m_b^{1S} = \pm 80\,\text{MeV}$ and $\pm 30\,\text{MeV}$.

Cuts on (q^2, m_X^2)	*Fract*	$\Delta_{\text{struct}} V_{ub}$	$\Delta_{\text{pert}} V_{ub}$	$\Delta_{m_b} V_{ub}$ $\pm 80/30\,\text{MeV}$	$\Delta_{1/m^3} V_{ub}$	ΔV_{ub}
Combined cuts						
$6\,\text{GeV}^2, 1.86\,\text{GeV}$	46%	−2%	4%	7%/2.5%	3%	8%/5%
$8\,\text{GeV}^2, 1.7\,\text{GeV}$	33%	−3%	6%	8%/3%	4%	9%/6%
$11\,\text{GeV}^2, 1.5\,\text{GeV}$	18%	−4%	13%	9%/3.5%	8%	14%/11%
Pure q^2 cuts						
$(m_B - m_D)^2, m_D$	0.17%	--	15%	19%/7%	18%	15%/12%

edge of the structure function, and using a realistic cut of $m_X^{\text{cut}} = 1.7\,\text{GeV}$, $|V_{ub}|$ can still be measured with uncertainties below the 10% level by choosing $q^2_{\text{cut}} \approx 8\text{GeV}^2$. With many different methods to measure $|V_{ub}|$, the question arises of which measurement to ultimately use. In my opinion, this question will most likely be answered once a full analysis is performed and the theoretical errors are investigated in conjunction with the experimental uncertainties. Depending on the details of the experimental efficiencies, some combination (most likely non-linear) of the three cuts on E_ℓ, m_X and q^2 will prove to be the most effective way to minimize the overall uncertainty. The cut on the hadronic invariant mass m_X will most likely be the best way to eliminate the $b \to c$ background, the cut on the leptonic invariant mass q^2 will limit the effect of the nonperturbative structure function and the cut on the lepton energy E_ℓ can help with experimental efficiencies.

A careful analysis, which includes a realistic structure function and conservative errors should be performed to find the ideal combination of cuts.

CONCLUSIONS

In this talk I reviewed the current status of determining the magnitude of the CKM matrix elements $|V_{ub}|$ and $|V_{cb}|$ from semileptonic B meson decays. For exclusive decays I reviewed theory and experiment and pointed towards a few improvements possible in the coming years. For inclusive decays, I reviewed how their decay spectra can be calculated (up to duality violations) using an operator product expansion, which at leading order reproduces the parton model results. Nonperturbative corrections are parameterized by matrix elements of local operators suppressed by powers of m_b. There are no corrections at $O(\Lambda/m_b)$, and at second order in the inverse heavy quark mass there are two matrix elements, commonly labeled by λ_1 and λ_2. While the value of λ_2 can be obtained from the $B-B^*$ mass splitting, the value of λ_1 is currently not known very well. Thus, in order to calculate the inclusive semileptonic observables precisely, we need to obtain values for the b quark mass, which enters the calculation, as well as the for the parameter λ_1.

For $B \to X_c \ell \bar{\nu}$, these parameters give rise to the dominant theoretical uncertainty in the determination of $|V_{cb}|$ and their precise determination is therefore crucial to lower this uncertainty. I have argued here that these parameters should be extracted from decay spectra itself, and then used to extract $|V_{cb}|$. I also put forward the idea that duality violations can be estimated using semileptonic decay spectra.

To measure $|V_{ub}|$ from the inclusive decay $B \to X_u \ell \bar{\nu}$ one has to deal with the large background from $b \to c$ transitions. Imposing kinematic cuts to suppress this background tends to destroy the convergence of the OPE and for both a cut on the lepton energy and a cut on the hadronic invariant mass an incalculable structure function is required, with corrections suppressed by Λ/m_b. This function can be extracted from the photon energy spectrum in $B \to X_s \gamma$, thus a precise measurement of this spectrum is very important. A cut on the leptonic invariant mass q^2 can be used to lower the effect of this structure function and is therefore crucial to minimize theoretical uncertainties. Ultimately, a dedicated analysis which combines the experimental and theoretical uncertainties will decide which combination of cuts will allow for the most precise determination of $|V_{ub}|$. I anticipate that such a study will eventually lead to a model independent measurement of $|V_{ub}|$ with uncertainties at the 5% level.

ACKNOWLEDGMENTS

First, I would like to thank the organizers for organizing such a pleasant meeting, and for dealing so well with the unforeseeable circumstances. I would also like to thank Zoltan Ligeti and Michael Luke for collaborations on some of the work presented here, and Thomas Mannel for comments on the manuscript. This work was supported by the Department of Energy, Contract DOE-FG03-97ER40546

REFERENCES

1. N. Isgur and M. B. Wise, Phys. Lett. B **232**, 113 (1989); Phys. Lett. B **237**, 527 (1990); E. Eichten and B. Hill, Phys. Lett. B **234**, 511 (1990). H. Georgi, Phys. Lett. B **240**, 447 (1990).
2. M. E. Luke, Phys. Lett. B **252**, 447 (1990).
3. A. Czarnecki, Phys. Rev. Lett. **76**, 4124 (1996); A. Czarnecki and K. Melnikov, Nucl. Phys. B **505**, 65 (1997).
4. A. F. Falk and M. Neubert, Phys. Rev. D **47**, 2965 (1993).
5. P. F. Harrison and H. R. Quinn [BABAR Collaboration], SLAC-R-0504
6. S. Hashimoto, A. S. Kronfeld, P. B. Mackenzie, S. M. Ryan and J. N. Simone, hep-ph/0110253.
7. C. G. Boyd, B. Grinstein and R. F. Lebed, Phys. Lett. B **353**, 306 (1995);
8. Nucl. Phys. B **461**, 493 (1996); Phys. Rev. D **56**, 6895 (1997); I. Caprini and M. Neubert, Phys. Lett. B **380**, 376 (1996); I. Caprini, L. Lellouch and M. Neubert, Nucl. Phys. B **530**, 153 (1998).
9. H. Kim, these proceedings and references therein.
10. R. Briere, these proceedings and references therein.
11. F. Parodi, these proceedings and references therein.
12. N. Uraltsev, Phys. Lett. B **501**, 86 (2001) and references therein.
13. A. F. Falk, M. E. Luke and M. J. Savage, Phys. Rev. D **53**, 2491 (1996); M. Gremm, A. Kapustin, Z. Ligeti and M. B. Wise, Phys. Rev. Lett. **77**, 20 (1996); A. F. Falk and M. E. Luke, Phys. Rev. D **57**, 424 (1998).
14. M. Gremm and A. Kapustin, Phys. Rev. D55 (1997) 6924. C. Bauer, Phys. Rev. D **57**, 5611 (1998) [Erratum-ibid. D **60**, 099907 (1998)]; C. W. Bauer and C. N. Burrell, Phys. Lett. B **469**, 248 (1999). Phys. Rev. D62 (2000) 114028;
15. I. I. Bigi, M. A. Shifman, N. G. Uraltsev and A. I. Vainshtein, Phys. Rev. D **50**, 2234 (1994); M. Beneke and V. M. Braun, Nucl. Phys. B **426**, 301 (1994).
16. M. E. Luke, A. V. Manohar and M. J. Savage, Phys. Rev. D **51**, 4924 (1995); P. Ball, M. Beneke and V. M. Braun, Phys. Rev. D **52**, 3929 (1995).
17. K. Melnikov and A. Yelkhovsky, Phys. Rev. D **59**, 114009 (1999); A. H. Hoang, Z. Ligeti and A. V. Manohar, Phys. Rev. Lett. **82**, 277 (1999); A. H. Hoang, Z. Ligeti and A. V. Manohar, Phys. Rev. D **59**, 074017 (1999); A. H. Hoang, Phys. Rev. D **61**, 034005 (2000); M. Beneke and A. Signer, Phys. Lett. B **471**, 233 (1999).
18. I. I. Bigi and N. Uraltsev, hep-ph/0106346 and references therein.
19. N. Isgur and M. B. Wise, Phys. Rev. D **42**, 2388 (1990).
20. Z. Ligeti and M. B. Wise, Phys. Rev. D **53**, 4937 (1996); Z. Ligeti, I. W. Stewart and M. B. Wise, Phys. Lett. B **420**, 359 (1998).
21. El-Khadra, A. X., Kronfeld, A. S., Mackenzie, P. B., Ryan, S. M., and Simone, J. N., Phys. Rev. D **64**, 014502 (2001); Kronfeld, A. S., these proceedings [hep-ph/0111376].
22. M. Neubert, Phys. Rev. D49 (1994) 4623; I.I. Bigi *et al.*, Int. J. Mod. Phys. A9 (1994) 2467.
23. C. W. Bauer, M. Luke and T. Mannel, hep-ph/0102089.
24. Z. Ligeti, M. E. Luke, A. V. Manohar and M. B. Wise, Phys. Rev. D **60**, 034019 (1999).
25. A.F. Falk, Z. Ligeti and M.B. Wise, Phys. Lett. B406 (1997) 225; I. Bigi, R. D. Dikeman and N. Uraltsev, Eur. Phys. J. C4 (1998) 453 .
26. A. K. Leibovich, I. Low and I. Z. Rothstein, Phys. Lett. B **486**, 86 (2000); I. Z. Rothstein, these proceedings [hep-ph/0111337].
27. C. W. Bauer, Z. Ligeti and M. Luke, Phys. Lett. B479 (2000) 395; hep-ph/0007054.
28. C. W. Bauer, Z. Ligeti and M. E. Luke, Phys. Rev. D **64**, 113004 (2001).

Radiative Penguins at Belle

Mikihiko Nakao representing the Belle collaboration

KEK, High Energy Accelerator Research Organization, Tsukuba, Ibaraki 305–0801, Japan

Abstract. We report the latest results from the Belle experiment on radiative and semi-leptonic electroweak decays of B meson. The analyses are based on full or partial set of 31.3 million $B\bar{B}$ events recorded at the the $\Upsilon(4S)$ resonance with the Belle detector at the KEKB e^+e^- storage ring. We study the $b \to s\gamma$ transition through an inclusive $B \to X_s\gamma$ measurement and through exclusive reconstruction of the final states $K^*\gamma$, $K_2^*(1430)\gamma$, $K^*\pi\gamma$ and $K\rho\gamma$. We search for the $b \to d\gamma$ transition in the decays $B \to \rho\gamma$ and set upper limits on their branching fractions. We search for the flavor-changing neutral current process $b \to s\ell^+\ell^-$ in the inclusive $X_s\ell^+\ell^-$ and exclusive $K^{(*)}\ell^+\ell^-$ final states. We observe the decay $B \to K\ell^+\ell^-$ for the first time; for other final states, we set stringent upper limits.

INTRODUCTION

In the Standard Model (SM), flavor changing neutral processes ($b \to s(d)\gamma$ and $b \to s(d)\ell^+\ell^-$ are prohibited at the tree level, and proceed only by a electroweak loop (penguin) diagram or further suppressed diagrams. The loop contains a t-quark and a W boson in the SM. If we assume there is new physics beyond the SM, the loop can also be mediated by a virtual heavy charged particles such as a charged Higgs or SUSY particles that interfere with the SM process. Such an interference can be observed as a difference in the branching fraction from the SM predictions [1]. The predicted branching fractions are so far calculated with better precision than existing measurements.

We have analyzed a data set of 29.5 fb^{-1} collected with the Belle detector [2] at the KEKB e^+e^- storage ring running at the $\Upsilon(4S)$ resonance. The analyses are based either on the full data set that contains 31.3 million $B\bar{B}$ events or partial data sets of 5.3 to 22.8 million $B\bar{B}$ events. Belle is a general purpose detector with a typical laboratory polar angular coverage between 17° to 150°. Charged tracks are recontructed with a 50 layer central drift chamber (CDC), and are then extrapolated and refitted with a three layer double sided silicon vertex detector (SVD) to provide precision track information for the decay vertex reconstruction. Particle identification, namely discrimination of kaons from pions, is provided by combining information from an array of silica aerogel Cherenkov counters (ACC) and a time-of-flight counter system (TOF), together with specific ionization (dE/dx) measurements from the CDC. Photons are measured with an electromagnetic calorimeter (ECL) of 8736 CsI(Tl) crystals. Electrons are identified by matching the charged track with the ECL, ACC and dE/dx information. These detectors are surrounded by a 1.5 T superconducting solenoid coil. An iron flux-return located outside of the coil is instrumented with resistive plate counters to detect K_L and muons (KLM).

CP618, *Heavy Flavor Physics: Ninth International Symposium*, edited by A. Ryd and F. C. Porter

RADIATIVE $b \to s(d)\gamma$ PROCESS

Inclusive $B \to X_s\gamma$ branching fraction measurement

The analysis is based on a dataset of 5.3 million $B\bar{B}$ events. Here we summarize what is already published in ref. [3]. The inclusive $B \to X_s\gamma$ candidates are reconstructed from a high energy photon, a kaon ($K^\pm$ or $K_S \to \pi^+\pi^-$) and one to four pions ($\pi^\pm$ or π^0, only up to one π^0). We apply a cut on $M_{X_s} < 2.05\,\mathrm{GeV}/c^2$, with which the $B\bar{B}$ background is reduced to a manageable level. We assume the M_{X_s} spectrum follows the prediction based on a spectator model [4] above $M_{X_s} > 1.15\,\mathrm{GeV}/c^2$, and a Breit-Wigner function for the K^* resonance below $1.15\,\mathrm{GeV}/c^2$. The main background source is from continuum $q\bar{q}$ events, where the high energy photon dominantly originates either from the initial state radiation process $e^+e^- \to q\bar{q}\gamma$ or from neutral meson (π^0, η, ...) decays. In order to reduce the continuum $q\bar{q}$ background and estimate the amount of background in the final sample, we introduce a new event shape variable, which we call the Super Fox-Wolfram (SFW). The SFW is a Fisher discriminant that consists of Fox-Wolfram like moments. The candidates are classified in the signal region and background-dominant (SFW sideband) region. We checked that the beam constrained mass (M_{bc}) does not correlate with the value of the SFW variable, and therefore we can use the SFW sideband data to model the background shape in the M_{bc} distribution. The background shape cannot be described by a threshold-type (ARGUS) function as seen in Fig. 1. We observe a signal of 106.5 ± 16.8 events, and obtain a branching fraction

$$\mathcal{B}(B \to X_s\gamma) = (3.36 \pm 0.53(\mathrm{stat}) \pm 0.42(\mathrm{sys})^{+0.50}_{-0.54}(\mathrm{th})) \times 10^{-4}, \qquad (1)$$

which is consistent with the SM prediction $(3.3 \pm 0.3) \times 10^{-4}$ [1] and previous measurements by CLEO [5] and ALEPH [6].

Radiative decays into photon + two-body final states

The exclusive $B \to K^*\gamma$ final state is reconstructed from a high energy photon with a $K^* \to K\pi$. The analysis of the $K^*\gamma$ final state is based on 11.3 million $B\bar{B}$ events. The K^* is reconstructed from $K^+\pi^-$, $K_S\pi^0$, $K^+\pi^0$ and $K_S\pi^+$. Continuum background is suppressed by a cut on a likelihood ratio, which is constructed from the SFW variable, B meson flight direction ($\cos\theta_B$) and K^* decay helicity angle ($\cos\theta_{hel}$). After applying the likelihood ratio cut, the $K^*\gamma$ final state is cleanly reconstructed with small background contributions only from continuum and $B \to K^*\pi\gamma$. Background contributions from other processes are negligibly small. From fits to the M_{bc} distribution (Fig. 2), we obtain the branching fractions

$$\begin{aligned} \mathcal{B}(B^0 \to K^{*0}\gamma) &= (4.96 \pm 0.67(\mathrm{stat}) \pm 0.45(\mathrm{sys})) \times 10^{-5}, & (2) \\ \mathcal{B}(B^+ \to K^{*+}\gamma) &= (3.89 \pm 0.93(\mathrm{stat}) \pm 0.41(\mathrm{sys})) \times 10^{-5}. & (3) \end{aligned}$$

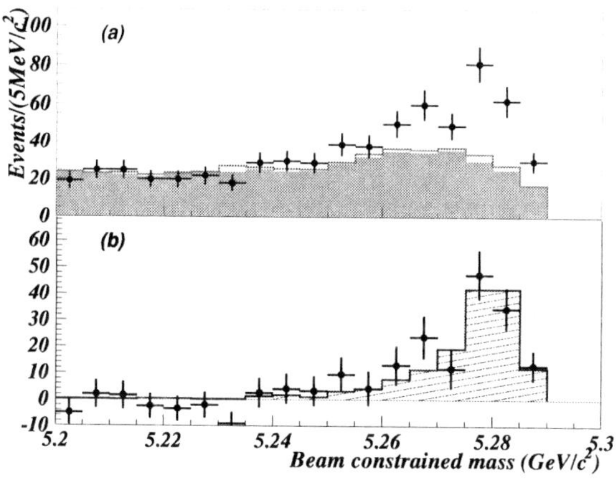

FIGURE 1. The M_{bc} distribution of $B \to X_s\gamma$: (a) data point on top of the continuum (solid histogram) and $B\bar{B}$ background (open histogram), (b) background subtracted data points compared with the Monte Carlo prediction (hatched histogram).

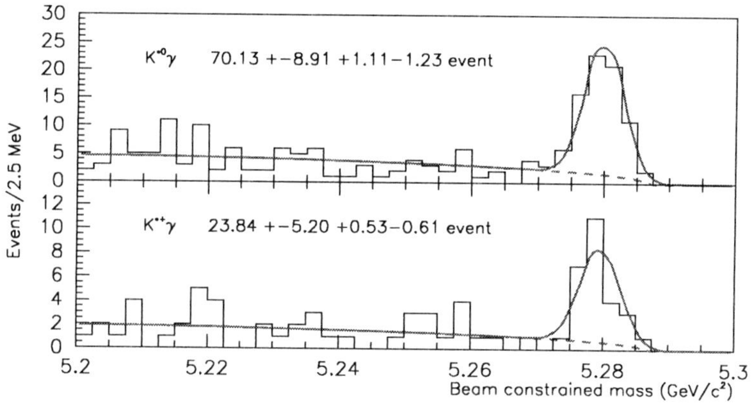

FIGURE 2. The M_{bc} distribution for the $B \to K^{*0}\gamma$ (left) and $B \to K^{*+}\gamma$ (right) candidates.

We also check the partial decay rate asymmetry between $B \to K^*\gamma$ and $\bar{B} \to \bar{K}^*\gamma$, where B ($\bar{B}$) stands for B^0 and B^+ ($\bar{B}$ and B^-). We find a small incorrectly tagged event fraction of 1.2%. From an inclusive high momentum K^* sample whose momentum spectrum is weighted to match that of $B \to K^*\gamma$, we find no detector induced asymmetry in the K^* reconstruction within a statistical error of 1%. From the $B \to K^*\gamma$ candidates,

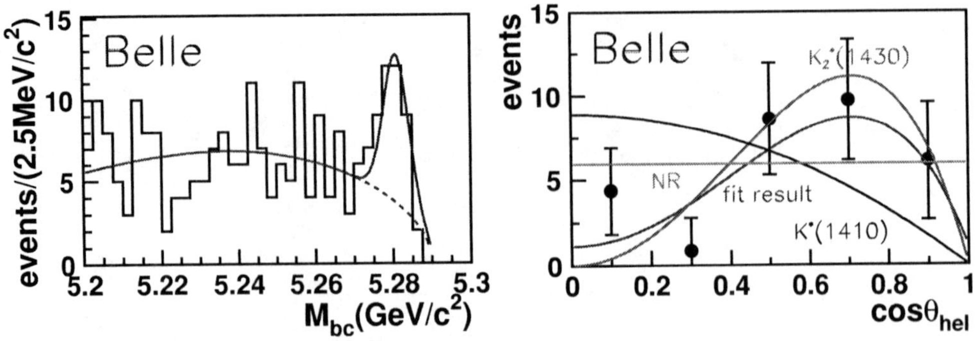

FIGURE 3. The M_{bc} (left) and $\cos\theta_{hel}$ (right) distribution for the $B \to K_2^*\gamma$ candidates.

we find its partial decay rate asymmetry is

$$A_{cp}(B \to K^*\gamma) \equiv \frac{\Gamma(\bar{B} \to \bar{K}^*\gamma) - \Gamma(B \to K^*\gamma)}{\Gamma(\bar{B} \to \bar{K}^*\gamma) + \Gamma(B \to K^*\gamma)} = (+2 \pm 11 \pm 1)\% \tag{4}$$

which is consistent with zero.

We then extend the $K\pi$ invariant mass range to higher resonances around 1.4 GeV$/c^2$, where $K_2^*(1430)$ and $K_1(1410)$ can contribute to the $K\pi$ final state. For this analysis, we use a sample of 22.8 million $B\bar{B}$ events. Since we cannot assume the spin state any more, the helicity angle is removed from the likelihood ratio calculation. After selecting candidates within a ± 125 MeV$/c^2$ mass window around $K_2^*(1430)$, we find a clear M_{bc} signal of $29.1 \pm 6.7^{+2.4}_{-1.9}$ events for the $B^0 \to K^+\pi^-\gamma$ mode (Fig. 3(left)). We estimate the feed-down contribution from other $b \to s\gamma$ processes to be 0.4 ± 0.3 events.

A fit to the helicity angle distribution is performed to disentangle the resonant components. The sample is divided into 5 bins of $\cos\theta_{hel}$ as shown in Fig. 3(right), in which the continuum background is subtracted in each bin. We find the K_2^* is the dominant component and obtain 20.1 ± 10.5 events from the fit. This leads to a branching fraction,

$$\mathcal{B}(B^0 \to K_2^*(1430)^0\gamma) = (1.26 \pm 0.66 \pm 0.10) \times 10^{-5}. \tag{5}$$

We find no significant $K^*(1410)$ component and set an upper limit,

$$\mathcal{B}(B \to K^*(1410)\gamma) < 8.0 \times 10^{-5} \quad (90\% \text{ C.L.}). \tag{6}$$

Radiative decays into photon + three-body final states

We extend the same analysis to the three body hadronic final states of radiative B decay on the sample of 22.8 million $B\bar{B}$ events. There are many kaonic resonances that decay into the $K\pi\pi$ final state ($K_1(1270)$, $K_1(1400)$, $K^*(1410)$, $K_2^*(1430)$, $K^*(1680)$, $K_3^*(1780)$, ...) through $K^*\pi$ and $K\rho$. All of these kaonic resonances have substantial

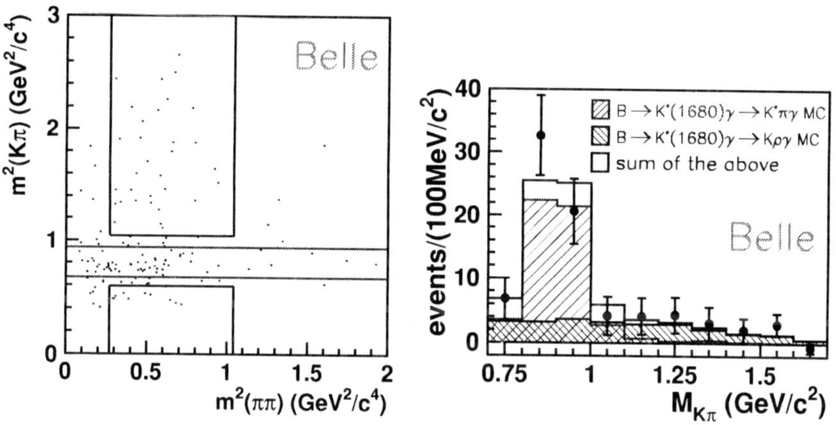

FIGURE 4. Dalitz plot (left, background included) and $K\pi$ invariant mass (right, background subtracted) for the $K\pi\pi$ final state of the $B \to K\pi\pi\gamma$ candidates.

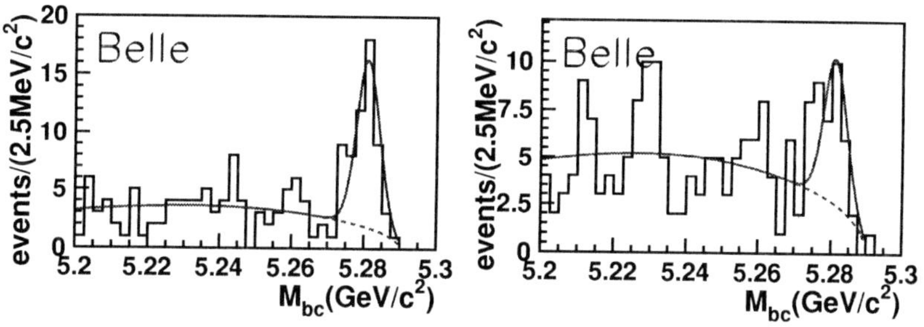

FIGURE 5. Beam constraint mass distribution for the $B \to K^*\pi\gamma$ (left) and $B \to K\rho\gamma$ candidates.

branching fractions to $K^*\pi$. We select $B^+ \to K^+\pi^+\pi^-\gamma$ candidates that have $M_{K\pi\pi} < 2.0\ \mathrm{GeV}/c^2$. Fig. 4 shows the $K\pi\pi$ Dalitz plot, in which events are concentrated around the K^* and ρ bands. We select events in the area indicated in the Dalitz plot. For the $B \to K^*\pi\gamma$ candidates, phase space overlaps with the region for $B \to K\rho\gamma$ events. Here we estimate the $K\rho\gamma$ component in the $K^*\pi\gamma$ candidates as indicated in Fig. 4(right) with an assumption of no interference, and subtract the $K\rho\gamma$ contribution. For the $K\rho\gamma$ analysis, we exclude the $K^*\pi\gamma$ region.

From the M_{bc} spectra (Fig. 5), we find $40.7 \pm 7.6^{+1.8}_{-2.7}$ events for the $K^*\pi\gamma$ signal after subtracting the $K\rho\gamma$ component, and $24.5 \pm 6.4^{+1.2}_{-2.3}$ events for the $K\rho\gamma$ signal. The corresponding branching fractions are obtained as

$$\mathcal{B}(B^+ \to (K^*\pi)^+\gamma;\ M_{K^*\pi} < 2.0\ \mathrm{GeV}) = (5.6 \pm 1.1 \pm 0.9) \times 10^{-5} \quad (7)$$
$$\mathcal{B}(B^+ \to (K\rho)^+\gamma;\ M_{K\rho} < 2.0\ \mathrm{GeV}) = (6.5 \pm 1.7^{+1.1}_{-1.2}) \times 10^{-5}. \quad (8)$$

where $(K^*\pi)^+$ includes both $K^{*0}\pi^+$ and $K^{*+}\pi^0$; $(K\rho)^+$ includes $K^+\rho^0$ and $K^0\rho^+$.

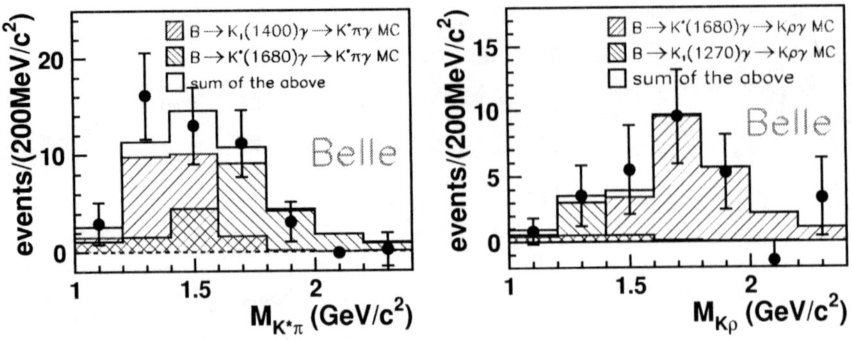

FIGURE 6. Background subtracted $K\pi\pi$ invariant mass distribution for the $B \to K^*\pi\gamma$ (left) and $B \to K\rho\gamma$ candidates.

The $K\pi\pi$ invariant mass spectra are shown in Fig. 6. Existence of multiple resonant states around the similar mass range does not allow us to uniquely determine the resonant decompositions. The overlaid Monte Carlo histograms indicate the mixture of two resonances that we use to estimate the reconstruction efficiencies. No single resonance seems to dominate the spectra. For the $K\rho\gamma$ channel, only $K_1(1270)$ can contribute in the low $M_{K\pi\pi}$ region below 1.4 GeV$/c^2$. As we do not observe a significant signal excess, we set an upper limit,

$$\mathcal{B}(B \to K_1(1270)\gamma) < 9.6 \times 10^{-5} \quad (90\% \text{ C.L.}). \tag{9}$$

For the $K^*\pi\gamma$ channel, upper limit is set in the following combination,

$$0.5 \times \mathcal{B}(B \to K_1(1270)\gamma) + \mathcal{B}(B \to K_1(1400)\gamma) + \mathcal{B}(B \to K^*(1410)\gamma) < 5.1 \times 10^{-5} \quad (90\% \text{ C.L.}). \tag{10}$$

The $B \to K^*\pi\gamma$ and $B \to K\rho\gamma$ branching fractions are both larger than sum of theoretically expected rates for radiative B decays to higher kaonic resonances [7]. If we assume the isospin invariance in the decay rate and sum up the two- and three-body final states of radiative B decays, the sum accounts for about half of the inclusive $B \to X_s\gamma$ branching fraction.

Search for CKM suppressed $B \to \rho\gamma$ decays

The $b \to d\gamma$ process is naively suppressed by a factor of $|V_{td}/V_{ts}|^2$. The final state is presumably similar to $b \to s\gamma$, except that the net strangeness should be zero. Therefore good kaon-pion separation is required to separate $b \to d\gamma$ from $b \to s\gamma$. The easiest route is to search for an exclusive channel $B \to \rho(,\omega)\gamma$, for which similar $b \to s\gamma$ final states are still the largest background sources.

We search for $B^0 \to \rho^0\gamma$ and $B^+ \to \rho^+\gamma$ using a sample of 11.3 million $B\bar{B}$ events. From a Monte Carlo study, we expect the $B \to K^*\gamma$ background can be well separated with the particle identification device, namely by using the ACC, and by vetoing the $\pi\pi$ combination that falls into the K^* mass region under a kaon hypothesis for one of the pions. Using the same likelihood ratio variables with a tigher cut, we find the M_{bc} spectrum is consistent with background, and set upper limits,

$$\mathcal{B}(B^0 \to \rho^0\gamma) < 10.6 \times 10^{-6} \quad (90\% \text{ C.L.}), \tag{11}$$

$$\mathcal{B}(B^+ \to \rho^+\gamma) < 9.9 \times 10^{-6} \quad (90\% \text{ C.L.}). \tag{12}$$

By taking a ratio between the $B \to \rho\gamma$ limit and measured $B \to K^*\gamma$ branching fraction, a large fraction of the systematic error cancels. Also assuming the isospin invariance, we set an upper limit on the ratio as,

$$\mathcal{B}(B \to \rho\gamma)/Br(B \to K^*\gamma) < 0.19 \quad (90\% \text{ C.L.}). \tag{13}$$

SEMI-LEPTONIC $b \to s\ell^+\ell^-$ PROCESS

Observation of $B \to K\ell^+\ell^-$ decay

The $b \to s\ell^+\ell^-$ process has been searched for by many experiments. Here we report the first observation of one of the exclusively reconstructed channels, $B \to K\ell^+\ell^-$, using a data sample of 31.3 million $B\bar{B}$ events.

The analysis proceeds as follows. We reconstruct the $B \to K^{(*)}\ell^+\ell^-$ candidates from an oppositly charged lepton pair and a kaon or K^* (K^+, $K_S \to \pi^+\pi^-$, $K^{*0} \to K^+\pi^-$ or $K_S\pi^0$, $K^{*+} \to K^+\pi^0$ or $K_S\pi^+$). Di-leptons from J/ψ and ψ' decays are vetoed; asymmetric veto windows are applied to remove radiative tails. The di-electrons from photon conversions inside the detector and $\pi^0 \to e^+e^-\gamma$ decays are removed by requiring $M_{e^+e^-} > 0.14$ GeV/c^2. Continuum background is suppressed by a likelihood ratio constructed from a Fisher discriminant and $\cos\theta_B$ (and the angle between the sphericity axis and beam axis for the di-electron channels). The Fisher discriminant is based on the energy flow in 9 cones around the sphericity axis and the normalized second Fox-Wolfram moment R_2. Another major background source is generic $B\bar{B}$ events in which the leptons come from the semi-leptonic decays $b \to c\ell\nu$. This background is suppressed by a cut on another likelihood ratio constructed from the missing energy in the detector and $\cos\theta_B$.

We extract the signal yields from the M_{bc} spectra (Fig. 7). We find $9.5^{+3.8+0.8}_{-3.1-1.0}$ events in the $B \to K\mu^+\mu^-$ channel with a significance of 4.7. We did not find a significant excess for other channels: $4.1^{+2.7+0.6}_{-2.1-0.8}$ events for Ke^+e^-, $6.3^{+3.7+1.0}_{-3.0-1.1}$ events for $K^*e^+e^-$ and $2.1^{+2.9+0.9}_{-2.1-1.0}$ events for $K^*\mu^+\mu^-$. The combinatoric backgrounds from continuum and $B\bar{B}$ are modelled by ARGUS functions in the M_{bc} spectrum, using a 400 fb^{-1} equivalent Monte Carlo sample. For the combined $B \to K\ell^+\ell^-$ mode, we observe $13.6^{+4.5}_{-3.8}$ signal events with a statistical significance of 5.3.

The background can contribute to the signal Gaussian function only when the muon identification fails twice for non-rare B meson decays with the same number of final state particles. Such background includes $B^+ \to J/\psi K^+ \to \mu^+\mu^-K^+$ identified as $K^+\mu^-\mu^+$,

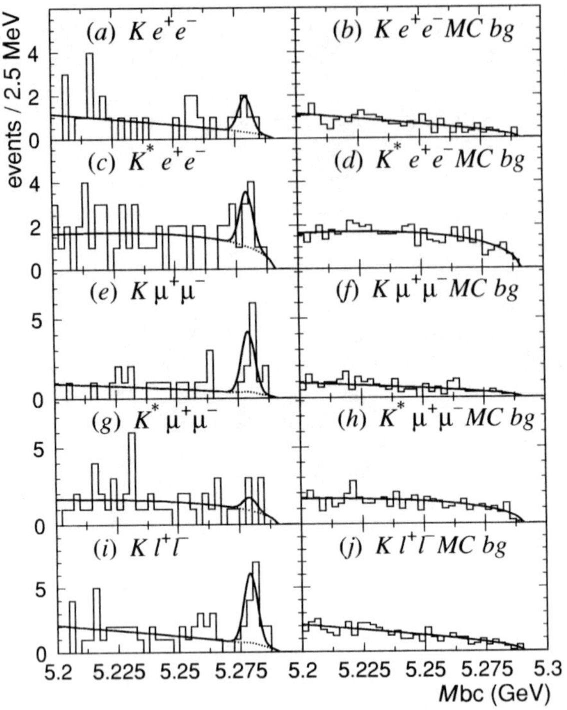

FIGURE 7. The M_{bc} distribution for the $B \to K^{(*)}\ell^+\ell^-$ candidates (open histogram) and fit results (curves) in the left column; corresponding Monte Carlo background distributions in the right column.

or $B \to K\pi^+\pi^-$ identified as $B \to K\mu^+\mu^-$. We estimate 0.27 background events for the $K\mu^+\mu^-$ channel using the measured muon fake rate and known branching fractions, and subtract this contribution from the signal yield. The corresponding background due to the fake electrons is negligibly small. The di-lepton mass spectrum (Fig. 8) is in a good agreement with a SM expectation.

The observed signal yields lead to the first measurements of branching fractions,

$$\mathcal{B}(B \to K\ell^+\ell^-) = (0.75^{+0.25}_{-0.21}(\text{stat}) \pm 0.09(\text{sys})) \times 10^{-6}, \tag{14}$$
$$\mathcal{B}(B \to K\mu^+\mu^-) = (0.99^{+0.40}_{-0.32}(\text{stat})^{+0.13}_{-0.14}(\text{sys})) \times 10^{-6}. \tag{15}$$

For other channels, we set upper limits,

$$\mathcal{B}(B \to Ke^+e^-) < 1.3 \times 10^{-6} \text{ (90\% C.L.)}, \tag{16}$$
$$\mathcal{B}(B \to K^*\mu^+\mu^-) < 3.1 \times 10^{-6} \text{ (90\% C.L.)}, \tag{17}$$
$$\mathcal{B}(B \to K^*e^+e^-) < 5.6 \times 10^{-6} \text{ (90\% C.L.)}. \tag{18}$$

We find these branching fractions are in a good agreement with the SM predictions [8], and the upper limits are consistent with the predictions, too.

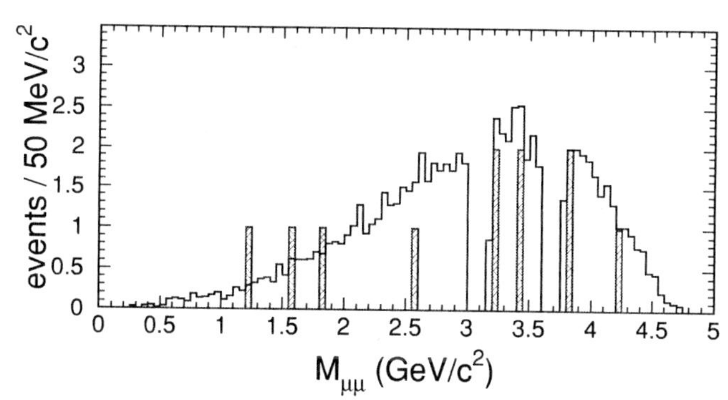

FIGURE 8. Dilepton invariant mass spectrum for the $B \to K\mu^+\mu^-$ candidates (hatched histogram) compared with the Monte Carlo signal distribution (open histogram).

Inclusive $B \to X_s\ell^+\ell^-$ search

We also apply the semi-inclusive analysis technique to the $b \to s\ell^+\ell^-$ process, by reconstructing $B \to X_s\ell^+\ell^-$ candidates from a lepton pair (e^+e^- or $\mu^+\mu^-$), a kaon (K^+ or K_S) and zero to four pions (up to one π^0). The data sample of 22.8 million $B\bar{B}$ events is used for this analysis.

The continuum background is suppressed by a cut on R_2 and a cosine of the angle between the thrust axis of the B candidate and the thrust axis of all the remaining tracks ($\cos\theta_{thr}$). In order to suppress the $B\bar{B}$ background, we form two likelihood ratios: the first one is constructucted from the missing energy and X_s mass, and the second one is constructed from $\cos\theta_B$ and a sum of two cosines of angles between the kaon and the leptons ($\cos\theta_{K\ell^+} + \cos\theta_{K\ell^-}$). We select the best candidate based on ΔE, and make a cut of $|\Delta E| < 30\ \mathrm{MeV}/c^2$.

The signal yield is extracted from a fit to the M_{bc} spectrum for each of $B \to X_s e^+e^-$ and $B \to X_s\mu^+\mu^-$ as shown in Fig. 9. We find no significant signal excess, and set upper limits,

$$\mathcal{B}(B \to X_s e^+e^-) < 10.2 \times 10^{-6} \quad (90\%\ \text{C.L.}), \qquad (19)$$

$$\mathcal{B}(B \to X_s\mu^+\mu^-) < 19.9 \times 10^{-6} \quad (90\%\ \text{C.L.}). \qquad (20)$$

Our results improve the limits by almost one order magnitude compared to the previously existing limits [9]. The limits are now close to the SM predictions for the branching fractions [10].

CONCLUSION

To conclude, we report the first observation of the decay $B \to K\ell^+\ell^-$, limits on other $b \to s\ell^+\ell^-$ channels and $B \to \rho\gamma$ decay, and the first measurements of $B \to K^*\pi\gamma$, $B \to K\rho\gamma$ among measurements of various inclusive and exclusive $b \to s\gamma$ decay channels. We

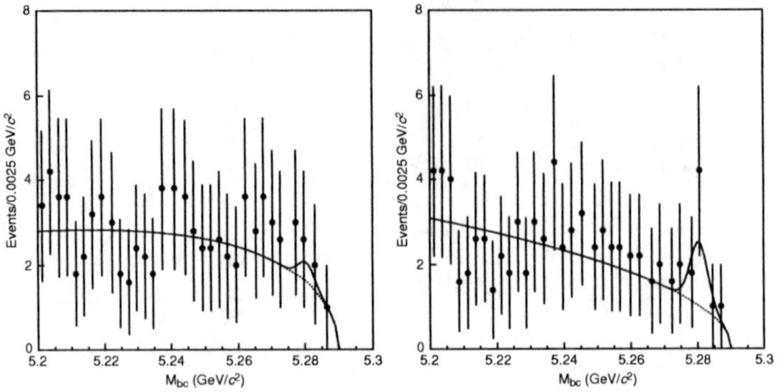

FIGURE 9. M_{bc} spectra for the $B \to X_s \ell^+ \ell^-$ candidates.

confirm the existence of radiative decay into a tensor + photon in the $B \to K_2^*(1430)\gamma$ channel. We find that the $B \to K^*\pi\gamma$ and $B \to K\rho\gamma$ branching fractions are quite large, and about half of the inclusive $B \to X_s\gamma$ branching fraction is accounted for by the two- and three-body radiative B decays. The observed branching fraction for $B \to K\ell^+\ell^-$ is consistent with the SM predictions; for other channels, a stringent test of the SM will soon be possible with several times larger data set which we anticipate in the coming half to one year of the Belle experiment.

ACKNOWLEDGMENTS

We wish to thank the KEKB accelerator group for the excellent operation of the KEKB accelerator.

REFERENCES

1. K. Chetyrkin, M. Misiak and M. Münz, *Phys. Lett.* **B400**, 206 (1997); Erratum ibid. **B425**, 414 (1998).
2. K. Abe *et al.* (Belle Collab.), *The Belle Detector, KEK Progress Report* 2000-4 (2000), to be published in *Nucl. Instr. Meth. A*.
3. K. Abe *et al.* (Belle Collab.), *Phys. Lett.* **B511**, 151 (2001).
4. A. L. Kagan, M. Neubert, *Eur. Phys. J.* **C7**, 5 (1999).
5. M. Alam *et al.* (CLEO Collab.), *Phys. Rev. Lett.* **74**, 2885 (1995).
6. R. Barate *et al.* (ALEPH Collab.), *Phys. Lett.* **B429**, 169 (1998).
7. S. Veseli and M. G. Olsson, *Phys. Lett.* **B367**, 309 (1996).
8. A. Ali, P. Ball, L. T. Handoko and G. Hiller, *Phys. Rev.* **D61**, 074024 (2000),
   C. Greub, A. Ioannissian and D. Wyler, *Phys. Lett.* **B346**, 149 (1995),
   D. Melikhov, N. Nikitin and S. Simula, *Phys. Lett.* **B410**, 290 (1997).
9. S. Glenn *et al.* (CLEO Collab.), *Phys. Rev. Lett.* **80**, 2289 (1998).
10. A. Ali, G. Hiller, L. T. Handoko and T. Morozumi, *Phys. Rev.* **D55**, 4105 (1997),
    F. Krüger and L. M. Sehgal, *Phys. Lett.* **B380**, 199 (1996).

Radiative Penguin Decays at BABAR

Anders Ryd representing the BABAR collaboration

California Institute of Technology, 356-48 Caltech, Pasadena, CA 91125, USA

Abstract. We report preliminary results based on a data sample of 20.7 fb $^{-1}$ recorded at the $\Upsilon(4S)$ resonance by the BABAR detector at the PEP-II energy asymmetric collider at the Stanford Linear Accelerator Center. We have measured the branching fraction $\mathcal{B}(B^0 \to K^{*0}\gamma) = (4.39 \pm 0.41 \pm 0.27) \times 10^{-5}$ and measured a charge asymmetry in the $B \to K^*\gamma$ decays consistent with zero: $A_{CP} = -0.035 \pm 0.076 \pm 0.012$. We also searched for the decay $B^0 \to \gamma\gamma$ and placed the 90% C.L. limit $\mathcal{B}(B^0 \to \gamma\gamma) < 1.7 \times 10^{-6}$. The search for the electroweak penguin decays $B \to K^{(*)}\ell^+\ell^-$ yielded the limits $\mathcal{B}(B \to K\ell^+\ell^-) < 0.6 \times 10^{-6}$ and $\mathcal{B}(B \to K^*\ell^+\ell^-) < 2.5 \times 10^{-6}$ at the 90% C.L.

INTRODUCTION

The decays $B \to K^*\gamma$, $B \to K^{(*)}\ell^+\ell^-$, and $B^0 \to \gamma\gamma$, reported on here, are examples of decays taking place via a flavor-changing neutral current (FCNC). In the Standard Model, FCNC transitions are forbidden at tree level and only takes place at higher order in amplitudes involving loops. An example of a quark level diagrams for a FCNC transition is shown in Figure 1 for the $b \to s\gamma$ transition. The decay $B \to K^*\gamma$, observed by CLEO [1], was the first observation of an electro-magnetic penguin decay of a B meson. CLEO also observed the inclusive decay $b \to s\gamma$ [2].

As these decays are suppressed in the Standard Model they provide an interesting window to look for physics beyond the Standard Model. New, heavy, particles as predicted *e.g.* by sypersymmetry can contribute to the rate of these decays by appearing virtually in the loop diagrams. The decay $B \to K^*\ell^+\ell^-$ is interesting as new physics can affect not only the rate, but also kinematic distributions, which in some cases can be predicted with less model dependence than the rate and allow for stringent tests of the Standard Model. For example, the lepton forward-backward asymmetry is expected to have a zero at $q^2 = m^2_{\ell^+\ell^-} \approx 3.0\ \mathrm{GeV}^2/c^4$, see *e.g.* Ref. [3].

In all of the analysis reported here we make use of the standard B reconstruction variables $\Delta E = \sum_i E_i^* - E^*_{\mathrm{beam}}$ and $m_{\mathrm{ES}} = \sqrt{E^{*2}_{\mathrm{beam}} - |\sum_i \mathbf{p}^*{}_i|^2}$, where a * denotes a quantity measured in the e^+e^- collision rest frame (CM), E_i and $\mathbf{p}_i$ are the energy and momentum respectively of the ith daughter of the B candidate, and E^*_{beam} is the beam energy in the the e^+e^- rest frame. ΔE peaks at zero for signal events and m_{ES} peaks at the B mass.

CP618, *Heavy Flavor Physics: Ninth International Symposium*, edited by A. Ryd and F. C. Porter

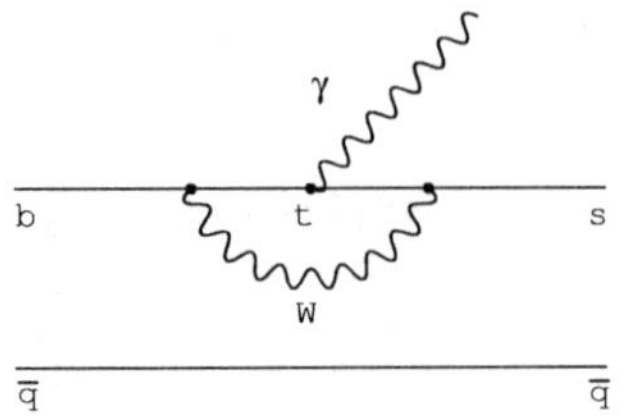

FIGURE 1. Quark level diagram contributing to $b \to s\gamma$.

DATA SAMPLE

The data sample used in these analyses was recorded with the BABAR detector from October 1999 to October 2000. Most of this time PEP-II ran on the $\Upsilon(4S)$ resonance and we recorded 20.7 fb^{-1} on the resonance. We also collected a 2.6 fb^{-1} sample at a center-of-mass energy 40 MeV below the $\Upsilon(4S)$ resonance. The off-resonance sample is useful for modeling non-$B\bar{B}$ backgrounds, in particular non-resonant $q\bar{q}$ production. The on-resonance sample corresponds to $(22.7 \pm 0.4) \times 10^6$ $B\bar{B}$ pairs. The BABAR detector is described in detail in Ref. [4].

STUDIES OF $B \to K^*\gamma$

We have reconstructed the decays $B^+ \to K^{*+}\gamma$ and $B^0 \to K^{*0}\gamma$ with $K^{*+} \to K^0_S\pi^+$, $K^+\pi^0$ and $K^{*0} \to K^0_S\pi^0$, $K^+\pi^-$. Events are selected that have a photon candidate with a CM energy between 2.30 and 2.85 GeV. Photon candidates are vetoed if combined with any other photon candidate in the event give a mass within 2σ of either the π^0 or η mass.

The photon candidate is combined with a K^* candidate to form the B candidate. The K^* candidate is required to have an invariant mass, $m_{K\pi}$, within 100 MeV/c^2 of the mean $K^*(892)$ mass. For the B candidate we calculate $\Delta E \equiv E^*_{K^*} + E^*_\gamma - E^*_{\text{beam}}$. We require that $-200 < \Delta E < 100$ MeV for modes without a π^0 and $-225 < \Delta E < 125$ MeV for modes with a π^0. The signal yields are extracted in m_{ES}. For the modes without a π^0 the magnitude of the photon momentum, $|p^*_\gamma| = E^*_\gamma$, is rescaled such that $\Delta E = 0$ when calculating m_{ES}. This rescaling removes the low tail in m_{ES} due to energy leakage in the calorimeter.

The main background in the analysis comes from non-resonant $q\bar{q}$ production. Most commonly, photons are either from initial state radiation or from decays of π^0 and η mesons. This background is suppressed by cutting on three variables that separate the background from the signal. The most powerful cut is $|\cos\theta^*_{\text{T}}| < 0.8$, where θ^*_{T} is the angle between the photon candidate and the thrust axis of the rest of the event, excluding the K^* daughter particles. The distribution of $|\cos\theta^*_{\text{T}}|$ is shown in Figure 2. We also require $|\cos\theta_{\text{H}}| < 0.75$, where θ_{H} is the decay angle of the kaon in the K^* rest frame. As the produced K^* mesons in the signal mode have either helicity ± 1, but not 0, the signal distribution in θ_{H} is given by a $\sin^2\theta_{\text{H}}$ distribution, whereas the background is approximately flat in $|\cos\theta_{\text{H}}|$. Similarly, we also require $|\cos\theta_B| < 0.75$, where θ_B is

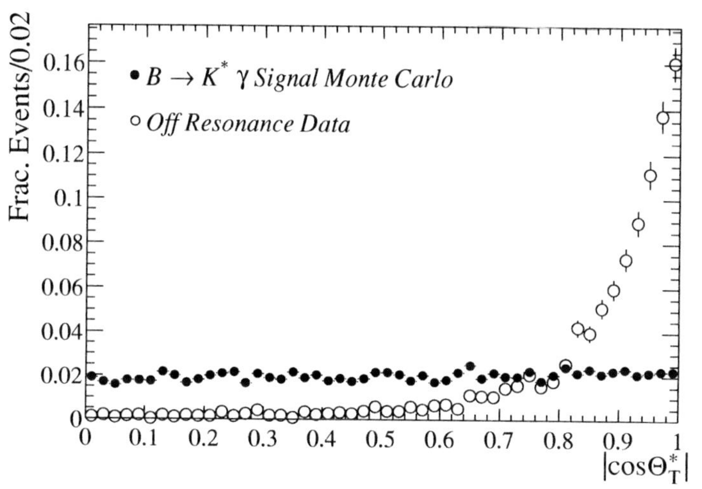

FIGURE 2. Distribution of $|\cos\theta^*_{\rm T}|$ for signal (solid dots) and off resonance data (open circles).

TABLE 1. The table summarizes efficiency, including K^* branching fractions, $(\varepsilon \times \mathcal{B})$; signal yields; cross feed background; and the measured branching fraction.

Mode	$\varepsilon \times \mathcal{B}$ (%)	Signal yield	Cross feed	$\mathcal{B}(B \to K^*\gamma)/10^{-5}$
$K^+\pi^-$	14.1	139.2 ± 13.1	1	$4.39 \pm 0.41 \pm 0.27$
$K^+\pi^0$	5.1	57.6 ± 10.4	3.8	$5.52 \pm 1.07 \pm 0.33$
$K^0_S\pi^0$	1.4	14.8 ± 5.6	1.4	$4.10 \pm 1.71 \pm 0.42$
$K^0_S\pi^+$	2.9	28.4 ± 6.4	1.7	$3.12 \pm 0.76 \pm 0.21$

the decay angle of the B candidate in the $\Upsilon(4S)$ frame.

In the four modes studied we obtain the signals as shown in Figure 3, and the results are summarized in Table 1. We also show the ΔE and $m_{K\pi}$ distributions in Figure 4, which shows that the signal is consistent with $B \to K^*(892)\gamma$. The largest systematic errors come from the photon detection efficiency and tracking efficiency. Recent predictions, Refs. [5] and [6], based on NLO calculations predicts somewhat larger branching fractions $\mathcal{B}(B \to K^*\gamma) = (7.1^{+2.5}_{-2.3}) \times 10^{-5}$ and $\mathcal{B}(B \to K^*\gamma) = (7.9^{+3.5}_{-3.0}) \times 10^{-5}$, respectively. The theoretical uncertainty is dominated by the $B \to K^*$ form factor.

Using the charge of either the K or K^* we measure the charge asymmetry,

$$A_{CP} \equiv \frac{\Gamma(\bar{B} \to \bar{K}^*\gamma) - \Gamma(B \to K^*\gamma)}{\Gamma(\bar{B} \to \bar{K}^*\gamma) + \Gamma(B \to K^*\gamma)}$$

and find $A_{CP} = -0.035 \pm 0.076 \pm 0.012$, consistent with the expectation of a rather small asymmetry (about -0.5%) in the Standard Model [5].

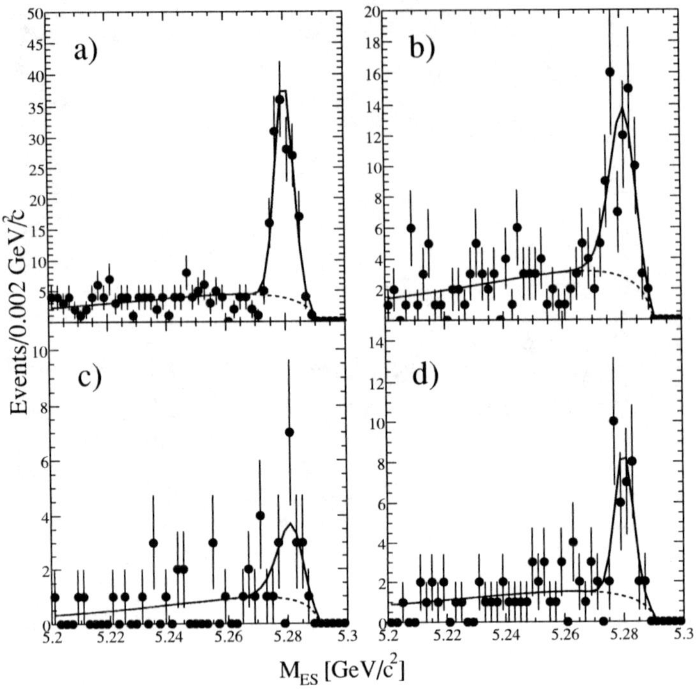

FIGURE 3. Signals in the $B \to K^*\gamma$ analysis where the K^* is reconstructed in the modes a) $K^+\pi^-$, b) $K^+\pi^0$, c) $K_S^0\pi^0$, and d) $K_S^0\pi^+$.

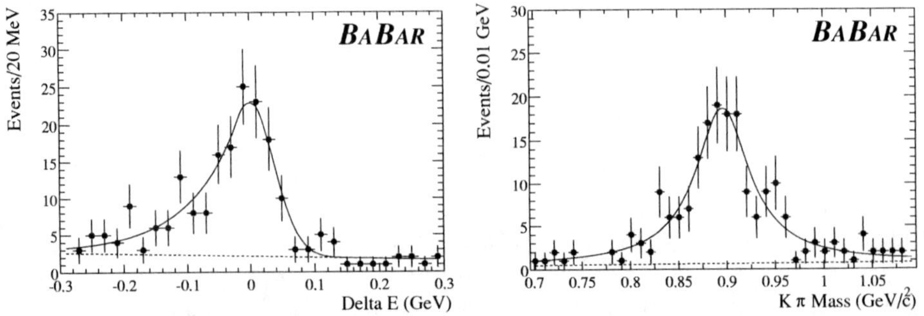

FIGURE 4. The ΔE (left) and $m_{K\pi}$ (right) distributions in the $K^+\pi^-$ mode in the $B \to K^*\gamma$ analysis.

SEARCH FOR $B^0 \to \gamma\gamma$

The decay $B^0 \to \gamma\gamma$ is highly suppressed in the Standard Model, expected branching fractions are $(0.1 - 2) \times 10^{-8}$ [7]. Figure 5 shows diagrams that contributes to the $B^0 \to \gamma\gamma$ decay.

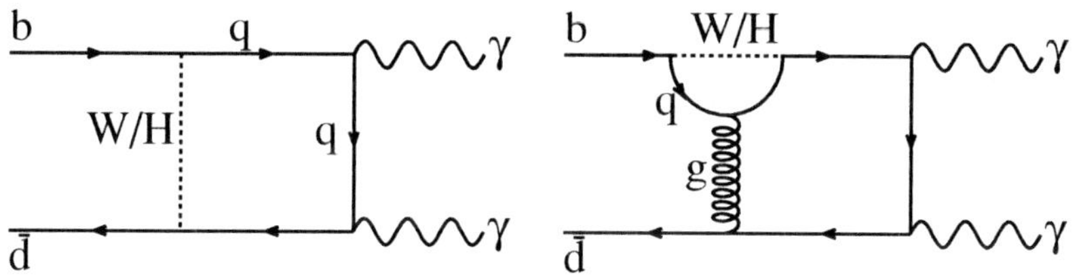

FIGURE 5. Quark level diagrams contributing to the $B^0 \to \gamma\gamma$ process.

The search for $B^0 \to \gamma\gamma$ was done blind, the signal region in ΔE versus $m_{\rm ES}$ were blind during the process of optimizing the event selection criteria. After unblinding, the signal yield is extracted by counting the number of events in the signal region. The data sample used in this analysis is somewhat smaller, 18 $\rm fb^{-1}$, than what is used in the other two analysis described in this presentation. Photon candidates in this analysis are rejected if they combined with any other photon candidate in the event are within 3σ of either the π^0 or η mass. This cut is somewhat tighter than that used in the $B \to K^*\gamma$ analysis as the expected S/B is much smaller. As in the $B \to K^*\gamma$ analysis the dominant backgrounds come from non-resonant $e^+e^- \to q\bar{q}$ production. The $q\bar{q}$ background is suppressed by requiring $|\cos\theta^*_{\rm T}| < 0.57$ and $|\cos\theta_B| < 0.81$. The ΔE vs. $m_{\rm ES}$ plane is shown in Figure 6. There is one event in the signal region. The expected background is 0.9 events based on extrapolations from the sideband. We place an upper limit at 90% C.L. assuming the observed event is signal, giving

$$\mathcal{B}(B^0 \to \gamma\gamma) < 1.7 \times 10^{-6}.$$

This result is a substantial improvement over previous measurements [8].

SEARCH FOR $B \to K^{(*)}\ell^+\ell^-$

The decays $B \to K\ell^+\ell^-$ and $B \to K^*\ell^+\ell^-$ are expected to have branching fractions of order $(0.3-0.6)\times 10^{-6}$ and $(1-2.5)\times 10^{-6}$ respectively [3, 9, 10, 11]. Figure 7 shows quark level diagrams contributing to these decays. In this analysis we searched for the following final states: $B^+ \to K^+\ell^+\ell^-$, $B^0 \to K^0_S\ell^+\ell^-$, $B^+ \to K^{*+}\ell^+\ell^-$, and $B^0 \to K^{0*}\ell^+\ell^-$; where $K^{*+} \to K^0_S\pi^+$, $K^{*0} \to K^+\pi^-$, and ℓ is either an electron or muon. This analysis was also carried out as a blind analysis; the signal region, as well as the near sideband, in ΔE vs. $m_{\rm ES}$ was blind during the optimization of the event selection criteria. The signal yield is extracted using an extended maximum likelihood fit in the ΔE versus $m_{\rm ES}$ plane.

Candidate events are reconstructed using electrons with a laboratory frame momentum greater than 0.5 GeV/c and muons with a momentum greater than 1.0 GeV/c. Charged kaon candidates are required to be positively identified using the DIRC, the BABAR Cherenkov particle identification detector, and dE/dx in the drift chamber and silicon vertex tracker. Our K^* candidates are required to have a mass, $m_{K\pi}$, within 75 MeV/c^2 of the mean $K^*(892)$ mass.

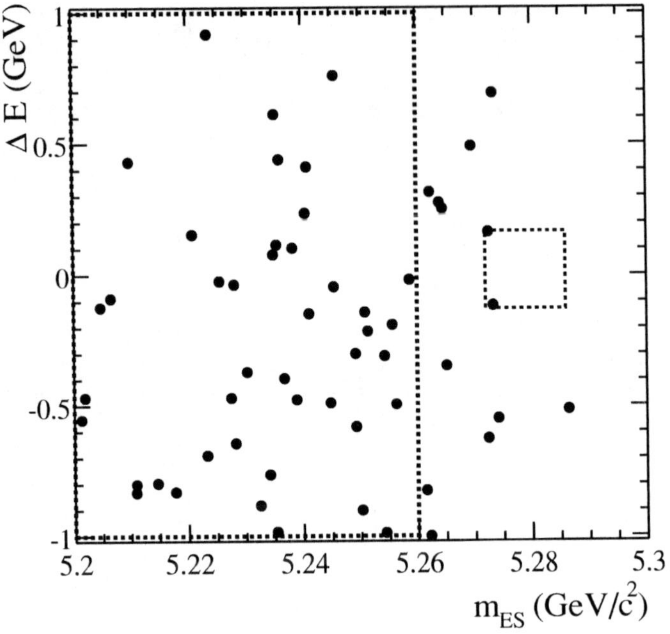

FIGURE 6. The unblinded signal region in the $B^0 \to \gamma\gamma$ analysis. There is one event in the signal regions; consistent with the background expectation of 0.9. events.

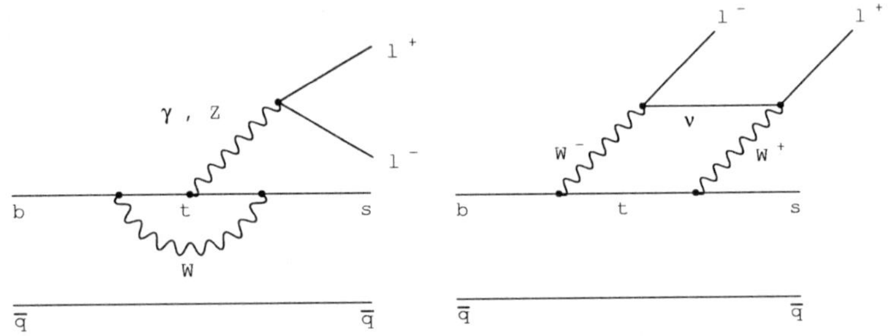

FIGURE 7. Quark level diagrams contributing to $B \to K^{(*)}\ell^+\ell^-$.

There are many backgrounds that have to be consider in this analysis. First, we have backgrounds that peak in the signal region. Most obviously, there is the $B \to J/\psi K^{(*)}$ and $B \to \psi(2S)K^{(*)}$ modes where the charmonium states decays to leptons. These decays have the same topology as the signal and are vetoed by a correlated cut in the dilepton mass and ΔE as illustrated in Figure 8. We also explicitly veto events in modes with a K^* where an additional random pion, from the other B decay, has been picked up to

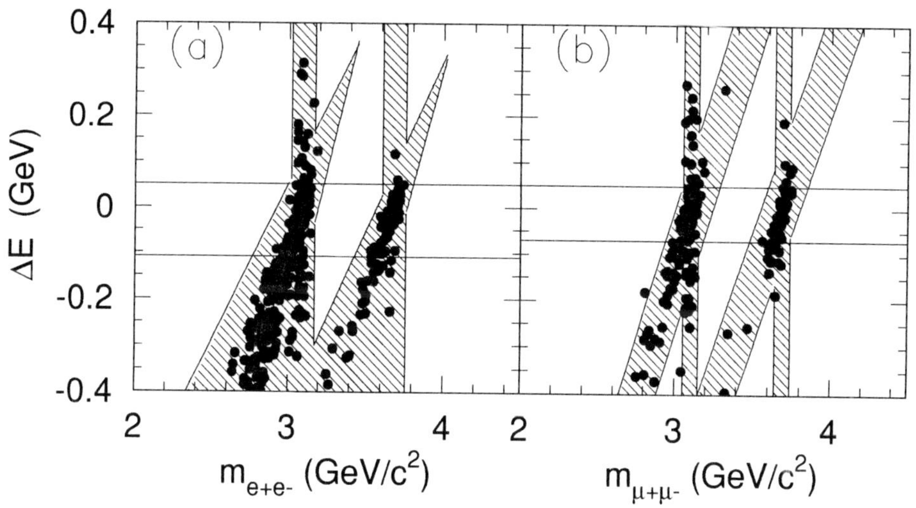

FIGURE 8. The charmonium vetoes in $B \to K^{(*)}\ell^+\ell^-$. We use a correlated cut in the dilepton mass, $m_{\ell\ell}$, and ΔE to remove the charmonium events not only in the signal region, but also in the ΔE sideband. This simplifies the description of the background shape in ΔE as otherwise there would be a large contamination from the radiative $B \to (J\psi, \psi(2S))K^{(*)}$ decays at low values of ΔE.

compensate for the energy lost by radiation of a lepton in a $B \to (J/\psi, \psi(2S))K$ decay.

Besides combinatorial backgrounds in non-resonant $q\bar{q}$ events, this analysis also have a substantial combinatorial background from $B\bar{B}$ events. These background sources are suppressed by cutting on two variables that are designed to suppress the combinatorial backgrounds from continuum and $B\bar{B}$ events respectively.

The backgrounds from continuum $e^+e^- \to q\bar{q}$ production are suppressed by a Fisher discriminant which combines $|\cos\theta^*_T|$, $|\cos\theta_B|$, R_2, and $m_{K\ell}$. R_2 is the ratio of second to zeroth Fox-Wolfram moment. The variable $m_{K\ell}$ uses the lepton with the charge consistent to be from a D decay with the strangeness given by the $K^{(*)}$. In decays of D mesons $m_{K\ell}$ will be below the D mass and hence provide useful separation between signal and background.

The combinatorial background from $B\bar{B}$ decays is suppressed using a likelihood which combines several variables. The most important variable is the missing energy in the event. Leptons in $B\bar{B}$ events are typically produced in semileptonic decays and hence have a large missing energy due to the undetected neutrinos in the event. We also use information from vertexing; if we pick up candidates from different B meson decays they do not come from a common vertex point and can be used to reject the $B\bar{B}$ background.

Figure 9 shows the unblinded ΔE versus $m_{\rm ES}$ plane and Fig. 10 shows the fit in the $m_{\rm ES}$ projections. The result of the fits is summarized in Table 2. The table also includes the results of the lepton-family-number violating modes.

Combining electron and muon channels we obtain

$$\mathcal{B}(B \to K\ell^+\ell^-) = (0.0 \pm 0.3(\text{stat.})) \times 10^{-6},$$

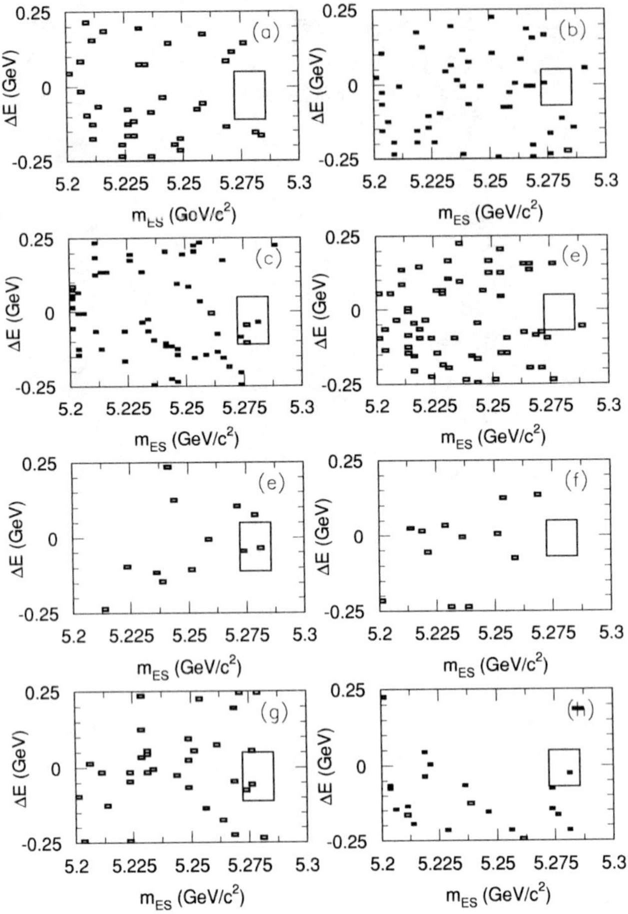

FIGURE 9. The signal region in the ΔE versus $m_{\rm ES}$ plane is shown in each of the modes. The region indicated by a box shows where the signal is expected. There is no evidence for a signal. The signal yield is extracted using a likelihood fit in the ΔE - $m_{\rm ES}$ regions shown in these plots.

$$\mathcal{B}(B \to K^*\ell^+\ell^-) = (0.7 \pm 1.1(\text{stat.})) \times 10^{-6}.$$

When combining the $B \to K^*e^+e^-$ and $B \to K^*\mu^+\mu^-$ modes we use the constraint $\mathcal{B}(B \to K^*e^+e^-)/\mathcal{B}(B \to K^*\mu^+\mu^-) = 1.2$ as given by Ref. [3]. As there is no evidence for a signal we set the upper limits

$$\mathcal{B}(B \to K\ell^+\ell^-) < 0.6 \times 10^{-6},$$

$$\mathcal{B}(B \to K^*\ell^+\ell^-) < 2.5 \times 10^{-6}.$$

These limits are now at the level of the predictions based on the Standard Model. Belle [12] claims a signal for $B \to K\ell^+\ell^-$ with a branching fraction $\mathcal{B}(B \to K\ell^+\ell^-) =$

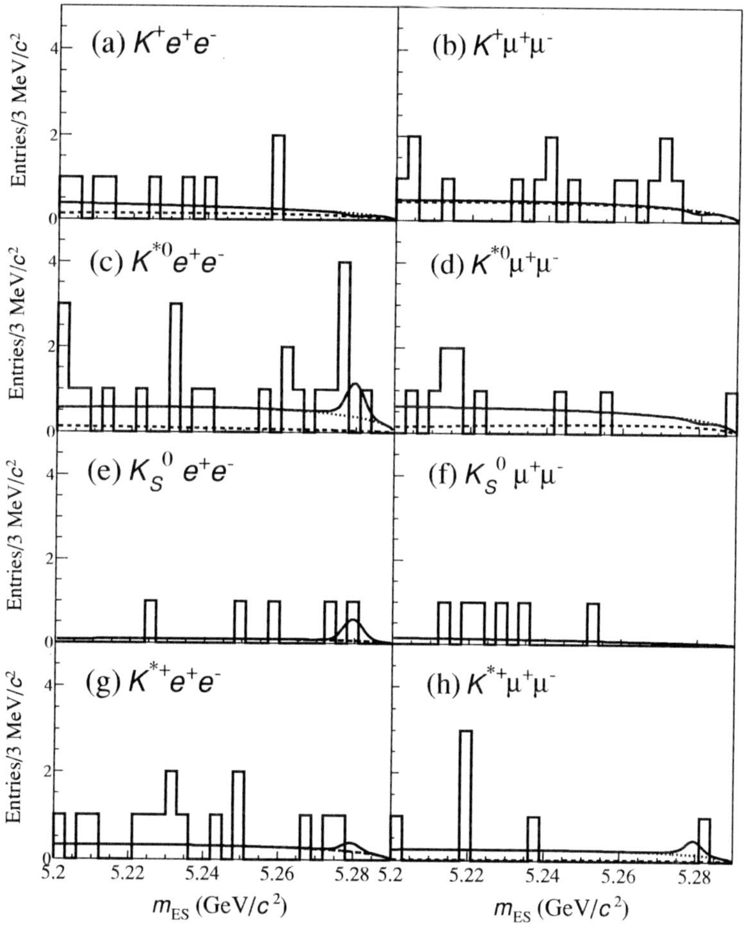

FIGURE 10. The fit is shown in an m_{ES} projection with $-0.11 < \Delta E < 0.05$ GeV for electrons and $-0.07 < \Delta E < 0.05$ GeV for muons.

$(0.75^{+0.25}_{-0.25}(\text{stat.}) \pm 0.09(\text{syst.})) \times 10^{-6}$, which is larger than our 90% C.L. limits. Future analysis of larger data samples will soon be able to settle the issue.

CONCLUSIONS

We have reported preliminary results on a first set of measurements from BABAR in electro-weak penguin decays. All results presented here provides significant improvements over previously published results. At the time of the completion of this contribution to the conference proceedings BABAR has a data sample that is almost three times the size used for the analysis presented here and by the summer of 2002 we expect to

TABLE 2. The results for the fits in the individual modes for $B \to K^{(*)}\ell^+\ell^-$ and the lepton-family-number violating decays $B \to K^{(*)}e^\pm\mu^\mp$. From left to right the columns are the signal yield; the signal efficiency, excluding branching fractions for K^* and K decays; systematic uncertainties on the efficiency and fit; and the upper limit including systematic uncertainties.

Mode	Signal yield	ε (%)	$(\Delta\mathcal{B}/\mathcal{B})_\varepsilon$	$(\Delta\mathcal{B}/\mathcal{B})_{\rm fit}$	$\mathcal{B}/10^{-6}$ at 90% C.L.
$B^+ \to K^+e^+e^-$	-0.2	17.5	±8.6	±10.6	0.9
$B^+ \to K^+\mu^+\mu^-$	-0.2	10.5	±8.6	±10.6	1.3
$B^0 \to K^{*0}e^+e^-$	2.5	10.2	±10.5	±10.6	5.0
$B^0 \to K^{*0}\mu^+\mu^-$	-0.3	8.0	±10.8	±10.6	3.6
$B^0 \to K^0e^+e^-$	1.3	15.7	±9.3	±10.6	4.7
$B^0 \to K^0\mu^+\mu^-$	0.0	9.6	±11.4	±10.6	4.5
$B^+ \to K^{*+}e^+e^-$	0.1	8.5	±11.4	±10.6	10.0
$B^+ \to K^{*+}\mu^+\mu^-$	1.0	5.8	±12.0	±10.6	17.5
$B^+ \to K^+e^\pm\mu^\mp$	-0.6	16.8	±8.6	±10.6	1.0
$B^0 \to K^{*0}e^\pm\mu^\mp$	0.6	11.9	±10.6	±10.6	2.7
$B^0 \to K^0e^\pm\mu^\mp$	0.8	14.6	±10.3	±10.6	3.3
$B^+ \to K^{*+}e^\pm\mu^\mp$	-0.4	9.3	±11.7	±10.6	8.7

have a sample of 100 fb^{-1}. Hence, in the near future we should expect many of the results presented here to be updated.

ACKNOWLEDGMENTS

I would like to thank all my colleagues at BABAR and PEP-II for making it possible to present these results.

REFERENCES

1. R. Ammar *et al.* (CLEO Collaboration), *Phys. Rev. Lett.*, **71**, 674 (1993).
2. M.S. Alam *et al.* (CLEO Collaboration), *Phys. Rev. Lett.*, **74**, 2885 (1995).
3. A. Ali, *et al.*, *Phys. Rev.*, **D61**, 074024 (2000).
4. B. Aubert *et al.* (BABAR Collaboration), submitted to Nucl. Instrum. Meth.
5. S.W. Bosch and G. Buchalla, hep-ph/0106081.
6. M. Beneke *et al.*, hep-ph/0106067.
7. G.G. Devidze and G.R. Jibuti, *Phys. Lett.*, **B429**, 48 (1998).
8. D.E. Groom, *et al.*, *Eur. Phys. J.*, **C15**, 1 (2000).
9. P. Colangelo, *et al.*, *Phys. Rev.*, **D53**, 3672 (1996).
10. D. Melikhov, N. Nikitin, and S. Simula, *Phys. Rev.*, **D57**, 6814 (1998).
11. T.M. Aliev *et al.*, *Phys. Lett.*, **B400**, 194 (1997).
12. M. Nakao, in these proceedings.

Extracting $|V_{ub}|$ Using the Radiative Decay Data

Ira Z. Rothstein

Dept. of Physics Carnegie Mellon University, Pittsburgh PA 15213, USA

Abstract. In this talk I review recent progress made in extracting $|V_{ub}|$, within a systematic expansion, from the cut electron energy and hadronic mass spectra of inclusive B meson decays utilizing the data from radiative decays. It is shown that an extraction is possible without modeling the B meson structure function. I discuss the issues involving the assumptions of local duality in various extractions. I also comment on the recent CLEO extraction of V_{ub}.

INTRODUCTION

It is an unfortunate fact that experimental cuts can take a nice clean theoretical prediction and turn it into a troublesome mess. A perfect example of this scenario arises in the extraction of V_{ub} from inclusive B decays. In principle this extraction should be straightforward. One measures the inclusive rate for semi-leptonic decays into non-charmed states and compares the result to the theoretical prediction for the total rate, which is under good theoretical control [1, 2]. Of course, the snag is that there is no simple way, at least at this time, to measure the total inclusive rate to charmless states. Thus, some cut must be applied to reject the charmed final states. Perhaps the simplest choice, from an experimentalist viewpoint, is to cut on the electron energy, rejecting all events with $E_l < (m_B^2 - m_D^2)/(2m_B)$. This is the oldest method for extracting V_{ub}. Unfortunately, the theoretical prediction for the integrated cut spectrum is rather complicated.

The problem arises from the fact that the cut introduces a new scale into the problem. Without the cut there are only two scales of relevant physics, namely, the mass m_B and the QCD scale Λ_{QCD}. Suppose we cut on a scaled kinematic variable, such as the electron energy $x_B = 2E_l/m_B$. If we cut near the endpoint, $x_B \simeq 1$, then the scale $(1-x)m_B$ can introduce a large dimensionless ratio into the calculation $1/(1-x)$. This can lead to power law, as well as logarithmic amplifications of what normally would be small effects.

The calculation of the total inclusive rate can be derived from first principles [1] within a systematic expansion in $\alpha_s(m_b)$ and Λ/m_b [1]. The aforementioned amplifications in the cut rate arise as corrections of the form $\Lambda/(m_b(1-x))$ and $\alpha_s Log^2(1-x)$. Physically, the reasons for these enhancements are clear. The non-perturbative corrections arise from the fact that near the endpoint the spectrum becomes very sensitive to the Fermi motion

[1] Even in the total rate for semi-leptonic decays one still has "mild" local-duality assumptions arising from the fact that the contour approaches the real axis at a point. Thus this calculation is technically not on the same theoretical footing as deep inelastic scattering

CP618, *Heavy Flavor Physics: Ninth International Symposium*, edited by A. Ryd and F. C. Porter

of the heavy quark. The logarithmic corrections are due to the exclusivity of the cut rate. In the limit where x approaches one there is no room for gluon radiation, thus leading to the usual infra-red divergences of the Bloch-Nordstrom type.

Of these two types of enhancements it is really the power corrections that make the extraction more difficult. The reason for this is that the effects of Fermi motion are incalculable. In the past the Fermi motion was modeled, leading to extractions of V_{ub} for which it was not really possible to make meaningful theorctical error estimates. This is not to say that those extractions will not lead to a number which in the end could turn out to be "correct", or that the old error band is totally unreasonable. However, given that we are now entering the age of precision CKM measurements, the theorist is obligated to make more systematic estimates of the errors. Fortunately, there is a way around having to model the Fermi motion. The relevant point is that the effect of the Fermi motion on the decay spectra are universal. Thus, if we can extract the necessary information about the end point spectrum from one decay we can use it to make a prediction for another decay spectrum.

THE CUT ELECTRON SPECTRUM

It can be shown from first principles that[3, 4], up to corrections of order Λ/m_b, the electron energy endpoint spectrum can be written as

$$\frac{d\Gamma}{dE} = \int_{2E-m_b}^{\bar{\Lambda}} dk_+ f(k_+) \frac{d\Gamma_p}{dE}(m_b^*), \tag{1}$$

where E is the charged lepton energy in semi-leptonic B decay. m_b^* is the effective mass which accounts for the residual momentum k_+, such that $m_b^* = m_b + k_+$ and the structure function, which accounts for the Fermi motion, is given by

$$f(k_+) = \langle B(v) \mid \bar{b}_v \delta(k_+ - iD_+) b_v \mid B(v) \rangle. \tag{2}$$

A similar expression can be given for the radiative decay spectrum. Thus, in principle, one could measure one end point spectrum, deconvolve (1), and then extract $f(k_+)$ to make a prediction for the shape of another spectrum. However, this would be a rather Herculean task which I would wish on neither friend nor foe. Fortunately, this analysis can be avoided by first taking the Mellin transform of the spectrum[2]. The point is that the Mellin transform turns the convolution in (1) into a product. Thus we may remove all dependence on the structure function[6] by first taking the ratio of the moments of two spectra and then taking the inverse Mellin transform, just as we do in relating deep-inelastic scattering to Drell-Yan processes. This is exactly what was done for the electron energy spectrum in ref. [7, 8]. In this reference it was shown that following this procedure $|V_{ub}|^2/|V_{ts}^* V_{tb}|^2$ may be extracted from the relation [7, 8]

[2] At tree level it can also be avoided without taking moments [5].

$$\frac{|V_{ub}|^2}{|V_{ts}^*V_{tb}|^2} = \frac{3\alpha C_7^{(0)}(m_b)^2}{\pi}(1+H_{mix}^{\gamma})\int_{x_B^c}^{1} dx_B \frac{d\Gamma}{dx_B} \times \left\{\int_{x_B^c}^{1} du_B W(u_B)\frac{d\Gamma^{\gamma}}{du_B}\right\}^{-1}, \qquad (3)$$

H_{mix}^{γ} represents the corrections due to interference coming from the operators O_2 and O_8 [5]

$$H_{mix}^{\gamma} = \frac{\alpha_s(m_b)}{2\pi C_7^{(0)}}\left[C_7^{(1)} + C_2^{(0)}\Re(r_2) + C_8^{(0)}\left(\frac{44}{9} - \frac{8\pi^2}{27}\right)\right], \qquad (4)$$

and x_B^c is the value of the cut. In Eq. (4), all the Wilson coefficients, evaluated at m_b, are "effective" as defined in [9], and $\Re(r_2) \approx -4.092 + 12.78(m_c/m_b - 0.29)$ [10]. The numerical values of the Wilson coefficients are: $C_2^{(0)}(m_b) \approx 1.11$, $C_7^{(0)}(m_b) \approx -0.31$, $C_7^{(1)}(m_b) \approx 0.48$, and $C_8^{(0)}(m_b) \approx -0.15$. The diagonal pieces from O_2 and O_8 are numerically insignificant. The function $W(u_B)$ is given by

$$W[u_B] = u_B^2 \int_{x_B^c}^{u_B} dx_B \left(1 - 3(1-x_B)^2 + \frac{\alpha_s}{\pi}(\frac{7}{2} - \frac{2\pi^2}{9} - \frac{10}{9}Log(1 - \frac{x_B}{u_B}))\right). \qquad (5)$$

The End Point Logs

Over the years there has been a concern over the perturbative part of the calculation discussed above [11]. This concern arises due to the fact that as the cut approaches the end-point, the perturbative corrections grow. This growth is due to the so called "Sudakov Logs", $\alpha_s Log^2(1-x_c)$. These logs appear are a consequence of the fact that near the end-point there is limited phase space available for radiation, and the fact is that you can't stop radiation; you can only hope to contain it. Indeed, it is well known that if one sums up these double logarithms, the rate dies off as $\Gamma \propto \exp[-\alpha_s Log^2(1-x_c)]$. The above mentioned concern, was that perhaps this exponentiation, and moreover, the exponentiation of sub-leading single logs would endanger the convergence of the perturbative expansion. Indeed, if we assume that the logs dominate the expansion at each order in perturbation theory, then we really should reorganize the calculation in terms of an expansion in the exponent. In fact, in moment space it was shown that [6, 12] the perturbative expansion could be written as

$$\frac{d\Gamma}{dN} \propto exp(Log(N)f(\alpha_s Log(N)) + g(\alpha_s Log(N)) + \alpha_s h(Log(N)) + ...). \qquad (6)$$

Thus the question of the convergence of the perturbative expansion becomes one of the nature of the series in the exponent. Not only should the series in the exponent converge (at least in an asymptotic sense), but it had also better be that the first term dropped in the exponent is much less than one. In [7], the logs were resummed including the subleading function g. There it was found that, after taking the ratio of the semi-leptonic and radiative decay rates, the net effect of resummation was only at $10-15\%$ level. Part of the reason for the smallness of this effect is that the f functions for the two processes are identical, therefore the leading double logs cancel in the ratio. The closed form expression for V_{ub} including the resummation is given by the same expression as above (3), with a new expression for the function W,

$$W[u_B] = u_B^2 \int_{x_B^c}^{u_B} K\left[x_B; \frac{4}{3\pi\beta_0} Log(1-\alpha_s\beta_0 l_{x_B/u_B})\right]. \tag{7}$$

$$\begin{aligned} K(x;y) &= 6\left\{\left[1+\frac{4\alpha_s}{3\pi}\left(1-\psi^{(1)}(4+y)\right)\right]\frac{1}{(y+2)(y+3)}\right. \\ &\quad \left. -\frac{\alpha_s}{3\pi}\left[\frac{1}{(y+2)^2}-\frac{7}{(y+3)^2}\right]-\frac{4\alpha_s}{3\pi}\left[\frac{1}{(y+2)^3}-\frac{1}{(y+3)^3}\right]\right\} \\ &\quad -3(1-x)^2. \end{aligned} \tag{8}$$

The function K has a non-integrable singularity in it due the fact that the sum in the exponent should be considered in terms of a prescription for a non-Borel summable series. Due to its asymptotic nature it can be well approximated by expanding the argument of the Log in a series and keeping as many terms as one wishes until the series starts to diverge. An excellent approximation for W is given by expanding the second argument of K[13],

$$\frac{4}{3\pi\beta_0} Log(1-\alpha_s\beta_0 l_{x_B/u_B}) \approx \alpha_s \frac{4}{3\pi} Log(1-x_B/u_B) - \alpha_s^2 \frac{25}{18\pi^2} Log^2(1-x_B/u_B). \tag{9}$$

While the use of this improved prediction may shift the central value slightly, it will not change the error bars.

THE HADRONIC MASS CUT

It is also possible to remove the background from charmed transitions by cutting on the hadronic invariant mass. While this choice presents a greater experimental challenge, it benefits from the fact that, unlike the electron spectrum, most of the $B \to X_u e\nu$ decays are expected to lie within the region $s_H < M_D^2$. Furthermore, it is believed that even though both the invariant mass region $s_H < M_D^2$ and electron energy regions $M_B/2 > E_e > (M_B^2 - M_D^2)/(2M_B)$ receive contributions from hadronic final states with invariant mass up to M_D, the cut mass spectrum will be less sensitive to local duality violations. This belief rests on the fact that the contribution of large mass states is kinematically suppressed for the electron energy spectrum in the region of interest. The expression for V_{ub} for the case of the hadronic mass cut is given by [14]

$$\frac{|V_{ub}|^2}{|V_{ts}|^2} = \frac{6\alpha C_7(m_b)^2(1+H_{mix}^\gamma)\,\delta\Gamma(c)}{\pi\,[I_0(c)+I_+(c)]}. \tag{10}$$

The expressions for $\Gamma(c)$, $I_0(c)$ and $I_+(c)$ can be found in [14]. The leading errors in this expression are again of order Λ/m_b. The effect of resummation of the end-point logs in this case was shown to be negligible [8].

SOME CAVEATS AND CONCLUDING REMARKS

We have emphasized the fact that we are now capable of extracting $| V_{ub} |$ in a systematic fashion. Which is to say that we have our errors under control. The leading errors are of order Λ/m_b and $\alpha_s(1-x)$. However, strictly speaking these are really the only errors we know how to quantify. As I emphasized the calculation is based upon certain assumptions about local duality. We don't know how to quantify these errors, all we know is when we do and do not expect local duality to be a valid assumption. Basically, the point is that we expect hadronic observables to well approximated in terms of partonic calculations when we are able to smear over resonances in a "sufficient manner"[3]. If the final state phase space is so restrictive that we only average over one or two resonances then, we begin to worry that our assumption is not well founded. It is interesting to note that in the charmed decay, the inclusive rate is saturated up to corrections of order Λ^2/m^2 by including only the D and D^* in the final states[15] in the small velocity limit [16]. Naively, one would think that pion emission would only be suppressed by p_π/f_π, but in the zero recoil limit the currents become generators of the effective theory and can only produce linear combinations of the D and D^*. While this is tantalizing, and gives one hope that local duality will work well even with small numbers of resonances, this case may be considered special.

None the less, the best choice of cuts, as far as local duality is concerned, is the hadronic mass, as it is believed to contain approximately 80% of the total rate, and if duality is going to work somewhere this is the place. Indeed, given that this is a relatively large percentage of the rate, compared to those expected in the electron energy cut $\simeq 20\%$ or leptonic mass cut $\simeq 20\%$ (which does need the radiative decay data [17, 18, 19]), we may hope to actually test the duality assumption. This may be done by simply varying the cut and seeing if the extracted value of V_{ub} remains fixed.

Let me pause at this point to make a few comments regarding some issues which were raised during the conference. First of all, it is often asked, "if we know that the hadronic mass cut really does include 80% of the rate, can't we get a relatively good extraction of V_{ub} without care about the radiative decays?". The answer to this is that we really don't know that the hadronic mass cut captures 80% of the rate. That number is a rough estimate which is made using an inspired guess for the wave function. The purpose of the number is simply to get a feeling for relative merits of the various cuts, but since the whole point of the exercise is to eliminate model dependence, these percentages should not be taken too seriously. Secondly, I would like to make a few comments about the recent CLEO extraction, as discussed in Roy Brieres' talk at this conference [20]. As I understand it, the extraction was performed by using a guess for the structure function, and varying the parameters of the model until a fit to the radiative decay spectrum was found. This fit was then used to make a prediction for the cut rate in the semi-leptonic decay. The robustness of this method was then "tested" by using two different parameterizations for the structure function. Now, I believe this is not the best way to do things. There are, no doubt, a large number of parameterizations for the structure

[3] Mathematically this assumption boils down to the fact that we are doing an operator product expansion in a kinematic region where its not really justified.

function which could fit the radiative decay data, and given that the function with which this parameterization is convoluted differs in the two different decay modes, there is no reason that the results should necessarily be identical in any parameterization. Of course, even in the method I described in this talk, one still needs to fit the radiative decay data, and this may indeed be well fit by using a model. The point is that in the method described above, all that is need is the *physical spectrum*, it doesn't matter what model you use to fit the data. In the end it may well indeed be that both methods give the same answer, but the present CLEO extraction as it stands, is not justified from first principles.

ACKNOWLEDGMENTS

I would like that thank my collaborators on this subject, Adam Leibovich and Ian Low. Thanks also go the organizers of this conference who were able to pull off a constructive and pleasant conference during difficult times. This work was supported in part by the Department of Energy under grant number DOE-ER-40682-143 and DE-AC02-76CH03000.

REFERENCES

1. Chay, J., Georgi, H., and Grinstein, B., *Phys. Lett.*, **B247**, 399–405 (1990).
2. Shifman, M. A., and Voloshin, M. B., *Sov. J. Nucl. Phys.*, **41**, 120 (1985).
3. Neubert, M., *Phys. Rev.*, **D49**, 3392–3398 (1994).
4. Bigi, I. I. Y., Shifman, M. A., Uraltsev, N. G., and Vainshtein, A. I., *Int. J. Mod. Phys.*, **A9**, 2467–2504 (1994).
5. Neubert, M., *Phys. Rev.*, **D49**, 4623–4633 (1994).
6. Korchemsky, G. P., and Sterman, G., *Phys. Lett.*, **B340**, 96–108 (1994).
7. Leibovich, A. K., Low, I., and Rothstein, I. Z., *Phys. Rev.*, **D61**, 053006 (2000).
8. Leibovich, A. K., Low, I., and Rothstein, I. Z., *Phys. Rev.*, **D62**, 014010 (2000).
9. Chetyrkin, K., Misiak, M., and Munz, M., *Phys. Lett.*, **B400**, 206–219 (1997).
10. Greub, C., Hurth, T., and Wyler, D., *Phys. Lett.*, **B380**, 385–392 (1996).
11. Falk, A. F., Jenkins, E., Manohar, A. V., and Wise, M. B., *Phys. Rev.*, **D49**, 4553–4559 (1994).
12. Akhoury, R., and Rothstein, I. Z., *Phys. Rev.*, **D54**, 2349–2362 (1996).
13. Leibovich, A. K., Low, I., and Rothstein, I. Z., *Phys. Lett.*, **B513**, 83–87 (2001).
14. Leibovich, A. K., Low, I., and Rothstein, I. Z., *Phys. Lett.*, **B486**, 86–91 (2000).
15. Boyd, C. G., Grinstein, B., and Manohar, A. V., *Phys. Rev.*, **D54**, 2081–2096 (1996).
16. Shifman, M. A., and Voloshin, M. B., *Sov. J. Nucl. Phys.*, **47**, 511 (1988).
17. Bauer, C. W., Ligeti, Z., and Luke, M. E., *Phys. Lett.*, **B479**, 395–401 (2000).
18. Bauer, C. W., Ligeti, Z., and Luke, M. E., *Phys. Rev.*, **D64**, 113004 (2001).
19. Bauer, C. W., *These proceedings* (2001).
20. Briere, R., *These proceedings* (2001).

New CLEO Results for $|V_{cb}|$ and $|V_{ub}|$

Roy A. Briere, representing the CLEO Collaboration

Carnegie Mellon University; 5000 Forbes Ave.; Pittsburgh, PA 15213

Abstract. We report recent measurements from CLEO of the first two moments of the photon energy spectrum for $b \to s\gamma$ decays and the hadronic recoil mass in $\bar{B} \to X_c \ell \bar{\nu}$. These physical quantities allow one to fix non-perturbative parameters occurring in calculations based on HQET and QCD. Predictions for semileptonic decay rates within this same framework depend in addition on the CKM matrix elements $V_{qq'}$ governing quark mixing. We can thus extract $|V_{cb}|$ from the inclusive semileptonic decay rate of B mesons, and $|V_{ub}|$ from the lepton endpoint spectrum of $\bar{B} \to X_u \ell \bar{\nu}$. Model dependence is reduced except for the assumption of quark-hadron duality. Finally, we update the classic measurement of $|V_{cb}|$ from $\bar{B} \to D^* \ell \bar{\nu}$ at zero recoil.

INTRODUCTION

Most of the ad-hoc parameters of the Standard Model are contained in the flavor sector. Mysteries of mixing, mass generation, and *CP* violation all meet here. *B* physics offers the possibility of direct measurements of two of the Cabibbo-Kobayashi-Maskawa (CKM) mixing-matrix elements, $|V_{cb}|$ and $|V_{ub}|$, as well as access to two others, $|V_{td}|$ and $|V_{ts}|$, via loop processes. We concentrate here on determinations of the former pair via semileptonic *B* decays.

Experimentally, we measure the number of certain semileptonic decays; we can convert this to a branching ratio via knowledge of $N_{B\bar{B}}$ (the number of *B* pairs present in our data sample) and finally to a partial width by using *B* lifetime measurements from elsewhere. Both these partial widths and certain other kinematic 'moments' may be calculated in a systematic Heavy Quark Effective Theory (HQET) [1, 2] and QCD expansion. The expressions depend on some a priori unknown non-perturbative parameters, most notably $\bar{\Lambda}, \lambda_1$ and λ_2. The kinematic moments will allow us to independently determine these parameters for use in a self-consistent way in other formulae.

Expressions for semileptonic decay rates also depend on CKM matrix elements. One can thus extract $|V_{cb}|$ from the inclusive semileptonic rate, or perform a more intricate extraction of $|V_{ub}|$ from the rate at the lepton endpoint of $b \to u$ semileptonic decays. The partial width calculations rely on quark-hadron duality [3]. That is, they assume that for sufficiently inclusive quantities, quark-gluon calculations can be used for observed hadronic processes. The chief issues are non-perturbative effects and the need to average over enough hadronic states, which is even more problematic for $|V_{ub}|$ from the endpoint. It is therefore desirable to compare these results to those obtained from previous methods; one such result is also updated here. This is our determination of $|V_{cb}|$ from $\bar{B} \to D^* \ell \bar{\nu}$ at zero-recoil as favored by more familiar HQET treatments.

The CLEO II detector [4] and the CLEO II.V upgrade with a silicon vertex detector [5] and new drift chamber gas [6] are described elsewhere. Overall, in CLEO II and II.V

CP618, *Heavy Flavor Physics: Ninth International Symposium,* edited by A. Ryd and F. C. Porter

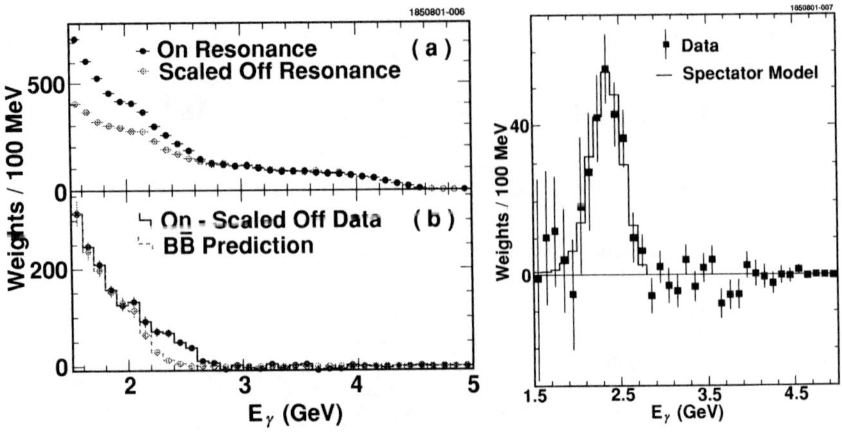

FIGURE 1. Left: Observed photon spectrum shown in a) with the scaled continuum background prediction from off-resonance data, and after continuum subtraction in b), where the $B\bar{B}$ background is now displayed. Right: Measured photon spectrum for $b \to s\gamma$ events.

data, the $\Upsilon(4S)$ to continuum luminosity ratio is about 2.1 and the effective $B\bar{B}$ cross section is 1.06 nb. The $b \to s\gamma$ and the $\bar{B} \to X_u \ell \bar{\nu}$ analyses both use the entire CLEO II and II.V datasets with a total luminosity of about 13.5 fb^{-1} and about 9.7×10^6 $B\bar{B}$ pairs, while the $\bar{B} \to X_c \ell \bar{\nu}$ and the $\bar{B} \to D^* \ell \bar{\nu}$ analyses use only the CLEO II portion with about 3.3×10^6 $B\bar{B}$ pairs.

THE $b \to s\gamma$ PHOTON SPECTRUM

This final state is distinguished by a high-energy photon. Our new analysis uses a lower cut of 2.0 GeV on E_γ, reducing model-dependence systematics. Continuum background dominates, but a factor of one hundred suppression is achieved by using event shape information, kinematics of detected leptons, and a pseudo-reconstruction technique (seeking the best $K(n\pi)\gamma$ combination). A neural network combines information to give a signal weight for each event. Both the raw photon spectrum indicating background sources and the final subtracted spectrum are displayed in Figure 1. Though $B\bar{B}$ background increases at the lowest energies, we still see sensible behavior albeit with larger errors.

Our focus here is extracting moments of the E_γ spectrum; for other analysis details, see [7]. This spectrum, naively a sharp line, has a width determined largely by the b quark Fermi motion and somewhat by the varying recoil mass (i.e., QCD effects). Two smaller sources of width are the small known B boost ($\beta \simeq 0.06$) and resolution smearing.

We determine moments directly from the data, accounting for energy-dependent efficiency, resolution, and boost smearing. As a check, we also take the Ali-Greub [8] or Kagan-Neubert models [9], propagated through our full GEANT-based detector simulation, and take moments of these models with parameters which best fit the data. All extractions are consistent, and variations are reflected in the systematic error.

We find, for $E_\gamma > 2.0$ GeV [10]:

$$\begin{aligned} < E_\gamma > &= (2.346 \pm 0.032 \pm 0.011)\,\text{GeV} \\ < E_\gamma^2 > - < E_\gamma >^2 &= (0.0226 \pm 0.0066 \pm 0.0020)\,\text{GeV}^2 \end{aligned}$$

where the brackets < ... > denote the average value.

The first moment, again with $E_\gamma > 2.0$ GeV, is calculated as [11, 12, 13]:

$$\begin{aligned} \langle E_\gamma \rangle = \frac{M_B}{2} [\,1 &- 0.385\frac{\alpha_s}{\pi} - 0.620\beta_0(\frac{\alpha_s}{\pi})^2 \\ &- \frac{\bar{\Lambda}}{M_B}(1 - .954\frac{\alpha_s}{\pi} - 1.175\beta_0(\frac{\alpha_s}{\pi})^2) \\ &- \frac{13\rho_1 - 33\rho_2}{12M_B^3} - \frac{\mathcal{T}_1 + 3\mathcal{T}_2 + \mathcal{T}_3 + 3\mathcal{T}_4}{4M_B^3} - \frac{\rho_2 C_2}{9M_D^2 M_B C_7} \\ &+ O(1/M_B^4)\,]\,. \end{aligned}$$

To lowest order, $\langle E_\gamma \rangle = \frac{1}{2}[M_B - \bar{\Lambda}]$. The parameter $\bar{\Lambda}$ measures the energy of the light degrees of freedom: the 'brown muck' surrounding the heavy-quark. There are further parameters appearing in general at second order in $1/M_B$; they are absent in this particular case, and are discussed later. Finally, the third-order parameters $\mathcal{T}_i$ and ρ_i are estimated as $O(0.5\,\text{GeV}^3)$ with variations included as systematics.

From the first moment alone, we find that $\bar{\Lambda} = (0.35 \pm 0.08 \pm 0.10)$ GeV [10]. Here, the entire experimental error is given first and the second error is due to the theoretical extraction. We avoid using second moments which give poorer determinations of the parameters and, in the case of hadronic moments below, have expressions which do not appear to converge as rapidly. Since teh value of $\bar{\Lambda}$ is not meaningful out of context, one must do all calculations consistently with respect to scheme and order. We choose $\overline{MS}$, $O(1/M_B^3)$, $O(\beta_0\alpha_s^2)$ everywhere.

HADRONIC MOMENTS IN $\bar{B} \to X_c \ell \bar{\nu}$

This analysis uses both e and μ with $1.5 < p_\ell < 2.5$ GeV/c and we reconstruct the neutrino properties via four-momentum balance with techniques developed at CLEO [14]. Great care is taken to account exactly once for all observed particles; cuts on charge balance, a multiple-lepton veto, and $(E_{miss}^2 - p_{miss}^2)$ help ensure that the missing four-momenta is due to a single missing neutrino. The resolution on missing momentum is $\sigma(p_{miss}) \simeq 110\,\text{MeV}/c$. After checking consistency with zero missing mass, we use $E_\nu = |p_{miss}|$ rather than E_{miss} since it has better resolution. We use B decay kinematics to determine $M_X^2 = M_B^2 + M_{\ell\bar{\nu}}^2 - 2E_B E_{\ell\bar{\nu}} + 2\vec{p}_B \cdot \vec{p}_{\ell\bar{\nu}}$ without needing to observe the hadrons or group them into those from the B vs. the $\bar{B}$ decay. The small final dot product, which averages zero, is ignored. Thus, four-momentum balance of the entire $B\bar{B}$ event measures the neutrino properties, leaving the hadronic recoil system unobserved. A typical exclusive mode analysis would instead observe the hadron(s) and have an unobserved neutrino. We expect 95% $b \to c\ell\bar{\nu}$ after continuum subtraction; the rest ($c \to s\bar{\ell}\nu$ secondaries, $b \to u\ell\bar{\nu}$) is subtracted via Monte Carlo simulation.

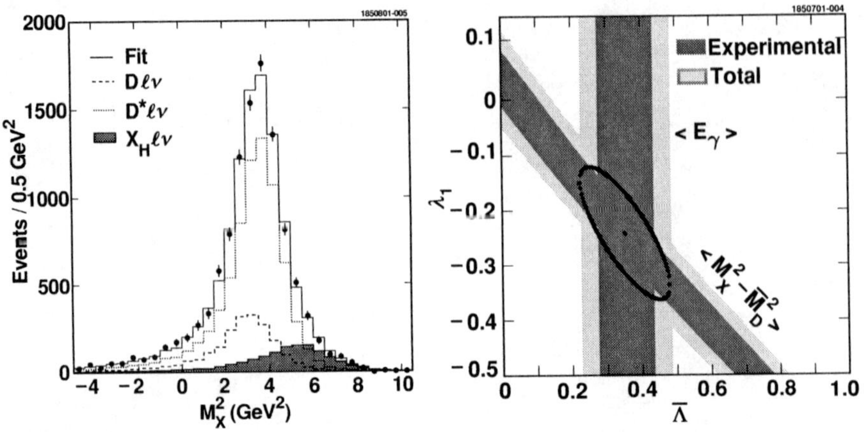

FIGURE 2. Left: Observed recoil mass in $\bar{B} \to X_c \ell \bar{\nu}$ showing data as points along with a fit to D, D^* and heavier charm meson contributions. Right: Constraints from measured $b \to s\gamma$ photon energy and $\bar{B} \to X_c \ell \bar{\nu}$ recoil mass first moments in the $\bar{\Lambda} - \lambda_1$ plane. The ellipse indicates $\Delta\chi^2 = 1$, including systematic errors.

Final results for the moments are calculated from the M_X^2 distributions corresponding to the mixture of $D\ell\bar{\nu}$, $D^*\ell\bar{\nu}$, and $X_H\ell\bar{\nu}$ spectra which best fit the data. We take moments of the generated M_X^2 distributions while fitting the data to reconstructed quantities passed through the full physics and detector simulation and analysis, hence accounting for the B boost, resolution and efficiency. Heavy states X_H beyond the D and D^* include D^{**} states modeled with ISGW2 [15] and non-resonant $D^{(*)}\pi$ treated with the Goity-Roberts prescription [16]. Their normalization is fixed by data. The fit distributions are shown in the left panel of Figure 2.

We finally arrive at [17]:

$$
\begin{aligned}
< M_X^2 - \bar{M}_D^2 > &= (0.251 \pm 0.023 \pm 0.062)\ \mathrm{GeV}^2 \\
< (M_X^2 - \bar{M}_D^2)^2 > &= (0.639 \pm 0.056 \pm 0.178)\ \mathrm{GeV}^4 \\
< (M_X^2 - < M_X^2 >)^2 > &= (0.576 \pm 0.048 \pm 0.163)\ \mathrm{GeV}^4
\end{aligned}
$$

where $\bar{M}_D$ denotes the spin-averaged D, D^* mass. The main systematics include the neutrino reconstruction efficiency and the models of the X_H states.

The theoretical expressions for the moments [18, 19, 20] use a consistent scheme and include (most of) the effects of the lepton energy cut. Unlike the mean photon energy discussed earlier, the second order HQET expansion parameters appear here. These are λ_1, related to the Fermi motion energy of the b quark, and λ_2, measuring the QCD hyperfine splitting; the latter is fixed from $m_{B^*} - m_B$ as measured by others. The third-order terms are treated as before.

Combining with the $b \to s\gamma$ result, we find [17]:

$$
\begin{aligned}
\bar{\Lambda} &= (0.35 \pm 0.07 \pm 0.10)\ \mathrm{GeV} \\
\lambda_1 &= (-0.236 \pm 0.071 \pm 0.078)\ \mathrm{GeV}^2
\end{aligned}
$$

The errors have the same meaning as for the $b \to s\gamma$ moments. The results are best viewed in the $\bar{\Lambda} - \lambda_1$ plane; see the right panel of Figure 2.

THE INCLUSIVE SEMILEPTONIC RATE AND $|V_{cb}|$

It is of course also possible to calculate the zeroth moment for semileptonic decays; this is simply $\Gamma_{sl} \equiv \Gamma(b \to c\ell\bar{\nu})$ [21, 22, 23]. The expression looks like a free-quark decay, akin to the classic muon decay rate, with a phase-space factor for finite m_c, augmented by QCD corrections and the HQET expansion. The actual formula used, gleaned from calculations in [19, 24, 25, 26, 27], may be found in [17].

Having consistently determined the lower-order HQET parameters with moments, we are poised to extract $|V_{cb}|$. The inclusive semileptonic rate is taken from a venerable CLEO result [28] using the tagged di-lepton method [29]. After subtracting 1% from the published value to correct for $\bar{B} \to X_u\ell\bar{\nu}$, we have: $\mathcal{B}(\bar{B} \to X_c\ell\bar{\nu}) = (10.39 \pm 0.46)\%$.

We convert to Γ_{sl} using $\tau_{B^\pm} = (1.548 \pm 0.032)$ ps [30], $\tau_{B^0} = (1.653 \pm 0.028)$ ps [30], and $f_{+-}/f_{00} = 1.04 \pm 0.08$ [31]. Combining, we finally arrive at [17]:

$$|V_{cb}| = (4.04 \pm 0.09 \pm 0.05 \pm 0.08) \times 10^{-2}$$

The largest errors are from (in order) the measurement of Γ_{sl}, the HQET parameters $\bar{\Lambda}, \lambda_1$, and the scale for α_s. This yields a precise (3.2%) determination, but given the global quark-hadron duality issues one should compare to results from other methods.

EXTRACTING $|V_{ub}|$ FROM THE $\bar{B} \to X_u\ell\bar{\nu}$ ENDPOINT

One might ask, why not do another expansion in $\bar{\Lambda}, \lambda_1, \lambda_2$ for the $\bar{B} \to X_u\ell\bar{\nu}$ partial width? One can do this for the fully inclusive width; for a discussion of subtleties, see [32]. But in experiments there are very large $b \to c$ backgrounds and we can therefore only measure the portion of the rate near the lepton momentum endpoint; that is, only above some p^ℓ_{min}. The $b \to X_c\ell\bar{\nu}, X_u\ell\bar{\nu}$ spectra are shown in the left half of Figure 3. The difficult calculation is the fraction of the leptons above a certain momentum cut. No only do we rely on local duality now, but terms of order $1/(M_B - 2p^\ell_{min})$ can enter, spoiling convergence. In more physical terms, we require the detailed shape and normalization of the spectrum near the endpoint.

Theory can profitably relate the endpoint $b \to u\ell\bar{\nu}$ rate to the observed $b \to s\gamma$ spectrum, since they are smeared by a common non-perturbative structure function [33, 34], up to corrections of order Λ_{QCD}/m_b (see [35] for a review). We can extract the structure function from $b \to s\gamma$ and then use this to predict the fraction of $b \to u\ell\bar{\nu}$ rate above the experimental lepton momentum cut. There is still an active debate concerning the details of the particular methodology we employ [36].

A neural net is used for continuum suppression and the signal region in lepton momentum comprises $2.2 < p_\ell < 2.6$ GeV/c. We have lowered our cut from 2.3 GeV/c to increase the rate. The data are shown in the right half of Figure 3. We observe good subtraction for $p_l > 2.6$ GeV/c, and extract $(1874 \pm 123 \pm 326)$ $\bar{B} \to X_u\ell\bar{\nu}$ events. This yields

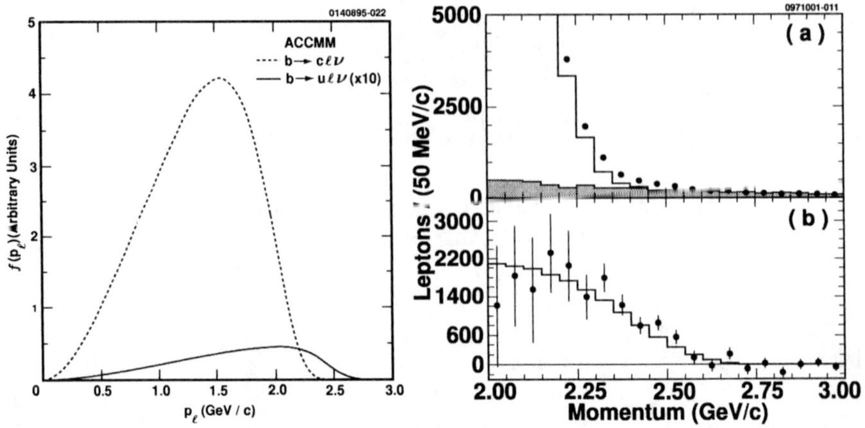

FIGURE 3. Left: typical prediction of the lepton spectra for $b \to c\ell\bar{\nu}$ and $b \to u\ell\bar{\nu}$ (note the x10 here!). Right: inclusive leptons in data near the endpoint. Plot a) gives the raw spectrum and shows the continuum (shaded) and $b \to c$ (open histogram) contributions; plot b) shows the extracted efficiency-corrected $b \to u\ell\bar{\nu}$ rate.

a partial branching ratio (before radiative corrections) of $\Delta\mathcal{B}_{ub}(2.2 - 2.6\,\mathrm{GeV}/c) = (2.35 \pm 0.15 \pm 0.45) \times 10^{-4}$. Systematics include variations of form factors and heavy charm states in Monte-Carlo modeling of $b \to c\ell\bar{\nu}$ backgrounds.

To extract $|V_{ub}|$, we start with an expression for the inclusive rate derived in the upsilon expansion [37]:

$$|V_{ub}| = [(3.06 \pm 0.08 \pm 0.08) \times 10^{-3}] \times [(\mathcal{B}_{ub}/0.001) \cdot (1.6\,\mathrm{ps}/\tau_B)]^{1/2}$$

The required $\mathcal{B}_{ub} \equiv B(\bar{B} \to X_u\ell\bar{\nu})$ is related to the observed rate $\Delta\mathcal{B}_{ub}$ in momentum window (p) by $\Delta\mathcal{B}_{ub}(p) = F_u(p)\,\mathcal{B}_{ub}$. The $b \to s\gamma$ spectrum will provide our prediction for $F_u(p)$, which is simply a properly normalized integral of the observed portion of the solid curve on the left of Figure 3. Using $b \to s\gamma$ data with $1.5 < E_\gamma < 2.8$ GeV, we fit the shape function [9] to various parameterizations. We use this to determine [38] $F_u(2.2 - 2.6\,\mathrm{GeV}/c) = 0.138 \pm 0.034$. Consistent results are obtained by the more model-dependent method of fitting the spectrum to parameters in the Ali-Greub spectator model [8] and feeding this information into the ACCMM model [39, 40].

Our *preliminary* result is:

$$|V_{ub}| = (4.09 \pm 0.14 \pm 0.66) \times 10^{-3}$$

This compares favorably with CLEO's exclusive $(\pi/\rho/\omega)\ell\bar{\nu}$ analyses [14, 41]:

$$|V_{ub}| = (3.25 \pm 0.14^{+0.21}_{-0.29} \pm 0.55) \times 10^{-3}$$

ZERO-RECOIL POINT OF $\bar{B} \to D^* \ell \bar{\nu}$

The newer, more inclusive methods above may have reduced model dependence in some sense, but quark-hadron duality is always involved. We now turn to a more traditional exclusive extraction of $|V_{cb}|$ from $\bar{B} \to D^* \ell \bar{\nu}$ decays.

This analysis measures the absolute rate as a function of q^2. Both $D^{*+} \ell \bar{\nu}$ and $D^{*0} \ell \bar{\nu}$ modes are reconstructed, with $0.8 < p_e < 2.4$ GeV/c and $1.4 < p_\mu < 2.4$ GeV/c. One usually replaces q^2 with $w = \vec{v}_B \cdot \vec{v}_D$, the product of B and D^* meson four-velocities. This new variable is just a particular linear transform, $w = a - bq^2$. HQET simplifies the analysis by relating the three form factors present to one universal Isgur-Wise function. In fact for $\bar{B} \to D^* \ell \bar{\nu}$, the point $w = 1$ (corresponding to maximum q^2) has no $O(1/M)$ corrections [42]. Physically, this occurs when $\vec{v}_B = \vec{v}_D$, that is at *zero-recoil* where the D^* is at rest relative to the B. This is the favorable place for a precise extraction of $|V_{cb}|$.

Background discrimination is accomplished by examining the angle, $\theta_{B-D^*\ell}$, between the B and the $D^*\ell$ system. This variable is similar to the familiar missing-mass. Since $\cos\theta_{B-D^*\ell}$ is calculated from four vectors it can be unphysical and its shape can be helpful in disentangling the many types of background. Determining the rate for each bin in w requires a detailed fit like the examples on the left of Figure 4. The right half of the figure shows both the fit rate and, after efficiency and kinematic factors, the extracted form-factor as a function of w. Inclusion of the D^{*0} mode adds significantly more efficiency at zero recoil as compared to using D^{*+} alone.

We determine the *preliminary* branching ratios:

$$\begin{aligned} \mathcal{B}(\bar{B}^0 \to D^{*+} \ell \bar{\nu}) &= (5.82 \pm 0.19 \pm 0.37)\% \\ \mathcal{B}(B^- \to D^{*0} \ell \bar{\nu}) &= (6.21 \pm 0.20 \pm 0.40)\% \end{aligned}$$

The intercept at zero-recoil and slope of the HQET form-factor (with the curvature related to the slope by dispersion relations [43, 44]) are *preliminarily* determined as:

$$\begin{aligned} F(1)|V_{cb}| &= (4.22 \pm 0.13 \pm 0.18) \times 10^{-2} \\ \rho^2 &= 1.61 \pm 0.09 \pm 0.21 \end{aligned}$$

Significant systematics include efficiency (especially for slow pions), the intermediate branching ratios, backgrounds, and (especially for ρ^2) the ratios of the three D^* form factors R_1 and R_2 [45]. Using $F(1) = 0.913 \pm 0.042$ [46], we extract:

$$|V_{cb}| = (4.62 \pm 0.14 \pm 0.20 \pm 0.21) \times 10^{-2}$$

where the errors are statistical, systematic, and theoretical, for a net 7% precision.

A comparison of recent results for the slope and intercept of the $\bar{B} \to D^* \ell \bar{\nu}$ form factor is shown in Figure 5. The differing shape of the CLEO ellipse is due to an interaction of the lepton momentum cut with variations of the form-factor ratios R_1 and R_2 within their errors. There is a disagreement at about the two sigma level. One difference in technique involves the $D^* X \ell \bar{\nu}$ background; CLEO includes these in the $\cos\theta_{B-D^*\ell}$ fits to the data, while LEP analyses use a model constrained to other LEP results on $\bar{B} \to D^* X \ell \bar{\nu}$.

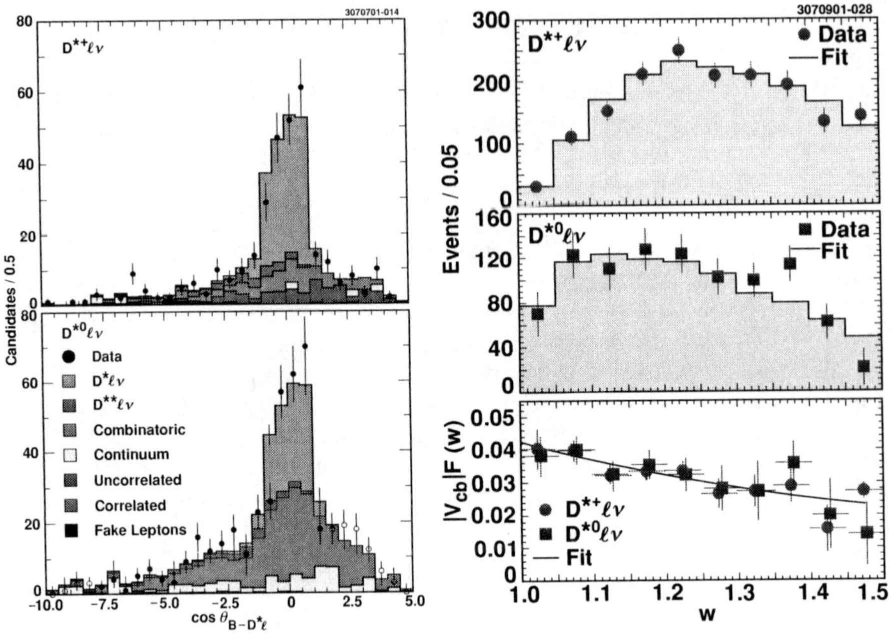

FIGURE 4. Left: Signal and backgrounds for $\bar{B} \to D^{*}\ell\bar{\nu}$ in the bin $1.10 < w < 1.15$ for both D^{*+} and D^{*0} modes. Note that backgrounds can be constrained from rates in the non-physical region of the $\cos\theta_{B-D^{*}\ell}$. Combinatoric background refers to fake D^{*}; (un)correlated background refers to cases where the real D^{*} and ℓ are (not) from the same B decay. Right: The top two panels show the fit yields in each w bin for each D^{*} mode, while the lower panel shows the extracted form factor with the final fit overlaid.

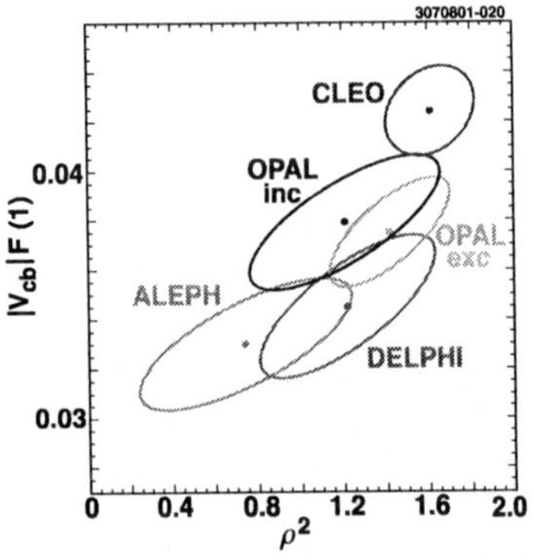

FIGURE 5. A graphical compilation of the present result along with published results from LEP experiments [47]. 'OPAL inc' denotes an analysis using partial reconstruction of $D^{*+}\ell\bar{\nu}$. Ellipses indicate $\Delta\chi^2 = 1$, including systematic errors.

THE FUTURE

There are several related analyses in progress at CLEO. These include $\bar{B} \to X_c \ell \bar{\nu}$ lepton spectrum moments ($< E_\ell >$) which will provide another band in the $\bar{\Lambda} - \lambda_1$ plane. Both low-background tagged (di-lepton) and higher statistics untagged analyses are being pursued. We also have more statistics for the exclusive $\bar{B} \to D^* \ell \bar{\nu}$ analysis.

Branching ratios and form-factor investigations for the $\bar{B} \to (\pi/\rho/\omega)\ell\bar{\nu}$ modes used for $|V_{ub}|$ are underway using neutrino reconstruction. We will further address inclusive measures of $|V_{ub}|$ with inclusive leptons, but now making full use of kinematics. Instead of singling out the lepton momentum, one can use quantities such as q^2 and the recoil mass [48]. We hope to accept a larger portion of rate while controlling background, thus reducing uncertainty on fraction of the $b \to u\ell\bar{\nu}$ rate observed.

Results for $|V_{cb}|$ from $D^{*+}e^+\bar{\nu}$ [49] and from $\mathcal{B}(\bar{B} \to Xe^-\bar{\nu})$ [50] have recently been presented by Belle. Future results from both Belle and BaBar will be of great interest.

CONCLUSION

We have measured the photon spectrum from $b \to s\gamma$ decays and the hadronic mass moments from $\bar{B} \to X_c\ell\bar{\nu}$. Using HQET and $\mathcal{B}(\bar{B} \to X_c\ell\bar{\nu})$, we extract $|V_{cb}|$ with more controlled theoretical systematics, but still subject to duality issues. New techniques relating studies of the lepton endpoint using a structure function constrained to $b \to s\gamma$ photon spectrum allow us to extract $|V_{ub}|$. Lastly, we update our $|V_{cb}|$ result using the $\bar{B} \to D^*\ell\bar{\nu}$ rate extrapolated to zero recoil. Ongoing analyses will extend all of this work further. Our emphasis is on using a variety of techniques, with differing systematics.

ACKNOWLEDGMENTS

It is a pleasure to thank my CESR and CLEO colleagues for their sustained efforts. Our research is supported by the NSF and DOE. Current information and results may be found at `http://w4.lns.cornell.edu/public/CLEO/`

REFERENCES

1. Isgur, N., and Wise, M. B., *Phys. Lett.* **B232**, 113–117 (1989); *ibid.* **B237**, 527–530 (1990).
2. Neubert, M., *Phys. Rept.* **245**, 259–396 (1994).
3. Bigi, I. I., "Flavour Dynamics – Central Mysteries of the Standard Model," in *Proceedings of the 30th International Conference on High Energy Physics, Vol. I*, edited by C. S. Lin and T. Yamanaka, World Scientific, Singapore, 2000, pp. 77–90.
4. CLEO Collaboration, Kubota, Y. et al., *Nucl. Instrum. Meth.* **A320**, 66–113 (1992).
5. Hill, T. S., *Nucl. Instrum. Meth.* **A418**, 32–39 (1998).
6. Briere, R. A., "Tracking in Helium-Based Gases: Present and Future *B* Factories," in *Proceedings of the Seventh International Symposium on Heavy Flavor Physics*, edited by C. Campagnari, World Scientific, Singapore, 1999, pp. 442–451.
7. Honscheid, K., these proceedings.

8. Ali, A., and Greub, C., *Phys. Lett.* **B361**, 146–154 (1995).
9. Kagan, A. L., and Neubert, M., *Eur. Phys. J.* **C7**, 5–27 (1999).
10. CLEO Collaboration, Chen, S. et al., CLNS 01/1751, submitted to *Phys. Rev. D*.
11. Bauer, C., *Phys. Rev.* **D57**, 5611–5619 (1998).
12. Ligeti, Z., Luke, M. E., Manohar, A. V., and Wise, M. B., *Phys. Rev.* **D60**, 034019 (1999).
13. Falk, A., and Ligeti, Z., private communication.
14. CLEO Collaboration, Alexander, J. P. et al., *Phys. Rev. Lett.* **77**, 5000–5004 (1996).
15. Isgur, N., Scora, D., Grinstein, B., and Wise, M. B., *Phys. Rev.* **D39**, 799–818 (1989); Scora, D., and Isgur, N., *Phys. Rev.* **D52**, 2783–2812 (1995).
16. Goity, J. L., and Roberts, W., *Phys. Rev.* **D51**, 3459–3477 (1995).
17. CLEO Collaboration, Cronin-Hennessy, D. et al. CLNS 01/1752, submitted to *Phys. Rev. Lett.*
18. Falk, A. F., Luke, M. E., and Savage, M. J., *Phys. Rev.* **D53**, 2491–2505 (1996); *ibid.* **D53**, 6316–6325 (1996).
19. Gremm, M., and Kapustin, A., *Phys. Rev.* **D55**, 6924–6932 (1997).
20. Falk, A. F., and Luke, M. E., *Phys. Rev.* **D57**, 424–430 (1998).
21. Voloshin, M. B., *Phys. Rev.* **D51**, 4934–4938 (1995).
22. Shifman, M. A., Uraltsev, N. G., and Vainshtein, A. I., *Phys. Rev.* **D51**, 2217–2223 (1995); [E: *ibid.* **D52**, 3149 (1995)].
23. Ball, P., Beneke, M., and Braun, V. M., *Phys. Rev.* **D52**, 3929–3948 (1995).
24. Bigi, I. I., Uraltsev, N. G., and Vainshtein, A.I., *Phys. Lett.* **B293**, 430–436 (1992) [E: *ibid.*, **B297**, 477 (1993)].
25. Bigi, I. I. Y., Shifman, M. A., Uraltsev, N. G., and Vainshtein, A. I., *Phys. Rev. Lett.* **71**, 496–499 (1993).
26. Jezabek, M., and Kuhn, J. H., *Nucl. Phys.* **B314**, 1–6 (1989).
27. Luke, M. E., Savage, M. J., and Wise, M. B., *Phys. Lett.* **B345**, 301–306 (1995).
28. CLEO Collaboration, Barish, B. et al., *Phys. Rev. Lett.* **76**, 1570–1574 (1996).
29. ARGUS Collaboration, Albrecht, H. et al., *Phys. Lett.* **B318**, 397–404 (1993).
30. Groom, D. E. et al., *Eur. Phys. J.* **C15**, 1–878 (2000).
31. CLEO Collaboration, Alexander, J. P. et al., *Phys. Rev. Lett.* **86-91**, 2737–2741 (2001).
32. Uraltsev, N., *Int. J. Mod. Phys.* **A14**, 4641–4652 (1999).
33. Neubert, M., *Phys. Rev.* **D49**, 4623–4633 (1994); Neubert, M., *Phys. Lett.* **B513**, 88–92 (2001).
34. Leibovich, A. K., Low, I., and Rothstein, I. Z., *Phys. Lett.* **B486**, 86–91 (2000); *ibid.* **B513**, 83–87 (2001).
35. Wise, M. B., "Recent Progress in Heavy Quark Physics", hep-ph/0111167, to appear in *Proceedings of the 20th International Symposium on Lepton and Photon Interactions at High Energies.*
36. Rothstein, I., these proceedings.
37. Hoang, A. H., Ligeti, Z., and Manohar, A. V., *Phys. Rev. Lett.* **82**, 277–280 (1999); Hoang, A. H., Ligeti, Z., and Manohar, A. V., *Phys. Rev.* **D59**, 074017 (1999).
38. De Fazio, F., and Neubert, M., *JHEP* **06**, 017 (1999).
39. Altarelli, G., Cabibbo, N., Corbo, G., Maiani, L., and Martinelli, G., *Nucl. Phys.* **B208**, 365 (1982).
40. Artuso, M., *Phys. Lett.* **B311**, 307 (1993).
41. CLEO Collaboration, Behrens, B. H. et al., *Phys. Rev.* **D61**, 052001 (2000).
42. Luke, M. E., *Phys. Lett.* **B252**, 447–455 (1990).
43. Boyd, C. G., Grinstein, B., and Lebed, R., *Phys. Rev.* **D56**, 6895–6911 (1997).
44. Caprini, I., Lellouch, L., and Neubert, M., *Nucl. Phys.* **B530**, 153–181 (1998).
45. CLEO Collaboration, Duboscq, J. E. et al., *Phys. Rev. Lett.* **76**, 3898–3902 (1996).
46. *The BaBar Physics Book*, edited by P. F. Harrison, and H. R. Quinn, H. R., SLAC-R-504 (1998).
47. LEP results are standardized by the LEP Heavy Flavor Working Group.
48. Falk, A. F. and Ligeti, Z., and Wise, M. B., *Phys. Lett.* **B406**, 225–231 (1997).
49. BELLE Collaboration, Abe, K. et al., BELLE 2001-19, hep-ex/0111060, submitted to *Phys. Lett. B*.
50. BELLE Collaboration, BELLE-CONF-0123 (2001).

Rare B Decays Beyond $B \to X_s\gamma$

Gustavo Burdman

Theoretical Physics Group, Lawrence Berkeley National Laboratory, Berkeley, CA 94720

Abstract. I discuss recent progress in our understanding of exclusive rare and semileptonic decays. I show the impact of HQET when combined with the predictions in the Large Energy Limit of QCD, focusing first on applications to $B \to K^*\ell^+\ell^+$. I also discuss the constraints on semileptonic form-factors that appear in HQET/LEET with the use of radiative decay data, and update these to include the effects of next-to-leading order contributions in $B \to K^*\gamma$, as well as the latest data.

INTRODUCTION

In the Standard Model (SM), Flavor Changing Neutral Currents (FCNCs) are forbidden at tree level. They can occur starting at one loop. For the $b \to s$ transitions, the diagrams involving the top quark dominate the short distance rate. This is essentially a consequence of non-decoupling in spontaneously broken gauge theories: heavy fermions in the loops give contributions that do not vanish as the mass increases but rather grow. Thus, FCNC processes at relatively low energies have the potential to explore high energy scales such as the weak scale or even beyond. Extensions of the SM, such as supersymmetry, technicolor, etc. would result in new contributions to FCNC loops, leading to deviations from the SM predictions. This sensitivity of FCNC processes to high energy scales, and therefore to new physics makes them of great interest. This is particularly true in B decays since it is possible that the third generation is involved in electroweak symmetry breaking [1]. However, B decays are affected by hadronic uncertainties that may obscure the interesting short distance processes. Such uncertainties result from the fact that hadronization is a non-perturbative problem only tractable from first principles by lattice gauge theory. The use of inclusive decay modes greatly circumvent this problems. There, the uncertainties are mostly not from hadronization (only relevant in the initial state) but from perturbative QCD. Exclusive modes, on the other hand, are largely affected by theoretical uncertainties from hadronic matrix elements of the short distance operators.

While lattice calculations progress toward greater precision and accuracy, we can ask the question: can rare B decays be used as tests of the one-loop structure of the SM with our current knowledge of hadronic matrix elements ? The comparison with the program of electroweak precision measurements, mostly performed at the Z pole, is interesting. That program relied on processes with large tree-level SM amplitudes

[1] For instance, this is the case in supersymmetry, where the stop plays a crucial role; as well as in topcolor models where top condensation is responsible for (at least partially) breaking electroweak symmetry.

CP618, *Heavy Flavor Physics: Ninth International Symposium*, edited by A. Ryd and F. C. Porter

(large SM background), but exquisite experimental precision made it possible to achieve sensitivity to one loop contributions since the accuracy of our theoretical knowledge matches this precision. On the other hand, rare B decays and FCNC processes in general, start at one loop but are affected by large theoretical uncertainties. These considerations are still valid even if we are talking about new physics entering at tree level. Very large data samples of $b \to s\gamma$ and $b \to s\ell^+\ell^-$ decays will be available soon. How good a test of the SM will these be? The inclusive $B \to X_s\gamma$ decay already constrains physics beyond the SM. Even more constraints will result once $b \to s\ell^+\ell^-$ modes begin to be measured.

Inclusive modes are theoretically cleaner. The uncertainties are, in principle, from perturbative QCD and the OPE for heavy quarks. In practice, there is the added problem that cuts are needed to make contact with experiment [1]. This introduces additional uncertainties. In addition, the experimental signals require a very clean environment. Exclusive decays are easier experimentally. But, as e mentioned earlier, they are affected by large hadronic uncertainties coming as unknown form-factors. These are the kinds of things the lattice will one day compute precisely. But for now (short of using models) we have to rely on symmetries and other related tricks in order to extract the short distance physics from these modes.

The factorization of short and long distance physics takes place in the effective hamiltonian at low energies. This is obtained by integrating out the heavy fields (e.g. the W, the top quark, the gluino, etc.) and has the form

$$\mathcal{H}_{eff} = -\frac{4G_F}{\sqrt{2}} V_{tb}^* V_{ts} \sum_{i=1}^{10} C_i(\mu)\, O_i(\mu) \tag{1}$$

Here, the basis $\{O_i\}$ constitute a complete set of operators leading to $b \to s$ transitions, whereas the Wilson coefficient functions $C_i(\mu)$ encode the short distance information coming from integrating out the heavy degrees of freedom. For instance in the SM the $C_i(\mu)$ come from loop integrals involving the W and the top quark and they will depend on their masses. If physics beyond the SM contributes to these decays it will shift the values of the Wilson coefficients with terms now depending on gluino, squark or technipion masses. The μ dependence of the Wilson coefficients is in principle canceled by that of the matrix elements of the operators. At leading order in α_s the operators mediating $b \to s$ transitions are

$$\mathcal{O}_7 = \frac{e}{16\pi^2} m_b \bar{s}_L \sigma_{\mu\nu} b_R\, F^{\mu\nu} \tag{2}$$

$$\mathcal{O}_9 = \frac{e^2}{16\pi^2} (\bar{s}_L \gamma_\mu b_L)(\bar{\ell}\gamma^\mu \ell) \tag{3}$$

$$O_{10} = \frac{e^2}{16\pi^2} (\bar{s}_L \gamma_\mu b_L)(\bar{\ell}\gamma^\mu \gamma_5 \ell)\,. \tag{4}$$

Among these, only $\mathcal{O}_7$ contributes to processes with the photon on-shell, such as $b \to s\gamma$. Mixing with other operators occurs due to the strong interactions. Most notably with the gluonic dipole operator $\mathcal{O}_8 = (g/16\pi^2) m_b \bar{s}_L T^a \sigma_{\mu\nu} b_R G^{a\mu\nu}$, and especially with $O_2 = (\bar{s}_L \gamma_\mu c_L)(\bar{c}_L \gamma^\mu b_L)$, which is generated by tree-level W exchange.

EXCLUSIVE DECAYS

The hadronic matrix elements can be parametrized in terms of form-factors. For the $B \to K\ell^+\ell^-$ decay the hadronic matrix elements of the operators O_7, O_9 and O_{10} can be written as

$$\langle K(k)|\bar{s}\sigma_{\mu\nu}q^\nu b|B(p)\rangle = i\frac{f_T}{m_B+m_K}\left\{(p+k)_\mu q^2 - q_\mu(m_B^2-m_K^2)\right\}, \quad (5)$$

$$\langle K(k)|\bar{s}\gamma_\mu b|B(p)\rangle = f_+(p+k)_\mu + f_- q_\mu, \quad (6)$$

with $f_T(q^2)$, and $f_\pm(q^2)$ unknown functions of $q^2=(p-k)^2=m^2_{\ell^+\ell^-}$. In the SU(3) limit $f_\pm$ in (6) are the same as the form-factors entering in the semileptonic decay $B\to\pi\ell\nu$. For $B\to K^*\ell^+\ell^-$ decays of the "semileptonic" matrix elements over vector and axial vector currents are

$$\langle K^*(k,\varepsilon)|\bar{q}\gamma_\mu b|B(p)\rangle = \frac{2V(q^2)}{m_B+m_V}\varepsilon_{\mu\nu\alpha\beta}\varepsilon^{*\nu}p^\alpha k^\beta \quad (7)$$

$$\begin{aligned}\langle K^*(k,\varepsilon)|\bar{q}\gamma_\mu\gamma_5 b|B(p)\rangle &= i2m_V A_0(q^2)\frac{\varepsilon^*\cdot q}{q^2}q_\mu + i(m_B+m_V)A_1(q^2)\left(\varepsilon^*_\mu - \frac{\varepsilon^*\cdot q}{q^2}q_\mu\right)\\ &\quad -iA_2(q^2)\frac{\varepsilon^*\cdot q}{m_B+m_V}\left((p+k)_\mu - \frac{m_B^2-m_V^2}{q^2}q_\mu\right). \quad (8)\end{aligned}$$

and for the FCNC magnetic dipole operator $\sigma_{\mu\nu}$

$$\begin{aligned}\langle K^*(k,\varepsilon)|\bar{q}\sigma_{\mu\nu}(1+\gamma_5)q^\nu b|B(p)\rangle &= i2T_1(q^2)\varepsilon_{\mu\nu\alpha\beta}\varepsilon^{*\nu}p^\alpha k^\beta\\ &\quad + T_2(q^2)\left\{\varepsilon^*_\mu(m_B^2-m_V^2) - (\varepsilon^*\cdot p)(p+k)_\mu\right\}\\ &\quad + T_3(q^2)(\varepsilon^*\cdot p)\left\{q_\mu - \frac{q^2}{m_B^2-m_V^2}(p+k)_\mu\right\} \quad (9)\end{aligned}$$

where ε_μ denotes the polarization four-vector of the K^*. In the $SU(3)$ limit these form-factors also describe $B\to\rho\ell\nu$, as well as $B_s\to\phi\ell^+\ell^-$ decays. The form-factors defined in eqns.(6)-(9) are not calculable in a controlled approximation with the notable exception of lattice gauge theory. We will see however, that symmetries and generally the behavior of hadrons in certain limits will allow us to gain important insight on these quantities. Notice that $T_1(0)=T_2(0)$ and T_3 does not contribute to the amplitude to the radiative decay into an on-shell photon, i.e. in $B\to K^*\gamma$. So now $Br(B\to K^*\gamma)\ \alpha\ |T_1(0)|^2$, where the short distance is basically the same as in inclusive decays[2].

Heavy Quark Symmetry and $B\to K^*\ell^+\ell^-$

In the HQL [2] $m_b \gg \Lambda_{\rm QCD}$ the form-factors $T_i(q^2)$ corresponding to the dipole operator are not independent of the semileptonic form-factors $V(q^2), A_i(q^2)$. Instead,

[2] With some caveats and exceptions we discuss later on. See also [9].

they obey the following relations [3, 4]

$$\begin{aligned}
T_1(q^2) &= \frac{m_B^2+q^2-m_V^2}{2m_B}\frac{V(q^2)}{m_B+m_V}+\frac{m_B+m_V}{2m_B}A_1(q^2), \\
\frac{m_B^2-m_V^2}{q^2}\left[T_1(q^2)-T_2(q^2)\right] &= \frac{3m_B^2-q^2+m_V^2}{2m_B}\frac{V(q^2)}{m_B+m_V}-\frac{m_B+m_V}{2m_B}A_1(q^2), \\
T_3(q^2) &= \frac{m_B^2-q^2+3m_V^2}{2m_B}\frac{V(q^2)}{m_B+m_V}+\frac{m_B^2-m_V^2}{m_Bq^2}m_VA_0(q^2) \\
&- \frac{m_B^2+q^2-m_V^2}{2m_Bq^2}\left[(m_B+m_V)A_1(q^2)-(m_B-m_V)A_2(q^2)\right].
\end{aligned} \tag{10}$$

In terms of the symmetries of the HQET, eqns. (10) result from the Heavy Quark *Spin* Symmetry (HQSS) that arises in the heavy quark limit due to the decoupling of the spin of the heavy quark [2]. These relations imply, for instance, that the $B\to\rho\ell\nu$ and $B\to\rho\ell^+\ell^-$ are described by the same form-factors. Additionally, if $SU(3)$ symmetry is assumed, these also are the $B\to K^*\ell^+\ell^-$ form-factors.

The Large Energy Limit

We now consider the Large Energy Limit (LEL) [5] for heavy-to-light transitions into a vector meson as the ones we are studying. As a result, one recovers the HQSS form-factor relations (10), but now there are additional new relations among the form-factors defined in (7-9). These read as [6]

$$V(q^2) = \left(1+\frac{m_V}{M}\right)\xi_\perp(M,E)\,, \tag{11}$$

$$A_1(q^2) = \frac{2E}{M+m_V}\xi_\perp(M,E)\,, \tag{12}$$

$$A_2(q^2) = \left(1+\frac{m_V}{M}\right)\left\{\xi_\perp(M,E)-\frac{m_V}{E}\xi_\parallel(M,E)\right\}, \tag{13}$$

$$A_0(q^2) = \left(1-\frac{m_V^2}{ME}\right)\xi_\parallel(M,E)+\frac{m_V}{M}\xi_\perp(M,E)\,, \tag{14}$$

and

$$T_1(q^2) = \xi_\perp(M,E)\,, \tag{15}$$

$$T_2(q^2) = \left(1-\frac{q^2}{M^2-m_V^2}\right)\xi_\perp(M,E)\,,\,, \tag{16}$$

$$T_3(q^2) = \xi_\perp(M,E)-\frac{m_V}{E}\left(1-\frac{m_V^2}{M^2}\right)\xi_\parallel(M,E)\,. \tag{17}$$

and will receive corrections that roughly go as $(\Lambda_{QCD})/E_h$. It is apparent from eqns. (11)-(17) that, in the LEL regime, the $B\to V\ell\ell'$ decays are described by only two form-factors: $\xi_\perp$ and $\xi_\parallel$, instead of the seven apriori independent functions in the general

Lorentz invariant ansatz of the matrix elements. Here, $\xi_\perp$ and $\xi_\|$ are functions of the heavy mass M and the hadronic energy E, and refer to the transverse and longitudinal polarizations, respectively.

This simplification leads to new relations among the form-factors. For instance, the ratio of the vector form-factor V to the axial-vector form-factor A_1,

$$R_V(q^2) \equiv \frac{V(q^2)}{A_1(q^2)} = \frac{(m_B+m_V)^2}{2E_V m_B}\,, \tag{18}$$

is independent of any of these unknown, non-perturbative functions $\xi_{\perp,\|}$ and is determined by purely kinematical factors. Here, $E_V = (m_B^2+m_V^2-q^2)/(2m_B)$ denotes the energy of the final light vector meson. A similar relation holds for T_1 and T_2, since they both are also proportional to the "transverse" form-factor $\xi_\perp$. As we will see below, these predictions have important consequences for observables at large recoil energies (low q^2).

Corrections to LEET

The predictions obtained in the LEL receive corrections from several sources. In the past year or so some of them have been extensively addressed in the literature. Here is a quick review.

Hard Corrections: These corrections result from the exchange of hard gluons. There are two kinds of them: factorizable and non-factorizable gluon exchange. The first kind corresponds to either the renormalization of the heavy-light currents which already appears in Ref. [2], or hard gluon exchange with the spectator quark. The latter can be computed in the Brodsky-Lepage formalism for exclusive processes at large momentum transfers [7]. This was done in Ref. [8], where it was found that the form-factor relations in (11)-(17) receive α_s corrections that are typically 10% or smaller. An interesting case is the ratio of vector to axial-vector form-factors in eqn. (18), which receives *no* α_s *corrections* to leading order in the $1/E$ expansion. This has an interesting explanation and we will come back to this point below.
The second kind of hard gluon corrections are the so called non-factorizable ones [9, 10]. These are mediated by diagrams where the gluon exchanged with the spectator comes from either the insertion of the operator O_8, or from the insertion of four-fermion operators $O_1 - O_6$ with the gluon attached to the quark loop. These corrections cannot be absorbed by form-factors or renormalization of the currents. Thus they are genuinely distinct hard corrections that do not occur in inclusive decays. In Refs. [9, 10] the effects of all these contributions are computed and are rather large in $B \to K^*\gamma$. The effective "exclusive" Wilson coefficient $C_7^{\rm excl.}$ is shifted by roughly 30%, whereas the effect in C_9 is smaller. The authors go on to compute the exclusive rate by making use of light-cone QCD sum rules for the form-factors, resulting in a rate which is approximately a factor of two larger than the experimental rate. This may be

signaling a problem with the form-factor calculation rather than one with the non-factorizable hard gluons.

Collinear Gluons: In Ref. [11] an effective field theory including collinear quarks and gluons is developed. This theory includes LEET but, unlike LEET, it has the correct infrared behavior. Sudakov logarithms are accounted for in this treatment. Concerning its application to exclusive decays, this "complete" LEET does not modify the form-factor relations implied by eqns.(11-17). A main ingredient of LEET is preserved when the interaction with collinear gluons is taken into account: energetic quarks still are two component spinors in the LEL. Thus the complete LEET including collinear gluons confirms that the LEET relations are valid in the large energy limit of QCD.

Power Corrections: Just as in HQET the $1/m_Q$ corrections are potentially a very important source of uncertainty, the LEL predictions are affected by $1/E$ corrections. Unlike in HQET, it is not clear that LEET is a framework where the $1/E$ corrections can be estimated. Even when collinear gluons are incorporated, it is difficult to separate degrees of freedom to be integrated out so that an effective field theory can emerge. Perhaps, we can still estimate the size of these corrections with this caveat in mind. More work is needed, and certainly once experimental tests of LEET predictions become available we hope to understand more about them.

Forward-Backward Lepton Asymmetry

The forward-backward asymmetry for leptons as a function of the dilepton mass squared $m^2_{\ell\ell} = q^2$ is defined as

$$A_{FB}(q^2) = \frac{\int_0^1 \frac{d^2\Gamma}{dxdq^2}dx - \int_{-1}^0 \frac{d^2\Gamma}{dxdq^2}dx}{\frac{d\Gamma}{dq^2}}, \tag{19}$$

where $x \equiv \cos\theta$, and θ is the angle between the ℓ^- and the $\bar{B}^0$ in the dilepton center-of-mass frame [3].

The asymmetry is proportional to the Wilson coefficient C_{10} and vanishes with it. Furthermore, it is proportional to a combination of $C_9^{\text{eff.}}$ and $C_7^{\text{eff.}}$ such that it has a zero in the physical region if the following condition is satisfied [12]

$$Re[C_9^{\text{eff.}}] = \frac{m_b}{q_0^2} C_7^{\text{eff.}} \left\{ \frac{T_1}{V}(m_B + m_{K^*}) - (m_B - m_{K^*}) \frac{T_2}{A_1} \right\}, \tag{20}$$

where q_0^2 is the position of the A_{FB} zero and all q^2-dependent quantities[4] are evaluated at q_0^2. In inclusive $B \to X_s \ell^+ \ell^-$, the zero of the asymmetry implies the relation:

---

[3] In Ref. [12] it is erroneously stated that θ is defined with respect to ℓ^+. But the expressions there correspond to the current definition. This explains the sign difference with respect to Ref. [10].

[4] Note that the sign difference in eqn.(20) with respect to a similar expression in Ref.[13] is due to a sign in the definition of V.

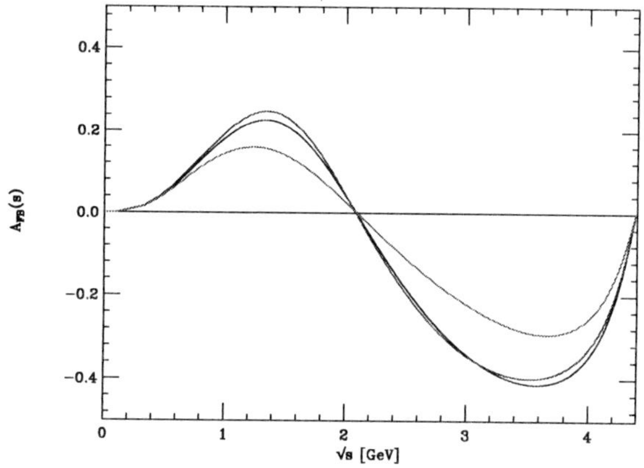

FIGURE 1. Differential forward-backward asymmetry vs. the dilepton mass squared for various model calculations.This includes all hard corrections, both factorizable and non-factorizable, so it is an update of Ref. [12], where the references for the models can be obtained.

$Re[C_9^{\text{eff.}}] = -2(m_B^2/q_0^2)C_7^{\text{eff.}}$. Thus, in principle the condition in (20) is affected by the presence of the hadronic form-factors making it a priori a more uncertain relation. However, by making use of the HQSS relations (10) for $T_1(q^2)$ and $T_2(q^2)$, the form-factor eqn. (20) simplifies to

$$\frac{Re[C_9^{\text{eff.}}]}{C_7^{\text{eff.}}} = -\frac{m_b}{q_0^2}\left\{\frac{2m_B k^2}{(m_B+m_{K^*})^2}R_V + \frac{(m_B+m_{K^*})^2}{2m_B R_V} + 2(m_B - E_{K^*})\right\} \tag{21}$$

where R_V is the ratio of vector to axial-vector form-factors defined in (18) and evaluated at q_0^2. Thus, the determination of the zero of $A_{FB}(q^2)$ in $B \to K^*\ell^+\ell^-$ gives a relation between the short distance Wilson coefficients $C_9^{\text{eff.}}$ and $C_7^{\text{eff.}}$ where the only uncertainty from form-factors is in the ratio R_V. As mentioned earlier, since (21) was derived using the heavy quark *spin* symmetry, it is expected to receive small corrections. In principle, information on $R_V(q^2)$ could be extracted from the $B \to \rho\ell\nu$ decay. But it turns out that this may not be necessary. If we plot the asymmetry vs. the dilepton mass in a variety of models we note that the asymmetry is agreed upon with exceptional accuracy! In Ref.[12] it was noted that this feature must emerge from a "factorization" of the soft physics in such a way that

$$V(q^2) \simeq A_{\text{kin.}} \times F_{\text{soft}}, \qquad A_1(q^2) \simeq B_{\text{kin.}} \times F_{\text{soft}}\ , \tag{22}$$

so that the soft physics cancels in the ratio R_V. It was first recognized in Ref. [13] that this is precisely what the LEET predicts: $R_V(q^2) = (m_B + m_{K^*})^2/(2m_B E_{K^*})$ as extracted from eqns. (11) and (12). Then, the condition for the vanishing of $A_{FB}(q^2)$ reads now

$$Re[C_9^{\text{eff.}}] = -2\frac{m_b m_B}{q_0^2}\, C_7^{\text{eff.}}\left(1 - \frac{m_{K^*}^2}{2m_B E}\right)\ . \tag{23}$$

Thus, in the LEL the position of the zero of $A_{FB}(q^2)$ in $B \to K^* \ell^+ \ell^-$ is predicted in terms of the short distance Wilson coefficients $C_9^{\mathrm{eff.}}$ and $C_7^{\mathrm{eff.}}$. The hard gluon corrections discussed above would shift the position of the zero by a calculable amount given in Ref.[10]. There it is pointed out that non-factorizable contributions change the value of q_0^2 from what it would be obtained if the "inclusive" Wilson coefficients are used by an amount around $(20-30)\%$. This is roughly $q_0^2 = (4.2 \pm 0.6)$ GeV^2. The plot in Fig.1 takes all these corrections into account.

Semileptonic Form-factors in the Large Energy Limit

We finally discuss another application of LEET. In Ref. [14] the LEET prediction for R_V in eqn.(18) was used in combination with the $B \to K^* \gamma$ and $B \to X_s \gamma$ data, and Heavy Quark Spin Symmetry in order to constrain the semileptonic form-factors V and A_1 at $q^2 = 0$. Here we update this analysis by incorporating the factorizable and non-factorizable hard corrections from Refs.[9, 10].
Since the NLO corrections in the exclusive mode are not canceled by the ones in the inclusive decay, we will not normalize the exclusive branching ratio to the inclusive one. Instead we make use of the exclusive data only in order to extract the $B \to K^* \gamma$ form-factor. The branching ratio reads

$$Br(B \to K^* \gamma) = \tau_B \frac{\alpha G_F^2}{32\pi^4} |V_{cb} V_{cs}^*|^2 m_b^2 m_B^3 \left(1 - \frac{m_{K^*}^2}{m_B^2}\right)^3 |A_7|^2 \, |T_1(0)|^2 \,, \tag{24}$$

where the form-factor $T_1(q^2)$ was defined in (9), and A_7 is the NLO effective Wilson coefficient for the exclusive mode, which differs from the one entering in inclusive decays. We use the running mass $m_b(m_b) = (4.2 \pm 0.2)$ GeV, and $|V_{cb} V_{cs}^*| = 0.04$. The value of A_7 was computed in Ref. [9] to be $A_7 = -0.4072 - i0.0256$. We make use of the world average for the neutral mode [15] $Br(B^0 \to K^{*0} \gamma) = (4.56 \pm 0.37) \times 10^{-5}$. We obtain $T_1(0) = 0.31 \pm 0.02$, where the error mainly reflects the experimental uncertainty, the uncertainty in the b quark mass and the scale dependence in A_7. This result assumes the SM for the calculation of the short distance coefficient A_7. However, this is not a very strong assumption given that the agreement of the inclusive calculation with the $B \to X_s \gamma$, and the fact that the new physics is likely to affect this rate and the exclusive one similarly. Armed with this extracted value of $T_1(0)$, now we can go to the HQSS relations and turn them into a relation between $V(0)$ and $A_1(0)$. This is

$$A_1(0) = \frac{2 m_B}{m_B + m_{K^*}} T_1(0) - \frac{m_B - m_{K^*}}{m_B + m_{K^*}} V(0) \,. \tag{25}$$

Eqn. (25) results in a constraint in the $(V(0), A_1(0))$ plane, which in Fig. 2 corresponds to the band descending from left to right. In addition, the expression (18) for R_V gives another constraint corresponding to a straight line going through the origin. This LEET prediction will surely receive corrections at next to leading order in the $1/E$ and $1/m_b$ expansions. These corrections have not been computed and we simply guess they are of order $m_{K^*}/m_B \simeq 0.17$, giving the cone in Fig. 2. We should notice, however, that the

LEET prediction for R_V does not receive hard gluon or collinear gluon corrections. We will come back to this point. The ellipses in Fig. 2 correspond to 68% and 95% C.L. intervals. The fit gives

$$V(0) = 0.36 \pm 0.04 , \qquad A_1(0) = 0.27 \pm 0.03 . \tag{26}$$

The results in eqn. (26) differ from Ref. [14] where the NLO corrections had not been taken into account. These lift the value of the Wilson coefficient by roughly 30% resulting in a lower value for $T_1(0)$, and through the fit in Fig. 2, in lower values for both $V(0)$ and $A_1(0)$. Also, here we use the most recent average of the neutral mode resulting in a thinner band than the one in Ref. [14].

We compare our findings for $V(0)$ and $A_1(0)$ with several model predictions. For illustration, we take the Bauer-Stech-Wirbel (BSW) model from Ref. [16] (cross), the modified version of the Isgur-Scora-Grinstein-Wise (ISGW2) model from Ref. [17] (diamond), a recent relativistic constituent quark model prediction by Melikhov and Stech (MS) [18] (star), a recent calculation in the Light Cone QCD Sum Rule (LCSR) formalism of Ref. [13] (diagonal cross) and the prediction by Ligeti and Wise (LW) from Ref.[19] (square). As it can be seen from Fig. 2, the latter is even more excluded now. As discussed in more detail in Ref.[14], this is likely to be due largely to the use of a monopole form-factor to extrapolate from the charm data to the maximum recoil in a B decay. Corrections to the heavy quark flavor symmetry, although large, are unlikely to be the sole explanation for such a discrepancy. A new finding in this update is the apparent exclusion of the MS and LCSR predictions for $V(0)$ and $A_1(0)$. These models, were in agreement with the fit in Ref. [14]. Of course, this exclusion refers to the predicted central values for the models. In some cases the predictions have large errors and modifications in the calculations may bring them into agreement with the fit.

A potentially large isospin violation splitting the neutral and charged modes was found in Ref.[20] in the context of QCD factorization. This would shift the value of A_7 in the neutral mode by a few percent, resulting in a similarly small shift in the value of $T_1(0)$ extracted from eqn.(24).

Finally, we comment on the leading order LEET prediction for R_V, eqn. (18). Making use of the HQSS relations (10), the transverse helicity amplitudes for a generic $B^- \to V^- \ell\ell'$ transition can be written as

$$H_\pm = \mathcal{F} \, (V \mp \frac{(m_B + m_V)^2}{2 m_B k_V} A_1) , \tag{27}$$

where $\mathcal{F}$ is a factor depending on the mode under consideration (e.g. Wilson coefficients, coupling constants, etc...) and k_V is the momentum of the vector meson. Thus, we see from the form of R_V in the large energy limit, that the "+" helicity vanishes $H_+ = 0$ in the LEL regime, up to residual terms of order $m_V^2/2E_V^2$. This is not a surprise: in the limit of an infinitely heavy quark decaying into a light quark, the helicity of the latter is "inherited" by the final vector meson. In the SM, the $(V-A)$ structure in semileptonic decays is reflected in the dominance of the H_- transverse helicity. On the other hand, the amplitude to flip the helicity of the fast outgoing light quark is suppressed by $1/E_h$. Quark models tend to have this concept built in, which explains the agreement on the position of the zero in Fig. 1. This is also the reason why

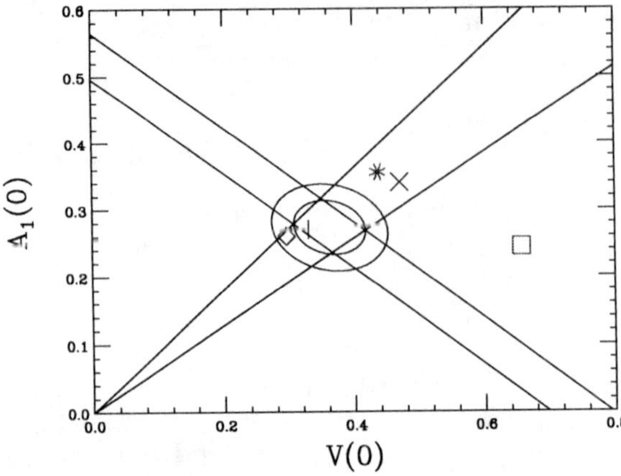

FIGURE 2. Constraints on the semileptonic form-factors $V(0)$ and $A_1(0)$ from $B^0 \to K^{*0}\gamma$ data plus HQSS (thicker band) together with the relation from the LEL (cone).The ellipses correspond to 68% and 95% confidence level intervals. Central values of model predictions are also shown and correspond to BSW [16] (vertical cross), ISGW2 [17] (diamond), MS [18] (star), LCSR [13] (diagonal cross) and LW [19] (square), respectively. This updates Ref. [14] to include hard corrections as computed in Ref. [9, 10], as well as the most recent data.

α_s corrections from hard gluon exchange between the spectator quark and the fast light quark do not affect eqn. (18): they are not helicity-changing at leading order in $1/E_h$. The same is true for the ratio of T_1 and T_2.

ACKNOWLEDGMENTS

I thank Gudrun Hiller, with whom part of the work reported here was done, for discussions and a careful reading of the and the organizers of HF9 at Caltech for such an excellent meeting, in spite of the difficult circumstances they had to face during its course. This work was supported by the Director, Office of Science, Office of High Energy and Nuclear Physics of the U.S. Department of Energy under Contract DE-AC0376SF00098.

REFERENCES

1. C. Bauer, these proceedings; I. Rothstein, these proceedings.
2. N. Isgur and M. B. Wise, Phys. Lett. **B232**, 113 (1989); Phys. Lett. **B237**, 527 (1990).
3. N. Isgur and M. B. Wise, Phys. Rev. **D42**, 2388 (1990).
4. G. Burdman and J. F. Donoghue, Phys. Lett. **B270**, 55 (1991).
5. M. Dugan and B. Grinstein, Phys. Lett. **B255**, 583 (1991).
6. J. Charles *et al.*, Phys. Rev. **D60**, 014001 (1999).
7. G. P. Lepage and S. J. Brodsky, Phys. Rev. **D22**, 2157 (1980). The first application to B decays is in A. Szczepaniak, E. M. Henley and S. J. Brodsky, Phys. Lett. B **243**, 287 (1990).
8. M. Beneke and T. Feldmann, Nucl. Phys. **B592**, 3 (2001).

9. S. W. Bosch and G. Buchalla, "The radiative decays $B \to V\gamma$ at next-to-leading order in QCD", hep-ph/0106081. Stefan Bosch, these proceedings.
10. M. Beneke, T. Feldmann and D. Seidel, Nucl. Phys.**B612**, 25 (2001).
11. C. W. Bauer, S. Fleming and M. E. Luke, Phys. Rev. **D63**, 014006 (2001); C. W. Bauer, S. Fleming, D. Pirjol and I. W. Stewart, Phys. Rev. **D63**, 114020 (2001).
12. G. Burdman, Phys. Rev. **D57**, 4254 (1998).
13. A. Ali, P. Ball, L. T. Handoko and G. Hiller, Phys. Rev.**D61**, 074024 (2000).
14. G. Burdman and G. Hiller, Phys. Rev. **D63**, 113008 (2001).
15. A. Ryd, for the Babar collaboration, these proceedings; M. Nakao, for the Belle collaboration, these proceedings; K. Honscheid, for the CLEO collaboration, these proceedings.
16. M. Wirbel, B. Stech and M. Bauer, Z. Phys. **C29**, 637 (1985).
17. N. Isgur and N. Scora, Phys. Rev. **D52**, 2783 (1995).
18. D. Melikhov and B. Stech, Phys. Rev. **D62**, 014006 (2000).
19. Z. Ligeti and M. Wise, Phys. ReV. **D60**, 117506 (1999).
20. A. L. Kagan and M. Neubert, "Isospin breaking in $B \to K^*\gamma$ decays", hep-ph/0110078.

Recent Results on Hadronic B Decay

Klaus Honscheid representing the CLEO collaboration

Ohio State University, 174 W 18th Avenue, Columbus, Ohio, 43210

Abstract. In this paper we present recent results obtained by the CLEO Collaboration on hadronic and rare decays of B mesons. Highlights include an improved branching fraction for inclusive electro-magnetic penguin decays, a measurement of the photon spectrum in these decays as well as the first observation color-suppressed B decays to final states with open-charm mesons.

INTRODUCTION

Acknowledging the fact that in this area of electronic publishing conference proceedings have lost most of their original meaning we will not present a detailed description of the results reported at HF9. Instead we will provide a brief summary combined with a list of electronic references where the interested reader can find current and detailed information on our work. The transparencies shown during the conference are available in pdf format [1].

The data analyzed in these studies were collected at the Cornell Electron Storage Ring (CESR) with the CLEO detector. Three different configurations were used. The CLEO II (1990-1995) and CLEO II.V (1995-1999) data sample consists of 9.1 fb^{-1} collected on the $\Upsilon(4S)$ resonance and 4.4 fb^{-1} taken in the e^+e^- continuum just below the resonance. The CLEO III data sample includes 6.9 fb^{-1} of $\Upsilon(4S)$ and 2.3 fb^{-1} of continuum data. The two data samples correspond to 9.7×10^6 and 7.4×10^6 $B\bar{B}$ events, respectively. Details about CESR and the CLEO be found following the hyperlinks given in [2] and [3].

RARE DECAYS OF B MESONS

$b \to s\gamma$ Branching Fraction and Photon Energy Spectrum

We have measured the branching fraction and photon energy spectrum for the radiative penguin process $b \to s\gamma$ [4]. The branching fraction is found to be

$$\mathcal{B}(b \to s\gamma) = (3.21 \pm 0.43 \pm 0.27^{+0.18}_{-0.10}) \times 10^{-4}$$

The errors are statistical, systematic, and from theory corrections. Analyzing the shape of photon energy spectrum we obtain the first and second moment for photon energies above 2 GeV

$$< E_\gamma >= 2.346 \pm 0.032 \pm 0.011 \; GeV$$

CP618, *Heavy Flavor Physics: Ninth International Symposium*, edited by A. Ryd and F. C. Porter

and

$$< E_\gamma^2 > - < E_\gamma >^2 = 0.0226 \pm 0.0066 \pm 0.0020\, GeV^2$$

The measured branching fraction can be compared to recent theoretical calculations by Chetyrkin, Misiak, Munz and by Kagan and Neubert which predict

$$\mathcal{B}(b \to s\gamma) = (3.29 \pm 0.33) \times 10^{-4}$$

The window for new physics in this decay mode is closing.

Improved Upper Limits on the FCNC Decays $B \to K\ell^+\ell^-$ and $B \to K^*(892)\ell^+\ell^-$

Using the CLEO II and II.V data samples we have searched for the flavor-changing neutral current decays $B \to K\ell^+\ell^-$ and $B \to K^*(892)\ell^+\ell^-$ [5]. For the latter decay we require that the dilepton mass exceed 0.5 GeV. In none of the decay modes an indication of a signal is observed. We obtain the following 90% confidence level upper limits

$$\mathcal{B}(B \to K\ell^+\ell^-) < 1.7 \times 10^{-6}$$

and

$$\mathcal{B}(B \to K^*(892)\ell^+\ell^-) < 3.3 \times 10^{-6}.$$

In order to compare with theoretical prediction we calculate the upper limit on the weighted average

$$0.65\mathcal{B}(B \to K\ell^+\ell^-) + 0.35\mathcal{B}(B \to K^*(892)\ell^+\ell^-) < 1.5 \times 10^{-6}.$$

We note that the weighted-average limit is only 50% above the Standard Model prediction.

Rare Hadronic B Decay

Preliminary results on hadronic rare B decays such as $B \to \pi\pi$ and $B \to K\pi$ have been reported using about half of the CLEO III data sample. The results are consistent with previously published CLEO measurements of these decay modes. For detailed information see the branching ratio tables in reference [1].

UNDERSTANDING HADRONIC B DECAY

Most hadronic decays of B mesons can be described by external and internal spectator decay diagrams. For charged B meson decays these two amplitudes interfere while for neutral B mesons they lead to separate final states. Most theoretical work attempting a description of these decays is based on factorization - the assumptions that the quarks from the virtual W and the meson remnant hadronize independently [6]. Recent work has provided a more solid foundation for this hypothesis [7]. Furthermore, two phenomenological parameter, a_1 and a_2, are introduced to absorb non-perturbative contributions to the external and internal spectator decay amplitudes, respectively. These parameter are expected to be process dependent but current experimental data can be described with universal values $a_1 \approx 1.1$ and $a_2 \approx 0.25$. In the following sub-sections we provide new information on experimental tests of the factorization hypothesis and on the parameter a_2.

Factorization

Experimental tests of the factorization hypothesis have been performed by comparing branching fractions as well as as angular correlations between hadronic and semileptonic B meson decays. Within the experimental precision (10 - 30%) and over the limited q^2 range probed so far the measurements agree with factorization predictions. Using the recent CLEO observation of $\mathcal{B}(\bar{B}^0 \to D^{*+}\pi^+\pi^-\pi^-\pi^0) = (1.72 \pm 0.14 \pm 0.24)\%$ [8], Ligeti, Luke and Wise compared the invariant 4π spectrum in $B \to D^*$ decays to $\tau^- \to \nu\pi^+\pi^-\pi^-\pi^0$ data [9]. Using factorization they find good agreement over the full accessible range up to $m_{4\pi}^2 < 2.9\,GeV^2$. This provides a good factorization test if it can be demonstrated that $B \to D^*4\pi$ transitions do not receive contributions from other decay diagrams. To answer this question, CLEO searched for the related decay mode $\bar{B}^0 \to D^{*0}\pi^+\pi^+\pi^-\pi^-$ which could proceed via D^{**} production or through an internal spectator decay. We find a branching ratio for this mode of $(0.30 \pm 0.07 \pm 0.06)\%$ [10]. We also observe a large $D^{**} \to D^{*0}\pi^+$ component indicating potential problems with the LLW factorization test. However, when we study the relative decay widths for $m_{4\pi}^2 < 2.9\,GeV^2$ we find that there is almost no contribution to this part of the 4 pion spectrum (90 % C.L.):

$$\frac{\Gamma(\bar{B}^0 \to D^{*0}\pi^+\pi^-\pi^+\pi^-)}{\Gamma(\bar{B}^0 \to D^{*+}\pi^+\pi^-\pi^+\pi^0)} < 0.13$$

The LLW factorization test is still valid.

Color Suppression

Since in internal spectator decays the quarks from the virtual W decay have to match the color of the quarks in the decaying hadron it is expected that the amplitude for this process is suppressed compared to external spectator decays. In the decays of charm mesons, the effect of this color suppression is obscured by effects of final state interactions or reduced by non-factorizable contributions. Color suppression is, however, believed to be operative in the B meson system. The only color-suppressed B meson decay modes that have been observed so far are final states containing charmonium states, e.g. $B \to \psi K$. We have used the CLEO II and CLEO II.V data samples to study the color-suppressed hadronic decays of neutral B mesons into the final states $D^{(*)0}\pi^0$

[11]. Using 9.67 million B-pairs we find

$$\mathcal{B}(\bar{B}^0 \to D^0\pi^0) = (2.74^{+0.36}_{-0.32} \pm 0.55) \times 10^{-4}$$

and

$$\mathcal{B}(\bar{B}^0 \to D^{*0}\pi^0) = (2.20^{+0.59}_{-0.52} \pm 0.79) \times 10^{-4}$$

The first error is statistical and the second systematic. The statistical significance of the $D^0\pi^0$ signal is 12.1 sigma (5.9 sigma for $D^{*0}\pi^0$). Combining these results with previous measurements of other $B \to D^{(*)}\pi$ final states we determine the strong interaction phase δ_I between the isospin 1/2 and 3/2 amplitudes in the $D\pi$ and $D^*\pi$ final states to be $\cos\delta_I = 0.89 \pm 0.08$ and $\cos\delta_I = 0.89 \pm 0.08$, respectively.

Comparing our results to models of hadronic B decays allows us to estimate $|a_2|_{B \to D^{(*)0}\pi^0}$ to ≈ 0.4 - significantly larger than the values for a_2 obtained from $B \to$ Charmonium and charged B decays. The model dependence of a_2 begins to show.

REFERENCES

1. Honscheid, K., *Recent Results on Rare and Hadronic B Decay* (2001), URL `http://www-physics.mps.ohio-state.edu/~klaus/research/talks/%hf9_short.pdf`.
2. *The Cornell Electron Storage Ring CESR* (2001), URL `http://www.lns.cornell.edu/public/CESR/`.
3. *The CLEO Experiment* (2001), URL `http://www.lns.cornell.edu/public/CLEO/`.
4. S. Chen *et al.* (CLEO Collaboration), *Phys. Rev. Lett.*, **87**, 251807 (2001).
5. S. Anderson *et al.* (CLEO Collaboration), *Phys. Rev. Lett.NS*, **87**, 181803 (2001).
6. M. Bauer, M. W., B. Stech, *Z. Phys. C*, **29**, 637 (1985).
7. Neubert, M., "Models of B Decay; these proceedings", 2001.
8. J. Alexander *et al.* (CLEO Collaboration), *Phys. Rev. D*, **64**, 092001 (2001).
9. Z. Ligeti, M. W., M. Luke, *Phys. Lett. B*, **507**, 142 (2001).
10. K.W. Edwards *et al.* (CLEO Collaboration), *Phys. Rev. D*, **65**, 012002 (2001).
11. T. Coan *et al.* (CLEO Collaboration), *Cornell CLNS*, **01/1755** (2001), (submitted to PRL).

Belle Results on Charmless B Decays

Hitoshi Ozaki representing the Belle collaboration

KEK, 1-1 Oho, Tsukuba, Ibaraki, 305-0801 Japan

Abstract. Recent results of charmless hadronic B decay analyses based on a 11 or 23 million $B\bar{B}$ sample will be presented. The data were collected with the Belle detector at the KEKB asymmetric e^+e^- collider.

INTRODUCTION

One of the most important goals of experiments at B-factories is to precisely measure the sides and angles of the unitarity triangle in the Cabbibo-Kobayashi-Masukawa matrix (Figure 1) and check its consistency. Any inconsistency is a clear signal of new physics beyond the Standard Model (SM). Charmless hadronic B decays play an important role in working towards this goal. For example, measurement of time-dependent asymmetry in the $B \to \pi\pi$ or $\rho\pi$ mode is a way to determine the ϕ_2 angle; measurements of branching fractions ($\mathcal{B}$) for the $B \to K\pi$, $\pi\pi$ constrain the ϕ_3 angle; $B \to \eta' K$, ϕK are alternatives to measure the ϕ_1 angle, which was measured in the summer of 2001 using charmonium + K decay modes [1].

In this talk, results of charmless hadronic B decay analyses based on a 11 or 23 million $B\bar{B}$ sample will be presented. The data were collected with the Belle detector [2] at the KEKB asymmetric e^+e^- collider [3].

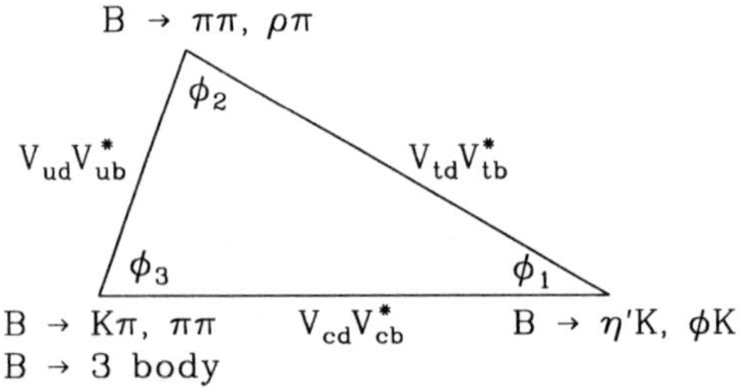

FIGURE 1. Unitarity triangle and related charmless hadronic decays.

ANALYSIS IN GENERAL

In all the decay modes presented here, the continuum process ($e^+e^- \to q\bar{q}$) is the dominant background. Since $B\bar{B}$ events are spherical while the continuum events are

CP618, *Heavy Flavor Physics: Ninth International Symposium*, edited by A. Ryd and F. C. Porter

jet-like, we apply cuts on various event shape variables (such as sphericity, thrust angle, Fox-Wolfram moments, and the production angle of B) to suppress the background.

B candidates are identified using two kinematical variables: beam constrained mass: $M_{BC} = \sqrt{E_{beam}^2 - p_B^2}$, and the energy difference: $\Delta E = E_B - E_{beam}$. Here E_{beam} is the beam energy, p_B and E_B are the momentum and energy of a reconstructed B candidate, respectively, where all variables are defined in the $\Upsilon(4S)$ rest frame.

The B-meson yield is extracted by a maximum-likelihood (ML) fit to the ΔE or M_{BC} distribution (simultaneous ML fits including other variables are not used).

K/π separation is performed by applying a cut on the likelihood ratio, $L_K/(L_\pi + L_K)$, where L_K (L_π) is a kaon (pion) likelihood computed from information from the particle identification devices: specific ionization loss in the central drift chamber, photo-electron yield in the aerogel Cherenkov counters, and time-of-flight [2].

RESULTS

$B \to \pi\pi,\ K\pi,\ KK$

These modes are of interest since their branching ratios constrain the ϕ_3 angle [4]; the $\pi\pi$ mode is the most promising mode to determine the ϕ_2 angle. They have already been observed by CLEO, but more precise measurements are needed to determine the ϕ_3 and ϕ_2 angles.

Figure 2 left shows the M_{BC} and ΔE distributions for the three decay modes with all-charged final states [5] [1]. From the fits, the yields and branching fractions are determined, which are summarized in Table 1 along with results from CLEO and BaBar. The results of the three groups are consistent, and the hierarchy $\mathcal{B}(B^0 \to K^+K^-) < \mathcal{B}(B^0 \to \pi^+\pi^-) < \mathcal{B}(B^0 \to K^+\pi^-)$ has been confirmed.

$B \to \eta' K,\ \eta K^*$

These modes are of interest since CLEO has reported branching fractions larger than theoretical predictions [8]. To make the situation clear it is necessary to confirm their results experimentally. These modes may also be used to determine the ϕ_1 angle.

We reconstruct η' and η in the following modes: $\eta' \to \eta\pi^+\pi^- \to \gamma\gamma\pi^+\pi^-$, $\eta' \to \rho^0\gamma \to \pi^+\pi^-\gamma$ [9], and $\eta \to \gamma\gamma$, $\pi^+\pi^-\pi^0$ [10].

Figure 2 right shows the M_{BC} and ΔE distributions [2]. The branching fractions obtained are summarized in Table 1 along with other experimental and theoretical results.

The results of the three groups are nearly consistent, confirming that the branching fractions are really larger than theoretical predictions.

---

[1] Inclusion of the charge conjugate modes is implied in this talk.

[2] Reprinted from ref. [9], with permission from Elsevier Science.

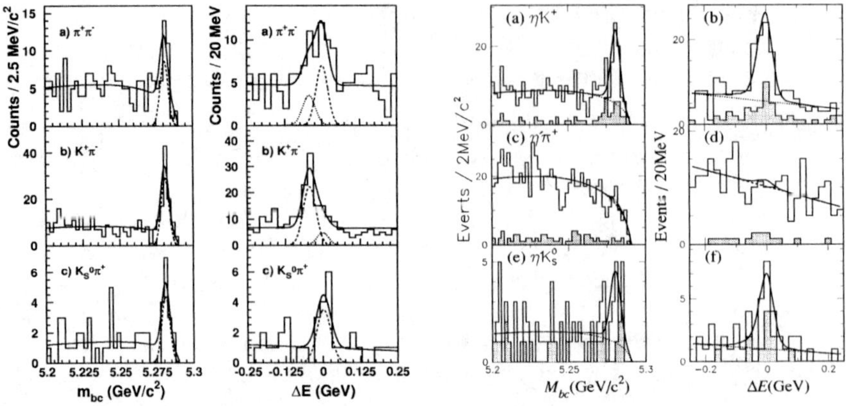

FIGURE 2. Left: M_{BC} and ΔE distributions for the $h\pi$ modes. The solid curves are the results of fits to data; dashed (dotted) curves represent signal (background) components. Right: M_{BC} and ΔE distributions for the $\eta' h$ modes. The shaded histograms are for the $\eta' \to \eta\pi^+\pi^-$ mode; open histograms are for the two modes combined; solid curves are the results of fits to data.

TABLE 1. Branching fractions obtained for the $hh^{(*)}$ modes. The first errors are statistical, and the second systematic. The numbers in [ ] represent statistical significances. Also shown are CLEO [6, 8, 15] and BaBar [7, 11, 16] results, and theoretical predictions [12].

Mode	$\mathcal{B}\,(10^{-6})$ (90% C.L.)			
	Belle	CLEO	BaBar	Theory
$\pi^+\pi^-$	$5.6^{+2.3}_{-2.0}\pm0.4$ [3.1]	$4.3^{+1.6}_{-1.4}\pm0.5$	$4.1\pm1.0\pm0.7$	
$\pi^+\pi^0$	<13.4	<12.7	<9.6	
$K^+\pi^-$	$19.3^{+3.4+1.5}_{-3.2-0.6}$ [7.8]	$17.2^{+2.5}_{-2.4}\pm1.2$	$16.7\pm1.6\pm1.3$	
$K^+\pi^0$	$16.3^{+3.5+1.6}_{-3.3-1.8}$ [7.2]	$11.6^{+3.0+1.4}_{-2.7-1.3}$	$10.8^{+2.1}_{-1.9}\pm1.0$	
$K^0\pi^+$	$13.7^{+5.7+1.9}_{-4.8-1.8}$ [3.5]	$18.2^{+4.6}_{-4.0}\pm1.6$	$18.2^{+3.3}_{-3.0}\pm2.0$	
$K^0\pi^0$	$16.0^{+7.2+2.5}_{-5.9-2.7}$ [3.9]	$14.6^{+5.9+2.4}_{-5.1-3.3}$	$8.2^{+3.1}_{-2.7}\pm1.2$	
K^+K^-	<2.7	<1.9	<2.5	
$K^+\bar{K}^0$	<5.0	<5.1	<2.4	
$\eta' K^+$	$79^{+12}_{-11}\pm9$ [12]	$80^{+10}_{-9}\pm7$	$70\pm8\pm5$	21 - 53
$\eta' K^0$	$55^{+19}_{-16}\pm8$ [5.4]	$89^{+18}_{-16}\pm9$	$42^{+13}_{-11}\pm4$	20 - 50
ηK^{*0}	$21.2^{+5.4}_{-4.7}\pm2.0$ [5.1]	$13.8^{+5.5}_{-4.6}\pm1.6$	$19.8^{+6.5}_{-5.6}\pm1.7$	2.0 - 8.2
ηK^{*+}	<49.9	$26.4^{+9.6}_{-8.2}\pm3.3$	<33.9	2.0 - 8.0
$\eta' \pi^+$	<7	<12	<12	1 - 3
$\eta\rho^0$	<5.5	<10		0.02 - 0.05
$\eta\rho^+$	<6.8	<15		4.2 - 11.6
ϕK^+	$11.2^{+2.2}_{-2.0}\pm1.4$ [8.5]	$5.5^{+2.1}_{-1.8}\pm0.6$	$7.7^{+1.6}_{-1.4}\pm0.8$	0.7 - 16
ϕK^0	$8.9^{+3.4}_{-2.7}\pm1.0$ [4.2]	<12.3	$8.1^{+3.1}_{-2.5}\pm0.8$	0.7 - 13
ϕK^{*0}	$13.0^{+6.4}_{-5.2}\pm2.1$ [3.6]	$11.5^{+4.5+1.8}_{-3.7-1.7}$	$8.6^{+2.8}_{-2.4}\pm1.1$	0.2 - 31
ϕK^{*+}	<19	<22.5	$9.7^{+4.2}_{-3.4}\pm1.7$	0.2 - 31

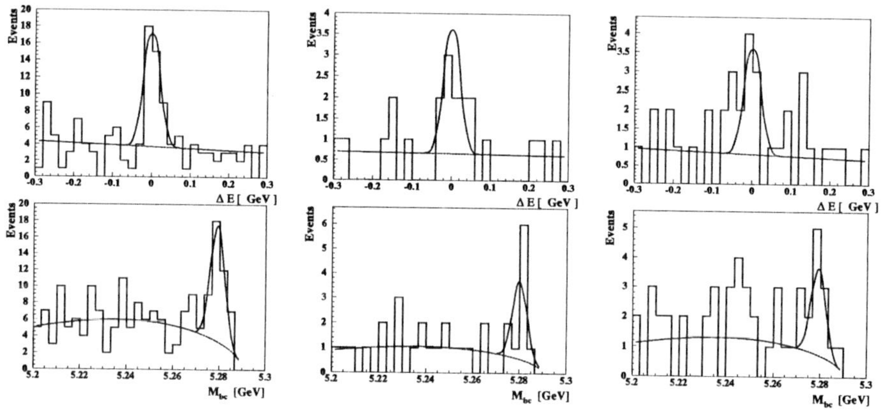

FIGURE 3. ΔE (top) and M_{BC} (bottom) distributions for the ϕK^+ (left), ϕK^0 (middle) and ϕK^{*0} (right) modes. The curves are the results of fits to data.

$B \to \phi K^{(*)}$

This decay is dominated by penguin ($b \to s$) diagrams in the SM. This mode is, therefore, sensitive to new physics as well as useful to extract information on V_{ts}. It can also be utilized to determine the ϕ_1 angle.

ϕ mesons are reconstructed in the K^+K^- decay mode [13]. Figure 3 shows the ΔE and M_{BC} distributions. The branching fractions obtained are summarized in Table 1.

Belle reported the first observation of a ϕK^+ signal at ICHEP2000 [14]. It is now confirmed with higher statistical significance. The ϕK^0 mode is also observed with a significance of 4.2σ. In addition, evidence for ϕK^{*0} is seen. The results of the three groups are nearly consistent, while theoretical predictions range widely, implying that some of the models are excluded.

$B^+ \to K^+\pi^-\pi^+,\ K^+K^-K^+$

It has been proposed that three-body decays can be used to determine the ϕ_3 angle [17]. We search for the three-body charmless decay $B^+ \to K^+\pi^-\pi^+$ [18]. Positive kaon identification is required for one of the tracks, and the other two tracks are required to be consistent with the pion hypothesis.

Figure 4a shows a Dalitz plot. The vertical band at ~ 3.5 (GeV/c^2)^2 is a background from $B^+ \to \bar{D}^0\pi^+ \to (K^+\pi^-)\pi^+$. The horizontal band at ~ 9.5 (14) (GeV/c^2)^2 is background from $J/\psi \to \mu^+\mu^-$ ($\psi(2S) \to \mu^+\mu^-$). These backgrounds are removed by applying cuts on the relevant invariant masses.

Figure 4b shows the ΔE distribution after background rejection. We see a clear peak corresponding to B-decays. From the fit to the ΔE distribution, the yield and branching fraction are determined to be 177 ± 20 events and

TABLE 2. Intermediate resonance contributions in the $K^+\pi^-\pi^+$ mode.

Mode	$\mathcal{B}_{B^+\to Rh^+} \times \mathcal{B}_{R\to h^+h^-}$ (10^{-6})	Significance
$K^*(892)^0\pi^+$	$11.1^{+2.5}_{-2.3}\pm1.4^{+2.0}_{-3.9}$	6.2
$K_X(1400)\pi^+$	$12.7^{+3.5}_{-3.4}\pm1.8^{+2.9}_{-5.8}$	4.1
$\rho^0(770)K^+$	< 9.6 (90% C.L.)	0.2
$f_0(980)K^+$	$11.7^{+2.5}_{-2.7}\pm1.5^{+4.1}_{-1.0}$	6.0
$f_X(1300)K^+$	$10.7^{+3.9}_{-3.5}\pm1.4^{+6.9}_{-2.8}$	3.2

$$\mathcal{B}(B^+ \to K^+\pi^-\pi^+) = (58.5 \pm 7.1 \pm 8.8) \times 10^{-6}.$$

This is the first observation of B^+ decaying into charmless $K^+\pi^-\pi^+$.

Intermediate resonances are searched for in the invariant masses of sub-systems. Shown in Figure 5 are the invariant masses of the $K^+\pi^-$ and $\pi^+\pi^-$ systems, where $M(\pi^+\pi^-) > 1.5$ GeV/c^2 and $M(K^+\pi^-) > 2$ GeV/c^2 are required, respectively, to remove the overlap regions in the Dalitz plot. A bump is seen at ~ 1.4 GeV/c^2 ($K_X(1400)$) along with a clear $K^*(892)^0$ peak. A clear $f_0(980)$ peak and a bump at ~ 1.3 GeV/c^2 ($f_X(1300)$) are also seen. [3]

To extract each resonance contribution, we perform a simple analysis: we divide the data into 7 regions in the Dalitz plot and fit the corresponding ΔE distributions with 5 decay modes simultaneously. Here $K_X(1400) = K_0^*(1430)$ and $f_X(1300) = f_0(1370)$ are assumed. No interference is taken into account in the fitting; its effect is evaluated separately using simulated events where interference is taken into account. Table 2 summarizes the results of the fit. The third errors represent the uncertainty coming from neglecting interference. The decays into $K^*(892)^0\pi^+$ and $f_0(980)K^+$ are the first observations. In particular, the latter is a new class of B modes, decays into a scalar and a pseudoscalar.

The same analysis is performed for the three kaon system, where positive kaon identification is required for all three tracks. Figure 6a shows a Dalitz plot. The vertical band at ~ 3.5 (GeV/c^2)^2 is a background from $B^+ \to \bar{D}^0K^+/\pi^+ \to (K^+K^-)K^+/\pi^+$. Figure 6b shows the ΔE distribution after background rejection. The yield and branching fraction are determined to be 162 ± 16 events and

$$\mathcal{B}(B^+ \to K^+K^-K^+) = (37.0 \pm 3.9 \pm 4.4) \times 10^{-6}.$$

Again this is the first observation of B^+ decaying into three kaons (approximately 20% is due to the contribution of $B^+ \to \phi K^+$).

[3] A peak at ~ 3.4 GeV/c^2 is χ_{c0}; see S. Suzuki's talk in this symposium.

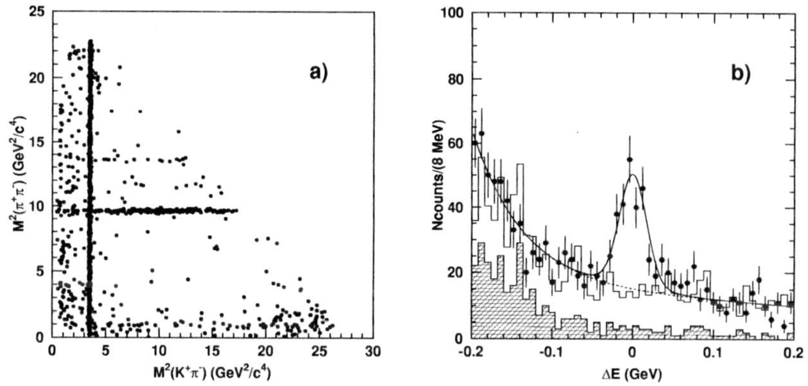

FIGURE 4. a) Dalitz plot, and b) ΔE distribution for the $K^+\pi^-\pi^+$ mode. The points with error bars are data, open histogram is the sum of off-resonance data and $B\bar{B}$ simulated events; hatched histogram shows the contribution of $B\bar{B}$ simulated events only; curve shows the fit to data.

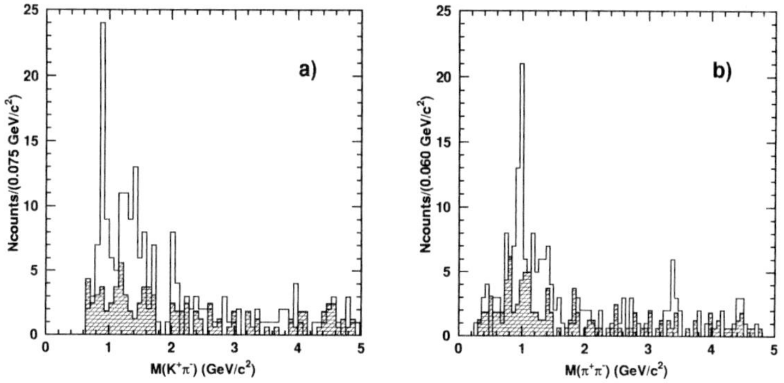

FIGURE 5. a) $K^+\pi^-$ mass spectrum for $M(\pi^+\pi^-) > 1.5$ GeV/c^2. b) $\pi^+\pi^-$ mass spectrum for $M(K^+\pi^-) > 2$ GeV/c^2. The open (hatched) histograms are for the signal region (ΔE sidebands).

$B^0 \to K^+\pi^-\pi^0$

We perform a similar analysis for the $K^+\pi^-\pi^0$ system [19]. The $\bar{D}^0 \to K^+\pi^-$ background is removed by applying a cut on the $K^+\pi^-$ invariant mass. The yield extracted by fitting is $105 \pm 23 \pm 15$ events with a significance of 4.8σ and the corresponding branching fraction is

$$\mathcal{B}(B^+ \to K^+\pi^-\pi^0) = (35.6^{+8.1}_{-7.7} \pm 5.2) \times 10^{-6}.$$

This is also the first observation of B^0 decaying into $K^+\pi^-\pi^0$. Intermediate resonance contributions are summarized in Table 3 along with results for three pion modes obtained

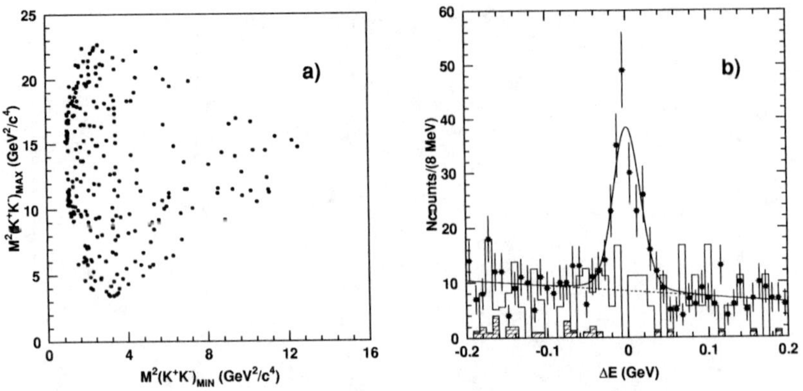

FIGURE 6. Same as Figure 4 but for the $K^+K^-K^+$ mode.

TABLE 3. Intermediate resonance contributions in the $K^+\pi^-\pi^0$ mode. Results for three pion modes are also shown.

Mode	$\mathcal{B}$ (10^{-6})	Significance
$K^*(892)^+\pi^-$	$26.0 \pm 8.3 \pm 3.5$	4.3
$\rho^- K^+$	$15.8^{+5.1+1.7}_{-4.6-3.0}$	4.2
$\rho^0\pi^0$	< 2.8 (90% C.L.)	-
$\rho^0\pi^+$	< 14.5 (90% C.L.)	1.9

in the same analysis.

Search for Direct CP Violation

Direct CP violation appears as a charge asymmetry between $\bar{B}$ and B decays:

$$\mathcal{A}_{\mathrm{CP}} = \frac{\mathcal{B}(\bar{B} \to \bar{f}) - \mathcal{B}(B \to f)}{\mathcal{B}(\bar{B} \to \bar{f}) + \mathcal{B}(B \to f)} \propto \sin\Delta\phi_S \sin\Delta\phi_W ,$$

where $\Delta\phi_S$ ($\Delta\phi_W$) is a strong (weak) phase difference between the two amplitudes. Under certain conditions, the asymmetry could be of order 10%.

We search for direct CP violation in the $K\pi$ and $\eta' K$ modes [20, 9]. The asymmetries obtained are summarized in Table 4. All the results are consistent with null asymmetry, hence we set 90% C.L. limits. Note that the systematic bias from charge asymmetry in the detectors is less than 1%, much smaller than the statistical errors at present.

TABLE 4. Results of searches for direct CP violation.

Mode	$\mathcal{A}_{CP}$	90% C.L.
$K^{\mp}\pi^{\pm}$	$0.044^{+0.186}_{-0.167}{}^{+0.018}_{-0.021}$	$-0.25 : 0.37$
$K^{\mp}\pi^{0}$	$-0.059^{+0.222}_{-0.196}{}^{+0.055}_{-0.017}$	$-0.40 : 0.36$
$K^0_S\pi^{\mp}$	$0.098^{+0.430}_{-0.343}{}^{+0.020}_{-0.063}$	$-0.53 : 0.82$
$K^{\mp}(\pi^{\pm}+\pi^0)$	$0.003^{+0.142}_{-0.126}{}^{+0.017}_{-0.014}$	$-0.22 : 0.25$
$\eta' K^{\mp}$	$0.06 \pm 0.15 \pm 0.01$	$-0.20 : 0.32$

SUMMARY

Recent results of charmless hadronic B decay analyses from Belle based on a 11 or 23 million $B\bar{B}$ sample have been presented.

- The following decay modes have been observed for the first time: $B \to K^+\pi^-\pi^+$, $K^+K^-K^+$, $K^+\pi^-\pi^0$, $K^*(892)^0\pi^+$, and $f_0(980)K^+$.
- Large $\mathcal{B}(B \to \eta' K)$ and $\mathcal{B}(B^0 \to \eta K^{*0})$ have been confirmed.
- $B^0 \to \phi K^0$ has been observed and evidence for $B^0 \to \phi K^{*0}$ has been seen.
- The hierarchy $\mathcal{B}(B^0 \to K^+K^-) < \mathcal{B}(B^0 \to \pi^+\pi^-) < \mathcal{B}(B^0 \to K^+\pi^-)$ has been confirmed.
- No evidence for direct CP violation has been seen.

We have collected data corresponding to 32 million $B\bar{B}$ events by the summer of 2001. Analyses based on this data sample are in progress.

REFERENCES

1. K. Abe *et al.* (Belle), *Phys. Rev. Lett.* **87**, 091802 (2001); B. Aubert *et al.* (BaBar), *Phys. Rev. Lett.* **87**, 091801 (2001).
2. K. Abe *et al.* (Belle) KEK Progress Report 2000-4 (2000).
3. KEKB B Factory Design Report, KEK Report 95-7 (1995), unpublished.
4. M. Neubert, hep-ph/0008072.
5. K. Abe *et al.* (Belle), *Phys. Rev. Lett.* **87**, 101801 (2001); 11 million $B\bar{B}$ sample.
6. D. Cronin-Hennessy *et al.* (CLEO), *Phys. Rev. Lett.* **85**, 515 (2000).
7. B. Aubert *et al.* (BaBar), SLAC-PUB-8838.
8. S. J. Richichi *et al.* (CLEO), *Phys. Rev. Lett.* **85**, 520 (2000).
9. K. Abe *et al.* (Belle), *Phys. Lett.* B **517**, 309 (2001); 11 million $B\bar{B}$ sample.
10. K. Abe *et al.* (Belle), Belle-CONF-0137; 23 $B\bar{B}$ million sample.
11. B. Aubert *et al.* (BaBar), SLAC-PUB-8914 and SLAC-PUB-8956.
12. Refs. [9, 10, 15] and references therein.
13. K. Abe *et al.* (Belle), Belle-CONF-0113; 23 million $B\bar{B}$ sample.
14. P. Chang (Belle), in Proc. of the 30th Int. Conf. on High Energy Phys. (ICHEP2000), edited by C. S. Lim and T. Yamanaka, World Scientific, Singapore (2001).
15. R. A. Briere *et al.* (CLEO), *Phys. Rev. Lett.* **86**, 3718 (2001).
16. B. Aubert *et al.* (BaBar), SLAC-PUB-8823.
17. R. E. Blanco *et al.* *Phys. Rev. Lett.* **86**, 2720 (2001).
18. K. Abe *et al.* (Belle), Belle-CONF-0114; 23 $B\bar{B}$ million sample.
19. K. Abe *et al.* (Belle), Belle-CONF-0115; 23 $B\bar{B}$ million sample.
20. K. Abe *et al.* (Belle), *Phys. Rev.* D **64**, 071101(R); 11 $B\bar{B}$ million sample.

Charmless Hadronic B Decays at BABAR

Carlo Dallapiccola representing the BABAR Collaboration

Dept. of Physics, University of Massachusetts, Amherst, MA 01003

Abstract. Results are presented for measurements of branching fractions and searches for direct CP violation in rare, charmless, hadronic B decays. The data were collected with the BABAR detector at the PEP-II asymmetric e^+e^- collider and consist of about 22.7 million $B\bar{B}$ pairs. The B mesons are reconstructed in two-body and quasi two-body final states with pions, kaons, η, η', ω, a_0, K^*, and ϕ resonances. Three-body final states involving ρ and K^* resonances are also studied.

INTRODUCTION

The study of B meson decays into charmless, hadronic final states plays an important role in the understanding of CP violation. In the Standard Model, all CP-violating phenomena are a consequence of a single complex phase in the Cabibbo-Kobayashi-Maskawa (CKM) quark-mixing matrix [1]. Significant CP-violating asymmetries have been observed in the K meson system [2] and, more recently, in B decays to charmonium [3]. Direct CP violation is due to the interference among decay amplitudes which differ in both weak and strong phases and has not yet been observed in the B system. Direct CP violation would be evident in a non-zero asymmetry of B decay rates:

$$\mathcal{A}_{CP} = \frac{\Gamma(\bar{B} \to \bar{f}) - \Gamma(B \to f)}{\Gamma(\bar{B} \to \bar{f}) + \Gamma(B \to f)}.$$

Charmless decays are manifestations of penguin and/or suppressed tree amplitudes. In many modes, the penguin (P) and tree (T) amplitudes are of similar size and substantial CP violation could arise from interference of these two terms[4]:

$$\mathcal{A}_{CP} = \frac{2|P||T|\sin\Delta\phi\sin\Delta\delta}{|P|^2 + |T|^2 + 2|P||T|\cos\Delta\phi\cos\Delta\delta},$$

where $\Delta\phi$ and $\Delta\delta$ are the differences in the weak and strong phases of the two amplitudes. The weak phase difference between the $b \to u$ tree amplitude T and the penguin amplitude P is the angle γ of the Unitarity Triangle, for the case of $b \to s$ penguins, or $\gamma + \beta$, for the case of $b \to d$ penguins. Thus, $\mathcal{A}_{CP}$ is sensitive to the CKM angle $\gamma \equiv \arg[-V_{ud}V_{ub}^*/V_{cd}V_{cb}^*]$ or $\alpha \equiv \arg[-V_{td}V_{tb}^*/V_{ud}V_{ub}^*]$, depending on the decay mode studied. Recent calculations based on effective theory and factorization predict asymmetries as large as 10% [5], but the uncertainties in the strong phases are large, which makes it difficult to extract the weak phase differences from the observed asymmetries. In those modes which receive substantial penguin contributions, the CP asymmetry may

CP618, *Heavy Flavor Physics: Ninth International Symposium*, edited by A. Ryd and F. C. Porter

be enhanced, to values of 30% or larger [6], due to the presence of new particles in the loops, such as charged Higgs or SUSY particles.

In addition to being an ideal system for the study of direct CP violation, charmless hadronic B^0 decays can be used to probe the time-dependent CP-violating asymmetries that arise from the interference between mixing and decay amplitudes. For the decays $B^0 \to \pi^+\pi^-$ and $B^0 \to \phi K_S^0$, for instance, the final states are CP eigenstates and the time-dependent rate asymmetries

$$\mathcal{A}_{CP}(\Delta t) \equiv \frac{\Gamma_{B^0}(\Delta t) - \Gamma_{\bar{B}^0}(\Delta t)}{\Gamma_{B^0}(\Delta t) + \Gamma_{\bar{B}^0}(\Delta t)}$$

are sensitive to the angles α and $\beta \equiv \arg[-V_{cd}V_{cb}^*/V_{td}V_{tb}^*]$, respectively, where Γ_{B^0} ($\Gamma_{\bar{B}^0}$) is the decay rate when the other B in the event is known to be a B^0 ($\bar{B}^0$) at its decay time, and Δt is the time difference for the two B decays.

ANALYSES

The data were collected with the *BABAR* detector [7] at the PEP-II e^+e^- collider at the Stanford Linear Accelerator Center. Most of the results presented in this paper are based on data taken in the 1999–2000 run. An integrated luminosity of 20.7 fb^{-1} was recorded at the $\Upsilon(4S)$ resonance, corresponding to 22.7 million $B\bar{B}$ pairs ("on-resonance"), with an additional 2.6 fb^{-1} about 40 MeV below this energy ("off-resonance") for the study of continuum backgrounds. The collider is operated with asymmetric beam energies, producing a boost ($\gamma\beta = 0.56$) of the $\Upsilon(4S)$ along the collision axis (z). The boost increases the momentum range of two-body B decay products from a narrow distribution centered near 2.6 GeV/c to a broad distribution extending from about 1.7 to 4.3 GeV/c.

A detailed description of the *BABAR* detector is presented in Ref. [7]. Charged particle momenta are measured in a tracking system consisting of a 5-layer, double-sided, silicon vertex tracker (SVT) and a 40-layer drift chamber (DCH) filled with a gas mixture of helium and isobutane, both operating within a 1.5 T superconducting solenoidal magnet. The typical decay vertex resolution for a fully reconstructed B decay is approximately 65 μm along the boost direction, z. Photons are detected in an electromagnetic calorimeter (EMC) consisting of 6580 CsI (Tl) crystals arranged in barrel and forward endcap subdetectors. The iron flux return (IFR) is segmented and instrumented with multiple layers of resistive plate chambers for the identification of muons and long-lived neutral hadrons.

Charged particle identification (PID) is provided by the average energy loss (dE/dx) in the tracking devices and by a unique, internally reflecting ring imaging Cherenkov detector (DIRC) covering the central region. A Cherenkov angle K–π separation of better than 4 standard deviations (σ) is achieved for tracks below 3 GeV/c momentum, decreasing to 2.5σ at the highest momenta in the final states considered here. Electrons and muons are identified with use of the EMC and IFR, respectively.

Candidate K_S^0 mesons are reconstructed in the final state $\pi^+\pi^-$ from pairs of oppositely-charged tracks that form a well-measured vertex and have an invariant mass within about 3.5σ of the nominal K_S^0 mass [8]. Additional selection requirements are placed on the decay length and angle between the flight and momentum vectors projected perpendicular to the beam axis.

Candidate π^0 mesons are formed from pairs of photons with invariant mass within about 3σ of the nominal π^0 mass. Only photons with energies above 30-50 MeV, with the exact value varying for different analyses, are used.

Two-Body and Quasi-Two-Body *B* Decays

The analyses of two-body and quasi-two-body charmless *B* decays have many aspects in common. Relatively loose selection criteria are placed on fully reconstructed *B* candidates in the decay mode of interest and the quantities which best discriminate between signal and background are used in an unbinned maximum likelihood fit that determines the signal yields. The yields are corrected by selection efficiencies determined from Monte Carlo simulation in order to produce branching fractions. In this paper, branching fractions, and, in some cases, *CP* asymmetries, are reported for the following decay modes:

- $\pi^+\pi^-$, $K^+\pi^-$, K^+K^-, $\pi^+\pi^0$, $K^+\pi^0$, $K^0\pi^+$, $\bar{K}^0K^+$, $K^0\pi^0$ and $K^0\bar{K}^0$;
- $\eta'K^+$, $\eta'\pi^+$, $\eta'K^0$, ηK^{*+}, ηK^{*0}, ωK^+, $\omega\pi^+$, ωK^0 and $\omega\pi^0$;
- ϕK^+, $\phi\pi^+$, ϕK^0, ϕK^{*0} and ϕK^{*+};
- $a_0^\pm\pi^\mp$.

The η' are reconstructed in the modes $\rho^0\gamma$, with $\rho^0 \to \pi^+\pi^-$, and $\eta\pi^+\pi^-$, with $\eta \to \gamma\gamma$. The ω are reconstructed in the final state $\pi^+\pi^-\pi^0$, and the ϕ in the final state K^+K^-. The K^{*0} are reconstructed in the mode $K^+\pi^-$ and the K^{*+} in the modes $K_S^0\pi^+$ and $K^+\pi^0$ (except for the ηK^{*+} analysis, which uses only $K^{*+} \to K_S^0\pi^+$ candidates). The $a_0^\pm$ resonance is identified in the mode $\eta\pi^\pm$.

In all cases, the *B* mesons are characterized by two kinematic observables that are only very weakly correlated. The energy-substituted mass variable, $m_{\rm ES}$, makes use of the fact that in the $\Upsilon(4S)$ frame the *B* meson energy $E_{\rm b}^*$ equals the beam energy. It is computed using only momenta measured in the lab frame and is given by $m_{\rm ES} = \sqrt{E_{\rm b}^2 - \mathbf{p}_B^2}$, where $E_{\rm b} = (s/2 + \mathbf{p}_i \cdot \mathbf{p}_B)/E_i$, $\sqrt{s}$ and E_i are the total energies of the e^+e^- system in the CM and lab frames, respectively, and $\mathbf{p}_i$ and $\mathbf{p}_B$ are the momentum vectors in the lab frame of the e^+e^- system and the *B* candidate, respectively. The calculation of $m_{\rm ES}$ only involves the three-momenta of the decay products, and is therefore independent of the masses assigned to them. The resolution on this quantity is dominated by the beam energy spread and is approximately $2.5\,{\rm MeV}/c^2$.

The second kinematic parameter ΔE is the difference between the energy of the *B* candidate and half the energy of the e^+e^- system. The resolution on ΔE is very mode dependent and varies in the range 25-45 MeV. Candidates with $m_{\rm ES} > 5.2\,{\rm GeV}/c^2$ and $|\Delta E| < 0.20\,{\rm GeV}$ are selected for use in the fit (the ΔE selection varies somewhat from mode to mode but is approximately $|\Delta E| < 0.20\,{\rm GeV}$ in all cases).

Charmless hadronic modes suffer from large backgrounds due to random combinations of tracks produced in the quark-antiquark ($q\bar{q}$) continuum. Monte Carlo simulation demonstrates that contamination from other B decays is negligible. The distinguishing feature of continuum backgrounds is their characteristic event shape resulting from the two-jet production mechanism. A variety of event shape variables defined in the center of mass system (CM) have been developed to exploit this difference.

One such variable is the angle θ_T between the thrust axis of the B candidate and the thrust axis of the rest of the event. This angle is small for continuum events, where the B candidate daughters tend to lie in the $q\bar{q}$ jets, and uniformly distributed for true $B\bar{B}$ events. Events are selected that satisfy $|\cos\theta_T| < 0.9$, or $|\cos\theta_T| < 0.8$, depending on the analysis.

Other topological variables used for background discrimination include the B polar angle θ_B (for $\Upsilon(4S)$ decays into two pseudoscalar B mesons, the θ_B distribution has a $\sin^2\theta_B$ dependence, whereas background is flat), the angle $\theta_{q\bar{q}}$ between the B-candidate thrust axis and the beam axis (follows a $1+\cos^2\theta_{q\bar{q}}$ distribution for background and is flat for signal), and the sum of track and neutral EMC cluster energies in concentric cones of 10° centered on the thrust axis of the B candidate (signal displays a more uniform distribution of energy flow in the cones than does continuum background). These discriminating variables are combined using either a neural network, NN, (the $a_0^{\pm}\pi^{\mp}$ analysis) or a Fisher discriminant $\mathcal{F}$, which is an optimized linear combination (all other analyses).

Signal yields are determined from unbinned maximum likelihood fits. Separate fits are performed for each of the various decay topologies listed above, where the likelihood for a given candidate j is given as:

$$\mathcal{L} = \exp\left(-\sum_{i=1}^{M} n_i\right) \prod_{j=1}^{N} \left[\sum_{i=1}^{M} n_i \mathcal{P}_i\left(\vec{x}_j; \vec{\alpha}_i\right)\right]. \tag{1}$$

The quantity $\mathcal{P}_i(\vec{x}_j;\vec{\alpha}_i)$ describes the probability for candidate j to belong to category i, based on its measured variables $\vec{x}_j$ and parameters $\vec{\alpha}_i$ that describe the expected distributions of these variables in each of the M categories. In the simplest case, the probabilities are summed over two categories ($M = 2$), signal and background. For modes in which the final state tags the decay flavor of the B (B^0 or $\bar{B}^0$) the yields are rewritten in terms of the sum $n_f + n_{\bar{f}}$ and the asymmetry $\mathcal{A}_i = (n_{\bar{f}} - n_f)/(n_{\bar{f}} + n_f)$, where $n_f\,(n_{\bar{f}})$ is the fitted number of events in the mode $B \to f\,(\bar{B} \to \bar{f})$. The event yields n_i and, where appropriate, asymmetries $\mathcal{A}_i$, in each category are determined by maximizing $\mathcal{L}$.

The probabilities $\mathcal{P}_i(\vec{x}_j;\vec{\alpha}_i)$ are evaluated as products of probability density functions (PDFs) for each of the independent input variables $\vec{x}_j$. Monte Carlo simulation is used to validate the assumption that the fit variables are uncorrelated. These input variables are $m_{\rm ES}$, ΔE, $\mathcal{F}$ (NN), and invariant masses of any intermediate resonance states (η', η, ω, ϕ, K^* and a_0^+). The ϕ and ω helicity angles are included for pseudoscalar-vector decays. For modes with at least one charged primary daughter, $B^{\pm} \to X^0 h^{\pm}$ or $B^0 \to h^{\pm} h'^{\mp}$ (h, $h' = \pi$ or K, and $X^0 = \eta'$, ω or ϕ) the π and K components are fit simultaneously.

The parameters for background $m_{\rm ES}$ and ΔE PDFs are determined from events in on-resonance ΔE sideband regions. The signal $m_{\rm ES}$ and ΔE PDF parameters are determined

TABLE 1. Summary of results for fitted signal yields (N_S), statistical significances (S), measured branching fractions ($\mathcal{B}$), 90% confidence levels (CL) for $\mathcal{B}$ measurements, and charge asymmetries ($\mathcal{A}_{CP}$).

Mode	N_S	S (σ)	$\mathcal{B}(10^{-6})$	$\mathcal{B}$ 90% CL	$\mathcal{A}_{CP}$
$\pi^+\pi^-$	$41\pm10\pm7$	4.7	$4.1\pm1.0\pm0.7$		
$K^+\pi^-$	$169\pm17\pm13$	15.8	$16.7\pm1.6\pm1.3$		$-0.19\pm0.10\pm0.03$
K^+K^-	$8.2^{+7.8}_{-6.4}\pm3.5$	1.3	$0.85^{+0.81}_{-0.66}\pm0.37$	<2.5	
$\pi^+\pi^0$	$37\pm14\pm6$	3.4	$5.1^{+2.0}_{-1.8}\pm0.8$	<9.6	
$K^+\pi^0$	$75\pm14\pm7$	8.0	$10.8^{+2.1}_{-1.9}\pm1.0$		$0.00\pm0.18\pm0.04$
$K^0\pi^+$	$59^{+11}_{-10}\pm6$	9.8	$18.2^{+3.3}_{-3.0}\pm2.0$		$-0.21\pm0.18\pm0.03$
$\bar{K}^0K^+$	$-4.1^{+4.5}_{-3.8}\pm2.3$	–	$-1.3^{+1.4}_{-1.0}\pm0.7$	<2.4	
$K^0\pi^0$	$17.9^{+6.8}_{-5.8}\pm1.9$	4.5	$8.2^{+3.1}_{-2.7}\pm1.2$		
$K^0\bar{K}^0$	$3.4^{+3.4}_{-2.4}\pm3.5$	1.5	$1.8^{+1.8}_{-1.2}\pm1.8$	<7.3	
ηK^{*0}	20.5 ± 6.0	5.4	$19.8^{+6.5}_{-5.6}\pm1.7$		
ηK^{*+}	14.3 ± 6.6	3.2	$22.1^{+11.1}_{-9.2}\pm3.3$	<33.9	
$\eta'_{\eta\pi\pi}K^+$	$49.5^{+8.1}_{-7.3}$	15	63^{+10}_{-9}		
$\eta'_{\rho\gamma}K^+$	$87.6^{+13.4}_{-12.5}$	11	80^{+12}_{-11}		
$\eta'K^+$		17	$70\pm8\pm5$		$-0.11\pm0.11\pm0.02$
$\eta'_{\eta\pi\pi}K^0$	$6.3^{+3.3}_{-2.5}$	4.7	28^{+15}_{-11}		
$\eta'_{\rho\gamma}K^0$	$20.8^{+7.4}_{-6.5}$	4.2	61^{+22}_{-19}		
$\eta'K^0$		5.9	$42^{+13}_{-11}\pm4$		
$\eta'_{\eta\pi\pi}\pi^+$	$5.7^{+3.8}_{-2.8}$	3.2	$7.1^{+4.8}_{-3.5}$		
$\eta'_{\rho\gamma}\pi^+$	$-0.9^{+7.8}_{-6.2}$	0.1	$-0.7^{+6.7}_{-5.3}$		
$\eta'\pi^+$		2.8	$5.4^{+3.5}_{-2.6}\pm0.8$	<12	
ωK^+	$6.4^{+5.6}_{-4.4}$	1.3	$1.4^{+1.3}_{-1.0}\pm0.3$	<4	
ωK^0	$8.1^{+4.6}_{-3.6}$	3.2	$6.4^{+3.6}_{-2.8}\pm0.8$	<13	
$\omega\pi^+$	$27.6^{+8.8}_{-7.7}$	4.9	$6.6^{+2.1}_{-1.8}\pm0.7$		$-0.01^{+0.29}_{-0.31}\pm0.03$
$\omega\pi^0$	$-0.9^{+5.0}_{-3.2}$	–	$-0.3\pm1.1\pm0.3$	<3	
ϕK^+	$31.4^{+6.7}_{-5.9}$	10.5	$7.7^{+1.6}_{-1.4}\pm0.8$		$-0.05\pm0.20\pm0.03$
ϕK^0	$10.8^{+4.1}_{-3.3}$	6.4	$8.1^{+3.1}_{-2.5}\pm0.8$		
$\phi K^{*+}_{K^+}$	$7.1^{+4.3}_{-3.4}$	2.7	$12.8^{+7.7}_{-6.1}\pm3.2$		
$\phi K^{*+}_{K^0}$	$4.4^{+2.7}_{-2.0}$	3.6	$8.0^{+5.0}_{-3.7}\pm1.3$		
ϕK^{*+}		4.5	$9.7^{+4.2}_{-3.4}\pm1.7$		$-0.43^{+0.36}_{-0.30}\pm0.06$
ϕK^{*0}	$20.8^{+5.9}_{-5.1}$	7.5	$8.7^{+2.5}_{-2.1}\pm1.1$		$0.00\pm0.27\pm0.03$
$\phi\pi^+$	$0.9^{+2.1}_{-0.9}$	0.6	$2.1^{+4.9}_{-2.1}\pm0.5$	<1.4	
$a_0^{\pm}\pi^{\mp}$	$18.1^{+8.7}_{-7.4}$	3.7	$6.2^{+3.0}_{-2.5}\pm1.1$	<11.5	

from fully reconstructed $B^+\to\bar{D}^0\pi^+$ and $B^+\to\bar{D}^0\rho^+$ ($\rho^+\to\pi^+\pi^0$) decays. Events in on-resonance m_{ES} sideband regions and Monte Carlo simulated signal decays are used to parameterize the Fisher discriminant PDFs for background and signal, respectively

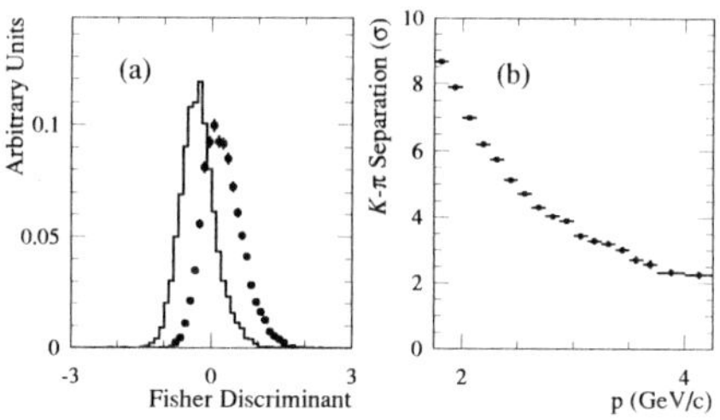

FIGURE 1. (a) The distributions of the Fisher discriminant for Monte Carlo simulated $B^0 \to h^+h'^-$ decays (histogram) and background events (points) in the $m_{\rm ES}$ sideband region $5.20 < m_{\rm ES} < 5.27\,{\rm GeV}/c^2$; (b) the K–π separation, in units of standard deviations, as a function of momentum, derived from the Cherenkov angle measurements of kaon and pion tracks in a $D^{*+} \to D^0\pi^+$ control sample, as described in the text.

(see Fig. 1(a)). Alternative parameterizations obtained from off-resonance data and Monte Carlo simulation are used as cross-checks and for determination of systematic uncertainties. The θ_c PDFs are derived from kaon and pion tracks in the momentum range of interest from approximately 42000 $D^{*+} \to D^0\pi^+$ ($D^0 \to K^-\pi^+$) decays. This control sample is used to parameterize the θ_c resolution σ_{θ_c} as a function of track polar angle. The resulting K–π separation, defined as $|\theta_c^K - \theta_c^\pi|/\sigma_{\theta_c}$, where θ_c^K (θ_c^π) is the expected Cherenkov angle for a kaon (pion), is shown as a function of momentum in Fig. 1(b).

The results [9] of the maximum likelihood fits are given in Table 1. For self-tagging modes in which significant signals are observed, the measured charge asymmetry $\mathcal{A}_{CP}$ is also given. Figures 2-3 show distributions of $m_{\rm ES}$ and ΔE for a few representative decay modes, where cuts have been placed on likelihood ratios to enhance the signal-to-background.

The analysis of all-charged, two-body decays $B^0 \to h^+h'^-$ ($h = \pi$ or K) has been updated [10] using additional data taken in 2001. This dataset amounts to $30.4\,{\rm fb}^{-1}$ collected on-resonance, which is about 50% more data than that which is used in the other measurements summarized in Table 1. The $B^0 \to K^+\pi^-$ charge asymmetry is measured to be

$$\mathcal{A}_{CP}(K^+\pi^-) = -0.07 \pm 0.08 \pm 0.02\,.$$

This measurement supersedes that given in Table 1. In addition, the flavor of the other B meson in the event is identified and the time difference between the two B decays is reconstructed, which allows for the first ever measurement of time-dependent CP-violating asymmetries in $B^0 \to \pi^+\pi^-$ decays [10]. That measurement is reported elsewhere in these proceedings [11].

Three-Body *B* Decays

The three-body decay modes $B^0 \to \pi^+\pi^-\pi^0$ and $B^+ \to K^{*0}\pi^+$, with $K^{*0} \to K^+\pi^-$, have also been studied. Unlike the analyses described above, a cut-based method has

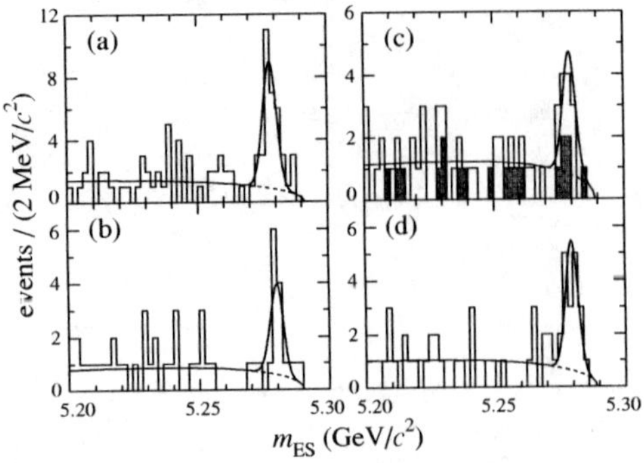

FIGURE 2. Projections onto the variable m_{ES}. The histograms show data for (a) $B^+ \to \phi K^+$; (b) $B^0 \to \phi K^0$; (c) $B^+ \to \phi K^{*+}$; (d) $B^0 \to \phi K^{*0}$ after a requirement on the signal probability $\mathcal{P}_{sig}/\Sigma\mathcal{P}_i$ with the PDF for m_{ES} excluded. In (c) the histogram is the sum of the two ϕK^{*+} channels while the shaded area is $K^{*+} \to K^0\pi^+$ alone. The solid (dashed) line shows the PDF projection of the full fit (background only).

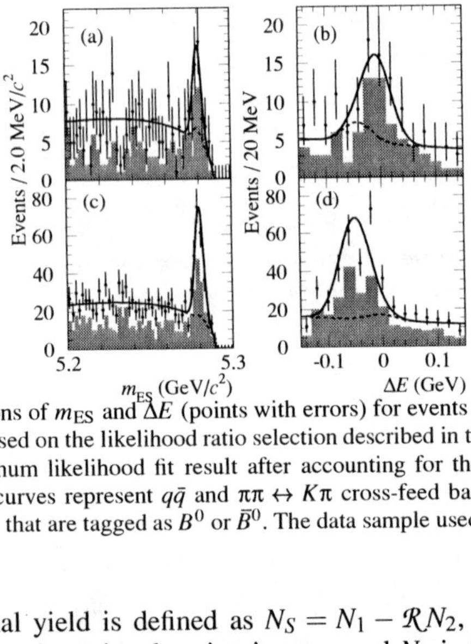

FIGURE 3. Distributions of m_{ES} and ΔE (points with errors) for events enhanced in signal (a), (b) $\pi\pi$ and (c), (d) $K\pi$ decays based on the likelihood ratio selection described in the text. Solid curves represent projections of the maximum likelihood fit result after accounting for the efficiency of the additional selection, while dashed curves represent $q\bar{q}$ and $\pi\pi \leftrightarrow K\pi$ cross-feed background. Shaded histograms show the subset of events that are tagged as B^0 or $\bar{B}^0$. The data sample used is 30.4 fb^{-1}.

been used. The signal yield is defined as $N_S = N_1 - \mathcal{R}N_2$, where N_1 is the number of observed candidates in a signal region in m_{ES}, and N_2 is the number of candidates observed in a sideband region. The quantity $\mathcal{R}$ is the ratio of background in the signal region to that in the sideband region.

Data for the $B^0 \to \pi^+\pi^-\pi^0$ final state can be represented on a Dalitz plot: we subdivide the Dalitz plot into distinct regions, each of which are chosen to be sensitive to a single resonance such as $\rho(770)$, $\rho(1450)$, etc. The regions are defined using selection criteria for the invariant mass of $\pi\pi$ combinations and the pair helicity angle, defined as the angle between the direction of one of the pions and the direction of the parent B candidate computed in the $\pi\pi$ rest frame. For the $B^+ \to K^{*0}\pi^+$ analysis, events are selected for which $K^+\pi^-$ combinations are consistent with the K^{*0} mass; no other regions of the Dalitz plot are explored. The results for these analyses are given in Table 2, where significant signals are observed for $B^0 \to \rho^\pm(770)\pi^\mp$ and $B^+ \to K^{*0}\pi^+$

TABLE 2. Summary of results for signal yields (N_S), statistical significances and measured branching fractions ($\mathcal{B}$). Upper limits are at 90% CL.

Mode	N_S	Stat. Sig. (σ)	$\mathcal{B}(10^{-6})$
$\rho^\pm(770)\pi^\mp$	$89 \pm 16 \pm 6$	5.0	$28.9 \pm 5.4 \pm 4.3$
$\rho^0(770)\pi^0$	$6.1 \pm 5.8 \pm 2.8$	1.0	< 10.6
$\pi^+\pi^-\pi^0$(NR)	$-4.2 \pm 7.3 \pm 3.8$	-	< 7.3
$K^{*0}\pi^+$	$34.8 \pm 7.6 \pm 1.1$	6.0	$15.5 \pm 3.4 \pm 1.8$

SUMMARY

We have made measurements of branching fractions and performed searches for direct *CP* violation in a large variety of rare, charmless, hadronic *B* decays. The *B* mesons were reconstructed in two-body and quasi two-body final states with pions, kaons, η, η', ω, a_0, K^*, and ϕ resonances. Three-body final states involving ρ and K^* resonances have also been studied. Summary of results are presented in Tables 1 and 2.

REFERENCES

1. M. Kobayashi and T. Maskawa, *Prog. Th. Phys.* **49**, 652 (1973). N. Cabibbo, *Phys. Rev. Lett.* **10**, 531 (1963);
2. J. H. Christenson, J. Cronin, V. Fitch, and R. Turlay, *Phys. Rev. Lett.* **13**, 138 (1964).
3. *BABAR* Collaboration, B. Aubert *et al.*, *Phys. Rev. Lett.* **87**, 091801 (2001); BELLE Collaboration, K. Abe *et al.*, *Phys. Rev. Lett.* **87**, 091802 (2001).
4. M. Bander, D. Silverman, and A. Soni, *Phys. Rev. Lett.* **43**, 242 (1979).
5. A. Ali, G. Kramer, and C.-D. Lü, *Phys. Rev. D* **59**, 014005 (1998); G. Kramer, W.F. Palmer, and H. Simma, *Nucl. Phys. B* **428**, 77 (1994).
6. I. Hinchliffe and N. Kersting, *Phys. Rev. D* **63**, 015003 (2001); X.-G. He, W.-S. Hou, and K.-C. Yang, *Phys. Rev. Lett.* **81**, 5738 (1998).
7. *BABAR* Collaboration, B. Aubert *et al.*, SLAC-PUB-8569, hep-ex/0105044, to appear in *Nucl. Instrum. and Methods*.
8. Particle Data Group, D.E. Groom *et al.*, *Eur. Phys. J. C* **15**, 1 (2000).
9. *BABAR* Collaboration, B. Aubert *et al.*, hep-ex/0111087, to appear in *Phys. Rev. D*. *BABAR* Collaboration, B. Aubert *et al.*, *Phys. Rev. Lett.* **87**, 221802 (2001); *BABAR* Collaboration, B. Aubert *et al.*, *Phys. Rev. Lett.* **87**, 151802 (2001); *BABAR* Collaboration, B. Aubert *et al.*, *Phys. Rev. Lett.* **87**, 151801 (2001).
10. *BABAR* Collaboration, B. Aubert *et al.*, hep-ex/0110062, to appear in *Phys. Rev. D*.
11. S. Prell, contribution to these proceedings.

Hadronic *B* Decays at *BABAR*

Enrico Robutti representing the *BABAR* collaboration

University and I.N.F.N. Genova, Via Dodecaneso, 33 – 16146 Genova, Italy

Abstract. The *BABAR* collaboration has performed a number of measurements on hadronic B decays to charmonium and open charm mesons. Preliminary results based on a sample of nearly 23 million $B\bar{B}$ pairs, collected between October 1999 and October 2000, are reviewed. These include measurements of exclusive branching fractions and ratios of branching fractions, as well as angular distributions.

INTRODUCTION

Hadronic decays account for a large fraction (between ~60% and ~75%) of B meson decays. They mainly proceed through $b \to c\bar{u}d$ and $b \to c\bar{c}s$ processes[1], whence consequent hadronization leads to the formation of charmed mesons and, to a lesser extent, charmonium states and charmed baryons.

Measurement of branching fractions, momentum and angular distributions for exclusive and inclusive decays of this kind is a key to a better understanding of the underlying dynamics at the quark level and a test for phenomenological models such as the factorization hypothesis and non-relativistic QCD. Moreover, many of these channels are of particular interest in the detection of *CP* violation effects.

The *BABAR* collaboration has performed a number of such measurements using a sample of $B\bar{B}$ pairs produced from $\Upsilon(4S)$ decays at the PEP II asymmetric e^+e^- storage ring at the Stanford Linear Accelerator Center. Results reported here are based on data collected with the *BABAR* detector [1] between October 1999 and October 2000, for an integrated luminosity of 20.7 fb^{-1}, corresponding to $(22.74 \pm 0.36) \times 10^6$ $B\bar{B}$ pairs, at the $\Upsilon(4S)$ resonance peak, and 2.6 fb^{-1} at 40 MeV below the peak.

EVENT SELECTION

Although results presented in the next sections are the outcome of several independent analyses, the techniques employed for event selection are often common. The main criteria used are briefly described in the following.

A detailed description of the *BABAR* detector can be found in [1]. After charged tracks and neutral cluster are reconstructed, a set of track multiplicity and event shape cuts selects multi-hadron events from $B\bar{B}$ decays while rejecting large part of the continuum background.

[1] Charge conjugate states are implied throughout this paper.

CP618, *Heavy Flavor Physics: Ninth International Symposium,* edited by A. Ryd and F. C. Porter

Reconstructed objects are then assigned particle types (non-exclusively) based on the information provided by all sub-detectors and on PID criteria with different levels of efficiency and purity.

"Composite" mesons are formed by combining sets of tracks and/or neutral clusters, whose mass is fixed to that of the particle type they are assigned to. A vertex constraint is usually applied before computing the invariant mass, which is then required to be compatible, within resolution, with the known meson mass. In some cases, additional criteria are applied in order to reject specific background modes or gain efficiency (*e.g.* bremsstrahlung recovery in $J/\psi \to e^+ e^-$, π^0 rejection in $\chi_{c1} \to J/\psi\gamma$). If the reconstructed meson is itself an intermediate state in the decay chain, its mass is constrained to the known mass before iterating the procedure.

Table 1 lists all decay modes used for meson reconstruction in all B decay channel considered in this paper.

Extraction of the B signal mainly relies on a pair of kinematical variables, exhibiting high discriminating power against combinatorial background and little correlation between each other:

$$\Delta E = E_B^* - \sqrt{s}/2\,; \tag{1}$$

$$m_{ES} = \sqrt{s/4 - p_B^{*2}}\,. \tag{2}$$

They make use of the reconstructed B energy and momentum in the center-of-mass frame, and of the total center of mass energy, known to a very good precision from the beam parameters. The signal is expected to cluster around $\Delta E = 0$ and $m_{ES} = m_B$.

Candidates are considered whose ΔE and m_{ES} values lie in a broad neighborhood of this point in a two dimensional plot ($|\Delta E| < \Delta E_{max}$, with ΔE_{max} tipically ~100 – 200 MeV, and 5.2 GeV/c^2 < m_{ES} < 5.3 GeV/c^2); a signal region is defined in a much smaller neighborhood, while large sidebands away from it are used for background shape evaluation. Where "cut and count" techniques are used (as opposed to likelihood fits) the background is fitted to an ARGUS[2] shape in m_{ES} and to a polynomial in ΔE in the sideband region: the extrapolated yield to the signal region is then subtracted from the total number of events found inside it.

TABLE 1. Decay channels used for the reconstruction of mesons.

***uds* mesons**	**charmonium**	**charmed mesons**
$\pi^0 \to \gamma\gamma$	$J/\psi \to e^+ e^-, \mu^+ \mu^-$	$D^+ \to K^- \pi^+ \pi^+, K_S \pi^+, K^- K^+ \pi^+$
$K_S^0 \to \pi^+ \pi^-, \pi^0 \pi^0$	$\psi(2S) \to e^+ e^-, \mu^+ \mu^-$	$D^0 \to K^- \pi^+, K^- \pi^+ \pi^0, K^- \pi^+ \pi^+ \pi^-, K_S \pi^+ \pi^-$
$K^{*+} \to K_S \pi^+, K^+ \pi^0$	$\psi(2S) \to J/\psi \pi^+ \pi^-$	$D^{*+} \to D^0 \pi^+, D^+ \pi^0$
$K^{*0} \to K_S \pi^0, K^+ \pi^-$	$\chi_{c1,2} \to J/\psi \gamma$	$D^{*0} \to D^0 \pi^0, D^0 \gamma$
$\phi \to K^+ K^-$		

[2] The ARGUS function is defined as $A(m_{ES}; m_0, \xi) = N \cdot m_{ES} \sqrt{1 - \left(\frac{m_{ES}}{m_0}\right)^2} \cdot e^{\xi(1-(m_{ES}/m_0)^2)}$.

BRANCHING FRACTIONS OF EXCLUSIVE *B* DECAYS

B Decays to Charmonium Mesons

Decays of B mesons to two-body final states containing a charmonium resonance constitute a very sensitive laboratory for the study of electroweak transitions, as well as the dynamics of strong interactions in heavy mesons systems. In particular, neutral B decays to these final states are expected to exhibit a significant CP asymmetry, the magnitude of which is cleanly related to the value of the angle β of the Unitarity Triangle [2].

We studied decay modes with a K or π meson accompanying the charmonium state. Due to the contributions of non-perturbative QCD interactions in the final state, theoretical estimates suffer from some degree of model dependence. Nevertheless isospin symmetry requires that the ratio of charged to neutral partial widths be unity for each type of decay, independently of the model used.

As seen in Table 1, charmonium states are reconstructed either through direct decay to $l^+ l^-$ (J/ψ and $\psi(2S)$) or through decay to a state containing a J/ψ, which in turn decays to $l^+ l^-$. The angular distribution of the lepton pair is exploited to reduce background when the light meson is a pseudoscalar, since in that case the (vector) charmonium state is longitudinally polarized.

ΔE and $m_{\rm ES}$ distributions are used to extract the signal yields for all channels studied, except for $B^0 \to J/\psi K_{\rm L}$. Fig. 1(*a*) shows the distributions for the case $B^0 \to J/\psi K_{\rm S}, K_{\rm S} \to \pi \pi^-$.

The $B^0 \to J/\psi K_{\rm L}$ case must to be treated differently since neither the $K_{\rm L}$ energy nor its momentum are measured. In this case the E_{K_L} energy is determined by constraining the B candidate mass to its known value, and the quantity $\Delta E_{K_L} \equiv E^*_{J/\psi} + E^*_{K_L} - E^*_{\rm beam}$

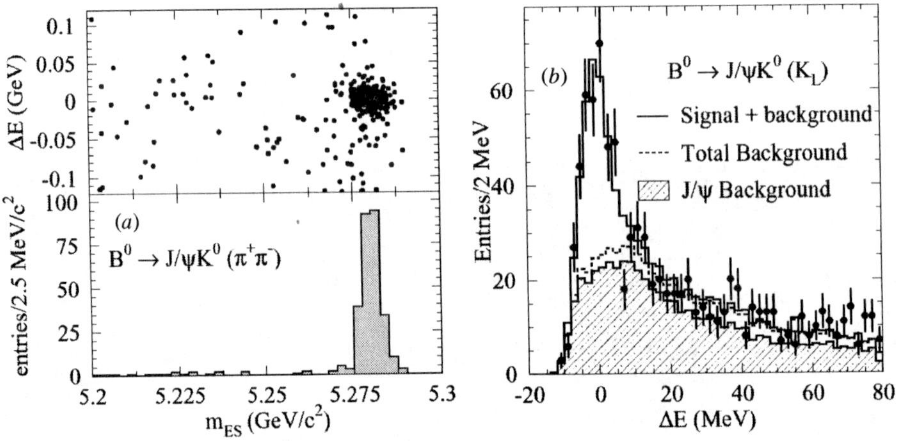

FIGURE 1. Signal evidence for two charmonium modes: (*a*) ΔE and $m_{\rm ES}$ distributions for $B^0 \to J/\psi K_{\rm S}$, $K_{\rm S} \to \pi^+ \pi^-$; (*b*) ΔE distribution for $B^0 \to J/\psi K_{\rm L}$: points are from real data, histograms from Monte Carlo events.

is plotted. The large (peaking) background coming from other J/ψ channels is modeled by a detailed Monte Carlo study. The ΔE_{K_L} distribution for our sample is shown in Fig. 1(*b*).

Table 2 summarizes the branching fraction values obtained for all channels. Most of them have a better precision than published world averages [3]. The resulting values of charged-to-neutral-mode ratios differ from 1 by at most 2σ.

TABLE 2. Measured branching fractions for exclusive *B* decays involving charmonium. The first error is statistical, the second systematic.

Mode		BR ($\times 10^{-4}$)	Mode	BR ($\times 10^{-4}$)
$B^0 \to J/\psi K^0$	$K^0_S \to \pi^+\pi^-$	$8.5 \pm 0.5 \pm 0.6$	$B^0 \to J/\psi \pi^0$	$0.20 \pm 0.06 \pm 0.02$
	$K^0_S \to \pi^0\pi^0$	$9.6 \pm 1.5 \pm 0.7$	$B^0 \to J/\psi \pi^+\pi^-$	$0.46 \pm 0.11 \pm 0.08$
	K^0_L	$6.8 \pm 0.8 \pm 0.8$	$B^0 \to \psi(2S) K^0$	$6.8 \pm 1.0 \pm 1.1$
	All	$8.3 \pm 0.4 \pm 0.5$	$B^+ \to \psi(2S) K^+$	$6.3 \pm 0.5 \pm 0.8$
$B^+ \to J/\psi K^+$		$10.1 \pm 0.3 \pm 0.5$	$B^0 \to \chi_{c1} K^0$	$5.4 \pm 1.4 \pm 1.1$
$B^0 \to J/\psi K^{*0}$		$12.4 \pm 0.5 \pm 0.9$	$B^+ \to \chi_{c1} K^+$	$7.5 \pm 0.8 \pm 0.8$
$B^+ \to J/\psi K^{*+}$		$13.7 \pm 0.9 \pm 1.1$	$B_0 \to \chi_{c1} K^{*0}$	$4.8 \pm 1.4 \pm 0.9$

B Decays to Open Charm Mesons

$B \to D^{(*)} \bar{D}^{(*)}$[3]

The Standard Model predicts sizeable *CP*-violating effects in the decays $B^0 \to D^{(*)+} D^{(*)-}$; in particular, time dependent asymmetries con be used to extract the value of sin 2β, as in the case of $B^0 \to J/\psi K_S$. An independent measurement of this quantity is especially important since several extensions to the Standard Model imply differences between the values extracted from the two different classes of processes. Charged *B* decays such as $B^\pm \to D^{*\pm} D^{*0}$ are also important since they provide calibration and control samples.

The rate of the Cabibbo-suppressed decays $B \to D^{(*)} \bar{D}^{(*)}$, can be estimated from the measured rate of the Cabibbo-favored decays $B \to D_s^{(*)} \bar{D}^{(*)}$, leading to values of the order of 0.1%. Previous measurements of branching fractions and upper limits for these modes were reported by CLEO [4] and ALEPH [5].

To search for signal in these channels the variable $\chi^2_{\text{Mass}} \equiv \Sigma\, [(m_i - m_i^{\text{PDG}}) / \sigma_{m_i}]^2$ was used, in addition to ΔE and m_{ES}, where m_i is the mass of the reconstructed $D^{(*)}$ candidate, σ_{m_i} its error, and m_i^{PDG} the corresponding PDG value and the sum is performed on all reconstructed D and D^* in the decay chain.

A clean signal is observed in the $B \to D^{*+} D^{*-}$ channel (Fig. 2); evidence for a peak is also seen in the m_{ES} plots for the $B \to D^{*+} D^-$ and $B^+ \to D^{*+} D^{*0}$ channels. Table 3 summarizes results for these 3 modes. We report a branching fraction only for $B^0 \to D^{*+} D^{*-}$: for the other two modes we quote the probability that the observed distribution be due to background fluctuation.

[3] Here and in the following, the symbol $D^{(*)}$ refers to either a D or a D^* state.

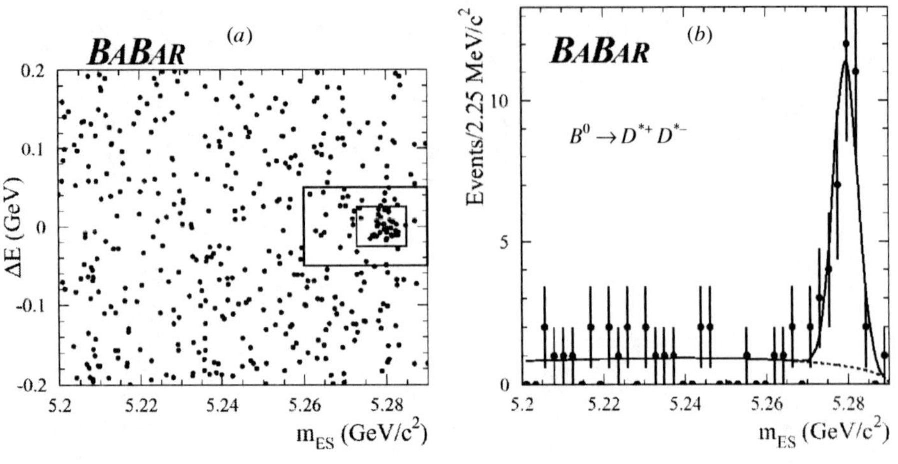

FIGURE 2. Evidence of signal for $B^0 \to D^{*+} D^{*-}$: (*a*) distribution in the m_{ES}-ΔE plane and (*b*) projection in m_{ES}.

TABLE 3. Yields and branching fractions for $B \to D^{(*)} \bar{D}^{(*)}$ modes.

Mode	N. signal events	Bkg.	BR [or prob. of bkg. fluctuation]
$B^0 \to D^{*+}D^{*-}$	38	6.2	$(8.0 \pm 1.6 \pm 1.2) \times 10^{-4}$
$B^0 \to D^{*+}D^{-}$	31	10.5	$[9.7 \times 10^{-7} (> 4.3\ \sigma)]$
$B^+ \to D^{*+}D^{*0}$	39	20.3	$[2.9 \times 10^{-6} (> 4.1\ \sigma)]$

$B \to D^{(*)} \bar{D}^{(*)} K$

Until 1994 it was believed that the $c\bar{s}$ pair in the process $b \to c\bar{c}s$ would hadronize dominantly as $D_s^{(*)+}$ mesons. If this conjecture is used in computing the total hadronic branching fraction of the B, it leads to an inconsistency with the measured value of the inclusive semileptonic branching fraction: their contributions sum up to about 80%, albeit with a large uncertainty [6]. In recent years, CLEO [7] and ALEPH [5] have reported evidence for a small number of completely reconstructed decays of the type $B \to D^{(*)} \bar{D}^{(*)} K$: this would point to a larger $b \to c\bar{c}s$ branching fraction, which would help solve the puzzle.

We have reconstructed $B \to D^{(*)} \bar{D}^{(*)} K$ events using all possible charge combinations. In the case of $B^0 \to D^{*-} D^{*0} K^+$, a partial reconstruction technique is adopted: the D^0 is reconstructed but not combined with the γ or π^0 to form the D^{*0}. The ΔE distribution peaks in this case at −154 MeV instead of 0. We found a significant number of events in 3 exclusive channels: m_{ES} distributions are shown in Fig. 3 for two of them, while resulting branching fractions are reported in Table 4. Moreover, several candidates have been observed in the semi-exclusive mode $B^0 \to D^{(*)} \bar{D}^{(*)} K_S$, which could be used for sin 2β measurements.

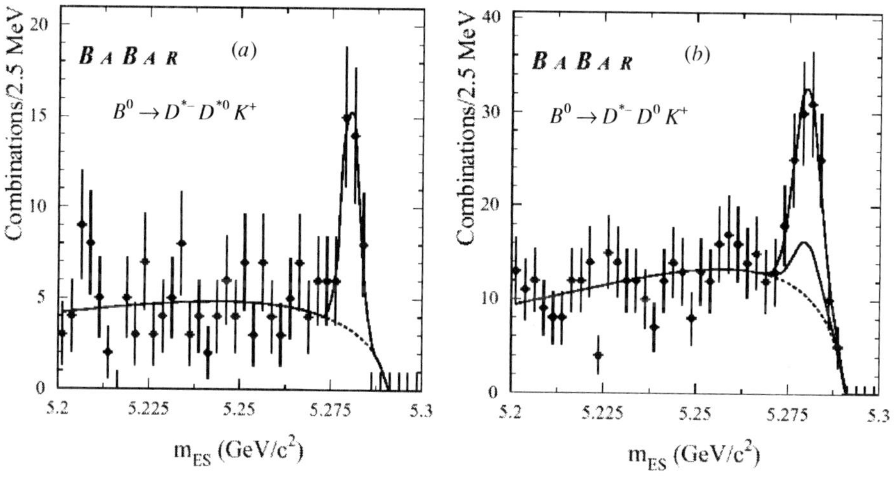

FIGURE 3. Evidence of signal for two exclusive $B \to D^{(*)} \bar{D}^{(*)} K$ modes: m_{ES} distribution for (*a*) $B^0 \to D^{*-} D^{*0} K^+$ and (*b*) $B^0 \to D^{*-} D^0 K^+$. The smaller peak in (*b*) represents the background contribution from $B^+ \to D^{*+} D^{*-} K^+$ decays, where the π^+ from the D^{*+} is not reconstructed.

TABLE 4. Yields and branching fractions for $B \to D^{(*)} \bar{D}^{(*)} K$ modes.

Mode	Signal (fit)	Bkg. (fit)	BR ($\times 10^{-3}$)
$B^+ \to D^{*-} D^{*+} K^+$	8.2 ± 3.5	1.7	3.4 ± 1.6 ± 0.9
$B^0 \to D^{*-} D^0 K^+$	29.6 ± 7.2	24.8	2.8 ± 0.7 ± 0.5
$B^0 \to D^{*-} D^{*0} K^+$	80.2 ± 15.3	20.6 ± 9.7	6.8 ± 1.7 ± 1.7
$B^0 \to D^{(*)-} D^{(*)+} K_S$	10.1 ± 3.7	3.4	

Measurement of Ratio of Branching Fractions with K/π Separation

In cases where two different final states differ only by the presence of either a charged K or π, the correct reconstruction of the decay depends on how well the two modes can be separated, based on PID information and kinematics. As one of the two decays is typically Cabibbo-suppressed with respect to the other, their branching fractions can differ by order of magnitudes, thus making the task hardly achievable by means of ordinary "cut and count" methods.

BABAR has measured two ratios of branching fractions of this kind, by using unbinned maximum likelihood fits on samples of reconstructed events where no K/π identification has been applied in the selection.

$$B(B^{\pm} \to J/\psi\, \pi^{\pm}) / B(B^{\pm} \to J/\psi\, K^{\pm})$$

Contribution from the tree diagrams alone would give a ratio of ~5% between the $B^{\pm} \to J/\psi\, \pi^{\pm}$ and $B^{\pm} \to J/\psi\; K^{\pm}$ branching fractions. A substantially different value of the measured ratio would point to significant interference with penguin diagrams. This could be the source of a sizeable direct CP-asymmetry [8].

A likelihood function based on p.d.f. for ΔE, m_{ES} and the bachelor track momentum for the π, K and background case was built and the number of events of each type fitted for. The result of the fit yields

$$B(B^{\pm} \to J/\psi\, \pi^{\pm}) \,/\, B(B^{\pm} \to J/\psi\, K^{\pm}) = (3.91 \pm 0.78 \pm 0.19)\ \%.$$

Fig. 4(a) shows the m_{ES} distribution for a sample where the $J/\psi\, \pi^{\pm}$ content has been enriched by applying tight PID cuts.

$B(B^{\pm} \to D^0\, K^{\pm}) /\, B(B^{\pm} \to D^0\, \pi^{\pm})$

Measurement of the $B^{\pm} \to D^0\, K^{\pm}$ branching fraction can in principle be used, in conjunction with other rarer B decays, to extract the value of the CKM angle γ in a theoretically clean way. A prediction for its ratio to the $B^{-} \to D^0\, \pi^{-}$ branching fraction can be obtained from $B(\tau^{-} \to K^{-} \nu_{\tau}) \,/\, B(\tau^{-} \to \pi^{-}\, \nu_{\tau})$, giving $(7.4 \pm 0.3)\%$. The first observation of the $B^{\pm} \to D^0\, K^{\pm}$ decay has been reported by CLEO [9].

In this case the likelihood function includes, along with the ΔE and m_{ES} p.d.f., a PID variable exploiting the K/π separation power provided by the DIRC. Contribution from both combinatorial and resonant background are fitted for. The result is:

$$B(B^{\pm} \to D^0\, K^{\pm}) \,/\, B(B^{\pm} \to D^0\, \pi^{\pm}) = (8.3 \pm 0.6 \pm 0.3)\ \%.$$

Fig. 4(b) shows the ΔE distribution for a sample where the $D^0\, K^{\pm}$ content has been enriched by applying tight PID cuts.

ANGULAR DISTRIBUTIONS

B decays to such states as $J/\psi\, K^{*0}$ and $D^{*+}\, D^{*-}$ are potentially sensitive to CP violation effects [10], [11]. Since the vector-vector final state has both a CP-even and a CP-odd component, a direct extraction of the CP-violating parameter ($\sin 2\beta$ in this case) would be affected by some dilution, unless an angular analysis is carried out.

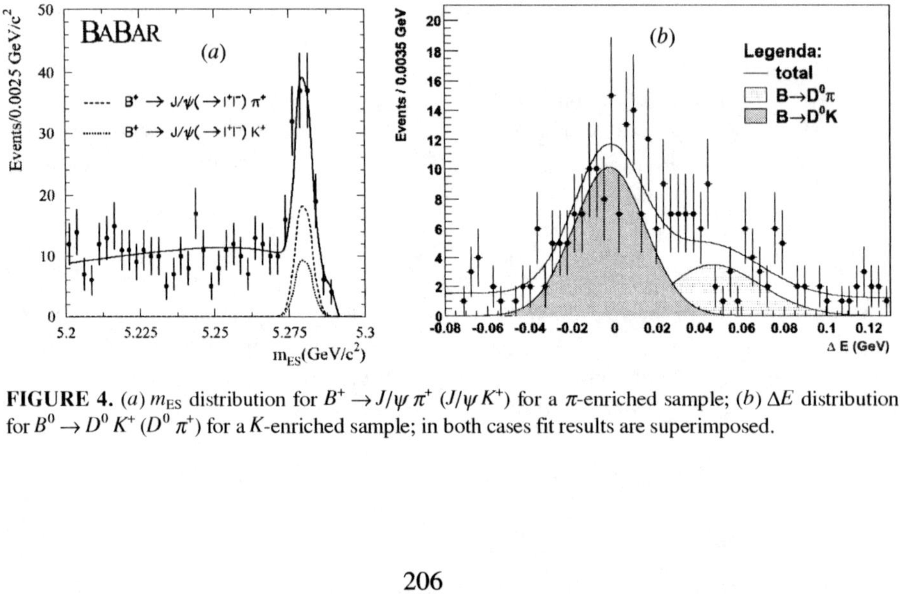

FIGURE 4. (a) m_{ES} distribution for $B^{+} \to J/\psi\, \pi^{+}$ ($J/\psi\, K^{+}$) for a π-enriched sample; (b) ΔE distribution for $B^0 \to D^0\, K^{+}$ ($D^0\, \pi^{+}$) for a K-enriched sample; in both cases fit results are superimposed.

In the transversity basis formalism [12], the angular distribution is described by three amplitudes A_0, $A_\parallel$, $A_\perp$ with CP eigenvalues +1, +1 and −1 respectively. The parameter

$$R_\perp \equiv \frac{|A_\perp|^2}{|A_0|^2 + |A_\parallel|^2 + |A_\perp|^2}; \tag{3}$$

describes the fraction of P-odd component: if this is neglected, the resulting dilution amounts to $1 - 2R_\perp$.

$B^0 \rightarrow J/\psi K^{*0}$

For the $B^0 \rightarrow J/\psi K^{*0}$ decay, an unbinned maximum likelihood fit to the data has been performed using a full angular distribution for both signal and background. Moduli and phases of the three amplitudes have been extracted from the fit, leading to

$$R_\perp = 0.160 \pm 0.032 \pm 0.014.$$

Fitted distributions are illustrated in Fig. 5(a). Two relative phases $\phi_\perp \equiv \arg(A_\perp/A_0)$ and $\phi_\parallel \equiv \arg(A_\parallel/A_0)$ are defined: if factorization holds, their value would be 0 or π. Our measurement of $\phi_\parallel = 2.50 \pm 0.20 \pm 0.08$, is inconsistent with this hypothesis. In general, all measurements significantly improve on previous ones by CLEO [13] and CDF [14].

$B^0 \rightarrow D^{*+} D^{*-}$

In the case of $B^0 \rightarrow D^{*+} D^{*-}$ decay, the angular distribution is integrated over two of the three angular variables, leaving a function with a single parameter, $R_\perp$. An unbinned maximum likelihood fit to the selected events yields

$$R_\perp = 0.22 \pm 0.18 \pm 0.03$$

The fitted distribution is shown in Fig. 5(b).

SUMMARY

Using a sample of 22.7 million $B\bar{B}$ pairs, *BABAR* has reconstructed a number of charged and neutral B decays to states containing charmonium or open charm mesons. Signals have been observed and branching fractions have been measured for several exclusive decays to $(c\bar{c})$ + light mesons, $D^{(*)}\bar{D}^{(*)}$, $D^{(*)}\bar{D}^{(*)}K$. The precision of these (preliminary) results is in most cases better than currently published world averages [3]. We also report measurements of the ratios of branching fractions $\mathrm{B}(B^\pm \rightarrow J/\psi\,\pi^\pm)$ / $\mathrm{B}(B^\pm \rightarrow J/\psi\,K^\pm)$ and $\mathrm{B}(B^\pm \rightarrow D^0 K^\pm)$ / $\mathrm{B}(B^\pm \rightarrow D^0\,\pi^\pm)$. Finally, angular distributions for $B^0 \rightarrow J/\psi K^{*0}$ and $B^0 \rightarrow D^{*+} D^{*-}$ have been studied and their parameters determined, with a substantial improvement over existing published results.

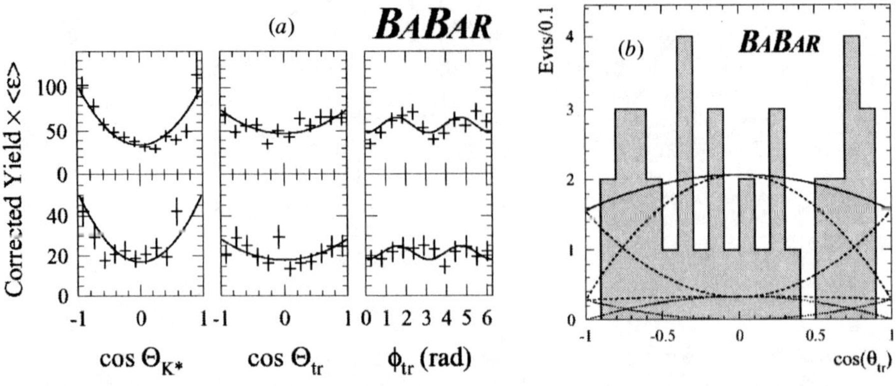

FIGURE 5. Fit results for angular distributions of (*a*) $B^0 \to J/\psi K^{*0}$ for channels without (top) and with (bottom) a π^0, and (*b*) $B^0 \to D^{*+} D^{*-}$. Solid curves are the fitted signal + background distributions.

REFERENCES

1. *BABAR* Collaboration, B. Aubert *et al.*, SLAC-PUB-8569, hep-ex/0105044, to appear in *Nucl. Instrum. and Methods*.
2. *BABAR* Collaboration, B. Aubert *et al.*, *Phys. Rev. Lett.* **86**, 2515 (2001).
   S. Prell, "*BABAR* measurements of *CP* violation, mixing and lifetimes of *B* mesons", this conference proceedings.
3. Particle Data Group, D. E. Groom *et al.*, *Eur. Phys. J.* **C15**, 1 (2000).
4. CLEO Collaboration, E. Lipeles *et al.*, *Phys. Rev.* **D 62**, 032005 (2000).
5. ALEPH Collaboration, R. Barate *et al.*, *Eur. Phys. J.* **C 4**, 387(1998).
6. T. E. Browder, "Hadronic *B* Decay", hep-ph/9802217 (1998).
7. CLEO Collaboration, CLEO CONF 97-26, EPS97, 337 (1997).
8. I. Dunietz, *Phys. Lett.* **B 316**, 561 (1993).
9. CLEO Collaboration, M. Athanas *et al.*, *Phys. Rev. Lett.* **80**, 5493 (1998).
10. I. I. Bigi, *Nuovo Cim.* **109 A**, 713 (1996).
11. Y. Grossman and M. Worah, *Phys. Lett.* **B 395**, 241 (1997)
    R. Fleischer, *Int. J. Mod. Phys.*, A **12**, 2459 (1997).
12. A. S. Dighe, I. Dunietz, H. J. Lipkin and J. L. Rosner, *Phys. Lett.* **B 369**, 144 (1996).
13. CLEO Collaboration, C. P. Jessop *et al.*, *Phys. Rev. Lett.* **79**, 4533 (1997).
14. CDF Collaboration, T. Affolder *et al.*, *Phys. Rev. Lett.* **85**, 4668 (2000).

Hadronic B Decays at Belle

Shiro Suzuki representing the Belle collaboration

Yokkaichi University, 1200 Kayo-cho, Yokkaichi, 512-8512 Japan

Abstract. Recent results on hadronic B decays from the Belle collaboration are reviewed. The results are based on the data sample with integrated luminosity of 21.3fb^{-1}, 22.8 million $B\bar{B}$ pairs in total. The first observations of $B^- \to \chi_{c0}K^-$, $B^0 \to D^{\pm}D^{*\mp}$, $B^- \to D_{CP}K^-$ and color suppressed modes $B^0 \to D^{(*)0}\pi^0$, $D^0\eta$, and $D^0\omega$ are reported.

INTRODUCTION

This summer, observation of a large CP violating asymmetry ($\sin 2\phi_1$) through $B^0\bar{B}^0$ mixing was reported by Belle and BaBar[1]. To shed light on the mechanism of CP violation in B decays, we need to determine $\sin 2\phi_1$ with much higher accuracy and under different systematic conditions. Also it is essential to measure the phase angles ϕ_2 and ϕ_3 as well as ϕ_1, and CKM matrix elements, V_{ub} in particular. To reliably measure the variables, we need a thorough understanding of hadronic interactions in B meson decays, such as factorization, color suppression and so on.

In this report, Belle's results on hadronic B decays with hidden and open charm are presented. The results are based on a 21.3 fb^{-1} data sample available in the summer of 2001, which contains 22.8 million $B\bar{B}$ events. The results are preliminary, unless otherwise stated.

B DECAYS WITH CHARMONIUM

As was reported by Belle and BaBar[1], $\sin 2\phi_1$ was measured to be large. We need to confirm this in other modes with different systematics. There are many B decay processes that include charmonia in the final state which provide $\sin 2\phi_1$ information besides the $J/\psi K_s$ and $J/\psi K_L$ final states, and we should examine the feasibility of using those processes in CP studies.

Extensive studies were made to measure two body decays of $\bar{B}^0$ and B^- with charmonium, J/ψ, $\psi(2S)$ and χ_{c1}. Some of them are sensitive to CP and some are calibration processes. The results are summarized in Table 1. The dominant systematic errors come from efficiency estimation. The uncertainty of the tracking efficiency is evaluated by comparing the ratio of the yields of $\eta \to \pi^+\pi^-\pi^0$ and $\eta \to \gamma\gamma$ for data and Monte Carlo, and found to be 2% per track. All the results are consistent with previous measurements. For several modes there is a significant improvement in precision compared to the previous measurements.

CP618, *Heavy Flavor Physics: Ninth International Symposium,* edited by A. Ryd and F. C. Porter

TABLE 1. Measured branching fractions for B decays with charmonium. The first errors are statistical and the second systematic. Averages from previous measurements are listed for comparison.

Decay modes	$\mathcal{B}(\times 10^{-4})$	**Previous**$(\times 10^{-4})$
$B^- \to J/\psi K^-$	$10.1 \pm 0.3 \pm 0.8$	10.0 ± 1.0[2]
$\bar{B^0} \to J/\psi K^0$	$7.7 \pm 0.4 \pm 0.7$	9.6 ± 0.9[2]
$B^- \to \psi(2S)K^-$, $\psi(2S) \to l^+l^-$	$6.7 \pm 0.6 \pm 0.7$	5.8 ± 1.0[2]
$B^- \to \psi(2S)K^-$, $\psi(2S) \to J/\psi\pi^+\pi^-$	$5.7 \pm 0.5 \pm 0.8$	5.8 ± 1.0[2]
$\bar{B^0} \to \psi(2S)K^0$, $\psi(2S) \to l^+l^-$	$6.0 \pm 1.1 \pm 0.7$	5.0 ± 1.3[3]
$\bar{B^0} \to \psi(2S)K^0$, $\psi(2S) \to J/\psi\pi^+\pi^-$	$7.2 \pm 1.1 \pm 1.1$	5.0 ± 1.3[3]
$B^- \to \chi_{c1}K^-$	$6.1 \pm 0.6 \pm 0.6$	10.0 ± 4.0[2]
$\bar{B^0} \to \chi_{c1}K^0$	$3.1 \pm 0.9 \pm 0.4$	$3.9^{+1.9}_{-1.4}$[2]
$B^- \to J/\psi\pi^-$	$0.52 \pm 0.07 \pm 0.07$	0.51 ± 0.15[2]
$\bar{B^0} \to J/\psi\pi^0$	$0.24 \pm 0.06 \pm 0.02$	$0.25^{+0.11}_{-0.09}$[2]

In the process of studying the non-resonant charmless three body modes $B \to \pi\pi K$ and KKK, appreciable signals for χ_{c0} were observed in the $\pi\pi$ and KK combinations. Resonant components D^0, K^{*0} in the $K\pi$ and ϕ in the KK were rejected by requiring the effective masses $M_{\pi\pi}$ and $M_{K\pi}$ be larger than 2 GeV. Continuum is suppressed by a thrust angle cut and a cut on a Fisher discriminant calculated using the virtual calorimeter method. The $B \to \chi_{c0}K$ process is suppressed in factorization models, and is only allowed by soft gluon emission. Here we can see a clear signal in $\pi\pi$ effective mass (fig.1). The peak around 3.7 GeV is feed across from $\psi(2S) \to \mu\mu$ with both μ's are misidentified as π's. The branching fraction determined in the $\chi_{c0} \to \pi\pi$ decay mode is

$$\mathcal{B}(B^+ \to \chi_{c0}K^+) = (6.0^{+2.1}_{-1.8} \pm 1.1) \times 10^{-4}.$$

This is remarkably large, and is comparable to $B^+ \to J/\psi K^+$.

We did not use the $\chi_{c0} \to KK$ mode in determining the branching fraction for the $B^+ \to \chi_{c0}K^+$, since we observed evidence for non-resonant $B^+ \to K^+K^+K^-$, and the K^+K^- invariant mass distribution could be distorted by interference with an amplitude not related to $B^+ \to \chi_{c0}K^+$.

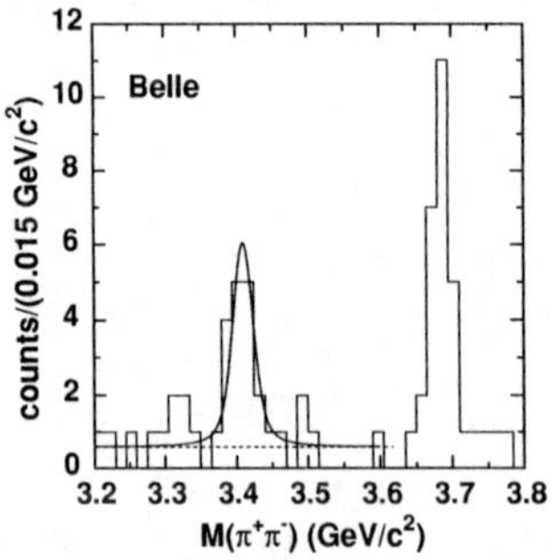

FIGURE 1. $\chi_{c0} \to \pi\pi$ signal found in the $B^+ \to K^+\pi^+\pi^-$ process.

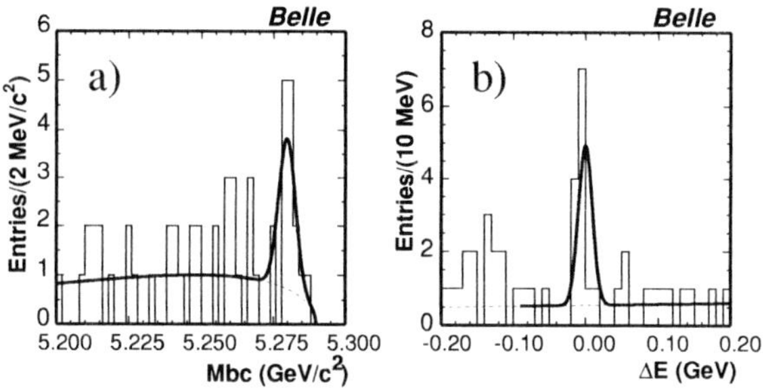

FIGURE 2. (a) M_{bc} and (b) ΔE for $B^0 \to D^{\pm}D^{*\mp}$ process.

B DECAYS WITH DOUBLE CHARM

$B \to D^{(*)}\bar{D}^{(*)}$ decays proceed via the Cabibbo suppressed $b \to c\bar{c}d$ transition and are sensitive to a time-dependent asymmetry proportional to $\sin(2\phi_1 + \delta_s)$, where δ_s is the possible strong phase. This asymmetry may have different systematics due to the relatively large contribution from a penguin diagram. The branching fraction is estimated to be around 0.1%, although a definitive observation is yet to be made.

Reconstruction of the $B \to D^{(*)}\bar{D}^{(*)}$ process is performed by detecting $D^{*+} \to D^0\pi^+$, and D^0 decays to $K^-\pi^+$, $K^-\pi^+\pi^0$ and $K^-\pi^+\pi^-\pi^+$ final states. We add the $D^0 \to K_s\pi^+\pi^-$ mode for the $D^{*+}D^{*-}$ study. We observe a $D^{\pm}D^{*\mp}$ signal of 11.2 ± 4.0 events with a statistical significance of 4.1σ (fig.2a). A consistent result is also obtained in the ΔE plot (fig.2b). We find that the branching fraction of the sum of $B^0 \to D^+D^{*-}$ and $B^0 \to D^-D^{*+}$ is

$$\mathcal{B}(B^0 \to D^{\pm}D^{*\mp}) = (1.04 \pm 0.38 \pm 0.22) \times 10^{-3}.$$

We have also confirmed the process $B^0 \to D^{*+}D^{*-}$. We observed 11.0 ± 3.7 events with a statistical significance of 5.0 σ. The branching fraction obtained for this process is

$$\mathcal{B}(B^0 \to D^{*+}D^{*-}) = (1.21 \pm 0.41 \pm 0.27) \times 10^{-3}.$$

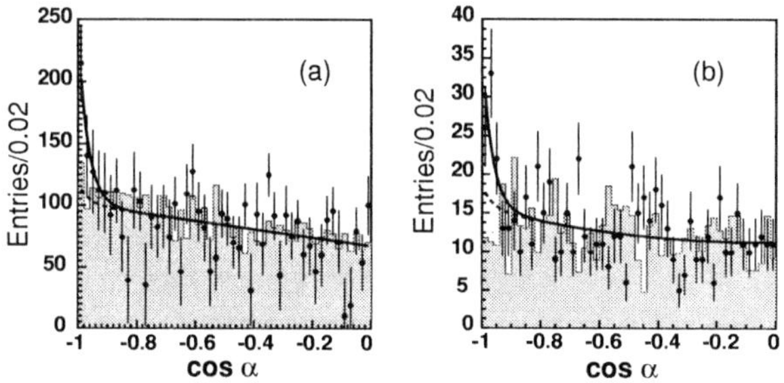

FIGURE 3. (a) $\cos\alpha$ for no-lepton tagged and (b) lepton tagged partially reconstructed DD^* events. Dots are data and histogram shows estimated $B\bar{B}$ background with Monte Carlo.

To compensate for the small detection efficiency of the D^0 and obtain much higher statistics, a partial reconstruction technique was used. This technique was originally developed to measure the $B \to D^*\pi$ process by the CLEO group and makes use of the small Q-value for $D^* \to D\pi$ decay[4]. The same procedure is applied to the $B \to D^+D^{*-}$ process reconstructing the D^+ and the slow pion from D^{*-}. Signatures of the signal are; the angle α between D^+ and slow π, which peaks around $\cos\alpha = -1$, and the helicity angle θ of the slow π in the D^* rest frame, which shows a $\cos^2\theta$ behavior due to helicity conservation. After subtracting continuum, we can see a peaking structure in $\cos\alpha = -1$ region in both the "no lepton-tagged" and "lepton-tagged" data samples. The histogram shows the estimated $B\bar{B}$ background. The helicity angle of slow pion for the events with $\cos\alpha < -0.9$ also shows the expected $\cos^2\theta$ behavior. The branching fraction for the sum of $B^0 \to D^+D^{*-}$ and D^-D^{*+} is

$$\mathcal{B}(B^0 \to D^{\pm}D^{*\mp}) = (1.84 \pm 0.46^{+0.68}_{-0.60}) \times 10^{-3},$$

consistent with the result of full reconstruction within errors.

ϕ_3 FROM $B \to D_{CP}K$

A direct CP asymmetry will be observed in $B^- \to D_{CP}K^-$ decays due to the interference of the $b \to c$ and $b \to u$ processes shown in fig.4. The angle ϕ_3 can be extracted from $B^- \to D_{1,2}K^-$ decays, where D_1 and D_2 are CP=+ and − eigenstates, respectively. If D^0-$\bar{D}^0$ mixing is negected, ϕ_3 can be expressed in terms of the observed quantities as,

$$\mathcal{A}_{1,2} = \frac{\mathcal{B}(B^- \to D_{1,2}K^-) - \mathcal{B}(B^+ \to D_{1,2}K^+)}{\mathcal{B}(B^- \to D_{1,2}K^-) + \mathcal{B}(B^+ \to D_{1,2}K^+)} = \frac{2r\sin\delta'\sin\phi_3}{1 + r^2 + 2r\cos\delta'\cos\phi_3},$$

$$R_{1,2} = R^{D_{CP}}_{K/\pi}/R^{D_{non-CP}}_{K/\pi} = 1 + r^2 + 2r\cos\delta'\cos\phi_3,$$

where $\mathcal{A}_{1,2}$ is a direct measure of CP asymmetry for $D_{1,2}$, r is the ratio of amplitudes $r = A(B^- \to D^0K^-)/A(B^- \to \bar{D}^0K^-)$ and δ' is the strong phase difference, $\delta' = \delta$ for CP = +1 and $\delta' = \delta + \pi$ for CP $= -1$. From this formula, ϕ_3 can be derived if we have enough statistics and knowledge of r and δ'. $R_{1,2}$ is the ratio of the $R_{K/\pi} = \mathcal{B}(B \to DK)/\mathcal{B}(B \to D\pi)$ for CP and non-CP eigenstates. Even if we cannot reach the necessary sensitivity for a direct derivation of ϕ_3 from $\mathcal{A}_{1,2}$, we can constrain the allowed region for ϕ_3 if $R_{1,2}$ is smaller than one.

Among the above quantities, $R^{D_{non-CP}}_{K/\pi}$ has been already measured to be $0.079 \pm 0.009 \pm 0.006$ by Belle[5]. After this measurement, we further analyzed this process with $D^0 \to$ "CP eigenstates" (fig.5). We obtained an improved $DK/D\pi$ signal for the $D^0 \to K^-\pi^+$ mode, and a $R^{D_{non-CP}}_{K/\pi}$ ratio consistent with the previous result. We observe a clear signal for $B^- \to D^0K^-$ in the CP=1 $D^0 \to K^+K^-$ mode with 12.3 ± 3.9 events. The statistical significance is 4.3σ. With the present statistics, we do not yet observe a clear signal in $D^0 \to \pi^+\pi^-$, but the $R_{K/\pi}$ ratio is consistent with other modes. The

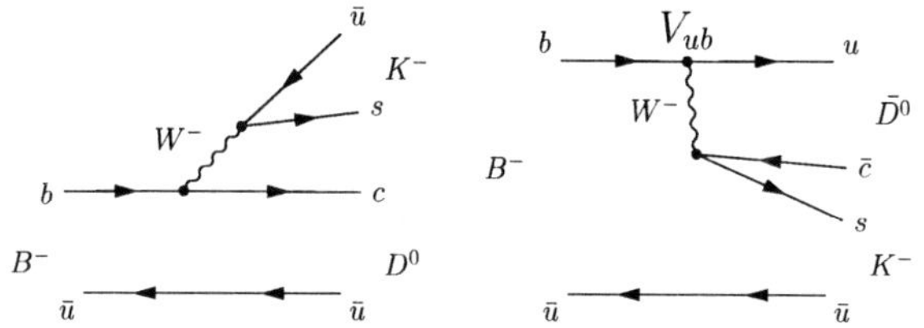

FIGURE 4. Two interfering diagrams in $B^- \to D_{CP}K^-$ process.

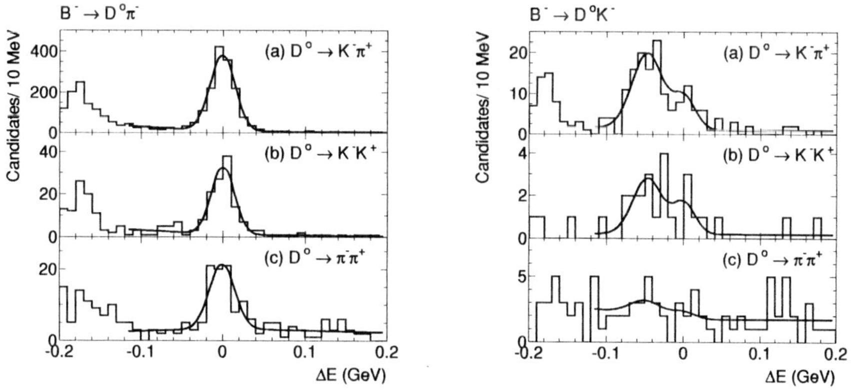

FIGURE 5. The ΔE distribution for $B^- \to D^0\pi^-$ and $B^- \to D^0K^-$ candidates for the various D^0 decay modes. $B^- \to D^0K^-$ signal makes a peak at $\Delta E \approx -49$MeV.

asymmetry parameters obtained are

$$\mathcal{A}_1 = 0.04^{+0.40}_{-0.35} \pm 0.15, \quad \text{and}$$
$$R_1 = 1.39 \pm 0.53 \pm 0.26.$$

This is the first attempt to measure CP asymmetry in the $B^- \to D^0K^-$ process. $\mathcal{A}_1$ is consistent with zero, and R_1 exceeds 1.0 including errors. At this moment we have no constraints on ϕ_3, but we have started seeing $B^- \to D_{CP}K^-$ processes which are sensitive to ϕ_3. Studies of other D^0 decays to CP eigenstates such as $K_s\pi^0$, $K_s\eta$, $K_s\eta'$, $K_s\phi$ and $K_s\omega$ are on going.

COLOR SUPPRESSED MODES

Another topic of interest in B decays with single charm are the newly observed color suppressed modes, $B^0 \to D^0h^0$ or $D^{*0}h^0$, where $h^0 = \pi^0$, η and ω. These decays proceed via internal spectator diagrams, and are expected to be suppressed relative to external diagrams (fig.6). Previously, CLEO has only given the upper limit of 1.2×10^{-4}[6]. Observation/measurement of these branching fractions will give us information about factorization models and final state interactions.

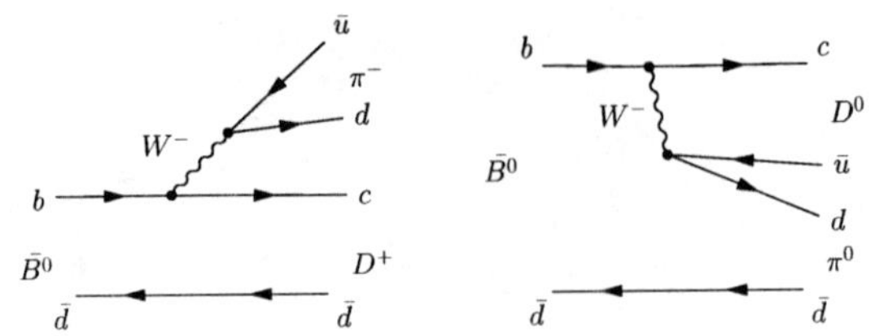

FIGURE 6. Tree and color suppressed diagrams for $B^0 \to D^+h^-$ and $B^0 \to D^0h^0$.

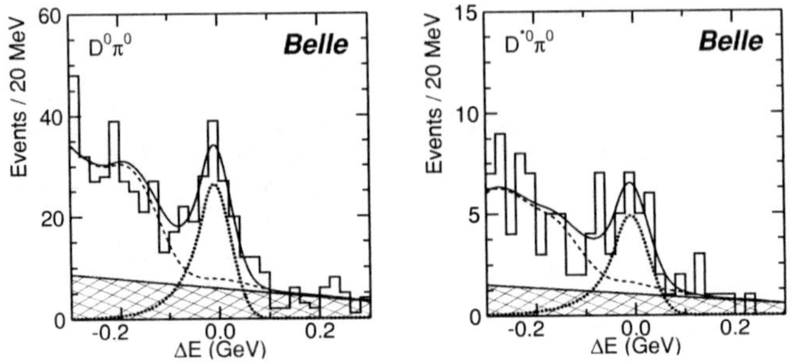

FIGURE 7. ΔE for $\bar{B}^0 \to D^0\pi^0$ and $\bar{B}^0 \to D^{*0}\pi^0$ processes.

$\bar{B}^0 \to D^0\pi^0$, $D^{*0}\pi^0$ decays

A careful search for $\bar{B}^0 \to D^0\pi^0$ and $\bar{B}^0 \to D^{*0}\pi^0$ processes has been carried out. Candidate π^0 mesons are reconstructed from pairs of photons within 16 MeV of the nominal π^0 mass. The energy of each photon is required to be larger than 50 MeV. Detected D^0 modes are $D^0 \to K^-\pi^+$, $K^-\pi^+\pi^0$ and $K^-\pi^+\pi^-\pi^+$, and the momentum of the π^0 from D^0 decay is required to be larger than 300MeV/c in the CM frame. The reconstructed D^0 mass is required to be within 2.5σ of the nominal mass. D^{*0}'s are detected in the $D^0\pi^0$ decay mode, and the cut on γ energy from the slow daughter π^0 is relaxed to 20 MeV. The large background contribution from continuum is suppressed by forming a Fisher discriminant, containing seven variables that quantify event topology. $B\bar{B}$ background from color favored $D^{(*)}(n\pi)^-$ and feed down from $D^{*0}h^0$ to D^0h^0 makes a large contribution with a complicated structure close to the signal region in the ΔE plot. The shape of the distribution is well understood by Monte Carlo.

In the fit we observe more than 100 events in $D^0\pi^0$, clearly separated from background (fig.7). We obtain a statistical significance of 9σ. Also we observe a clean signal in $D^{*0}\pi^0$ with a significance of 4σ. These are the first measurements of color suppressed modes. The measured branching fractions are;

$$\mathcal{B}(\bar{B}^0 \to D^0\pi^0) = (3.1 \pm 0.4 \pm 0.5) \times 10^{-4} \quad \text{and}$$
$$\mathcal{B}(\bar{B}^0 \to D^{*0}\pi^0) = (2.8^{+0.8}_{-0.7}\,^{+0.5}_{-0.6}) \times 10^{-4}.$$

These are larger than the CLEO limit, and also appreciably larger than the values expected from factorization.

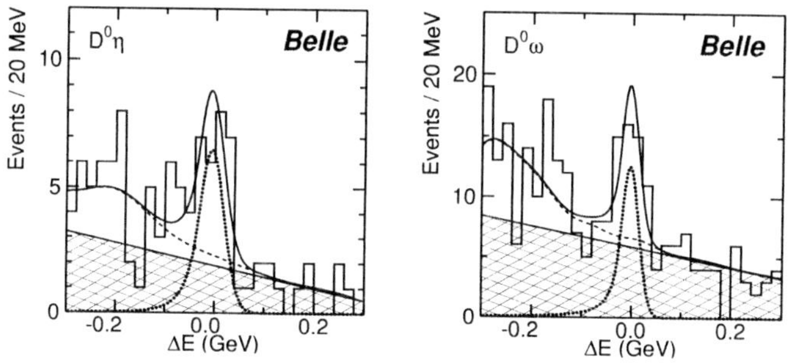

FIGURE 8. ΔE for $\bar{B^0} \to D^0\eta$ and $\bar{B^0} \to D^0\omega$ processes.

$\bar{B^0} \to D^0\eta, D^0\omega$ decays

Signals are also observed in $B^0 \to D^0\eta$ and $D^0\omega$. The η candidates are detected in the $\pi^+\pi^-\pi^0$ mode as well as the 2γ mode. The ω is reconstructed in $\pi^+\pi^-\pi^0$. For η candidates in the $\gamma\gamma$ mode, the γ energy is required to be larger than 100 MeV, and the reconstructed mass should be within 10.4 MeV of the nominal value. This value is 3.4 MeV for the three pion mode. Events are rejected if one of the γ's can be combined with any other γ to form a π^0 mass. We observe a signal with 4.2σ significance (fig.8). Similarly, a 4.4σ signal is observed in $D^0\omega$. The measured branching fractions are;

$$\mathcal{B}(\bar{B^0} \to D^0\eta) = (1.4^{+0.5}_{-0.4} \pm 0.3) \times 10^{-4} \quad \text{and}$$
$$\mathcal{B}(\bar{B^0} \to D^0\omega) = (1.8^{+0.5}_{-0.6} \pm 0.3) \times 10^{-4}.$$

Again, the branching fractions obtained for $D^0\eta$ and $D^0\omega$ are higher than the factorization prediction.

Fairly clean signals are seen in $D^{*0}\eta$ and $D^{*0}\omega$. Since their significance is just above 3σ, we quote only upper limit;

$$\mathcal{B}(\bar{B^0} \to D^{*0}\eta) < 4.6 \times 10^{-4} \quad \text{and}$$
$$\mathcal{B}(\bar{B^0} \to D^{*0}\omega) < 7.9 \times 10^{-4}$$

both at 90% confidence level.

$B \to D_S\pi, D_SK$ PROCESSES

$B^0 \to D_s^+\pi^-$ which proceeds via a tree level $b \to u$ transition is sensitive to $|V_{ub}|$. $B^0 \to D_s^- K^+$ proceeds via W exchange and should be highly suppressed. We have tried to measure these processes using the decay modes to $\phi\pi^+$, $K^{*0}K^+$ and K_sK^+. Super Fox-Wolfram moments and $\cos\theta_B$ are used to suppress the continuum. We have some hints of a signal (fig.9), but the significance is not high enough to quote a branching fraction. We determine the upper limits;

$$\mathcal{B}(B^0 \to D_s^+\pi^-) < 1.1 \times 10^{-4} \quad \text{and}$$

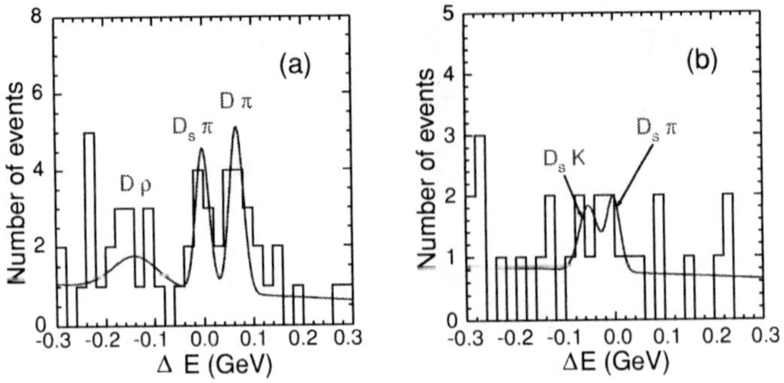

FIGURE 9. ΔE distributions for (a) $B^0 \to D_s^+\pi^-$ and (b) $B^0 \to D_s^- K^+$.

$$\mathcal{B}(B^0 \to D_s^- K^+) \;<\; 0.7 \times 10^{-4}$$

both at 90% confidence level.

SUMMARY

In the course of studying the origin of CP violation, we have succeeded in making the first observation of the color suppressed decay modes $B^0 \to D^{(*)0}h^0$, $B^0 \to DD^*$ in full and partial reconstruction, $B^- \to D_{CP}K^-$, and $B^+ \to \chi_{c0}K^+$. Color suppressed modes have much larger branching fractions than expected in factorization models, and remain problematic. We have made a good step toward new measurements of ϕ_1 with different systematics in the DD^* and D^*D^* modes, and with non-golden mode charmonium states. We have started to observe $B^- \to D_{CP}K^-$ which is the beginning of a program of the mesurement of ϕ_3, and made good progress in the determination of $|V_{ub}|$ in the $B \to D_s\pi$, D_sK modes.

REFERENCES

1. K. Abe et al. (Belle collaboration), *Phys. Rev. Lett.* **87**, 091802 (2001); B. Aubert et al. (BaBar collaboration), *Phys. Rev. Lett.* **87**, 091801 (2001).
2. Particle Data Group, D.E. Groom et al., *Eur. Phys. J.* **C15** 1 (2000) and 2001 partial update for edition 2002(URL:http://pdg.lbl.gov).
3. S.J. Richichi et al. (CLEO collaboration), *Phys. Rev.* **D63**, 031103 (2001).
4. G. Brandenburg et al. (CLEO collaboration), *Phys. Rev. Lett.* **80**, 2762 (1998).
5. K. Abe et al. (Belle collaboration), *Phys. Rev. Lett.* **87**, 111801 (2001).
6. B. Nemati et al. (CLEO collaboration), *Phys. Rev.* **D57**, 5363 (1998).

Aspects of QCD Factorization[1]

Matthias Neubert

Newman Laboratory of Nuclear Studies, Cornell University, Ithaca, NY 14853, USA

Abstract. The QCD factorization approach provides the theoretical basis for a systematic analysis of nonleptonic decay amplitudes of B mesons in the heavy-quark limit. After recalling the basic ideas underlying this formalism, several tests of QCD factorization in the decays $B \to D^{(*)}L$, $B \to K^*\gamma$, and $B \to \pi K, \pi\pi$ are discussed. It is then illustrated how factorization can be used to obtain new constraints on the parameters of the unitarity triangle.

INTRODUCTION

In many years of intense experimental and theoretical investigations the flavor sector of the Standard Model has been explored in great detail by studying mixing and weak decays of B mesons and kaons. CP violation has been observed in K–$\bar{K}$ mixing (1964), $K \to \pi\pi$ decays (1999), and most recently in the interference of mixing and decay in $B \to J/\psi K$ (2001). There is now compelling evidence that the Cabibbo–Kobayashi–Maskawa (CKM) mechanism accounts for the dominant source of CP violation in low-energy hadronic weak interactions. Most notably, the discovery of a large CP asymmetry in the B system has established that CP is not an approximate symmetry of Nature. Rather, the smallness of CP-violating effects in kaon (and charm) physics reflects the hierarchy of CKM matrix elements.

Measurements of $|V_{cb}|$ and $|V_{ub}|$ in semileptonic B decays and of the magnitude and phase of V_{td} in K–$\bar{K}$ mixing, $B_{d,s}$–$\bar{B}_{d,s}$ mixing, and $B \to J/\psi K$ decays has helped to determine the parameters of the unitarity triangle $V_{ub}^* V_{ud} + V_{cb}^* V_{cd} + V_{tb}^* V_{td} = 0$ with good accuracy. The current values obtained at 95% confidence level are $\bar{\rho} = 0.21 \pm 0.12$, $\bar{\eta} = 0.38 \pm 0.11$ for the coordinates of the apex of the (rescaled) triangle, and $\sin 2\beta = 0.74 \pm 0.15$, $\sin 2\alpha = -0.14 \pm 0.57$, $\gamma = (61 \pm 16)^\circ$ for its angles [1]. These studies have established the existence of a CP-violating phase in the top sector of the CKM matrix, i.e., $\mathrm{Im}(V_{td}^2) \neq 0$. The next step in testing the CKM paradigm must be to explore the CP-violating phase in the bottom sector, i.e., $\gamma = \arg(V_{ub}^*) \neq 0$. In the Standard Model the two phases are, of course, related to each other. However, there is still plenty of room for New Physics to affect the magnitude of flavor violations in both mixing and weak decays (see, e.g., [2]). In particular, the present upper bound on γ is derived from the experimental limit on B_s–$\bar{B}_s$ mixing, which has not yet been seen experimentally and could well be affected by New Physics.

---

[1] This talk was delivered on 11 September 2001, a few hours after terrorist attacks hit various targets in the U.S. Our thoughts and sympathy are with the many innocent victims of this tragedy.

CP618, *Heavy Flavor Physics: Ninth International Symposium*, edited by A. Ryd and F. C. Porter

FIGURE 1. Tree and penguin topologies in charmless hadronic B decays.

Common lore says that measurements of γ are difficult. Several "theoretically clean"[2] determinations of this phase have been suggested (see, e.g., [3, 4]), which are extremely challenging experimentally. Likewise, "clean" measurements of $\alpha = \pi - \beta - \gamma$ [5, 6] are very difficult. It is more accessible experimentally to probe γ (and α) via the sizeable tree–penguin interference in charmless hadronic decays such as $B \to \pi K$ and $B \to \pi\pi$. The basic decay topologies contributing to these modes are shown in Figure 1. Experiment shows that the tree-to-penguin ratios in the two cases are roughly $|T/P|_{\pi K} \approx 0.2$ and $|P/T|_{\pi\pi} \approx 0.3$, indicating a sizeable amplitude interference. It is important that the relative weak phase between the two amplitudes can be probed not only via CP asymmetry measurements ($\sim \sin\gamma$), but also via measurements of CP-averaged branching fractions ($\sim \cos\gamma$). Extracting information about CKM parameters from the analysis of nonleptonic B decays is a challenge to theory, since it requires some level of control over hadronic physics, including strong-interaction phases. Such challenges, combined with the importance of the issue, is what triggers theoretical progress.

QCD FACTORIZATION

Hadronic weak decay amplitudes simplify greatly in the heavy-quark limit $m_b \gg \Lambda_{\rm QCD}$. This statement should not surprise those who have followed the dramatic advances in our theoretical understanding of B physics in the past decade. Many areas of B physics, from spectroscopy to exclusive semileptonic decays to inclusive rates and lifetimes, can now be systematically analyzed using heavy-quark expansions. Yet, the more complicated exclusive nonleptonic decays have long resisted any theoretical progress. The technical reason is that, whereas in most other applications of heavy-quark expansions one proceeds by integrating out heavy fields (leading to local operator product expansions), in the case of nonleptonic decays the large scale m_b enters as the energy carried by light fields. Therefore, in addition to hard and soft subprocesses collinear degrees of freedom become important. This complicates the understanding of hadronic decay amplitudes

---

[2] In this area of flavor physics many practitioners would consider a method to be "theoretically clean" only if it exclusively relies on elementary geometry (amplitude triangles) and, perhaps, isospin symmetry. We adopt the rationale followed in most other branches of high-energy physics and call a method theoretically clean if it relies on systematic expansions in small parameters. The methods discussed later in this talk are theoretically clean in this wider sense.

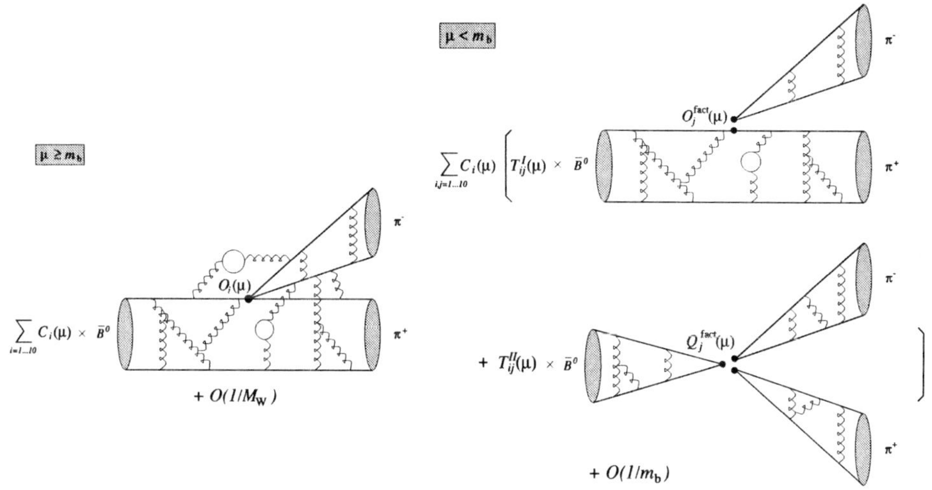

FIGURE 2. Factorization of short- and long-distance contributions in hadronic B decays. Left: Factorization of short-distance effects into Wilson coefficients of the effective weak Hamiltonian. Right: Factorization of hard "nonfactorizable" gluon exchanges into hard-scattering kernels (QCD factorization).

using the language of effective field theory. (Yet, very significant progress towards an effective field-theory description of nonleptonic decays has been made recently with the establishment of a "collinear–soft effective theory" [7]. The reader is referred to these papers for more details on this important development.)

The importance of the heavy-quark limit is based on the physical idea of color transparency [8, 9, 10]. A fast-moving light meson (such as a pion) produced in a point-like source (a local operator in the effective weak Hamiltonian) decouples from soft QCD interactions. More precisely, the couplings of soft gluons to such a system can be analyzed using a multipole expansion, and the first contribution (from the color dipole) is suppressed by a power of $\Lambda_{\rm QCD}/m_b$. The QCD factorization approach provides a systematic, model-independent implementation of this idea [11, 12]. It gives rigorous results in the heavy-quark limit, which are valid to leading power in $\Lambda_{\rm QCD}/m_b$ but to all orders of perturbation theory. Having obtained control over nonleptonic decays in the heavy-quark limit is a tremendous advance. We are now able to talk about power corrections to a well-defined and calculable limiting case, which captures a substantial part of the physics in these complicated processes.

The workings of QCD factorization can best be illustrated with the cartoons shown in Figure 2. The first graph shows the well-known concept of an effective weak Hamiltonian obtained by integrating out the heavy fields of the top quark and weak gauge bosons from the Standard Model Lagrangian. This introduces new effective interactions mediated by local operators $O_i(\mu)$ (typically four-quark operators) multiplied by calculable running coupling constants $C_i(\mu)$ called Wilson coefficients. This reduction in complexity (nonlocal heavy particle exchanges $\rightarrow$ local effective interactions) is exact up to corrections suppressed by inverse powers of the heavy mass scales. The resulting picture at scales at or above m_b is, however, still rather complicated, since gluon exchange is possible between any of the quarks in the external meson states. Additional simplifications occur when the renormalization scale μ is lowered below the scale m_b. Then color transparency comes to play and implies systematic cancellations of soft and collinear gluon exchanges. As a result, all "nonfactorizable" exchanges, i.e., gluons connecting the light meson at the "upper" vertex to the remaining mesons, are dominated by virtualities of order m_b and can be calculated. Their effects are absorbed into a new set of running couplings $T_{ij}^{I,II}(\mu)$ called hard-scattering kernels, as shown in the two graphs on the right-hand side. What remains are "factorized" four-quark and six-quark operators $O_j^{\rm fact}(\mu)$ and $Q_j^{\rm fact}(\mu)$, whose matrix elements can be expressed in terms of form factors, decay constants and light-cone distribution amplitudes. As before, the reduction in complexity (local four-quark operators $\to$ "factorized" operators) is exact up to corrections suppressed by inverse powers of the heavy scale, now set by the mass of the b quark.

The factorization formula is valid in all cases where the meson at the "upper" vertex is light, meaning that its mass is much smaller than the b-quark mass. The second term in the factorization formula (the term involving "factorized" six-quark operators) gives a power-suppressed contribution when the final-state meson at the "lower" vertex is a heavy meson (i.e., a charm meson), but its contribution is of leading power if this meson is also light. Aspects of this power counting will be discussed in more detail later.

Factorization is a property of decay amplitudes in the heavy-quark limit. Comparing the magnitude of "nonfactorizable" effects in kaon, charm and beauty decays, there can be little doubt about the relevance of the heavy-quark limit to understanding nonleptonic processes [13]. Yet, for phenomenological applications it is important to explore the structure of at least the leading power-suppressed corrections. While no complete classification of such corrections has been given to date, several classes of power-suppressed terms have been analyzed and their effects estimated. These estimates (with conservative errors) have been implemented in the phenomenological applications to be discussed later in this talk. Specifically, the corrections that have been analyzed are "chirally-enhanced" power corrections [11], weak annihilation contributions [12, 14], and power corrections due to nonfactorizable soft gluon exchange [15, 16, 17]. With the exception of the "chirally-enhanced" terms, no unusually large power corrections (i.e., corrections exceeding the naive expectation of 5–10%) have been identified so far. Nevertheless, it is important to refine and extend the estimates of power corrections. Fortunately, the QCD factorization approach has a wide range of applicability and makes many testable predictions. Ultimately, therefore, the data will give us conclusive evidence on the relevance of power-suppressed effects. Many tests can, in fact, already be done using existing data. Several examples will now be discussed in detail.

Tests of Factorization in $B \to D^{(*)}L$ Decays

In B decays into a heavy–light final state, when the light meson is produced at the "upper" vertex, the factorization formula assumes its simplest form. Then only the form factor term (the first graph on the right-hand side in Figure 2) contributes at leading power. This is also the place where QCD factorization is best established theoretically. In [12], the systematic cancellation of soft and collinear singularities was demonstrated explicitly at two-loop order. The proof of these cancellations has recently been extended to all orders in perturbation theory [18]. In order to complete a rigorous proof of factorization one would still have to show that the hard-scattering kernels are free of endpoint singularities stronger than $1/x$ or $1/(1-x)$ as one of the quarks in the light meson becomes a soft parton. It has been demonstrated that the kernels tend to a constant (modulo logarithms) at the endpoints in the so-called "large-β_0 limit" of QCD, i.e., to order $\beta_0^{n-1}\alpha_s^n$ for arbitrary n in perturbation theory [17]. However, it is an open question whether such a smooth behavior persists in higher orders of full QCD.

Let us first consider the decays $\bar{B}^0 \to D^{(*)+}L^-$, where L denotes a light meson. In this case the flavor content of the final state is such that the light meson can only be produced at the "upper" vertex, so factorization applies. One finds that process-dependent "nonfactorizable" corrections from hard gluon exchange, though present, are numerically very small. All nontrivial QCD effects in the decay amplitudes are then described by a quasi-universal coefficient $|a_1(D^{(*)}L)| = 1.05 \pm 0.02 + O(\Lambda_{\rm QCD}/m_b)$ [12]. For a given decay channel this coefficient can be determined experimentally from the ratio [8]

$$\frac{\Gamma(\bar{B}^0 \to D^{*+}L^-)}{d\Gamma(\bar{B}^0 \to D^{*+}l^-\nu)/dq^2|_{q^2=m_L^2}} = 6\pi^2 |V_{ud}|^2 f_L^2\, |a_1(D^{(*)}L)|^2\,.$$

Using CLEO data one obtains $|a_1(D^*\pi)| = 1.08 \pm 0.07$, $|a_1(D^*\rho)| = 1.09 \pm 0.10$, and $|a_1(D^*a_1)| = 1.08 \pm 0.11$, in good agreement with theory. This is a first indication that power corrections in these modes are under control, but more precise data are required for a firm conclusion. Other tests of factorization in B decays to heavy–light final states have been discussed in [12, 19, 20].

Recently, the experimental observation of unexpectedly large rates for color-suppressed decays such as $\bar{B}^0 \to D^{0(*)}\pi^0$ [21, 22] has attracted some attention. QCD factorization does not allow us to calculate the amplitudes for these processes in a reliable way. It predicts that these amplitudes are power-suppressed with respect to the corresponding $\bar{B}^0 \to D^{+(*)}\pi^-$ amplitudes, but only by one power of $\Lambda_{\rm QCD}/m_c$. Specifically, the prediction is that a certain ratio of isospin amplitudes approaches unity in the heavy-quark limit: $A_{1/2}/(\sqrt{2}A_{3/2}) = 1 + O(\Lambda_{\rm QCD}/m_c)$ [12]. Considering that charm is not a particularly heavy quark, we find that this scaling law is respected by the experimental data, which give $A_{1/2}/(\sqrt{2}A_{3/2}) = (0.70 \pm 0.11)\, e^{\pm i(27\pm 7)^\circ}$ for $B \to D\pi$ and $(0.72 \pm 0.08)\, e^{\pm i(21\pm 8)^\circ}$ for $B \to D^*\pi$ [13]. Assuming the hierarchy $m_b \gg m_c > \Lambda_{\rm QCD}$, a rough theoretical estimate of the amplitude ratio, $A_{1/2}/(\sqrt{2}A_{3/2}) \sim 0.75\, e^{-15^\circ i}$, had been obtained prior to the observation of the color-suppressed decays [12]. It anticipated the correct order of magnitude of the deviation from the heavy-quark limit.

Tests of Factorization in $B \to K^*\gamma$ Decays

The QCD factorization approach not only applies to nonleptonic decays, but also to other exclusive processes such as $B \to V\gamma$ and $B \to V\, l^+l^-$ (where $V = K^*, \rho, \ldots$ is a vector meson) [23, 24]. The resulting factorization formula is similar (but simpler) to that for B decays into two light mesons. Therefore, the study of exclusive radiative transitions not only extends the range of applicability of the method, but also provides a new testing ground for the factorization idea.

Interestingly, the analysis of isospin-breaking effects in radiative B decays gives a direct probe of power corrections to the factorization formula. Experimentally, it is found that [25, 26, 27]

$$\Delta_{0-} \equiv \frac{\Gamma(\bar{B}^0 \to \bar{K}^{*0}\gamma) - \Gamma(B^- \to \bar{K}^{*-}\gamma)}{\Gamma(\bar{B}^0 \to \bar{K}^{*0}\gamma) + \Gamma(B^- \to \bar{K}^{*-}\gamma)} = 0.11 \pm 0.07 \,,$$

indicating (albeit with a large error) that isospin-breaking effects could be as large as 10% at the level of the decay amplitudes. Such effects are absent in the heavy-quark limit. A detailed theoretical analysis of the leading power-suppressed contributions leads to the prediction $\Delta_{0-} = (8.0^{+2.1}_{-3.2})\% \times (0.3/T_1^{B\to K^*})$ [28], where $T_1^{B\to K^*}$ is a tensor form factor, whose value is expected to be close to 0.3. By far the largest contribution to the result comes from an annihilation contribution involving the $(V-A)\otimes(V+A)$ penguin operator O_6 in the effective weak Hamiltonian. Therefore, the quantity Δ_{0-} is a sensitive probe of the magnitude and sign of the ratio $C_6/C_{7\gamma}$ of Wilson coefficients.

The above discussion shows that in the Standard Model one indeed expects a sizeable isospin breaking in the $B \to K^*\gamma$ decay amplitudes, in agreement with the current central experimental value. If the agreement persists as the data become more precise, this would not only test the penguin sector of the effective weak Hamiltonian, but also provide a quantitative test of factorization at the level of power corrections.

Tests of Factorization in $B \to \pi K, \pi\pi$ Decays

The factorization formula for B decays into two light mesons is more complicated because of the presence of the two types of contributions shown in the graphs on the right-hand side in Figure 2. The finding that these two topologies contribute at the same power in $\Lambda_{\rm QCD}/m_b$ is nontrivial [14] and relies on the heavy-quark scaling law $F^{B\to L}(0) \sim m_b^{-3/2}$ for heavy-to-light form factors. Whereas this scaling law has been obtained from several independent studies (see, e.g., [29, 30, 31]), it is not as rigorously established as the corresponding scaling law for heavy-to-heavy form factors. In the QCD factorization approach the kernels $T_{ij}^I(\mu)$ are of order unity, whereas the kernels $T_{ij}^{II}(\mu)$ contribute first at order α_s. Numerically, the latter ones give corrections of about 10–20% with respect to the leading terms. This is consistent with being of the same power but down by a factor of α_s. Therefore, the scaling laws that form the basis of the QCD factorization formula appear to work well empirically.

The factorization formula for B decays into two light mesons can be tested best by using decays that have negligible amplitude interference. In that way any sensitivity to the value of the weak phase γ is avoided. For a complete theoretical control over charmless hadronic decays one must control the magnitude of the tree topologies, the magnitude of the penguin topologies, and the relative strong-interaction phases between trees and penguins. It is important that these three key features can be tested separately. Once these tests are conclusive (and assuming they are successful), factorization can be used to constrain the parameters of the unitarity triangle. (Of course, alternative schemes such as pQCD [32] and "charming penguins" [33] must face the same tests.)

Magnitude of the Tree Amplitude. The magnitude of the leading $B \to \pi\pi$ tree amplitude can be probed in the decays $B^\pm \to \pi^\pm\pi^0$, which to an excellent approximation do not receive any penguin contributions. The QCD factorization approach makes an absolute prediction for the corresponding branching ratio [14],

$$\mathrm{Br}(B^\pm \to \pi^\pm\pi^0) = \left[5.3^{+0.8}_{-0.4}\,(\text{pars.}) \pm 0.3\,(\text{power})\right] \cdot 10^{-6} \times \left[\frac{|V_{ub}|}{0.0035}\,\frac{F_0^{B\to\pi}(0)}{0.28}\right]^2,$$

which compares well with the experimental result $(5.6 \pm 1.5) \times 10^{-6}$ (see Table 7 in [14] for a compilation of the experimental data on charmless hadronic B decays). The theoretical uncertainties quoted are due to input parameter variations and to the modeling of the leading power corrections. An additional large uncertainty comes from the present error on $|V_{ub}|$ and the semileptonic $B \to \pi$ form factor. The sensitivity to these quantities can be eliminated by taking the ratio

$$\frac{\Gamma(B^\pm \to \pi^\pm\pi^0)}{d\Gamma(\bar{B}^0 \to \pi^+ l^- \bar{\nu})/dq^2|_{q^2=0}} = 3\pi^2 f_\pi^2 \underbrace{|a_1^{(\pi\pi)} + a_2^{(\pi\pi)}|^2}_{1.33^{+0.20}_{-0.11}\,(\text{pars.}) \pm 0.07\,(\text{power})} = (0.68^{+0.11}_{-0.06})\,\mathrm{GeV}^2.$$

This prediction includes a sizeable ($\sim 25\%$) contribution of the hard-scattering term in the factorization formula (the lower graph on the right-hand side in Figure 2). Unfortunately, this ratio has not yet been measured experimentally.

Magnitude of the T/P Ratio. The magnitude of the leading $B \to \pi K$ penguin amplitude can be probed in the decays $B^\pm \to \pi^\pm K^0$, which to an excellent approximation do not receive any tree contributions. Combining it with the measurement of the tree amplitude just described, a tree-to-penguin ratio can be determined via the relation

$$\varepsilon_{\text{exp}} = \left|\frac{T}{P}\right| = \tan\theta_C\,\frac{f_K}{f_\pi}\left[\frac{2\mathrm{Br}(B^\pm \to \pi^\pm\pi^0)}{\mathrm{Br}(B^\pm \to \pi^\pm K^0)}\right]^{\frac{1}{2}} = 0.223 \pm 0.034\,.$$

The quoted experimental value of this ratio is in good agreement with the theoretical prediction $\varepsilon_{\text{th}} = 0.24 \pm 0.04\,(\text{pars.}) \pm 0.04\,(\text{power}) \pm 0.05\,(V_{ub})$ [14], which is independent of form factors but proportional to $|V_{ub}/V_{cb}|$. This is a highly nontrivial test of the QCD factorization approach. Recall that when the first measurements of charmless hadronic decays appeared several authors remarked that the penguin amplitudes were much larger than expected based on naive factorization models. We now see that QCD factorization reproduces naturally (i.e., for central values of all input parameters) the correct magnitude of the tree-to-penguin ratio. This observation also shows that there is no need to supplement the QCD factorization predictions in an ad hoc way by adding enhanced phenomenological penguin amplitudes, such as the "nonperturbative charming penguins" introduced in [33]. In their most recent paper [34], the advocates of charming penguins parameterize the effects of these animals in terms of a nonperturbative "bag

TABLE 1. Direct CP asymmetries in $B \to \pi K$ decays

	Experiment [35, 36, 37, 38]	Theory: Beneke et al. [14]	Keum et al. [32]	Ciuchini et al. [33]
$A_{CP}(\pi^+K^-)$ (%)	-4.8 ± 6.8	5 ± 9	-18	$\pm(17 \pm 6)$
$A_{CP}(\pi^0K^-)$ (%)	-9.6 ± 11.9	7 ± 9	-15	$\pm(18 \pm 6)$
$A_{CP}(\pi^-\bar{K}^0)$ (%)	-4.7 ± 13.9	1 ± 1	-2	$\pm(3 \pm 3)$

parameter" $\hat{B}_1 = (0.13 \pm 0.02)\, e^{i(188 \pm 82)^\circ}$ fitted to the data on charmless decays. By definition, this parameter contains the contribution from the perturbative charm loop, which is calculable in QCD factorization. Using the factorization approach as described in [14] we find that $\hat{B}_1^{\rm fact} = (0.09\,^{+0.03+0.04}_{-0.02-0.02})\, e^{i(185 \pm 3 \pm 21)^\circ}$, where the errors are due to input parameter variations and the estimate of power corrections. The perturbative contribution to the central value is 0.08; the remaining 0.01 is mainly due to weak annihilation. We conclude that, within errors, QCD factorization can account for the "charming penguin bag parameter", which is, in fact, dominated by short-distance physics.

Strong Phase of the T/P Ratio. QCD factorization predicts that (most) strong-interaction phases in charmless hadronic B decays are parametrically suppressed in the heavy-quark limit, i.e., $\sin\phi_{\rm st} = O[\alpha_s(m_b), \Lambda_{\rm QCD}/m_b]$. This implies small direct CP asymmetries since, e.g., $A_{\rm CP}(\pi^+K^-) \simeq -2|\frac{T}{P}|\sin\gamma \sin\phi_{\rm st}$. The suppression results as a consequence of systematic cancellations of soft contributions, which are missed in phenomenological models of final-state interactions. In many other schemes the strong-interaction phases are predicted to be much larger, and therefore larger CP asymmetries are expected. Table 1 shows that first experimental data provide no evidence for large direct CP asymmetries in $B \to \pi K$ decays. However, the errors are still too large to draw a definitive conclusion that would allow us to distinguish between different theoretical predictions.

Remarks on Sudakov Logarithms

In recent years, Li and collaborators have proposed an alternative scheme for calculating nonleptonic B decay amplitudes based on a perturbative hard-scattering approach [32]. From a conceptual point of view, the main difference between QCD factorization and this so-called pQCD approach lies in the latter's assumption that Sudakov form factors effectively suppress soft-gluon exchange in diagrams such as those shown in the graphs on the right-hand side in Figure 2. As a result, the $B \to \pi$ and $B \to K$ form factors are assumed to be perturbatively calculable. This changes the counting of powers of α_s. In particular, the nonfactorizable gluon exchange diagrams included in the QCD factorization approach, which are crucial in order to cancel the scale and scheme-dependence in the predictions for the decay amplitudes, are formally of order α_s^2 in the pQCD scheme and consequently are left out. Thus, to the considered order there are no loop graphs that could give rise to strong-interaction phases in that scheme. (However, in [32] large phases are claimed to arise from on-shell poles of massless propagators in tree diagrams. These phases are entirely dominated by soft physics. Hence, the prediction of large direct CP asymmetries in the pQCD approach rests on assumptions that are strongly model dependent.)

The assumption of Sudakov suppression in hadronic B decays is questionable, because the relevant "large" scale $Q^2 \sim m_b\Lambda_{\mathrm{QCD}} \sim 1\,\mathrm{GeV}^2$ is in fact not large for realistic b-quark masses. Indeed, one finds that the pQCD calculations are very sensitive to details of the $p_\perp$ dependence of the wave functions [39]. This sensitivity to infrared physics invalidates the original assumption of an effective suppression of soft contributions. The argument just presented leaves open the conceptual question whether Sudakov logarithms are relevant in the asymptotic limit $m_b \to \infty$. This question has not yet been answered in a satisfactory way.

NEW CONSTRAINTS ON THE UNITARITY TRIANGLE

The QCD factorization approach, combined with a conservative estimate of power corrections, offers several new strategies to derive constraints on CKM parameters. This has been discussed at length in [14], to which we refer the reader for details. Some of these strategies will be illustrated below. Note that the applications of QCD factorization are not limited to computing branching ratios. The approach is also useful in combination with other ideas based on flavor symmetries and amplitude relations. In this way, strategies can be found for which the residual hadronic uncertainties are simultaneously suppressed by three small parameters, since they vanish in the heavy-quark limit ($\sim \Lambda_{\mathrm{QCD}}/m_b$), the limit of SU(3) flavor symmetry ($\sim (m_s - m_q)/\Lambda_{\mathrm{QCD}}$), and the large-$N_c$ limit ($\sim 1/N_c$).

Determination of γ *with minimal theory input.* Some years ago, Rosner and the present author have derived a bound on γ by combining measurements of the ratios $\varepsilon_{\mathrm{exp}} = |T/P|$ and $R_* = \frac{1}{2}\Gamma(B^\pm \to \pi^\pm K^0)/\Gamma(B^\pm \to \pi^0 K^\pm)$ with the fact that for an arbitrary strong phase $-1 \le \cos\phi_{\mathrm{st}} \le 1$ [40]. The model-independent observation that $\cos\phi_{\mathrm{st}} = 1$ up to second-order corrections to the heavy-quark limit can be used to turn this bound into a determination of γ (once $|V_{ub}|$ is known). The resulting constraints in the $(\bar\rho,\bar\eta)$ plane, obtained under the conservative assumption that $\cos\phi_{\mathrm{st}} > 0.8$ (corresponding to $|\phi_{\mathrm{st}}| < 37^\circ$) are shown in the left-hand plot in Figure 3 for several illustrative values of the ratio R_*. Note that for $0.8 < R_* < 1.1$ (the range preferred by the Standard Model) the theoretical uncertainty reflected by the widths of the bands is smaller than for any other constraint on $(\bar\rho,\bar\eta)$ except for the one derived from the $\sin 2\beta$ measurement. With present data the Standard Model is still in good shape, but it will be interesting to see what happens when the experimental errors are reduced.

Determination of $\sin 2\alpha$*.* With the help of QCD factorization it is possible to control the "penguin pollution" in the time-dependent CP asymmetry in $B \to \pi^+\pi^-$ decays, defined such that $S_{\pi\pi} = \sin 2\alpha \cdot [1 + O(P/T)]$. This is illustrated in the right-hand plot in Figure 3, which shows the constraints imposed by a measurement of $S_{\pi\pi}$ in the $(\bar\rho,\bar\eta)$ plane. It follows that even a result for $S_{\pi\pi}$ with large experimental errors would imply a useful constraint on the unitarity triangle. A first, preliminary measurement of the

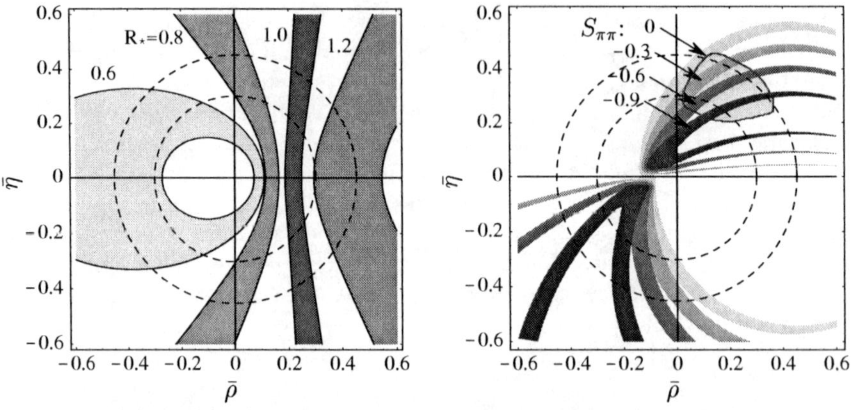

FIGURE 3. Left: Allowed regions in the $(\bar{\rho},\bar{\eta})$ plane corresponding to $\varepsilon_{\mathrm{exp}} = 0.22$ and different values of the ratio R_* as indicated. The widths of the bands reflect the total theoretical uncertainty. The current experimental values are $\varepsilon_{\mathrm{exp}} = 0.22 \pm 0.03$ and $R_* = 0.71 \pm 0.14$. Right: Allowed regions in the $(\bar{\rho},\bar{\eta})$ plane corresponding to different values of the mixing-induced CP asymmetry $S_{\pi\pi}$. The widths of the bands reflect the total theoretical uncertainty. The corresponding bands for positive values of $S_{\pi\pi}$ are obtained by a reflection about the $\bar{\rho}$ axis. The bounded light area is the allowed region obtained from the standard global fit of the unitarity triangle [1].

asymmetry has been presented by the BaBar Collaboration this summer. Their result is $S_{\pi\pi} = 0.03^{+0.53}_{-0.56} \pm 0.11$ [38].

Global Fit to $B \to \pi K, \pi\pi$ Branching Ratios. Various ratios of CP-averaged $B \to \pi K, \pi\pi$ branching fractions exhibit a strong dependence on γ and $|V_{ub}|$, or equivalently, on the parameters $\bar{\rho}$ and $\bar{\eta}$ of the unitarity triangle. From a global analysis of the experimental data in the context of the QCD factorization approach it is possible to derive constraints in the $(\bar{\rho},\bar{\eta})$ plane in the form of regions allowed at various confidence levels. The results are shown in Figure 4. The best fit of the QCD factorization theory to the data yields an excellent χ^2/n_{dof} of less than 0.5. (We should add at this point that we disagree with the implementation of our approach presented in [34] and, in particular, with the numerical results labeled "BBNS" in Table II of that paper, which led the authors to the premature conclusion that the "theory of QCD factorization ... is insufficient to fit the data". Even restricting $(\bar{\rho},\bar{\eta})$ to lie within the narrow ranges adopted by these authors, we can find parameter sets for which QCD factorization fits the data with a good χ^2/n_{dof} of less than 1.5.)

The results of this global fit are compatible with the standard CKM fit using semileptonic decays, K–$\bar{K}$ mixing and B–$\bar{B}$ mixing ($|V_{ub}|$, $|V_{cb}|$, ε_K, Δm_d, Δm_s, $\sin 2\beta$), although the fit prefers a slightly larger value of γ and/or a smaller value of $|V_{ub}|$. The combination of the results from rare hadronic B decays with $|V_{ub}|$ from semileptonic decays excludes $\bar{\eta} = 0$ at 95% CL, thus showing first evidence for the existence of a CP-violating phase in the bottom sector. In the near future, when the data become more precise, this will provide a powerful test of the CKM paradigm.

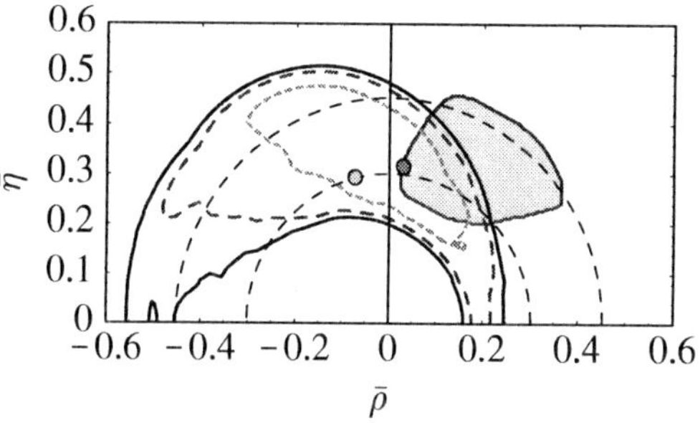

Decay Mode	Best Fit	Exp. Average
$B^0 \to \pi^+\pi^-$	4.6	4.4 ± 0.9
$B^\pm \to \pi^\pm\pi^0$	5.3	5.6 ± 1.5
$B^0 \to \pi^\mp K^\pm$	17.9	17.2 ± 1.5
$B^\pm \to \pi^0 K^\pm$	11.3	12.1 ± 1.7
$B^\pm \to \pi^\pm K^0$	17.7	17.2 ± 2.5
$B^0 \to \pi^0 K^0$	7.1	10.3 ± 2.5

FIGURE 4. 95% (solid), 90% (dashed) and 68% (short-dashed) confidence level contours in the $(\bar{\rho},\bar{\eta})$ plane obtained from a global fit of QCD factorization results to the CP-averaged $B \to \pi K, \pi\pi$ branching fractions. The dark dot shows the overall best fit, whereas the light dot indicates the best fit for the default choice of all theory input parameters. The table compares the best fit values for the various CP-averaged branching fractions (in units of 10^{-6}) with the world average data.

OUTLOOK

The QCD factorization approach provides the theoretical framework for a systematic analysis of hadronic and radiative exclusive *B* decay amplitudes based on the heavy-quark expansion. This theory has already passed successfully several nontrivial tests, and will be tested more thoroughly with more precise data. A new effective field-theory language appropriate to QCD factorization is emerging in the form of the collinear–soft effective theory. Ultimately, the developments reviewed in this talk may lead to theoretical control over a vast variety of exclusive *B* decays, giving us new constraints on the unitarity triangle.

ACKNOWLEDGMENTS

I wish to thank the organizers of *Heavy Flavours 9* for their efforts to run this conference successfully under the difficult circumstances created by the terrorist attacks of 11 September 2001. This work was supported in part by the National Science Foundation.

REFERENCES

1. A. Höcker, H. Lacker, S. Laplace and F. Le Diberder, *Eur. Phys. J. C* **21**, 225 (2001) [hep-ph/0104062]; for an updated analysis, see `http://www.slac.stanford.edu/~laplace/ckmfitter.html`.
2. A. L. Kagan and M. Neubert, *Phys. Lett. B* **492**, 115 (2000) [hep-ph/0007360];
Y. Grossman, M. Neubert and A. L. Kagan, *JHEP* **9910**, 029 (1999) [hep-ph/9909297].
3. D. Atwood, I. Dunietz and A. Soni, *Phys. Rev. Lett.* **78**, 3257 (1997) [hep-ph/9612433].
4. I. Dunietz and R. G. Sachs, *Phys. Rev. D* **37**, 3186 (1988) [Erratum: *ibid. D* **39**, 3515 (1988)];
I. Dunietz, *Phys. Lett. B* **427**, 179 (1998) [hep-ph/9712401].
5. M. Gronau and D. London, *Phys. Rev. Lett.* **65**, 3381 (1990).

6. A. E. Snyder and H. R. Quinn, *Phys. Rev. D* **48**, 2139 (1993).
7. C. W. Bauer, S. Fleming, D. Pirjol and I. W. Stewart, *Phys. Rev. D* **63**, 114020 (2001) [hep-ph/0011336];
C. W. Bauer and I. W. Stewart, *Phys. Lett. B* **516**, 134 (2001) [hep-ph/0107001];
C. W. Bauer, D. Pirjol and I. W. Stewart, hep-ph/0109045.
8. J. D. Bjorken, *Nucl. Phys. Proc. Suppl.* **11**, 325 (1989).
9. M. J. Dugan and B. Grinstein, *Phys. Lett. B* **255**, 583 (1991).
10. H. D. Politzer and M. B. Wise, *Phys. Lett. B* **257**, 399 (1991).
11. M. Beneke, G. Buchalla, M. Neubert and C. T. Sachrajda, *Phys. Rev. Lett.* **83**, 1914 (1999) [hep-ph/9905312].
12. M. Beneke, G. Buchalla, M. Neubert and C. T. Sachrajda, *Nucl. Phys. B* **591**, 313 (2000) [hep-ph/0006124].
13. M. Neubert and A. A. Petrov, *Phys. Lett. B* **519**, 50 (2001) [hep-ph/0108103].
14. M. Beneke, G. Buchalla, M. Neubert and C. T. Sachrajda, *Nucl. Phys. B* **606**, 245 (2001) [hep-ph/0104110].
15. A. Khodjamirian, *Nucl. Phys. B* **605**, 558 (2001) [hep-ph/0012271].
16. C. N. Burrell and A. R. Williamson, *Phys. Rev. D* **64**, 034009 (2001) [hep-ph/0101190].
17. T. Becher, M. Neubert and B. D. Pecjak, *Nucl. Phys. B* **619**, 538 (2001) [hep-ph/0102219].
18. C. W. Bauer, D. Pirjol and I. W. Stewart, *Phys. Rev. Lett.* **87**, 201806 (2001) [hep-ph/0107002].
19. Z. Ligeti, M. Luke and M. B. Wise, *Phys. Lett. B* **507**, 142 (2001) [hep-ph/0103020].
20. M. Diehl and G. Hiller, *JHEP* **0106**, 067 (2001) [hep-ph/0105194].
21. E. von Toerne (CLEO Collaboration), talk at the International Europhysics Conference on High-Energy Physics, Budapest, Hungary, July 2001.
22. K. Abe *et al.* [Belle Collaboration], hep-ex/0107048.
23. M. Beneke, T. Feldmann and D. Seidel, *Nucl. Phys. B* **612**, 25 (2001) [hep-ph/0106067].
24. S. W. Bosch and G. Buchalla, *Nucl. Phys. B* **621**, 459 (2002) [hep-ph/0106081].
25. T. E. Coan *et al.* [CLEO Collaboration], *Phys. Rev. Lett.* **84**, 5283 (2000) [hep-ex/9912057].
26. Y. Ushiroda [Belle Collaboration], hep-ex/0104045.
27. J. Nash [BaBar Collaboration], talk at the 20th International Symposium on Lepton and Photon Interactions, Rome, Italy, July 2001.
28. A. L. Kagan and M. Neubert, hep-ph/0110078.
29. V. L. Chernyak and I. R. Zhitnitsky, *Nucl. Phys. B* **345**, 137 (1990).
30. A. Ali, V. M. Braun and H. Simma, *Z. Phys. C* **63**, 437 (1994) [hep-ph/9401277].
31. E. Bagan, P. Ball and V. M. Braun, *Phys. Lett. B* **417**, 154 (1998) [hep-ph/9709243];
P. Ball and V. M. Braun, *Phys. Rev. D* **58**, 094016 (1998) [hep-ph/9805422].
32. Y. Keum, H. Li and A. I. Sanda, *Phys. Lett. B* **504**, 6 (2001) [hep-ph/0004004]; *Phys. Rev. D* **63**, 054008 (2001) [hep-ph/0004173];
Y. Keum and H. Li, *Phys. Rev. D* **63**, 074006 (2001) [hep-ph/0006001].
33. M. Ciuchini, E. Franco, G. Martinelli and L. Silvestrini, *Nucl. Phys. B* **501**, 271 (1997) [hep-ph/9703353];
M. Ciuchini, E. Franco, G. Martinelli, M. Pierini and L. Silvestrini, *Phys. Lett. B* **515**, 33 (2001) [hep-ph/0104126].
34. M. Ciuchini, E. Franco, G. Martinelli, M. Pierini and L. Silvestrini, hep-ph/0110022.
35. S. Chen *et al.* [CLEO Collaboration], *Phys. Rev. Lett.* **85**, 525 (2000) [hep-ex/0001009].
36. K. Abe *et al.* [Belle Collaboration], *Phys. Rev. D* **64**, 071101 (2001) [hep-ex/0106095].
37. B. Aubert *et al.* [BaBar Collaboration], *Phys. Rev. Lett.* **87**, 151802 (2001) [hep-ex/0105061].
38. B. Aubert *et al.* [BaBar Collaboration], hep-ex/0107074.
39. S. Descotes and C. T. Sachrajda, hep-ph/0109260.
40. M. Neubert and J. L. Rosner, *Phys. Lett. B* **441**, 403 (1998) [hep-ph/9808493];
M. Neubert, *JHEP* **9902**, 014 (1999) [hep-ph/9812396].

A Novel PQCD Approach in Charmless B-meson Decays

Y.-Y. Keum [1]

Institute of Physics, Academia Sinica, Nankang Taipei 151, Taiwan (R.O.C.)

Abstract. We discuss a novel pertubative QCD approach on the exclusive non-leptonic two body B-meson decays. We briefly review its ingredients and some important theoretical issues on the factorization approaches. We show numerical results which is compatible with recent experimantal data for the charmless B-meson decays.

INTRODUCTION

The aim of the study on weak decay in B-meson is two folds: (1) To determine precisely the elements of CKM matrix and to explore the origin of CP-violation in low energy scale, (2) To understand strong interaction physics related to the confinements of quarks and gluons within hadrons.

Both tasks complement each other. An understanding of the connection between quarks and hadron properties is a necessary prerequeste for a precise deterination of CKM matrix elements and CP-violating phases, so called KM-phase.

The theoretical description of hadronic weak decays is difficult since nonperturbative QCD interactions are involved. This makes a difficult to interpret correctly data from asymmetric B-factories and to seek the origin of CP violation. In the case of B-meson decays into two light mesons, we can explain roughly branching ratios by using the factorization approximation [1, 2]. Since B-meson is quite heavy, when it decays into two light mesons, the final-state mesons are moving so fast that it is difficult to exchange gluons between final-state mesons. So we can express the amplitude in terms of the product of weak decay constant and transition form factors by the factorization (color-transparancy) argument. In this approach we can not calculate non-factorizable contributions and annihilation contributions even though which is not dominant. Because of this weakness, asymmetry of CP violation can not be predicted precisely.

Recently two different QCD approaches beyond naive and general factorization assumption [1, 2, 3, 4] was proposed: (1) QCD-factorization in heavy quark limit [5, 6] in which non-factorizable terms and a_i are calculable in some cases. (2) A Novel PQCD approach [7, 8, 9] including the resummation effects of the transverse momentum carried by partons inside meson. In this talk, I discuss some important theoretical issues in the PQCD factorization and numerical results for charmless B-decays.

---

[1] Work supported by National Science Council of R.O.C. under the Grant No. NSC-90-2811-M-002

CP618, *Heavy Flavor Physics: Ninth International Symposium*, edited by A. Ryd and F. C. Porter

INGREDIENTS OF PQCD

Factorization in PQCD. The idea of pertubative QCD is as follows: When heavy B-meson decays into two light mesons, the hard process is dominant. Since two light mesons fly so fast with large momentum, it is reasonable assumptions that the final-state interaction is not important for charmless B-decays and hard gluons are needed to boost the resting spectator quark to get large momentum and finally to hadronize a fast moving final meson. So the dominant process is that one hard gluon is exchanged between specator quark and other four quarks.

Let's start with the lowest-order diagram of $B \to K\pi$. The soft divergences in the $B \to \pi$ form factor can be factorized into a light-cone B meson wave function, and the collinear divergences can be factorized into a pion distribution amplitude. The finite pieces of them is absorbed into the hard part. Then in the natural way we can factorize amplitude into two pieces: $G \equiv H(Q,\mu) \otimes \Phi(m,\mu)$ where H stands for hard part which is calculable with a perturbative way, and Φ is wave functions which belong to the non-perturbative physics.

PQCD adopt the three scale factorization theorem [10] based on the perturbative QCD formalism by Brodsky and Lepage [11], and Botts and Sterman [12], with the inclusion of the transverse momentum components which was carried by partons inside meson.

We have three different scales: electroweak scale: M_W, hard interaction scale: $t \sim O(\sqrt{(}\bar{\Lambda}m_b))$, and the factorization scale: $1/b$ where b is the conjugate variable of parton transverse momenta. The dynamics below $1/b$ is completely non-perturbative and can be parameterized into meson wave funtions which is universal and process independent. In our analysis we use the results of light-cone distribution amplitudes (LCDAs) by Ball [13, 14] with light-cone sum rule.

The ampltitude in PQCD is expressed as

$$A \sim C(t) \times H(t) \times \Phi(x) \times \exp\left[-s(P,b) - 2\int_{1/b}^{t} \frac{d\mu}{\mu}\, \gamma_q(\alpha_s(\mu))\right] \tag{1}$$

where $C(t)$ are Wilson coefficients, $\Phi(x)$ are meson LCDAs and variable t is the factorized scale in hard part.

Sudakov Suppression Effects. When we include $k_\perp$, the double logarithms $\ln^2(Pb)$ are generated from the overlap of collinear and soft divergence in radiative corrections to meson wave functions, where P is the dominant light-cone component of a meson momentum. The resummation of these double logarithms leads to a Sudakov form factor $exp[-s(P,b)]$ in Eq.(1), which suppresses the long distance contributions in the large b region, and vanishes as $b > 1/\Lambda_{QCD}$.

This suppression renders $k_\perp^2$ flowing into the hard amplitudes of order

$$k_\perp^2 \sim O(\bar{\Lambda}M_B)\,. \tag{2}$$

The off-shellness of internal particles then remain of $O(\bar{\Lambda}M_B)$ even in the end-point region, and the singularities are removed. This mechanism is so-called Sudakov suppression.

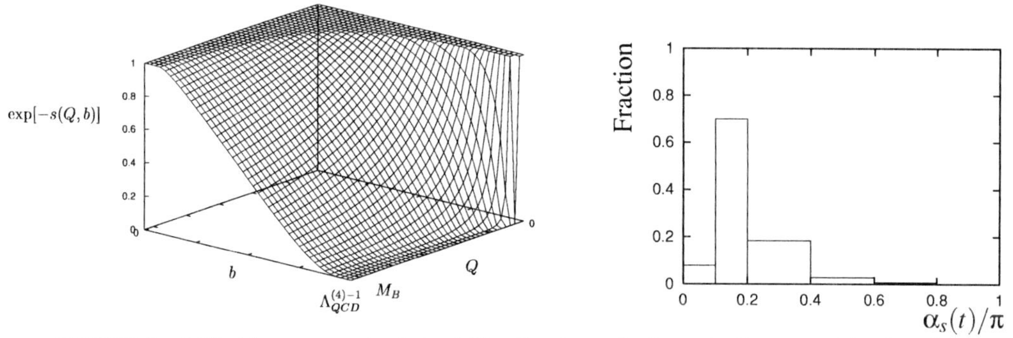

FIGURE 1. (a)Sudakov suppression factor (b)Fractional contribution to the $B \to \pi$ transition form factor $F^{B\pi}$ as a function of $\alpha_s(t)/\pi$.

Du *et al.* have studied the Sudakov effects in the evaluation of nonfactorizable amplitudes [15]. If equating these amplitudes with Sudakov suppression included to the parametrization in QCDF, it was observed that the corresponding cutoffs are located in the reasonable range proposed by Beneke *et al.* [6]. Sachrajda *et al.* have expressed an opposite opinion on the effect of Sudakov suppression in [16]. However, their conclusion was drawn based on a very sharp B meson wave function, which is not favored by experimental data.

Here I would like to commnent on the negative opinions on the large $k_\perp^2 \sim O(\bar{\Lambda}M_B)$. It is easy to understand the increase of $k_\perp^2$ from $O(\bar{\Lambda}^2)$, carried by the valence quarks which just come out of the initial meson wave functions, to $O(\bar{\Lambda}M_B)$, carried by the quarks which are involved in the hard weak decays. Consider the simple deeply inelastic scattering of a hadron. The transverse momentum $k_\perp$ carried by a parton, which just come out of the hadron distribution function, is initially small. After infinite many gluon radiations, $k_\perp$ becomes of $O(Q)$, when the parton is scattered by the highly virtual photon, where Q is the large momentum transfer from the photon. The evolution of the hadron distribution function from the low scale to Q is described by the Dokshitzer-Gribov-Lipatov-Altarelli-Parisi (DGLAP) equation [17, 18]. The mechanism of the DGLAP evolution in DIS is similar to that of the Sudakov evolution in exclusive B meson decays. The difference is only that the former is the consequence of the single-logarithm resummation, while the latter is the consequence of the double-logarithm resummation.

By including Sudakov effects, all contributions of the $B \to \pi$ form factor comes from the region with $\alpha_s/\pi < 0.3$ [8] as shown in Figure 1. It indicate that our PQCD results are well within the perturbative region.

Threshold Resummation. The other double logarithm is $\alpha_s \ln^2(1/x)$ from the end point region of the momentum fraction x [19]. This double logarithm is generated by the corrections of the hard part in Figure 2. This double logarithm can be factored out of the hard amplitude systematically, and its resummation introduces a Sudakov factor $S_t(x) = 1.78[x(1-x)]^c$ with $c = 0.3$ into PQCD factorization formula. The Sudakov factor from threshold resummation is universal, independent of flavors of internal quarks, twists and topologies of hard amplitudes, and decay modes.

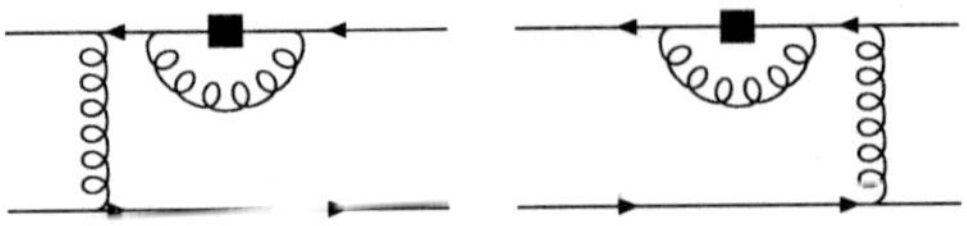

FIGURE 2. The diagrams generate double logarithm corrections for the threshold resummation.

Threshold resummation[19] and $k_\perp$ resummation [20, 12, 21] arise from different subprocesses in PQCD factorization and suppresses the end-point contributions, making PQCD evaluation of exclusive B meson decays reliable. If excluding resummation effects, the PQCD predictions for the $B \to K$ form factors are infrared divergent. If including only $k_\perp$ resummation, the PQCD predictions are finite. However, the two-parton twist-3 contributions are still huge, so that the $B \to K$ form factors have an unreasonably large value $F^{BK} \sim 0.57$ at maximal recoil. The reason is that the double logarithms $\alpha_s \ln^2 x$ have not been organized. If including both resummations, we obtain the reasonable result $F^{BK} \sim 0.35$. These studies indicate the importance of resummations in PQCD analyses of B meson decays. In conclusion, if the PQCD analysis of the heavy-to-light form factors is performed self-consistently, there exist no end-point singularities, and both twist-2 and twist-3 contributions are well-behaved.

Power Counting Rule in PQCD. The power behaviors of various topologies of diagrams for two-body nonleptonic B meson decays with the Sudakov effects taken into account has been discussed in details in [22]. The relative importance is summarized below:

$$\text{emission} : \text{annihilation} : \text{nonfactorizable} = 1 : \frac{2m_0}{M_B} : \frac{\bar{\Lambda}}{M_B}, \tag{3}$$

with m_0 being the chiral symmetry breaking scale. The scale m_0 appears because the annihilation contributions are dominated by those from the $(V-A)(V+A)$ penguin operators, which survive under helicity suppression. In the heavy quark limit the annihilation and nonfactorizable amplitudes are indeed power-suppressed compared to the factorizable emission ones. Therefore, the PQCD formalism for two-body charmless nonleptonic B meson decays coincides with the factorization approach as $M_B \to \infty$. However, for the physical value $M_B \sim 5$ GeV, the annihilation contributions are essential. In Table 1 we can easily check the relative size of the different topology in Eq.(3) by the peguin contribution for W-emission ($f_\pi F^P$), annihilation($f_B F_a^P$) and non-factorizable(M^P) contributions.

Note that all the above topologies are of the same order in α_s in PQCD. The nonfactorizable amplitudes are down by a power of $1/m_b$, because of the cancellation between a pair of nonfactorizable diagrams, though each of them is of the same power as the factorizable one. I emphasize that it is more appropriate to include the nonfactorizable

TABLE 1. Amplitudes for the $B_d^0 \to K^+\pi^-$ decay where F (M) denotes factorizable (non-factorizable) contributions, P (T) denotes the penguin (tree) contributions, and a denotes the annihilation contributions. Here we adopted $\phi_3 = 80^0$, $R_b = 0.38$.

Amplitudes	Left-handed gluon exchange	Right-handed gluon exchange	Total
$Re(f_\pi F^T)$	$7.07 \cdot 10^{-2}$	$3.16 \cdot 10^{-2}$	$1.02 \cdot 10^{-1}$
$Im(f_\pi F^T)$	–	–	–
$Re(f_\pi F^P)$	$-5.52 \cdot 10^{-3}$	$-2.44 \cdot 10^{-3}$	$-7.96 \cdot 10^{-3}$
$Im(f_\pi F^P)$	–	–	–
$Re(f_B F_a^P)$	$4.13 \cdot 10^{-4}$	$-6.51 \cdot 10^{-4}$	$-2.38 \cdot 10^{-4}$
$Im(f_B F_a^P)$	$2.73 \cdot 10^{-3}$	$1.68 \cdot 10^{-3}$	$4.41 \cdot 10^{-3}$
$Re(M^T)$	$7.06 \cdot 10^{-3}$	$-7.17 \cdot 10^{-3}$	$-1.11 \cdot 10^{-4}$
$Im(M^T)$	$-1.10 \cdot 10^{-2}$	$1.35 \cdot 10^{-2}$	$2.59 \cdot 10^{-3}$
$Re(M^P)$	$-3.05 \cdot 10^{-4}$	$3.07 \cdot 10^{-4}$	$2.17 \cdot 10^{-6}$
$Im(M^P)$	$4.50 \cdot 10^{-4}$	$-5.29 \cdot 10^{-4}$	$-7.92 \cdot 10^{-5}$
$Re(M_a^P)$	$2.03 \cdot 10^{-5}$	$-1.37 \cdot 10^{-4}$	$-1.16 \cdot 10^{-4}$
$Im(M_a^P)$	$-1.45 \cdot 10^{-5}$	$-1.27 \cdot 10^{-4}$	$-1.42 \cdot 10^{-4}$

contributions in a complete formalism. The factorizable internal-W emisson contributions are strongly suppressed by the vanishing Wilson coefficient a_2 in the $B \to J/\psi K^{(*)}$ decays [23], so that nonfactorizable contributions become dominant. In the $B \to D\pi$ decays, there is no soft cancellation between a pair of nonfactorizable diagrams, and nonfactorizable contributions are significant [23].

In QCDF the factorizable and nonfactorizable amplitudes are of the same power in $1/m_b$, but the latter is of next-to-leading order in α_s compared to the former. Hence, QCDF approaches FA in the heavy quark limit in the sense of $\alpha_s \to 0$. Briefly speaking, QCDF and PQCD have different counting rules both in α_s and in $1/m_b$. The former approaches FA logarithmically ($\alpha_s \propto 1/\ln m_b \to 0$), while the latter does linearly ($1/m_b \to 0$).

IMPORTANT THEORETICAL ISSUES

End Point Singularity and Form Factors. If calculating the $B \to \pi$ form factor $F^{B\pi}$ at large recoil using the Brodsky-Lepage formalism [11, 24], a difficulty immediately occurs. The lowest-order diagram for the hard amplitude is proportional to $1/(x_1 x_3^2)$, x_1 being the momentum fraction associated with the spectator quark on the B meson side. If the pion distribution amplitude vanishes like x_3 as $x_3 \to 0$ (in the leading-twist, *i.e.*, twist-2 case), $F^{B\pi}$ is logarithmically divergent. If the pion distribution amplitude is a constant as $x_3 \to 0$ (in the next-to-leading-twist, *i.e.*, twist-3 case), $F^{B\pi}$ even becomes linearly divergent. These end-point singularities have also appeared in the evaluation of the nonfactorizable and annihilation amplitudes in QCDF mentioned above.

TABLE 2. Branching ratios of $B \to \pi\pi$ and $K\pi$ decays with $\phi_3 = 80^0$, $R_b = 0.38$. Here we adopted $m_0^\pi = 1.3$ GeV and $m_0^K = 1.7$ GeV. Unit is 10^{-6}.

Decay Channel	CLEO	BELLE	BABAR	World Av.	PQCD
$\pi^+\pi^-$	$4.3^{+1.6}_{-1.4} \pm 0.5$	$5.6^{+2.3}_{-2.0} \pm 0.4$	$4.1 \pm 1.0 \pm 0.7$	4.4 ± 0.9	$7.0^{+2.0}_{-1.5}$
$\pi^+\pi^0$	$5.6^{+2.6}_{-2.3} \pm 1.7$	$7.8^{+3.8+0.8}_{-3.2-1.2}$	$5.1^{+2.0}_{-1.8} \pm 0.8$	5.6 ± 1.5	$3.7^{+1.3}_{-1.1}$
$\pi^0\pi^0$	< 5.7	–	–	–	$0.3 + 0.1$
$K^0\pi^\pm$	$18.2^{+4.6}_{-4.0} \pm 1.6$	$13.7^{+5.7+1.9}_{-4.8-1.8}$	$18.2^{+3.3}_{-3.0} \pm 2.0$	17.3 ± 2.7	$16.4^{+3.3}_{-2.7}$
$K^\pm\pi^\mp$	$17.2^{+2.5}_{-2.4} \pm 1.2$	$19.3^{+3.4+1.5}_{-3.2-0.6}$	$16.7 \pm 1.6 \pm 1.3$	17.3 ± 1.5	$15.5^{+3.1}_{-2.5}$
$K^\pm\pi^0$	$11.6^{+3.0+1.4}_{-2.7-1.3}$	$16.3^{+3.5+1.6}_{-3.3-1.8}$	$10.8^{+2.1}_{-1.9} \pm 1.0$	12.1 ± 1.7	$9.1^{+1.9}_{-1.5}$
$K^0\pi^0$	$14.6^{+5.9+2.4}_{-5.1-3.3}$	$16.0^{+7.2+2.5}_{-5.9-2.7}$	$8.2^{+3.1}_{-2.7} \pm 1.2$	10.4 ± 2.7	8.6 ± 0.3

When we include small parton transverse momenta $k_\perp$, we have

$$\frac{1}{x_1\, x_3^2 M_B^4} \quad \longrightarrow \quad \frac{1}{(x_3 M_B^2 + k_{3\perp}^2)\,[x_1 x_3 M_B^2 + (k_{1\perp} - k_{3\perp})^2]} \tag{4}$$

and the end-point singularity is smeared out.

In PQCD, we can calculate analytically space-like form factors for $B \to P,V$ transition and also time-like form factors for the annihilation process [22, 25].

Strong phases. While stong phases in FA and QCDF come from the Bander-Silverman-Soni (BSS) mechanism[26] and from the final state interaction (FSI), the dominant strong phase in PQCD come from the factorized annihilation diagram[7, 8, 9]. It has been argued that the two sources of strong phases in the FA and QCDF approaches are in fact strongly suppressed by the charm mass threshold and by the end-point behavior of meson wave functions.

Dynamical Penguin Enhancement vs Chiral Enhancement. As explained before, the hard scale is about 1.5 GeV. Since the RG evolution of the Wilson coefficients $C_{4,6}(t)$ increase drastically as $t < M_B/2$, while that of $C_{1,2}(t)$ remain almost constant, we can get a large enhancement effects from both wilson coefficents and matrix elements in PQCD.

In general the amplitude can be expressed as

$$Amp \sim [a_{1,2} \pm a_4 \pm m_0^{P,V}(\mu) a_6] \cdot < K\pi|O|B > \tag{5}$$

with the chiral factors $m_0^P(\mu) = m_P^2/[m_1(\mu) + m_2(\mu)]$ for pseudoscalr meson and $m_0^V = m_V$ for vector meson. To accommodate the $B \to K\pi$ data in the factorization and QCD-factorization approaches, one relies on the chiral enhancement by increasing the mass m_0 to as large values about 3 GeV at $\mu = m_b$ scale. So two methods accomodate large branching ratios of $B \to K\pi$ and it is difficult for us to distinguish two different methods in $B \to PP$ decays. However we can do it in $B \to PV$ because there is no chiral factor in LCDAs of the vector meson.

We can test whether dynamical enhancement or chiral enhancement is responsible for the large $B \to K\pi$ branching ratios by measuring the $B \to \phi K$ modes. In these modes

TABLE 3. Branching ratios of $B \to \phi K^{(*)}$ decays with $\phi_3 = 80^0$, $R_b = 0.38$. Here we adopted $m_0^{\pi} = 1.3$ GeV and $m_0^K = 1.7$ GeV. Unit is 10^{-6}.

Decay Channel	CLEO	BELLE	BABAR	PQCD
$\phi K^{\pm}$	$5.5^{+2.1}_{-1.8} \pm 0.6$	$11.2^{+2.2}_{-2.0} \pm 0.14$	$7.7^{+1.6}_{-1.4} \pm 0.8$	$10.2^{+3.9}_{-2.1}$
ϕK^0	< 12.3	$8.9^{+3.4}_{-2.7} \pm 1.0$	$8.1^{+3.1}_{-2.5} \pm 0.8$	$9.6^{+3.7}_{-2.0}$
$K^{*0}\pi^{\pm}$	$7.6^{+3.5}_{-3.0} \pm 1.6$	$19.4^{+4.2}_{-3.9} \pm 2.1^{+3.5}_{-6.8}$	$15.5 \pm 3.4 \pm 1.5$	$12.2^{+2.4}_{-2.0}$
$K^{*\pm}\pi^{\mp}$	22^{+8+4}_{-6-5}	$-$	$-$	$9.6^{+2.0}_{-1.6}$

penguin contributions dominate, such that their branching ratios are insensitive to the variation of the unitarity angle ϕ_3. According to recent works by Cheng *at al.* [27], the branching ratio of $B \to \phi K$ is $(2-7) \times 10^{-6}$ including 30% annihilation contributions in QCD-factorization approach. However PQCD predicts 10×10^{-6} [22, 28].

Fat Imaginary Penguin in Annihilation. There is a falklore that annihilation contribution is negligible compared to W-emission one. In this reason annihilation contribution was not included in the general factorization approach and the first paper on QCD-factorization by Beneke et al. [5]. In fact there is a suppression effect for the operators with structure $(V-A)(V-A)$ because of a mechanism similar to the helicity suppression for $\pi \to \mu\nu_{\mu}$. However annihilation from the operators $O_{5,6,7,8}$ with the structure $(S-P)(S+P)$ via Fiertz transformation survive under the helicity suppression and can get large imaginary value. The real part of factorized annihilation contribution becomes small because there is a cancellation between left-handed gluon exchanged one and right-handed gluon exchanged one as shown in Table 1. This mostly pure imaginary value of annihilation is a main source of large CP asymmetry in $B \to \pi^+\pi^-$ and $K^+\pi^-$. In Table 5 we summarize the CP asymmetry in $B \to K(\pi)\pi$ decays.

NUMERICAL RESULTS

Branching ratios and Ratios of CP-averaged rates. The PQCD approach allows us to calculate the amplitudes for charmless B-meson decays in terms of ligh-cone distribution amplitudes upto twist-3. We focus on decays whose branching ratios have already been measured. We take allowed ranges of shape parameter for the B-meson wave funtion as $\omega_B = 0.36 - 0.44$ which accomodate to reasonable form factors, $F^{B\pi}(0) = 0.27 - 0.33$ and $F^{BK}(0) = 0.31 - 0.40$. We use values of chiral factor with $m_0^{\pi} = 1.3 GeV$ and $m_0^K = 1.7 GeV$. Finally we obtain branching ratios for $B \to K(\pi)\pi$ [22, 29], $K\phi$ [22, 28] and $K^*\pi$, which is well agreed with present experimental data (see Table 2 and 3).

In order to reduce theoretical uncertainties from decay constant of B-meson and from light-cone distribution amplitudes, we consider rates of CP-averaged branching ratios, which is presented in Table 4. While the first ratio is hard to be explained by QCD factorization approach with $\phi_3 < 90^o$, our prediction can be reached to 0.30.

TABLE 4. Ratios of CP-averaged rates in $B \to K\pi, \pi\pi$ decays with $\phi_3 = 80^0$, $R_b = 0.38$. Here we adopted $m_0^\pi = 1.3$ GeV and $m_0^K = 1.7$ GeV.

Quatity	CLEO	PQCD	BBNS
$\frac{Br(\pi^+\pi^-)}{Br(\pi^\pm K^\mp)}$	0.25 ± 0.10	$0.30 - 0.69$	$0.5 - 1.9$
$\frac{Br(\pi^\pm K^\mp)}{2Br(\pi^0 K^0)}$	0.59 ± 0.27	$0.78 - 1.05$	$0.9 - 1.4$
$\frac{2\,Br(\pi^0 K^\pm)}{Br(\pi^\pm K^0)}$	1.27 ± 0.47	$0.77 - 1.60$	$0.9 - 1.3$
$\frac{\tau(B^+)}{\tau(B^0)}\frac{Br(\pi^\mp K^\pm)}{Br(\pi^\pm K^0)}$	1.00 ± 0.30	$0.70 - 1.45$	$0.6 - 1.0$

CP Asymmetry of $B \to \pi\pi, K\pi$. Because we have a large imaginary contribution from factorized annihilation diagrams in PQCD approach, we predict large CP asymmetry ($\sim$ 20%) in $B^0 \to \pi^+\pi^-$ decays and about -15% CP violation effects in $B^0 \to K^+\pi^-$. The detail prediction is given in Table 5. The precise measurement of direct CP asymmetry (both magnitude and sign) is a crucial way to test factorization models which have different sources of strong phases. Our predictions for CP-asymmetry on $B \to K(\pi)\pi$ have a totally opposite sign to those of QCD factorization.

Determination ϕ_3 in $B \to \pi\pi, K\pi$. Some years ago, many authors[30, 31, 32] have derived a bound on ϕ_3 from rates of branching ratios on $B \to K(\pi)\pi$. The ratio R_K is

TABLE 5. CP-asymmetry in $B \to K\pi, \pi\pi$ decays with $\phi_3 = 40^0 \sim 90^0$, $R_b = 0.38$. Here we adopted $m_0^\pi = 1.3$ GeV and $m_0^K = 1.7$ GeV.

$A_{CP}(\%)$	Experiment (CLEO,BELLE,BABAR)	Theory PQCD	BBNS	CFMPS ($\lvert A_{CP}\rvert$)
π^+K^-	-4.8 ± 6.8 (BaBar) $-7 \pm 8 \pm 2$	$-12.9 \sim -21.9$	5 ± 9	17 ± 6
π^0K^-	-9.6 ± 11.9	$-10.0 \sim -17.3$	7 ± 9	18 ± 6
$\pi^-\bar{K}^0$	-4.7 ± 13.9	$-0.6 \sim -1.5$	1 ± 1	3 ± 3
$\pi^+\pi^-$	-25 ± 48	$15.0 \sim 20.0$	-6 ± 12	58 ± 29

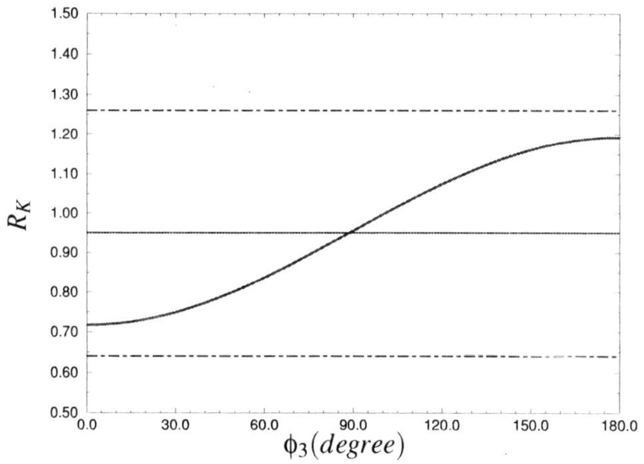

FIGURE 3. Dependence of the ratio R_K on ϕ_3. The dashed (dotted) lines correspond to the bounds (central value) of the data.

given by

$$R_K \equiv \frac{Br(B^0 \to K^{\pm}\pi^{\mp})}{Br(B^{\pm} \to K^0\pi^{\pm})} = 1.0 + \frac{2\,\lambda^2\,R_b}{a_K}\,cos\phi_3 \qquad (6)$$

where $\lambda = 0.22$, $R_b = 0.41 \pm 0.07$ and $a_K = (a_4 + 2a_6 r_K)/a_1$. As shown in Figure 3, arbitrary ϕ_3 is allowed due to large experimantal uncertainties in present data (0.95 ± 0.30). We expect more precise measurement of R_K to determine ϕ_3 in future within 2-3 years.

SUMMARY AND OUTLOOK

In this talk I have discuss ingredients of PQCD approach and some important theoretical issues with numerical results by comparing exparimental data. A new PQCD factorization approach provides a useful theoretical framework for a systematic analysis on non-leptonic two-body B-meson decays. This approach explain sucessfully present experimental data upto now and will be tested more thoroughly with more precise data in near future.

ACKNOWLEDGMENTS

It is a great pleasure to thank the organizers for the invitation to this exciting workshop and S.-H. Yang for his hospitality. I wish to acknowlege the fruitful collaboration with

H. Y. Cheng, H.-N. Li, A. I. Sanda and joyful discussions with other members of PQCD working group.

REFERENCES

1. Bauer, M., Stech, B. and Wirbel, M., *Z. Phys.* **C 29**, 637 (1985).
2. Bauer, M., Stech, B. and Wirbel, M., *Z. Phys.* **C 34**, 103 (1987).
3. Ali, A., Kramer, G. and Lu, C.-D., *Phys. Rev.* **D 58**, 094009 (1998).
4. Chen, Y.-H., Cheng, H.-Y., Tseng, B. and Yang, K.-C., *Phys. Rev.* **D 60**, 094014 (1999).
5. Beneke, M., Buchalla, G., Neubert, M. and Sachrajda, C. T., *Phys. Rev. Lett.* **83**, 1914 (1999).
6. Beneke, M., Buchalla, G., Neubert, M. and Sachrajda, C. T., *Nucl. Phys.* **B 591**, 313 (2000).
7. Keum, Y.-Y., Li, H.-N. and Sanda, A. I., *Phys. Lett.* **B 504**, 6 (2001).
8. Keum, Y.-Y., Li, H.-N. and Sanda, A. I., *Phys. Rev.* **D 63**, 074006 (2001).
9. Keum, Y.-Y. and Li, H.-N., *Phys. Rev.* **D 63**, 054008 (2001).
10. Chang, C. H. and Li, H.-N., *Phys. Rev.* **D 55**, 5577 (1997).
11. Lepage, G. P. and Brodsky, S., *Phys. Rev.* **D 22**, 2157 (1980).
12. Botts, J. and Sterman, G., *Nucl. Phys.* **B 225**, 62 (1989).
13. Ball, P., *JHEP* **9809**, 005 (1998).
14. Ball, P., *JHEP* **9901**, 010 (1999).
15. Du, D.-S., Huang, C.-S., Wei, Z.-T. and Yang, M.-Z., *Phys. Lett.* **B 520**, 50 (2001).
16. Descotes-Genon, S. and Sachrajda, C. T., *hep-ph/0109260*.
17. Gribov, V. N. and Lipatov, L. N., *Sov. J. Nucl. Phys.* **15**, 428 (1972).
18. Altarelli, G. and Parisi, G., *Nucl. Phys.* **B 126**, 298 (1977).
19. Li, H.-N., *hep-ph/0102013*.
20. Collins, J. C. and Soper, D. E., *Nucl. Phys.* **B 193**, 381 (1981).
21. Li, H.-N. and Sterman, G., *Nucl. Phys.* **B 381**, 129 (1992).
22. Chen, C.-H., Keum, Y.-Y. and Li, H.-N., *Phys. Rev.* **D 64**, 112002 (2001).
23. Yeh, T.-W. and Li, H.-N., *Phys. Rev.* **D 56**, 1615 (1997).
24. Szczepaniak, A., Henley, E. M. and Brodsky, S., *Phys. Lett.* **B 243**, 287 (1990).
25. Kurimoto, T., Li, H.-N. and Sanda, A. I., *Phys. Rev.* **D 65**, 014007 (2002).
26. Bander, M., Silverman, D. and Soni, A., *Phys. Rev. Lett.* **43**, 242 (1979).
27. Cheng, H.-Y. and Yang, K.-C., *Phys. Rev.* **D 64**, 074004 (2001).
28. Mishima, S., *Phys. Lett.* **B 521**, 252 (2001).
29. Lu, C. D., Ukai, K. and Yang, M. Z., *Phys. Rev.* **D 63**, 074009 (2001).
30. Mannel, T. and Fleischer, R., *Phys. Rev.* **D 57**, 2752 (1998).
31. Buras, A. J. and Fleischer, R., *Euro. Phys. J.* **C 16**, 97 (2000).
32. Neubert, M. and Rosner, J. L., *Phys. Lett.* **B 441**, 403 (1998).

Testing Factorization

Gudrun Hiller [1]

Stanford Linear Accelerator Center, Stanford University, Stanford, CA 94309, USA

Abstract. We briefly review the status of factorization in b-decays. We discuss several experimental tests of its nature and stress their importance. We show that decays into mesons which have small decay constants or spin greater than one ('designer mesons') offer a variety of new opportunities.

INTRODUCTION

Major theory issues in b-physics are 1) Is the CKM description of CP violation correct or are there other sources of CP violation? and 2) Is the Standard Model the correct effective theory up to scales of order $\sim$ 1 TeV? The latter could be probed for example with flavor-changing neutral current (FCNC) decays such as $b \to s\gamma$, $b \to s\ell^+\ell^-$ and $b \to sq\bar{q}$. Both questions can be addressed in b-physics in a unique way, which has stimulated many experimental and theoretical activities. However, there is another one: 3) Is our understanding of non-perturbative QCD good enough to answer the above questions? which is part of the whole picture. Among effects due to hadronization, those related to the factorization of matrix elements of hadronic 2-body decays are of peak importance. Their understanding and knowledge of the limitations in their theoretical description will become more urgent in the near future since the 'QCD background' limits the potential of precision tests of the Standard Model. Fortunately, there are many decay modes, observables and facilities where this can be further explored and checked.

FACTORIZATION CONCEPT

Naive factorization is a working hypothesis [1], which allows one to express the matrix elements of hadronic 2-body decays in terms of known objects, with the product (decay constant $\times$ form factor). Diagrammatically, it amounts to cutting the amplitude across the W-boson line and resembles the description of semileptonic decays. The picture of color transparency (CT) [2] justifies this for some decays when the meson emitted from the weak vertex is fast in the rest frame of the decaying parent. An example is $\bar{B}^0 \to D^+\pi^-$. The CT explanation however must fail e.g. for $\bar{B}^0 \to D^-\pi^+$. If $1/N_c$ counting arguments [3] are at work, then factorization would still hold here. There exist no general theory for all 2-body decays, but they have been classified into color allowed, suppressed, heavy-light, etc. QCD based approaches [4, 5] differ in their treatment of α_s

[1] Work supported by the Department of Energy, Contract DE-AC03-76SF00515

CP618, *Heavy Flavor Physics: Ninth International Symposium,* edited by A. Ryd and F. C. Porter

TABLE 1. Examples of 'designer' mesons.

X	a_0	b_1	π	a_2	a_0	π_2	ρ_3	χ_{c0}	K_0^*	K_2^*	D_2^*
m_X [MeV]	985	1230	1300	1318	1474	1670	1691	3415	1412	1426	2459
J^{PC}	0^{++}	1^{+-}	0^{-+}	2^{++}	0^{++}	2^{-+}	3^{--}	0^{++}	0^+	2^+	2^+

and $1/m_b$ power corrections. Thus it is important to test experimentally where (naive) factorization holds and where corrections arise to understand its dynamical origin (CT, $1/N_c$, ...).

Status

Factorization has been tested with tree level dominated modes, i.e. where possible New Physics effects are tiny. Currently, the factorization concept in color allowed B-decays rules:

- Heavy-to-light $\bar{B}^0 \to D^{(*)+}(\pi,\rho,a_1)^-$ decays can be described by one universal coefficient $|a_1| = 1.1 \pm 0.1$, see e.g. [4]; factorization is ok up to the $O(10\%)$ level.
- In $\bar{B}^0 \to D^{(*)+}X^-$ decays, where $X = 4\pi, \omega\pi$ the factorization hypothesis can be tested as a function of the hadronic mass m_X of the emitted $n\pi$ state [6]. Its 'decay constant' $\langle X|\bar{d}\Gamma u|0\rangle$ is obtained from hadronic τ-data. CT holds if X is fast, so one expects corrections to factorization for growing m_X, but no kinematical dependence in the $1/N_c$ approach. Current data indicate no factorization breaking but further experimental studies should be pursued.
- In $\bar{B}^0 \to D^-_{(s)}\pi^+$ decays there is no CT explanation of factorization. The factorizable amplitude is proportional to $\mathcal{A}_{fac} \sim V_{ub}F^{B\to\pi}(m^2_{D(s)})f_{D_s}$ and quantitative tests need good control over all parameters on the r.h.s. Experimentally, to date an upper bound exists, $\mathcal{B}(\bar{B}^0 \to D_s^-\pi^+) < 1.1 \cdot 10^{-4}$ @90% C.L. [7] which is ok with factorization.
- In heavy-to-heavy $B \to D^{(*)}D_s^{(*)}$ decays factorization holds within errors [8]; confirmed by study including penguins [9]. The dominant uncertainty in these analyses comes from the decay constants, e.g. $f_{D_s} = 280 \pm 48$ MeV from $D_s \to \mu\nu_\mu$ [10].

Significant improvement in precision is required to isolate factorization breaking effects in the above decays. Alternatively, one can study factorization in those decays, where the corrections to factorization are not hidden behind a large factorizable contribution [11, 12]. Then less precision is required, although for the price of less events.

NEW WAYS

Recently, it has been proposed to explore factorization with final states whose coupling to the W-boson is suppressed either because the spin > 1 or the decay constant is

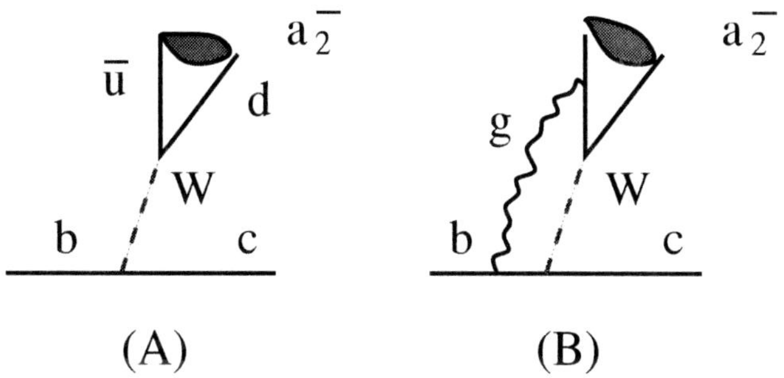

FIGURE 1. Examples of a diagram which does not (A) and does (B) contribute to $b \to c a_2^-$.

suppressed [11]. Examples of such 'designer mesons' are given in Table 1. The relevant property in the context of factorization tests is that amplitudes in *naive* factorization are suppressed. In the case of a vanishing decay constant, they even vanish e.g. $\mathcal{B}_{naive}(\bar{B}^0 \to D^+ a_2^-) = 0$, see Fig. 1 (A). However, corrections e.g. induced by hard gluon exchange, see Fig. 1 (B) circumvent suppression, because the a_2^- is now produced from a non-local vertex, which allows for higher spins. Other mechanisms that avoid the suppression include annihilation topologies, interactions with the spectator and those induced by charm [13]. Thus non-factorizable effects can be highlighted by the choice of specific 'designer' final states.

Decay constants

Decay constants for scalars S with momentum p are defined as $\langle S(p)|\bar{q}\gamma_\mu q'|0\rangle = -i f_S p^\mu$. For $q = q'$ the decay constant vanishes by charge conjugation e.g. $f_{a_0^0}, f_{\chi_{c0}} = 0$. The decay constant of the charged a_0 is proportional to $m_d - m_u$ and thus isospin suppressed and small compared to e.g. $f_\pi = 131$ MeV. Analogous arguments apply to the axial vector b_1. Also a 1^+ single charm meson $D_{J=1}(j = 3/2)$ with vanishing decay constant in the heavy quark limit is predicted [14]. The decay constants of the mesons in Table 1 are only poorly known. The current theory spread as compiled in [11] reads as

$$\begin{aligned} f_{a_0(980)} &= 0.7 - 2.5\,\mathrm{MeV} \\ f_{\pi(1300)} &= 0.5 - 7.2\,\mathrm{MeV} \\ f_{K_0^*} &= 33 - 46\ \ \mathrm{MeV} \end{aligned} \tag{1}$$

As expected, $f_{K_0^*} > f_{a_0}$ due to larger quark mass splitting. No estimate of the b_1 (longitudinal) decay constant has been reported. The decay constants of $a_0, \pi(1300)$, b_1, K_0^* mesons could be determined in hadronic τ-decays. Estimates of the corresponding branching ratios are given in Table 2. Note that the bound $\mathcal{B}(\tau \to (\pi(1300) \to 3\pi)\nu_\tau) < 1 \cdot 10^{-4}$ [10] implies $f_{\pi(1300)} < 8.4$ MeV.

TABLE 2. Theory estimates of decay constants as complied in [11] and the corresponding $\tau \to \nu_\tau X$ branching ratios.

X	$a_0(980)$	$a_0(1450)$	$\pi(1300)$	$K_0^*(1430)$
f_X [MeV]	1.1	0.7	7.2	42
$\mathcal{B}(\tau \to \nu_\tau X)$	$3.8 \cdot 10^{-6}$	$3.7 \cdot 10^{-7}$	$7.3 \cdot 10^{-5}$	$7.7 \cdot 10^{-5}$

Flavor selection

Not every decay into a designer final state is suppressed. Instead one needs flavor selection criteria to find the designer modes. It is crucial that the spectator does not end up in the designer. An example where this condition does not hold is the decay $B^- \to D^0 a_0^-$. It proceeds via two different topologies. The color allowed one has a suppressed amplitude, because the a_0^- is emitted from the weak vertex. The color suppressed contribution to the amplitude however produces the a_0^- from the spectator. Since form factors for designer mesons are not anomalous, this topology escapes suppression. Examples of modes which satisfy the criteria and do have a suppressed factorizable contribution are given in Table 3, (for the full listing and details see [11]). There are many decays of $B^0, B^\pm, B_s$ mesons and Λ_b, Ω_b baryons, many final states, which cover a wide range of masses (every particle can be replaced by another one with the same flavor content and W-coupling features), many topologies (tree, annihilation, penguin annihilation,...), and many classifications (color allowed, color suppressed). Heavy-light, light-light and decays into charmonium should factorize according to CT while light-heavy and heavy-heavy not. Note that baryons differ in quark content and in particular in annihilation topologies from mesons, i.e. they cannot be fully annihilated by 4-Fermi operators. They also offer more degrees of freedom accessible to experiments such as polarization and have no background from the decay of the CP conjugate parent into the same final state (this is a problem for e.g. $\bar{B}^0 \to \pi^+ a_0^-$, because $B^0 \to \pi^+ a_0^-$ is unsuppressed).

Existing data

Experimental information on some designer modes from Table 3 is already available. The Belle collaboration has reported recently [15]

$$
\begin{aligned}
\mathcal{B}(B^+ \to \chi_{c0} K^+) &= (6.0^{+2.1}_{-1.8} \pm 1.1) \cdot 10^{-4} \\
R = \frac{\mathcal{B}(B^+ \to \chi_{c0} K^+)}{\mathcal{B}(B^+ \to J/\Psi K^+)} &= 0.60^{+0.21}_{-0.18} \pm 0.05 \pm 0.08
\end{aligned} \tag{2}
$$

Because it has a small radius, CT is expected to work for charmonium despite the fact it is not light. However, problems of factorization with color suppressed decays are no surprise, since radiative corrections come in without color suppression and are large [4, 11]. Indeed, eq. (2) represents an $O(1)$ violation of naive factorization since $R_{naive} = 0$.

TABLE 3. Some color allowed, color suppressed and baryon decays which satisfy the flavor selection criteria specified in text, adapted from [11]. The magnitude of the amplitudes is given in powers of the Wolfenstein parameter $\lambda \simeq 0.22$.

example decay	**factorizing contribution**			**annihilation**	
	quark level	**tree**	**penguin**	**tree**	**penguin**
$\bar{B}^0 \to D^+ a_0^-$	$\bar{d}b \to \bar{d}(c\bar{u}d)$	λ^2		λ^2	
$B^- \to \pi^0 D_2^{*-}$	$\bar{u}b \to \bar{u}(u\bar{c}d)$	λ^4		λ^4	
$\bar{B}_s \to D_s^+ D_2^{*-}$	$\bar{s}b \to \bar{s}(c\bar{c}d)$	λ^3	λ^3		λ^3
$B^- \to \pi^- \bar{K}_2^{*0}$	$\bar{u}b \to \bar{u}(d\bar{d}s)$		λ^2	λ^4	λ^2
$B^- \to K^- K_2^{*0}$	$\bar{u}b \to \bar{u}(s\bar{s}d)$		λ^3	λ^3	λ^3
$\bar{B}^0 \to \pi^0 D_2^{*0}$	$\bar{d}b \to \bar{d}(d\bar{u}c)$	λ^2		λ^2	
$B^- \to K^- \chi_{c0}$	$\bar{u}b \to \bar{u}(s\bar{c}c)$	λ^2	λ^2	λ^4	λ^2
$\bar{B}_s \to K^0 a_2^0$	$\bar{s}b \to \bar{s}(d\bar{u}u)$	λ^3	λ^3		λ^3
$\Lambda_b \to \quad n D_2^{*0}$	$udb \to ud(c\bar{u}d)$	λ^2		λ^2	
$\Lambda_b \to \Lambda_c D_{sJ}^-, \Lambda\chi_{c0}$	$udb \to ud(c\bar{c}s)$	λ^2	λ^2	λ^4	λ^2
$\Omega_b \to \Omega_c a_0^-, \Xi^- D_2^{*0}$	$ssb \to ss(c\bar{u}d)$	λ^2			

The branching ratio into a light-light final state has been measured by Belle $\mathcal{B}(B^+ \to (K_X(1400) \to K^+\pi^-)\pi^+) = (12.7^{+3.5+1.8+2.9}_{-3.4-1.8-5.8}) \cdot 10^{-6}$ [16]. This is comparable in magnitude with the one into the corresponding unsuppressed mode, $\mathcal{B}(B^+ \to K^0\pi^+) = (13.7^{+5.7+1.9}_{-4.8-1.8}) \cdot 10^{-6}$ (Belle) [17] and $\mathcal{B}(B^+ \to K^0\pi^+) = (18.2^{+3.3}_{-3.0} \pm 2.0) \cdot 10^{-6}$ (BaBar) [18]. Both decays are dominated by QCD penguins. In particular, they receive large contributions from scalar penguins $(\bar{q}(1-\gamma_5)b)(\bar{s}(1+\gamma_5)q)$, which are parametrically enhanced by factors

$$r_K = \frac{2m_K^2}{m_b m_s}, \qquad r_{K_0^*} = \frac{2m_{K_0^*}^2}{m_b m_s}$$

for pseudoscalar K and scalar K_0^* mesons, respectively. Since the penguin enhancement $r_{K_0^*}/r_K = m_{K_0^*}^2/m_K^2$ compensates for the decay constant suppression $f_{K_0^*}/f_K \sim 1/4$, the hypothesis $K_X(1400) = K_0^*$ is consistent with factorization [11, 19]. Note that both contributions remain finite in the chiral limit – for the Goldstone bosons because $m_K \to 0$ in the same limit and for the scalars because $f_{K_0^*} \sim m_s$ which multiplies $r_{K_0^*}$ vanishes. Measurement of the branching ratio into the tensor would be much more exciting since in naive factorization $\mathcal{B}_{naive}(B^\pm \to K_2^*\pi^\pm) = 0$. Thus, one would directly probe the factorization breaking corrections, an issue in light-light decays that is controversial between perturbative QCD [5] and QCD factorization [4, 20] (Some problems in the pQCD approach have been pointed out recently in Ref. [21]). Experimentally, angular analysis is required to discriminate the nearby kaon resonances $K_0^*, K^*(1410), K_2^*$.

Quantitatively: color allowed $B \to D^{(*)}X$ decays $X = a_0, a_2, b_1, \pi(1300), \pi_2, \rho_3$ and K_0^*, K_2^*

Generically we have the branching ratios $\mathcal{B}(B \to D^{(*)}(\pi,\rho,a_1)) \sim 10^{-3}$. Assuming $O(10)\%$ corrections to factorization arising from $1/N_c^2$ and/or Λ_{QCD}/m_b one expects for the $I = 1$ designer mesons $\mathcal{B}(B \to D^{(*)}X) \sim 10^{-5} - 10^{-6}$. The branching ratios can be calculated in QCD factorization [4] from evaluation of the matrix element [11]

$$\langle D^{(*)+}X^-|\mathcal{H}_{eff}|B\rangle \sim a_1(\mu) f_X + \frac{\alpha_s(\mu)}{4\pi} C_2(\mu) \frac{C_F}{N_c} \int_0^1 du F(u) \varphi(u;\mu). \tag{3}$$

The first term corresponds to the expression in naive factorization. It vanishes if $f_X = 0$. In terms of light cone distribution amplitudes (DA) $\varphi(u;\mu)$, where u denotes the momentum fraction carried by the quark in X, the decay constant is given as

$$f_X = \int_0^1 du \varphi(u).$$

However, a small or vanishing zeroth moment $f_X \simeq 0$ does not imply that the DA is small or vanishing. The contribution from hard gluon exchange, which is given by the second term in eq. (3), thus escapes the suppression mechanism [11]. From charge conjugation, the following symmetry properties for meson DA's hold up to isospin breaking

$$\begin{aligned} \pi, \pi(1300), \pi_2, \rho_3 \ &: \ \varphi(u) = +\varphi(1-u) \\ a_0, a_2, b_1, K_0^*, K_2^* \ &: \ \varphi(u) = -\varphi(1-u) \end{aligned} \tag{4}$$

The leading coefficients in the expansion of the DA in Gegenbauer polynomials $C_n^{3/2}$

$$\varphi(u;\mu) = f^{\varphi} 6u(1-u) \left[B^0 + \sum_{n=1}^{\infty} B_n(\mu) C_n^{3/2}(2u-1) \right]$$

are estimated for mesons X with Fock space normalization techniques á la pion as

$$\begin{aligned} |f^{\varphi} B_1|_{a_0,b_1,a_2,K_0^*,K_2^*} &\approx 75 \text{ MeV} \\ |f^{\varphi} B_2|_{\pi(1300),\pi_2,\rho_3} &\approx 50 \text{ MeV} \end{aligned} \tag{5}$$

at the renormalization scale $\mu = m_b = 4.4$ GeV ($B_0 = 1$ and $f^{\varphi} = f_X$ for spin$(X) \leq 1$, otherwise $B_0 = 0$) [11]. Resulting branching ratios are given in Table 4. Similar ones are expected for $B_s \to D_s X$ decays accessible at the Tevatron and LHC. The QCD factorization result shows an enhancement over the one obtained in the naive factorization approach. However, the branching ratios are still much smaller than the ones of the corresponding 'non-designer' modes like $B \to D\pi$, but within experimental reach. Since QCD factorization picks up hard α_s contributions which are enhanced in the designer decays, their amplitude is more sensitive to the renormalization scale and gains large strong phases. Current dominant uncertainties are due to the decay constants, which could be measured in hadronic τ decays, see Table 2. Furthermore, information on the DA's of the neutral $a_0, a_2, \pi(1300), \pi_2$ could be obtained from $\gamma\gamma^*$ collisions in $e^+e^- \to e^+e^- X$ processes similar to the analysis performed by CLEO for π, η, η' [22].

TABLE 4. Branching ratios obtained in QCD factorization for two choices of the renormalization scale μ and in the naive factorization approach, taken from [11]. †With $f_{b_1} = 0$.

mode	naive factorization	QCD factorization	
		$\mu = m_b$	$\mu = m_b/2$
$\bar{B}^0 \to D^+ a_0(980)$	$1.1 \cdot 10^{-6}$	$2.0 \cdot 10^{-6}$	$4.0 \cdot 10^{-6}$
$\bar{B}^0 \to D^+ a_0(1450)$	$8.6 \cdot 10^{-8}$	$5.8 \cdot 10^{-7}$	$2.1 \cdot 10^{-6}$
$\bar{B}^0 \to D^+ a_2, D^+ b_1{}^{\dagger}$	0	$3.5 \cdot 10^{-7}$	$1.7 \cdot 10^{-6}$
$\bar{B}^0 \to D^+ \pi(1300)$	$9.1 \cdot 10^{-6}$	$9.3 \cdot 10^{-6}$	$9.6 \cdot 10^{-6}$
$\bar{B}^0 \to D^+ \pi_2, D^+ \rho_3$	0	$1.4 \cdot 10^{-9}$	$8.1 \cdot 10^{-9}$
$\bar{B}^0 \to D^+ K_0^*(1430)$	$2.0 \cdot 10^{-5}$	$2.0 \cdot 10^{-5}$	$2.1 \cdot 10^{-5}$
$\bar{B}^0 \to D^+ K_2^*$	0	$1.9 \cdot 10^{-8}$	$9.2 \cdot 10^{-8}$
$\bar{B}^0 \to D^{*+} a_0(980)$	$1.0 \cdot 10^{-6}$	$1.8 \cdot 10^{-6}$	$3.7 \cdot 10^{-6}$
$\bar{B}^0 \to D^{*+} a_0(1450)$	$7.9 \cdot 10^{-8}$	$5.2 \cdot 10^{-7}$	$1.9 \cdot 10^{-6}$
$\bar{B}^0 \to D^{*+} a_2 D^{*+} b_1{}^{\dagger}$	0	$2.9 \cdot 10^{-7}$	$1.5 \cdot 10^{-6}$
$\bar{B}^0 \to D^{*+} \pi(1300)$	$8.3 \cdot 10^{-6}$	$8.4 \cdot 10^{-6}$	$8.4 \cdot 10^{-6}$
$\bar{B}^0 \to D^{*+} \pi_2, D^{*+} \rho_3$	0	$5.7 \cdot 10^{-10}$	$3.2 \cdot 10^{-9}$
$\bar{B}^0 \to D^{*+} K_0^*(1430)$	$1.8 \cdot 10^{-5}$	$1.9 \cdot 10^{-5}$	$1.9 \cdot 10^{-5}$
$\bar{B}^0 \to D^{*+} K_2^*$	0	$1.5 \cdot 10^{-8}$	$7.7 \cdot 10^{-8}$

Measure strong phases as a byproduct of $2\beta + \gamma$

Time dependent measurements in $B \to D^{\pm}\pi^{\mp}$ decays allow the extraction of CKM angles $2\beta + \gamma$ together with strong phases without model assumptions on strong dynamics [23]. Unfortunalty, the effect scales with a tiny $\sim 1\%$ asymmetry. Choosing instead $B \to D^{\pm}X^{\mp}$ decays, where $X = a_0, a_2, b_1, \pi(1300), \pi_2, \rho_3$ the CKM hierarchy of the amplitudes is compensated by the decay constants and large [24], for example for the a_0

$$\frac{\mathcal{A}(B^0 \to D^+ a_0^-)}{\mathcal{A}(\bar{B}^0 \to D^+ a_0^-)} \simeq \frac{f_D}{f_{a_0}} \frac{V_{cd}V_{ub}^*}{V_{cb}V_{ud}^*} \sim O(1). \tag{6}$$

Therefor designer mesons are competitive with $B \to D^{\pm}\pi^{\mp}$ decays and help to resolve ambiguities because different and large strong phases are involved.

CONCLUSIONS

Factorization successfully works in color allowed b-decays within present errors. This includes $B \to DD_s$ decays where the color transparency explanation is absent. More data are needed to see to what accuracy it works and perhaps to understand why it works. I discussed new strategies to explore factorization with modes where the factorizable contribution is suppressed such that the corrections show up isolated. This is similar to pure annihilation decays $B^0 \to K^+K^-, B_s \to \pi^+\pi^-, \pi^0\pi^0$ and $B \to$ *baryon baryon*. These measurements could be carried out at operating and future high luminosity e^+e^--facilities [25]. A dedicated factorization study requires improved knowledge of non-perturbative input such as decay constants and distribution amplitudes. Their determination should be part of such an experimental program. A qualitatively and quantitatively accurate description of hadronic matrix elements in B-decays is particularly important for the extraction of the CKM unitarity angles γ and α. Finally, 'designer' final states and modes are also promising to study CP violation and to search for New Physics effects in rare FCNC processes. Steps in this direction have already been undertaken [24, 26, 27].

ACKNOWLEDGMENTS

It is a pleasure to thank the organizers for the invitation to this exciting symposium and Dr. Diehl for joyful collaboration on parts of the work presented here. I am grateful to David E. Kaplan for useful comments on the manuscript.

REFERENCES

1. M. Bauer, B. Stech and M. Wirbel, Z. *Phys.* C **34**, 103 (1987).
2. J. D. Bjorken, *Nucl. Phys. Proc. Suppl.* **11**, 325 (1989);
   M. J. Dugan and B. Grinstein, *Phys. Lett.* B **255**, 583 (1991);
   H. D. Politzer and M. B. Wise, *Phys. Lett.* B **257**, 399 (1991).
3. A. J. Buras, J. M. Gerard and R. Ruckl, *Nucl. Phys.* B **268** (1986) 16.
4. M. Beneke *et al.*, *Nucl. Phys.* B **591**, 313 (2000) [hep-ph/0006124].
5. Y. Y. Keum, H. N. Li and A. I. Sanda, *Phys. Rev.* D **63**, 054008 (2001) [hep-ph/0004173], and references therein.
6. Z. Ligeti, M. E. Luke and M. B. Wise, *Phys. Lett.* B **507**, 142 (2001) [hep-ph/0103020].
7. Belle Collaboration, contributed paper for 2001 Summer Conferences, BELLE-CONF-0125.
8. Z. Luo and J. L. Rosner, *Phys. Rev.* D **64**, 094001 (2001) [hep-ph/0101089].
9. C. S. Kim, Y. Kwon, J. Lee and W. Namgung, hep-ph/0108004.
10. D. E. Groom *et al.* [Particle Data Group Collaboration], *Eur. Phys. J.* C **15**, 1 (2000).
11. M. Diehl and G. Hiller, *JHEP* **0106**, 067 (2001) [hep-ph/0105194], and references therein.
12. P. F. Harrison and H. R. Quinn [BABAR Collaboration], SLAC-R-0504.
13. M. Ciuchini *et al.*, *Nucl. Phys.* B **501**, 271 (1997) [hep-ph/9703353];
    S. J. Brodsky and S. Gardner, hep-ph/0108121, to appear in Phys. Rev. D.
14. A. Le Yaouanc *et al.*, *Phys. Lett.* B **387**, 582 (1996) [hep-ph/9607300].
15. K. Abe *et al.* [Belle Collaboration], KEK Preprint 2001-132, Belle Preprint 2001-12.
16. K. Abe *et al.* [Belle Collaboration], hep-ex/0107051.
17. K. Abe *et al.* [BELLE Collaboration], *Phys. Rev. Lett.* **87**, 101801 (2001) [hep-ex/0104030].
18. B. Aubert *et al.* [BABAR Collaboration], *Phys. Rev. Lett.* **87**,151802 (2001) [hep-ex/0105061].
19. V. Chernyak, *Phys. Lett.* B **509**, 273 (2001) [hep-ph/0102217].
20. M. Beneke *et al.*, *Nucl. Phys.* B **606**, 245 (2001) [hep-ph/0104110].
21. S. Descotes-Genon and C. T. Sachrajda, hep-ph/0109260.
22. J. Gronberg *et al.* [CLEO Collaboration], *Phys. Rev.* D **57**, 33 (1998) [hep-ex/9707031].
23. I. Dunietz, *Phys. Lett.* B **427**, 179 (1998) [hep-ph/9712401], and references therein.
24. M. Diehl and G. Hiller, *Phys. Lett.* B **517**, 125 (2001) [hep-ph/0105213].
25. "Physics at a 10**36 asymmetric B factory," SLAC-PUB-8970.
26. S. Laplace and V. Shelkov, hep-ph/0105252.
27. B. Aubert *et al.* [BABAR Collaboration], hep-ex/0107075.

$B \to V\gamma$ at NLO from QCD Factorization[1]

Stefan W. Bosch

Max-Planck-Institut für Physik, Föhringer Ring 6, D-80805 Munich, Germany

Abstract. We discuss the exclusive radiative B-meson decays $B \to K^*\gamma$ and $B \to \rho\gamma$ in a model-independent manner. The analysis is based on the heavy-quark limit of QCD. This allows a factorization of perturbatively calculable contributions to the $B \to V\gamma$ matrix elements from non-perturbative form factors and universal light-cone distribution amplitudes. These results allow us to compute *exclusive* $b \to s(d)\gamma$ decays systematically beyond the leading logarithmic approximation. We present results for these decays complete to next-to-leading order in QCD and to leading order in the heavy-quark limit. Phenomenological implications for various observables of interest are discussed, including direct CP violation and isospin breaking effects.

INTRODUCTION

The radiative transitions $b \to s(d)\gamma$ are among the most valuable probes of flavour physics. Although they are rare decays the Cabibbo-favoured $b \to s\gamma$ modes are experimentally accessible already at present. The inclusive branching fraction has been measured to be

$$B(B \to X_s\gamma) = (3.23 \pm 0.42) \cdot 10^{-4} \tag{1}$$

combining the results of [2, 3, 4]. The branching ratios for the exclusive channels have been determined by CLEO [5], and more recently also by BABAR [6] and BELLE [7] to give the averaged values:

$$B(B^0 \to K^{*0}\gamma) = (4.51 \pm 0.54) \cdot 10^{-5} \tag{2}$$

$$B(B^+ \to K^{*+}\gamma) = (3.86 \pm 0.59) \cdot 10^{-5}. \tag{3}$$

On the theoretical side, the flavour-changing neutral current (FCNC) reactions $b \to s(d)\gamma$ are characterized by their high sensitivity to New Physics and by the particularly large impact of short-distance QCD corrections [8, 9, 10, 11]. Considerable efforts have therefore been devoted to achieve a full calculation of the inclusive decay $b \to s\gamma$ at next-to-leading order (NLO) in renormalization group (RG) improved perturbation theory [12, 13, 14] (see [15] for recent reviews).

Whereas the inclusive mode can be computed perturbatively, using the fact that the b-quark mass is large and employing the heavy-quark expansion, the treatment of the exclusive channel $B \to K^*\gamma$ is in general more complicated. In this case bound state effects are essential and need to be described by nonperturbative hadronic quantities

---

[1] Based on work together with Gerhard Buchalla [1]

CP618, *Heavy Flavor Physics: Ninth International Symposium*, edited by A. Ryd and F. C. Porter

(form factors). The basic mechanisms at next-to-leading order were already discussed previously for the $B \to V\gamma$ amplitudes [16]. However, hadronic models were used to evaluate the various contributions, which did not allow a clear separation of short- and long-distance dynamics and a clean distinction of model-dependent and model-independent features.

In this talk we present the results of [1] where a systematic analysis of the exclusive radiative decays $B \to V\gamma$ ($V = K^*$, ρ) in QCD, based on the heavy quark limit $m_b \gg \Lambda_{QCD}$, was performed. We quote factorization formulas for the evaluation of the relevant hadronic matrix elements of local operators in the weak Hamiltonian. A similar subject was treated in [17, 18]. Factorization holds in QCD to leading power in the heavy quark limit. This result relies on arguments similar to those used previously to demonstrate QCD factorization for hadronic two-body modes of the type $B \to \pi\pi$ [19, 20].

This framework allows us to separate perturbatively calculable contributions from the nonperturbative form factors and universal meson light-cone distribution amplitudes (LCDA) in a systematic way. This includes the treatment of loop effects from light quarks, in particular up and charm. Such loop effects are straightforwardly included for the inclusive decays $b \to s(d)\gamma$. For the exclusive modes, however, the effects from virtual charm and up quarks have so far been considered to be uncalculable "long-distance" contributions and have never been treated in a model independent fashion.

Finally, power counting in Λ_{QCD}/m_b implies a hierarchy among the possible mechanisms for $B \to V\gamma$ transitions. This allows us to identify leading and subleading contributions. For example, weak annihilation contributes only at subleading power in the heavy quark limit.

Within this approach, higher order QCD corrections can be consistently taken into account. We give the $B \to V\gamma$ decay amplitudes at next-to-leading order (NLO). Furthermore numerical values for CP-asymmetries and isospin-violating ratios are given. After including NLO corrections the largest uncertainties still come from the $B \to V$ form factors, which are at present known only with limited precision ($\sim \pm 15\%$), mostly from QCD sum rule calculations [21]. The situation should improve in the future with the help of both lattice QCD [22] and analytical methods based on the heavy-quark and large-energy limits [23, 24, 25].

BASIC FORMULAS

The effective Hamiltonian for $b \to s\gamma$ transitions reads

$$\mathcal{H}_{eff} = \frac{G_F}{\sqrt{2}} \sum_{p=u,c} \lambda_p^{(s)} \left[C_1 Q_1^p + C_2 Q_2^p + \sum_{i=3,\ldots,8} C_i Q_i \right] \tag{4}$$

where $\lambda_p^{(s)} = V_{ps}^* V_{pb}$. The relevant operators are given by

$$Q_1^p = (\bar{s}p)_{V-A}(\bar{p}b)_{V-A} \tag{5}$$

$$Q_2^p = (\bar{s}_i p_j)_{V-A}(\bar{p}_j b_i)_{V-A} \tag{6}$$

$$Q_7 = \frac{e}{8\pi^2} m_b \bar{s}_i \sigma^{\mu\nu}(1+\gamma_5) b_i F_{\mu\nu} \tag{7}$$

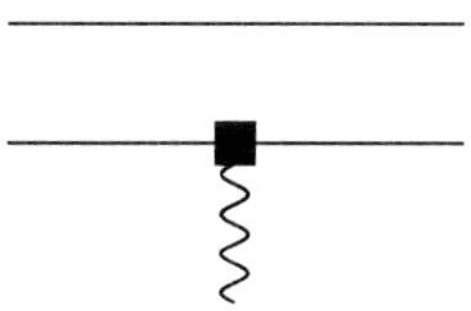

FIGURE 1. Contribution of the magnetic penguin operator Q_7 described by $B \to V$ form factors. All possible gluon exchanges between the quark lines are included in the form factors and have not been drawn explicitly.

$$Q_8 \quad = \quad \frac{g}{8\pi^2} m_b \bar{s}_i \sigma^{\mu\nu}(1+\gamma_5) T^a_{ij} b_j G^a_{\mu\nu} \tag{8}$$

Note that the numbering of $Q^p_{1,2}$ is reversed with respect to the convention of [26]. Throughout this work we neglect the contribution from the QCD penguin operators $Q_{3...6}$, which enter at $O(\alpha_s)$ and are further suppressed by very small Wilson coefficients. The effective Hamiltonian for $b \to d\gamma$ is obtained from (4–8) by the replacement $s \to d$.

The most difficult step in computing the decay amplitudes is the evaluation of the hadronic matrix elements of the operators in (4). A systematic treatment can be given in the heavy-quark limit. In this case the following factorization formula is valid

$$\langle V\gamma(\varepsilon)|Q_i|\bar{B}\rangle = \left[F^{B\to V}(0)\, T^I_i + \int_0^1 d\xi\, dv\, T^{II}_i(\xi, v)\, \Phi_B(\xi)\, \Phi_V(v)\right] \cdot \varepsilon \tag{9}$$

where ε is the photon polarization 4-vector. Here $F^{B\to V}$ is a $B \to V$ transition form factor, and Φ_B, Φ_V are leading twist light-cone distribution amplitudes of the B meson and the vector meson V, respectively. These quantities are universal, nonperturbative objects. They describe the long-distance dynamics of the matrix elements, which is factorized from the perturbative, short-distance interactions expressed in the hard-scattering kernels T^I_i and T^{II}_i. The QCD factorization formula (9) holds up to corrections of relative order Λ_{QCD}/m_b.

In the leading logarithmic approximation (LO) and to leading power in the heavy-quark limit, Q_7 gives the only contribution to the amplitude of $\bar{B} \to V\gamma$ and the factorization formula (9) is trivial. The matrix element is simply expressed in terms of the standard form factor, T^I_7 is a purely kinematical function and the spectator term T^{II}_7 is absent. An illustration is given in Fig. 1. The matrix element reads

$$\langle V(k,\eta)\gamma(q,\varepsilon)|Q_7|\bar{B}\rangle = -\frac{e}{2\pi^2} m_b c_V F_V \left[\varepsilon^{\mu\nu\lambda\rho}\varepsilon_\mu \eta_\nu k_\lambda q_\rho + i(\varepsilon\cdot\eta\, k\cdot q - \varepsilon\cdot k\, \eta\cdot q)\right] \tag{10}$$

where $c_V = 1$ for $V = K^*$, ρ^- and $c_V = 1/\sqrt{2}$ for $V = \rho^0$. The $\bar{B} \to V$ form factor F_V is evaluated at momentum transfer $q^2 = 0$. Our phase conventions coincide with those of [21, 27].

The matrix elements of Q_1 and Q_8 start contributing at $O(\alpha_s)$. In this case the factorization formula becomes nontrivial. The diagrams for the hard-scattering kernels T^I_i are shown in Fig. 2 for Q_1 and in Fig. 3 for Q_8. These diagrams were computed in [13] to get the virtual corrections to the inclusive matrix elements of Q_1 and Q_8.

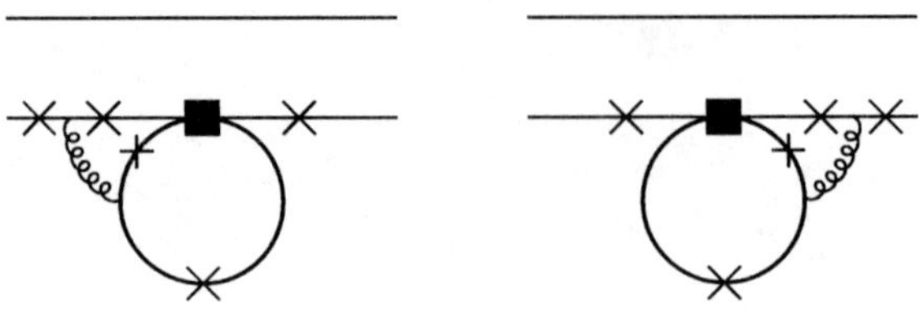

FIGURE 2. $O(\alpha_s)$ contribution to the hard-scattering kernels T_1^I from four-quark operators Q_1. The crosses indicate the places where the emitted photon can be attached.

FIGURE 3. $O(\alpha_s)$ contribution to the hard-scattering kernels T_8^I from chromomagnetic penguin operator Q_8.

In our case they determine the kernels T_1^I and T_8^I. As required for the consistency of the factorization formula these corrections must be dominated by hard scales of order m_b and hence must be infrared finite. This is indeed the case. Re-interpreted as the perturbative hard-scattering kernels for the exclusive process, the results from [13] imply

$$\langle Q_{1,8}\rangle^I = \langle Q_7\rangle \frac{\alpha_s C_F}{4\pi} G_{1,8} \tag{11}$$

where $C_F = (N^2-1)/(2N)$, with $N=3$ the number of colours, and

$$G_1(s_c) = -\frac{104}{27}\ln\frac{\mu}{m_b} + g_1(s_c) \tag{12}$$

$$G_8 = \frac{8}{3}\ln\frac{\mu}{m_b} + g_8 \tag{13}$$

The finite part $g_1(s_c)$ of the Q_1-contribution depends via $s_c = \frac{m_c^2}{m_b^2}$ on the mass of the quark running in the loop in Fig. 2.

We now turn to the mechanism where the spectator participates in the hard scattering.

The non-vanishing contributions to T_i^{II} are shown in Fig. 4. To find the correction for $\langle Q_1\rangle$ we compute the first diagram in Fig. 4 We obtain

$$\langle Q_1\rangle^{II} = \langle Q_7\rangle \frac{\alpha_s(\mu_h) C_F}{4\pi} H_1(s_c) \tag{14}$$

with

$$H_1(s) = -\frac{2\pi^2}{3N}\frac{f_B f_V^\perp}{F_V m_B^2}\int_0^1 d\xi \frac{\Phi_{B1}(\xi)}{\xi}\int_0^1 dv\, h(\bar{v},s)\Phi_\perp(v) \tag{15}$$

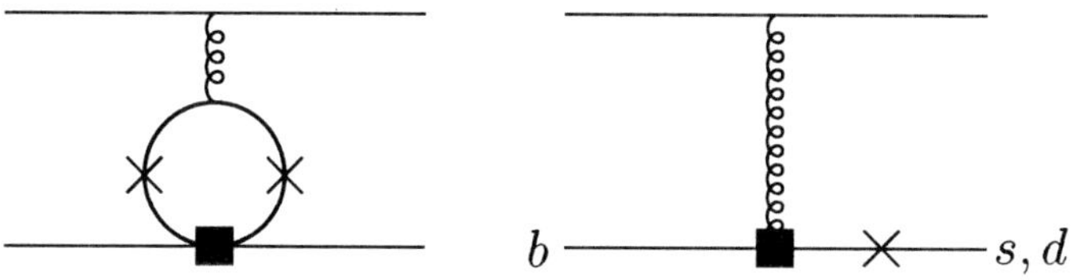

FIGURE 4. $O(\alpha_s)$ and leading power contribution to the hard-scattering kernels T_i^{II} from four-quark operators Q_i (left) and from Q_8.

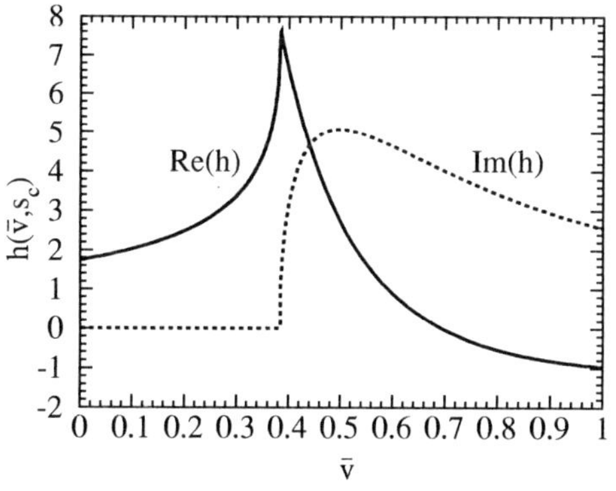

FIGURE 5. The hard-scattering kernel $h(\bar{v}, s_c)$ as a function of $\bar{v}$.

The first negative moment of the light-cone distribution amplitude $\Phi_{B1}(\xi)$ can be parametrized by m_B/λ_B where $\lambda_B = O(\Lambda_{QCD})$. The hard-scattering kernel $h(\bar{v}, s_c)$ is displayed in Fig. 5. It is real for $\bar{v} \leq 4s$ and develops an imaginary part for the light-cone momentum fraction of the anti-quark in the vector meson $\bar{v} > 4s$.

The correction to $\langle Q_8 \rangle$ from the hard spectator interaction comes from the second diagram in Fig. 4. One finds

$$\langle Q_8 \rangle^{II} = \langle Q_7 \rangle \frac{\alpha_s(\mu_h) C_F}{4\pi} H_8 \tag{16}$$

where

$$H_8 = \frac{4\pi^2}{3N} \frac{f_B f_V^{\perp}}{F_V m_B^2} \int_0^1 d\xi \frac{\Phi_{B1}(\xi)}{\xi} \int_0^1 dv \frac{\Phi_{\perp}(v)}{v} \tag{17}$$

Here the LCDA $\Phi_{\perp}(v)$ is of leading power for a transversly polarized vector meson. The last convolution integral in (17) can be performed explicitely and leads to a combination of Gegenbauer moments of $\Phi_{\perp}$ [28, 21].

There are further mechanisms that can in principle contribute to $\bar{B} \to V\gamma$ decays. One possibility is weak annihilation. In this case the leading-power projection onto the meson V vanishes because the trace over an odd number of Dirac matrices is zero. A non-vanishing result arises from the subleading-power projections onto $\Phi_{\parallel}$ and $g_{\perp}$.

Despite its power suppression, the dominant annihilation amplitude can be computed within QCD factorization. This is because the colour-transparency argument applies to the emitted, highly energetic vector meson in the heavy-quark limit [20, 29].

Since weak annihilation is a power correction, we will content ourselves with the lowest order result ($O(\alpha_s^0)$) for our estimates below. In particular, we shall include the annihilation effects from operators $Q_{1,2}$ to estimate isospin-breaking corrections in $B \to \rho\gamma$ decays. The reason for including this class of power corrections is that they come with a numerical enhancement from the large Wilson coefficients $C_{1,2}$ ($C_1 \approx 3|C_7|$) and are not CKM suppressed. Instead, a CKM suppression of annihilation effects occurs for $B \to K^*\gamma$ and these contributions are thus very small in this case.

RESULTS

Finally, we can combine these results and write, adding the up- and charm- quark contribution,

$$A(\bar{B} \to V\gamma) = \frac{G_F}{\sqrt{2}} \left[\sum_{p=u,c} \lambda_p^{(s)} a_7^p(V\gamma) \right] \langle V\gamma|Q_7|\bar{B}\rangle \tag{18}$$

where, at NLO, the factorization coefficients $a_7^p(V\gamma)$ are given as

$$\begin{aligned} a_7^p(V\gamma) = C_7 \; &+ \; \frac{\alpha_s(\mu)C_F}{4\pi}\left(C_1(\mu)G_1(s_p) + C_8(\mu)G_8\right) \\ &+ \; \frac{\alpha_s(\mu_h)C_F}{4\pi}\left(C_1(\mu_h)H_1(s_p) + C_8(\mu_h)H_8\right) \end{aligned} \tag{19}$$

Here the NLO expression for C_7 [14] has to be used, while the leading order values are sufficient for C_1 and C_8. Numerically we obtain for central values of all input parameters, at $\mu = m_b$, and displaying separately the size of the various correction terms:

$$\begin{aligned} a_7^c(K^*\gamma) &= \underset{C_7^{LO}}{-0.3221} \;\; \underset{\Delta C_7^{NLO}}{+0.0113} \;\; \underset{T_{1,8}^{I}\text{-contribution}}{-0.0820 - 0.0147i} \;\; \underset{T_{1,8}^{II}\text{-contribution}}{-0.0144 - 0.0109i} \\ &= -0.4072 - 0.0256i. \end{aligned} \tag{20}$$

We note a sizable enhancement of the leading order value, dominated by the T^I-type correction. A complex phase is generated at NLO, where the T^I-corrections and the hard-spectator interactions (T^{II}) yield comparable effects.

The net enhancement of a_7 at NLO leads to a corresponding enhancement of the branching ratios, for fixed value of the form factor. This is illustrated in Fig. 6, where we show the residual scale dependence for $B(\bar{B} \to \bar{K}^{*0}\gamma)$ and $B(B^- \to \rho^-\gamma)$ at leading and next-to-leading order. As expected, the inclusion of the hard-vertex corrections (T^I) reduces the scale dependence coming from the Wilson coefficients. The scale dependence of the complete NLO result is "deteriorated" because the type-II contributions appear at $O(\alpha_s)$ for the first time and therefore introduce a completely new scale dependence.

For the decay $\bar{B} \to \rho\gamma$ both sectors of the effective Hamiltonian have the same order of magnitude. The amplitude for the CP-conjugated mode $B \to \rho\gamma$ is obtained by replacing

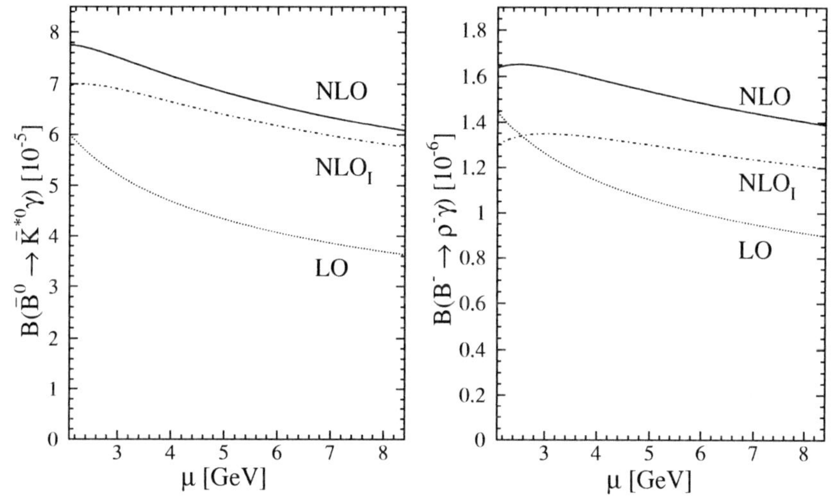

FIGURE 6. Dependence of the branching fractions $B(\bar{B}^0 \to \bar{K}^{*0}\gamma)$ and $B(B^- \to \rho^-\gamma)$ on the renormalization scale μ. The dotted line shows the LO, the dash-dotted line the NLO result including type-I corrections only and the solid line shows the complete NLO result.

$\lambda_p^{(d)} \to \lambda_p^{(d)*}$. We may then consider the CP asymmetry

$$\mathcal{A}_{CP}(\rho\gamma) = \frac{\Gamma(B \to \rho\gamma) - \Gamma(\bar{B} \to \rho\gamma)}{\Gamma(B \to \rho\gamma) + \Gamma(\bar{B} \to \rho\gamma)} \tag{21}$$

It is substantial for the $\rho\gamma$ modes and much less dependent on the form factors. Here the largest theoretical uncertainty comes from the scale dependence. This is to be expected because the direct CP asymmetry is proportional to the perturbative strong phase difference, which arises at $O(\alpha_s)$ for the first time. Unknown power corrections could have some impact on the prediction.

A further interesting observable is the charge averaged isospin breaking ratio

$$\Delta(\rho\gamma) = \frac{\Gamma(B^+ \to \rho^+\gamma)}{4\Gamma(B^0 \to \rho^0\gamma)} + \frac{\Gamma(B^- \to \rho^-\gamma)}{4\Gamma(\bar{B}^0 \to \rho^0\gamma)} - 1 \tag{22}$$

Within our approximations, isospin breaking is generated by weak annihilation. Isospin breaking was already discussed in [30], partially including NLO corrections.

$B \to \rho\gamma$ also depends sensitively on fundamental CKM parameters, such as $|V_{ub}/V_{cb}|$ and γ, and can thus in principle serve to constrain the latter quantities once measurements become available. This is further illustrated in Fig. 7, where the dependence on γ is shown for $\mathcal{A}_{CP}(\rho\gamma)$ and $\Delta(\rho\gamma)$, respectively. We remark that our sign of $\Delta(\rho\gamma)$ differs from the one found in [30].

Another application of our results concerns an estimate of U-spin breaking effects in $B \to V\gamma$ decays [31, 32, 33]. Let us state here only that U-spin breaking effects can be sizeable. For further details we refer to [1].

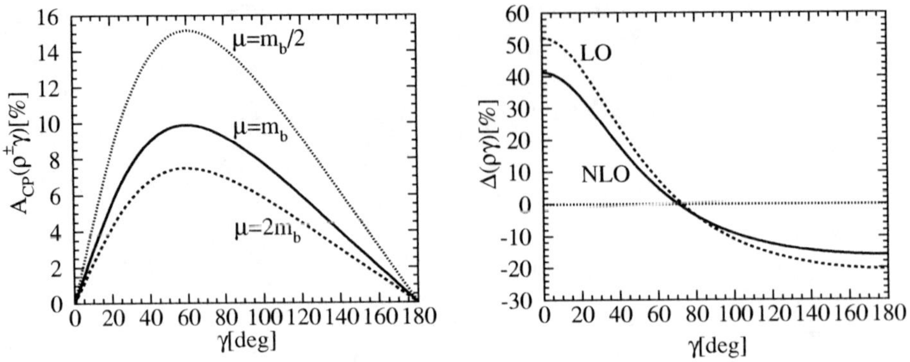

FIGURE 7. Left: The CP asymmetry $\mathcal{A}_{CP}(\rho\gamma)$ as a function of the CKM angle γ for three values of the renormalization scale $\mu = m_b/2$, m_b and $2m_b$. Right: The isospin-breaking asymmetry $\Delta(\rho\gamma)$ as a function of the CKM angle γ at leading and next-to-leading order.

CONCLUSIONS

In this talk we have presented a systematic and model-independent framework for the exclusive radiative decays $B \to V\gamma$ based on the heavy-quark limit. This allowed us to compute the decay amplitudes for these modes consistently at next-to-leading order in QCD.

An important conceptual aspect of this analysis is the interpretation of loop contributions with charm and up quarks, which come from leading operators in the effective weak Hamiltonian. We have argued that these effects are calculable in terms of perturbative hard-scattering functions and universal meson light-cone distribution amplitudes. They are $O(\alpha_s)$ corrections, but are leading power contributions in the framework of QCD factorization. This picture is in contrast to the common notion that considers charm and up-quark loop effects as generic, uncalculable long-distance contributions. Non-factorizable long-distance corrections may still exist, but they are power-suppressed.

Another important feature of the NLO calculation are the strong interaction phases, which are calculable at leading power. They play a crucial role for CP violating observables. We have seen that weak-annihilation amplitudes are power-suppressed, but can be numerically important for $B \to \rho\gamma$ because they enter with large coefficients. These effects also turn out to be calculable and were included in our phenomenological discussion at leading order in QCD.

Finally, a numerical analysis leads to our predictions for the branching ratios $B(\bar{B}^0 \to \bar{K}^{*0}\gamma) \sim 7.09 \cdot 10^{-5}$ and $B(B^- \to \rho^-\gamma) \sim 1.58 \cdot 10^{-6}$, and for the CP asymmetries $\mathcal{A}_{CP}(K^*\gamma) \sim -0.5\%$ and $\mathcal{A}_{CP}(\rho\gamma) \sim 10\%$ for central values of the input parameters. The uncertainties for the branching ratios are of $O(30\ldots35\%)$ where the form factors F_{K^*} and F_ρ taken from [21] clearly dominate this uncertainty. This situation however can be systematically improved. In particular, our approach allows for a consistent perturbative matching of the nonperturbative form factor to the short-distance part of the amplitude.

ACKNOWLEDGMENTS

I would like to thank the organizers for inviting me to such an interesting symposium. It is a great pleasure to thank Gerhard Buchalla for the extremely enjoyable collaboration. For the work presented here, I also gratefully acknowledge financial support from the Studienstiftung des deutschen Volkes and thank the CERN Theory Division for the kind hospitality.

REFERENCES

1. S. W. Bosch and G. Buchalla, hep-ph/0106081.
2. S. Chen *et al.* [CLEO Collaboration], hep-ex/0108032; K. Honscheid, this conference.
3. R. Barate *et al.* [ALEPH Collaboration], *Phys. Lett. B* **429** (1998) 169.
4. K. Abe *et al.* [Belle Collaboration], *Phys. Lett. B* **511**, 151 (2001). M. Nakao [Belle collaboration], this conference.
5. T. E. Coan *et al.* [CLEO Collaboration], *Phys. Rev. Lett.* **84** (2000) 5283.
6. S. Spanier [BaBar Collaboration], SLAC-PUB-8987 *Presented at the Kaon 2001 Conference, Pisa, Italy, June 12 - June 17, 2001.* A. Ryd [BaBar collaboration], this conference.
7. M. Nakao [Belle collaboration], this conference.
8. S. Bertolini, F. Borzumati and A. Masiero, *Phys. Rev. Lett.* **59** (1987) 180.
9. A. Ali, C. Greub and T. Mannel, DESY-93-016, *To be publ. in Proc. of ECFA Workshop on the Physics of a B Meson Factory, Eds. R. Aleksan, A. Ali, 1993.*
10. A. J. Buras, M. Misiak, M. Münz and S. Pokorski, *Nucl. Phys. B* **424** (1994) 374.
11. M. Ciuchini *et al.*, *Phys. Lett. B* **316** (1993) 127. M. Ciuchini, E. Franco, L. Reina and L. Silvestrini, *Nucl. Phys. B* **421** (1994) 41.
12. K. Adel and Y. Yao, *Phys. Rev. D* **49** (1994) 4945. C. Greub and T. Hurth, *Phys. Rev. D* **56** (1997) 2934. A. J. Buras, A. Kwiatkowski and N. Pott, *Nucl. Phys. B* **517** (1998) 353.
13. C. Greub, T. Hurth and D. Wyler, *Phys. Rev. D* **54** (1996) 3350. A. J. Buras, A. Czarnecki, M. Misiak and J. Urban, *Nucl. Phys. B* **611** (2001) 488.
14. K. Chetyrkin, M. Misiak and M. Münz, *Phys. Lett. B* **400** (1997) 206. Erratum-ibid. B **425** (1998) 414.
15. M. Misiak, hep-ph/0002007. C. Greub, hep-ph/9911348.
16. H. H. Asatrian, H. M. Asatrian and D. Wyler, *Phys. Lett. B* **470** (1999) 223. C. Greub, H. Simma and D. Wyler, *Nucl. Phys. B* **434** (1995) 39 [Erratum-ibid. B **444** (1995) 447].
17. M. Beneke, T. Feldmann and D. Seidel, *Nucl. Phys. B* **612** (2001) 25.
18. A. Ali and A. Y. Parkhomenko, hep-ph/0105302.
19. M. Beneke, G. Buchalla, M. Neubert and C. T. Sachrajda, *Phys. Rev. Lett.* **83** (1999) 1914.
20. M. Beneke, G. Buchalla, M. Neubert and C. T. Sachrajda, *Nucl. Phys. B* **591** (2000) 313.
21. P. Ball and V. M. Braun, *Phys. Rev. D* **58** (1998) 094016.
22. J. M. Flynn and C. T. Sachrajda, hep-lat/9710057.
23. J. Charles *et al.*, *Phys. Rev. D* **60** (1999) 014001.
24. M. Beneke and T. Feldmann, *Nucl. Phys. B* **592** (2001) 3.
25. G. Burdman and G. Hiller, *Phys. Rev. D* **63**, 113008 (2001).
26. G. Buchalla, A. J. Buras and M. E. Lautenbacher, *Rev. Mod. Phys.* **68** (1996) 1125.
27. A. Ali, V. M. Braun and H. Simma, *Z. Phys. C* **63** (1994) 437.
28. P. Ball and V. M. Braun, *Phys. Rev. D* **54** (1996) 2182.
29. B. Grinstein and D. Pirjol, *Phys. Rev. D* **62** (2000) 093002.
30. A. Ali, L. T. Handoko and D. London, *Phys. Rev. D* **63** (2000) 014014.
31. M. Gronau and J. L. Rosner, *Phys. Lett. B* **500** (2001) 247. M. Gronau, *Phys. Lett. B* **492** (2000) 297.
32. T. Hurth and T. Mannel, *Phys. Lett. B* **511**, 196 (2001)
33. R. Fleischer, *Phys. Lett. B* **459** (1999) 306.

Penguins in B Decays

Michael Gronau

Department of Physics, Technion-Israel Institute of Technology
Technion City, 32000 Haifa, Israel

Abstract. We report on recent progress in studying two aspects of B physics, in which penguin amplitudes play an important role:

1. Bounds on *penguin pollution* in $B^0(t) \to \pi^+\pi^-$ constraining the CKM parameters ρ and η, and a lower bound on $B^0 \to \pi^0\pi^0$ improving precision in $\sin 2\alpha$.
2. A suggestion for measuring the photon polarization in *electroweak penguin* decay, $B \to K_1(1400)\gamma$, providing a test of the Standard Model and a probe for new physics.

INTRODUCTION

When being asked to choose a topic for my talk at this conference, I responded without hesitation by making the above choice. The topic of penguins in B decays is quite broad and covers a large variety of aspects, some of which are discussed by other speakers at this conference. The two particular aspects to which I will address my talk are almost as old as the entire field of B physics. Let me remind you how penguins entered heavy flavor physics. *Gluonic penguin diagrams* were introduced twenty five years ago [1] when analyzing QCD effects in hadronic K meson decays. Shortly afterwards penguin amplitudes were shown to play an important role in direct CP violation, first studied in K decays [2] and soon afterwards in B decays [3]. A couple of years later it was realized that intermediate heavy fermions, such as a heavy top quark, imply sizable *electroweak penguin amplitudes* governing radiative K [4] and B [5] decays. These historical remarks lead naturally to my two topics.

Penguin pollution in $B^0(t) \to \pi^+\pi^-$

Calculations of direct CP violation in B decays, due to interference between tree and penguin amplitudes, involve theoretical uncertainties in nonperturbative hadronic matrix elements of weak operators and uncertainties in final state interaction phases [6]. This poses a difficulty in interpreting a measurement of the time-dependent CP asymmetry in $B^0(t) \to \pi^+\pi^-$ in terms of a fundamental CKM phase [7, 8]. This problem, known as the problem of *penguin pollution*, is dealt with in the next section. I will show first that a crude asymmetry measurement and an approximate knowledge of the ratio of penguin to tree amplitudes improve significantly our present knowledge of CKM parameters.

I will then discuss the cleanest way of resolving the penguin pollution [9], which is based on applying isospin symmetry to the system of all three decays $B^0 \to \pi^+\pi^-, B^+ \to$

CP618, *Heavy Flavor Physics: Ninth International Symposium,* edited by A. Ryd and F. C. Porter

$\pi^+\pi^0$ and $B^0 \to \pi^0\pi^0$. The most challenging experimental task in this method is measuring decay rates into two neutral pions while distinguishing between B^0 and $\bar{B}^0$ decays. I will report on recent theoretical progress made in order to overcome this difficulty.

Photon polarization in radiative B decays

In the Standard Model radiative B decays have one unambiguous signature which has not yet been tested. Namely, in decays of B^- and $\bar{B}^0$ (containing a b quark) the emitted photon is left-handed polarized, while in B^+ and B^0 it is right-handed. This prediction of *maximal parity violation*, holds to within a percent and, in principle, can serve as a *precision test* of the Standard Model. Deviations from this prediction are sensitive probes of new physics. In the third section I will survey several suggestions for studying photon helicity effects in radiative B and Λ_b decays, focusing on a particular method, which was suggested very recently. A measurement of the photon polarization in $B \to K\pi\pi\gamma$, $m(K\pi\pi) = 1400$ MeV, through decay particle angular distributions, will be shown to be feasible at currently operating B factories.

Finally, the last section contains several concluding remarks.

BOUNDS ON PENGUIN POLLUTION IN $B^0 \to \pi^+\pi^-$

CP asymmetry in $B^0(t) \to \pi^+\pi^-$

The weak phase $\alpha \equiv \arg(-V_{tb}^* V_{td}/V_{ub}^* V_{ud}) = \pi - \beta - \gamma$ occurs in the time-dependent rate of $B^0(t) \to \pi^+\pi^-$ and would dominate its asymmetry if only a *tree* amplitude T contributed. In reality this process involves a second amplitude P due to *penguin* operators which carries a different weak phase than the dominant tree amplitude,

$$A(B^0 \to \pi^+\pi^-) = |T|e^{i\delta_T} e^{i\gamma} + |P|e^{i\delta_P} \quad . \tag{1}$$

The two terms contain CKM factors $V_{ub}^* V_{ud}$, $V_{cb}^* V_{cd}$ and weak phases γ and 0, respectively. This leads to a generalized form of the time-dependent asymmetry, which includes in addition to the $\sin(\Delta mt)$ term a $\cos(\Delta mt)$ term due to direct CP violation [7]

$$\mathcal{A}(t) = C_{\pi\pi} \cos(\Delta mt) + S_{\pi\pi} \sin(\Delta mt) \ , \tag{2}$$

$$C_{\pi\pi} \equiv a_{\rm dir} = \frac{1 - |\lambda_{\pi\pi}|^2}{1 + |\lambda_{\pi\pi}|^2} \ , S_{\pi\pi} \equiv \sqrt{1 - a_{\rm dir}^2} \sin 2(\alpha + \Delta\alpha) = \frac{2{\rm Im}(\lambda_{\pi\pi})}{1 + |\lambda_{\pi\pi}|^2} \ , \tag{3}$$

where

$$\lambda_{\pi\pi} \equiv e^{-2i\beta} \frac{A(\overline{B}^0 \to \pi^+\pi^-)}{A(B^0 \to \pi^+\pi^-)} \quad . \tag{4}$$

In the absence of the penguin amplitude one would have $C_{\pi\pi} = \Delta\alpha = 0, S_{\pi\pi} = \sin 2\alpha$.

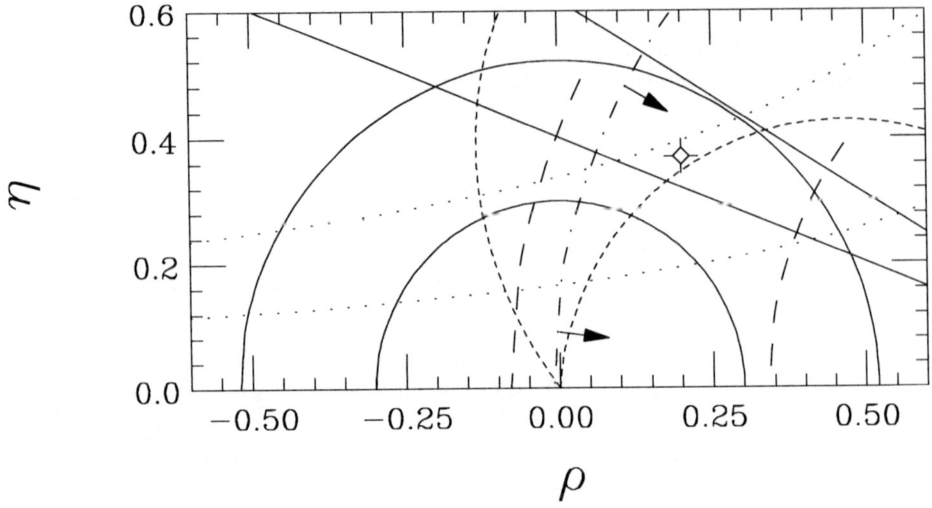

FIGURE 1. Constraints on parameters of the CKM matrix. Solid circles denote limits on $|V_{ub}/V_{cb}| = 0.090 \pm 0.025$ from charmless b decays. Dashed arcs denote limits from B^0–$\overline{B}^0$ mixing. Dot-dashed arc denotes limit from B_s–$\overline{B}_s$ mixing. Dotted hyperbolae are associated with limits on CP-violating K^0–$\overline{K}^0$ mixing. Limits of $\pm 1\sigma$ from CP asymmetries in $B^0 \to J/\psi K_S$, $\sin(2\beta) = 0.79 \pm 0.10$, are shown by the solid rays. The small dashed lines represent constraints due to 1σ bounds $-0.53 \le S_{\pi\pi} \le 0.59$, with $0.21 \le |P/T| \le 0.34$. The plotted point lies in the middle of the allowed region.

The time-dependent asymmetry measurement provides two equations for $C_{\pi\pi}$ and $S_{\pi\pi}$ in terms of $|P/T|$, $\delta \equiv \delta_P - \delta_T$ and α. This is insufficient for a determination of α. Knowledge of $|P/T|$ would, in principle, enable this determination up to discrete ambiguities. A crude estimate [10], $|P/T| = 0.3 \pm 0.1$, was obtained several years ago by applying flavor SU(3) to the ratio of $B \to \pi\pi$ and $B \to K\pi$ branching ratios first measured by CLEO [11]. A more precise evaluation including SU(3) breaking, from averaging recent CLEO, Belle and BaBar branching ratios [12], yields [13] $|P/T| = 0.276 \pm 0.064$. In a QCD factorization approach, where absolute hadronic weak amplitudes including strong phases are calculated, one finds [14] $|P/T| = 0.285 \pm 0.076$.

The BaBar Collaboration reported recently the first measurements of $C_{\pi\pi}$ and $S_{\pi\pi}$ [15]

$$C_{\pi\pi} = -0.25^{+0.45}_{-0.47} \pm 0.14 \ , \quad S_{\pi\pi} = 0.03^{+0.53}_{-0.56} \pm 0.11 \ . \tag{5}$$

This asymmetry measurement is still very crude. Nonetheless, to anticipate the significance of future improvements, we have studied recently [13] the implication of present 1σ bounds, $-0.53 \le S_{\pi\pi} \le 0.59$, on CKM parameters. Assuming that δ is small [6, 14], one has

$$S_{\pi\pi} \simeq \sin[2(\alpha + \Delta\alpha)] \ , \quad \tan\alpha = \frac{\eta}{\eta^2 - \rho(1-\rho)} \ , \quad \tan\Delta\alpha = \frac{\eta|P/T|}{\sqrt{\rho^2+\eta^2} + \rho|P/T|} \ . \tag{6}$$

Using $0.21 \le |P/T| \le 0.34$ we found in [13] that *the above 1σ $S_{\pi\pi}$ bounds exclude more than half of the (ρ, η) parameter space [16] allowed by all other constraints on*

CKM parameters. Results are shown in Fig. 1. The exclusion plot due to $S_{\pi\pi}$ is rather striking in view of the large uncertainties assumed here for $S_{\pi\pi}$ and $|P/T|$ which imply an uncertainty $\Delta\alpha$ in α as large as about $\pm 20°$. The assumption of a small δ can be tested by improving limits on $C_{\pi\pi}$.

Combining $B^0 \to \pi^+\pi^-$ with $B^+ \to \pi^+\pi^0$, $B^0 \to \pi^0\pi^0$

The isospin method [9] requires measuring also the time-integrated rates of $B^+ \to \pi^+\pi^0$, $B^0 \to \pi^0\pi^0$ and their charge-conjugates. The three $B \to \pi\pi$ amplitudes obey an isospin triangle relation,

$$A(B^0 \to \pi^+\pi^-)/\sqrt{2} + A(B^0 \to \pi^0\pi^0) = A(B^+ \to \pi^+\pi^0) \ , \tag{7}$$

while a similar relation holds for the charge-conjugate processes. One uses the different isospin properties of the penguin ($\Delta I = 1/2$) and tree ($\Delta I = 1/2, 3/2$) contributions and the well-defined weak phase (γ) of the tree amplitude. The electroweak penguin amplitude is very small [17], and can be dealt with as function as V_{td}/V_{ub} in the isospin symmetry limit [18]. Laying the two isospin triangles such that they have a common side, $A(B^+ \to \pi^+\pi^0) = A(B^- \to \pi^-\pi^0)$, the angle between $A(B^0 \to \pi^+\pi^-)$ and $A(\bar{B}^0 \to \pi^+\pi^-)$ is $2\Delta\alpha$ which then determines α from the asymmetry (2).

While this method is the cleanest theoretically, it suffers from the experimental difficulty associated with the two neutral pion mode. In fact, this introduces three kinds of practical complications:

1. $B^0 \to \pi^0\pi^0$ is argued to be *color-suppressed* and, although there exists no reliable calculation for this branching ratio, it is customarily assumed to be much smaller than the other two branching ratios.
2. One requires neutral B *flavor tagging* in order to distinguish between $B^0 \to \pi^0\pi^0$ and $\bar{B}^0 \to \pi^0\pi^0$.
3. Neutral pions have a somewhat *lower detection efficiency* than charged pions.

In the subsequent discussion we will show how to overcome the first two obstacles, which lead one to ask the following question: Assuming that one has only an upper bound on the sum of B^0 and $\bar{B}^0$ decay branching ratios to $\pi^0\pi^0$, can one put an upper limit on $|\Delta\alpha|$? This question was addressed a few years ago, and a partial answer was given under the assumption that the sum of rates of $B^+ \to \pi^+\pi^0$ and its charge conjugate is known. An upper bound, based on right angle isospin triangles, was given in terms of the ratio of charged-to-neutral B lifetimes [19] and the ratio of charge-averaged branching ratios $\mathcal{B}(B \to \pi^0\pi^0)/\mathcal{B}(B^\pm \to \pi^\pm\pi^0)$ [20]:

$$|\sin(\Delta\alpha)| \leq \sqrt{\frac{r_\tau \mathcal{B}(B \to \pi^0\pi^0)}{\mathcal{B}(B^\pm \to \pi^\pm\pi^0)}} \ , \quad r_\tau \equiv \frac{\tau_{B^+}}{\tau_{B^0}} = 1.068 \pm 0.016 \ . \tag{8}$$

A slight improvement, involving the direct CP asymmetry in $B^0 \to \pi^+\pi^-$, $a_{\rm dir}$, as well as an independent bound assuming the knowledge of $\mathcal{B}(B \to \pi^+\pi^-)$ instead of $\mathcal{B}(B^\pm \to \pi^\pm\pi^0)$, were suggested in [21].

Although these two bounds are somehow related to the isospin triangles, neither of them involves all three $B \to \pi\pi$ processes, implying that the saturation of these bounds may be inconsistent with the closure of the triangles. Thus, the real question is what is the maximum value of $|\Delta\alpha|$, consistent with the closure, for given $\mathcal{B}(B \to \pi^+\pi^-)$ and $\mathcal{B}(B^\pm \to \pi^\pm\pi^0)$ and for an upper bound on $\mathcal{B}(B \to \pi^0\pi^0)$. The correct answer to this question was found recently [22],

$$\cos(2\Delta\alpha) \geq \frac{(B^{+-}/2 \quad B^{00} + B^{+0}/r_\tau)^2 - B^{+-}B^{+0}/r_\tau}{\sqrt{1 - a_{\rm dir}^2}B^{+-}B^{+0}/r_\tau} , \tag{9}$$

where B^{ij} are corresponding charge-averaged branching ratios. This bound is stronger than Eq. (8) and the bound [21], as demonstrated in [22]. A crucial difference between Eq. (9) and the earlier bounds is that (9) includes also a lower bound on B^{00}, following from the triangle construction:

$$B^{00} \geq B^{+0}/r_\tau + B^{+-}/2 - \sqrt{(1+\sqrt{1-a_{\rm dir}^2})B^{+-}B^{+0}/r_\tau} \geq (\sqrt{B^{+0}/r_\tau} - \sqrt{B^{+-}/2})^2 . \tag{10}$$

The advantage of the two bounds Eqs. (9) and (10) over (8) and [21] was demonstrated in [22] when using the present world averaged branching ratios in units of 10^{-6} [12]

$$B^{+-} = 4.4 \pm 0.9 , \quad B^{+0} = 5.6 \pm 1.5 , \quad B^{00} < 5.7 \text{ (90\% C.L.)} . \tag{11}$$

In order to illustrate the future potential power of these bounds in reducing the error in α due to penguin pollution, we list below values of the three branching ratios with corresponding errors, which were measured and which can be measured at B factories with higher integrated luminosities. Errors in B^{ij} scale down as $1/\sqrt{\text{luminosity}}$. For illustration purpose, we will take the future central value of B^{+0} to be less than 1σ above its present central value.

luminosity:	30 fb^{-1}	120 fb^{-1}	500 fb^{-1}
B^{+-}	4.4 ± 0.9	4.4 ± 0.4	4.4 ± 0.2
B^{+0}	5.6 ± 1.5	7.0 ± 0.8	7.0 ± 0.4
B^{00}	< 5.7	< 1.4	< 0.4 or seen
B^{00}	$\geq 0.78 \pm 0.62$	$\geq 1.35 \pm 0.38$	$\geq 1.35 \pm 0.19$

The last line in the table gives the lower bounds on B^{00} obtained from Eq. (10) for the corresponding values of B^{+-} and B^{+0}.

Thus, while an upper bound on B^{00} can be obtained from a direct measurement, useful lower bounds follow from measuring the other two branching ratios. If B^{00} is not very small, which does not seem unlikely in view of the present values of B^{+-} and B^{+0}, one may be able to restrict its values from above and below to a narrow range. Consequently, the uncertainty in measuring α becomes small. Assuming, for instance, that one finds $1.2 \leq B^{00} \leq 1.3$, permitted by the lower bound derived for 500 fb^{-1}, one obtains from Eq. (9) $|\Delta\alpha| < 9°$. In comparison, the bound (8) implies only $|\Delta\alpha| < 26°$. We stress that this demonstration of a rather precise determination of α (where the uncertainty follows only from penguin pollution) assumes no separation between B^0 and $\bar{B}^0$ decays to $\pi^0\pi^0$. Neutral B flavor tagging will reduce the uncertainty further.

THE PHOTON POLARIZATION IN $b \to s\gamma$

The present agreement between experiment and the Standard Model (SM) prediction for the rate of inclusive $B \to X_s\gamma$ is reasonable, at a level of 20% [23]. However, one basic feature, the left-handedness of the emitted photon in $b \to s\gamma$, has never been tested. The photon is predominantly left-handed, since the recoil s quark which couples to a W is left-chiral. In several extensions of the SM, including left-right symmetric [24] and supersymmetric models [25], in which decay amplitudes involve $W_L - W_R$ mixing and scalar exchange, the photon can acquire a large right-handed component without affecting substantially the inclusive rate.

Formally, the effective weak Hamiltonian for radiative b decays contains two Wilson coefficients, C_{7L} and C_{7R}, multiplying operators, $\mathcal{O}_{7L}$ and $\mathcal{O}_{7R}$, describing left and right handed emitted photons,

$$\mathcal{H}_{\rm rad} = -\frac{4G_F}{\sqrt{2}} V_{tb}V_{ts}^* \left(C_{7L}\mathcal{O}_{7L} + C_{7R}\mathcal{O}_{7R}\right) , \quad \mathcal{O}_{7L,R} \equiv \frac{e}{16\pi^2} m_b \bar{s}\sigma_{\mu\nu} \frac{1 \pm \gamma_5}{2} b F^{\mu\nu} . \tag{12}$$

The photon polarization in inclusive $b \to s\gamma$ is

$$\lambda_\gamma \equiv \frac{|C_R|^2 - |C_L|^2}{|C_R|^2 + |C_L|^2} . \tag{13}$$

In the SM, where $C_{7R}/C_{7L} = m_s/m_b$, the polarization in exclusive decays is $\lambda_\gamma = -1$ within a percent, also when modified by long distance hadronic effects [26]. This prediction can provide precision tests of the SM and sensitive probes for new physics.

Several ways of carrying out such measurements were proposed in the past. They require very high luminosity B factories or new experimental facilities. We will describe very briefly these early suggestions, and will focus our attention on a recent proposal which is feasible at currently operating B factories.

CP asymmetry in $B^0(t) \to X^{CP}_{s(d)}\gamma$

Consider the time-dependent rate of [27] $B^0(t) \to X^{CP}_{s(d)}\gamma$, where $X^{CP}_s = K^{*0} \to K_S\pi^0$ or $X^{CP}_d = \rho^0 \to \pi^+\pi^-$. The time-dependent CP asymmetry follows from interference between B^0 and $\bar{B}^0$ decay amplitudes into a common state of definite photon polarization, and is proportional to C_{7R}/C_{7L}. For instance, in the SM the asymmetry in $B^0(t) \to f$, $f = K^{*0}\gamma \to (K_S\pi^0)\gamma$, is given by

$$\mathcal{A}(t) \equiv \frac{\Gamma(B^0(t) \to f) - \Gamma(\bar{B}^0(t) \to f)}{\Gamma(B^0(t) \to f) + \Gamma(\bar{B}^0(t) \to f)} = \frac{2A_LA_R}{A_L^2 + A_R^2} \sin 2\beta \sin(\Delta m t) , \tag{14}$$

where $A_{L(R)}$ is the amplitude for a left (right) handed photon in $\bar{B} \to \bar{K}^*\gamma$. In the SM one expects $A_R/A_L \leq 0.05$ in the presence of long distance effects, whereas in extensions of the SM this ratio may be much larger [27].

Angular distribution in $\bar{B} \to \bar{K}^*\gamma \to \bar{K}\pi e^+e^-$

Consider the decay distribution in this process as function of the angle ϕ between the $\bar{K}\pi$ and e^+e^- planes, where the photon can be virtual [28] or real, converting in the beam pipe to an electron-positron pair [29]. The e^+e^- plane acts as a polarizer, the distribution in ϕ is isotropic for purely circular polarization, and the angular distribution is sensitive to interference between left and right polarization. One finds

$$\frac{d\sigma}{d\phi} \propto 1 + \xi \frac{A_L A_R}{A_L^2 + A_R^2} \cos(2\phi + \delta) , \tag{15}$$

where the parameters ξ and δ are calculable and involve hadronic physics.

Forward-backward asymmetry in $\Lambda_b \to \Lambda\gamma \to p\pi\gamma$

The forward-backward asymmetry of the proton with respect to the Λ_b in the Λ rest-frame is proportional to the photon polarization λ_γ [30]. Using polarized Λ_b's from extremely high luminosity e^+e^- Z factories, one can also measure the forward-backward asymmetry of the Λ momentum with respect to the Λ_b boost axis [31]. This asymmetry is proportional to the product of the Λ_b and photon polarizations.

Angular distribution in $B \to K_1(1400)\gamma \to K\pi\pi\gamma$

In order to measure the photon polarization λ_γ in radiative B decays through the recoil hadron distribution, one requires that the hadrons consist of at least three particles. A hadronic quantity which is proportional to λ_γ must be parity odd. The pseudoscalar quantity, which contains the smallest number of hadron momenta, is a triple product. The idea is then to measure an expectation value $\langle \vec{p}_\gamma \cdot (\vec{p}_1 \times \vec{p}_2) \rangle$, where $\vec{p}_1$ and $\vec{p}_2$ are momenta of two of the hadrons. Since the triple product is also time-reversal odd, a nonzero expectation value requires a phase due to final state interactions. While in general such a phase would be incalculable, there are special cases where the decay occurs through two isospin-related intermediate resonance states, and the phase can be calculated simply in terms of Breit-Wigner forms [32].

Consider the decays $B^+ \to K_1^+(1400)\gamma$ and $B^0 \to K_1^0(1400)\gamma$, where K_1^+ and K_1^0 are observed through

$$K_1^+(1400) \to \left\{ \begin{array}{c} K^{*+}\pi^0 \\ K^{*0}\pi^+ \end{array} \right\} \to K^0\pi^+\pi^0 , \quad K_1^0(1400) \to \left\{ \begin{array}{c} K^{*+}\pi^- \\ K^{*0}\pi^0 \end{array} \right\} \to K^+\pi^-\pi^0 . \tag{16}$$

Two Breit-Wigner amplitudes interfere due to intermediate K^{*+} and K^{*0}, $\mathcal{B}(K_1 \to K^*\pi) = 0.94 \pm 0.06$ [16]. Decay to ρK will be neglected at this point, $\mathcal{B}(K_1 \to \rho K) = 0.03 \pm 0.03$. The two K^* amplitudes are related by isospin; therefore phases other those related to the Breit-Wigner phase cancel. The decay $K_1 \to K^*\pi$ is dominated by an S wave and involves a small D waves, where the D/S ratio of rates is $D/S = 0.04 \pm 0.01$ [16]. Using Lorentz invariance, it is straightforward to write down the decay amplitude for $B \to (K\pi\pi)_{K_1}\gamma$, and to calculate the decay distribution [32],

$$\frac{d\Gamma}{ds_{13}ds_{23}d\cos\theta} \propto |\vec{J}|^2(1+\cos^2\theta) + \lambda_\gamma 2\mathrm{Im}\left(\hat{n}\cdot(\vec{J}\times\vec{J}^*)\right)\cos\theta\,, \tag{17}$$

where

$$\begin{aligned}\vec{J} = \quad & \vec{p}_1\left[\left((1-\frac{m_K^2-m_\pi^2}{m_{K^*}^2})(1-\kappa(p_{K_1}\cdot p_1 - m_\pi^2)) - 2\kappa p_1\cdot p_2\right)B(s_{23}) - 2B(s_{13})\right] \\ & -(p_1 \leftrightarrow p_2)\,, \end{aligned}\tag{18}$$

$$B(s) = \left(s - m_{K^*}^2 - im_{K^*}\Gamma_{K^*}\right)^{-1}\,, \quad s_{ij} = (p_i+p_j)^2\,. \tag{19}$$

κ parametrizes the D wave contribution, $\kappa = [0.38 + 8.66|A_D/A_S|e^{i(\delta_D-\delta_S)}][1 + 0.71|A_D/A_S|e^{i(\delta_D-\delta_S)}]^{-1}\mathrm{GeV}^{-2}$, $|A_D/A_S|^2 = 0.04 \pm 0.01$, $\delta_D - \delta_S = (260 \pm 20)^\circ$ [16]. p_1 and p_2 are the two pion momenta, p_3 is the K momentum, and θ is the angle between the normal to the decay plane $\hat{n} \equiv (\vec{p}_1 \times \vec{p}_2)/|\vec{p}_1 \times \vec{p}_2|$ and $-\vec{p}_\gamma$, all measured in the K_1 rest frame. A useful definition of the normal is in terms of the slow and fast pion momenta, $(\vec{p}_{\mathrm{slow}} \times \vec{p}_{\mathrm{fast}})/|\vec{p}_{\mathrm{slow}} \times \vec{p}_{\mathrm{fast}}|$. The angle between this norml and $-\vec{p}_\gamma$ will be denoted by $\tilde{\theta}$.

The decay distribution exhibits an up-down asymmetry of the photon momentum with respect to the K_1 decay plane. The up-down asymmetry is proportional to the photon polarization. When integrating over the entire Dalitz plot one finds

$$\mathcal{A}_{\mathrm{up-down}} \equiv \frac{\int_0^{\pi/2}\frac{d\Gamma}{d\cos\tilde{\theta}}d\cos\tilde{\theta} - \int_{\pi/2}^{\pi}\frac{d\Gamma}{d\cos\tilde{\theta}}d\cos\tilde{\theta}}{\int_0^{\pi}\frac{d\Gamma}{d\cos\tilde{\theta}}d\cos\tilde{\theta}} = (0.34 \pm 0.05)\lambda_\gamma\,. \tag{20}$$

The uncertainty follows from experimental errors in the ρK and in the D wave amplitudes. In the SM, where $\lambda_\gamma \approx -1$, the asymmetry is $34 \pm 5\%$ and the polarization signature is unambiguous: *In B^- and $\bar{B}^0$ decays the photon prefers to be emitted in the hemisphere of $\vec{p}_{\mathrm{slow}} \times \vec{p}_{\mathrm{fast}}$, while in B^+ and B^0 it is more likely to be emitted in the opposite hemisphere.*

Is this measurement feasible at currently operating B factories? A 3σ measurement of a 34% up-down asymmetry requires about 80 reconstructed $B \to K_1(1400)\gamma \to K\pi\pi\gamma$ events, including charged and neutral B and $\bar{B}$ decays. Assuming $\mathcal{B}(B \to K_1\gamma) = 0.7 \times 10^{-5}$ [33] and including K_1 and K^* branching ratios to the relevant charge states, one finds that this number of reconstructed events can be obtained from a total of 2×10^7 $B\bar{B}$ pairs, including charged and neutrals. This number has already been produced at e^+e^- colliders. Since we ignored experimental efficiencies, resolution and background, one may have to wait a year or so before obtaining the required number of events.

The region of $K\pi\pi$ invariant mass around 1400 MeV contains also two other resonances, a spin 2 positive-parity $K_2^*(1430)$ which has already been observed in radiative B decays [34, 35], and a vector state $K_1^*(1410)$, both of which decay to $K^*\pi$. (A nonresonant contribution in a narrow bin around $m(K\pi\pi) = 1400$ MeV is expected to be very small.) The K_2^* decays involve a smaller up-down asymmetry with the same sign as K_1 [32] (although the integrated asymmetry vanishes), while the decay of K_1^* is up-down symmetric. Whereas the overall polarization signature is unchanged, the integrated up-down asymmetry would be diluted relative to the asymmetry from $K_1(1400)$ if all three resonance contributions would be added. It is therefore useful to isolate the K_1 from the other two resonances. This can be achieved by applying to the data an angular decay distribution characterizing an axial vector particle.

CONCLUDING REMARKS

- While the study of CP asymmetry in $B^0 \to \pi^+\pi^-$ in terms of $\sin 2\alpha$ is complicated by a penguin amplitude, even crude limits on the asymmetry may exclude a large part of the presently allowed CKM parameter space.
- A lower bound on the charge-averaged branching ratio of $B \to \pi^0\pi^0$ from measured $B \to \pi^+\pi^-$ and $B^\pm \to \pi^\pm\pi^0$ may reduce the uncertainty of measuring $\sin 2\alpha$, without carrying out the complete isospin analysis.
- The photon polarization in $b \to s\gamma$, predicted to be left-handed in the Standard Model, can be measured through angular decay distributions in $B \to K\pi\pi\gamma$ around $m(K\pi\pi) = 1400$ MeV.

I expect that in a year these measurements will lead to interesting and useful results.

ACKNOWLEDGMENTS

I am grateful to D. Atwood, Y. Grossman, D. London, D. Pirjol, J. L. Rosner, A. Ryd, N. Sinha, R. Sinha, and A. Soni for enjoyable collaborations on work discussed in this talk. I wish to thank SLAC and the Aspen Center for Physics where this talk was prepared. This work was supported in part by the Fund for the Promotion of Research at the Technion, by the Israel Science Foundation founded by the Israel Academy of Sciences and Humanities, and by the U. S. – Israel Binational Science Foundation through Grant No. 98-00237.

REFERENCES

1. M. A. Shifman, A. I. Vainstein, and V. I. Zakharov, *Nucl. Phys.* B **120**, 316 (1977); J. R. Ellis, M. K. Gaillard, D. V. Nanopoulos and S. Rudaz, *Nucl. Phys.* B **131**, 285 (1977); (E) *ibid.* B **132**, 541 (1978).
2. J. R. Ellis, M. K. Gaillard, D. V. Nanopoulos, *Nucl. Phys.* B **109**, 213 (1976); F. J. Gilman, and M. B. Wise, *Phys. Lett.* B **83**, 83 (1979).

3. M. Bander, D. Silverman, and A. Soni, *Phys. Rev. Lett.* **43**, 242 (1979).
4. T. Inami and C. S. Lim, *Prog. Theor. Phys.* **65**, 297 (1981).
5. B. A. Campbell and P. J. O'Donnell, *Phys. Rev.* D **25**, 1989 (1982).
6. See talks by Y. Y. Keum, M. Neubert and H. R. Quinn at this conference.
7. M. Gronau, *Phys. Rev. Lett.* **63**, 1451 (1989).
8. D. London and R. Peccei, *Phys. Lett.* B **223**, 257 (1989); B. Grinstein, *Phys. Lett.* B **229**, 280 (1989).
9. M. Gronau and D. London, *Phys. Rev. Lett.* **65**, 3381 (1990).
10. A. Dighe, M. Gronau, and J. L. Rosner, *Phys. Rev. Lett.* **79**, 4333 (1997).
11. CLEO Collaboration, R. Godang *et al.*, *Phys. Rev. Lett.* **80**, 3456 (1998).
12. CLEO Collaboration, D. Cronin-Hennessy *et al.*, *Phys. Rev. Lett.* **85**, 515 (2000); S. J. Richichi *et al.*, *Phys. Rev. Lett.* **85**, 520 (2000); Belle Collaboration, K. Abe *et al.*, *Phys. Rev. Lett.* **87**, 101801 (2001); BaBar Collaboration, B. Aubert *et al.*, *Phys. Rev. Lett.* **87**, 151802 (2001).
13. M. Gronau and J. L. Rosner, *Phys. Rev.* D **65**, 013004 (2002); See also Z. Luo and J. L. Rosner, hep-ph/0108024.
14. M. Beneke, G. Buchalla, M. Neubert, and C. T. Sachrajda, *Nucl. Phys.* B **606**, 245 (2001).
15. BaBar Collaboration, B. Aubert *et al.*, hep-ex/0107074, submitted to XX International Symposium on Lepton and Photon Interactions at High Energies, Rome, Italy, July 23-28, 2001; see also hep-ex/0110062.
16. Particle Data Group, D. E. Groom *et al.*, *Eur. Phys. J.* C **15**, 1 (2000).
17. N. G. Deshpande and X. G. He, *Phys. Rev. Lett.* **74**, 26, 4099(E) (1995).
18. M. Gronau, D. Pirjol, and T. M. Yan, *Phys. Rev.* D **60**, 034021 (1999); A. J. Buras and R. Fleischer, *Eur. Phys. J.* C **11**, 93 (1999).
19. K. Osterberg, talk presented at the International Europhysics Conference on High-Energy Physics, Budapest, Hungary, 12–18 July 2001.
20. Y. Grossman and H. R. Quinn, *Phys. Rev.* D **58**, 017504 (1998).
21. J. Charles, *Phys. Rev.* D **59**, 054007 (1999).
22. M. Gronau, D. London, N. Sinha, and R. Sinha, *Phys. Lett.* B **514**, 315 (2001).
23. For experimental reports at this conference, see talks by K. Honscheid, M. Nakao and A. Ryd. For two recent theoretical reviews, see M. Misiak, hep-ph/0105312, talk at the XXXVIth Rencontres de Moriond, Les Arcs, March 10–17, 2001; T. Hurth, hep-ph/0106050, talk at the Eighth International Conference on Supersymmetries in Physics, Geneva, Switzerland, June 26 – July 1, 2000.
24. K. Fujikawa and A. Yamada, *Phys. Rev.* D **49**, 5890 (1994); K. S. Babu, K. Fujikawa, and A. Yamada, *Phys. Lett.* B **333**, 196 (1994); P. Cho and M. Misiak, *Phys. Rev.* D **49**, 5894 (1994).
25. See *e.g.* L. Everett, G. L. Kane, S. Rigolin, L. T. Wang, and T. T. Wang, hep-ph/0112129 and references therein.
26. B. Grinstein and D. Pirjol, *Phys. Rev.* D **62**, 093002 (2000).
27. D. Atwood, M. Gronau, and A. Soni, *Phys. Rev. Lett.* **79**, 185 (1997).
28. D. Melikhov, N. Nikitin, and S. Simula, *Phys. Lett.* B **442**, 381 (1998); F. Krüger, L. M. Sehgal, N. Sinha, and R. Sinha, *Phys. Rev.* D **61**, 114028 (2000); C. S. Kim, Y. G. Kim, C. D. Lü, and T. Morozumi, *Phys. Rev.* D **62**, 034013 (2000).
29. Y. Grossman and D. Pirjol, JHEP **0006**, 029 (2000).
30. T. Mannel and S. Recksiegel, *Acta Phys. Polonica* **B28**, 2489 (1997).
31. G. Hiller and A. Kagan, hep-ph/0108074.
32. M. Gronau, Y. Grossman, D. Pirjol, and A. Ryd, *Phys. Rev. Lett.* **88**, 051802 (2002). See also T. Islam and A. Weinstein, CLEO Internal Note, July 1999 (unpublished).
33. T. Altomari, *Phys. Rev.* D **37**, 677 (1988); A. Ali, T. Ohl, and T. Mannel, *Phys. Lett.* B **298**, 195 (1993); D. Atwood and A. Soni, Zeit. Phys. C **64**, 241 (1994); S. Veseli and M. G. Olsson, *Phys. Lett.* B **367**, 309 (1996); D. Ebert, R. N. Faustov, V. O. Galkin, and H. Toki, *Phys. Lett.* B **495**, 309 (2000); *Phys. Rev.* D **64**, 054001 (2001).
34. CLEO Collaboration, T. E. Coan *et al.*, *Phys. Rev. Lett.* **84**, 5283 (2000).
35. Belle Collaboration, K. Abe *et al.*, hep-ex/0107065, submitted to XX International Symposium on Lepton and Photon Interactions at High Energies, Rome, Italy, July 23-28, 2001.

Extraction of γ

Robert Fleischer

Deutsches Elektronen-Synchrotron DESY, Notkestr. 85, D–22607 Hamburg, Germany

Abstract. After a brief look at the well-known standard approaches to determine the angle γ of the unitarity triangle, we focus on two kinds of strategies, employing $B \to \pi K$ modes, and U-spin-related B decays. Interesting "puzzles", which may already be indicated by the present B-factory data, are pointed out, and the importance of the extraction of hadronic parameters, which are provided by these strategies as by-products, is emphasized.

SETTING THE STAGE

The recent observation of CP violation in the B system by the BaBar and Belle collaborations [1] manifests the beginning of a new era in the exploration of particle physics. One of the central goals of the B-factories is to overconstrain the unitarity triangle of the Cabibbo–Kobayashi–Maskawa (CKM) matrix as much as possible through independent measurements both of its sides and of its angles α, β and γ [2].

A particularly important element in this stringent test of the Kobayashi–Maskawa mechanism of CP violation is the direct determination of the angle γ. The corresponding experimental values may well be in conflict with the indirect results provided by the fits of the unitarity triangle, yielding at present $\gamma \sim 60^\circ$ [3]. Moreover, we may encounter discrepancies between various different approaches to determine γ directly. In such exciting cases, the data may shed light on the physics lying beyond the Standard Model.

Key Problem in the Determination of γ

At leading order of the well-known Wolfenstein expansion of the CKM matrix, all matrix elements are real, apart from

$$V_{td} = |V_{td}|e^{-i\beta} \quad \text{and} \quad V_{ub} = |V_{ub}|e^{-i\gamma}. \tag{1}$$

In non-leptonic B-meson decays, we may obtain sensitivity on γ through interference effects between different CKM amplitudes. Making use of the unitarity of the CKM matrix, it can be shown that at most two different weak amplitudes contribute to a given non-leptonic decay $B \to f$, so that we may write

$$A(B \to f) = |A_1|e^{i\delta_1} + e^{+i\gamma}|A_2|e^{i\delta_2}, \quad A(\overline{B} \to \overline{f}) = |A_1|e^{i\delta_1} + e^{-i\gamma}|A_2|e^{i\delta_2}, \tag{2}$$

CP618, *Heavy Flavor Physics: Ninth International Symposium,* edited by A. Ryd and F. C. Porter

where γ enters through V_{ub} and the $|A_{1,2}|e^{i\delta_{1,2}}$ denote CP-conserving strong amplitudes. Consequently, the corresponding direct CP asymmetry takes the following form:

$$\mathcal{A}_{\rm CP} = \frac{|A(B\to f)|^2 - |A(\overline{B}\to \overline{f})|^2}{|A(B\to f)|^2 + |A(\overline{B}\to \overline{f})|^2} = \frac{2|A_1||A_2|\sin(\delta_1-\delta_2)\sin\gamma}{|A_1|^2+2|A_1||A_2|\cos(\delta_1-\delta_2)\cos\gamma+|A_2|^2}. \quad (3)$$

Measuring such a CP asymmetry, we may in principle extract γ. However, due to hadronic uncertainties, which affect the strong amplitudes

$$|A|e^{i\delta} \sim \sum_k \underbrace{C_k(\mu)}_{\text{pert. QCD}} \times \underbrace{\langle \overline{f}|Q_k(\mu)|\overline{B}\rangle}_{\text{“unknown”}}, \quad (4)$$

a reliable determination of γ is actually a challenge.

Major Approaches to Determine γ

In order to deal with the problems arising from the hadronic matrix elements, we may employ one of the following approaches:

- The most obvious one – and, unfortunately, also the most challenging from a theoretical point of view – is to try to calculate the $\langle \overline{f}|Q_k(\mu)|\overline{B}\rangle$. In this context, interesting progress has recently been made through the development of the “QCD factorization” [4, 5] and perturbative hard-scattering (or “PQCD”) [6] approaches, as discussed by Neubert and Keum, respectively, at this symposium. As far as the determination of γ is concerned, $B\to\pi K$, $\pi\pi$ modes play a key rôle.
- Another avenue we may follow is to use decays of neutral B_d- or B_s-mesons, where interference effects between B^0_q–$\overline{B^0_q}$ mixing ($q\in\{d,s\}$) and decay processes arise. There are fortunate cases, where hadronic matrix elements cancel:
  - Decays of the kind $B_d\to D^{(*)\pm}\pi^{\mp}$, allowing a clean extraction of $\phi_d+\gamma$ [7], where the B^0_d–$\overline{B^0_d}$ mixing phase $\phi_d=2\beta$ can be fixed through $B_d\to J/\psi K_S$.
  - Decays of the kind $B_s\to D_s^{(*)\pm}K^{(*)\mp}$, allowing a clean extraction of $\phi_s+\gamma$ [8], where the B^0_s–$\overline{B^0_s}$ mixing phase ϕ_s is negligibly small in the Standard Model, and can be probed through $B_s\to J/\psi\phi$ modes.
- An important tool to eliminate hadronic uncertainties in the extraction of γ is also provided by certain amplitude relations. There are two kinds of such relations:
  - Exact relations between $B^\pm\to K^\pm\{D^0,\overline{D^0},D^0_\pm\}$ amplitudes [9], where $D^0_\pm$ denotes the CP eigenstates of the neutral D system. This approach is realized in an ideal way in the $B^\pm_c\to D^\pm_s\{D^0,\overline{D^0},D^0_\pm\}$ system [10]. Unfortunately, B_c-mesons are not as accessible as “ordinary” B_u- or B_d-mesons.
  - Amplitude relations, which are implied by the flavor symmetries of strong interactions, i.e. isospin or $SU(3)$ [11, 12]. In the corresponding strategies to determine γ, we have to deal with $B_{(s)}\to\pi\pi,\pi K,KK$ modes.

In the following discussion, we shall focus on the latter kind of strategies, involving the $B \to \pi K$ system, and the U-spin-related[1] decays $B_d \to \pi^+\pi^-$, $B_s \to K^+K^-$ and $B_d \to \pi^\mp K^\pm$, $B_s \to \pi^\pm K^\mp$. These approaches are particularly promising from a practical point of view: BaBar, Belle and CLEO-III can probe γ nicely through $B \to \pi K$ modes, whereas the U-spin strategies, requiring also the measurement of B_s-meson decays, are interesting for Tevatron-II and can be fully exploited at BTeV and the LHC experiments.

EXTRACTION OF γ FROM $B \to \pi K$ DECAYS

Let us first point out some interesting features of the $B \to \pi K$ system. Because of the small ratio $|V_{us}V_{ub}^*/(V_{ts}V_{tb}^*)| \approx 0.02$, these decays are dominated by QCD penguin topologies, despite their loop suppression. Due to the large top-quark mass, we have also to care about electroweak (EW) penguins. In the case of $B^+ \to \pi^+K^0$ and $B_d^0 \to \pi^-K^+$, these topologies contribute only in color-suppressed form and are hence expected to play a minor rôle, whereas they contribute also in color-allowed form to $B_d^0 \to \pi^0K^0$ and $B^+ \to \pi^0K^+$ and may here even compete with tree-diagram-like topologies.

Using the isospin flavor symmetry of strong interactions, we may derive the following amplitude relations:

$$\sqrt{2}A(B^+ \to \pi^0K^+) + A(B^+ \to \pi^+K^0) = \sqrt{2}A(B_d^0 \to \pi^0K^0) + A(B_d^0 \to \pi^-K^+)$$

$$= -\left[|T+C|e^{i\delta_{T+C}}e^{i\gamma} + P_{\rm ew}\right] \propto \left[e^{i\gamma} + q_{\rm ew}\right], \qquad (5)$$

where T and C denote the strong amplitudes of color-allowed and color-suppressed tree-diagram-like topologies, respectively, $P_{\rm ew}$ is due to color-allowed and color-suppressed EW penguins, δ_{T+C} is a CP-conserving strong phase, and $q_{\rm ew}$ denotes the ratio of the EW to tree-diagram-like topologies. A relation with an analogous phase structure holds also for the "mixed" $B^+ \to \pi^+K^0$, $B_d^0 \to \pi^-K^+$ system. Because of these relations, the following combinations of $B \to \pi K$ decays were considered in the literature to probe γ:

- The "mixed" $B^\pm \to \pi^\pm K$, $B_d \to \pi^\mp K^\pm$ system [13]–[16].
- The "charged" $B^\pm \to \pi^\pm K$, $B^\pm \to \pi^0 K^\pm$ system [17]–[19].
- The "neutral" $B_d \to \pi^0 K$, $B_d \to \pi^\mp K^\pm$ system [19, 20].

Interestingly, already CP-averaged $B \to \pi K$ branching ratios may lead to non-trivial constraints on γ [14, 17]. In order to *determine* this angle, also CP-violating rate differences have to be measured. To this end, we introduce the following observables [19]:

$$\left\{ \begin{array}{c} R \\ A_0 \end{array} \right\} \equiv \left[\frac{\mathrm{BR}(B_d^0 \to \pi^-K^+) \pm \mathrm{BR}(\overline{B_d^0} \to \pi^+K^-)}{\mathrm{BR}(B^+ \to \pi^+K^0) + \mathrm{BR}(B^- \to \pi^-\overline{K^0})}\right] \frac{\tau_{B^+}}{\tau_{B_d^0}} \qquad (6)$$

---

[1] U spin is an $SU(2)$ subgroup of $SU(3)_{\rm F}$, relating down and strange quarks to each other.

$$\left\{ \begin{array}{c} R_{\rm c} \\ A_0^{\rm c} \end{array} \right\} \equiv 2 \left[\frac{\mathrm{BR}(B^+ \to \pi^0 K^+) \pm \mathrm{BR}(B^- \to \pi^0 K^-)}{\mathrm{BR}(B^+ \to \pi^+ K^0) + \mathrm{BR}(B^- \to \pi^- \overline{K^0})} \right] \tag{7}$$

$$\left\{ \begin{array}{c} R_{\rm n} \\ A_0^{\rm n} \end{array} \right\} \equiv \frac{1}{2} \left[\frac{\mathrm{BR}(B_d^0 \to \pi^- K^+) \pm \mathrm{BR}(\overline{B_d^0} \to \pi^+ K^-)}{\mathrm{BR}(B_d^0 \to \pi^0 K^0) + \mathrm{BR}(\overline{B_d^0} \to \pi^0 \overline{K^0})} \right], \tag{8}$$

where the $R_{(\rm c,n)}$ are ratios of CP-averaged branching ratios and the $A_0^{(\rm c,n)}$ represent CP-violating observables.

If we employ the $SU(2)$ flavor symmetry, which implies (5), and make plausible dynamical assumptions, concerning mainly the smallness of certain rescattering processes, we obtain parametrizations of the following kind [16, 19] (for alternative ones, see [18]):

$$R_{(\rm c,n)}, A_0^{(\rm c,n)} = \text{functions}\left(q_{(\rm c,n)}, r_{(\rm c,n)}, \delta_{(\rm c,n)}, \gamma\right). \tag{9}$$

Here $q_{(\rm c,n)}$ denotes the ratio of EW penguins to "trees", $r_{(\rm c,n)}$ is the ratio of "trees" to QCD penguins, and $\delta_{(\rm c,n)}$ the strong phase between "trees" and QCD penguins for the mixed, charged and neutral $B \to \pi K$ systems, respectively. The EW penguin parameters $q_{(\rm c,n)}$ can be fixed through theoretical arguments: in the mixed system [13]–[15], we have $q \approx 0$, as EW penguins contribute only in color-suppressed form; in the charged and neutral $B \to \pi K$ systems, $q_{\rm c}$ and $q_{\rm n}$ can be fixed through the $SU(3)$ flavor symmetry without dynamical assumptions [17]–[20]. The $r_{(\rm c,n)}$ can be determined with the help of additional experimental information: in the mixed system, r can be fixed through arguments based on factorization [4, 13, 15] or U-spin, as we will see below, whereas $r_{\rm c}$ and $r_{\rm n}$ can be determined from the CP-averaged $B^\pm \to \pi^\pm \pi^0$ branching ratio by using only the $SU(3)$ flavor symmetry [11, 17]. The uncertainties arising in this program from $SU(3)$-breaking effects can be reduced through the QCD factorization approach [4, 5], which is moreover in favour of small rescattering processes. For simplicity, we shall neglect such FSI effects below; more detailed discussions can be found in [21].

Since we are in a position to fix the parameters $q_{(\rm c,n)}$ and $r_{(\rm c,n)}$, we may determine $\delta_{(\rm c,n)}$ and γ from the observables given in (9). This can be done separately for the mixed, charged and neutral $B \to \pi K$ systems. It should be emphasized that also CP-violating rate differences have to be measured to this end. Using just the CP-conserving observables $R_{(\rm c,n)}$, we may obtain interesting constraints on γ. In contrast to $q_{(\rm c,n)}$ and $r_{(\rm c,n)}$, the strong phase $\delta_{(\rm c,n)}$ suffers from large hadronic uncertainties. However, we can get rid of $\delta_{(\rm c,n)}$ by keeping it as a "free" variable, yielding minimal and maximal values for $R_{(\rm c,n)}$:

$$\left. R_{(\rm c,n)}^{\rm ext} \right|_{\delta_{(\rm c,n)}} = \text{function}\left(q_{(\rm c,n)}, r_{(\rm c,n)}, \gamma\right). \tag{10}$$

Keeping in addition $r_{(\rm c,n)}$ as a free variable, we obtain another – less restrictive – minimal value for $R_{(\rm c,n)}$:

$$\left. R_{(\rm c,n)}^{\rm min} \right|_{r_{(\rm c,n)}, \delta_{(\rm c,n)}} = \text{function}\left(q_{(\rm c,n)}, \gamma\right) \sin^2\gamma. \tag{11}$$

These extremal values of $R_{(\rm c,n)}$ imply constraints on γ, since the cases corresponding to $R_{(\rm c,n)}^{\rm exp} < R_{(\rm c,n)}^{\rm min}$ and $R_{(\rm c,n)}^{\rm exp} > R_{(\rm c,n)}^{\rm max}$ are excluded. The present experimental status is

TABLE 1. Present experimental status of the observables $R_{(\mathrm{c,n})}$. For the evaluation of R, we have used $\tau_{B^+}/\tau_{B^0_d} = 1.060 \pm 0.029$.

	CLEO [22]	BaBar [23]	Belle [24]
R	1.00 ± 0.30	0.97 ± 0.23	1.50 ± 0.66
R_c	1.27 ± 0.47	1.19 ± 0.35	2.38 ± 1.12
R_n	0.59 ± 0.27	1.02 ± 0.40	0.60 ± 0.29

summarized in Table 1. We observe that both the CLEO and the Belle data point towards $R_\mathrm{c} > 1$ and $R_\mathrm{n} < 1$, whereas the central values of the BaBar collaboration are close to one, with a small preferrence of $R_\mathrm{c} > 1$.

In Fig. 1, we show the dependence of (10) and (11) on γ for the neutral $B \to \pi K$ system; the charged $B \to \pi K$ curves look very similar [20]. Here the crossed region below the R_min curve, which is described by (11), is excluded. On the other hand, the shaded region is the allowed range (10) for R_n, arising in the case of $r_\mathrm{n} = 0.17$. This figure allows us to read off immediately the allowed range for γ for a given value of R_n. In order to illustrate this feature, let us assume that $R_\mathrm{n} = 0.6$ has been measured, which would be in accordance with the central values of CLEO and Belle in Table 1. In this case, the R_min curve implies $0° \leq \gamma \leq 19° \vee 97° \leq \gamma \leq 180°$. If we use additional information on r_n, we may put even stronger constraints on γ. For $r_\mathrm{n} = 0.17$, we obtain, for instance, the allowed range $134° \leq \gamma \leq 180°$. It is interesting to note that the R_min curve is only effective for $R_\mathrm{n} < 1$. Assuming $R_\mathrm{c} = 1.3$ to illustrate implications of the CP-averaged charged $B \to \pi K$ branching ratios, (11) is not effective and r_c has to be fixed in order to constrain γ. Using $r_\mathrm{c} = 0.21$, we obtain $84° \leq \gamma \leq 180°$. Although it is too early to draw definite conclusions, it is important to emphasize that the second quadrant for γ,

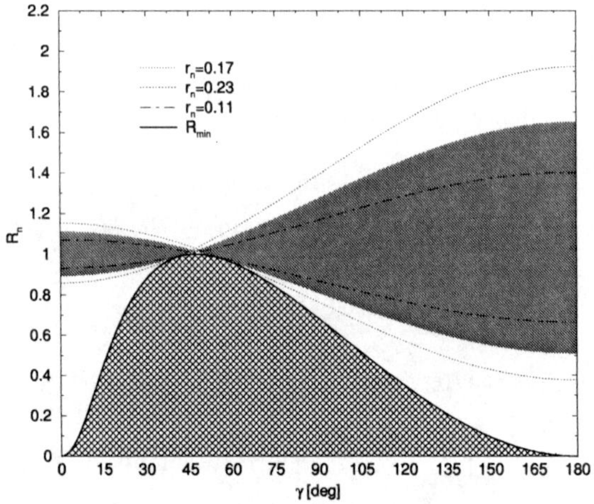

FIGURE 1. The dependence of the extremal values of R_n on γ for $q_\mathrm{n} = 0.68$.

i.e. $\gamma \geq 90^\circ$, would be preferred in these cases. Interestingly, such a situation would be in conflict with the standard analysis of the unitarity triangle [3], yielding $\gamma \sim 60^\circ$. Here the stringent present experimental lower bound on ΔM_s implies $\gamma < 90^\circ$.

Another "puzzle" may arise from CP-conserving strong phases, which can also be constrained through the observables $R_{(\mathrm{c,n})}$ [20]. Interestingly, the CLEO and Belle data point towards $\cos\delta_{\mathrm{c}} > 0$ and $\cos\delta_{\mathrm{n}} < 0$, whereas we would expect equal signs for these quantities. Moreover, $\cos\delta_{\mathrm{n}} < 0$ would be in conflict with factorization.

Consequently, the present CLEO and Belle data point towards a "puzzling" situation, whereas no such discrepancies are indicated by the results of the BaBar collaboration. It is of course too early to draw any definite conclusions. However, if future data should confirm such a picture, it may be an indication for new physics or large flavor-symmetry-breaking effects [20]. Further studies are desirable to distinguish between these cases. Since $B \to \pi K$ modes are governed by penguin processes, they actually represent sensitive probes for new physics [25].

Due to the recent theoretical progress in the description of $B \to \pi K, \pi\pi$ decays, the theoretical uncertainties of $r_{\mathrm{c,n}}$ and $q_{\mathrm{c,n}}$ can be reduced to the level of [5]

$$O\left(\frac{1}{N_{\mathrm{C}}} \times \frac{m_s - m_d}{\Lambda_{\mathrm{QCD}}} \times \frac{\Lambda_{\mathrm{QCD}}}{m_b}\right) = O\left(\frac{1}{N_{\mathrm{C}}} \times \frac{m_s - m_d}{m_b}\right), \tag{12}$$

and confidence into dynamical assumptions related to rescattering effects can be gained. Making more extensive use of QCD factorization, approaches complementary to the ones discussed above, which rely on a "minimal" input from theory, are provided. As a first step, we may use that the CP-conserving strong phase δ_{c} is predicted to be very small in QCD factorization, so that $\cos\delta_{\mathrm{c}}$ governing R_{c} is close to one. As a second step, information on γ can be obtained from the predictions for the branching ratios and the observables $R_{(\mathrm{c,n})}$. Finally, the information from all CP-averaged $B \to \pi K, \pi\pi$ branching ratios can be combined into a single global fit for the allowed region in the $\bar{\rho}$–$\bar{\eta}$ plane [5, 26]. For these approaches, it is of course crucial that the corrections entering in the QCD factorization formalism at the $\Lambda_{\mathrm{QCD}}/m_b$ level can be controlled reliably. As argued in a recent paper [27], non-perturbative contributions with the same quantum numbers as penguin topologies with internal charm- and up-quark exchanges may play an important rôle in this context. The issue of $\Lambda_{\mathrm{QCD}}/m_b$ corrections in phenomenological analyses will certainly continue to be a hot topic in the future.

U-SPIN STRATEGIES

Let us now focus on strategies to extract γ from pairs of B-meson decays, which are related to each other through the U-spin flavor symmetry of strong interactions. In order to deal with non-leptonic B decays, U-spin offers an important tool, and first approaches to extract CKM phases were already pointed out in 1993 [28]. However, the great power of the U-spin symmtery to determine weak phases and hadronic parameters was noticed just recently in the strategies proposed in [29]–[32]. Since these methods involve also decays of B_s-mesons, B experiments at hadron colliders are required to implement them in practice. At Tevatron-II, we will have first access to the corresponding modes and interesting results are expected [33]. In the era of BTeV and the LHC, the U-spin strategies can then be fully exploited [34], as emphasized by Stone at this symposium. In the following discussion, we shall focus on two particularly promising approaches, using the $B_d \to \pi^+\pi^-$, $B_s \to K^+K^-$ [30] and $B_d \to \pi^\mp K^\pm$, $B_s \to \pi^\pm K^\mp$ [31] systems.

Extraction of β and γ from $B_d \to \pi^+\pi^-$, $B_s \to K^+K^-$ Decays

Looking at the corresponding Feynman diagrams, we observe that $B_s \to K^+K^-$ is obtained from $B_d \to \pi^+\pi^-$ by interchanging all down and strange quarks. The structure of the corresponding decay amplitudes is given as follows [30]:

$$A(B_d^0 \to \pi^+\pi^-) = \mathcal{C}\left[e^{i\gamma} - de^{i\theta}\right] \tag{13}$$

$$A(B_s^0 \to K^+K^-) = \lambda\,\mathcal{C}'\left[e^{i\gamma} + \left(\frac{1-\lambda^2}{\lambda^2}\right)d'e^{i\theta'}\right], \tag{14}$$

where $\mathcal{C}$, $\mathcal{C}'$ are CP-conserving strong amplitudes, and $de^{i\theta}$, $d'e^{i\theta'}$ measure, sloppily speaking, ratios of penguin to tree amplitudes. Using these general parametrizations, we obtain the following expressions for the direct and mixing-induced CP asymmetries:

$$\mathcal{A}_{\rm CP}^{\rm dir}(B_d \to \pi^+\pi^-) = \text{function}(d,\theta,\gamma) \tag{15}$$

$$\mathcal{A}_{\rm CP}^{\rm mix}(B_d \to \pi^+\pi^-) = \text{function}(d,\theta,\gamma,\phi_d = 2\beta) \tag{16}$$

$$\mathcal{A}_{\rm CP}^{\rm dir}(B_s \to K^+K^-) = \text{function}(d',\theta',\gamma) \tag{17}$$

$$\mathcal{A}_{\rm CP}^{\rm mix}(B_s \to K^+K^-) = \text{function}(d',\theta',\gamma,\phi_s \approx 0). \tag{18}$$

Consequently, we have four observables, depending on six "unknowns". However, since $B_d \to \pi^+\pi^-$ and $B_s \to K^+K^-$ are related to each other by interchanging all down and strange quarks, the U-spin flavor symmetry of strong interactions implies

$$d'e^{i\theta'} = d\,e^{i\theta}. \tag{19}$$

Using this relation, the four observables (15)–(18) depend on the four quantities d, θ, $\phi_d = 2\beta$ and γ, which can hence be determined. It should be emphasized that no dynamical assumptions about rescattering processes have to be made in this approach, which is an important conceptual advantage in comparison with the $B \to \pi K$ strategies discussed above. The theoretical accuracy is hence only limited by U-spin-breaking effects. Theoretical considerations allow us to gain confidence into (19), which does not receive U-spin-breaking corrections in factorization [30]. Moreover, there are general relations between observables of U-spin-related decays, allowing experimental insights into U-spin breaking [29, 30, 35, 36].

The U-spin arguments can be minimized, if we employ the B_d^0–$\overline{B_d^0}$ mixing phase $\phi_d = 2\beta$ as an input, which can be determined straightforwardly through $B_d \to J/\psi K_S$. The observables $\mathcal{A}_{\rm CP}^{\rm dir}(B_d \to \pi^+\pi^-)$ and $\mathcal{A}_{\rm CP}^{\rm mix}(B_d \to \pi^+\pi^-)$ allow us then to eliminate the strong phase θ and to determine d as a function of γ. Analogously, $\mathcal{A}_{\text{CP}}^{\text{dir}}(B_s \to K^+K^-)$ and $\mathcal{A}_{\text{CP}}^{\text{mix}}(B_s \to K^+K^-)$ allow us to eliminate the strong phase θ' and to determine d' as a function of γ. The corresponding contours in the γ–d and γ–d' planes can be fixed in a *theoretically clean* way. Using now the U-spin relation $d' = d$, these contours allow the determination both of the CKM angle γ and of the hadronic quantities d, θ, θ'; for a detailed illustration, see [30].

This approach is very promising for Tevatron-II and the LHC era, where experimental accuracies for γ of $O(10^\circ)$ [33] and $O(1^\circ)$ [34] may be achieved, respectively. It should be emphasized that not only γ, but also the hadronic parameters d, θ, θ' are of particular interest, as they can be compared with theoretical predictions, thereby allowing valuable insights into hadron dynamics. For strategies to probe γ and constrain hadronic penguin parameters using a variant of the $B_d \to \pi^+\pi^-$, $B_s \to K^+K^-$ approach, where the latter decay is replaced through $B_d \to \pi^\mp K^\pm$, the reader is referred to [37].

Extraction of γ from $B_{(s)} \to \pi K$ Decays

Another interesting pair of decays, which are related to each other by interchanging all down and strange quarks, is the $B_d^0 \to \pi^- K^+$, $B_s^0 \to \pi^+ K^-$ system [31]. In the strict U-spin limit, the corresponding decay amplitudes can be parametrized as follows:

$$A(B_d^0 \to \pi^- K^+) = -P\left(1 - re^{i\delta}e^{i\gamma}\right) \tag{20}$$

$$A(B_s^0 \to \pi^+ K^-) = P\sqrt{\varepsilon}\left(1 + \frac{1}{\varepsilon}re^{i\delta}e^{i\gamma}\right), \tag{21}$$

where P denotes a CP-conserving complex amplitude, $\varepsilon \equiv \lambda^2/(1-\lambda^2)$, r is a real parameter, and δ a CP-conserving strong phase. At first sight, it appears as if γ, r and δ could be determined from the ratio of the CP-averaged rates and the two CP asymmetries provided by these modes.[2] However, because of the relation

$$\begin{aligned} |A(B_d^0 \to \pi^- K^+)|^2 - |A(\overline{B_d^0} \to \pi^+ K^-)|^2 &= 4r\sin\delta\sin\gamma \\ &= -\left[|A(B_s^0 \to \pi^+ K^-)|^2 - |A(\overline{B_s^0} \to \pi^- K^+)|^2\right], \end{aligned} \tag{22}$$

we have actually only two independent observables, so that the three parameters γ, r and δ cannot be determined. To this end, the overall normalization P has to be fixed, requiring a further input. Assuming that rescattering processes play a minor rôle and that color-suppressed EW penguins can be neglected as well, the isospin symmetry implies

$$P = A(B^+ \to \pi^+ K^0). \tag{23}$$

In order to extract γ and the hadronic parameters, it is useful to introduce observables R_s and A_s by replacing $B_d \to \pi^\mp K^\pm$ through $B_s \to \pi^\pm K^\mp$ in (6). Using (20), (21) and (23),

---

[2] Note that these observables are independent of P.

we then obtain

$$R_s = \varepsilon + 2r\cos\delta\cos\gamma + \frac{r^2}{\varepsilon} \tag{24}$$

$$A_s = -2r\sin\delta\sin\gamma = -A_0. \tag{25}$$

Together with the parametrization for R as sketched in (9), these observables allow the determination of all relevant parameters. The extraction of γ and δ is analogous to the "mixed" $B_d \to \pi^\mp K^\pm$, $B^\pm \to \pi^\pm K$ approach discussed above. However, now the advantage is that the U-spin counterparts $B_s \to \pi^\pm K^\mp$ of $B_d \to \pi^\mp K^\pm$ allow us to determine also the parameter r without using arguments related to factorization [31]:

$$r = \sqrt{\varepsilon\left[\frac{R + R_s - 1 - \varepsilon}{1+\varepsilon}\right]}. \tag{26}$$

On the other hand, we still have to make dynamical assumptions concerning rescattering and color-suppressed EW penguin effects. The theoretical accuracy is further limited by $SU(3)$-breaking effects. An interesting consistency check is provided by the relation $A_s = -A_0$, which is due to (22). A variant of this approach using the CKM angle β as an additional input was proposed in [38].

CONCLUSIONS

There are many strategies to determine γ, suffering unfortunately in several cases from experimental problems. The approaches discussed above, employing penguin processes, are on the other hand very promising from a practical point of view and exhibit further interesting features. As a by-product, they also allow us to determine strong phases and other hadronic parameters, allowing comparisons with theoretical predictions. Moreover, these strategies are sensitive probes for the physics lying beyond the Standard Model, which may lead to discrepancies in the extraction of γ or the hadronic quantities. Let us hope that signals for new physics will actually emerge this way.

REFERENCES

1. B. Aubert *et al.* [BaBar Collaboration], *Phys. Rev. Lett.* **87**, 091801 (2001) [hep-ex/0107013]; K. Abe *et al.* [Belle Collaboration], *Phys. Rev. Lett.* **87**, 091802 (2001) [hep-ex/0107061].
2. For recent reviews, see Y. Nir, WIS-18-01-AUG-DPP [hep-ph/0109090]; A. J. Buras, TUM-HEP-402-01 [hep-ph/0101336]; J. L. Rosner, EFI-2000-47 [hep-ph/0011355]; R. Fleischer, DESY-00-170 [hep-ph/0011323].
3. For recent analyses, see A. Höcker *et al.*, *Eur. Phys. J.* C **21**, 225 (2001) [hep-ph/0104062]; M. Ciuchini *et al.*, *JHEP* **0107**, 013 (2001) [hep-ph/0012308]; A. Ali and D. London, *Eur. Phys. J.* C **18**, 665 (2001) [hep-ph/0012155].
4. M. Beneke *et al.*, *Phys. Rev. Lett.* **83**, 1914 (1999) [hep-ph/9905312].
5. M. Beneke *et al.*, *Nucl. Phys.* B **606**, 245 (2001) [hep-ph/0104110].
6. H.-n. Li and H.L. Yu, *Phys. Rev.* D **53**, 2480 (1996) [hep-ph/9411308]; Y.Y. Keum, H.-n. Li and A. I. Sanda, *Phys. Lett.* B **504**, 6 (2001) [hep-ph/0004004];

Phys. Rev. D **63**, 054008 (2001) [hep-ph/0004173];
Y.Y. Keum and H.-n. Li, *Phys. Rev.* D **63**, 074006 (2001) [hep-ph/0006001].
7. R.G. Sachs, EFI-85-22 (unpublished);
I. Dunietz and R. G. Sachs, *Phys. Rev.* D **37**, 3186 (1988) [E: *Phys. Rev.* D **39**, 3515 (1988)];
I. Dunietz, *Phys. Lett.* B **427**, 179 (1998) [hep-ph/9712401];
M. Diehl and G. Hiller, *Phys. Lett.* B **517**, 125 (2001) [hep-ph/0105213].
8. R. Aleksan, I. Dunietz and B. Kayser, *Z. Phys.* C **54**, 653 (1992);
R. Fleischer and I. Dunietz, *Phys. Lett.* B **387**, 361 (1996) [hep-ph/9605221];
A. F. Falk and A. A. Petrov, *Phys. Rev. Lett.* **85**, 252 (2000) [hep-ph/0003321];
D. London, N. Sinha and R. Sinha, *Phys. Rev. Lett.* **85**, 1807 (2000) [hep-ph/0005248].
9. M. Gronau and D. Wyler, *Phys. Lett.* B **265**, 172 (1991);
D. Atwood, I. Dunietz and A. Soni, *Phys. Rev. Lett.* **78**, 3257 (1997) [hep-ph/9612433];
M. Gronau, *Phys. Rev.* D **58**, 037301 (1998) [hep-ph/9802315].
10. R. Fleischer and D. Wyler, *Phys. Rev.* D **62**, 057503 (2000) [hep-ph/0004010].
11. M. Gronau, J. L. Rosner and D. London, *Phys. Rev. Lett.* **73**, 21 (1994) [hep-ph/9404282].
12. O. F. Hernández *et al.*, *Phys. Lett.* B **333**, 500 (1994) [hep-ph/9404281];
M. Gronau *et al.*, *Phys. Rev.* D **50**, 4529 (1994); *Phys. Rev.* D **52**, 6356 (1995) [hep-ph/9504326].
13. R. Fleischer, *Phys. Lett.* B **365**, 399 (1996) [hep-ph/9509204].
14. R. Fleischer and T. Mannel, *Phys. Rev.* D **57**, 2752 (1998) [hep-ph/9704423].
15. M. Gronau and J. L. Rosner, *Phys. Rev.* D **57**, 6843 (1998) [hep-ph/9711246].
16. R. Fleischer, *Eur. Phys. J.* C **6**, 451 (1999) [hep-ph/9802433].
17. M. Neubert and J. L. Rosner, *Phys. Lett.* B **441**, 403 (1998) [hep-ph/9808493];
Phys. Rev. Lett. **81**, 5076 (1998) [hep-ph/9809311].
18. M. Neubert, *JHEP* **9902**, 014 (1999) [hep-ph/9812396].
19. A. J. Buras and R. Fleischer, *Eur. Phys. J.* C **11**, 93 (1999) [hep-ph/9810260].
20. A. J. Buras and R. Fleischer, *Eur. Phys. J.* C **16**, 97 (2000) [hep-ph/0003323].
21. L. Wolfenstein, *Phys. Rev.* D **52**, 537 (1995);
J. F. Donoghue *et al.*, *Phys. Rev. Lett.* **77**, 2178 (1996) [hep-ph/9604283];
M. Neubert, *Phys. Lett.* B **424**, 152 (1998) [hep-ph/9712224];
J. M. Gérard and J. Weyers, *Eur. Phys. J.* C **7**, 1 (1999) [hep-ph/9711469];
A. J. Buras, R. Fleischer and T. Mannel, *Nucl. Phys.* B **533**, 3 (1998) [hep-ph/9711262];
A. F. Falk, A. L. Kagan, Y. Nir and A. A. Petrov, *Phys. Rev.* D **57**, 4290 (1998) [hep-ph/9712225];
R. Fleischer, *Phys. Lett.* B **435**, 221 (1998) [hep-ph/9804319].
22. D. Cronin-Hennessy *et al.* [CLEO Collaboration], *Phys. Rev. Lett.* **85**, 515 (2000).
23. B. Aubert *et al.* [BaBar Collaboration], *Phys. Rev. Lett.* **87**, 151802 (2001) [hep-ex/0105061].
24. K. Abe *et al.* [Belle Collaboration], *Phys. Rev. Lett.* **87**, 101801 (2001) [hep-ex/0104030].
25. R. Fleischer and T. Mannel, TTP-97-22, [hep-ph/9706261];
D. Choudhury, B. Dutta and A. Kundu, *Phys. Lett.* B **456**, 185 (1999) [hep-ph/9812209];
X. He, C. Hsueh and J. Shi, *Phys. Rev. Lett.* **84**, 18 (2000) [hep-ph/9905296];
R. Fleischer and J. Matias, *Phys. Rev.* D **61**, 074004 (2000) [hep-ph/9906274];
Y. Grossman, M. Neubert and A. L. Kagan, *JHEP* **9910**, 029 (1999) [hep-ph/9909297];
J. Matias, UAB-FT-514 [hep-ph/0105103].
26. For sophisticated fits, see H. Lacker, these proceedings, and
`http://www.slac.stanford.edu/~laplace/ckmfitter.html`.
27. M. Ciuchini *et al.*, *Phys. Lett.* B **515**, 33 (2001) [hep-ph/0104126].
28. I. Dunietz, FERMILAB-CONF-93/90-T (1993).
29. R. Fleischer, *Eur. Phys. J.* C **10**, 299 (1999) [hep-ph/9903455].
30. R. Fleischer, *Phys. Lett.* B **459**, 306 (1999) [hep-ph/9903456].
31. M. Gronau and J. L. Rosner, *Phys. Lett.* B **482**, 71 (2000) [hep-ph/0003119].
32. P. Z. Skands, *JHEP* **0101**, 008 (2001) [hep-ph/0010115].
33. M. Tanaka [CDF Collaboration], *Nucl. Instrum. Meth.* A **462**, 165 (2001).
34. P. Ball *et al.*, CERN-TH-2000-101 [hep-ph/0003238].
35. R. Fleischer, *Phys. Rev.* D **60**, 073008 (1999) [hep-ph/9903540].
36. M. Gronau, *Phys. Lett.* B **492**, 297 (2000) [hep-ph/0008292].
37. R. Fleischer, *Eur. Phys. J.* C **16**, 87 (2000) [hep-ph/0001253].
38. C. Chiang and L. Wolfenstein, *Phys. Lett.* B **493**, 73 (2000) [hep-ph/0004255].

Charm Decay Results from CLEO

David Cinabro representing the CLEO collaboration

Wayne State University, Department of Physics and Astronomy, 666 West Hancock, Detroit, Michigan, 48202, USA

Abstract. I summarize the latest charm decay results from CLEO.

INTRODUCTION

Charm physics is a rich field with many paths of inquiry that often take unexpected turns. CLEO combining the cleanliness of the e^+e^-, with the precision of a high quality silicon vertex detector in a very well understood tracking system, and a large data set has produced many unexpected results recently in this field. In these proceedings I summarize recent CLEO results on the width of the D^{*+}, searches for direct CP-violation and mixing in $D^0 \to K^+K^-$ and $\pi^+\pi^-$, observation of the wrong sign decay $D^0 \to K^+\pi\pi\pi$, measurements in $D^+ \to \bar{K}^{*0}\ell\nu$, and a number of charm baryon results.

The CLEO II and its II.V upgraded detector have been described in detail elsewhere [1]. The analyses described below are based on 14/fb of e^+e^- collisions taken with a center of mass energy around 10.6 GeV. Most of the data (9/fb) were taken in the II.V configuration and the high precision analyses use only that subset.

THE WIDTH OF D^{*+}

A measurement of $\Gamma(D^{*+})$ opens an important window on the non-perturbative strong physics involving heavy quarks. The basic framework of the theory is well understood, however, there is still much speculation - predictions for the width range from 15 keV to 150 keV [2]. We know the D^{*+} width is dominated by strong decays. The level splitting in the B sector is not large enough to allow real strong transitions. Therefore, a measurement of the width of the D^{*+} gives unique information about the strong coupling constant in heavy-light meson systems. This width only depends on g, a universal strong coupling between heavy vector and pseudoscaler mesons to the pion, since the small contribution of the electromagnetic decay can be neglected, yielding

$$\Gamma(D^{*+}) = \frac{2g^2}{12\pi f_\pi^2} p_{\pi^+}^3 + \frac{g^2}{12\pi f_\pi^2} p_{\pi^0}^3, \tag{1}$$

where f_π is the pion decay constant and the momenta are for the indicated particle in D^{*+} decay in the D^{*+} rest frame [3].

CP618, *Heavy Flavor Physics: Ninth International Symposium,* edited by A. Ryd and F. C. Porter

TABLE 1. The data sample, results of the fits, and simulation biases. The uncertainties are only statistical.

	Sample		
Parameter	Nominal	Tracking	Kinematic
Candidates	11496	368	3284
N_{signal}	11207 ± 109	353 ± 20	3151 ± 57
$N_{\text{background}}$	289 ± 31	15 ± 7	133 ± 16
f_{mis} (%)	5.3 ± 0.5	NA	NA
σ_{mis} (keV)	508 ± 39	NA	NA
Q_0 (keV)	5853 ± 2	5854 ± 10	5850 ± 4
Γ_0 (keV)	98.9 ± 4.0	106.0 ± 19.6	108.1 ± 5.9
$\Gamma_{\text{fit}} - \Gamma_{\text{generated}}$ (keV)	2.7 ± 2.1	1.7 ± 6.4	4.3 ± 3.1
D^{*+} Width (keV)	96.2 ± 4.0	104 ± 20	103.8 ± 5.9

Our nominal sample follows the selection of $D^{*+} \to \pi^+_{\text{slow}} D^0 \to K^- \pi^+ \pi^+_{\text{slow}}$ candidates used in our $D^0 - \bar{D}^0$ mixing analysis[4]. Our reconstruction method takes advantage of the small CESR beam spot and the kinematics and topology of the $D^{*+} \to \pi^+_{\text{slow}} D^0 \to \pi^+_{\text{slow}} K^- \pi^+$ decay chain. The K^- and π^+ are required to form a common vertex[5]. The resultant D^0 candidate momentum vector is then projected back to the CESR luminous region to determine the D^0 production point. The CESR luminous region has a Gaussian width $\sim 10\ \mu$m vertically and $\sim 300\ \mu$m horizontally. This procedure determines an accurate D^0 production point for D^0's moving out of the horizontal plane. Then the π^+_{slow} track is refit constraining its trajectory to intersect the D^0 production point. This improves the resolution on the energy release, $Q = M(K^- \pi^+ \pi^+_{\text{slow}}) - M(K^- \pi^+) - m_{\pi^+}$, by more than 30% over simply forming the appropriate invariant masses of the tracks. Our resolution, σ_Q is typically 150 keV and agrees well with the prediction of the Monte Carlo.

We select two subsamples of the D^{*+} candidates. One selects candidates with the best measured tracks by making very tight cuts on tracking parameters. A second alternative is to choose data based on specific kinematic properties of the D^{*+} decay that minimize the dependence of the width of the D^{*+} on detector mismeasurements. In all three samples the width is extracted with an unbinned maximum likelihood fit to the energy release distribution. These three different approaches yield consistent values for the width of the D^{*+} giving us confidence that our simulation accurately models our data.

Table 1 summarizes the statistics in our three samples. The tracking and kinematic samples are subsets of the nominal sample. The two subsets contain 94 common candidates. We consider the pairs of Q and σ_Q for $D^{*+} \to \pi^+_{\text{slow}} D^0 \to K^- \pi^+ \pi^+_{\text{slow}}$ where σ_Q is given for each candidate by propagating the tracking errors in the kinematic fit of the charged tracks. The underlying signal shape of the Q distribution is assumed to be given by a P-wave Breit-Wigner with central value Q_0. For each candidate the signal shape is convolved with a resolution Gaussian with width σ_Q, determined by the tracking errors, as a model of our resolution. The fits also includes a background contribution with a fixed shape derived from our simulation, and modeled with a third order polynomial. We allow a small fraction of the signal, f_{mis}, to be parametrized by a single Gaussian resolution function of width σ_{mis}. This shape is included in the fit to model tracking

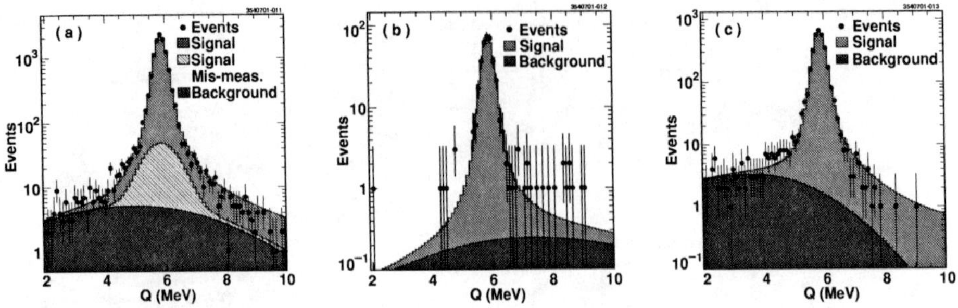

FIGURE 1. Fits to the three data samples: a) nominal; b) tracking; and c) kinematic. The different contributions to the fits are shown by different colors or patterns.

mishaps which our simulation predicts to be at the 5% level in the nominal sample and negligible in both the tracking and kinematic selected samples.

Figure 1 displays the fits to the three data sample. The results of the fits are summarized in Table 1.

The most important contribution systematic uncertainty is the variation of the results as a function of the kinematic parameters of the D^{*+} decay (± 16 keV on the width, ± 8 keV on Q_0). The next most important contribution comes from any mismodeling of σ_Q's dependence on the kinematic parameters (± 11 keV) We take into account correlations among the less well measured parameters of the fit by fixing each parameter at $\pm 1\sigma$ from their central fit values, repeating the fit, and adding in quadrature the variation in the width of the D^{*+} and Q_0 from their central values (± 8 keV). We have also checked that our simulation accurately models the line shape of other narrow resonances visible in our data. Notably the decay $\Lambda^0 \to p\pi^-$ when we select the π^- to have a momentum in the range of the π^+_{slow} in the D^{*+} decay, has a visible width which agrees to a few percent between data and simulation. An extra and dominant source of uncertainty on Q_0 is the energy scale of our measurements. We evaluate this uncertainty by studying $K_s \to \pi^+\pi^-$ decays in our data, and we find a shift in the fit value of Q_0 of -4 keV. We evaluate uncertainties in the energy scale by varying an overall momentum scale to change the $K_s \to \pi^+\pi^-$ mass by ± 30 keV, the uncertainty on that mass [6], and applying the statistical errors we obtain from the calculations of the pion momentum corrections.

We have measured the width of the D^{*+} by studying the distribution of the energy release in $D^{*+} \to D^0\pi^+$ followed by $D^0 \to K^-\pi^+$ decay. We obtain

$$\Gamma(D^{*+}) = 96 \pm 4 \pm 22 \text{ keV}, \tag{2}$$

where the first uncertainty is statistical and the second is systematic. It corresponds to a strong coupling[2]

$$g = 0.59 \pm 0.01 \pm 0.07. \tag{3}$$

We also measure the mean value for the energy release in $D^{*+} \to D^0\pi^+$ decay

$$Q_0 = 5842 \pm 2 \pm 12 \text{ keV}, \tag{4}$$

where the first error is statistical and second is systematic. Combining this with the mass of the charged pion, 139.570 MeV with an uncertainty less than 1 keV [6], we calculate

$$m_{D^*(2010)^+} - m_{D^0} = 145.412 \pm 0.002 \pm 0.012\ MeV. \quad (5)$$

STUDIES WITH $D^0 \to K^+K^-$ AND $\pi^+\pi^-$

The SU(3) flavor symmetry predicts $\Gamma\left(D^0 \to K^+K^-\right)/\Gamma\left(D^0 \to \pi^+\pi^-\right) = 1$, while the previously measured value is 2.80 ± 0.20 [6]. This deviation is most likely caused by large final state interactions. A measure of *CP* violation in these decays, the direct *CP* violation asymmetry, is proportional to the amount of *CP* violation in the decays and the sine of the strong phase difference. I describe our measurement of the ratio of partial widths, $\Gamma\left(D^0 \to K^+K^-\right)/\Gamma\left(D^0 \to \pi^+\pi^-\right)$, and present our search for direct *CP* violation in these decays. In the absence of *CP* violation, the *D* meson mass eigenstates $D_{1,2}$ are also *CP* eigenstates. The decay of a D^0 to a *CP* eigenstate has a purely exponential lifetime characteristic of the associated mass eigenstate. Therefore, in the limit of no *CP* violation $y_{CP} = y = \Delta\Gamma/2\Gamma$, and $\Delta\Gamma$ is the width difference between the two mass eigenstates. We can measure y_{CP} simply by measuring the ratio of lifetimes of the D^0 decaying to a *CP* eigenstate (τ_{CP+}) and a *CP* neutral state such as $K^-\pi^+$ (τ). Then $y_{CP} = \tau/\tau_{CP+} - 1$. We have used $\tau = (\tau_{CP+} + \tau_{CP-})/2$, and assumed that the lifetime difference is small so that the $K^+\pi^-$ lifetime distribution can be fit with a single exponential.

The events are selected by searching for the decay chain $D^{*+} \to D^0\pi_s^+$, with subsequent decays of the D^0 to K^+K^-, $\pi^+\pi^-$, or $K^-\pi^+$ [7]. The charge of the slow pion, π_s^+, from the D^{*+} decay is a tag of the initial D^0 flavor. All pairs of oppositely-charged tracks of good quality are used to form D^0 candidates assuming four particle assignments: K^+K^-, $K^+\pi^-$, π^+K^-, and $\pi^+\pi^-$. The same technique as described above for the $\Gamma(D^{*+})$ analysis is used to improve the resolution giving a Q distribution width of approximately 190 keV. Particle identification using specific ionization is not required since the different mass hypotheses are separated by greater than 8.5 standard deviations.

The partial width measurements are obtained from binned maximum likelihood fits to the Q distribution of the D^{*+} decay. To obtain $R_{\pi\pi} = \Gamma\left(D^0 \to \pi^+\pi^-\right)/\Gamma\left(D^0 \to K^-\pi^+\right)$ we fit the Q distributions for the ratio of signal yields between the $\pi\pi$ and $K\pi$ channels, and for the normalization of the background, where we have used the signal shape and background parameters determined from the $K\pi$ data and Monte Carlo samples, respectively. For the KK channel we add an additional component from pseudoscalar-vector decay (PV) background, primarily $D^0 \to K^-\rho^+$, where the shape is taken from Monte Carlo and the normalization is allowed to float. We observe approximately 20,000 $K^-\pi^+$, 1900 K^+K^-, and 710 $\pi^+\pi^-$ events.

The systematic uncertainty due to the fitting shapes is assessed by performing a series of fits using different assumptions for the background and also several fits to the D^0 mass distribution. We vary the bin sizes, Q fit range, Q signal region and candidate D^0 mass requirement. We also estimate systematic uncertainties associated with some of the event selection requirements by doing the analysis without those requirements. We use the $K\pi$ data sample to study the effect of any mismodeling in the simulation of the fragmentation and the detector acceptance. The final results are $R_{KK} = \Gamma\left(D^0 \to K^+K^-\right)/\Gamma\left(D^0 \to K^-\pi^+\right) = 0.1040 \pm 0.0033 \pm 0.0027$ and $R_{\pi\pi} = \Gamma\left(D^0 \to \pi^+\pi^-\right)/\Gamma\left(D^0 \to K^-\pi^+\right) = 0.0351 \pm 0.0016 \pm 0.0017$. We can combine the results, accounting for cancellations and correlations among the uncertainties to calculate $R_{KK}/R_{\pi\pi} = 2.96 \pm 0.16(\text{stat}) \pm 0.15(\text{syst})$.

We can use the same procedure to search for the direct CP asymmetries

$$A_{CP} = \frac{\Gamma(D^0 \to f) - \Gamma(\overline{D^0} \to f)}{\Gamma(D^0 \to f) + \Gamma(\overline{D^0} \to f)}.$$

The K^+K^- and $\pi^+\pi^-$ data are separated into D^0 and $\overline{D^0}$ samples based on the charge of the slow pion. However, we still normalize by the entire $K\pi$ sample to eliminate possible bias from any asymmetry in $D^0 \to K^-\pi^+$ decay. The sources of possible systematic error for the CP asymmetry measurement are the shapes used for fitting and a charge dependent slow pion acceptance. To assess the systematic uncertainty from the fitting shapes we perform fits in which we vary the candidate D^0 mass window, remove the vertex confidence level requirement, vary the width of the $K\pi$ signal region and the Q fit region, alter the number of bins, and split the $K\pi$ sample into two according to the charge of the associated slow pion and fit the two samples separately. We have considered artificial asymmetries induced by having a detector made of matter, tracking bias between negative and positive tracks, and that the center of the luminous region was not exactly at the center of the detector coupled with a small forward-backward asymmetry induces an acceptance asymmetry. Summing all of the systematic uncertainties in quadrature and applying the corrections summarized above we measure $A_{CP}^{KK} = (0.0 \pm 2.2 \pm 0.8)\%$ and $A_{CP}^{\pi\pi} = (1.9 \pm 3.2 \pm 0.8)\%$. We see no evidence of direct CP violation in these decays.

In the limit of no CP violation in the D meson sector y_{CP} is equivalent to y. We use the same data sample described above, using the decay length and momentum to determine the proper decay time. We apply a cut on Q, and fit the candidate D^0 mass spectrum to a signal plus small background. The fit values are converted into a mass dependent probability for signal and background and are used as an input to the lifetime fits. For the signal portion of the probability distribution function for the lifetime fits we constrain the candidate D^0 mass to a fixed value, which gives us a better measurement of y_{CP}. The mass constraint introduces a systematic bias in the lifetime measurement, which cancels for y_{CP} which only depends on the ratio of lifetimes. The resolution on the D^0 decay point is typically 40 μm in each dimension. The resolution in t is typically $\sigma_t = 0.4$ in units of D^0 lifetimes. We determine the proper decay time in the three dimensions separately, and combine them to arrive at the best estimate of t and σ_t.

We fit the lifetime distribution using an unbinned likelihood method. The signal probability distribution function consists of an exponential convolved with a resolution function, composed of the sum of three parts. For most events the calculated covariance matrix for the D^0 daughters is assumed to be correct to within a global scale factor, with a Gaussian resolution function of width σ_t. A few percent of the events have one or more particles that have undergone a hard scatter, rendering the extrapolated vertex errors virtually meaningless. We model the contribution from these events with a single Gaussian whose normalization and width are allowed to float in the fit. For a very small fraction of events the vertex location is extremely mismeasured. These events have a

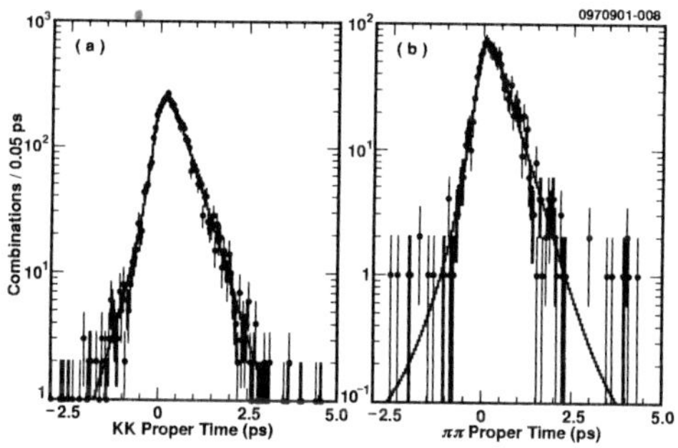

FIGURE 2. The proper time distribution for all $D^0 \to K^+K^-$ (left) and $D^0 \to \pi^+\pi^-$ (right) candidates included in the fit. The curves are the fit results discussed in the text.

nearly flat distribution in lifetime. We model this contribution with a broad Gaussian, assigning a fixed width of 8 ps. The normalization of this contribution is allowed to float in the fit. The signal PDF is multiplied, on an event-by-event basis, by the mass-dependent signal probability from the D^0 candidate mass fit. The background lifetime distribution contains two pieces: a prompt piece and a piece with non-zero lifetime. The component with non-zero lifetime comes from partially reconstructed charm decays. We model this component with a single exponential where the lifetime is another parameter of the fit. We expect and find the fitted value of the background lifetime to be consistent with the D^0 lifetime. The relative amount of background with and without lifetime is also allowed to float in the fit. Both sorts of background are convolved with a resolution function that is modeled in the same manner as the signal, but with an independent set of parameters. The fit results are shown in Figure 2. Small corrections to the lifetimes are computed by comparing the generated and measured values in a Monte Carlo analysis on a fully simulated sample, including backgrounds, corresponding to roughly ten times the data sample.

We estimate systematic uncertainties from slightly different energy release in the three decays with the Monte Carlo. We study the effects of background shape mismodeling by varying the amount and composition of the background. We study the effect of our treatment of the proper time outlier events, which we have modeled with a wide Gaussian, by varying its width. We investigate the bias introduced by constraining all of the events to the same D^0 mass by removing this constraint. Summing all of the listed systematic uncertainties in quadrature, including the Monte Carlo statistics, we obtain the final results $y_{CP}^{KK} = -0.019 \pm 0.029 \pm 0.016$, $y_{CP}^{\pi\pi} = 0.005 \pm 0.043 \pm 0.018$ and combining the two results we obtain $y_{CP} = -0.012 \pm 0.025 \pm 0.014$, which is consistent with zero.

THE WRONG SIGN DECAY $D^0 \to K^+\pi\pi\pi$

We have continued to search for wrong sign decays of the D^0, mediated either by mixing or Double-Cabibbo-Suppressed-Decays. The cleanliness of our underlying signal and

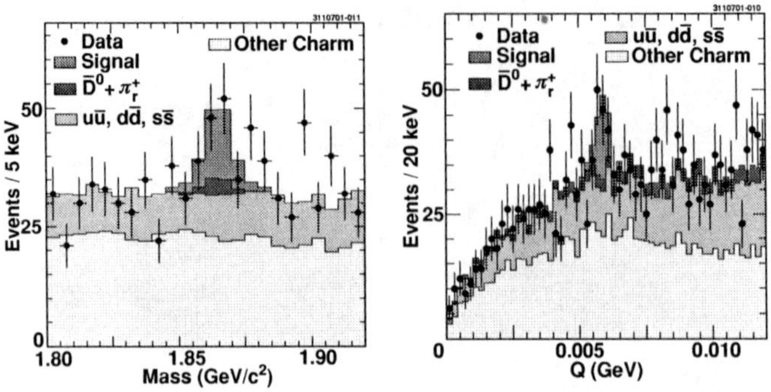

FIGURE 3. The projections of the wrong sign $D^0 \to K3\pi$ 2D fit.

our technique of taking advantage of the small CESR luminous region have given us the ability to see signals even in the most inhospitable modes. For example the all charged decay $D^0 \to K^-3\pi$ has an energy release resolution of 210 keV and a signal to noise of 35:1 [8]. We observe over 13000 candidates in this mode. The wrong sign mode, $D^0 \to K^+3\pi$, is overwhelmed by combinatoric background. Another difficulty in this mode is the complicated nature of resonance sub-structure that can contribute to the final state. We evaluate the relative efficiency for the right and wrong sign decays using the observed distribution of data events to weight our simulations calculation of the efficiency for the two modes. This procedure is limited by the statistics in the wrong sign sample. We extract this signal in a two dimensional fit to the wrong sign candidate Q versus $K3\pi$ mass using the right signal to give the wrong sign signal shape. Figure 3 shows the projections of wrong sign signal. Systematic uncertainties are dominated by the background which is evaluated by repeating the fit for the signal in restricted regions of the 2D space, and changing assumptions about the background shape. We measure

$$R_{\rm WS} \equiv \frac{\mathcal{B}(D^0 \to K^+\pi^-\pi^+\pi^-)}{\mathcal{B}(D^0 \to K^-\pi^+\pi^+\pi^-)} = (0.0041^{+0.0012}_{-0.0011} \pm 0.0004 \times (1.07 \pm 0.10), \qquad (6)$$

which is in good agreement with other observations of the wrong sign rate in D^0 decays.

$D^+ \to \bar{K}^{*0}\ell\nu$

In this decay the hardronic current depends on three form factors: $A_1(q^2)$; $A_2(q^2)$; and $V(q^2)$. The calculation of these form factors is theoretically difficult and their measurement can aid in the extraction of V_{ub} in charmless, semileptonic B decays. We measure the ratio of the semileptonic decay rate to $D^+ \to K^-\pi^+\pi^+$, R_ℓ^+, and combine with results from E791[9] on the form factor ratios to calculate the form factors.

Preliminarily we find $R_\ell^+ = 0.74 \pm 0.04 \pm 0.05$ where the dominant systematics came from the $D^+ \to K^-\pi^+\pi^+$ branching ratio and shape of the signal and background shapes used to extract the signal in a neutrino reconstruction analysis. We calculate $A_1(0) = 0.69 \pm 0.07$, $A_2(0) = 0.48 \pm 0.08$, and $V(0) = 1.25 \pm 0.15$.

CHARM BARYON STUDIES

Space limits me to only briefly mention our recent charm baryon results. We have measured the lifetime of the Ξ_c^+ at $503 \pm 47 \pm 18$ fs [10]. We have measured the mass and width of the Σ_c^0 and Σ_c^{++}. We find, all in MeV, $M(\Sigma_c^{++}) - M(\Lambda_c^+) = 167.4 \pm 0.1 \pm 0.2$, $\Gamma(\Sigma_c^{++}) = 2.3 \pm 0.2 \pm 0.3$, $M(\Sigma_c^0) - M(\Lambda_c^+) = 167.2 \pm 0.1 \pm 0.2$, and $\Gamma(\Sigma_c^0) = 2.5 \pm 0.2 \pm 0.3$ [11]. Preliminarily we have measured the form factor ratio f_1/f_2 in $\Lambda_c \to \Lambda e\nu$ decay to be $-0.31 \pm 0.06 \pm 0.06$ in agreement with a model from Korner and Kramer [12]. Preliminarily we have observed the decay $\Omega_c \to \Omega e\nu$ and measure $\mathcal{B}(\Omega_c \to \Omega \ell \nu)\sigma(e^+e^- \to \Omega_c X) = 42.2 \pm 14.1 \pm 11.9$ fb.

CONCLUSION

Charm physics presents an unusual challenge both to experiment and theory. Precision measurements in the charm sector remains one of the best hunting ground for physics beyond the standard model. The recent results presented here show that CLEO is leading the way in all sectors, searches for charm mixing, semileptonic decays, and baryons of charm physics.

ACKNOWLEDGMENTS

I would like to thank Dave Hitlin, Alan Weinstein, and the other members of the local organizing committee for such a smoothly run conference in the face of such difficult circumstances. CLEO gratefully acknowledge the effort of the CESR staff in providing us with excellent luminosity and running conditions. This work was supported by the National Science Foundation, the U.S. Department of Energy, and the Natural Sciences and Engineering Research Council of Canada.

REFERENCES

1. Y. Kubota *et al.*, *Nucl. Instrum. Methods Phys. Res. A* **320**, 66 (1992); T.S. Hill, *Nucl. Instrum. Methods Phys. Res. A* **418**, 32 (1998); D. Peterson, *Nucl. Phys.* **B** (Proc. Suppl.) **54B**, 31 (1997).
2. V.M.Belyaev *et al. Phys. Rev.* D 51, 6177 (1995) contains a recent survey summarizing and referencing previous theoretical work. P. Singer *Acta Phys. Polon.* B30 3849 (1999), J. L. Goity and W. Roberts JLAB-THY-00-45 (hep-ph/0012314), K. O. E. Henriksson, *et al. Nuc. Phys.* A 686, 355 (2001), and M. Di Pierro and E. Eichten hep-ph/0104208 appear since that survey.
3. M. Wise, *Phys. Rev.* D **45**, R2188 (1992), G. Burdman and J. F. Donoghue, *Phys. Lett.* B **280**, 287 (1992), and T. Yan *et al.*, *Phys. Rev.* D **46**, 1148 (1992) [Erratum-ibid. D **55**, 5851 (1992)]. D. Becirevic and A. Le Yaouanc, JHEP **9903**, 021 (1999) contains a recent review referencing previous theoretical work.

4. R. Godang *et al.* (CLEO Collaboration), *Phys. Rev. Lett.* **84**,5038 (2000).
5. The analysis is described in full detail in A. Anastassov *et al.*, (CLEO Collaboration), CLNS 01/1741, CLEO 01-13, accepted by PRD.
6. D. E. Groom *et al.* (Particle Data Group), *Eur. Phys. J.* **C 15**, 1 (2000).
7. These analyses are described in full detail in S. E. Csorna *et al.*, (CLEO Collaboration), CLNS 01/1763, CLEO 01-21, submitted to PRD.
8. This analysis is described in full detail in S. A. Dytman *et al.*, (CLEO Collaboration), CLNS 01/1748, CLEO 01-15, submitted to PRL.
9. R. Zaliznyak, Ph.D. thesis, Stanford University (1998).
10. A. H. Mahmood *et al.*, (CLEO Collaboration), CLNS 01/1761, CLEO 01-20, submitted to PRL.
11. M. Artuso *et al.*, (CLEO Collaboration), CLNS 01/1759, CLEO 01-19, submitted to PRL.
12. J. G. Korner and M. Kramer, *Phys. Lett.* **B 275**, 496 (1992).

Results on Mixing in the D^0 System from BABAR

Monika Grothe representing the BABAR collaboration

University of California, Santa Cruz Institute for Particle Physics, Santa Cruz CA 95064, USA

Abstract. With 12.3 fb^{-1} collected by the BABAR experiment in 2001, the mixing parameter $y = \Delta\Gamma/2\Gamma$ is determined from the ratio of the D^0 lifetimes measured in the $D^0 \to K^-\pi^+$ and in the $D^0 \to K^-K^+$ decay modes. The preliminary result $y = (-1.0 \pm 2.2(\text{stat.}) \pm 1.7(\text{syst.}))$ % is obtained. Also presented is the status of measuring the mixing parameters y and $x^2 = (\Delta M/\Gamma)^2$ from a simultaneous fit to the time evolution of the decay time distributions of Cabibbo-favored right-sign ($D^0 \to K^-\pi^+$) and doubly Cabibbo-suppressed wrong-sign ($D^0 \to K^+\pi^-$) decays. The wrong-sign decay rate, R_{WS} = (# WS decays) / (# RS decays) = $(0.38 \pm 0.04(\text{stat.}) \pm 0.02(\text{syst.}))$ % is obtained from the fit to 23 fb^{-1} of BABAR data taken in 2000.

INTRODUCTION

If CP conservation holds in the D^0 system, the CP-even and CP-odd eigenstates are mass eigenstates with masses M_+, M_- and widths Γ_+, Γ_-. The mixing parameters $x = 2(M_+ - M_-)/(\Gamma_+ + \Gamma_-)$ and $y = (\Gamma_+ - \Gamma_-)/(\Gamma_+ + \Gamma_-)$ measure the difference of these masses and lifetimes, respectively. Since in the Standard Model D - $\bar{D}$ mixing is doubly Cabibbo-suppressed and vanishes in the $SU(3)$ flavor limit both parameters are expected to be small (10^{-3}) [1]. If D - $\bar{D}$ mixing were found to be large, this might be a sign either for large $SU(3)$ flavor breaking ($y \gtrsim x$) or for new physics ($x \gg y$) [2].

The mixing parameter y can be determined from the lifetime of D^0 mesons[1] that decay into final states of specific CP symmetry [3]. A final state that is an equal mixture of CP-even and CP-odd is produced by the Cabibbo-favored decay $D^0 \to K^-\pi^+$. If y is small, the lifetime distribution of D^0 mesons decaying into this final state can be approximated as an exponential with lifetime $\tau_{K\pi} = 1/\Gamma$ where $\Gamma = (\Gamma_+ + \Gamma_-)/2$. A CP-even final state is produced by the singly Cabibbo-suppressed decay $D^0 \to K^-K^+$. The decay time distribution of D^0 mesons that decay into K^-K^+ is exponential with a lifetime $\tau_{KK} = 1/\Gamma_+$. This lifetime can be compared to $\tau_{K\pi}$ to obtain y:

$$y = \frac{\tau_{K\pi}}{\tau_{KK}} - 1 \,. \quad (1)$$

The mixing parameters x^2 and y can be determined simultaneously from the time evolution of the wrong-sign (WS) decay $D^0 \to K^+\pi^-$. The WS decay rate has contributions from the doubly Cabibbo-suppressed decay, described by a pure exponential, and from

[1] In the following, the charge conjugates of the D^0 meson and its decay modes are always implied as well.

CP618, *Heavy Flavor Physics: Ninth International Symposium*, edited by A. Ryd and F. C. Porter

D - $\bar{D}$ mixing, described by the same exponential modified by a coefficient quadratic in the decay time t. A third term arises from the interference of the two, where the exponential has a coefficient linear in t. The time evolution of the WS decay rate is thus described by [4]:

$$\Gamma(t) \propto \exp(-t)\,[\,R + \sqrt{R}\,y'\,t + 1/4\,(\,x'^2 + y'^2\,)\,t^2\,]. \tag{2}$$

Here, t is given in units of the lifetime of the D^0. The parameters x' and y' are related to the mixing parameters x, y by a rotation:

$$x' = x\cos\delta + y\sin\delta, \qquad y' = y\cos\delta - x\sin\delta. \tag{3}$$

The phase of the rotation is the strong phase between the doubly Cabibbo-suppressed contribution and the one from mixing.

In the following, two results are presented. Firstly, a preliminary measurement is described of y from the difference of the D^0 lifetime determined in the $K^-\pi^+$ and K^-K^+ decay channels. Secondly, a preliminary measurement of the WS decay rate of the D^0 is discussed, extracted from a full fit to the time evolution of the combined right-sign and wrong-sign D^0 decay rates.

The measurements are based on data collected with the *BABAR* detector at the PEP-II asymmetric e^+e^- collider. Data taken on and off the $\Upsilon(4S)$ resonance are used. Their center-of-mass is boosted along the beam axis with a nominal Lorentz boost of $\beta\gamma = 0.56$. The size of the interaction point (IP) transverse to the beam direction is typically 6 μm in the vertical direction and 120 μm in the horizontal direction.

Candidates for D^0 particles are identified through the decay $D^{*+} \to D^0\pi^+$. Charged particles are detected and their momenta are measured by a combination of a 40-layer drift chamber (DCH) and a five-layer, double-sided, silicon vertex tracker (SVT), both operating in a 1.5 T solenoidal magnetic field. A ring-imaging Cherenkov detector (DIRC) is used for charged particle identification. A detailed description of the *BABAR* detector is available in [5].

MEASUREMENT OF y

The result is obtained from a sample of 12.4 fb^{-1} of 2001 *BABAR* data that were reconstructed with the most advanced tracking alignment parameters and reconstruction algorithms.

Event Selection

The widths Γ_- and Γ_+ are determined by fitting the decay time distributions of independent samples of $D^0 \to K^-\pi^+$ and $D^0 \to K^-K^+$ decays. The D^0 candidates for each sample are identified by means of the charged particles in their final state. The decay $D^{*+} \to D^0\pi^+$ and $K^\pm$ particle identification are used to suppress backgrounds.

D^0 candidates are selected by searching for pairs of tracks of opposite charge and combined invariant mass near the D^0 mass m_D. Each track is required to contain a minimum number of SVT and DCH hits in order to ensure reconstruction quality. The two daughter tracks of the D^0 candidate are fitted to a common vertex. The χ^2 probability of this vertex fit has to be better than 1%.

Each $K^{\pm}$ candidate among the D^0 daughter tracks is required to pass a likelihood-based particle identification algorithm. This algorithm is based on the measurement of the Cherenkov angle by the DIRC for momenta $p \gtrsim 0.6\,\mathrm{GeV}/c$ and on the energy loss (dE/dx) measured by SVT and DCH for momenta $p \lesssim 0.6\,\mathrm{GeV}/c$. An average efficiency for $K^{\pm}$ identification of approximately 80% is reached for tracks within the DIRC acceptance while the $\pi^{\pm}$ misidentification probability amounts to about 2%.

The decay $D^{*+} \rightarrow D^0\pi^+$ can be distinguished by a π^+ of low momentum, commonly referred to as the slow pion (π_s). To increase acceptance, π_s candidate tracks do not have to contain DCH hits. To improve momentum resolution, a vertex fit is used to constrain each π_s candidate track to pass through the intersection of the D^0 trajectory and the IP. If the χ^2 probability of this vertex fit is less than 1%, the D^* candidate is discarded.

The D^* candidates peak at a value of $\delta m \approx 145.4\,\mathrm{MeV}/c^2$, where δm is the difference in the reconstructed D^* and D^0 masses. Backgrounds are reduced by rejecting events with a value of δm that deviates more than a given margin from the peak. The size of this margin corresponds to approximately three standard deviations and varies between 1 and $2.5\,\mathrm{MeV}/c^2$, depending on the quality of the π_s track.

To remove background from B meson decays, each D^* candidate has to have a momentum p^* greater than $2.6\,\mathrm{GeV}/c$ in the center-of-mass. This condition is also effective at removing combinatorial background that tends to accumulate at lower momenta.

After all selection requirements, about 45,000 D^0 candidates in the decay channel $K^-\pi^+$ and about 4,000 in the K^-K^+ decay channel are retained. The mass and δm distribution of the selected events are shown in Fig. 1. The relative size of the remaining background is about 2% and 5% for the $K^-\pi^+$ and K^-K^+ samples, respectively, when measured inside a $\pm 20\,\mathrm{MeV}/c^2$ window.

Lifetime Determination

The flight length and its measurement error are determined for each D^0 candidate by a global, three-dimensional, multiple-vertex fit that includes the D^0 daughter tracks, the π_s track, and the IP envelope. This fit does not include explicit constraints on the D^* or D^0 masses. The value listed by the Particle Data Group (PDG) for the D^0 mass ($m_D = 1.8654$ GeV$/c^2$ [6]) and the momentum of the D^0 obtained with the vertex fit are used to calculate the boost of the D^0 and obtain the proper decay time.

The lifetimes of the D^0 candidate samples are extracted by means of an unbinned maximum likelihood fit. The likelihood function is divided into two distinct decay time distribution functions, one for the signal and one for the background. The signal function is the convolution of an exponential and a resolution function. The resolution function consists of the sum of two Gaussian distributions with zero mean and with widths that are proportional to the measurement error (typically 180 fs) of the decay time of each D^0 candidate. The parameters in the fit associated with the signal are the lifetime, the proportionality factors for the widths of the two Gaussians, and the fraction of signal

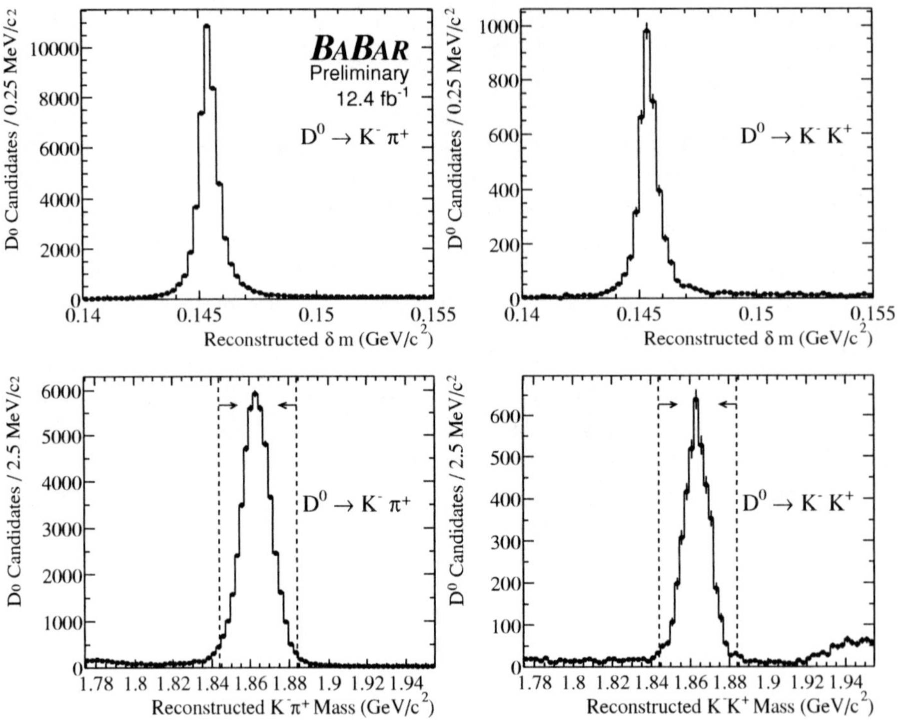

FIGURE 1. The reconstructed δm and D^0 mass distributions after event selection for the $K^-\pi^+$ and K^-K^+ decay modes. The δm plots include candidates both inside and outside the δm selection requirement which fall within the m_D window indicated in the lower plots.

that is assigned to the second Gaussian.

Like the signal likelihood function, the background function consists of the convolution of a resolution function and a lifetime distribution. The background lifetime distribution is determined as the sum of an exponential distribution and a delta function at zero, the latter corresponding to those sources of background that originate at the IP. The resolution function consists of the sum of three Gaussian distributions. The first two are chosen to match the resolution function of the signal. The third is given a width independent of the decay time error and accounts for outliers produced by long-lived particles or reconstruction errors. The additional fit parameters associated with the background include the fraction assigned to zero lifetime sources, the background lifetime, the width of the third Gaussian, and the fraction of background assigned to the third Gaussian.

The reconstructed mass of each D^0 candidate provides the likelihood of this candidate to be part of the signal. This likelihood is based on a separate fit of the reconstructed D^0 mass distribution. This fit includes a resolution function composed of a Gaussian with an asymmetric tail and a linear portion to describe the background. The slope of the background part is constrained with D^0 candidates in the δm sideband ($151 < \delta m <$

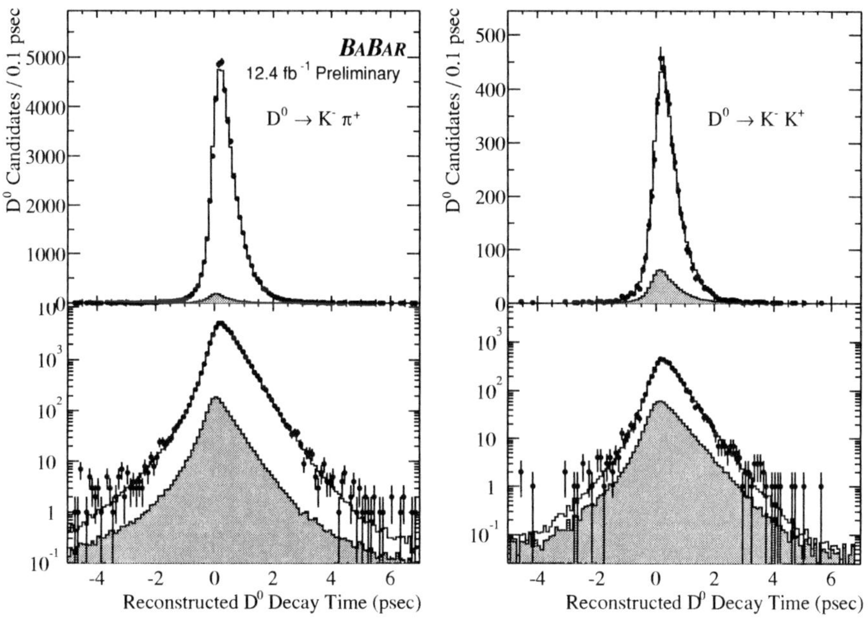

FIGURE 2. The fit to the reconstructed D^0 decay time distribution in the two D^0 decay modes and for all events including the D^0 mass sidebands. The white histogram represents the result of the unbinned maximum likelihood fit described in the text. The gray histogram is the portion assigned by the fit to background.

$159\,\mathrm{MeV}/c^2$).

The results of the lifetime fits are shown in Fig. 2. Typical values for the fitted parameters are a background lifetime similar to the D^0 lifetime and a width of the third Gaussian that is several times larger than the typical decay time error. The proportionality factors associated with the two Gaussians in the resolution function correspond to a root-mean-square of approximately 1.2.

To ensure that the analysis was performed in an objective manner, the D^0 lifetime and y values were blinded throughout the analysis process. This blinding was performed by adding to each of the $\tau_{K\pi}$ and τ_{KK} fit results an offset chosen from a random Gaussian distribution of width 10 fs. The values of the two offsets and the positive ($\tau_i > 0$) sides of the lifetime distribution of the data and of the fit result (Fig. 2) were concealed. The value of y was unblinded only after the analysis method and systematic uncertainties were finalized and the result was committed for public release.

Systematic Errors and Result

Many systematic uncertainties cancel because y is measured from the ratio of lifetimes. The uncertainties that do not cancel are associated mostly with backgrounds. These are tested by varying each event selection requirement within its uncertainty and

TABLE 1. Individual contributions to the systematic uncertainty in y, in %

Event Selection and Background	Reconstruction and Vertexing	Detector Alignment	Quadrature Sum of All Contributions
1.7	0.4	0.3	1.7

recording the subsequent change Δ_i in the measured value of y. The quadrature difference $(\delta\Delta_i)^2 = |\sigma_0^2 - \sigma_i^2|$ is used as an estimate of the statistical error $\delta\Delta_i$ in Δ_i, where σ_0 (σ_i) is the statistical error in y before (after) the ith systematic check. Only systematic checks with $\Delta_i > \delta\Delta_i$ are included in the sum $\sum \Delta_i^2 - (\delta\Delta_i)^2$. The square root of this sum (1.7%) provides an estimate of the systematic uncertainty originating from the event selection and backgrounds.

Biases in tracking reconstruction are investigated by studying Monte Carlo samples, which, within statistics, show no reconstruction bias. In addition, a variety of vertexing techniques, including constraining the D^0 mass and using separate D^* and D^0 vertex fits are employed as systematic checks. A systematic uncertainty of 0.4% is assigned from this source.

Detector misalignment is another potential source of systematic uncertainty. Residual internal misalignment of the SVT, even as small as a few microns, can produce significant variations in the apparent D^0 lifetime.

This source is studied with the help of $e^+e^- \to \gamma\gamma \to 4\pi^\pm$ events in which the four charged tracks are known to originate from the IP. By selecting oppositely charged pairs of these tracks with opening angles similar to two-prong D^0 decays, it is possible to measure the apparent IP position as a function of D^0 trajectory and to calculate a correction to the D^0 lifetime obtained by the fit. For the data samples used in this analysis, a correction of +5 fs is found, with negligible statistical error and a systematic uncertainty of ±5 fs. This type of correction nearly cancels in the lifetime ratio and introduces little systematic uncertainty in y.

The systematic uncertainties in y are summarized in Table 1. Adding the contributions of all systematic checks in quadrature, the preliminary result for y is obtained:

$$y = (-1.0 \pm 2.2 \pm 1.7)\ \% \,, \tag{4}$$

where the first error is statistical and the second systematic. This result is consistent with the Standard Model expectation of zero and with results from E791 [7], FOCUS [8], Belle [9] and CLEO [10].

An important test of the soundness of the analysis is a D^0 lifetime that agrees with the PDG value of 412.6 ± 2.8 fs [6]. A corrected value of $\tau_{K\pi} = 412 \pm 2$ fs is found. The systematic uncertainty is approximately 6 fs and is dominated by detector alignment effects.

MEASUREMENT OF THE WRONG-SIGN D^0 DECAY RATE

Wrong-sign (WS) decays of the type $D^0 \to K^+\pi^-$ are selected by requiring that the slow pion and the kaon have the same charge. The measurement is based on 23 fb^{-1} of *BABAR*

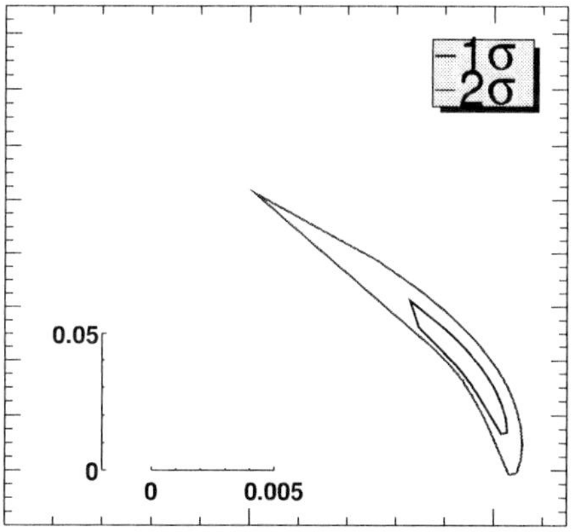

FIGURE 3. The 1 and 2 σ contours of the statistical error in the x'^2-y'-plane (y' on vertical axis) obtained from the simultaneous fit to the time evolution of RS and WS decay rates. A length scale is given for the axes but no axis labels and thus no central values for x'^2, y'.

data taken in 2000. Vertex reconstruction, event selection and systematic error analysis are similar to the ones employed in the measurement of y, described above. After all event selection requirements a sample of about 200 WS candidate events is retained.

A simultaneous unbinned maximum likelihood fit to the decay time distribution of right-sign (RS) and wrong-sign D^0 decay candidates is performed. The fit function consists of a lifetime distribution function convoluted with a resolution function. For RS events, a pure exponential is assumed as functional form of their lifetime distribution while for WS events the functional form follows Eq. 2. The simultaneous fit of RS and WS candidates exploits the high statistics of the RS sample to obtain an accurate description of the resolution function and of the type and size of the background contributions.

The 1 and 2 σ contours of the statistical error of the fit result in the y'-x'^2-plane (Fig. 3) exhibit a strong correlation between the two parameters that extends into the unphysical region $x'^2 < 0$. Note that a length scale is given for the axes but no axis labels and thus no central values for x'^2, y'. The analysis of systematic errors shows that the dominating systematic error source in the 2000 *BABAR* data is the internal alignment of the SVT. It will soon be possible to extract the central values of x'^2 and y' from more data reconstructed with the most advanced tracking alignment constants.

The fit function contains as one parameter the WS decay rate, defined as:

$$R_{WS} = \text{(number of wrong-sign decays) / (number of right-sign decays)}$$

The simultaneous fit to the RS and WS candidate events yields the following result for the WS decay rate:

$$R_{WS} = (\, 0.38 \pm 0.04 \pm 0.02)\ \%,$$

where the first error is statistical, the second systematic. This result is compatible, within its errors, with the results obtained by E791 [11], ALEPH [12], CLEO [13],

FOCUS [14], Belle [9] and is the most precise experimental measurement of R_{WS} currently available.

CONCLUSION

With 12.4 fb^{-1} of data taken by the *BABAR* detector in 2001, the preliminary result for y is obtained:

$$y = (-1.0 \pm 2.2 \pm 1.7)\ \%\ , \tag{5}$$

where the first error is statistical and the second systematic. This result is consistent with the Standard Model expectation of zero and is consistent with published values from E791 [7] and FOCUS [8] and preliminary results from Belle [9] and CLEO [10]. A reduction of the statistical error by half is feasible in the near future by considering additional decay channels ($D^0 \to \pi^-\pi^+$) and by including the 23 fb^{-1} of *BABAR* data collected in 2000.

The extraction of the mixing parameters x^2 and y by measuring the time evolution of the wrong-sign ($D^0 \to K^+\pi^-$) decay rate in a sample of 35 fb^{-1} of combined 2000 and 2001 *BABAR* data will be possible soon.

Already with the 23 fb^{-1} of *BABAR* data collected in 2000, the most precise existing measurement of the wrong-sign decay rate $R_{WS} = (\#\ WS\ decays)/(\#\ RS\ decays)$ is achieved:

$$R_{WS} = (0.38 \pm 0.04\ (\text{stat.}) \pm 0.02\ (\text{syst.}))\ \%\ . \tag{6}$$

This result is consistent with results by E791 [11], ALEPH [12], CLEO [13], FOCUS [14] and a preliminary result by Belle [9].

REFERENCES

1. F. Buccella, M. Lusignoli, and A. Publiese, Phys. Lett. B **379** (1996) 249; F. Buccella *et al.*, *Phys. Rev. D* **51**, 3478 (1995).
2. Z. Ligeti, D-$\bar{D}$ mixing, these proceedings.
3. E. Golowich and S. Pakvasa, *Phys. Lett. B* **505**, 94 (2001).
4. Y. Nir, CP-violation in and beyond the Standard Model, Lectures given in the XXVII SLAC Summer Institute on Particle Physics, 1999, hep-ph9911321.
5. The BABAR Collaboration, B. Aubert *et al.*, to appear in *Nucl. Instrum. Methods*, hep–ex/0105044.
6. D.E. Groom *et al.*, *Eur. Phys. Jour. C* **15**, 1 (2000).
7. The E791 Collaboration, E.M. Aitala *et al.*, *Phys. Rev. Lett.* **83**, 32 (1999).
8. The FOCUS Collaboration, J.M. Link *et al.*,]it Phys. Lett. B **485**, 62 (2000).
9. B.D. Yabsley, Proceedings of the International Europhysics Conference on High Energy Physics (EPS HEP 2001), JHEP PRHEP-hep2001/059, Belle–CONF–0131;
   J. Tanaka, D-$\bar{D}$ mixing results from Belle, these proceedings.
10. The CLEO Collaboration, to appear in the proceedings of the 4th International Conference on B Physics and CP Violation (BCP 4), Japan (2001) hep–ex/0104008.
11. The E791 Collaboration, E.M. Aitala *et al.*, *Phys. Rev.* **D57**, 13 (1998).
12. The ALEPH Collaboration, R. Barate *et al.*, *Phys. Lett. B* **436**, 211 (1998).
13. The CLEO Collaboration, R. Godang *et al.*, *Phys. Rev. Lett.* **84**, 5038 (2000).
14. The FOCUS Collaboration, S. Bianco *et al.*, *Nucl. Phys., Proc. Suppl.* **99**, 191 (2001).

Search for D^0-$\overline{D}^0$ Mixing with Belle

Junichi Tanaka representing the Belle collaboration

Department of Physics, University of Tokyo, Tokyo 113-0033, Japan

Abstract. The search for D^0-$\overline{D}^0$ mixing provides an important window on physics beyond the Standard Model. We present results from Belle on the lifetime difference parameter y_{CP}, the rate R_{ws} of the doubly Cabibbo-suppressed decay $D^0 \to K^+\pi^-$, and a measurement of D^0 decaying to $K_L^0\pi^0$ and $K_S^0\pi^0$. All results are preliminary.

INTRODUCTION

The search for D^0-$\overline{D}^0$ mixing provides an important window on physics beyond the Standard Model (SM). Since the predicted rate of mixing in the SM is very small ($x, y \sim O(10^{-3})$) [1], large mixing could indicate a contribution from non-SM processes, *e.g.* non-SM particles in the box diagrams.

The D^0-$\overline{D}^0$ mixing parameters, $y \equiv (\Gamma_H - \Gamma_L)/(\Gamma_H + \Gamma_L)$ and $x \equiv 2(M_H - M_L)/(\Gamma_H + \Gamma_L)$, can be explored by measuring the lifetime difference of the D^0 mesons decaying into the final state $K^-\pi^+$ and the CP even eigenstate K^-K^+. The parameter y_{CP}, defined by

$$y_{CP} \equiv \frac{\hat{\Gamma}(\text{CP even}) - \hat{\Gamma}(\text{CP odd})}{\hat{\Gamma}(\text{CP even}) + \hat{\Gamma}(\text{CP odd})} \approx \frac{\tau(K^-\pi^+)}{\tau(K^-K^+)} - 1,$$

is related to y and x by the expression $y_{CP} = y\cos\phi - \frac{A_{mix}}{2}x\sin\phi$, where ϕ is a CP violating weak phase due to the interference of decays with and without mixing, and A_{mix} is related to CP violation in mixing. E791, FOCUS, and CLEO have measured y_{CP} to be $(0.8 \pm 2.9 \pm 1.0)\%$ [2], $(3.42 \pm 1.39 \pm 0.74)\%$ [3], and $(-1.1 \pm 2.5 \pm 1.4)\%$ [4], respectively. It is interesting that the FOCUS result is non-zero by more than two standard deviations. Since non-SM processes are expected to manifest themselves as a large value of x, such a large value for y_{CP} could be interpreted as a failure of the SM prediction [6]. On the other hand, CLEO gives results [5] for D^0-$\overline{D}^0$ mixing through the doubly Cabibbo suppressed decay $D^0 \to K^+\pi^-$, $y' = (-2.5^{+1.4}_{-1.6} \pm 0.3)\%$, $x' = (0.0 \pm 1.5 \pm 0.2)\%$ and $A_{mix} = 0.23^{+0.63}_{-0.80} \pm 0.01$, where $y' = y\cos\delta_{K\pi} - x\sin\delta_{K\pi}$ and $x' = x\cos\delta_{K\pi} + y\sin\delta_{K\pi}$; the parameter $\delta_{K\pi}$ is the strong phase difference between the doubly Cabibbo suppressed decay $D^0 \to K^+\pi^-$ and the Cabibbo allowed decay $\overline{D}^0 \to K^+\pi^-$ ($\delta_{K\pi} = 0$ in the $SU(3)$ limit). The FOCUS (y_{CP}) and CLEO (x', y', and A_{mix}) results could be consistent if there is a large $SU(3)$ breaking effect in $D^0 \to K^\pm\pi^\mp$ decays [7].

In this paper, we describe three measurements related to D^0-$\overline{D}^0$ mixing, and show results using the data collected by the Belle detector [8] at the KEKB collider [9].

CP618, *Heavy Flavor Physics: Ninth International Symposium*, edited by A. Ryd and F. C. Porter

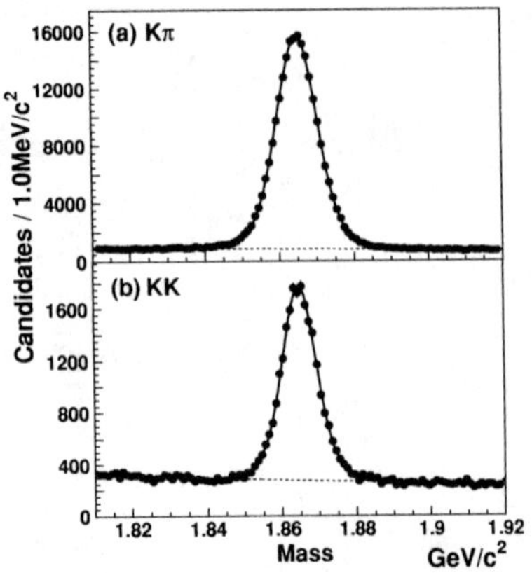

FIGURE 1. The invariant mass distributions for (a) $D^0 \to K^-\pi^+$ and (b) $D^0 \to K^-K^+$ candidates. The results of the fit to two Gaussians (signal) and a linear function (background) are superimposed. The dotted line indicates the background

LIFETIME DIFFERENCE

We have measured the lifetime difference parameter y_{CP} using $D^0 \to K^-\pi^+$ and $D^0 \to K^-K^+$ decays. The decay vertex ($\vec{x}_{\rm dec}$) of the charm meson is determined using both tracks that form the charm meson candidate. The production vertex ($\vec{x}_{\rm pro}$) is obtained by extrapolating the D^0 flight path to the interaction point. The projected decay length (ℓ) in 3-dimensional space and the proper-time (t) are obtained from $\ell = (\vec{x}_{\rm dec} - \vec{x}_{\rm pro}) \cdot \vec{p}_D/|\vec{p}_D|$ and $t = \ell m_D/|\vec{p}_D|$, respectively, where $\vec{p}_D$ and m_D are the momentum vector of the reconstructed charm meson and the PDG [10] average mass.

Figure 1 shows invariant mass distributions for $D^0 \to K^-\pi^+$ and $D^0 \to K^-K^+$ candidates, superimposed with the result of a fit to two Gaussians (for signal) plus a linear function (for background). We find 214260 ± 562 $D^0 \to K^-\pi^+$ and 18306 ± 189 $D^0 \to K^-K^+$ signal events, as estimated by the area of the two Gaussians.

We extract the value of y_{CP} by combining likelihood functions for $D^0 \to K^-\pi^+$ and $D^0 \to K^-K^+$ decays,

$$L_{y_{CP}} = L_{K^-\pi^+} \cdot L_{K^-K^+},$$

and expressing the lifetime for $D^0 \to K^-K^+$ as

$$\tau(K^-K^+) = \tau(K^-\pi^+)/(1 + y_{CP})$$

in an unbinned maximum likelihood fit. This method allows us to properly estimate correlated systematic errors. The definition of the likelihood functions $L = L_{K^-\pi^+}$ and $L_{K^-K^+}$ is given by [11].

We have measured [12]:

$$y_{CP} = (-0.5 \pm 1.0(\text{stat.})^{+0.7}_{-0.8}(\text{syst.}))\%,$$

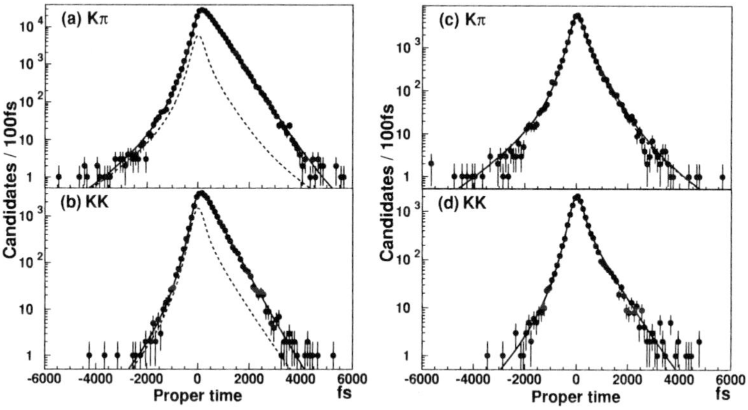

FIGURE 2. Proper-time distributions and fit results for the (a)(c) $D^0 \to K^-\pi^+$, and (b)(d) $D^0 \to K^-K^+$ decay modes in the D^0 mass signal and sideband regions, respectively. The solid line is the result of the fit. The dotted line indicates the background contribution.

using about 23 $\mathrm{fb}^{-1}$of data, and including a correction based on Monte Carlo simulation (MC). This result is consistent with zero within 1σ. Figures 2 (a)–(d) show the proper-time distributions and fit results for $D^0 \to K^-\pi^+$ and $D^0 \to K^-K^+$ decays. Figures 2 (a)(b) show the results of the fit in the D^0-mass signal region ($< \pm 3\sigma$) while Figures 2 (c)(d) show the results of the fit in the D^0-mass sideband region ($> \pm 3\sigma$), where σ is the weighted average of the standard deviations of two Gaussians in the D^0 candidate mass distributions. The solid lines show the fit and the dotted lines show the background contribution in the fit. We verify that the background shape is properly estimated using the sideband region where the background contribution is dominant. Because y_{CP} is measured as a ratio of two lifetimes, the correlated uncertainties cancel. The dominant contributions are reconstruction bias of the proper-time obtained from the MC simulation, fit bias due to the presence of the background events estimated by changing PID information [12].

DOUBLY CABIBBO-SUPPRESSED DECAY

We are studying wrong-sign decays $D^0 \to K^+\pi^-$ to measure x' and y' from the proper-time distribution of wrong-sign decays. We can see a clear peak in the Q $(= m_{D^*} - m_{D^0} - m_\pi)$ distribution for wrong-sign decays as shown in Figure 3 and have measured the time-integrated ratio R_{ws} between wrong-sign and right-sign processes [13] to be

$$R_{ws} \equiv \frac{\Gamma(D^0 \to K^+\pi^- \oplus D^0 \to \overline{D}^0 \to K^+\pi^-)}{\Gamma(D^0 \to K^-\pi^+)} = (0.30 \pm 0.06(\mathrm{stat.}) \pm 0.08(\mathrm{syst.}))\%,$$

using about 10 $\mathrm{fb}^{-1}$of data. This is consistent with other recent measurements [10]. We are still studying the various background components and their distributions in (Q, m_{D^0}) to extract the x' and y' parameters.

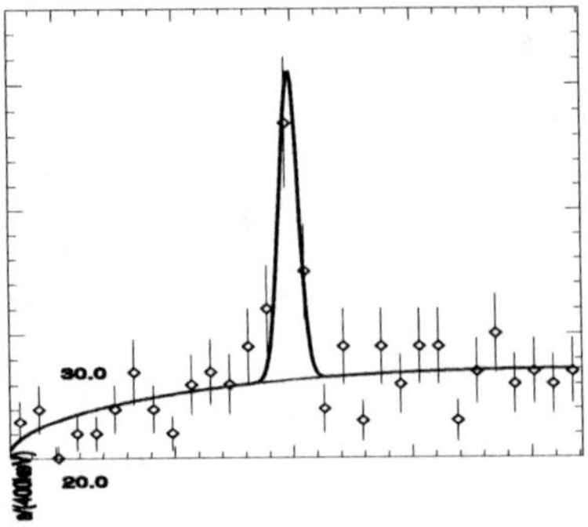

FIGURE 3. Q distribution of wrong-sign decay $D^0 \to K^+\pi^-$.

$D^0 \to K_L^0\pi^0$ AND $K_S^0\pi^0$ DECAYS

We have measured the rate asymmetry $\mathcal{A}$ for D^0 decaying into $K_L^0\pi^0$ and $K_S^0\pi^0$ final states. This asymmetry can be used to disentangle the Cabibbo favored $D^0 \to \overline{K}^0\pi^0$ and doubly Cabibbo suppressed $D^0 \to K^0\pi^0$ amplitudes, and contributes to the important goal of constraining the strong phase difference $\delta_{K\pi}$ between $D^0 \to K^-\pi^+$ and $D^0 \to K^+\pi^-$. We reconstruct $D^0 \to K_L^0\pi^0$ using a D^0 mass constraint and observe a $D^{*+} \to D^0\pi^+$ signal in the D^* mass plots as shown in Figure 4. We measured the rate asymmetry $\mathcal{A}$ to be [14]

$$\mathcal{A} \equiv \frac{\Gamma(D^0 \to K_S^0\pi^0) - \Gamma(D^0 \to K_L^0\pi^0)}{\Gamma(D^0 \to K_S^0\pi^0) + \Gamma(D^0 \to K_L^0\pi^0)} = 0.06 \pm 0.05(\text{stat.}) \pm 0.05(\text{syst.}),$$

using about 23 $\text{fb}^{-1}$of data; an asymmetry of order 5% ($\tan^2\theta_c$) is expected [15]. We are currently studying ways to reduce the systematic error of our technique, to allow a significant measurement of the sign and magnitude of $\mathcal{A}$.

SUMMARY

We have presented three interesting results related to D^0-$\overline{D}^0$ mixing. In particular, our y_{CP} result is consistent with zero within 1σ. All results are preliminary. With the increasing Belle data sample, a better understanding of D^0-$\overline{D}^0$ mixing is expected in the near future.

ACKNOWLEDGMENTS

We wish to thank the KEKB accelerator group for the excellent operation of the KEKB accelerator. We acknowledge support from the Ministry of Education, Culture, Sports,

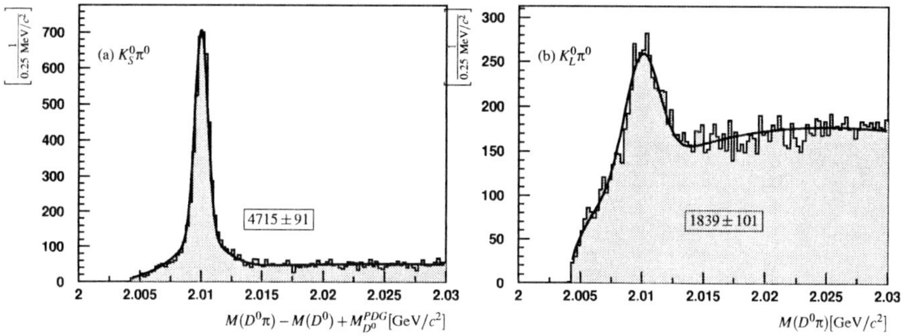

FIGURE 4. D^* mass plots for (a) $D^0 \to K_S^0\pi^0$ and (b) $D^0 \to K_L^0\pi^0$ candidates.

Science and Technology of Japan and the Japan Society for the Promotion of Science; the Australian Research Council and the Australian Department of Industry, Science and Resources; the Department of Science and Technology of India; the BK21 program of the Ministry of Education of Korea and the SRC program of the Korea Science and Engineering Foundation; the Polish State Committee for Scientific Research under contract No.2P03B 17017; the Ministry of Science and Technology of Russian Federation; the National Science Council and the Ministry of Education of Taiwan; and the U.S. Department of Energy.

REFERENCES

1. H. Nelson, hep-ex/9908021 and references therein.
2. E.M. Aitala *et al.* (E791) *Phys. Rev. Lett.* **83**, 32 (1999).
3. J.M. Link *et al.* (FOCUS) *Phys. Lett. B* **485**, 62 (2000).
4. D. Cronin-Hennessy *et al.* (CLEO) hep-ex/0102006.
5. R. Godang *et al.* (CLEO) *Phys. Rev. Lett.* **84**, 5038 (2000).
6. I.I. Bigi, N.G. Uraltsev, *Nucl. Phys. B* **592**, 92 (2001).
7. S. Bergmann, Y. Grossman, Z. Ligeti, Y. Nir and A.A. Petrov, *Phys. Lett. B* **486**, 418 (2000).
8. K. Abe *et al.* (Belle) KEK Progress Report 2000-4 (2000), to be published in Nucl. Inst. and Meth. A.
9. KEKB B Factory Design Report, KEK Report 95-7 (1995), unpublished; Y. Funakoshi *et al.*, Proc. 2000 European Particle Accelerator Conference, Vienna (2000).
10. Particle Data Group, D.E. Groom *et al.*, *Eur. Phys. J.* **C15**, 1 (2000) and 2001 off-year partial update for the 2002 edition available on the PDG WWW pages (URL: http://pdg.lbl.gov/).
11. K. Abe *et al.* (Belle) BELLE-CONF-0131, submitted as a contribution paper to 2001 Lepton Photon Conference.
12. K. Abe *et al.* (Belle) hep-ex/0111026 submitted to PRL. This value is different with that presented at HF9 conference. There are two small systematic corrections to y_{CP}: one for background and the other for multiple scattering. After additional MC study, the first was found to cancel (its systematic error was estimated by changing PID) and the second was more precisely determined.
13. Y. Ban's talk (Belle) at the International Conference on Flavor Physics (ICFP 2001).
14. K. Abe *et al.* (Belle) hep-ex/0107078.
15. E. Golowich and S. Pakvasa, *Phys. Lett. B* **505**, 94 (2001).

$D^0 - \overline{D}^0$ Mixing

Zoltan Ligeti

Ernest Orlando Lawrence Berkeley National Laboratory
University of California, Berkeley, CA 94720

Abstract. The main challenge in the Standard Model calculation of the mass and width difference in $D^0 - \overline{D}^0$ mixing is to estimate the size of $SU(3)$ breaking effects. We consider the possibility that phase space effects may be the dominant source of $SU(3)$ breaking. In particular, we explore whether the Standard Model mass and width differences may satisfy $\Delta\Gamma/2\Gamma \sim 1\%$ and $\Delta\Gamma \gg \Delta m$, in which case the sensitivity of D mixing to new physics would be substantially reduced [1, 2].

INTRODUCTION

It is a common assertion that the Standard Model prediction for mixing in the $D^0 - \overline{D}^0$ system is very small, making this process a sensitive probe of new physics. Two physical parameters that characterize $D^0 - \overline{D}^0$ mixing are

$$x \equiv \frac{\Delta M}{\Gamma}, \qquad y \equiv \frac{\Delta\Gamma}{2\Gamma}, \tag{1}$$

where ΔM and $\Delta\Gamma$ are the mass and width differences of the two neutral D meson mass eigenstates, and Γ is their average width. The $D^0 - \overline{D}^0$ system is unique among the neutral mesons in that it is the only one whose mixing proceeds via intermediate states with down-type quarks. The mixing is very slow in the Standard Model, because the third generation plays a negligible role due to the smallness of $|V_{ub}V_{cb}|$ and the relative smallness of m_b, and so the GIM cancellation is very effective [3, 4, 5, 6].

The current experimental upper bounds on x and y are on the order of a few times 10^{-2}, and are expected to improve significantly in the coming years. To regard a future discovery of nonzero x or y as a signal for new physics, we would need high confidence that the Standard Model predictions lie significantly below the present limits. As we will show, in the Standard Model x and y are generated only at second order in $SU(3)$ breaking, so schematically

$$x, y \sim \sin^2\theta_C \times [SU(3)\ \text{breaking}]^2, \tag{2}$$

where θ_C is the Cabibbo angle. Therefore, predicting the Standard Model values of x and y depends crucially on estimating the size of $SU(3)$ breaking. Although y is expected to be determined by Standard Model processes, its value nevertheless affects significantly the sensitivity to new physics of experimental analyses of D mixing [2].

At present, there are three types of experiments which measure x and y. Each is actually sensitive to a combination of x and y, rather than to either quantity directly. First,

CP618, *Heavy Flavor Physics: Ninth International Symposium*, edited by A. Ryd and F. C. Porter

there is the D^0 lifetime difference to CP even and CP odd final states [7, 8, 9, 10, 11], which to leading order measures

$$y_{CP} = \frac{\Gamma(CP\,\text{even}) - \Gamma(CP\,\text{odd})}{\Gamma(CP\,\text{even}) + \Gamma(CP\,\text{odd})} \simeq \frac{\hat{\tau}(D \to \pi^+ K^-)}{\hat{\tau}(D \to K^+ K^-)} - 1 = y\cos\phi - x\sin\phi\,\frac{A_m}{2}, \quad (3)$$

where the D mass eigenstates are $|D_{L,S}\rangle = p\,|D^0\rangle \pm q\,|\overline{D}^0\rangle$, $A_m = |q/p|^2 - 1$, and $\phi = \arg(q/p)$ is a possible CP violating phase of the mixing amplitude. The experimental results are

$$y_{CP} = \begin{cases} 0.8 \pm 3.1\% & \text{E791 [7]} \\ 3.4 \pm 1.6\% & \text{FOCUS [8]} \\ -1.1 \pm 2.9\% & \text{CLEO [9]} \\ -0.5 \pm 1.3\% & \text{BELLE [10]} \\ -1.0 \pm 2.8\% & \text{BABAR [11]} \end{cases}, \quad (4)$$

which yield a world average of $y_{CP} \simeq 0.65 \pm 0.85\%$ at present. Second, one can measure the time dependence of doubly Cabibbo suppressed decays, such as $D^0 \to K^+\pi^-$ [12], which is sensitive to the three quantities

$$(x\cos\delta + y\sin\delta)\cos\phi, \qquad (y\cos\delta - x\sin\delta)\sin\phi, \qquad x^2 + y^2, \quad (5)$$

where δ is the strong phase between the Cabibbo allowed and doubly Cabibbo suppressed amplitudes. A similar study for $D^0 \to K^-\pi^+\pi^0$ would allow the strong phase difference to be extracted simultaneously from the Dalitz plot analysis [13]. Third, one can search for D mixing in semileptonic decays [14], which is sensitive to $x^2 + y^2$.

In a large class of models, the best hope to discover new physics in D mixing is to observe the CP violating phase, $\phi_{12} = \arg(M_{12}/\Gamma_{12})$, which is very small in the Standard Model. However, if $y \gg x$, then the sensitivity of any physical observable to ϕ_{12} is suppressed, since $A_m \propto x/y$ and $\phi \propto (x/y)^2$, even if new physics makes a large contribution to M_{12} and ϕ_{12} [2]. It is also clear from Eq. (5) that if y is significantly larger than x, then δ must be known very precisely for experiments to be sensitive to new physics in the terms linear in x and y.

$SU(3)$ ANALYSIS OF $D^0 - \overline{D}^0$ MIXING

We now prove that $D^0 - \overline{D}^0$ mixing arises only at second order in $SU(3)$ breaking effects. The proof is valid when $SU(3)$ violation enters perturbatively. This would not be the case, for example, if D transitions were dominated by intermediate states or single resonances close to threshold. Other than in these exceptional situations, treating $SU(3)$ violation perturbatively seems to us to be a mild assumption.

The quantities M_{12} and Γ_{12} which determine x and y depend on matrix elements with the general structure

$$\langle \overline{D}^0 |\, \mathcal{H}_w \mathcal{H}_w \,| D^0 \rangle, \quad (6)$$

where in this section we let $\mathcal{H}_w$ denote specifically the $\Delta C = -1$ part of the weak Hamiltonian. Let D be the field operator that creates a D^0 meson and annihilates a $\overline{D}^0$.

Then the matrix element may be written as

$$\langle 0|D\,\mathcal{H}_w\mathcal{H}_w D\,|0\rangle\,. \tag{7}$$

Let us focus on the $SU(3)$ flavor group theory properties of this expression.

Since the operator D is of the form $\bar{c}u$, it transforms in the fundamental representation of $SU(3)$, which we will represent with a lower index, D_i. We use a convention in which the correspondence between matrix indices and quark flavors is $(1,2,3)=(u,d,s)$. The only nonzero element of D_i is $D_1=1$. The $\Delta C=-1$ part of the weak Hamiltonian has the flavor structure $(\bar{q}_i c)(\bar{q}_j q_k)$, so its matrix representation is written with a fundamental index and two antifundamentals, H^{ij}_k. This operator is a sum of irreducible representations contained in the product $3\times\overline{3}\times\overline{3}=\overline{15}+6+\overline{3}+\overline{3}$. In the limit in which the third generation is neglected, H^{ij}_k is traceless, so only the $\overline{15}$ (symmetric on i and j) and 6 (antisymmetric on i and j) representations appear. That is, the $\Delta C=-1$ part of $\mathcal{H}_w$ may be decomposed as $\frac{1}{2}(\mathcal{O}_{\overline{15}}+\mathcal{O}_6)$, where

$$\begin{aligned}
\mathcal{O}_{\overline{15}} &= (\bar{s}c)(\bar{u}d)+(\bar{u}c)(\bar{s}d)+s_1(\bar{d}c)(\bar{u}d)+s_1(\bar{u}c)(\bar{d}d)\\
&\quad -s_1(\bar{s}c)(\bar{u}s)-s_1(\bar{u}c)(\bar{s}s)-s_1^2(\bar{d}c)(\bar{u}s)-s_1^2(\bar{u}c)(\bar{d}s)\,,\\
\mathcal{O}_6 &= (\bar{s}c)(\bar{u}d)-(\bar{u}c)(\bar{s}d)+s_1(\bar{d}c)(\bar{u}d)-s_1(\bar{u}c)(\bar{d}d)\\
&\quad -s_1(\bar{s}c)(\bar{u}s)+s_1(\bar{u}c)(\bar{s}s)-s_1^2(\bar{d}c)(\bar{u}s)+s_1^2(\bar{u}c)(\bar{d}s)\,,
\end{aligned} \tag{8}$$

and $s_1=\sin\theta_C\approx 0.22$. The matrix representations $H(\overline{15})^{ij}_k$ and $H(6)^{ij}_k$ have nonzero elements

$$\begin{array}{lll}
H(\overline{15})^{ij}_k: & H^{13}_2=H^{31}_2=1\,, & H^{12}_2=H^{21}_2=s_1\,,\\
& H^{13}_3=H^{31}_3=-s_1\,, & H^{12}_3=H^{21}_3=-s_1^2\,,\\
H(6)^{ij}_k: & H^{13}_2=-H^{31}_2=1\,, & H^{12}_2=-H^{21}_2=s_1\,,\\
& H^{13}_3=-H^{31}_3=-s_1\,, & H^{12}_3=-H^{21}_3=-s_1^2\,.
\end{array} \tag{9}$$

We introduce $SU(3)$ breaking through the quark mass operator $\mathcal{M}$, whose matrix representation is $M^i_j=\mathrm{diag}(m_u,m_d,m_s)$. Although $\mathcal{M}$ is a linear combination of the adjoint and singlet representations, only the 8 induces $SU(3)$ violating effects. It is convenient to set $m_u=m_d=0$ and let $m_s\neq 0$ be the only $SU(3)$ violating parameter. All nonzero matrix elements built out of D_i, H^{ij}_k and M^i_j must be $SU(3)$ singlets.

We now prove that $D^0-\overline{D}^0$ mixing arises only at second order in $SU(3)$ violation, by which we mean second order in m_s. First, we note that the pair of D operators is symmetric, and so the product D_iD_j transforms as a 6 under $SU(3)$. Second, the pair of $\mathcal{H}_w$'s is also symmetric, and the product $H^{ij}_k H^{lm}_n$ is in one of the representations which appears in the product

$$\begin{aligned}
\left[(\overline{15}+6)\times(\overline{15}+6)\right]_S &= (\overline{15}\times\overline{15})_S+(\overline{15}\times 6)+(6\times 6)_S \\
&= (\overline{60}+\overline{24}+15+15'+\overline{6})+(42+24+15+\overline{6}+3)+(15'+\overline{6})\,.
\end{aligned} \tag{10}$$

A straightforward computation shows that only three of these representations actually appear in the decomposition of $\mathcal{H}_w\mathcal{H}_w$. They are the $\overline{60}$, the 42, and the $15'$ (actually twice, but with the same nonzero elements both times). So we have product operators of the form

$$DD = \mathcal{D}_6\,, \qquad \mathcal{H}_w\mathcal{H}_w = O_{\overline{60}} + O_{42} + O_{15'}\,, \tag{11}$$

where the subscript denotes the representation of $SU(3)$.

Since there is no $\overline{6}$ in the decomposition of $\mathcal{H}_w\mathcal{H}_w$, there is no $SU(3)$ singlet which can be made with $\mathcal{D}_6$, and no $SU(3)$ invariant matrix element of the form (7) can be formed. This is the well known result that $D^0-\overline{D}^0$ mixing is prohibited by $SU(3)$ symmetry.

Now consider a single insertion of the $SU(3)$ violating spurion $\mathcal{M}$. The combination $\mathcal{D}_6\mathcal{M}$ transforms as $6\times 8 = 24+\overline{15}+6+\overline{3}$. Note that there is still no invariant to be made with $\mathcal{H}_w\mathcal{H}_w$. It follows that $D^0-\overline{D}^0$ mixing is not induced at first order in $SU(3)$ breaking.

With two insertions of $\mathcal{M}$, it becomes possible to make an $SU(3)$ invariant. The decomposition of $\mathcal{D}\mathcal{M}\mathcal{M}$ is

$$\begin{aligned} 6\times(8\times 8)_S &= 6\times(27+8+1) \\ &= (60+\overline{42}+24+\overline{15}+\overline{15}'+6)+(24+\overline{15}+6+\overline{3})+6\,. \end{aligned} \tag{12}$$

There are three elements of the 6×27 part which can give invariants with $\mathcal{H}_w\mathcal{H}_w$. Each invariant yields a contribution proportional to $s_1^2 m_s^2$. As promised, $D^0-\overline{D}^0$ mixing arises only at second order in the $SU(3)$ violating parameter m_s.

ESTIMATING THE SIZE OF $SU(3)$ BREAKING

There is a vast literature on estimating x and y within and beyond the Standard Model; for a compilation of results, see Ref. [15]. Roughly speaking, there are two approaches, neither of which give reliable results because m_c is in some sense intermediate between heavy and light.

"Inclusive" approach. The inclusive approach is based on the operator product expansion (OPE), first applied to $D^0-\overline{D}^0$ mixing by Georgi [3] and later extended by other authors [4, 16]. In the $m_c \gg \Lambda$ limit, where Λ is a scale characteristic of the strong interactions, such as m_ρ or $4\pi f_\pi$, ΔM and $\Delta\Gamma$ can be expanded in terms of matrix elements of local operators [3, 4, 16]. The use of the OPE relies on local quark-hadron duality, and on Λ/m_c being small enough to allow a truncation of the series after the first few terms. The charm mass may not be large enough for these to be good approximations, especially for nonleptonic D decays. An observation of y of order 10^{-2} could be ascribed to a breakdown of the OPE or of duality [16], but such a large value of y is certainly not a generic prediction of OPE analyses.

In the limit $m_c \gg \Lambda$, one can write

$$\Gamma_{12} = \frac{1}{2m_D}\,\mathrm{Im}\,\langle\overline{D}^0|\,i\int d^4x\,T\left\{\mathcal{H}_w^{\Delta C=1}(x)\,\mathcal{H}_w^{\Delta C=1}(0)\right\}|D^0\rangle\,, \tag{13}$$

where $\mathcal{H}_w^{\Delta C=1}$ is the $|\Delta C|=1$ effective Hamiltonian. In the OPE, the time ordered product in Eq. (13) can be expanded in local operators of increasing dimension; the higher dimension operators are suppressed by powers of Λ/m_c.

TABLE 1. Enhancement of ΔM and $\Delta\Gamma$ relative to the box diagram at various orders in the OPE. Λ denotes a hadronic scale around $4\pi f_\pi \sim 1\,\text{GeV}$, and $\beta_0 = 11 - 2n_f/3 = 9$.

Ratio	4-quark	6-quark	8-quark
$\Delta M/\Delta M_{\text{box}}$	1	$\Lambda^2/m_s m_c$	$(\alpha_s/4\pi)(\Lambda^2/m_s m_c)^2$
$\Delta\Gamma/\Delta M$	m_s^2/m_c^2	$\alpha_s/4\pi$	$\beta_0\,\alpha_s/4\pi$

The leading contribution comes from 4-quark operators in the $|\Delta C| = 2$ effective Hamiltonian, corresponding to the short distance box diagram. The result is of the form

$$\Delta M_{\text{box}} \sim \frac{2X_D}{3\pi^2}\frac{m_s^4}{m_c^2}, \qquad \Delta\Gamma_{\text{box}} \sim \frac{4X_D}{3\pi}\frac{m_s^6}{m_c^4}, \tag{14}$$

where $X_D = V_{cs}^2 V_{cd}^2 G_F^2 m_D B_D f_D^2$. Eq. (14) then lead to the estimates

$$x_{\text{box}} \sim \text{few} \times 10^{-5}, \qquad y_{\text{box}} \sim \text{few} \times 10^{-7}. \tag{15}$$

$\Delta\Gamma_{\text{box}}$ is proportional to m_s^6. This factor comes from three sources: (i) m_s^2 from an $SU(3)$ violating mass insertion on each quark line in the box graph; (ii) m_s^2 from an additional mass insertion on each line to compensate the chirality flip from the first insertion; (iii) m_s^2 to lift the helicity suppression for the decay of a scalar meson into a massless fermion pair. The last factor of m_s^2 is absent from ΔM; this is why at leading order in the OPE, $y_{\text{box}} \ll x_{\text{box}}$. Higher order terms in the OPE are important, because the chiral suppressions can be lifted by quark condensates instead of by mass insertions, allowing ΔM and $\Delta\Gamma$ to be proportional to m_s^2. This is the minimal suppression required by $SU(3)$ symmetry, as we proved earlier.

The order of magnitudes of the resulting contributions are explained in Ref. [1], and are summarized in Table 1. One finds that the dominant contributions to x are from 6- and 8-quark operators, while the dominant contribution to y is from 8-quark operators. With some assumptions about the hadronic matrix elements, the resulting estimates are

$$x \sim y \sim 10^{-3}. \tag{16}$$

It is a general feature of OPE based analyses that $x \gtrsim y$. We emphasize that at this time these methods are useful for understanding the order of magnitude of x and y, but not for obtaining reliable quantitative results. To turn the estimates presented here into a systematic computation of x and y would require the calculation of almost two dozen nonperturbative matrix elements.

"Exclusive" approach. A long distance analysis of D mixing is complementary to the OPE. Instead of assuming that the D meson is heavy enough for duality to hold between the partonic rate and the sum over hadronic final states, here one assumes that D transitions are dominated by a small number of exclusive processes, which are examined explicitly. This is particularly interesting for studying $\Delta\Gamma$, which depends on real final states in D decays. However, the D is not light enough that its decays are dominated by a few final states. Since there are cancellations between states within a given $SU(3)$ multiplet, one needs to know the contribution of each state with high precision. In the absence of sufficiently precise data on many decay rates and on strong phases, one is forced to use some assumptions. While most studies find $x, y \lesssim 10^{-3}$, Refs. [17, 18, 19] obtain x and y at the 10^{-2} level by arguing that $SU(3)$ violation is actually of order unity, but the source of the large $SU(3)$ breaking is not made explicit.

For a long distance analysis, it is useful to express the width difference directly in terms of observable decay rates,

$$y = \sum_n \eta_{\rm CKM}(n)\, \eta_{CP}(n)\, \cos\delta_n \sqrt{\mathcal{B}(D^0 \to n)\, \mathcal{B}(\overline{D}^0 \to n)}\,, \tag{17}$$

where δ_n is the strong phase difference between the $D^0 \to n$ and $\overline{D}^0 \to n$ amplitudes. In decays to many-body final states, the strong phases may have different values in different regions of the Dalitz plot, in which case the sum is supplemented by an integral over the Dalitz plot for each final state. The CKM factor is $\eta_{\rm CKM} = (-1)^{n_s}$, where n_s is the number of s and $\bar{s}$ quarks in the final state. The factor $\eta_{CP} = \pm 1$ is determined by the CP transformation of the final state, $CP|f\rangle = \eta_{CP}|\bar{f}\rangle$, which is well-defined since $|f\rangle$ and $|\bar{f}\rangle$ are in the same $SU(3)$ multiplet.

In practice, we cannot use Eq. (17) to get a reliable estimate of y, since the doubly Cabibbo suppressed rates have large errors, and there are very little data on strong phase differences. To proceed further, we would be forced to introduce model dependent assumptions about the amplitudes and/or their strong phases. For example, in two-body D decays to charged pseudoscalars ($\pi^+\pi^-$, π^+K^-, $K^+\pi^-$, K^+K^-), the $SU(3)$ violation can enter through the decay rates or the strong phase difference. We know experimentally that in some of these rates the $SU(3)$ breaking is sizable; for example $\mathcal{B}(D^0 \to K^+K^-)/\mathcal{B}(D^0 \to \pi^+\pi^-) \simeq 2.8$ [20]. Such effects were the basis for the claim in Ref. [17] that $SU(3)$ is simply inapplicable to D decays. In contrast, we know very little about the strong phase δ which vanishes in the $SU(3)$ limit; Ref. [21] presented a model calculation resulting in $\cos\delta \gtrsim 0.8$, but it is also possible to obtain much larger values for δ [18]. Using Eq. (17), the value of y_a corresponding to the U-spin doublet of charged π and K is

$$y_{\pi K} = \mathcal{B}(D^0 \to \pi^+\pi^-) + \mathcal{B}(D^0 \to K^+K^-) - 2\cos\delta \sqrt{\mathcal{B}(D^0 \to K^-\pi^+)\, \mathcal{B}(D^0 \to K^+\pi^-)}\,. \tag{18}$$

The experimental central values, allowing for D mixing in the doubly Cabibbo suppressed rates, yield $y_{\pi K} \simeq (5.76 - 5.29\cos\delta) \times 10^{-3}$ [2]. For small δ there is an almost perfect cancellation even though the ratios of the individual rates significantly violate $SU(3)$. In the "exclusive" approach, x is obtained from y by use of a dispersion relation, and one generally finds $x \sim y$.

At this stage, one cannot use the exclusive approach to predict either x or y. Any estimate of their sizes depends on computing $SU(3)$ breaking effects. While this problem is not tractable in general, one source of $SU(3)$ breaking in y, from final state phase space, can be calculated with only minimal and reasonable assumptions. We will estimate these effects in the next section.

$SU(3)$ BREAKING FROM PHASE SPACE

We now turn to the contributions to y from on-shell final states. There is a contribution to the D^0 width difference from every common decay product of D^0 and $\overline{D}^0$. In the $SU(3)$ limit, these contributions cancel when one sums over complete $SU(3)$ multiplets in the final state. The cancellations depend on $SU(3)$ symmetry both in the decay matrix elements and in the final state phase space. While there are certainly $SU(3)$ violating corrections to both of these, it is difficult to compute the $SU(3)$ violation in the matrix elements in a model independent manner.[1] However, with some mild assumptions about the momentum dependence of the matrix elements, the $SU(3)$ violation in the phase space depends only on the final particle masses and can be computed. In this section we estimate the contributions to y solely from $SU(3)$ violation in the phase space.[2] We will find that this source of $SU(3)$ violation can generate y of the order of a percent.

The mixing parameter y may be written in terms of the matrix elements for common final states for D^0 and $\overline{D}^0$ decays, Let us concentrate on final states F which transform within a single $SU(3)$ multiplet R. Assuming CP symmetry in D decays, which in the Standard Model and almost all scenarios of new physics is an excellent approximation, relates $\langle\overline{D}^0|\mathcal{H}_w|n\rangle$ to $\langle D^0|\mathcal{H}_w|\overline{n}\rangle$. Since $|n\rangle$ and $|\overline{n}\rangle$ are in a common $SU(3)$ multiplet, they are determined by a single effective Hamiltonian. Hence the contribution of the states $n \in F_R$ to y is

$$y = \frac{1}{\Gamma}\langle\overline{D}^0|\,\mathcal{H}_w\left\{\eta_{CP}(F_R)\sum_{n\in F_R}|n\rangle\rho_n\langle n|\right\}\mathcal{H}_w\,|D^0\rangle = \frac{\sum_{n\in F_R}\langle\overline{D}^0|\,\mathcal{H}_w|n\rangle\rho_n\langle n|\mathcal{H}_w\,|D^0\rangle}{\sum_{n\in F_R}\Gamma(D^0\to n)}, \tag{19}$$

where ρ_n is the phase space available to the state n. In the $SU(3)$ limit, all the ρ_n are the same for $n \in F_R$, and the quantity in braces above is an $SU(3)$ singlet. Since the ρ_n depend only on the known masses of the particles in the state n, incorporating the true values of ρ_n in the sum is a calculable source of $SU(3)$ breaking.

We begin by computing $y_{F,R}$ for D decays to states $F = PP$ consisting of a pair of pseudoscalar mesons such as π, K, η. We neglect $\eta - \eta'$ mixing throughout this analysis, and we have checked that this simplification has a negligible effect on the numerical results. Since PP is symmetric in the two mesons, it must transform as an element of $(8\times 8)_S = 27 + 8 + 1$. It is straightforward to construct the $SU(3)$ invariants, and compute $y_{F,R}$ from them. As an example, for the PP pair in an 8 there are invariants with $\mathcal{H}_w$ in a $\overline{15}$, $A_8^{\overline{15}}(PP_8)_i^k H_k^{ij} D_j$, and with $\mathcal{H}_w$ in a 6, $A_8^6(PP_8)_i^k H_k^{ij} D_j$. In this particular case, the product $H_k^{ij} D_j$ with (ij) symmetric (the $\overline{15}$) is proportional to $H_k^{ij} D_j$ with (ij) antisymmetric (the 6), and the linear combination $A_8 \equiv A_8^{\overline{15}} - A_8^6$ is the only one which

---

[1] The $SU(3)$ breaking in matrix elements may be modest even in cases such as $D\to K^+K^-$ and $D\to \pi^+\pi^-$, for which the ratio of measured rates appears to be very far from the $SU(3)$ limit [22].

[2] The phase space difference alone can explain the large $SU(3)$ breaking between the measured $D\to K^*\ell\bar{\nu}$ and $D\to\rho\ell\bar{\nu}$ rates, assuming no $SU(3)$ breaking in the form factors [23]. Recently it was shown that the lifetime ratio of the D_s and D^0 mesons may also be explained this way [24].

TABLE 2. Values of $y_{F,R}$ for two-body final states. This represents the value which y would take if elements of F_R were the only channel open for D^0 decay.

Final state representation		$y_{F,R}/s_1^2$	$y_{F,R}$ (%)
PP	8	−0.0038	−0.018
	27	−0.00071	−0.0034
PV	8_S	0.031	0.15
	8_A	0.032	0.15
	10	0.020	0.10
	$\overline{10}$	0.016	0.08
	27	0.040	0.19
$(VV)_{s\text{-wave}}$	8	−0.081	−0.39
	27	−0.061	−0.30
$(VV)_{p\text{-wave}}$	8	−0.10	−0.48
	27	−0.14	−0.70
$(VV)_{d\text{-wave}}$	8	0.51	2.5
	27	0.57	2.8

appears. Thus we find that $y_{PP,8}$ is proportional to

$$y_{PP,8} \propto s_1^2 \Big[\frac{1}{2}\Phi(\eta,\eta) + \frac{1}{2}\Phi(\pi^0,\pi^0) + \frac{1}{3}\Phi(\eta,\pi^0) + \Phi(\pi^+,\pi^-) + \Phi(K^+,K^-) \tag{20}$$
$$- \frac{1}{6}\Phi(\eta,K^0) - \frac{1}{6}\Phi(\eta,\overline{K}^0) - \Phi(K^+,\pi^-) - \Phi(K^-,\pi^+) - \frac{1}{2}\Phi(K^0,\pi^0) - \frac{1}{2}\Phi(\overline{K}^0,\pi^0)\Big],$$

where $\Phi(P_1,P_2)$ is the phase space integral for the decay into mesons P_1 and P_2. In the $SU(3)$ limit when all Φ's are equal, Eq. (20) vanishes. In a two-body decay, $\Phi(P_1,P_2)$ is proportional to $|\vec{p}\,|^{2\ell+1}$, where $\vec{p}$ and ℓ are the spatial momentum and orbital angular momentum of the final state particles. For $D^0 \to PP$, the decay is into an s wave. It is straightforward to compute the required phase space factors from the known pseudoscalar masses. The results are shown in Table 2. These effects are no larger than one finds in the inclusive analysis, since as in the parton picture, the final states are far from threshold.

Next we turn to final states of the form PV, consisting of a pseudoscalar and a vector meson. Note that three-body final states $3P$ can resonate through PV, and so are partially included here. In this case there is no symmetry between the mesons, so in principle all representations in the combination $8 \times 8 = 27 + 10 + \overline{10} + 8_S + 8_A + 1$ can appear. For simplicity, we take the quark content of the ϕ and ω respectively to be $\bar{s}s$ and $(\bar{u}u + \bar{d}d)/\sqrt{2}$, and consider only the combination which appears in the $SU(3)$ octet. Reasonable variations of the $\phi - \omega$ mixing angle have a negligible effect on our results. Both because one of the particles is more massive, and because the decay is now into a p wave, the phase space dependence is stronger than for the PP final state, see Table 2. For any representation of the final state, the effects are less than one percent.

For the VV final state, decays into s, p and d waves are all possible. Bose symmetry and the restriction to zero total angular momentum together imply that only the symmetric $SU(3)$ combinations appear. Because some VV final states, such as ϕK^*, lie near the

TABLE 3. Values of $y_{F,R}$ for three- and four-body final states.

Final state representation		$y_{F,R}/s_1^2$	$y_{F,R}$ (%)
$(3P)_{s\text{-wave}}$	8	−0.48	−2.3
	27	−0.11	−0.54
$(3P)_{p\text{-wave}}$	8	−1.13	−5.5
	27	−0.07	−0.36
$(3P)_{\text{form-factor}}$	8	−0.44	−2.1
	27	−0.13	−0.64
$4P$	8	3.3	16
	27	2.2	9.2
	27′	1.9	11

D threshold, the inclusion of vector meson widths is quite important. Our model for the resonance line shape is a Lorentz invariant Breit-Wigner normalized on $0 < m < \infty$,

$$f(m; m_R, \Gamma_R) = N(m_R, \Gamma_R) \frac{m^2 \Gamma_R^2}{(m^2 - m_R^2)^2 + m^2 \Gamma_R^2}, \tag{21}$$

where m_R and Γ_R are the mass and width of the vector meson, and m^2 is the square of its four-momentum in the decay. The results for s, p, and d wave decays are shown in Table 2. With these heavier final states and with the higher partial waves, effects at the level of a percent are quite generic. The vector meson widths turn out to be quite important; if they were neglected, the results in the p- and d-wave channels would be larger by approximately a factor of three. The finite widths soften the $SU(3)$ breaking which otherwise would be induced by a sharp phase space boundary. We have checked that our results are not very sensitive to variations in the line shape used to model the vector meson widths. Again, $4P$ and PPV final states can resonate through VV, so they are partially included here.

As we go to final states with more particles, the combinatoric possibilities begin to proliferate. We will consider the final states $3P$ and $4P$, and for concreteness require that the pseudoscalars be found in a totally symmetric 8 or 27 representation of $SU(3)$. This assumption is convenient, because the phase space integration is much simpler if it can be performed symmetrically. These final states should be representative; we have no reason to believe that this choice selects final state multiplets for which phase space effects are particularly enhanced or suppressed. Note that $3 \times (\overline{15} + 6)$ contains no representation larger than a 27. The results are shown in Table 3.

In contrast to the two-body case, for three-body final states the momentum dependence of the matrix elements is no longer fixed by the conservation of angular momentum. The simplest assumption is to take a momentum independent matrix element, with all three final state particles in an s wave. We have also considered other matrix elements; for example, if one of the mesons has angular momentum $\ell = 1$ in the D^0 rest frame (balanced by the combination of the other two mesons). Alternatively, we could imagine introducing a mild "form factor suppression," with a weight such as $\Pi_{i \neq j}(1 - m_{ij}^2/Q^2)^{-1}$, where $m_{ij}^2 = (p_i + p_j)^2$, and $Q = 2\,\text{GeV}$ is a typical resonance mass. The resulting $y_{F,R}$

TABLE 4. Total D^0 branching fractions to classes of final states, rounded to the nearest 5% [20].

Final state	PP	PV	$(VV)_{s\text{-wave}}$	$(VV)_{d\text{-wave}}$	$3P$	$4P$
Branching fraction	5%	10%	5%	5%	5%	10%

values corresponding to each of these hypothesis are show in in Table 3.

Finally, we have studied the final state with four pseudoscalars, with the mesons in an overall symmetric 8 or a symmetric 27. We take a momentum independent matrix element. There are actually two symmetric 27 representations; we call the 27 the representation of the form $R^{ij}_{kl} = [M^i_m M^m_k M^j_n M^n_l + \text{symmetric} - \text{traces}]$ and the $27'$ the one of the form $R^{ij}_{kl} = [M^i_m M^m_n M^n_k M^j_l + \text{symmetric} - \text{traces}]$. The results are again summarized in Table 3. Here the partial contributions to y are very large, of the order of 10%. This is not surprising, since $4P$ final states containing more than one strange particle are close to D threshold, and the ones containing no pions are kinematically inaccessible. There is no reason to expect $SU(3)$ cancellations to persist effectively in this regime.

Formally, one could construct y from the individual $y_{F,R}$ by weighting them by their D^0 branching ratios,

$$y = \frac{1}{\Gamma} \sum_{F,R} y_{F,R} \Big[\sum_{n \in F_R} \Gamma(D^0 \to n) \Big] . \tag{22}$$

However, the data on D decays are neither abundant nor precise enough to disentangle the decays to the various $SU(3)$ multiplets, especially for the three- and four-body final states. Nor have we computed $y_{F,R}$ for all or even most of the available representations. Instead, we can only estimate individual contributions to y by assuming that the representations for which we know $y_{F,R}$ to be typical for final states with a given multiplicity, and then to scale to the total branching ratio to those final states. The total branching ratios of D^0 to two-, three- and four-body final states can be extracted from Ref. [20]. The results are presented in Table 4, where we round to the nearest 5% to emphasize the uncertainties in these numbers. Close to half of all D^0 decays are accounted for in this table; the rest are decays to other modes such as PPV, decays to states with $SU(3)$ singlet mesons, decays to higher resonances, semileptonic decays, and other suppressed processes. Based on data in the channel $\overline{K}^{0*}\rho^0$, the VV final state is dominantly CP even, consistent with an equal distribution between s and d wave decays (although favoring a small s wave enhancement).

We estimate the contribution to y from a given type of final state by taking the product of the typical $y_{F,R}$ found in our calculation with the approximate branching ratios given in Table 4. While in most cases the contributions are small, of the order of 10^{-3} or less, we observe that D^0 decays to nonresonant $4P$ states naturally contribute to y at the percent level. The reason for such unusually large $SU(3)$ violating effects in y is that approximately 10% of D^0 decays are to final states for which the complete $SU(3)$ multiplets are not kinematically accessible.

We have not considered all possible final states which might give large contributions to y. In particular, the branching ratio for $D^0 \to K^- a_1^+$ is $(7.3 \pm 1.1)\%$ [20], even though this final state is quite close to D threshold. Unfortunately, the identities of the $SU(3)$ partners of the $a_1(1260)$, which has $J^{PC} = 1^{++}$, are not well established. While it is natural to identify the $K_1(1400)$ as the corresponding strange axial vector meson, and the $f_1(1285)$ as the analogue of the ω, there is no natural candidate for the $\bar{s}s$ analogue of the ϕ. The size of y_{PV^*} is quite sensitive to this choice, as well as to the value taken for the poorly measured width of the a_1. If we take the $\bar{s}s$ state to be the $f_1(1420)$, and $\Gamma(a_1) = 400\,\text{MeV}$, we find $y_{PV^*,8_S} = 1.8\%$. If instead we take the $f_1(1510)$, we find $y_{PV^*,8_S} = 1.7\%$. With $\Gamma(a_1) = 250\,\text{MeV}$, these numbers become 2.5% and 2.4%, respectively. Although it is clear that percent level contributions to y are possible from $SU(3)$ violation in this channel, the data are still too poor to draw firm conclusions.

On the basis of this analysis, in particular as applied to the $4P$ final state, we would conclude that y on the order of a percent would be completely natural. Anything an order of magnitude smaller would require significant cancellations which would only be expected if they were enforced by the OPE, that is, if the charm quark were heavy enough that the "inclusive" approach were applicable. The hypothesis underlying the present analysis is that this is not the case.

CONCLUSIONS

The motivation most often cited in searches for $D^0 - \overline{D}^0$ mixing is the possibility of observing a signal from new physics which may dominate over the Standard Model contribution. But to look for new physics in this way, one must be confident that the Standard Model prediction does not already saturate the experimental bound. Previous analyses based on short distance expansions have consistently found $x, y \sim 10^{-3}$, while naive estimates based on known $SU(3)$ breaking in charm decays allow an effect an order of magnitude larger. Since current experimental sensitivity is at the level of a few percent, the difference is quite important.

We have performed a general $SU(3)$ analysis of the contributions to y. We proved that if $SU(3)$ violation may be treated perturbatively, then $D^0 - \overline{D}^0$ mixing in the Standard Model is generated only at second order in $SU(3)$ breaking effects. Within the exclusive approach, we identified an $SU(3)$ breaking effect, $SU(3)$ violation in final state phase space, which can be calculated with minimal model dependence. We found that phase space effects alone can provide enough $SU(3)$ breaking to induce $y \sim 10^{-2}$. Large effects in y appear for decays to final states close to D threshold, where an analytic expansion in $SU(3)$ violation is no longer possible.

The implication of our results for the Standard Model prediction for x is less apparent. While analyses based on the "inclusive" approach generally yield $x \gtrsim y$, it is not clear what the "exclusive" approach predicts. The effect of $SU(3)$ breaking in phase space in x is softer than in y, so one would expect $x < y$ from our analysis. Thus if $x > y$ is found experimentally, it may still be an indication of a new physics contribution to x, even if y is also large. On the other hand, if $y > x$ then it will be hard to find signals of new physics, even if such contributions dominate ΔM. The linear sensitivity to new physics in the analysis of the time dependence of $D^0 \to K^+\pi^-$ is from $x' = x\cos\delta + y\sin\delta$ and $y' = y\cos\delta - x\sin\delta$ instead of x and y. If $y > x$, then δ would have to be known precisely for these terms to be sensitive to new physics in x.

There remain large uncertainties in the Standard Model predictions of x and y, and values near the current experimental bounds cannot be ruled out. Thus, it will be hard to find a clear indication of physics beyond the Standard Model in $D^0 - \overline{D}^0$ mixing. We believe that at this stage the only robust potential signal of new physics in $D^0 - \overline{D}^0$ mixing is CP violation, for which the Standard Model prediction is very small. Unfortunately, if y is larger or much larger than x, then the observable CP violation in $D^0 - \overline{D}^0$ mixing is necessarily small, even if new physics dominates x. Therefore, searching for new physics and CP violation in $D^0 - \overline{D}^0$ mixing should aim at precise measurements of both x and y, and at more complicated analyses involving the extraction of the strong phase in the time dependence of doubly Cabibbo suppressed decays.

ACKNOWLEDGMENTS

It is a pleasure to thank Sven Bergmann, Adam Falk, Yuval Grossman, Yossi Nir, and Alexey Petrov for very enjoyable collaborations on the subjects discussed in this talk. This work was supported in part by the Director, Office of Science, Office of High Energy and Nuclear Physics, Division of High Energy Physics, of the U.S. Department of Energy under Contract DE-AC03-76SF00098.

REFERENCES

1. A.F. Falk, Y. Grossman, Z. Ligeti and A.A. Petrov, hep-ph/0110317, to appear in *Phys. Rev.* D.
2. S. Bergmann, Y. Grossman, Z. Ligeti, Y. Nir and A.A. Petrov, *Phys. Lett.* B486, 418 (2000).
3. H. Georgi, *Phys. Lett.* B297, 353 (1992).
4. T. Ohl, G. Ricciardi and E. H. Simmons, *Nucl. Phys.* B403, 605 (1993).
5. A. Datta and M. Khambakhar, *Zeit. Phys.* C27, 515 (1985).
6. J. Donoghue, E. Golowich, B. Holstein and J. Trampetic, *Phys. Rev.* D33, 179 (1986).
7. E.M. Aitala *et al.*, E791 Collaboration, *Phys. Rev. Lett.* 83, 32 (1999).
8. J.M. Link *et al.*, FOCUS Collaboration, *Phys. Lett.* B485, 62 (2000).
9. D. Cronin-Hennessy *et al.*, CLEO Collaboration, hep-ex/0102006.
10. K. Abe *et al.*, BELLE Collaboration, hep-ex/0111026.
11. B. Aubert *et al.*, BABAR Collaboration, hep-ex/0109008.
12. R. Godang *et al.*, CLEO Collaboration, *Phys. Rev. Lett.* 84, 5038 (2000).
13. G. Brandenburg *et al.*, CLEO Collaboration, *Phys. Rev. Lett.* 87, 071802 (2001).
14. E.M. Aitala *et al.*, E791 Collaboration, *Phys. Rev.* D57, 13 (1998).
15. H.N. Nelson, hep-ex/9908021.
16. I. Bigi and N. Uraltsev, *Nucl. Phys.* B592, 92 (2000).
17. L. Wolfenstein, *Phys. Lett.* B164, 170 (1985).
18. P. Colangelo, G. Nardulli and N. Paver, *Phys. Lett.* B242, 71 (1990).
19. T.A. Kaeding, *Phys. Lett.* B357, 151 (1995).
20. D.E. Groom *et al.*, Particle Data Group, *Eur. Phys. J.* C15, 1 (2000).
21. A.F. Falk, Y. Nir and A.A. Petrov, *JHEP* 12, 019 (1999).
22. M.J. Savage, *Phys. Lett.* B257, 414 (1991).
23. Z. Ligeti, I.W. Stewart and M.B. Wise, *Phys. Lett.* B420, 359 (1998).
24. S. Nussinov and M.V. Purohit, hep-ph/0108272.

Beyond the Standard Model in B Decays: Three Topics

Alexander L. Kagan [1]

Physics Department, University of Cincinnati, Cincinnati OH 45221, USA

Abstract. Three new results are discussed: (a) A non-vanishing amplitude for the 'wrong sign' kaon decay $B \to J/\Psi \overline{K}$ or its CP conjugate is shown to be a necessary condition for obtaining different CP asymmetries in $B \to J/\Psi K_{S,L}$. A significant effect would require a scale of new physics far below the weak scale, all but ruling out this possibility. (b) The leading isospin breaking contributions to the $B \to K^*\gamma$ decay amplitudes can be calculated in QCD factorization, providing a sensitive probe of the penguin sector of the effective weak Hamiltonian. New physics models which reverse the predicted Standard Model amplitude hierarchy could be ruled out with more precise data. (c) A slowly falling $g^*g\eta'$ form factor can be ruled out using the ARGUS η' spectrum in $\Upsilon(1S)$ decays. Thus the origin of the anomalously large $B \to \eta' X_s$ rate remains unknown, perhaps requiring the intervention of New Physics.

INTRODUCTION

Three recent developments with potential implications for new physics and B decays are discussed. Recently, the BaBaR and Belle collaborations presented new measurements of the time-dependent CP asymmetries in $B \to J/\Psi K_S$ and $B \to J/\Psi K_L$ [1, 2]. In the Standard Model the two CP asymmetries are predicted to have the same magnitude but opposite sign. Although the measured values are equal within errors the CP asymmetry in the $B \to J/\Psi K_L$ channel is somewhat larger, which raises the question of whether $|a_{\rm CP}(B \to \psi K_L)| \neq |a_{\rm CP}(B \to \psi K_S)|$ is possible. The conditions which must be fulfilled for this to happen are explained below [3]. A model-independent analyis implies that the associated scale of new physics interactons would have to lie at a prohibitively low scale of a few GeV in order to obtain a significant effect.

Experimental measurements of the exclusive $B \to K^*\gamma$ branching ratios have been reported by the CLEO, Belle and BaBar Collaborations, with the results (averaged over CP-conjugate modes):

$$10^5\,{\rm Br}(\bar{B}^0 \to \bar{K}^{*0}\gamma) = \begin{cases} 4.55^{+0.72}_{-0.68} \pm 0.34 & [4] \\ 4.96 \pm 0.67 \pm 0.45 & [5] \\ 4.37 \pm 0.40 \pm 0.26 & [6] \end{cases}$$

[1] Work supported by the Department of Energy, Grant No. DOE DE FG02-84ER-40153

CP618, *Heavy Flavor Physics: Ninth International Symposium*, edited by A. Ryd and F. C. Porter

$$10^5\,\mathrm{Br}(B^- \to \bar{K}^{*-}\gamma) \;=\; \begin{cases} 3.76^{+0.89}_{-0.83} \pm 0.28 & [4] \\ 3.89 \pm 0.93 \pm 0.41 & [5] \\ 3.92 \pm 0.62 \pm 0.21 & [6] \end{cases}$$

The average branching ratios for the two modes are $(4.53 \pm 0.36) \cdot 10^{-5}$ and $(3.87 \pm 0.47) \cdot 10^{-5}$. When corrected for the difference in the B-meson lifetimes, $\tau_{B^-}/\tau_{\bar{B}^0} = 1.068 \pm 0.016$ [7], these results imply

$$\Delta_{0-} \equiv \frac{\Gamma(\bar{B}^0 \to \bar{K}^{*0}\gamma) - \Gamma(B^- \to \bar{K}^{*-}\gamma)}{\Gamma(\bar{B}^0 \to \bar{K}^{*0}\gamma) + \Gamma(B^- \to \bar{K}^{*-}\gamma)} = 0.11 \pm 0.07. \tag{1}$$

Although there is no significant deviation of this quantity from zero, the fact that all three experiments see a tendency for a larger neutral decay rate raises the question of whether the Standard Model can account for isospin-breaking effects of order 10% in the decay amplitudes.

Recently it has been shown that in the heavy-quark limit the decay amplitudes for these processes can be calculated in a model-independent way using a QCD factorization approach [8, 9], which is similar to the scheme developed for the analysis of nonleptonic two-body decays of B mesons [10]. To leading order in Λ/m_b one finds that the amplitudes for the decays $\bar{B}^0 \to \bar{K}^{*0}\gamma$ and $B^- \to K^{*-}\gamma$ coincide. Here we report on subsequent work [11], in which the QCD factorization approach was used to estimate the leading isospin-breaking contributions for the $B \to K^*\gamma$ decay amplitudes in the Standard Model. These are due to annihilation graphs which enter at order Λ/m_b. Because of their relation to matrix elements of penguin operators, we will see that isospin-breaking effects in $B \to K^*\gamma$ decays are sensitive probes of physics beyond the Standard Model.

Finally, the CLEO collaboration and more recently, as we heard at this workshop, the BaBaR collaboration have measured very large rates for fast η' production in $B \to \eta' X_s$ decays:

$$\mathcal{BR}(B \to \eta' X_s)_{p_{\eta'} > 2\ \mathrm{GeV}} = \begin{cases} 6.2 \pm 1.6 \pm 1.3^{+0}_{-1.5} \times 10^{-4}; & \text{CLEO [12]}, \\ 6.8^{+.7}_{-1.0} \pm 1^{+0}_{-.5} \times 10^{-4}; & \text{BaBar [13]}. \end{cases} \tag{2}$$

The experimental cut on $p_{\eta'}$ is beyond the kinematic limit for most $b \to c$ decays. A majority of the events lie at large recoil mass, consistent with a three-body or higher multiplicity decay. The η' yield from $b \to c$ transitions is dominated by intermediate charmonia decays, which contribute only $\approx 1.1 \times 10^{-4}$ [14] to the branching ratio. Charmless η' production proceeding via the quark content of the η' has been estimated using factorization [15, 16], giving a contribution to the branching ratio of $\approx 1 \times 10^{-4}$ with a quasi two-body recoil spectrum that is peaked at low energies in conflict with the observed spectrum.

The surprisingly large η' yield led Atwood and Soni [14] to propose that it is associated with the gluonic content of the η' via the subprocess $b \to s(g^* \to \eta' g)$. Making the key assumption that the form factor remains constant up to $q^2 \sim m_b^2$, where q is the virtual gluon's momentum, Atwood and Soni showed that the large η' yield could easily be reproduced in the Standard Model. Moreover, they observed that the three-body decay leads to an η' recoil spectrum that is consistent with observation. Hou and Tseng [17] argued that the factor of α_s implicit in H should be evaluated at the scale of momentum transfer through the $g^* g\eta'$ vertex. However this would only introduce a mild logarithmic suppression of the form factor versus q^2 and therefore still lead to a large η' yield.

Assymptotically, perturbative QCD (pQCD) predicts that the leading form factor contributions should fall like $1/q^2$. The question is in what region of q^2 does this behaviour set in? We will see [16] that the η' spectrum measured in $\Upsilon(1S) \to \eta' X$ decays by the ARGUS Collaboration [18] rules out form factors which fall slowly in the range $q^2 \leq m_b^2$. However, a rapidly falling form factor [16] representative of pQCD predictions [21, 22] is consistent with the data. The corresponding $b \to sg\eta'$ branching ratio in the Standard Model is about a factor of 20 smaller than the measured values in Eq. (2).

CAN THE CP ASYMMETRIES IN $B \to J/\Psi K_{S,L}$ BE DIFFERENT?

The relevant quantities are [23]

$$\lambda_{S,L} \equiv \frac{q_B}{p_B} \frac{\bar{A}_{S,L}}{A_{S,L}}, \tag{3}$$

where

$$\bar{A}_{S,L} \equiv A(\bar{B} \to \psi K_{S,L}), \qquad A_{S,L} \equiv A(B \to \psi K_{S,L}). \tag{4}$$

The neutral B and K meson mass eigenstates are defined in the usual way in terms of flavor eigenstates,

$$|B_{L,H}\rangle = p_B|B\rangle \pm q_B|\bar{B}\rangle, \qquad |K_{S,L}\rangle = p_K|K\rangle \pm q_K|\bar{K}\rangle. \tag{5}$$

The time-dependent CP asymmetries are given by

$$a_{\rm CP}(B \to \psi K_{S,L}) = -2\frac{{\rm Im}\lambda_{S,L}}{1+|\lambda_{S,L}|^2} \sin\Delta m_B t + \frac{1-|\lambda_{S,L}|^2}{1+|\lambda_{S,L}|^2} \cos\Delta m_B t. \tag{6}$$

In the limit of no direct CP violation ($|\lambda_{S,L}| = 1$), as is the case in the Standard Model, the asymmetries reduce to the familiar form $-{\rm Im}\lambda_{S,L} \sin\Delta m_B t$.

We need to rewrite $\lambda_{S,L}$ in terms of the decay amplitudes into kaon flavor eigenstates,

$$\bar{A}_K \equiv A(\bar{B} \to \psi K), \qquad \bar{A}_{\bar{K}} \equiv A(\bar{B} \to \psi \bar{K}), \tag{7}$$

$$A_K \equiv A(B \to \psi K), \qquad A_{\bar{K}} \equiv A(B \to \psi \bar{K}). \tag{8}$$

Allowing for the possibility that the 'wrong-sign' kaon amplitudes $\bar{A}_K$ and $A_{\bar{K}}$ receive new physics contributions (they are negligible in the Standard Model), one obtains

$$\lambda_{S,L} = \pm\lambda \left(\frac{1 \pm a}{1 \pm b}\right), \tag{9}$$

where $\lambda \equiv q_B q_K \bar{A}_{\bar{K}} / p_B p_K A_K$, and a and b are proportional to ratios of wrong-sign to right-sign kaon amplitudes

$$a \equiv \frac{p_K \bar{A}_K}{q_K \bar{A}_{\bar{K}}}, \qquad b \equiv \frac{q_K A_{\bar{K}}}{p_K A_K}. \tag{10}$$

In the Standard Model and, more generally, in any model in which the wrong-sign amplitudes are negligibly small ($a = b = 0$) this reduces to $\lambda_{S,L} = \pm\lambda$ so that $\mathrm{Im}(\lambda_S + \lambda_L) = 0$. The $\sin \Delta m_b t$ term therefore has equal magnitude but opposite sign for the two CP asymmetries.

From the general relation $\lambda_S + \lambda_L = \lambda 2(a-b)/(1-b^2)$ we learn that a necessary and sufficient condition for $\lambda_S \neq \lambda_L$ to be satisfied is the presence of non-vanishing wrong-sign amplitudes with $a \neq b$. As an example, if each right-sign and wrong-sign amplitude is dominated by a single contribution, then $|a| \approx |b|$. If in addition $\mathrm{Re}\lambda \sim \mathrm{Im}\lambda \sim O(1)$, as in the Standard Model, and $\mathrm{Arg}[a] \sim \mathrm{Arg}[b] \sim O(1)$, we obtain $\mathrm{Im}(\lambda_S + \lambda_L) \sim |a| \sim |b|$. Thus the difference in CP asymmetries is of order the ratio of wrong-sign to right-sign kaon amplitudes.

Constraints on New Physics Scenarios

Consider the wrong-sign decay $B^0 \to J/\Psi\overline{K}$. As the final state does not contain a d quark, the decay must proceed via annihilation of the B meson. It must therefore be mediated by six-quark operators, with an effective Hamiltonian

$$\mathcal{H}_{\mathrm{eff}} \sim \frac{g}{M^5} b s c \bar{c} \bar{d} \bar{d}, \tag{11}$$

where g is a dimensionless coupling, M is the scale of new physics (the color indices and Dirac structure of the operators have been suppressed). An estimate of the amplitude in the factorization approximation yields

$$A_{\bar{K}} \sim \frac{g}{M^5} f_B f_K f_\Psi m_B m_\Psi (\varepsilon_\Psi \cdot p_K). \tag{12}$$

For purposes of comparison, we note that in the Standard Model the right-sign amplitude in the factorization approximation is given by

$$A_K = \frac{G_F}{\sqrt{2}} V_{cb} a_2 f_\Psi F_1 m_\Psi (\varepsilon_\Psi \cdot p_K), \tag{13}$$

where F_1 is the $B \to K$ form factor, and a_2 is a function of the current-current operator Wilson coefficients C_1 and C_2 arising from W exchange. The observed $B \to J/\Psi K_S$ amplitude is reproduced if $a_2 \approx .25$.

We can get an order of magnitude upper bound on the quantity g/M^2 by integrating out the charm quarks at one-loop, yielding a contribution to the Hamiltonian mediating charmless hadronic B decays of order

$$\frac{1}{16\pi^2} \frac{g}{M^2} b s \bar{d} \bar{d}. \tag{14}$$

Upper bounds on the strenghths of such operators can be obtained by considering their contributions to the rare decays $B^\pm \to \pi^\pm K_S$ or, more specifically, to the wrong-sign kaon decays $B^\pm \to \pi^\pm \bar{K}^0$ [24] in the factorization approximation, yielding $g/M^2 \lesssim 10^{-5}\ \mathrm{GeV}^{-2}$. In turn, equating our estimate for $A_{\bar{K}}$ in Eq. (12) with the factorization model expression for A_K in Eq. (13) (for $a_2 \approx .25$) implies that a scale of new physics M of a few GeV is required in order to obtain significant wrong-sign kaon amplitudes.

We know of only one scenario with such a potentially low scale for new flavor-changing interactions: supersymmetric models with a light bottom squark $\tilde{b}$ of mass 2-5.5 GeV and light gluinos of mass 12-16 GeV [25], which have been proposed to enhance the b quark production cross section at hadron colliders. Among the new operators which can arise at tree-level are several of the form $\bar{d}b\tilde{b}^*\tilde{b}$. Stringent upper bounds on their strenghths from rare B decays have been obtained in [26]. Interactions of the form given in Eq. (11) would be generated from these operators if the R-parity violating Yukawa couplings mediating $\tilde{b} \to \bar{c}\bar{d}$ and $\tilde{b} \to \bar{c}\bar{s}$ decays were also present. Unfortunately, an upper bound of order 10^{-5} on the product of these two couplings from box-graph contributions to $K - \bar{K}$ mixing implies that the wrong-sign kaon amplitudes would be negligibly small. Although this does not constitute a no-go theorem, it appears that the possibility of significantly different CP asymmetries in $B \to J/\Psi K_{S,L}$ decays is all but ruled out by the requirement of such a new interaction scale.

ISOSPIN VIOLATION IN $B \to K^*\gamma$

In the Standard Model, the effective weak Hamiltonian for $b \to s\gamma$ transitions is

$$\mathcal{H}_{\rm eff} = \frac{G_F}{\sqrt{2}} \sum_{p=u,c} \lambda_p^{(s)} \Big(C_1 Q_1^p + C_2 Q_2^p + \sum_{i=3,\dots,8} C_i Q_i \Big) , \qquad (15)$$

where $\lambda_p^{(s)} = V_{ps}^* V_{pb}$, $Q_{1,2}^p$ are the current–current operators, $Q_{3,\dots,6}$ are the local four-quark QCD penguin operators, and Q_7 and Q_8 are the electro-magnetic and chromo-magnetic dipole operators. (We adopt the conventions of [10]; in particular, $C_1 \approx 1$ is the largest coefficient.) The Wilson coefficients C_i and the matrix elements of the renormalized operators Q_i depend on the renormalization scale μ.

At leading power in Λ/m_b the $B \to K^*\gamma$ decay amplitude is given by

$$i\mathcal{A}_{\rm lead} = \frac{G_F}{\sqrt{2}} \lambda_c^{(s)} a_7^c \langle \bar{K}^*(k,\eta)\gamma(q,\varepsilon)|Q_7|\bar{B}\rangle , \qquad (16)$$

where the next-to-leading order (NLO) result for the coefficient $a_7^c = C_7 + \dots$ can be found in [8]. The ellipses denote $O(\alpha_s)$ hard spectator interaction contributions.

The leading isospin-breaking effects arise from the diagrams shown in Fig. 1. Their contributions to the decay amplitudes can be parametrized as $\mathcal{A}_q = b_q \mathcal{A}_{\rm lead}$, where q is the flavor of the spectator antiquark in the $\bar{B}$ meson. We neglect NLO terms of order $\alpha_s C_{3,\dots,6}$ while retaining terms of order $\alpha_s C_{1,8}$. This is justified, because the penguin coefficients $C_{3,\dots,6}$ are numerically very small. Also, it is a safe approximation to neglect

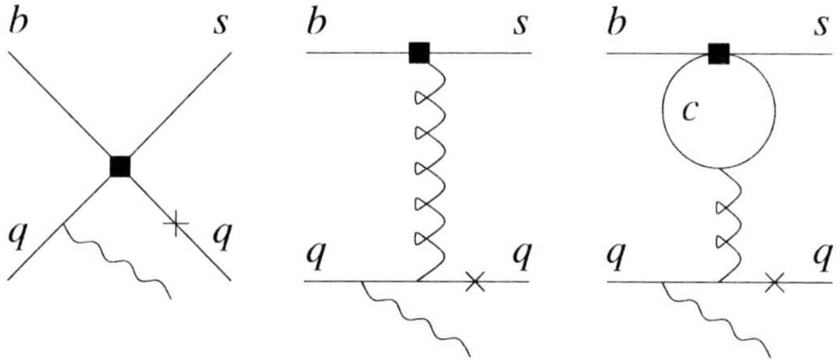

FIGURE 1. Spectator-dependent contributions from local 4-quark operators (left), the chromomagnetic dipole operator (center), and the charm penguin (right). Crosses denote alternative photon attachments.

terms of order $\alpha_s \lambda_u^{(s)}/\lambda_c^{(s)}$. It then suffices to evaluate the contributions of the 4-quark operators shown in the first diagram at tree level.

The QCD factorization approach gives an expression for the coefficients b_q in terms of convolutions of hard-scattering kernels with light-cone distribution amplitudes for the K^* and B mesons. The result can be written as

$$b_q = \frac{12\pi^2 f_B Q_q}{m_b T_1^{B\to K^*} a_7^c} \left(\frac{f_{K^*}^{\perp}}{m_b} K_1 + \frac{f_{K^*} m_{K^*}}{6\lambda_B m_B} K_2 \right), \tag{17}$$

where $T_1^{B\to K^*}$ is a form factor in the decomposition of the $B \to K^*$ matrix element of the tensor current, and m_b denotes the running b-quark mass. λ_B is a hadronic parameter which enters the first inverse moment of the relevant B meson distribution amplitude. Heavy-quark scaling laws imply that b_q scales like Λ/m_b. However, because of the large numerical factor $12\pi^2$ in the numerator the values of b_q turn out to be larger than anticipated in [8, 9].

The dimensionless quantities $K_1[C_1, C_5, C_6, C_8]$ and $K_2[C_1, C_2, C_3, C_4]$ are linear functions of the Wilson coefficients. The latter are multiplied by convolution integrals of hard scattering kernels with meson distribution amplitudes. The integrals associated with the four-quark QCD penguin operators and current-current operators exist for any reasonable choice of the distribution amplitudes, which shows that the QCD factorization approach holds at subleading power for their matrix elements, to the order we are working. However, the convolution integral associated with the chromomagnetic dipole operator suffers from a logarithmic end point singularity, indicating that at subleading power factorization breaks down for this matrix element. We regulate the singularity by introducing a cutoff. A large uncertainty is assigned to this estimate since this contribution must be dominated by soft physics.

To leading-order in small quantities the theoretical expression for the isospin-breaking parameter (see Eq. (1)) is $\Delta_{0-} = \mathrm{Re}(b_d - b_u)$. A dominant uncertainty in the prediction for Δ_{0-} comes from the tensor form factor $T_1^{B\to K^*}$, recent estimates of which range from $0.32^{+0.04}_{-0.02}$ [27] to 0.38 ± 0.06 [28]. On the other hand, a fit to the $B \to K^*\gamma$ branching fractions yields the lower value 0.27 ± 0.04 [9]. To good approximation the result for Δ_{0-} is inversely proportional to the value of the form factor. We take $T_1^{B\to K^*} = 0.3$ (at

$\mu = m_b$) as a reference value. Values for the remaining input parameters together with their uncertainties can be found in Ref. [11].

Combining all sources of uncertainty we obtain the Standard Model result

$$\Delta_{0-} = (8.0^{+2.1}_{-3.2})\% \times \frac{0.3}{T_1^{B\to K^*}}. \quad (18)$$

The three largest contributions to the error from input parameter variations are due to λ_B ($^{+1.0}_{-2.5}\%$), the divergent integral in the matrix element of Q_8 ($\pm 1.2\%$), and the decay constant f_B ($\pm 0.8\%$). The perturbative uncertainty is about $\pm 1\%$. Our result is in good agreement with the current central experimental value of Δ_{0-} including its sign, which is predicted unambiguously. By far the most important source of isospin-breaking is due to the four-quark penguin operator Q_6, whose contribution to Δ_{0-} is about 9% (at $\mu \approx m_b$). The other terms are much smaller. In particular, the contribution of the chromo-magnetic dipole operator, for which factorization does not hold, is less than 1% in magnitude and therefore numerically insignificant. Hence, the important isospin-breaking contributions can be reliably calculated using QCD factorization. It follows from our result that these effects *mainly test the magnitude and sign of the ratio* $\mathrm{Re}(C_6/C_7)$ *of penguin coefficients.*

Because of their relation to matrix elements of penguin operators, isospin-breaking effects in $B \to K^*\gamma$ decays are sensitive probes of physics beyond the Standard Model. In particular, scenarios in which the sign of Δ_{0-} is flipped could be ruled out in the near future with more precise data. For simplicity lets restrict ourselves to new physics models which do not enlarge the Standard Model operator basis, so that this would amount to flipping the sign of $\mathrm{Re}(C_6/C_7)$. As a specific example, we consider the minimal supersymmetric standard model (MSSM) with minimal flavor violation, and with $tan\beta$ enhanced contributions to $B \to X_s\gamma$ taken into account beyond leading-order [29]. In this scenario new contributions to $Q_{3,\ldots,6}$, and Q_8 are too small to have a significant effect. For low $\tan\beta$, $\mathrm{Re}C_7(m_b)$ is negative as in the Standard Model. However, for $\tan\beta \gtrsim 27$ $\mathrm{Re}C_7(m_b)$ can take on both positive or negative values with positive values becoming more probable as $tan\beta$ increases [30]. Therefore, isospin breaking could rule out significant regions of MSSM parameter space at large $tan\beta$. The sign of $\mathrm{Re}C_7$ can also be flipped in supersymmetric models with non-minimal flavor violation (independently of $tan\beta$) via gluino-down squark loop graphs, without affecting the sign of C_6. A more detailed treatment of new physics effects, including the possibility of an enlarged operator basis will be presented elsewhere.

CONSTRAINTS ON THE $G^*G\eta'$ FORM FACTOR FROM $\Upsilon(1S)$ DECAYS

The effective $\eta' g^* g$ coupling can be written as

$$H(q^2)\varepsilon_{\alpha\beta\mu\nu}q^\alpha k^\beta \varepsilon_1^\mu \varepsilon_2^\nu, \quad (19)$$

where $q = p_b - p_s$ is the virtual gluon's momentum, k is the 'on-shell' gluon's momentum, and $H(q^2)$ is the $g^*g\eta'$ transition form factor. The QCD axial anomaly determines

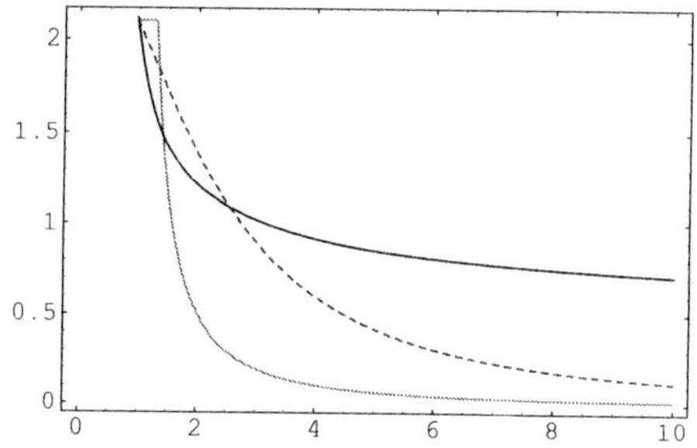

FIGURE 2. $H(q^2)$ vs. q^2 [GeV^2] for three form factors, as described in the text: (a) the slowly falling form factor (blue), (b) a rapidly falling form factor representative of perturbative QCD calculations (green), (c) an intermediate example (red-dashed)

the form factor at small momentum transfer, i.e., in the $q^2 \to 0$ limit. An estimate of this limit from the decay rate for $J/\psi \to \eta'\gamma$ gives [14] $H(0) \approx 1.8$ GeV^{-1}.

As we have discussed, the crucial issue that needs to be addressed in determining the $b \to sg\eta'$ decay rate is the dependence of H on q^2. A simple model for the $gg\eta'$ vertex [16] in which a pseudoscalar current is coupled perturbatively to two gluons through quark loops yields a form factor which can be parametrized as

$$H(q^2) = -\frac{H_0 m_{\eta'}^2}{q^2 - m_{\eta'}^2}. \tag{20}$$

The dependence of H_0 on q^2 is subleading, but it insures the absence of a pole at $q^2 = m_{\eta'}^2$. To first approximation it can be modeled by a constant H_0 to be identified with the low energy coupling extracted from $J/\psi \to \eta'\gamma$, e.g., $H_0 \approx 1.8$ GeV^{-1}. The above parametrization agrees well with recent pQCD calculations of the form factor [21, 22] (in which hard amplitudes involving quark and gluon exchanges are convoluted with the η' quark and gluon wave functions), particularly if $H_0 \approx 1.7$ GeV^{-1}.

Below we will consider three representative choices for $H(q^2)$: (a) The slowly falling form factor of Ref. [17], $H(q^2) = \sqrt{N_f}\alpha_s(q^2)\cos\theta/(\pi f_{\eta'}) \approx 2.1\alpha_s(q^2)/\alpha_s(m_{\eta'}^2)$ GeV^{-1} (θ is the pseudoscalar mixing angle). It gives BR$(b \to sg\eta') \approx 6.8 \times 10^{-4}$ for $p_{\eta'} > 2$ GeV. (b) The rapidly falling form factor of Eq. (20), with $H_0 \approx 1.7$ GeV^{-1}. At low q^2 it is matched onto the value of the previous form factor at $m_{\eta'}^2$. The result is BR$(b \to sg\eta') \approx 3 \times 10^{-5}$ for $p_{\eta'} > 2$ GeV. (c) An intermediate purely phenomenological form factor $H(q^2) \propto 1/(q^2 + M^2)$ with $M = 2.2$ GeV, which gives BR$(b \to sg\eta') \approx 4.4 \times 10^{-4}$ for $p_{\eta'} > 2$ GeV. The three form factors are plotted in Fig. 2.

The $g^*g\eta'$ form factor induces the decay ${}^3S_1^{[1]} \to gg(g^* \to \eta' g)$ of the dominant color-singlet $\Upsilon(1S)$ Fock state. Moreover, the region of q^2 relevance for fast η' production, e.g., $E_{\eta'} > .7m_\Upsilon/2$, has a large overlap with the region of q^2 relevant for fast η' production in $b \to sg\eta'$ decays. Therefore, the η' spectrum in $\Upsilon(1S)$ decays could constrain the $g^*g\eta'$ form factor, and at the same time tell us if the subprocess $b \to sg\eta'$ can account for the η' yield in B decays [20].

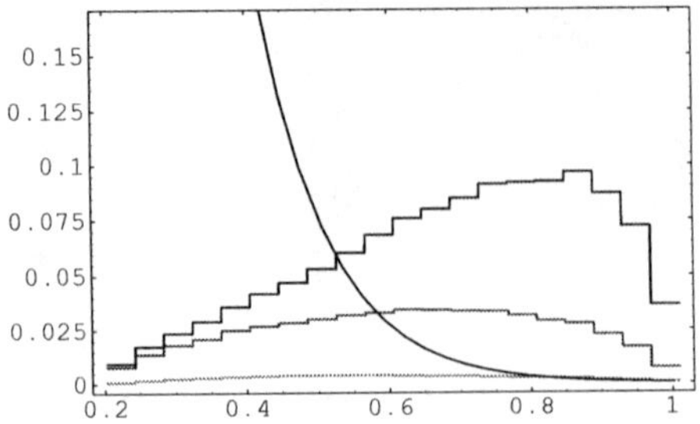

FIGURE 3. $d\text{BR}(\Upsilon(1S) \to \eta' X)/dz$ vs. z, for the three $g^* g \eta'$ form factors (a) Slowly falling (blue), (b) pQCD (green) (c) intermediate (red), with the model-dependent ARGUS fit (black), as described in the text

The ARGUS collaboration observed inclusive η' production in $\Upsilon(1S)$ decays [18]. Unfortunately, the statistics were insufficient to subtract the continuum contribution from the η' spectrum, so that modeling of the two components was required. Their fit to the $\Upsilon(1S)$ spectrum gives $d\text{BR}/dz(\Upsilon(1S) \to \eta' X) \approx 16.8\beta e^{-10.6z}$, where $z \equiv 2E_{\eta'}/m_\Upsilon$. However, since the total number of $\Upsilon(1S)$ produced is known, it is possible to obtain a model-independent upper bound on the $\Upsilon(1S) \to \eta' X$ branching ratio for each bin in z by assuming that the η' yield in each bin is entirely due to $\Upsilon(1S)$ decays.[2] Of particular interest is the bound for the highest energy bin ($.7 < z < 1.0$) for which we obtain $\text{BR}(\Upsilon(1S) \to \eta' X)_{z>.7} < 6.4 \times 10^{-4}$. The model-dependent fit gives a slightly larger result, 8.8×10^{-4}.

We have calculated the η' spectrum, $d\Gamma(^3S_1^{[1]} \to ggg\eta')/dz$ for each of the three $g^* g\eta'$ form factors considered above at leading-order in QCD using a four-body phase space Monte Carlo [20]. Relativistic corrections have not been taken into account. The spectra have been normalized with respect to the leading-order three gluon decay width, $\Gamma(^3S_1^{[1]} \to ggg)$ to obtain estimates of the differential branching ratio $d\text{BR}(\Upsilon(1S) \to \eta' X)/dz$. The results are compared to the ARGUS fit in Fig. 3. Only the rapidly falling perturbative QCD form factor is consistent with the ARGUS data at large energies, e.g., $z > .6$. Comparison of the theoretical branching ratio estimates for the three form factors with the model-independent ARGUS upper bound (for $z > .7$) is even more conclusive. We obtain

$$\frac{\Gamma(\Upsilon(1S) \to \eta' X)^{\text{theory}}_{z>.7}}{\Gamma(\Upsilon(1S) \to \eta' X)^{\text{exp}}_{z>.7}} \gtrsim \begin{cases} 41; & \text{(a) slowly falling,} \\ 1; & \text{(b) pQCD,} \\ 13; & \text{(c) intermediate.} \end{cases} \tag{21}$$

Note that the bulk of the η' yield is expected to originate from long-distance fragmentation of the three gluon configuration and should therefore be quite soft, in agreement with the ARGUS data.

---

[2] I am grateful to Axel Lindner and Dietrich Wegener for supplying the raw η' yields and efficiencies from Ref. [19].

We have checked that significant contributions from decays of higher Fock states of the $\Upsilon(1S)$ induced by the $g^*g\eta'$ coupling, e.g., ${}^1S_0^{[8]}, {}^3P_J^{[8]} \to gg\eta'$, would only harden the η' spectra. It is also extremely unlikely that relativistic corrections, next-to-leading order QCD corrections, and other theoretical refinements would be sufficient to eliminate the large excesses given in Eq. (21).

CONCLUSION

We have seen that a necessary condition for obtaining significantly different time-dependent CP asymmetries in $B \to J/\Psi K_{S,L}$ is a non-vanishing amplitude for the 'wrong sign' kaon decay $B \to J/\Psi\overline{K}$ or its CP conjugate. We note that it may be possible to test directly for the presence of these amplitudes by searching for wrong-sign $B^0 \to J/\Psi(\bar{K}^* \to K^+\pi^-)$ decays. A model-independent analysis shows that the scale of new physics needs to be prohibitively low, of order a few GeV, making this possibility extremely unlikely. The only potential example we have found, in the framework of supersymmetric models with an ultra-light bottom squark and light gluinos, would lead to gross violation of $K - \bar{K}$ mixing constraints.

We have seen that isospin breaking effects in $B \to K^*\gamma$ decays can be calculated model-indepndently in the QCD factorization framework. For the Standard Model, the decay rate for $\bar{B}^0 \to \bar{K}^{*0}\gamma$ is predicted to be about 10–20% larger than that for $B^- \to K^{*-}\gamma$, in agreement with the measured central values. The direction of this amplitude hierarchy is particularly sensitive to the sign of the Wilson coefficient ratio $\mathrm{Re}(C_6/C_7)$. As an application, more precise data which will be available in the not too-distant future could rule out models in which the sign of $\mathrm{Re}C_7$ is flipped relative to the Standard Model.

Finally, we have seen that slowly falling $g^*g\eta'$ form factors which can explain the large $B \to \eta' X_s$ rate are ruled out by the ARGUS $\Upsilon(1S) \to \eta' X$ data. However, the rapidly falling form factor predicted by perturbative QCD calculations is compatible. The corresponding fast η' yield in the Standard Model due to the subprocess $b \to sgg\eta'$ is a factor of 20 smaller than the measured central values. An enhanced $b \to sg$ chromomagnetic dipole operator, motivated by the low semileptonic branching ratio and charm multiplicity in B decays, would improve agreement with experiment [16].

ACKNOWLEDGMENTS

It is a pleasure to thank my collaborators Yixiong Chen, Yuval Grossman, Matthias Neubert, and Zoltan Ligeti. I am grateful to the organizers for arranging a very stimulating symposium, under very trying circumstances. I would also like to thank Alexey Petrov, Soeren Prell, and Zack Sullivan for useful discussions.

REFERENCES

1. K. Abe *et al.* [Belle Collaboration], *Phys. Rev. Lett.* **87**, 091802 (2001).
2. B. Aubert *et al.* [BaBar Collaboration], *Phys. Rev. Lett.* **87**, 091801 (2001); B. Aubert *et al.* [BaBar Collaboration], BaBar-PUB-01/03, SLAC-PUB-9060 [arXiv:hep-ex/0201020].
3. Y. Grossman, A.L. Kagan, Z. Ligeti, in preparation.
4. T. E. Coan *et al.* [CLEO Collaboration], *Phys. Rev. Lett.* **84**, 5283 (2000) [ArXiv:hep-ex/9912057].
5. Y. Ushiroda [Belle Collaboration], in the *Proceedings of the 4th Int. Conference on B Physics and CP Violation*, February 2001, Ise-Shima, Japan, KEK Preprint 2001-13, KUNS-1717 [ArXiv:hep-ex/0104045].
6. J. Nash [BaBar Collaboration], in the *Proceedings of the* 20th *Int. Symposium on Lepton and Photon Interactions*, Rome, Italy, July 2001. [ArXiv:hep-ex/0201002].
7. K. Osterberg, in the *Proceedings of the Int. Europhysiscs Conference on High Energy Physics, Budapest*, Hungary, July 2001.
8. S. W. Bosch and G. Buchalla, *Nucl.Phys.* **B621**, 459 (2002) [ArXiv:hep-ph/0106081]; Also see S. W. Bosch, in these proceedings.
9. M. Beneke, Th. Feldmann, D. Seidel, *Nucl.Phys.* **B612**, 25 (2001) [ArXiv:hep-ph/0106067].
10. M. Beneke, G. Buchalla, M. Neubert, C. T. Sachrajda, *Phys. Rev. Lett.* **83**, 1914 (1999) [ArXiv:hep-ph/9905312]; *Nucl. Phys.* **B591**, 313 (2000) [ArXiv:hep-ph/0006124]; *Nucl. Phys.* **B606**, 245 (2001) [ArXiv:hep-ph/0104110].
11. A. L. Kagan and M. Neubert, CLNS-01/1756 [ArXiv:hep-ph/0110078]; Also see M. Neubert, in these proceedings.
12. T. E. Browder *et al.* [CLEO Collaboration], *Phys. Rev. Lett.* **81**, 1786 (1998), [ArXiv:hep-ex/9804018].
13. C. Dallapiccola, [BaBar Collaboration], talk given at this symposium.
14. D. Atwood and A. Soni, *Phys. Lett.* **B405**, 150 (1997).
15. A. Datta, X. -G. He, S. Pakvasa, *Phy. Lett.* **B419**, 369 (1998).
16. A. L. Kagan and A. A. Petrov, [ArXiv:hep-ph/9707354]; Also see A. L. Kagan, in the *Proceeedings of the 7th Int. Symposium on Heavy Flavor Physics*, Santa-Barbara, CA, July 1997 [ArXiv:hep-ph/9806266]
17. W. Hou and B. Tseng, *Phys. Rev. Lett.* **80**, 434 (1998).
18. H. Albrecht *et al.* [ARGUS Collaboration] *Z. Phys. C* **58**, 199 (1993).
19. A. Zimmermann, Diplom Thesis, University of Dortmund, April 1992.
20. Y. Chen and A. L. Kagan, in preparation.
21. T. Muta and M. -Z. Yang, *Phys. Rev.* **D61**, 054007 (2000), [ArXiv:hep-ph/9909484].
22. A. Ali and A. Ya. Parkhomenko, accepted in *Phys. Rev.* **D** [ArXiv:hep-ph/0012212].
23. For a review see Y. Nir, *Lectures at XXVII SLAC Summer Institute on Particle Physics*, IASSNS-HEP-99-96 [arXiv:hep-ph/9911321].
24. Y. Grossman, A. L. Kagan, M. Neubert, *JHEP* **10**, 029, 1999 [arXiv:hep-ph/9909297].
25. E. Berger *et al.*, *Phys. Rev. Lett.* **86**, 4231 (2001) [ArXiv:hep-ph/0012001]; M. Carena, S. Heinemeyer, C. E. M. Wagner, G. Weiglein, *Phys. Rev. Lett.* **86**, 4463 (2001) [ArXiv:hep-ph/0008023]; Also see E. Berger, in these proceedings and references therein.
26. T. Becher, S. Braig, A. L. Kagan, M. Neubert, in the *Proceedings of the Int. Europhysics Conference on High-Energy Physics*, Budapest, Hungary, July 2001 [ArXiv:hep-ph/0112129].
27. L. Del Debbio, J. M. Flynn, L. Lellouch, J. Nieves, *Phys. Lett.* **B416**, 392 (1998) [ArXiv:hep-lat/9708008].
28. P. Ball, V. M. Braun, Y. Koike, K. Tanaka, *Nucl. Phys.* **B529**, 323 (1998) [ArXiv:hep-ph/9802299]; P. Ball and V. M. Braun, *Phys. Rev.* **D58**, 094016 (1998) [ArXiv:hep-ph/9805422].
29. G. Degrassi, P. Gambino, G. F. Giudice, *JHEP* **12**, 009 (2000) [ArXiv:hep-ph/0009337]; M. Carena, D. Garcia, U. Nierste, C. E. Wagner, *Phys. Lett.* **B499** 141 (2001) [ArXiv:hep-ph/0010003]; D. A. Demir and K. A. Olive, *Phys. Rev.* **D65**, 034007 (2002).
30. M. Boz and N. K. Pak, [ArXiv:hep-ph/0201199].

Charm Lifetimes and Mixing

Harry W. K. Cheung

Fermilab, P.O. Box 500, Batavia, IL 60510-0500

Abstract. A review of the latest results on charm lifetimes and D-mixing is presented. The e^+e^- collider experiments are now able to measure charm lifetimes quite precisely, however comparisons with the latest results from fixed-target experiments show that possible systematic effects could be evident. The new D-mixing results from the B-factories have changed the picture that is emerging. Although the new world averaged value of y_{CP} is now consistent with zero, there is still a very interesting and favoured scenario if the strong phase difference between the Doubly-Cabibbo-suppressed and the Cabibbo-flavoured $D^0 \to K\pi$ decay is large.

MOTIVATION FOR THE STUDY OF CHARM LIFETIMES

The study of charm lifetimes is essentially a study of strong interactions [1], and in particular provide a test of the theoretically challenged part of the Standard Model, namely non-perturbative QCD. It is hoped that experimental results in charm lifetimes (possibly combined with other charm results), can give some guidance as to what is needed to theoretically describe strong interactions at all energy scales. This is important not only to improve our theoretical understanding of strong interactions, but also because the theoretical tools used to calculate lifetimes are the same or similar to those used in other areas, for example to extract V_{cs} and V_{cd} in charm decays; to calculate the b-particle lifetimes; and to extract other Standard Model parameters or decay constants in heavy flavor physics.

The other motivation is more mundane. Theoretical calculations are used to calculate decay rates whereas experimentally one measures branching fractions. One needs precise particle lifetimes to convert measured branching ratios to decay rates so one can compare to theory and to extract Standard Model parameters. Lifetimes are also important as an experimental tool since the correctness of the measured lifetime will test techniques or probe systematic effects in other areas where lifetimes and lifetime resolution is important. For example in D and B_s Δm mixing measurements or in $\Delta\Gamma$ measurements for D and B mesons.

COMPARISON OF EXPERIMENTS

Table 1 shows a comparison of the experiments for which new results in charm lifetimes and mixing have been presented recently. These include both fixed-target experiments and experiments at e^+e^- colliders. This is significant as the two types of experiments are quite different and will thus have different systematics. Therefore a comparison of

CP618, *Heavy Flavor Physics: Ninth International Symposium,* edited by A. Ryd and F. C. Porter

TABLE 1. Comparison of experiments with recent results on charm lifetimes and mixing

	Fixed Target			e^+e^- Collider		
Experiment	E791	SELEX	FOCUS	CLEO	BaBar	BELLE
Beam	hadronic		photon	off-resonance e^+e^-		
Charm	$\sim 10^5$	$\sim 10^4$	$\sim 10^6$		$> 10^6$	
σ_t (fs)	~ 40	~ 20	~ 40	~ 140	~ 160	
Method	Uses vertex detachment cut Uses 3-D decay length			Needs average IP position Uses $\sim$2-D decay length		

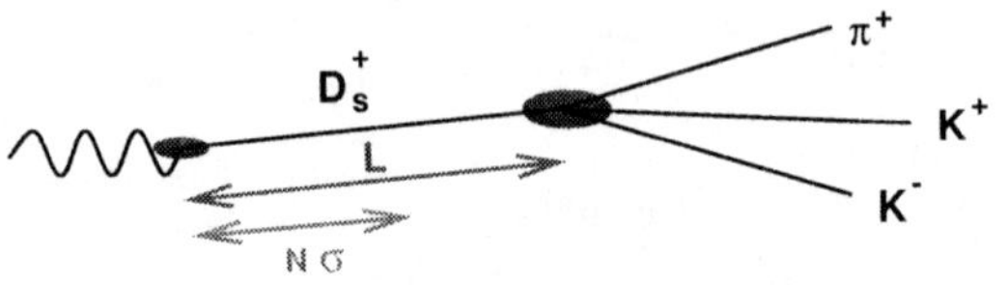

FIGURE 1. Illustration of decay length L and vertex separation

the results between the two types of experiments will be an important check of any systematic effects.

Typically fixed-target experiments have excellent vertex and proper time resolutions in 3-dimensions but large (non-charm) backgrounds. These backgrounds are eliminated by selecting decay vertices that are well separated from the production vertex, *e.g.* $L > N\sigma_L$ (see Fig.1), and optionally also outside of target material. This means that short-lived decays are preferentially eliminated and the proper time distribution would look like that given in Fig.2(a) requiring large non-uniform acceptance corrections as illustrated in Fig.2(b). This problem is avoided by using the reduced proper time, $t' = (L - N\sigma_L)/\beta\gamma c$, which starts the clock at the minimum allowed proper time. The lifetime follows the same exponential wherever one chooses to start, so the reduced proper time distribution will follow an exponential with the true lifetime. This is illustrated in Figures 1(c) and (d). The acceptance correction obtained using Monte Carlo simulations can be checked with data by using K_s^0 decays reconstructed with the vertex detector as these have a well measured lifetime. Absorption corrections are typically small and can similarly be checked in data.

In fixed-target experiments there is usually a compromise between systematics due to backgrounds and systematics due to the acceptance correction. The latter can become larger if one uses other types of vertexing cuts to eliminate more background at the expense of introducing a (larger) lifetime dependence in the acceptance.

In contrast, the e^+e^- experiments operating near B threshold have much less background but have poorer vertex resolutions and hence poorer proper time resolutions. The average interaction point is normally used for constraining the location of the production point but its position is usually only known well in 2-dimensions or even only 1-dimension. The proper time resolution can be large compared to the charm particle lifetime under study. Due to proper time smearing from the poor resolution one has to take into account this smearing on an event-by-event basis. This necessarily requires

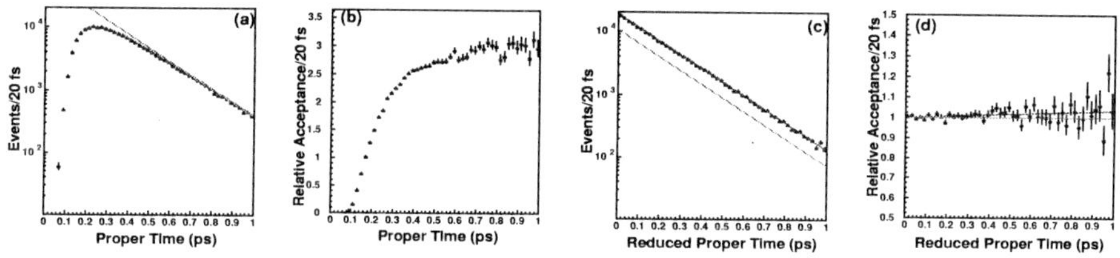

FIGURE 2. (a) Proper time distribution for a MC sample when a vertex separation requirement is made to reduce backgrounds; (b) resulting acceptance correction function; (c) reduced proper time distribution showing an exponential where the offset line shows the slope of the true generated lifetime; (d) acceptance as a function of reduced proper time.

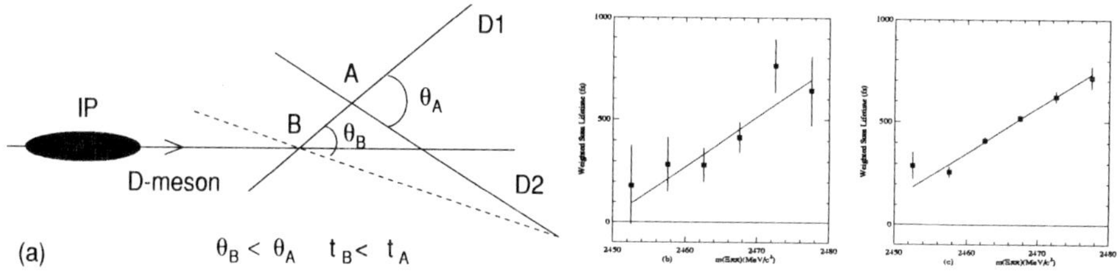

FIGURE 3. (a) Illustration of altering a track slope so that track *D*2 is at intersection point *B* instead of the unbiased initial reconstruction point *A*; (b) Correlation between mass and lifetime as seen by CLEO in data; and (c) in MC.

a more complicated event-by-event likelihood analysis where one has to parameterize the time and mass resolutions as well as background lifetime distributions. These resolutions can be known well as they can be obtained from data but they are not usually parameterizable by a simple function. The resolution can sometimes be improved by using additional constraints like forcing the reconstructed decay secondaries to come from a single point or using the average IP position. However this can also lead to fit biases and subsequent corrections. An example of this is illustrated in Fig.3(a) where constraining a decay track to come from point *B* instead of *A* will decrease the openning angle and hence also the reconstructed mass, it would also decrease the lifetime. This produces a correlation between the reconstructed mass and lifetime as seen in CLEO for $\Xi_c^+ \to \Xi^- \pi^+ \pi^+$ decays in both data, Fig.3(b), and MC Fig.3(c) [2]. Systematic concerns are therefore usually related to the fit method, resolutions and fit biases.

RESULTS ON LIFETIMES

Both BaBar and BELLE would like to demonstrate that they have excellent understanding of their vertexing and lifetime resolutions by measuring the lifetimes of the charm particles. In fact they already have enough data to get charm lifetimes with a precision comparable to the current world averages. BELLE has led the way with preliminary lifetime measurements for the D^0, D^+ and D_s^+ mesons [3]. A summary of the most recent

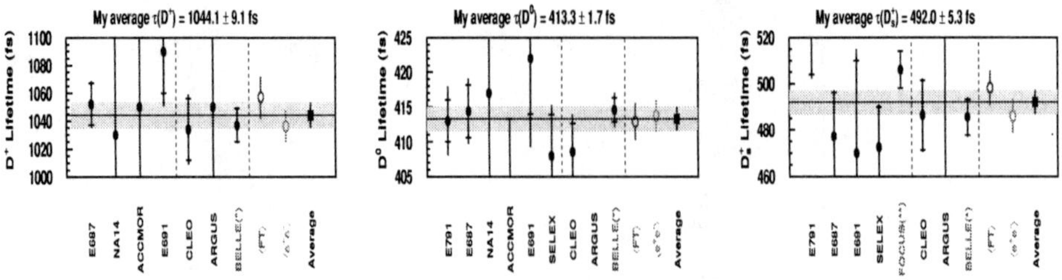

FIGURE 4. Summaries of recent D-meson lifetime measurements and their averages.

TABLE 2. World averaged lifetimes including also preliminary results.

Quantity	World Average	Quantity	World Average
$\tau(D^+)$ (fs)	1044.1 ± 9.1	$\tau(\Lambda_c^+)$ (fs)	200.2 ± 3.2
$\tau(D^0)$ (fs)	413.3 ± 1.7	$\tau(\Xi_c^+)$ (fs)	433 ± 19
$\tau(D_s^+)$ (fs)	492.0 ± 5.3	$\tau(\Xi_c^0)$ (fs)	106.0^{+9}_{-8}

charm lifetime results are given in Fig.4. Together with the published results [4, 5, 6] I have included the preliminary results shown in the figure in my new world averages given in Table 2. For the D_s^+ FOCUS preliminary result [7] which does not yet include a systematic uncertainty I have taken the total uncertainty to be $\sqrt{2}$ times the statistical error. The fixed-target and e^+e^- averages are also separately shown and agree fairly well suggesting no additional unaccounted-for systematics. The e^+e^- averages are currently dominated by the preliminary BELLE measurements.

The ratio $\tau(D_s^+)/\tau(D^0)$ continues to be of interest in determining the importance of the suppression of W-exhange and W-annihilation contributions in D-meson decays. There are now three direct measurements of this ratio giving an average of 1.171 ± 0.018 which agrees well with just taking the ratio of the two world average lifetimes: 1.190 ± 0.014. The ratio is much larger than 1.07 which is the maximum expected size for no W-exchange/W-annihilation contributions [8]. An accurately measured value for this ratio can be used to determine phenomenlogically the relative size of W-exchange/W-annihilation contributions [9].

There are new preliminary FOCUS lifetime results for the Λ_c^+, Ξ_c^+ and Ξ_c^0 baryons [10, 11] and a CLEO preliminary lifetime result for Ξ_c^+ [2]. These are shown together with published results in Fig.5. As previously, for the preliminary FOCUS lifetime result for Ξ_c^0 which does not include a systematic uncertainty I have taken the total uncertainty to be $\sqrt{2}$ times the statistical error in determining the world average.

There are two items of note in the new results. The first is that the CLEO published lifetime for Λ_c^+ appears to disagree with the fixed-target average value when the new FOCUS preliminary number is included. This could point to a systematic problem starting to appear which might be related to the short-lived nature of the Λ_c^+ decay. An example of a possible effect is the mass-lifetime correlation seen in CLEO which introduces very large MC correlations to the lifetime. It is known that the size and type of the resonance

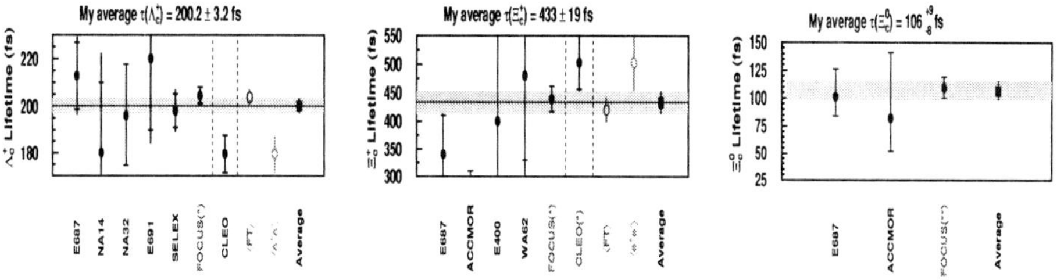

FIGURE 5. Summaries of recent charm baryon lifetime measurements and their averages.

sub-structure of a decay like $\Lambda_c^+ \to pK^-\pi^+$ can be important to the acceptance corrections, and may also be important in the mass-lifetime correlations since the resonance sub-structure alters the angular distributions of the daughter tracks. Uncertainties related to the resonance sub-structure can thus introduce additional systematic studies. Any systematic problems between the fixed-target and e^+e^- results should be kept in mind as the BaBar and BELLE lifetime and mixing results become more precise.

The other interesting feature of the new results is that the ratio $\tau(\Xi_c^+)/\tau(\Lambda_c^+)$ is now 2.16 ± 0.11 compared to the PDG2000 value of $1.60^{+0.31}_{-0.22}$. Many of the calculations favour a value of 1.2–1.6 though with large uncertainties [12]. It would be interesting to see if one can feed back into the calculations this new information and others like the $\tau(D_s^+)/\tau(D^0)$ ratio to get a better understanding, and to see if it affects the prediction for other related quantities like the ratio $\tau(\Lambda_b^0)/\tau(B^0)$. Both the mentioned results suggest that the W-exchange contribution is much more important than have so far been assumed in the calculations.

D-MIXING REVIEW

The parameters we use to describe D-mixing can best be defined by the relevant equations relating the states with definite mass and lifetimes: $|D_H(t)\rangle = e^{-im_H t}e^{-\Gamma_H t/2}|D_H\rangle$, and $|D_L(t)\rangle = e^{-im_L t}e^{-\Gamma_L t/2}|D_L\rangle$ to the observed $|D^0\rangle$ and $|\overline{D^0}\rangle$ states: $|D_H\rangle = p|D^0\rangle + q|\overline{D^0}\rangle$ and $|D_L\rangle = p|D^0\rangle - q|\overline{D^0}\rangle$. So for example the doubly-Cabibbo-suppressed (DCS) decay rate $\Gamma(D^0 \to K^+\pi^-) = |\langle K^+\pi^-|T|D^0(t)\rangle|^2$. We will assume CP conservation in charm decays and use the following approximations for charm: $|q/p| = 1$ and $|x|, |y|, R_{DCS} << 1$. Where $x = \Delta m/\Gamma$, $\Delta m = m_H - m_L$, $\Gamma = (\Gamma_H + \Gamma_L)/2$, $y = \Delta\Gamma/2\Gamma = (\Gamma_{CPeven} - \Gamma_{CPodd})/(\Gamma_{CPeven} + \Gamma_{CPodd}) = y_{CP}$, and $R_{DCS} = |\langle K^+\pi^-|T|D^0\rangle/\langle K^+\pi^-|T|\overline{D^0}\rangle|^2$. The ratio of the "wrong-sign" $D^0 \to K^+\pi^-$ to "right-sign" $D^0 \to K^-\pi^+$ decays is given by

$$R_{WS}(t) = \left[R_{DCS} + (y\cos\delta - x\sin\delta)t\sqrt{R_{DCS}} + \frac{(x^2+y^2)}{4}t^2\right]e^{-t}$$

$$R_{WS}(t) = \left[R_{DCS} + y't\sqrt{R_{DCS}} + \frac{(x'^2+y'^2)}{4}t^2\right]e^{-t}$$

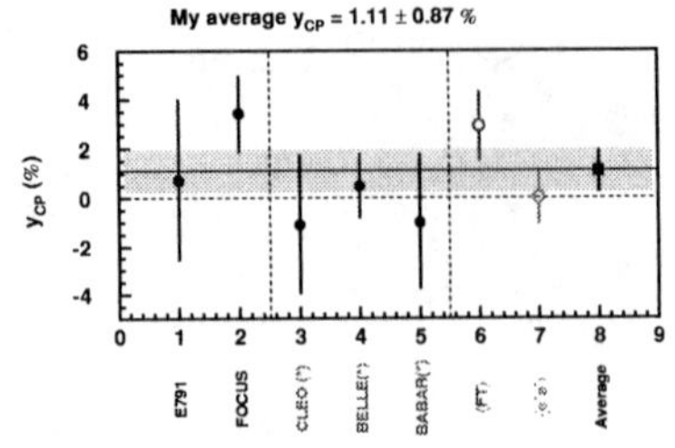

FIGURE 6. Summary of recent y_{CP} measurements.

where $y' \equiv y\cos\delta - x\sin\delta$ and $x' \equiv x\cos\delta + y\sin\delta$ and δ is the strong phase difference between the Cabibbo-favoured and the DCS decay.

Information on the charm mixing parameters can be obtained in several ways:

1. Measure the lifetime difference between CP-even, CP-odd and flavour specific states to give y_{CP}.
2. Measure wrong-sign semileptonic decays which do not require good lifetime resolution but only give information on (x^2+y^2) and cannot separate x and y.
3. Measure wrong-sign hadronic decays like $D^0 \to K^+\pi^-$ which require a lifetime study and high S/B and can give information on both x^2 and y separately. However there is an additional complication of an unknown strong phase difference, e.g. $\delta_{K\pi}$ between $D^0 \to K^+\pi^-$ and $D^0 \to K^-\pi^+$ contributions.

The published results are summarized in Fig.7(a) which include the y_{CP} measurement from FOCUS [15], the limits on (x',y') from CLEO allowing and not allowing for CP violation [14], and the E791 limits from semileptonic decays [13].

The preliminary measurements from CLEO [16], BELLE [3] and BaBar [17] for y_{CP} are given in Fig.6. Including these produces a world average of $1.11 \pm 0.87\%$ which is quite consistent with zero[1]. However the situation is still interesting due to in part to a preliminary result from FOCUS for the allowed region in x' and y' [19]. The 95% confidence level allowed regions for the FOCUS and CLEO measurements are given in Fig.7(b) in the (x,y) space assuming the strong phase difference $\delta_{K\pi} = 0$. The slightly larger (lighter) region for CLEO is when CP-conservation is not assumed. Also shown are the allowed region in y from the combined lifetime difference measurements assuming no CP-violation, and the circular allowed region from wrong-sign semileptonic decay limits from E791 [13]. The smaller circular line gives the expected size of the allowed region from FOCUS wrong-sign semileptonic decays [20].

Even though I do not have the likelihood contour of the CLEO allowed region in order to combine the FOCUS and CLEO results, it can be seen that the combined allowed region is beginning to exclude zero, especially if no CP-violation is assumed. The other

---

1 At "press time" the y_{CP} value of $+0.5 \pm 1.0^{+0.8}_{-0.9}$ shown by BELLE at the conference was superceeded by a new value of $-0.5 \pm 1.0^{+0.7}_{-0.8}$ contained in their new preprint [18]. The same data sample was used but the analysis contained updated MC corrections. This moves the world average value down to 0.63 ± 0.85.

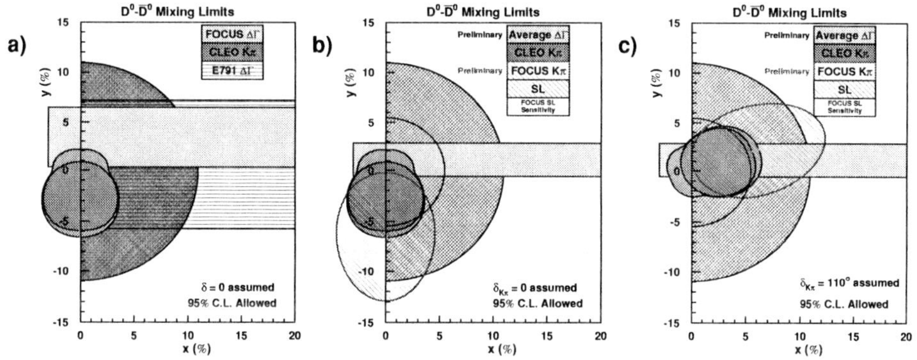

FIGURE 7. Summary of D-mixing results showing 95% confidence level allowed regions in (x,y) (a) for published results assuming $\delta_{K\pi} = 0$; and including preliminary results assuming (b) $\delta_{K\pi} = 0$ and (c) $\delta_{K\pi} = 110°$.

point is that the combined FOCUS and CLEO $K\pi$ result for the allowed region does not agree well with the world average allowed y_{CP} range when one assumes $\delta_{K\pi} = 0$. This could indicate one of three things: it is just a statistical fluctuation; the systematic uncertainties are underestimated; or the most interesting is that $\delta_{K\pi}$ is large and non-zero. Knowing the value of $\delta_{K\pi}$ is crucially important. For example in the absence of a theoretically favoured value, the experimentally preferred value is about 110° as shown in Fig.7(c). This would be a really interesting situation since the favoured scenario is with y near zero and a large value of x of $\approx 3\%$[2]. This is the most likely expected signature of new physics beyond the Standard Model since these can produce sizable non-zero values in Δm but not usually in $\Delta\Gamma$.

CONCLUSIONS

Charm lifetime measurements continue to be an interesting way to study non-perturbative strong interaction physics and evaluate possible systematic problems for measurements that require good lifetime resolutions. The larger than expected values of $\tau(D_s^+)/\tau(D^0)$ and $\tau(\Xi_c^+)/\tau(\Lambda_c^+)$ indicates that W-exchange is much more important that normally thought which could have implications on other theoretical predictions, like $\tau(\Lambda_b^+)/\tau(B^0)$.

The D-mixing situation is still very interesting as the data now favour a y value near zero and a large value of x of $\approx 3\%$. This could be a signature of new physics beyond the Standard Model. However its requires the strong phase difference $\delta_{K\pi} \approx 110°$ which is unexpected theoretically. The possibility that this is a statistical fluctuation or

[2] Note that the new lower world average value of y_{CP} does not significantly change these favoured values.

a systematic underestimation can be greatly clarified by new precise D-mixing results from BaBar and BELLE and we look forward eagerly for these results in the future.

ACKNOWLEDGMENTS

I wish to thank Monika Grothe and Jun'ichi Tanaka for providing the BaBar and BELLE results respectively, and for providing prompt replies to my questions.

REFERENCES

1. G. Bellini, I. I. Y. Bigi and P. J. Dornan, *Phys. Rept.*, **289**, 1 (1997).
2. M. Artuso *et al.*, *"Measurement of the Ξ_c^+ Lifetime"*, CLEO-CONF-01-03, Jul. (2001) hep-ex/0107040.
3. J. Tanaka, *these proceedings*; and K. Abe *et al.*, *"Precise Measurements of Charm Meson Lifetimes and the Search for $D^0 - \overline{D^0}$ Mixing at Belle"*, BELLE-CONF-0131 (2001).
4. D. E. Groom *et al.*, *Eur. Phys. Jour.*, **C15**, 1 (2000); and partial update for edition 2002.
5. A. Kushnirenko, *et al.*, *Phys. Rev. Lett.* **85**, 5243–5246 (2001).
6. M. Iori *et al.*, *"Measurement of the $D_s^\pm$ Lifetime"*, FERMILAB-PUB-01-086-E, May (2001) hep-ex/0106005.
7. H. W. K. Cheung, FOCUS collaboration, *"Preliminary D_s^+ Lifetime from FOCUS"*, presented at the APS Centennial Meeting, Altanta, Georgia, U.S.A., March 20-26 1999.
8. I. I. Bigi and N. G. Uraltsev, *Z. Physik.*, **C62**, 623 (1994).
9. Harry W. K. Cheung, *"Review of Charm Lifetimes"*, FERMILAB-CONF-99-344, Jul. (1999) hep-ex/9912021. In proceedings of the 8th International Symposium on Heavy Flavour Physics, Southampton, U.K., 25-29 July 1999, PRHEP-hf8/022.
10. E. W. Vaandering, FOCUS collaboration, *"FOCUS Spectroscopy and Lifetimes"*, presented at the 9th Intern. Conf. on Hadron Spectroscopy, Protvino, Russia, Aug. 25–Sep. 1 (2001) `http://www.hep.vanderbilt.edu/~vondo/focus-public/hadron_2001.ps`
11. J. M. Link *et al.*, *"A New Measurement of the Ξ_c^+ Lifetime"*, Submitted to Phys. Lett. B, FERMILAB-PUB-01-296-E, Oct. (2001) hep-ex/0107040.
12. B. Guberina, R. Ruckl and J. Trampetic, *Z. Phys.* **C33** 297 (1986); M. A. Shifman and M. B. Voloshin, *Sov. Phys. JETP* **64**, 698 (1986); B. Blok and M. A. Shifman, TPI-MINN-93-55-T, UMN-TH-1227-93, TECHNION-PH-93-41, Nov. (1991) hep-ph/9311331; I. I. Bigi, UND-HEP-96-BIG06, Nov. (1996) hep-ph/9612293; B. Guberina and B. Melic, *Eur. Phys. J.* **C2**, 697 (1998).
13. E. M. Aitala *et al.*, *Phys. Rev. Lett.* **77** 2384 (1996).
14. R. Godang *et al.*, *Phys. Rev. Lett.* **84** 5038 (2000).
15. J. M. Link *et al.*, *Phys. Lett.* **B485** 62 (2000).
16. D. Cronin-Hennessy *et al.*, *"Mixing and CP Violation in the Decay of Neutral D Mesons at CLEO"*, CLEO-CONF-01-1 Feb. (2001) hep-ex/0102006.
17. M. Grothe, *these proceedings*; and B. Aubert *et al.*, *"Search for a Lifetime Difference in D^0 Decays"*, BABAR-CONF-01/29, SLAC-PUB-8983, Sep. (2001) hep-ex/0109008.
18. K. Abe *et al.*, *"A Measurement of Lifetime Difference in D^0 Meson Decays"*, KEK preprint 2001-143, BELLE preprint 2001-18, Nov. (2001) hep-ex/0111026.
19. J. M. Link, FOCUS Collaboration, *"$D^0 - \overline{D^0}$ Mixing in FOCUS"*, Jun (2001). To appear in the proceedings of 36th Rencontres de Moriond on Electroweak Interactions and Unified Theories, Les Arcs, France, Mar. 10-17 (2001) hep-ex/0106093; and J. M. Link *et al.*, Phys. Rev. Lett. **86** 2955.
20. D. Kim, FOCUS collaboration, *"Recent Results on CP Lifetime Differences of Neutral D Mesons"*, presented at the Frontiers in Contemporary Physics - II, Mar. 5-10 (2001). `http://web.hep.uiuc.edu/home/doriskim/talks/fcp01/kim_doris_tr.pdf`

Open and Hidden Charm Spectroscopy and Decays: an Overview

Roberto Mussa

INFN - Sezione di Torino, V.P.Giuria 1, 10125, Torino(Italy)

Abstract. An overview of charmonium and charm hadron spectroscopy is given, with a special emphasis on spin splittings, total widths, and transitions between excited states. Preliminary results from E835 and recent results from CLEO,BES and E831 are shown. After a decade of high statistics data taking, major problems remain unsolved and new results are needed.

INTRODUCTION

In recent years, the charm sector has regained attention, for many reasons:

- The failure of QCD predictions on prompt J/ψ and ψ' production at the Tevatron energies [1] triggered a number of efforts to produce a more rigorous field theoretical approach to QCD [2].
- Standard Model predictions, by dooming the D^0 system as fruitless for CP violation studies, promote it as a sensitive probe of New Physics beyond SM [3].
- The extraction of more precise values of the CKM parameters from heavy flavors is limited by systematic errors on the hadronic matrix elements.

Charm bound state dynamics is here discussed by comparing the interactions of the charm quark with its antiquark (charmonium), with light antiquarks (D-mesons) and with a pair of light quarks (C-baryons). Despite the fact that spectral patterns differ, a common feature arises: the narrow width of the excited states. This fact has two implications: it provides a clear signature of the heavy flavor, and increases the chances to use electromagnetic transitions to probe the structure of the strongly bound systems.

CHARMONIUM

Due to its clean signature, the charmonium system is the ideal arena to challenge our understanding of QCD dynamics at the boundary between perturbative and non-perturbative regime. Traditionally, the $c\bar{c}$ spectrum was studied with potential models. In recent years, theory efforts were focused on defining limits and methods for the application of non relativistic QCD (NRQCD) to the charmonium system [4, 5, 6], and on extending chiral lagrangians to heavy mesons [7]. With the increasing power of computational techniques, Lattice QCD methods are finally coming to age [8].

CP618, *Heavy Flavor Physics: Ninth International Symposium,* edited by A. Ryd and F. C. Porter

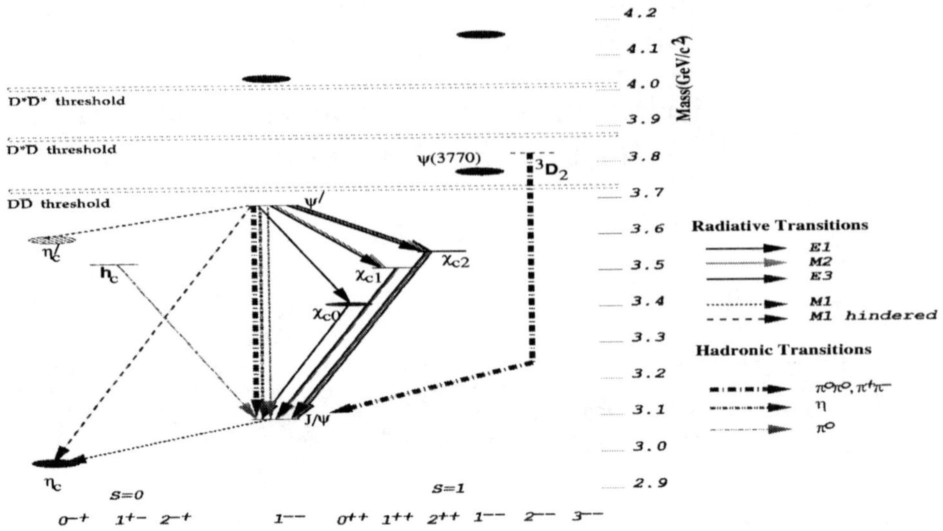

FIGURE 1. Charmonium spectrum and transitions between states.

Since charmonium discovery, the main road for the experimental study of these states was traced by e^+e^- colliders, both in formation, and in $\gamma\gamma$ fusion. In 1984, the CERN experiment R704 pioneered the study of $c\bar{c}$ narrow states in $p\bar{p}$ annihilation. The three techniques complement each other:

- e^+e^- annihilations give direct access only to vector $c\bar{c}$ states, and are sensitive to all decay channels;the use of crystal calorimeters greatly extends the sensitivity of these experiments to the $c\bar{c}$ states reachable through radiative transitions;
- Central production in $e^+e^- \to \gamma\gamma e^+e^-$ allows to access states with positive C-parity and is again sensitive to all decay channels;
- $p\bar{p}$ annihilations allow direct access to all the $c\bar{c}$ states, but the huge hadronic background limits the study to EM decay channels. The coupling to J=0 and S=0 states was forbidden by massless QCD. The helicity selection rule is indeed badly violated in the case of χ_{c0} and η_c states, studied during the last decade.

Mass spectroscopy

The $c\bar{c}$ bound states are charge conjugation and parity eigenstates. As in positronium, for every value of angular momentum L$\neq$ 0, we can identify a triplet of states with S=1, and C=(-1)$^{L+1}$ and a singlet with S=0 and C=(-1)^{L} (see Figure 1). Parity depends only on the angular momentum: $P = (-1)^{L+1}$. The number of expected narrow states below $D\bar{D}$ threshold is limited to eight. An additional pair of narrow states between $D\bar{D}$ and $D\bar{D}^*$ thresholds is also expected , as their quantum numbers ($J^{PC} = 2^{-+}, 2^{--}$) forbid the OZI allowed open charm decays. Two of the narrow states below open charm threshold, namely the $^1S_0(2s)$ (η_c') and the $^1P_1$ (h_c) were observed by only one experiment and need confirmation.

The spectrum of radial excitations is the first information that became available, soon after the J/ψ and Υ discoveries. The pattern of excited states is reasonably well described by an interaction kernel composed by a one gluon exchange vector term, dominant at short distances, and a long range scalar term, linearly increasing with the radial distance. However, in the case of charmonium the structure is complicated by the presence of a significant relativistic mixing between S wave and D wave vector states. Despite this, the splitting between ground state and first radial excitation in $c\bar{c}$ (589.1±0.1 MeV/c^2) is only 5% larger than in $b\bar{b}$ (563.0±0.4 MeV/c^2).

The tensor structure of the binding potential is investigated by studying the pattern of spin dependent splittings. The mass splittings between triplet P states (χ_c) have been known with excellent precision since a decade, but the theory predictions still fail by a large amount. The fine splitting is described by the mass formula:

$$M\left\{\begin{array}{c}\chi_{c0}\\ \chi_{c1}\\ \chi_{c2}\end{array}\right\} = M_{cog} + h_{LS}\left\{\begin{array}{c}-2\\ -1\\ 1\end{array}\right\} + h_T\left\{\begin{array}{c}-2\\ 1\\ -\frac{1}{5}\end{array}\right\}$$

Both the spin orbit term h_{LS} and the tensor term h_T should scale with the inverse of the heavy quark mass; table 1 shows a comparison between the relevant parameters of charmonium and bottomonium P-states.

TABLE 1. Splittings between χ_c and χ_b states

	$c\bar{c}(n=1)$	$b\bar{b}(n=1)$	$b\bar{b}(n=2)$
$\Delta M_{21} = M(\chi_2) - M(\chi_1)$ (in MeV)	45.6±0.2	19.9±0.8	13.0±0.8
$\Delta M_{10} = M(\chi_1) - M(\chi_0)$ (in MeV)	95.3±0.4	32.8±1.2	23.4±0.8
$\rho(\chi) = \Delta M_{21}/\Delta M_{10}$	0.470±0.003	0.61±0.04	0.56±0.04
h_T (in MeV)	20.2±0.1	6.3±0.4	4.7±0.3
h_{LS} (in MeV)	34.5±0.1	13.8±0.3	9.3±0.3

Lattice QCD[9, 10] still fails to fit the hyperfine and χ_c splittings : despite light quark loop effects are held responsible for this disagreement, recent unquenched lattice calculations [11] do not show any improvement.

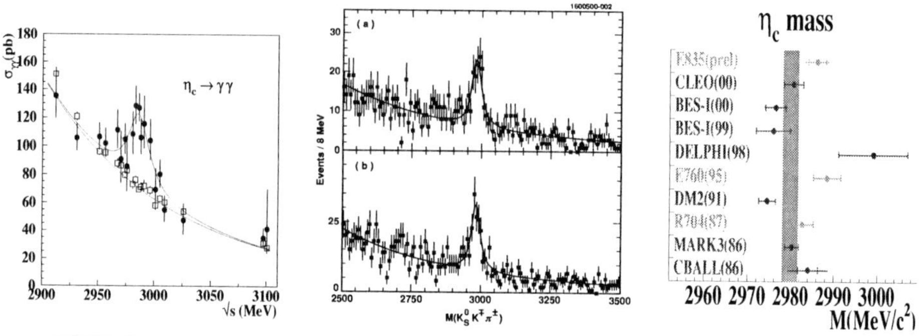

FIGURE 2. η_c mass plots by E835(left) and CLEO(center). Summary of all measures (right).

The precision on the 1S hyperfine splittings is limited by the experimental difficulty to extract the η_c signal. Recently, this state was indeed observed with the three existing techniques (see Figure 2):

a) in e^+e^-, at BES, from the radiative decay of ψ'[12] or J/ψ[13].
b) in $\bar{p}p \to \eta_c \to \gamma\gamma$, by E835; the $\gamma\gamma$ decay channel, however, is affected by a large and steep feeddown from $\pi^o\pi^o$ and $\pi^o\gamma$ continuum.
c) in $e^+e^- \to e^+e^-\gamma\gamma$, where the initial $\gamma\gamma$ flux has a steep dependence on $\sqrt{s}$, by CLEO[14] and DELPHI[17].

A few MeV systematic discrepancy between the results from different experimental techniques is not resolved at this time.

The hyperfine splitting between triplet and singlet P states yields information on the long range behaviour of the vector potential. The state h_c has been observed[18] by E760 in the reaction $p\bar{p} \to h_c \to J/\psi\pi^0$, and awaits confirmation. E835 analysis on Ldt = 97 pb^{-1} is in progress. Given a series of modifications to the machine parameters and data taking conditions, a lengthy set of cross checks on the energy measurements is crucial.

The hyperfine splitting between excited S states $\Delta M(2S) = M(\psi') - M(\eta_c')$ was measured by Crystal Ball[19]: the experiment saw a bump in the inclusive photon spectrum from $1.8{*}10^6\psi'$ decays. E835 has extensively searched [20] for the η_c' in the reaction $p\bar{p} \to (\eta_c') \to \gamma\gamma$, without confirming it. BES,CLEO-c and B-factories have a reasonable chance to find this state in the near future.

Hints for a possible evidence of the 2^{--} state above $D\bar{D}$ threshold at M=3836$\pm$13 MeV/c^2 were seen by the FNAL experiment E771 [21], in the reaction $\pi Li \to J/\psi\pi^+\pi^- + X$.

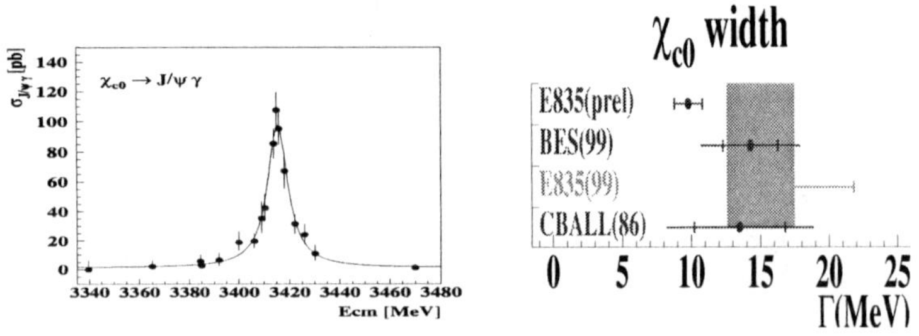

FIGURE 3. χ_{c0} resonance scan by E835 (left). χ_{c0} width measurements (right)

Radiative and Hadronic Decay Widths

The widths of the J=1 states below threshold range between 100 keV and 1 MeV. In this case electromagnetic and hadronic decays have comparable magnitude. States with J=0,2 have widths ranging between 2 and 25 MeV, due to their allowed decay to two hard gluons. The most recent measurements by CLEO[14] and E835[15] of the η_c width suggest a value close to 20 MeV, larger than the PDG value $\Gamma(\eta_c)$= $13.2^{+3.8}_{-3.2}$[22].

On the contrary, the first 10% measurement of the χ_{c0} width, $\Gamma(\chi_{c0}) = 9.8 \pm 1.0 \pm 0.1$, done by E835[16] in 2000, shifts the central PDG value 30% below the current average.

Below open charm threshold, charmonium states' decays can be grouped in seven different classes, as summarized in Table 2. Most partial widths of charmonium states are known with a precision not better than 10%, barely sufficient to distinguish among different theoretical approaches.

TABLE 2. Charmonium decays (LH stands for *light hadrons)*.

Decay mode	η_c	J/ψ	χ_{c0}	χ_{c1}	χ_{c2}	ψ'
a - $(c\bar{c}) \to gg, gg^* \to LH$	} 100%	0	} 99%	} 73%	} 87%	0
b - $(c\bar{c}) \to ggg \to LH$		} 70%				
c - $(c\bar{c}) \to \gamma gg \to \gamma + LH$	0		0	0	0	} 7%
d - $(c\bar{c}) \to \gamma^* \to LH$	0	17%	0	0	0	
e - $(c\bar{c}) \to e^+e^- + \mu^+\mu^-, \gamma\gamma$	$3 \cdot 10^{-4}$	12%	$\sim 10^{-4}$	0	$\sim 10^{-4}$	2%
f - $(c\bar{c}) \to \gamma + (c\bar{c})$	0	1%	0.6%	27%	13%	30%
g - $(c\bar{c}) \to LH + (c\bar{c})$	0	0	?	?	?	58%

Electromagnetic transitions between charmonium states probe the long range behaviour of the wavefunction. Relativistic corrections are expected to give first order effects on $\psi' \to \chi_c\gamma$ and on $\psi', J/\psi \to \eta_c\gamma$ transitions. E1 Radiative widths have been known at $\sim$ 15% level during the last decade. Equal contributions to the statistical error came from the branching ratios (measured by Crystal Ball) and from the total width (E760). The study of angular distributions allows to access the suppressed M2 amplitudes through the M2-E1 interference terms in radiative decays. A recent result from E835[23] hints at a 2σ discrepancy between theory and experiment on the χ_c radiative decay. A global refitting of all the radiative transition rates, correctly accounting for correlations between different measurements can be found in ref.[24]. Table 3 summarizes the current experimental situation. The new fit value of $\Gamma_{rad}(\chi_{c0})$ incorporates the recent E835 fit of the χ_{c0} total width.

TABLE 3. Radiative decays between charmonium states. Partial widths (with relative errors) from ref.[22] are compared with the results of a new fit from ref.[24]

Transition	E_γ,MeV	Γ_{rad}^{PDG},keV	$\frac{\delta\Gamma_{tot}}{\Gamma_{tot}} \oplus \frac{\delta BR}{BR}$	$\frac{\delta\Gamma_{rad}}{\Gamma_{rad}}$	Γ_{rad}^{newfit},keV	M2/E1
$\psi' \to \chi_{c2}\gamma$	127.5	22±3	11% ⊕ 10%	= 15%	19±3	(8±5)%
$\psi' \to \chi_{c1}\gamma$	171.3	24±4	11% ⊕ 9%	= 14%	23±4	(13±9)%
$\psi' \to \chi_{c0}\gamma$	260.8	26±4	11% ⊕ 10%	= 15%	19±4	0
$\chi_{c2} \to J/\psi\gamma$	429.6	270±32	9% ⊕ 8%	= 12%	374±62	-(9±4)%
$\chi_{c1} \to J/\psi\gamma$	389.2	240±41	16% ⊕ 6%	= 17%	282±52	(0±2)%
$\chi_{c0} \to J/\psi\gamma$	303.6	98±32	17% ⊕ 27%	= 32%	118±32	0
$\psi' \to \eta_c\gamma$	638	0.78±0.19	11%⊕ 21%	= 24%		0
$J/\psi \to \eta_c\gamma$	115	1.13±0.35	6%⊕ 31%	= 31%		0

Very few hadronic transitions between charmonia were observed so far, namely $\psi' \to J/\psi\pi\pi, J/\psi\eta, J/\psi\pi^0$, and the unconfirmed $h_c \to J/\psi\pi^0$. The analysis of a high statistics sample of $\psi' \to J/\psi\pi^+\pi^-$ decays [25] recently yielded the evidence of a 18% D-wave contribution in the dipion. Due to the isospin violation, $c\bar{c} \to c\bar{c}\pi^0$ transitions (BR=9.7±2.1*10^{-4}) are suppressed by more than one order of magnitude with respect to $c\bar{c} \to c\bar{c}\eta$(BR=2.2±0.4%). χ_c transitions to $J/\psi + \pi, \pi\pi$ are suppressed by C-parity conservation, which forbids purely hadronic transitions. E835 is sensitive to $\Gamma(\chi_{c1,2} \to J/\psi + LH)$ at $\sim 10^{-4}$ level, in $\chi_{c1,2}$ decays. No $(c\bar{c}) \to \eta_c + LH$ transitions have been detected so far.

BES has recently published papers concerning exclusive light meson decay channels for J/ψ and ψ' states: these show that baryon-antibaryon and vector-pseudoscalar decays are consistent with the rule [26]:

$$\frac{BR(\psi' \to X_h)}{BR(J/\psi \to X_h)} = \left[\frac{\alpha_s(\psi')}{\alpha_s(J/\psi)}\right]^3 \frac{BR(\psi' \to e^+e^-)}{BR(J/\psi \to e^+e^-)} = 11.6 \pm 2.2\%$$

However, since many years, some exclusive channels ($\rho\pi$ being the most dramatic one) were found to violate the rule. The systematic study of these regularities and anomalies will eventually be of great help in the understanding of charmonium dynamics.

Recent observations of hadronic exclusive decays of χ_c states with charged final products were reported by BES[12]. e^+e^- experiments can only measure the products BR($J/\psi, \psi' \to \gamma R_{c\bar{c}}$)*BR($R_{c\bar{c}} \to X_h$), therefore only the ratios (e.g. BR($R_{c\bar{c}} \to J/\psi\gamma$)/BR($R_{c\bar{c}} \to p\bar{p}$)) will not depend on BR($J/\psi, \psi' \to \gamma R_{c\bar{c}}$), which is currently the dominant source of systematic error on the hadronic branching ratios from η_c, χ_c states. On the contrary, the $p\bar{p}$ experiments only allowed us to measure the products BR($R_{c\bar{c}} \to p\bar{p}$)*BR($R_{c\bar{c}} \to J/\psi\gamma$). The measurement of BR($R_{c\bar{c}} \to p\bar{p}$) at BES has a special interest, because it finally allows to complement $p\bar{p}$ and e^+e^- data, and remove the dependence from the radiative BR's.

CHARMED HADRONS

From the experimental point of view, while charmonium can be formed, D-mesons and c-baryons can only be produced , and the measurement of the widths of narrow states strongly depend on the ability to deconvolute the detector resolution on final product energies and momenta. The extraction of a clean signal with respect to the combinatorial background relies upon three factors: low multiplicity of the final state, good particle identification and good vertex resolution. In the last decade, most results on charm hadron spectroscopy came either from photoproduction experiments (E687 and E831 at FNAL), or from e^+e^- colliders (LEP and CESR).

D-meson spectroscopy

If charmonium is the positronium of QCD , we can say that D-mesons are the analogue of the hydrogen atom. In the case of the excited D-mesons , the charge conjugation symmetry is no longer useful, and the mass asymmetry between the constituents suggests to split the total angular momentum $\vec{J} = \vec{s_q} + \vec{s_c} + \vec{L}$ in the angular momentum $\vec{j_q} = \vec{s_q} + \vec{L}$ of the light orbiting quark and the spin $\vec{s_c}$ if the heavy, quasi-static, quark.

Theoretical predictions on the dynamics of excited heavy-light mesons have been made in the framework of the Heavy Quark Effective Theory (HQET) or the low energy chiral effective theory. Both approaches do not aim to give accurate results: lattice gauge theory is the ultimate candidate for more detailed predictions.

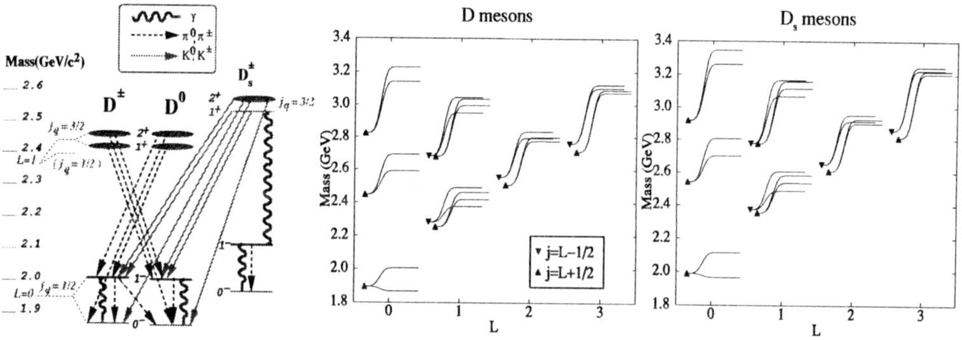

FIGURE 4. D-meson experimental spectrum and theory expectations (from ref.[27])

The hyperfine splitting $M(D^*) - M(D)$ is expected to depend on the inverse mass of the heavy quark, and should not depend on the light quark flavor. The agreement with experiment is at 3% level, as shown in Table 4.

TABLE 4. $1/M_Q$ dependence of the hyperfine splitting.

	$c\bar{u}$	$c\bar{d}$	$c\bar{s}$	$b\bar{u}, b\bar{d}$	$b\bar{s}$
$M(1^-)$-$M(0^-)$,MeV/c^2	142.12 ± 0.07	140.64 ± 0.10	143.8 ± 0.4	45.78 ± 0.35	47.0 ± 2.6

The four possible P states are grouped in two doublets because of parity conservation in the hadronic decay:

- The one with $j_q^P = 3/2^-$ $(J^P = 1^+, 2^+)$ can decay to $D\pi$ only via the D-wave , and will have narrower width (about 20 MeV). Parity conservation forbids $D(1^+) \rightarrow D(0^-)\pi$.
- The one with $j_q^P = 1/2^-$ $(J^P = 0^+, 1^+)$ decays to $D\pi$ via S-wave, therefore will be much wider (about 200 MeV). Parity conservation allows only $D(1^+) \rightarrow D(1^-)\pi$ and $D(0^+) \rightarrow D(0^-)\pi$.

A mixing between the two 1^+ states is expected. The same rules apply also on the D_s states, after replacing the pion with the kaon. The observation of broad structures is obstacled by the large combinatorial background. Recently, E831 showed that a broad 0^+ bump, overlapping the narrow $D^{**}(2^+)$ peak, can provide a better fit to the $D^+\pi^-$ mass distribution.

It is worth to notice that the flavor independence (within 2%) of the splitting between 1^- and 2^+ states is valid also for the heavy quarkonia (see Table 5), in agreement with

TABLE 5. Flavor independence of the tensor-vector splitting.

	$c\bar{u}$	$c\bar{d}$	$c\bar{s}$	$c\bar{c}$	$b\bar{b}$
M(2$^+$)-M(1$^-$), in MeV/c^2	452 ± 2	449±4	461±2	458.3±0.1	452.3±0.6

the $m_Q \to \infty$ limit [28]. These are the states having the highest level of simmetry with respect to $SU(2)_{spin}$, in the two cases L=0 and 1.

The orbital excitation of the D^* state was seen only by DELPHI [29], and was not confirmed by OPAL [30] and CLEO [31]. The state was seen in the channel $D^*\pi^+\pi^-$ at $M = 2637 \pm 2 \pm 6 MeV/c^2$, but theoretical predictions expect it about 50 MeV above. The same signal is being searched for by E831.

D-meson decays

The dynamics of hadronic transitions between D mesons, involving slow pions or kaons, is fully in the non perturbative QCD regime. The hyperfine splitting $M(D^*) - M(D)$ is very close to the π mass. Therefore all the D^* states have very little phase space for the hadronic decay, and are expected to be very narrow, below 100 keV. Recently, CLEO has published the first measurement of the D^{*+} width [32]:$\Gamma(D^{*\pm}) = (96 \pm 4_{stat} \pm 22_{syst}) keV$. This remarkable result is unlikely to be repeated in the near future on the widths of the D_s^* and the D^{0*}.

In the framework of the chiral quark model, all the partial widths $\Gamma(M \to M' + \pi, K)$ are related through Clebsch-Gordan coefficients, and are proportional to the effective coupling constant $g_A^8 = 0.82 \pm 0.09$, as obtained [27] from the measurement of the D^* width. All the available measurements in the D-meson sector are consistent with this result. The lattice calculation of g_A^8 is $\sim 30\%$ below this value.

Radiative decays between D-mesons are detectable only for those states whose width is very narrow: the three vector D^* and the $D_s(1^+)$, limited by the phase space of the $D_s(1^+) \to D^*K$ transition. By combining $BR(D^{*\pm} \to D^\pm\gamma)$ with the above measurement of $\Gamma(D^{*\pm})$ we obtain the first measurement of the M1 radiative width, $\Gamma(D^{*\pm} \to D^\pm\gamma) = (1.6 \pm 0.5)$ keV, statistically consistent with $\Gamma(J/\psi \to \eta_c\gamma)$= 1.13±0.35 keV.

Charmed Baryon Spectra and Decays

In recent years there has been great progress in the spectroscopy of excited charm baryons, from both e^+e^- colliders at Υ energies (ARGUS,CLEO), and photoproduction experiments (E687,E831 at Fermilab). The most recent discoveries were done by CLEO and concern the spectroscopy of charmed strange baryons. As in the case of D-mesons, the assignment of quantum numbers is tightly related to the decay channel by which a state has been observed.

In the baryon spectrum, we classify the states according to the symmetry of the light quark pair:

- In the Λ_c (cud), the light quark pair has spin S=0 and isospin I=0, therefore $\vec{J}(\Lambda_c) = \vec{l} + \vec{s_c}$. Three states have been observed so far.
- The Σ_c states $(cqq, qq = uu, ud, dd)$ come in triplets, as the light quark pair has S=1 and I=1; in the two lower triplets observed so far, the mass splitting is due to the relative orientation of the charm quark's and light diquark's spins.
- In the case of the Ξ_c baryon doublets $(csq, q = u, d)$, the nomenclature is misleading: in this case the semi-light diquark sq can be both spin symmetric and antisymmetric. The state with a symmetric light diquark is identified by a prime, Ξ_c'. Two Ξ_c and two Ξ_c' doublets have been observed so far.

An updated summary of mass splittings and widths concerning the charmed non-strange baryon states is given in Table 6. In the Λ_c sector, the observation of the first doublet of orbitally excited states allows to compare both orbital and fine splittings with their analogues in the strange hyperon sector. For what concerns the orbital motion, the ratio between the $M(\frac{3}{2}^-) - M(\frac{1}{2}^+)$ splitting and the $M(2^+) - M(1^-)$ meson splitting is constant, and roughly equal to 3/4 [35]:

$$\frac{\Lambda_c(\frac{3}{2}^-) - \Lambda_c(\frac{1}{2}^+)}{D(2^+) - D^*(1^-)} = \frac{342}{455} = 0.752 \quad ; \quad \frac{\Lambda(\frac{3}{2}^-) - \Lambda(\frac{1}{2}^+)}{K(2^+) - K^*(1^-)} = \frac{404}{538} = 0.759$$

Recently, both CLEO[34] and E831 [33] have measured the masses and narrow widths of the $\Sigma_c(2455)$ resonances, as well as the masses of the three $\Sigma_c^*(2520)$ states. CLEO has also put an upper limit (at 17 MeV) for the width of the $\Sigma_c^{*+}(2520)$[36].

The spin-orbit splitting $\Delta_{LS} = M(\frac{3}{2}^-) - M(\frac{1}{2}^-)$ between the two excited states scales with the inverse mass of the heavy quark, as $\Delta_{LS}(cdu) = 32.8 \pm 1.6$ MeV/c^2 and $\Delta_{LS}(sdu) = 113.5 \pm 4.0$ MeV/c^2. The hyperfine splitting $\Delta_{hf}^B = M(\frac{3}{2}^+) - M(\frac{1}{2}^+)$ between the two lower Σ_c states is due to the coupling of the spin of the charm quark with the spin of the vector diquark, in analogy to the hyperfine splitting Δ_{hf}^M between vector and pseudoscalar mesons. In this case, due to the symmetry of the diquark wave func-

TABLE 6. Masses and widths of Λ_c, Σ_c baryons

state	J^P	M-M(Λ_c^+),MeV	Γ,MeV	ref.
$\Sigma_c^0(2455)$	$\frac{1}{2}^+$	167.32±0.15	2.4±0.5	[22, 33, 34]
$\Sigma_c^+(2455)$	$\frac{1}{2}^+$	166.4±0.4	<4.6 (90%CL)	[22, 33, 34]
$\Sigma_c^{++}(2455)$	$\frac{1}{2}^+$	167.67±0.15	2.5±0.5	[22, 33, 34]
$\Sigma_c^{*0}(2520)$	$\frac{3}{2}^+$	232.6±1.3	17.9±5.3	[22, 34]
$\Sigma_c^{*+}(2520)$	$\frac{3}{2}^+$	231.0±2.3	<17 (90%CL)	[34]
$\Sigma_c^{*++}(2520)$	$\frac{3}{2}^+$	234.5±1.4	13.0±5.2	[22, 34]
$\Lambda_c^+(2593)$	$\frac{1}{2}^-$	308.9±0.6	$3.6^{+2.0}_{-1.3}$	[22]
$\Lambda_c^+(2625)$	$\frac{3}{2}^-$	341.7±0.6	<1.9 (90%CL)	[22]

tion, the comparison can be extended to the light baryons. The ratio $\Delta^B_{hf}/\Delta^M_{hf}$ shows little dependence on the quark mass, as shown in table 7.

TABLE 7. Hyperfine splitting in mesons and baryons

ΔM	$q\bar{q}$,qqq	$s\bar{q}$,sqq	$c\bar{q}$,cqq	$c\bar{s}$,csq
Δ^B_{hf},MeV	(Δ-N) 294	(Σ)191	(Σ_c) 66	(Ξ_c) 69
Δ^M_{hf},MeV	($\rho-\pi$) 635	(K^*-K) 397	(D^*-D) 142	($D_s^*-D_s$) 144
ratio	0.446	0.481	0.465	0.480

The first Ξ_c excited states were observed by CLEO in 1995. Recently, CLEO reported the observation of other two doublets:

- the Ξ'_c, which can only decay to $\Xi_c\gamma$ [37]; its width is then extremely narrow.
- the $\Xi'_c(2645)(\frac{3}{2}^+)$, which decays to the ground state Ξ_c mostly through the double cascade $\Xi'_c(2645) \to \pi\Xi_c(2576) \to \pi\pi\Xi_c$[38]. HQET explains the suppression of the direct decay to $\pi\Xi_c$ by totally decoupling the heavy quark motion from the light diquark degrees of freedom.

TABLE 8. Masses and widths of excited Ξ_c baryons

state	J^P	M-M(Ξ_c^+),MeV	Γ,MeV	ref.
$\Xi_c'^0$	$\frac{1}{2}^+$	112.5±3.2		[37]
$\Xi_c'^+$	$\frac{1}{2}^+$	107.8±3.3		[37]
$\Xi_c'^0(2645)$	$\frac{3}{2}^+$	178.2±1.8	<5.5 (90%CL)	[22]
$\Xi_c'^+(2645)$	$\frac{3}{2}^+$	181.1±2.0	<3.1 (90%CL)	[22]
$\Xi_c^{*0}(2815)$	$\frac{3}{2}^-$	352.7±2.5	<6.5 (90%CL)	[38]
$\Xi_c^{*+}(2815)$	$\frac{3}{2}^-$	348.6±1.8	<3.5 (90%CL)	[38]

All the widths of the strange c-baryons observed so far are very narrow (see Table 8). The $\Xi_c(\frac{1}{2}^-)$ and all the excited states of the $\Omega_c(ssc)$ are still unobserved.

CONCLUSIONS

In this overview, I summarized the recent experimental results in the charm sector, and the unsolved issues concerning spectroscopy and decay of heavy hadrons.

In spectroscopy, the precision of experimental data, when available, is ahead of theory; the missing narrow states of charmonium, if confirmed, can shed light on many issues. While lattice experts are focusing on the rôle of light quark pairs in the still unexplained spin splittings, a few simple experimental patterns, consistent with valence quark approximations and with the $m_Q \to \infty$ limit, arise from the comparison of all bound systems containing charm quarks. The experimental pattern of charmed mesons and baryons up to the first radial excitation is approaching completion. The narrow width is the most powerful tag of bound states containing a charm quark: when this signature is missing, like for the $j_q = 1/2^-$ D-mesons, the identification is much harder.

On the other side, the experimental knowledge on decay widths is not yet sufficient, to challenge theoretical predictions. The hidden charm decay patterns are by far more instructive than open charm ones (both annihilations and hadronic transitions are OZI suppressed, and provide access to pure glue dynamics) but larger statistics are needed, to stimulate theory calculations at higher order. EM transitions probe details of the wavefunctions in all bound systems containing a charm quark.

In the near future, new experimental results in the charm sector are expected from BES, E831, E835, and the B-factories. The conversion of the CESR ring in a charm factory [39] will provide an unvaluable source of interesting new high statistics data.

REFERENCES

1. Abe, F., *et al.*, *Phys. Rev. Lett.*, **69**, 3704–3708 (1992).
2. Bodwin, G. T., Braaten, E., and Lepage, G. P., *Phys. Rev.*, **D51**, 1125–1171 (1995).
3. Link, J. M., *et al.* (2001), hep-ex/0109022.
4. Brambilla, N., Pineda, A., Soto, J., and Vairo, A., *Phys. Lett.*, **B470**, 215 (1999).
5. Fleming, S., Rothstein, I. Z., and Leibovich, A. K., *Phys. Rev.*, **D64**, 036002 (2001).
6. Hoang, A. H., Manohar, A. V., and Stewart, I. W., *Phys. Rev.*, **D64**, 014033 (2001).
7. Casalbuoni, R., *et al.*, *Phys. Rept.*, **281**, 145–238 (1997).
8. Bali, G. S., *Phys. Rept.*, **343**, 1–136 (2001).
9. Ali Khan, A., *et al.*, *Nucl. Phys. Proc. Suppl.*, **94**, 325–328 (2001).
10. Bernard, C. W., *Nucl. Phys. Proc. Suppl.*, **94**, 159–176 (2001).
11. Stewart, C., and Koniuk, R., *Phys. Rev.*, **D63**, 054503 (2001).
12. Bai, J. Z., *et al.*, *Phys. Rev.*, **D60**, 072001 (1999).
13. Bai, J. Z., *et al.*, *Phys. Rev.*, **D62**, 072001 (2000).
14. Brandenburg, G., *et al.*, *Phys. Rev. Lett.*, **85**, 3095–3099 (2000).
15. Ambrogiani, M., *et al.*, in preparation.
16. Bagnasco, S., *et al.*, in preparation.
17. Abreu, P., *et al.*, *Phys. Lett.*, **B441**, 479–490 (1998).
18. Armstrong, T. A., *et al.*, *Phys. Rev. Lett.*, **69**, 2337–2340 (1992).
19. Gaiser, J., *et al.*, *Phys. Rev.*, **D34**, 711 (1986).
20. Ambrogiani, M., *et al.*, *Phys. Rev.*, **D64**, 052003 (2001).
21. Antoniazzi, L., *et al.*, *Phys. Rev.*, **D50**, 4258–4264 (1994).
22. Groom, D. E., *et al.*, *Eur. Phys. J.*, **C15**, 1–878 (2000).
23. Ambrogiani, M., *et al.* (2001), Fermilab-PUB-01-334-E, to be published on *Phys. Rev.D*
24. Patrignani, C., *Phys. Rev.*, **D64**, 034017 (2001).
25. Bai, J. Z., *et al.*, *Phys. Rev.*, **D62**, 032002 (2000).
26. Bai, J. Z., *et al.*, *Phys. Rev.*, **D63**, 032002 (2001).
27. Di Pierro, M., and Eichten, E., *Phys. Rev.*, **D64**, 114004 (2001).
28. Isgur, N. (2000), *nucl-th/0007008*.
29. Abreu, P., *et al.*, *Phys. Lett.*, **B426**, 231–242 (1998).
30. Abbiendi, G., *et al.*, *Eur. Phys. J.*, **C20**, 445–454 (2001).
31. Rodriguez, J. L. (1998), *hep-ex/9901008*.
32. Anastassov, A., *et al.* (2001), *hep-ex/0108043*.
33. Link, J. M., *et al.*, *Phys. Lett.*, **B488**, 218–224 (2000).
34. Artuso, M., *et al.* (2001), *hep-ex/0110071*.
35. Lewis, R., Mathur, N., and Woloshyn, R. M., *Phys. Rev.*, **D64**, 094509 (2001).
36. Ammar, R., *et al.*, *Phys. Rev. Lett.*, **86**, 1167–1170 (2001).
37. Jessop, C. P., *et al.*, *Phys. Rev. Lett.*, **82**, 492–496 (1999).
38. Alexander, J. P., *et al.*, *Phys. Rev. Lett.*, **83**, 3390–3393 (1999).
39. Shipsey, I., these proceedings.

Hadronic Decays of Charm

Kevin Stenson

Dept of Physics & Astronomy, VU Station B351807, Vanderbilt University, Nashville, TN 37235

Abstract. Recent hadronic charm decay results from fixed-target experiments are presented. New measurements of the $D^0 \to K^-K^+K^-\pi^+$ branching ratio are shown as are recent results from Dalitz plot fits to $D^+ \to K^-K^+\pi^+$, $\pi^+\pi^-\pi^+$, $K^-\pi^+\pi^+$, $K^+\pi^-\pi^+$ and $D_s^+ \to \pi^+\pi^-\pi^+$, $K^+\pi^-\pi^+$. These fits include measurements of the masses and widths of several light resonances as well as strong evidence for the existence of two light scalar particles, the $\pi\pi$ resonance σ and the $K\pi$ resonance κ.

INTRODUCTION

Hadronic decays of charm are rich in information about QCD. For instance, the suppression of $D^0 \to \pi^+\pi^-$ relative to $D^0 \to K^-K^+$ proved the importance of final state interactions in charm decays. Also, hadronic decays give rise to the $2.5\times$ difference between the D^+ and D^0 lifetimes. Hadronic decays can provide information on relative strengths of decay diagrams (spectator, exchange, annihilation, etc.). Spectator diagrams are believed to be responsible for the vast majority of the charm decay rate. In a spectator diagram, the charm quark decays while the other quark in the meson is a spectator. By contrast, exchange and annihilation diagrams require a connection between the charm quark and the other quark in the meson and are therefore suppressed. Determining the contributions of these diagrams is an interesting open question in charm physics.

More recently, charm has been used to investigate the light resonances which are products of charm decays. Although very high statistics scattering experiments have been performed for many years to investigate these resonances, many parameters are still virtually unknown. Charm offers a unique way to investigate these resonances by nature of its low background and well defined initial state (pseudoscalar meson).

Experimentally, hadronic decays can be investigated by comparing branching ratios and by analyzing the resonant structure of multibody decays. The results presented here come from the Fermilab experiments E791 and FOCUS. E791 (FOCUS) ran in 1991 (1996–7) with a 500 GeV/c^2 π^- beam (180 GeV photon beam) on five (four) targets.

BRANCHING RATIO MEASUREMENTS

Calculations of charm *quark* decay rates via the weak interaction have been possible for many years. Unfortunately, only charm *hadron* decay rates (which are affected by the strong force) can be measured by experiments. The strong force is even more intimately involved when the charm particle decays into hadrons. Thus, deviations from the naïve weak prediction for a given decay can provide insight into the nature of the strong force.

CP618, *Heavy Flavor Physics: Ninth International Symposium*, edited by A. Ryd and F. C. Porter

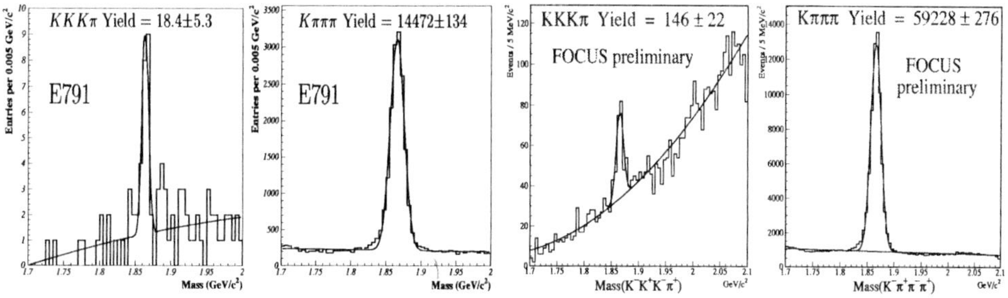

FIGURE 1. E791 and FOCUS signals from $D^0 \to K^-K^+K^-\pi^+$ and $D^0 \to K^-\pi^+\pi^-\pi^+$ decays.

$D^0 \to K^-K^+K^-\pi^+$ branching ratio

The decays $D^0 \to K^-K^+K^-\pi^+$ and $D^0 \to K^-\pi^+\pi^-\pi^+$ are both Cabibbo favored decays. A Cabibbo-favored hadronic $D^0(c\bar{u})$ decay produces one s quark from the charm, one $\bar{u}$ spectator quark, and a u and $\bar{d}$ quark from the virtual W. If both decay modes were entirely non-resonant, forming the $K^-\pi^+\pi^-\pi^+$ final state would require popping $u\bar{u}d\bar{d}$ from the vacuum while $K^-K^+K^-\pi^+$ would require $u\bar{u}s\bar{s}$ and the branching ratio between $D^0 \to K^-K^+K^-\pi^+$ and $D^0 \to K^-\pi^+\pi^-\pi^+$ could be used to determine the $s\bar{s}$ suppression relative to $d\bar{d}$. Multi-body charm decays generally proceed through resonances, however, which complicates the issue. Even in this case, one can note that the $D^0 \to K^-\pi^+\pi^-\pi^+$ decay can occur through resonances using only the four quarks from the decay, *e.g.* $\overline{K}^*(892)^0(s\bar{d})$ $\rho(770)^0(u\bar{u})$, while the $D^0 \to K^-K^+K^-\pi^+$ decay requires either an $s\bar{s}$ pair from the vacuum or a final state interaction which couples $\pi\pi$ to $K\overline{K}$. The recent E791 result contains some of these speculations [1]. Signals for these two modes from E791 and FOCUS are shown in Fig. 1. E791 finds $\mathrm{BR}\left(\frac{D^0 \to K^-K^+K^-\pi^+}{D^0 \to K^-\pi^+\pi^-\pi^+}\right) = 0.0054 \pm 0.0016 \pm 0.0008$, significantly higher than the E687 measurement of $0.0028 \pm 0.0007 \pm 0.0001$ [1, 2]. The preliminary FOCUS result is much closer to the E687 result at 0.00306 ± 0.00047 (statistical error only).

DALITZ PLOT ANALYSES

Multibody decays of charm particles can occur through various strong resonances which can interfere with each other. In a three-body decay, a "Dalitz" plot can be constructed to show the effect of these resonances and their interferences by plotting the squared invariant mass of two combinations of the final state particles against each other. In the absence of interference, how a resonance appears on the Dalitz will depend on its mass and width (a relativistic Breit-Wigner) as well as on its spin (Legendre polynomials). Interference effects can significantly alter these shapes. Fitting a Dalitz plot to a fully coherent sum of resonances allows one to extract information about how much each resonance contributes and how each resonance interferes with other resonances. By performing a coherent Dalitz plot analysis one can extract information about weak decays and the effects of the strong force on the weak decays.

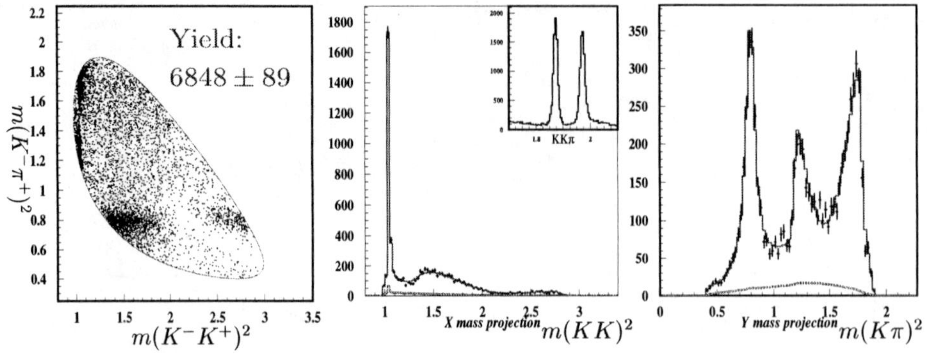

FIGURE 2. Preliminary $D^+ \to K^-K^+\pi^+$ FOCUS Dalitz plot and projections. Projections show the data background, data signal+background, and the fitted result.

TABLE 1. Preliminary FOCUS Dalitz plot fit results for the decay $D^+ \to K^-K^+\pi^+$. (Statistical errors only.)

Decay mode	Fraction (%)	Phase (°)
$K^*(892)K^+$	22.0 ± 1.1	0 (fixed)
$a_0(980)\pi^+$	27.8 ± 4.8	146 ± 5
$\phi(1020)\pi^+$	27.8 ± 0.9	244 ± 6
$f_2(1270)\pi^+$	0.7 ± 0.2	12 ± 7
$f_0(1370)\pi^+$	5.9 ± 1.2	60 ± 6
$K^*(1410)K^+$	8.8 ± 1.9	135 ± 6
$K_0^*(1430)K^+$	69.3 ± 6.3	63 ± 4
$\phi(1680)\pi^+$	1.5 ± 0.5	-70 ± 9
sum	163.8	

$D^+ \to K^-K^+\pi^+$ decays

FOCUS has a high statistics sample of the singly Cabibbo suppressed decay $D^+ \to K^-K^+\pi^+$. The Dalitz plot as well as projections along each axis for this mode are shown in Fig. 2. In the Dalitz plot and on the $m^2(KK)$ projection, a very clear $\phi(1020)$ can be seen. The $K^*(892)$ across the Dalitz plot and along the $m^2(K\pi)$ projection is also clear. Both contributions show a $\cos\theta$ modulation of the amplitude due to the spin-1 nature of the resonances. Distortions due to interferences are also visible. Preliminary fit results from a fully coherent analysis of the data are tabulated in Table 1. These results support the presence of significant amounts of $K^*(892)$ and $\phi(1020)$. The large contribution from the $K_0^*(1430)$ partly explains the broad enhancement at high $m^2(K\pi)$. The existance of many resonances with very different phases indicates significant interferences, as does the fact that the fit fraction sum is much greater than 100%. These interferences explain the obvious distortions seen in the Dalitz plot. Work is currently underway to investigate direct *CP* violation by comparing the Dalitz plots of the D^+ and D^- decays. While for two-body decays a simple branching ratio is sufficient, for multi-body modes a Dalitz plot analysis is necessary to extract all of the information on direct *CP* violation.

FIGURE 3. Diagram of annihilation (left) and resonance (right) contributions to $D_s^+ \to \pi^+\pi^-\pi^+$.

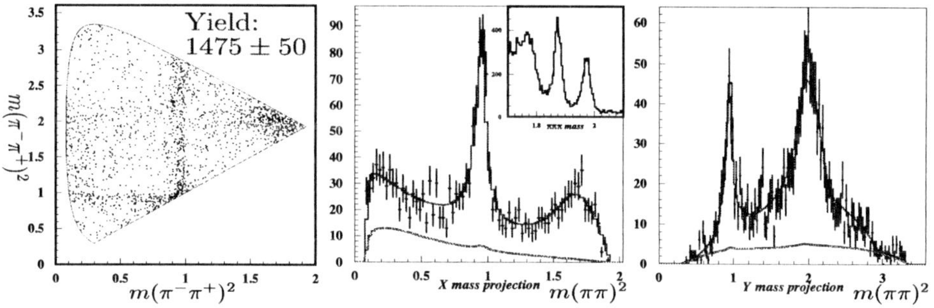

FIGURE 4. Preliminary $D_s^+ \to \pi^+\pi^-\pi^+$ FOCUS Dalitz plot and projections. Projections show the data background, data signal+background, and the fitted result.

$D_s^+ \to \pi^+\pi^-\pi^+$ decays

Although $D_s^+ \to \pi^+\pi^-\pi^+$ is Cabibbo favored, it can only occur via a spectator diagram if it uses a resonance which couples to both $K\overline{K}$ and $\pi\pi$ or via final state interactions. It can also proceed via an annihilation diagram. These possibilities are sketched in Fig. 3. Since $\rho(770)$ does not couple to $K\overline{K}$, any $\rho(770)$ would indicate an annihilation diagram contribution or final state interactions. A resonance known to couple to $K\overline{K}$ and $\pi\pi$ is the $f_0(980)$. This mysterious state has been proposed as a four-quark state and $K\overline{K}$ molecule among other things. The presence of the $a_0(980)$ further complicates the understanding of this unique resonance. The $f_0(980)$ mass is below the $K\overline{K}$ threshold but is broad enough to have a significant branching fraction to $K\overline{K}$ even with limited phase space.

The FOCUS Dalitz plot and projections for $D_s^+ \to \pi^+\pi^-\pi^+$ are shown in Fig. 4. Clear $f_0(980)$ bands are visible in the Dalitz plot and projections. A concentration at $m^2(\pi^+\pi^-) \sim 2$ GeV/c^2 is also visible. The results of a preliminary fit to this distribution as well as E791 results [3] are shown in Table 2. Both results clearly show $f_0(980)$ dominance and no significant $\rho(770)$. Thus, this decay proceeds almost entirely through resonance modes with no evidence of an annihilation diagram contribution.

Using data, E791 and FOCUS have made measurements of the poorly measured scalars contributing to the $D_s^+ \to \pi^+\pi^-\pi^+$ decay. Although the PDG [4] lists $f_0(980)$ mass and width measurements from 1973 to the present, with 13 measurements in 1999 alone, there is still no consensus on either value. The I=0, J=0 states betweeen 1200 and 1500 MeV/c^2 are even more murky.

E791 finds a mass and width of the $f_0(1370)$ of $1434 \pm 18 \pm 9$ MeV/c^2 and $172 \pm 32 \pm 6$ MeV/c^2, respectively. FOCUS uses $S_0(1475)$ for a scalar around 1475 MeV/c^2, as seen in an E687 analysis. The preliminary mass and width are found to be 1473 ± 8 MeV/c^2 and 112 ± 17 MeV/c^2, respectively; quite comparable to the E791 result for $f_0(1370)$.

TABLE 2. E791 [3] and preliminary FOCUS Dalitz plot fit results for the decay $D_s^+ \to \pi^+\pi^-\pi^+$. Errors on FOCUS results are statistical only.

	E791		FOCUS (preliminary)	
Decay mode	**Fraction (%)**	**Phase (°)**	**Fraction (%)**	**Phase (°)**
$f_0(980)\pi^+$	$56.5\pm4.3\pm4.7$	0 (fixed)	94.4 ± 2.5	0 (fixed)
non-resonant	$0.5\pm1.4\pm1.7$	$181\pm94\pm51$	25.5 ± 4.4	246 ± 4
$\rho(770)\pi^+$	$5.8\pm2.3\pm3.7$	$109\pm24\pm5$		
$f_2(1270)\pi^+$	$19.7\pm3.3\pm0.6$	$133\pm13\pm28$	9.8 ± 1.2	140 ± 6
$\rho(1450)\pi^+$	$4.4\pm2.1\pm0.2$	$162\pm26\pm17$	4.1 ± 0.7	188 ± 14
$S_0(1475)\pi^+$			17.4 ± 2.2	250 ± 4
$f_0(1370)\pi^+$	$32.4\pm7.7\pm1.9$	$198\pm19\pm27$		
sum	119.3		151.2	

E791 finds slightly better fits using the WA92 coupled channel Breit-Wigner formalism to describe the $f_0(980)$. A standard relativistic Breit-Wigner is proportional to $1/(m^2 - m_r^2 + im_r\Gamma_r(m^2))$ for a resonance r with a mass and width of m_r and Γ_r, respectively and at a two-body mass-squard of m^2. In the $f_0(980)$ WA92 coupled channel formula, Γ_r is replaced with $\Gamma_r^\pi + \Gamma_r^K$ where $\Gamma_r^\pi(m^2) = g_\pi\sqrt{m_{\pi\pi}^2/4 - m_\pi^2}$ and $\Gamma_r^K(m^2) = g_K\left(\sqrt{m_{\pi\pi}^2/4 - m_{K^+}^2} + \sqrt{m_{\pi\pi}^2/4 - m_{K^0}^2}\right)$. This framework yields $M_{f_0(980)} = 977\pm3\pm2$ MeV/c^2, $g_\pi = 0.09\pm0.01\pm0.01$, and $g_K = 0.02\pm0.04\pm0.03$ from the E791 data. Fitting with a standard Breit-Wigner results in minor changes: $M_{f_0(980)} = 975\pm3\pm2$ MeV/c^2 and $\Gamma_{f_0(980)} = 44\pm2\pm2$ MeV/c^2. FOCUS finds the K-matrix framework works quite well in dealing with the coupled channel nature of the $f_0(980)$ [5, 6]. In this framework, transformed variables are used: $m_0^2 = m_r^2 + (\gamma_{KK}/\gamma_{\pi\pi})^2(|\rho_{KK}(m_r)|/\rho_{\pi\pi}(m_r))m_r\Gamma_r$ and $\Gamma_0 = m_r\Gamma_r/(m_0\rho_{\pi\pi}(m_r)\gamma_{\pi\pi}^2)$ where $\rho_{\pi\pi}$ and ρ_{KK} are phase space terms and $\gamma_{\pi\pi}$ and γ_{KK} are coupling constants normalized to $\gamma_{\pi\pi}^2 + \gamma_{KK}^2 = 1$. In this framework, the preliminary FOCUS fits return $M_{f_0(980)} = 963\pm6$ MeV/c^2, $\Gamma_{f_0(980)} = 297\pm92$ MeV/c^2, and $\gamma_{KK}^2/\gamma_{\pi\pi}^2 = 2.09\pm0.53$ which translates to $M_{f_0(980)} = 982\pm30$ MeV/c^2 and $\Gamma_{f_0(980)} = 89\pm32$ MeV/c^2 for a standard Breit Wigner (errors are statistical only).

$D^+ \to \pi^+\pi^-\pi^+$

E791 has published results for a coherent Dalitz plot analysis of $D^+ \to \pi^+\pi^-\pi^+$ [7]. In their initial fit to the Dalitz plot using all known resonances, the fit quality was very poor with a confidence level of 10^{-5}. By including a low mass scalar particle (σ) the fit was significantly improved and yielded a confidence level of 75%, providing strong evidence for the elusive light scalar. The results from both fits are tabulated in Table 3. From the fit, the σ parameters were determined to be $M_\sigma = 478^{+24}_{-23}\pm17$ MeV/c^2 and $\Gamma_\sigma = 324^{+42}_{-40}\pm21$ MeV/c^2. Many checks were made to validate the existence of the σ in this decay. These checks included fitting with a vector and tensor state, and fitting with no phase variation. The fit with a phase-varying scalar particle was clearly preferred.

TABLE 3. E791 Dalitz plot fit results for the decay $D^+ \to \pi^+\pi^-\pi^+$ [7]. Parameter errors from "without σ" fits are statistical only.

	Without σ: CL = 0.001%		With σ: CL = 75%	
Decay mode	**Fraction (%)**	**Phase (°)**	**Fraction (%)**	**Phase (°)**
$\rho(770)\pi^+$	20.8 ± 2.4	0 (fixed)	$33.6 \pm 3.2 \pm 2.2$	0 (fixed)
non-resonant	38.6 ± 9.7	150 ± 12	$7.8 \pm 6.0 \pm 2.7$	$57 \pm 20 \pm 6$
$f_0(980)\pi^+$	7.4 ± 1.4	152 ± 16	$6.2 \pm 1.3 \pm 0.4$	$165 \pm 11 \pm 3$
$f_2(1270)\pi^+$	6.3 ± 1.9	103 ± 16	$19.4 \pm 2.5 \pm 0.4$	$57 \pm 8 \pm 3$
$f_0(1370)\pi^+$	10.7 ± 3.1	143 ± 10	$2.3 \pm 1.5 \pm 0.8$	$105 \pm 18 \pm 1$
$\rho(1450)\pi^+$	22.6 ± 3.7	46 ± 15	$0.7 \pm 0.7 \pm 0.3$	$319 \pm 39 \pm 11$
$\sigma\pi^+$			$46.3 \pm 9.0 \pm 2.1$	$206 \pm 8 \pm 5$
sum	106.4		116.3	

$D^+ \to K^-\pi^+\pi^+$

The Cabibbo favored decay $D^+ \to K^-\pi^+\pi^+$ provides a very high statistics mode in which to study charm decays. Previous analyses of this decay [8, 9] have identified two mysteries in this decay. The first mystery is why there is a dominant non-resonant contribution; unique in charm decays. The second mystery is why a good fit to this Dalitz plot seems to be impossible to achieve. The E791 data sample of 15,090 events (94% signal) can be used to shed light on these mysteries. Fitting the data using all known resonances results in a large non-resonant contribution and a very poor fit (confidence level of 10^{-11}), similar to past attempts. Given the evidence for the σ in the $D^+ \to \pi^+\pi^-\pi^+$ mode, an additional scalar (κ) was added to the $D^+ \to K^-\pi^+\pi^+$ fit. This provides a dramatic reduction in the non-resonant contribution (from 91% to 13%) and a much improved fit (confidence level of 95%). The preliminary results of both fits are shown in Table 4. The preliminary κ parameters are found to be $M_\kappa = 797 \pm 19 \pm 42$ MeV/c^2 and $\Gamma_\kappa = 410 \pm 43 \pm 85$ MeV/c^2. Attempts to explain the data in other ways (*e.g.* including a vector state, tensor state, non-phase-varying state, structured non-resonant contribution) have been inadequate. Preliminary meausrements of the $K_0^*(1430)$ parameters ($M_{K_0^*(1430)} = 1459 \pm 7 \pm 6$ MeV/c^2 and $\Gamma_{K_0^*(1430)} = 175 \pm 12 \pm 12$ MeV/c^2) have also been extracted in this analysis [10].

TABLE 4. Preliminary E791 Dalitz plot fit results for the decay $D^+ \to K^-\pi^+\pi^+$. Parameter errors from "without κ" fits are statistical only.

	Without κ: CL = 10^{-11}		With κ: CL = 95%	
Decay mode	**Fraction (%)**	**Phase (°)**	**Fraction (%)**	**Phase (°)**
NR	90.9 ± 2.6	0 (fixed)	$13.0 \pm 5.8 \pm 2.6$	$349 \pm 14 \pm 8$
$\overline{K}^*(892)\pi^+$	13.8 ± 0.5	54 ± 2	$12.3 \pm 1.0 \pm 0.9$	0 (fixed)
$\overline{K}_0^*(1430)\pi^+$	30.6 ± 1.6	54 ± 2	$12.5 \pm 1.4 \pm 0.4$	$48 \pm 7 \pm 10$
$\overline{K}_2^*(1430)\pi^+$	0.4 ± 0.1	33 ± 8	$0.5 \pm 0.1 \pm 0.2$	$306 \pm 8 \pm 6$
$\overline{K}^*(1680)\pi^+$	3.2 ± 0.3	66 ± 3	$2.5 \pm 0.7 \pm 0.2$	$28 \pm 13 \pm 15$
$\kappa\pi^+$			$47.8 \pm 12.1 \pm 3.7$	$187 \pm 8 \pm 17$
sum	138.9		88.6	

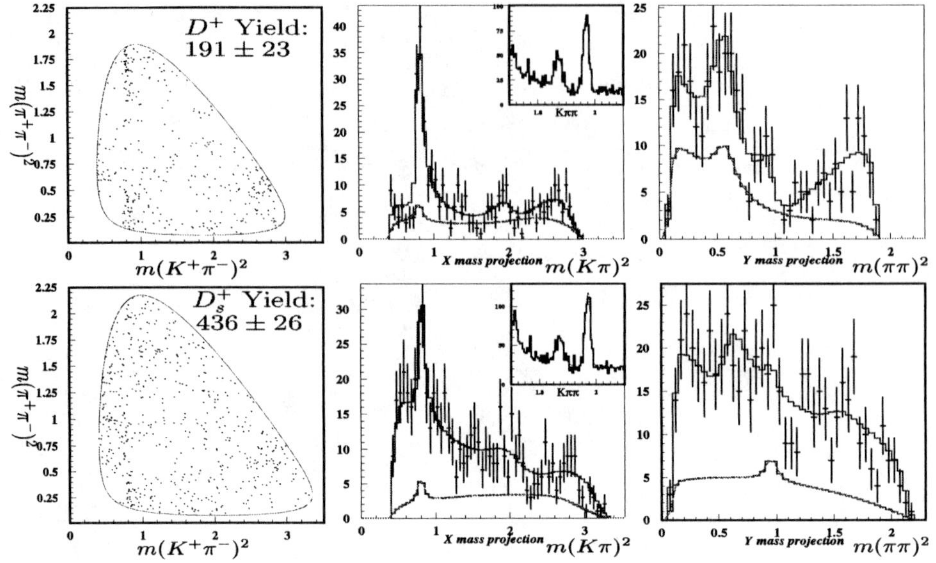

FIGURE 5. FOCUS Dalitz plots and projections for $D^+ \to K^+\pi^-\pi^+$ and $D_s^+ \to K^+\pi^-\pi^+$. Projections show the data background, data signal+background, and the fitted result.

TABLE 5. Preliminary FOCUS Dalitz plot fit results for $D^+ \to K^+\pi^-\pi^+$ and $D_s^+ \to K^+\pi^-\pi^+$. Statistical errors only.

	$D^+ \to K^+\pi^-\pi^+$		$D_s^+ \to K^+\pi^-\pi^+$	
Decay mode	**Fraction (%)**	**Phase (°)**	**Fraction (%)**	**Phase (°)**
$\rho(770)K^+$	51 ± 10	0 (fixed)	40 ± 4	0 (fixed)
non-resonant	9 ± 5	-6 ± 16	18 ± 4	34 ± 7
$K^*(892)\pi^+$	43 ± 7	208 ± 16	22 ± 3	163 ± 7
$f_0(980)K^+$	9 ± 5	73 ± 31		
$f_2(1270)K^+$			2 ± 1	33 ± 21
$K^*(1410)\pi^+$	12 ± 8	133 ± 23	14 ± 5	-10 ± 7
$K_0^*(1430)\pi^+$			14 ± 6	68 ± 7
$K_2^*(1430)\pi^+$	6 ± 3	48 ± 27		
$\rho(1450)K^+$	10 ± 5	247 ± 15	8 ± 2	219 ± 14
$K^*(1680)\pi^+$	22 ± 10	2 ± 20		
sum	162		118	

$D^+, D_s^+ \to K^+\pi^-\pi^+$

FOCUS has obtained preliminary results from Dalitz plot fits to the doubly Cabibbo suppressed decay $D^+ \to K^+\pi^-\pi^+$ and the singly Cabibbo suppressed decay $D_s^+ \to K^+\pi^-\pi^+$. The Dalitz plots and projections are shown in Fig. 5 and the fit results in Table 5. This is the first fit to the $D_s^+ \to K^+\pi^-\pi^+$ Dalitz plot. Both fit results indicate a rich resonance structure, dominated by $\rho(770)$.

TABLE 6. Fitted masses and widths for various scalars involved in charm decays. E791 κ and $K_0^*(1430)$ results are preliminary. The FOCUS results are preliminary and errors shown are only statistical.

	E791		FOCUS	
Resonance	**Mass (MeV/c^2)**	**Γ (MeV/c^2)**	**Mass (MeV/c^2)**	**Γ (MeV/c^2)**
σ	$478^{+24}_{-23}\pm 17$	$324^{+42}_{-40}\pm 21$		
κ	$797\pm 19\pm 42$	$410\pm 43\pm 85$		
$f_0(980)$	$975\pm 3\pm 2$	$44\pm 2\pm 2$	982 ± 30	89 ± 32
$f_0(1370)/S_0(1475)$	$1434\pm 18\pm 9$	$172\pm 32\pm 6$	1473 ± 8	112 ± 17
$K_0^*(1430)$	$1459\pm 7\pm 6$	$175\pm 12\pm 12$		

CONCLUSION

Hadronic decays of charm provides an environment to study many aspects of high energy physics including measuring the contributions of various Feynman diagrams, studying the effect of final state interactions, searching for *CP* violation, and measuring the mass and width of light resonances. The relatively large decay rate of $D_s^+ \to \pi^+\pi^-\pi^+$ decay has been found to be due to resonances which couple simultaneously to $K\overline{K}$ and $\pi\pi$ rather than annihilation diagrams. Strong evidence for, and precise measurements of, two particles which have existed on the fringe for many years (σ and κ) are presented. Using the clean charm environment to measure light resonance parameters is a new and interesting use of charm hadronic decays. Table 6 summarizes all of the measured light resonance values described in these proceedings. The future holds the prospect of even more precise measurements of light resonances, a search for direct *CP* violation from $D^+ \to K^-K^+\pi^+$ decays, and much more information from many decay modes.

REFERENCES

1. E. M. Aitala *et. al.*, *Phys. Rev.*, **D64**, 112003 (2001).
2. P. L. Frabetti *et. al.*, *Phys. Lett.*, **B354**, 486 (1995).
3. E. M. Aitala *et. al.*, *Phys. Rev. Lett.*, **86**, 765 (2001).
4. D. E. Groom *et. al.*, *Eur. Phys. J.*, **C15**, 1 (2000).
5. S. Malvezzi *et. al.*, "D-Meson Dalitz Fit from Focus", in *Intersections of Particle and Nuclear Physics: 7th Conference*, edited by Z. Parsa and W. J. Marciano, American Institute of Physics, 2000, vol. CP549.
6. S. U. Chung *et. al.*, *Ann Physik.*, **4**, 404 (1995).
7. E. M. Aitala *et. al.*, *Phys. Rev. Lett.*, **86**, 770 (2001).
8. J. C. Anjos *et. al.*, *Phys. Rev.*, **D46**, 56 (1993).
9. P. L. Frabetti *et. al.*, *Phys. Lett.*, **B331**, 217 (1994).
10. C. Göbel *et. al.*, "Light Meson Physics from Charm Decays at Fermilab E791", in *IX International Conference on Hadron Spectroscopy*, edited by D. Amelin, American Institute of Physics, 2001.

A Survey of Charm Hadroproduction Results

James S. Russ

Physics Department, Carnegie Mellon University, Pittsburgh, PA 15213 USA

Abstract. Charm hadroproduction experiments at fixed target energies were intended to explore perturbative QCD at the charm mass scale. High- statistics studies with π^- beams suggested strong non-perturbative effects at large x_F. Recent results from proton and Σ^- beams show further systematics of non-perturbative behavior. This review summarizes the systematics of these effects as developed for different charm hadrons and different beam particles.

INTRODUCTION

Charm hadroproduction studies in fixed target experiments have been carried out for the past 2 decades. Having two heavy flavor species (c,b) to compare allows us in principle to test the flavor-independence of hadroproduction predicted by QCD. However, the original goal of testing ideas of perburbative QCD at the charm quark mass scale has proved to be elusive. Thorny theoretical issues have been recognized by many authors. [1] Since the actual observables are the color-singlet final state hadrons, there may be significant non-perturbative hadronization effects: (a) color-drag effects between either the c or $\bar{c}$ quark and the outgoing colored fragment from the beam or target hadron; and (b) other hadronization effects, including rescattering of the separating charmed and anti-charmed hadrons because of the limited p_T range of fixed target experiments.

The consensus is that the original goal of testing theory is unattainable. Instead, the intent of the latest round of charm hadroproduction experiments is to systematize the non-perturbative factors at work by comparing production of several charm species at different energies by several different beam hadrons. One can test:

- hadronization effects in the p_T spectrum
- hadronization and k_T effects in charm pair production
- color-drag effects and x_F dependence versus charm type
- $c\bar{c}$ asymmetries for various beam hadrons

This report summarizes the current understanding of non-perturbative features of the data on charm hadroproduction. I should note that the strength of non-perturbative physics in charm photoproduction at fixed target energies is much weaker than in hadroproduction. The photoproduction data are predominantly diffractive, i.e., are dominated by quark pair production, rather than through the partonic interactions of the photon. Such a distinction from fixed target hadroproduction suggests that many of the non-perturbative features of hadroproduction stem from the the effects listed above. We shall try to systematize what is known.

CP618, *Heavy Flavor Physics: Ninth International Symposium*, edited by A. Ryd and F. C. Porter

GENERAL THEORETICAL FEATURES OF CHARM HADROPRODUCTION

The conventional picture of Heavy Quark production assumes factorization, i.e., a separation between the quarks produced in the hard scattering process and the hadron fragments that are left in the beam and target hadrons. Factorization allows one to use parton ideas from Deep Inelastic Scattering to analyze the quark production process at an unphysical high mass scale where the calculation is understood, then evolve the distributions to the m_Q-scale. The NLO parton calculation includes single gluon emission effects in the production process and was first analyzed by Nason, Dawson and Ellis. [2, 3] To use the NLO picture one chooses: (a) the renormalization scale for the parton-level process; (b) the factorization scale for the parton distributions (generally the same as (a)); and (c) the effective quark constituent mass. One adds parton transverse momentum k_T, attributable to confinement, by hand. The NLO calculation leaves the LO prediction for single-quark distributions in x_F or p_T unchanged, except in scale. NLO effects mainly modify $Q\overline{Q}$ pair distributions, chiefly the azimuthal angle distribution between Q and $\overline{Q}$.

The NLO calculation leaves us with quark-level distributions. Observables are usually related to hadrons, and non-perturbative fragmentation mechanisms have to be introduced. As a consequence of the heavy quark Q's fragmenting into a hadron, distributions in x_F and p_T for charm hadrons are expected to be softer than those for the quarks themselves. NLO calculations are presented as computational packages with adjustable scale parameters and use various parton distribution functions. [1] Typically hadronization is treated by applying the empirical Peterson function developed for $e^+ e^-$ production to the quark-level distributions.

FEATURES OF CHARM HADROPRODUCTION DATA

As an illustration of the general features of the data, consider the Fermilab E791 (500 GeV/c π^-) x_F and p_T^2 distributions for D^0 mesons [+ c.c.] covering the central and forward production regions shown in Fig. 1. [4]

One sees that the overall shape of both distributions matches the quark-level calculation. Normalization here is to the data; calculation falls well below data. The x_F behaviour is smooth, with some experimental suggestion of a slower decrease at large x_F than the NLO calculation. The hadronized distribution using the Petersen function is softer than the data.

The p_T^2 distribution shows a clear break in slope from the gaussian shape in the forward region to a power-law form for $p_T^2 \geq 2\ (\mathrm{GeV}/c)^2$. The form is suggested by the $O(\alpha_s^3)$ calculation, [1] but both the position of the break and the power are determined from the data, not by calculation.

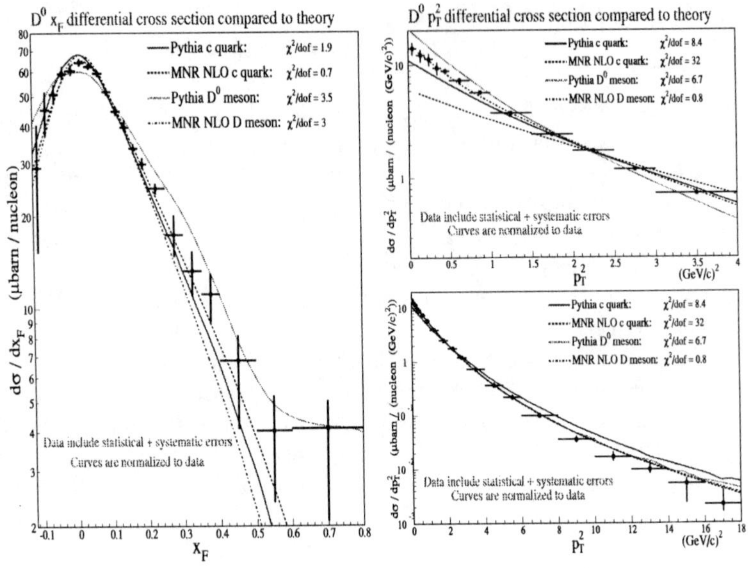

FIGURE 1. E791 D^0 + c.c. x_F and p_T^2 distributions (500 GeV/c π^-)

Particle-Antiparticle Asymmetries in Charm Hadroproduction

Both LO and NLO pictures of charm hadroproduction predict very small asymmetries in any single-particle distribution comparison of particle and antiparticle. Therefore, it was a major surprise when the E791 high-statistics data confirmed earlier suggestions of significant asymmetry between D^+ and D^- production, as shown in the left panel of Fig. 2. [5] However, the right panel shows that interpretation is not simple. $D^{*\pm}$ distributions, ostensibly having the same valence quark effects as the $D^\pm$, show NO strong asymmetry, as we will discuss later. [6]

Because these data cannot be explained by perturbative QCD, the questions raised focus on other physics: (a) why do some but not all charm hadrons show large asymmetry? (b) is there a beam-hadron dependence of the effect? and (c) why is it x_F dependent? These questions can be framed within two possible pictures of large non-perturbative effects: the Quark-Gluon-String Model (QGSM) originally included in the PYTHIA Monte Carlo [7] and the Intrinsic Charm Model (ICM) [8].

The QGSM recognizes that the low- p_T environment of fixed target production may permit color-field interactions between the c- or $\bar{c}$ -quark and the fragments of the beam hadron to produce a colorless hadron. This color-drag effect will *increase* the x_F of the charm hadron when it occurs, compared to the perturbative value. When the charm quark is well-separated from the beam fragments at large p_T, the effect should be smaller.

The ICM, having rather different kinematics, came from QED analyses by Brodsky and co-workers. In the ICM the valence quark distribution of the beam hadron includes an intrinsic c- $\bar{c}$ component with some probability. Therefore, in soft collisions the beam hadron may dissociate, leaving the c- or $\bar{c}$ quark carrying a large fraction of the beam

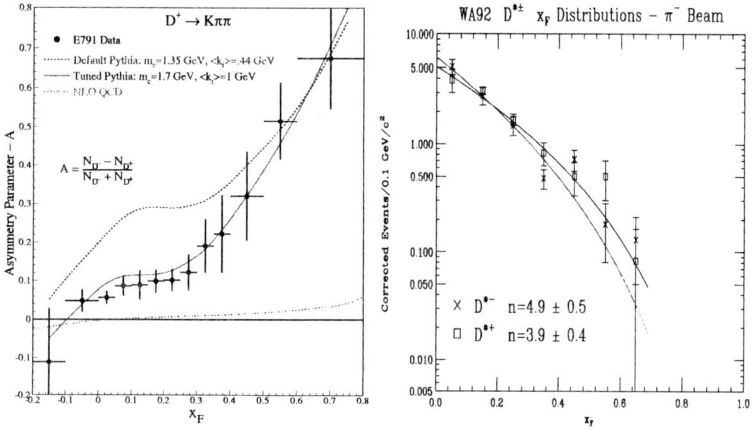

FIGURE 2. (left) E791(500 GeV/c π^-) $D^\pm$ production asymmetry; (right) WA92(350 GeV/c π^- beam) $D^{*\pm}$ asymmetry

momentum. This leads to an excess cross section at low p_T, large x_F.

SYSTEMATICS OF NON-PERTURBATIVE EFFECTS IN CHARM HADROPRODUCTION

The strong asymmetries seen in hadroproduction have shifted emphasis in from testing pQCD to systematizing the non- perturbative effects. Experimental studies focus on comparing the x_F distributions of particle and antiparticle. One defines a *leading* charm particle (antiparticle) as one which shares one or more valence quarks with the beam hadron. Both QGSM and ICM mechanisms increase the yields of a leading charm particle and/or antiparticle at large x_F.

For the observed set of charm hadrons produced by π^-, Σ^-, and proton beams, we have the following structure of leading charm particles or antiparticles in these pictures:

TABLE 1. Leading Charm Hadrons for π^-, Σ^- and p beams

π^-		Σ^-		p	
L	NL	L	NL	L	NL
D^0	$\overline{D}^0$	*none*	D^0, $\overline{D}^0$	$\overline{D}^0$	D^0
D^-, D^{*-}	D^+, D^{*+}	D^-, D^{*-}	D^+, D^{*+}	D^-, D^{*-}	D^+, D^{*+}
none	$D_s^\pm$	D_s^-	D^+	*none*	$D_s^\pm$
Λ_c^+, $\overline{\Lambda_c^+}$	*none*	Λ_c^+	$\overline{\Lambda_c^+}$	Λ_c^+	$\overline{\Lambda_c^+}$

In the following sections we shall see what the data for fixed target charm production by these three beam hadrons say about the leading systematics.

Comparison of $D^{\pm}$ and $D^{*\pm}$ Meson Production

The D* mesons are the S=1 hyperfine excitations of the D mesons and have the same valence quark content. Any leading behaviour should be similar for the two systems, yet we saw in Fig. 2 that the $D^{*\pm}$ behavior is clearly quite different from that of the $D^{\pm}$ family for a π^- beam. What about production by different beam hadrons? SELEX has PRELIMINARY data on D* production by Σ^-, shown at left in Fig. 3.

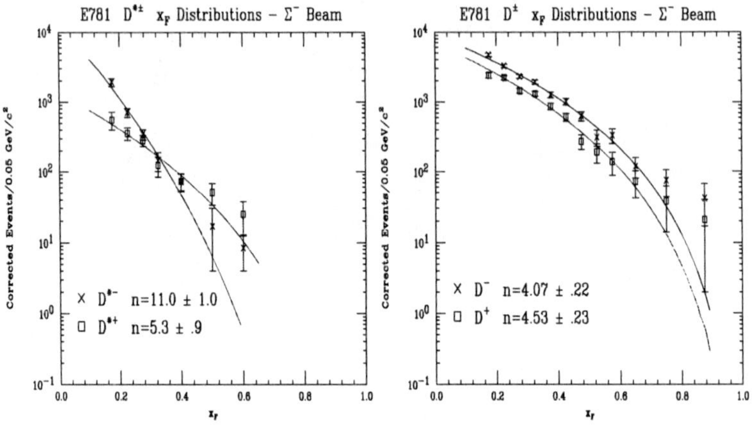

FIGURE 3. SELEX (600 GeV Σ^- beam) x_F distributions for (left) D^{*-} and D^{*+}; (right) D^- and D^+

These results are remarkable for the steep x_F fall of the *leading* D^{*-} events. Even though the integral yield for $x_F \geq 0.15$ favors the leading hadron, the asymmetry changes sign for $x_F \geq .3$. This is quite different from the π^- production case.

For $D^{\pm}$ production, shown at right in Fig. 3, we get a different kind of surprise. There is no shape distinction between the leading D^- and the non-leading D^+, only an x_F-independent normalization preference for leading particle production.

What should one conclude? The valence d-quark structure is different in the π^- and Σ^-, but the impact of having a single quark or diquark seems to be quite different for the S=0 and S=1 mesons. Adding proton data (single d quark) from future SELEX analysis may be instructive.

$D_S^{\pm}$ and D^0 Meson Production

Consider next D_S production. For the π^- beam, neither D_S^+ nor D_S^- is leading. In that case E791 reports a very small asymmetry with no particular x_F structure, consistent with Table 1. [9] In contrast, the Σ^- beam has an s quark that makes the D_S^- leading. Preliminary SELEX Σ^- data are shown at left in Fig. 4. Strong leading behavior is clearly seen, unlike the nominally-comparable case of the D^-. Thus, D_S production is consistent with expectations from Table 1 for both π^- and Σ^- beams.

The remaining meson to be considered is the D^0. The production distributions here are contaminated by feed-down from the dominant decays of $D^{*\pm,0}$ mesons to final states

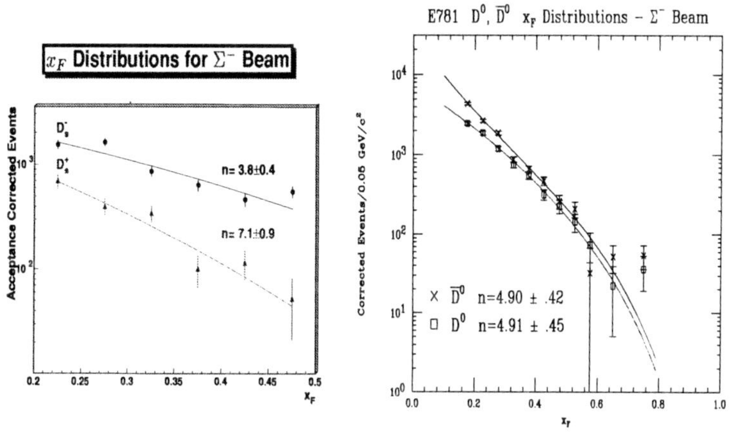

FIGURE 4. SELEX (600 GeV Σ^- beam) x_F distributions for (left) $D_S^{\pm}$; (right) D^0 and $\overline{D}^0$

with D^0s. SELEX has made the first effort to look at the x_F dependence for D^0 and $\overline{D}^0$ separately, shown at right in Fig. 4. Feeddown from the steep D^{*-} distribution for the Σ^- beam can be seen in these data as a sharpening of the $\overline{D}^0$ distribution for $x_F \leq 0.3$. The D^{*+} x_F shape, on the other hand, is very similar to what we see here for D^0. If one fits both D^0 and $\overline{D}^0$ distributions for $x_F \geq 0.3$ to $(1 - \mathrm{x_F})^n$, the n-values for the two particles are very similar, as shown in Fig. 4. The curve for $\overline{D}^0$ in the figure includes a contribution with a fixed $(1 - \mathrm{x_F})^{11}$ behavior to account for the D^{*-} contribution at small x_F. The overall conclusion is that Σ^- production shows no leading behavior for the D^0 system, consistent with Table 1. The upcoming π and proton beam results from SELEX, where leading effects are expected, will be important to see if the D^0 system is predicted correctly by non-perturbative effects.

Λ_c Production

Finally we turn our attention to charm baryon production. Recall that the early measurements of Λ_c^+ production by pions near $x_F \sim 0$ showed comparable yields for charmed baryon and antibaryon. E791 has reported a good- statistics measurement of the asymmetry in the central production region showing approximately equal production of Λ_c^+ and $\overline{\Lambda_c}^-$, consistent with the expectations of Table 1. [10] The solid lines in the left panel of Fig. 5 are default Pythia predictions.

SELEX presents x_F distributions from all 3 beam hadrons for Λ_c^+ and $\overline{\Lambda_c}^-$ separately, as seen in the right panel of Fig. 5. For a *baryon* beam, antibaryon production is strongly suppressed, and the suppression grows stronger at large x_F. These findings again agree with the expectations of Table 1. In terms of the asymmetry parameter A, both baryon beams have A $\sim$ 1 for $x_F \geq 0.5$.

The Λ_c^+ p_T distribution from the Σ^- beam is shown at left in Fig. 6. The same break in

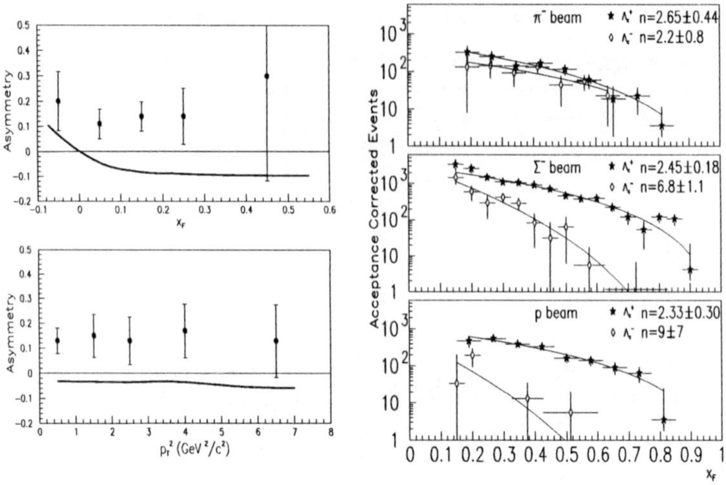

FIGURE 5. (left)E791 (500 GeV π^-) Λ_c^+ asymmetry vs x_F; (right) SELEX (600 GeV Σ^-) Λ_c^+ and $\overline{\Lambda_c}^-$ x_F distributions

slope previously seen in D^0 data (Fig. 1) is observed here, suggesting that it is a feature of c-quark production and is independent of non-perturbative issues like hadronization or color-drag effects. SELEX has looked at the x_F distributions in the diffractive regime ($p_T \leq 1$ GeV/c^2) and in the power-law regime ($p_T^2 \geq 2$ GeV/c^2). These two distributions are shown at right in Fig. 6.

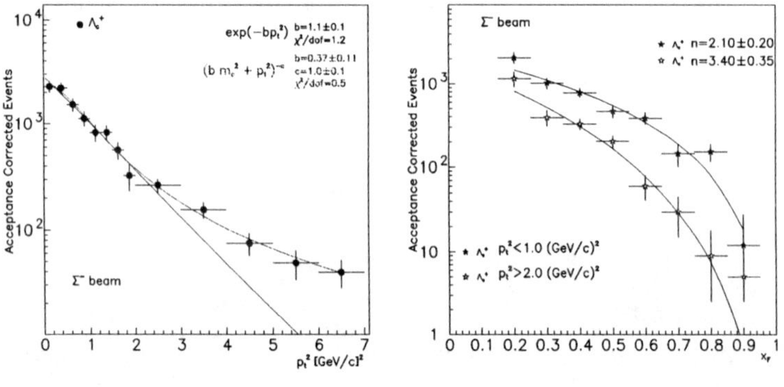

FIGURE 6. (left) SELEX (600 GeV Σ^-) Λ_c^+ p_T^2 distribution ; (right) x_F distributions for $p_T^2 \leq 1(\mathrm{GeV/c})^2$ and $p_T^2 \geq 2(\mathrm{GeV/c})^2$

The x_F distribution falls faster for the large p_T data than does the one at low p_T. This is consistent with QGSM effects. The c-quark at large p_T is more isolated from the beam fragments and experiences less color drag than one at low p_T. SELEX has also looked for ICM-type effects in these data, viz., a sharpening of the p_T distribution to favor small p_T in the large x_F sample. No such effect is observed.

PRODUCTION SUMMARY AND STATUS

Not all of the channels listed in Table 1 have been discussed here. SELEX is continuing analysis to fill in the table. Results can be expected within the coming year. We see the following pattern in the currently-available data:

1. D_S, D^0, and Λ_c^+ production by π^- and Σ^- beams follows the pattern in Table 1, showing leading effects where predicted.
2. D^+ and D^{*+} + c.c. production by π^- and Σ^- beams shows strong non-perturbative effects, but they are not at all consistent with model predictions.

The surprises in item 2 come from production differences in S=0 and S=1 mesons of the same valence quark content. The other charm hadrons follow the non-perturbative model rather well. Can we understand this?

There are several Russian groups who have attempted to deal with all charm hadroproduction data in a self-consistent fashion. They treat non-perturbative effects by using different charm quark fragmentation descriptions along with including color-drag effects invoking detailed valence quark configurations in the parton distribution functions. There are many parameters in the models, but Likhoded and coworkers [11] and Piskounova and coworkers [12] use e^+e^- and HERA ep data as well as hadroproduction data to fit parameters and predict the high-energy results. It will be a real challenge to the models to confront the complete set of high energy data that will be available soon. An important question is whether the information gleaned from the model fits can be related to a more fundamental picture of the non-perturbative interactions in charm hadroproduction.

We end this section by recalling the original reason to study heavy quark (HQ) production: testing flavor-independence in HQ processes. Are charm cross sections at a perturbative scale the same as b cross sections? What is the right way to make the comparison? The collider data from Tevatron Run II may offer an interesting new look into these issues if one compares charm and beauty production at m_T scales far above m_b.

AND NOW SOMETHING COMPLETELY DIFFERENT

Soon after charmonium states were discovered, theoretical calculation of ccq baryon states were done and mass spectra predicted using a Coulomb-like potential. Further refinements were added, but the absence of experimental stimulus has stymied progress. The current situation is summarized in Ref. [13].

From the experimental viewpoint, one expects that a ccq baryon will have the cc pair in a symmetric spin state, like the $J/\psi(2S)$ at 3.69 GeV/c^2. The lowest ccq mass according to Richard's summary is expected to be in the range 3.63-3.75 GeV/c^2. Decay modes have not been explored. The likely products are a charm baryon plus meson, with a lifetime comparable to or shorter than the Λ_c^+ lifetime of 200 fs (half the c-quark lifetime).

SELEX has searched for the decay $ccu^{++} \rightarrow \Lambda_c^+ + (K^-\pi^+\pi^+)$. The $(K^-\pi^+\pi^+)$ vertex is formed topologically, with no particle ID requirements. This 3-prong vertex must be distinct both from the primary vertex (vertex separation significant of 1-4 σ) and the Λ_c^+ decay vertex (similar requirement). The results are shown in the left panel of Fig. 7. One sees an interesting bump (3.3 σ) at a mass of 3.79 GeV/c^2. The resolution of the gaussian fit to the data is consistent with experimental simulation. The lifetime appears to be very short, but not zero.

Background evaluations are shown, along with the signal channel, in the right panel of Fig. 7. We switched mass assignments of the 3 mesons, requiring $(K^+\pi^-\pi^+)$ (top right). We looked at overall neutral systems which *cannot* be (ccq) systems. Fig. 7 shows these neutral states with the meson system corresponding to $(K^-\pi^+\pi^-)$ (lower left) or $(K^+\pi^-\pi^-)$ (lower right). Only one bin in the neutral system shows any significant fluctuation up from the nominal background shape.

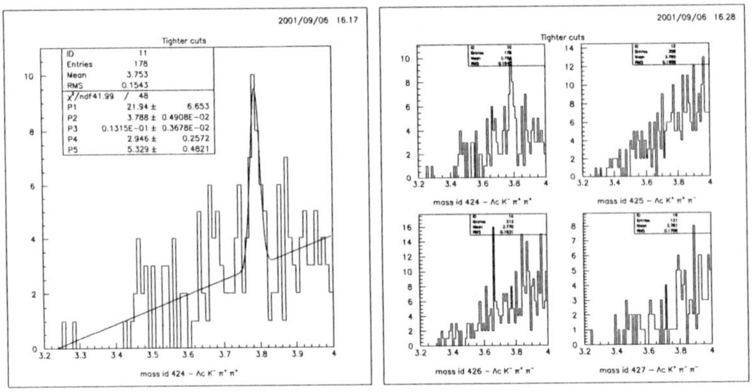

FIGURE 7. (left) SELEX ccu^{++} search ; (right) mass distributions for ccu^{++} signal and background channels as discussed in text

These data suggest a possible ccu^{++} state at 3.79 GeV/c^2, but the statistics are limited. The yield of this state, if real, into this mode is about 1% of the yield of $\Lambda_c^+ \rightarrow pK^-\pi^+$. It would be very exciting to see a confirmation of such an effect from, e.g., FOCUS, which has a very different production environment and different background.

REFERENCES

1. Stefano Frixione, Michelangelo L. Mangano, Paolo Nason, and Giovanni Ridolfi. *Adv. Ser. Direct. High Energy Phys.*, 15:609–706, 1998.
2. P. Nason, S. Dawson, and R. K. Ellis. *Nucl. Phys.*, B327:49–92, 1989.
3. P. Nason, S. Dawson, and R. K. Ellis. *Nucl. Phys.*, B335:260, 1990.
4. E. M. Aitala *et al. Phys. Lett.*, B462:225–236, 1999.
5. E. M. Aitala *et al. Phys. Lett.*, B371:157–162, 1996.

6. M. I. Adamovich *et al. Nucl. Phys.*, B547:3–14, 1999.
7. T. Sjostrand. *Comput. Phys. Commun.*, 82:74–90, 1994.
8. R. Vogt, S. Brodsky, and P. Hoyer. *Nucl. Phys.*, B383:643–684, 1992.
9. E. M. Aitala *et al. Phys. Lett.*, B411:230–236, 1997.
10. E. M. Aitala *et al. Phys. Lett.*, B495:42–48, 2000.
11. A. K. Likhoded and S. R. Slabospitsky. 2000.
12. O. I. Piskounova. *Nucl. Phys. Proc. Suppl.*, 93:144–147, 2001.
13. S. Fleck and J. M. Richard. *Prog. Theo. Phys.*, 82:760–774, 1989.

Heavy Flavour Production at HERA

Christoph Grab
representing the H1 and ZEUS collaborations

Institute for Particle Physics, ETHZ, 8093 Zürich, Switzerland

Abstract. Selected recent measurements of inelastic heavy flavour production in lepton-proton collisions performed at HERA by the experiments H1 and ZEUS are reviewed. The experimental results, covering the region in four-momentum transfer squared Q^2 from photoproduction ($Q^2 \approx 0$) to deep-inelastic scattering at high Q^2 are compared in detail with theoretical predictions, based on QCD calculations in both leading order and next-to-leading order perturbative approaches. While in the charm sector no large deviations from the expectations were observed within the theoretical uncertainties, in the beauty sector the absolute normalisation of the QCD predictions are significantly below the data.

Results on inelastic leptoproduction of J/ψ mesons are directly compared with predictions based on a NRQCD/factorisation approach. A search for colour octet processes predicted within this approach is conducted in many regions of phase space, but no unambiguous evidence has been found.

INTRODUCTION

The main interests in studying heavy flavours (HF) at the HERA $e-p$ collider are threefold: a) investigating the production mechanism allows detailed comparisons with theoretical predictions, in particular perturbative QCD calculations, and extends the studies to non-perturbative phenomena; b) exploiting HF as a probing tool provides access to the parton density functions in the proton, the photon or the pomeron; c) they serve as a probe for non-standard physics processes, such as, e.g., signs of SUSY or instanton searches through flavour democracy.

At HERA, heavy quarks are produced predominantly by the *photon gluon fusion* process $\gamma g \rightarrow c\bar{c}(b\bar{b})$ where a real or virtual photon emitted by the electron[1] and a gluon from the proton generate a $c\bar{c}(b\bar{b})$ pair. In this case, colour is transferred from the proton so that the proton breaks up (inelastic scattering). In general, elastic scattering may also occur, where no colour is transferred, and the proton can remain intact (diffractive interactions); however this topic is not covered here.

The heavy quarks hadronise and are then either detected as "open states", i.e., in final states with charmed or beauty hadrons, or alternatively as "hidden states", such as J/Ψ .

The kinematics of HF production is described by three independent variables, the centre-of-mass energy $\sqrt{s}$, the photon four-momentum squared $q^2 = -Q^2$ and either

[1] Hereafter, a reference to electron implies a reference to either electron or positron. Most of the results shown were obtained in positron-proton collisions.

CP618, *Heavy Flavor Physics: Ninth International Symposium,* edited by A. Ryd and F. C. Porter

one of the scaling variables $y = (q \cdot P / l \cdot P)$, the inelasticity of the ep-interaction, or Bjorken-x, $x = Q^2/(2P \cdot l)$. Here P and l denote the four-momentum of the proton and the electron, respectively. The $\gamma - p$ centre-of-mass energy squared is given by $W_{\gamma p}^2 = W^2 \approx y \cdot s - Q^2$.

The results presented here are based upon data recorded by the H1 [1] and ZEUS [2] collaborations during 1996 through 2000, where the e and p energies were 27.5 GeV and 820 GeV (920 GeV after 1998), yielding an $e - p$ centre-of-mass energy $\sqrt{s} = 300$ GeV (318 GeV), respectively,

OPEN CHARM PRODUCTION

Production mechanisms and theories

The description of inelastic open heavy quark production is based on perturbative QCD (pQCD), detailed here for the charm case, however also valid for the beauty case. In leading order (LO) the direct process of photon-gluon fusion (BGF of $O(\alpha\alpha_s)$, i.e., $\gamma g \to c\bar{c}$) is the dominant contribution. In photoproduction, resolved photon interactions ($\sim O(\alpha/\alpha_s) \times O(\alpha_s^2)$, i.e., $gg \to c\bar{c}, q\bar{q} \to c\bar{c}$) contribute as well and lead to the notion of a photon structure function. However, beyond LO *only the sum* of direct and resolved processes is a well-defined quantity.

The QCD calculations in NLO exist in several schemes: massive and massless ones. Both assume the scale to be hard enough (i.e., α_s small enough) to render pQCD feasible and the factorisation theorem valid. The *massive approach* [3, 4] is a fixed order (in α_s) pQCD calculation (FO=fixed order) with a massive charm quark, and assumes the number of active flavours in the proton to be three ($N_f = 3$) plus the gluon. The charm quark is only produced at the perturbative level. The massive calculations are considered reliable in the low p_T-regime and they agree within the theoretical uncertainties with measurements done, e.g., in $p\bar{p}$-experiments. However, this approach breaks down for large scales (see below) $p_T \gg m_c$ because no large-p_T resummation is done (initial- and final-state collinear divergences are $\sim ln(p_T^2/m_c^2)$).

In the *massless or resummed approach* [5, 6] (RS), charm is assumed to be an active flavour in the proton and the photon, and hence additional processes may occur ($cg \to cg$, $cq \to cq$, ...). Within this approach the final state collinear divergences are absorbed into the fragmentation functions. The large logarithms (high-p_T) are resummed by evolving the parton density functions (PDF) and fragmentation functions (FF) to the relevant scale, either by a coherence treatment [5] or equivalently [7] by a perturbative fragmentation function approach (PFF) [6]. The massless approach is expected to be applicable for $p_T \gg m_c$, however it is anticipated to break down for $p_T \leq m_c$. Both massive and massless schemes have been successfully applied in photoproduction. In DIS, the FO has been in use so far [4, 8].

Recently, for the photoproduction regime a combined approach has become available, a so-called "matched calculation" (FONLL) [9]. This combination asymptotically turns into the FO-model in the low p_T-limit and into the RS-model in the high p_T-limit. The *variable flavour number scheme* (VFNS) [10] adjusts the number of active partons N_f

TABLE 1. ZEUS: measured fragmentation fractions f_c of excited D, D_s mesons, compared with numbers from e^+e^-, in [%].

$f(c \to D_1^0) =$	$1.46 \pm 0.18^{+0.33}_{-0.27} \pm 0.06$	ZEUS
	1.8 ± 0.3	CLEO
$f(c \to D_2^{*0}) =$	$2.00 \pm 0.58^{+1.40}_{-0.48} \pm 0.41$	ZEUS
	1.9 ± 0.3	CLEO
$f(c \to D_{s1}^+) =$	$1.24 \pm 0.18^{+0.08}_{-0.06} \pm 0.14$	ZEUS
	$1.6 \pm 0.4 \pm 0.3$	OPAL

according to the relevant scale. It is applicable only in DIS for inclusive quantities such as σ_{tot}, F_2^c etc.

The available Monte Carlo simulation programs are based on the LO processes (BGF and resolved in some cases). Traditionally, they are based on DGLAP-evolution equations and leading-log parton showers, such as the implementations AROMA, HERWIG and PYTHIA. A new program, CASCADE [11], is based on the CCFM-evolution equation [12], which assumes an angular ordered initial gluon cascade in the evolution, and hence uses unintegrated k_T-dependent parton density functions.

In comparing measured distributions with QCD predictions, the various components of the calculations have to be deconvoluted to access the particular component of interest. The typical cross section $d\sigma \sim g(x) \otimes f_\gamma(x) \otimes \hat{\sigma}_{BGF} \otimes D_{NP}(z)$ contains a) the gluon parton density ($g(x)$) extracted from combined global data sets, including HERA measurements; b) the photon parton density ($f_\gamma(x)$) taken from e^+e^- data and WWA [13], c) the parton level cross sections ($\hat{\sigma}_{BGF}$) calculated in QCD; and e) non-perturbative fragmentation effects ($D_{NP}(z)$).

Studies of fragmentation

The question of the universality of non-perturbative fragmentation effects has been addressed by ZEUS in a study of production ratios of various D-meson states and comparison with the numbers observed at e^+e^--colliders. The measurements [14, 15] concentrated on the relative fragmentation fractions f_c of excited D, D_s mesons in the decay chains $D_1(2420)^0$, $D_2^*(2460)^0 \to D^{*\pm}\pi^{\mp}$ and $D_{s1}^{\pm}(2536) \to D^{*\pm}K_S^0$. Based on a sample of ≈ 30000 D^* mesons, they find the same fractions f_c in ep as in e^+e^-, as indicated in Table 1. Another study compares the strangeness production rate in D_s and D^* mesons. The ratio of the production cross sections of D_s and D^* has been found by ZEUS [16] to be $\frac{\sigma_{ep \to D_s X}}{\sigma_{ep \to D^* X}} = 0.41 \pm 0.07^{+0.03}_{-0.05}$, to be compared with the number obtained in e^+e^--collisions of 0.43 ± 0.04. Within the LUND model, this ratio can be converted into the strangeness suppression factor γ_S. The findings are $\gamma_S = 0.27 \pm 0.05$ by ZEUS, and 0.26 ± 0.03 by e^+e^-. A third measurement compares vector (D^*) to pseudoscalar (D^0) meson production. A measurement of $P_V = \frac{V}{V+P} = \frac{D^*}{D^*+D^0}$, under the assumption $\sigma(D^{*\pm}) = \sigma(D^{*0})$, yields a value of $P_V = 0.546 \pm 0.045 \pm 0.028$ by ZEUS [17]. This can be compared with corresponding values measured at LEP of 0.57 ± 0.05 (OPAL) or

0.595 ± 0.045 (ALEPH).

All these measurements support the hypothesis that the charm fragmentation process is universal, e.g., the results from e^+e^- can be applied to ep-processes.

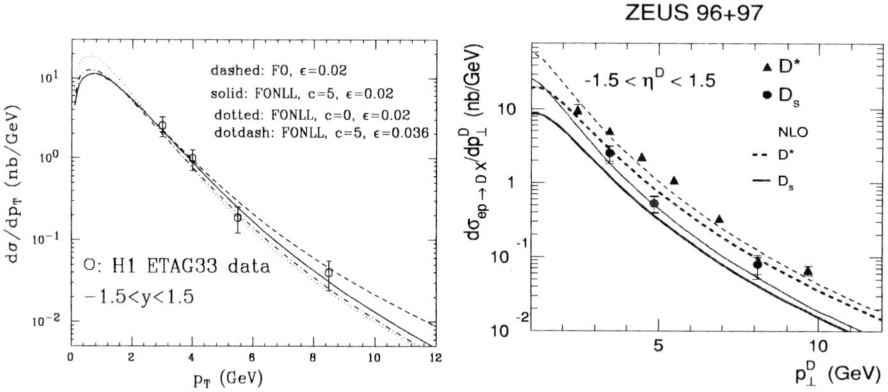

FIGURE 1. $d\sigma/dp_t$ distribution in photoproduction data; (left) H1 D^* results compared with FO and FONLL predictions; (right) ZEUS D^*, D_s results compared with FO predictions with standard (thick lines) or extreme (thin lines) parameter settings.

D-meson Photoproduction

In photoproduction D mesons are identified in the H1 and ZEUS detectors by the decay chain $D^{*+} \to D^0\pi^+$, $D^0 \to K^-\pi^+$or $K^-\pi^+\pi^-\pi^+$ (ZEUS) and $D_s \to \phi\pi$, in the *visible* kinematic region with $p_t(D) > 2.5$ GeV (3.0 GeV in ZEUS analysis) and rapidity $|\hat{y}(D)| < 1.5$.

The visible differential distributions in $d\sigma/dp_t$ for low $Q^2 < 0.01$ GeV^2 for H1 [18] are shown in Fig. 1(left). Overlaid are the predictions by the FO NLO QCD calculations.[2] H1 data [18] are pretty well described by both FO and FONLL [9] with $f_c = 0.27$. The shape of the ZEUS data [16] in Fig. 1(right) is reproduced by NLO predictions both with standard parameters (solid line) and with extreme parameter choices ($m_c = 1.2, \mu_r = 0.5\mu_T$), but the normalisation is too low in the first case. The shape of the rapidity distributions predicted by NLO do not describe the measured data well, with the deviations being maximal in the forward direction. This fact is independently confirmed by a ZEUS measurement based on tagged charm jets [19].

The question is how much of the observed differences are due to fragmentation and how much originate from higher order corrections? An adjustment of the Peterson fragmentation parameter alone does not result in a proper description of the data. There remain discrepancies in the forward region, the clarification of which requires more

---

[2] The renormalisation scale μ_r, the hadron and photon factorization scales μ_{Fh} and $\mu_{F\gamma}$ are customarily given relative to the charm quark mass m_c. In the case of FOPT in photoproduction, the standard values are: $\mu = m_T = \sqrt{m_c^2 + p_t^2}$; and $\mu_r = \mu; \mu_{Fh} = \mu_{F\gamma} = 2\mu$.

studies. The visible range is found to be fairly insensitive to the choice of PDF parametrisations within FO calculations, which are most distinctive, unfortunately, in the hard-to-access low-p_t region for the proton and in the large-$\hat{y}$ region for the photon.

D-meson Production in Deep-Inelastic Scattering

The inclusive production of $D^{*\pm}(2010)$ mesons in deep-inelastic e^+p scattering has been studied both by H1 [20] and ZEUS [21].

Using about 19 pb^{-1} of data taken in 1996 and 1997 in the kinematic region $1 < Q^2 < 100$ GeV^2, H1 measured the production cross section in the visible range $0.05 < y < 0.7$, $|\eta_{D^*}| < 1.5$ and $p_{t\,D^*} > 1.5$ GeV and determined $\sigma_{\rm vis}(e^+p \to e^+D^{*\pm}X) = (8.50 \pm 0.42(\text{stat.})^{+1.21}_{-1.00}(\text{syst.}) \pm 0.65(\text{model}))$ nb.

The FO NLO pQCD calculation [4, 8] predicts for $\sigma_{\rm vis}$ values of $5.2-7.0$ nb depending on the assumptions on the charm quark mass and the fragmentation parameter ε_c. The LO CASCADE program yields a cross section of $\sigma_{\rm vis} \sim 8.0-10.8$ nb.

Furthermore, single and double differential cross sections have been measured and compared to both QCD calculations, as shown in Fig. 2. A slightly better agreement with the data is found with the CCFM model when using the initial gluon distribution fitted to the inclusive F_2-data, in particular at large η-values for low p_t. Note, that only part of the differences between the two predictions is due to the model itself, as, e.g., the GRV-HO parametrisation renders lower normalisation values.

From such differential cross section measurements, the charm structure function F_2^c has been extracted as a function of Q^2 for different values of x, the relative proton momentum of the charm quark:

$$\frac{d^2\sigma^{c\bar{c}}}{dxdQ^2} = \frac{2\pi\alpha^2}{Q^4x}\left[1+(1-y)^2\right]F_2^{c\bar{c}}(x,Q^2)\,. \tag{1}$$

The results are shown in Fig. 3 for both the measurements of the ZEUS and H1 collaborations. The data show a rise of F_2^c with Q^2 with increasing steepness towards small-x values. Similar to the inclusive structure function F_2 the scaling violations are largest for small values of x. However, the steep rise at small x is even larger than predicted by NLO QCD calculations. Charm contributes very significantly - up to 50% - to the scaling violations of F_2. The ratio F_2^c/F_2 is found to be largest at low x and amounts to 10% at low Q^2 and even exceeds 25% for $Q^2 > 25$ GeV^2.

The Q^2-dependence of the cross section measured by ZEUS [21] is compared to the NLO predictions up to very high Q^2-values in Fig. 4(left). Within the large errors, no deviations are visible. Furthermore, in Fig. 4 (right) the Q^2-dependences of $\sigma(e^+p \to D^*X)$ and $\sigma(e^-p \to D^*X)$ are separately shown [22]. The ratio appears - unexpectedly - to slightly increase towards high Q^2 and towards large x (not shown). However, both $\sigma(e^-p)$ and $\sigma(e^+p)$ are described by NLO QCD within errors.

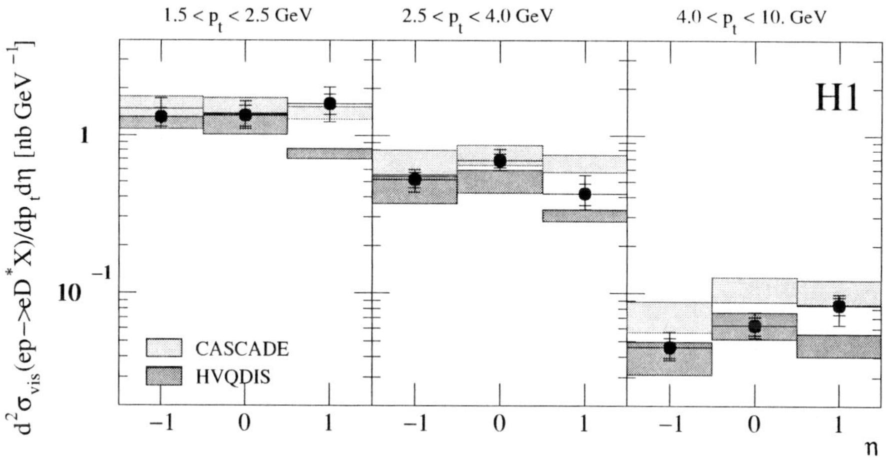

FIGURE 2. Double differential inclusive cross section $\mathrm{d}^2\sigma/\mathrm{d}p_t\mathrm{d}\eta$ in bins of p_{tD^*} Inner and outer error bars correspond to the statistical and the total errors. The expectation of the NLO QCD with GRV98-HO parton densities is indicated by the darker band, while the lighter band shows the LO CASCADE predictions with the initial gluon distribution fitted to the inclusive F_2 data. The upper and lower bounds of both calculations correspond to ($m_c = 1.3$ GeV, $\varepsilon_c = 0.035$) and ($m_c = 1.5$ GeV, $\varepsilon_c = 0.10$), respectively.

INELASTIC CHARMONIUM PRODUCTION

One of the main interests in the leptoproduction of J/Ψ mesons is a clarification of its production process. Studies of charmonium production allow one to probe the strong interaction physics in the transition region between the perturbative QCD regime and non-perturbative phenomenological models, where long range or soft interactions are considerable. The different production processes may be classified in terms of *elastic* and *inelastic* processes, depending on whether the proton stays intact or breaks up. Experimentally, the separation is usually achieved by means of the *inelasticity* variable $z \equiv \frac{q \cdot P_\psi}{q \cdot P}$, where P_ψ is the four-momentum of the J/ψ meson. In the proton rest-frame, z is the relative energy of the J/ψ with respect to the photon, $z \approx \frac{E^*_\psi}{E^*_\gamma}$, with values of 1 for purely elastic, < 0.9 for inelastic and typically < 0.25 for resolved γp-interactions. The J/ψ mesons are experimentally identified by their decays to lepton pairs. Herein, only the inelastic cases are considered.

Predictions for the direct inelastic production of J/Ψ are based on the photon-gluon fusion mechanism, where the photon participates in the hard interaction with a gluon from the proton to form a $c\bar{c}$ pair. The hard interaction is calculable in perturbative QCD. To achieve a colourless J/ψ state, at least one further gluon must be involved in the interaction.

In the theoretically favoured NRQCD/factorization approach [23], the cross section is described by $\sigma(\gamma p \to J/\psi X) = \sum \langle O[n] \rangle \cdot \hat{\sigma}(\gamma p \to c\bar{c}[n]X)$, where $\langle O[n] \rangle$ are the NRQCD long distance matrix elements and $\hat{\sigma}$ the short distance coefficients, calculable in pQCD, with n being the quantum numbers of the colour state, ${}^{2S+1}L_J$.

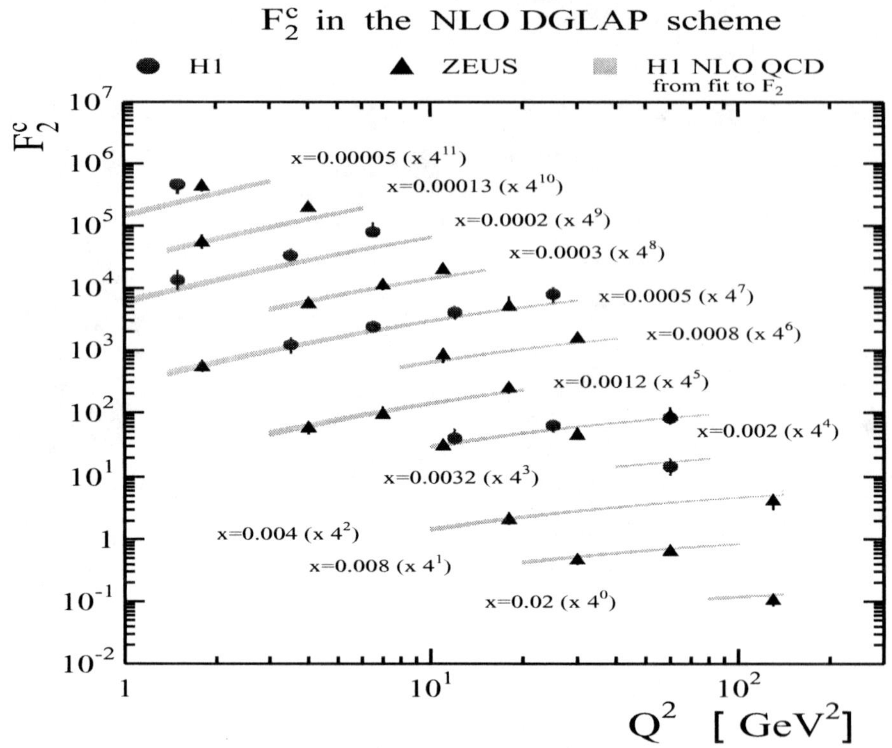

FIGURE 3. F_2^c as derived from the inclusive $D^{*\pm}$ meson production cross sections as a function of Q^2 for different values of x for both H1 and ZEUS data. The curves represent the predictions of the NLO DGLAP evolution based on the parton densities in the proton obtained by the fit to the inclusive F_2.

This NRQCD-approach contains both the traditional colour-singlet model [24, 25] (CSM), where the radiation of a gluon from a $c(\bar{c})$ line to form a colourless $c\bar{c}$ state is part of the hard interaction, and the colour-octet model [26] (COM) where the transition from the coloured $c\bar{c}$ state produced in the hard sub-process to the colour-singlet final state proceeds non-perturbatively. Data from FNAL [27] suggested the existence of sizable CO contributions. Since these CO-contributions $\langle O[8] \rangle$ are believed to be universal, they could be determined from a fit to Tevatron J/ψ data ($^1S_0, ^3P_J$) and applied at HERA. Thus a measure of these COM-contributions at HERA provides a very sensitive test for the consistency of the model.

Up to now, the CSM calculations suffered from large NLO corrections; however recently NLO calculations for CSM in γp for HERA [28] became available, which considerably reduces the uncertainties in the normalisation of the prediction.

In the following cross section measurements are shown for both the medium z-range ($z > 0.35$) and for the low z-range. For comparison with NLO calculations [28] in the CSM, cross-section determinations are restricted to the kinematic region $p_t(J/\psi) >$ 1 GeV.

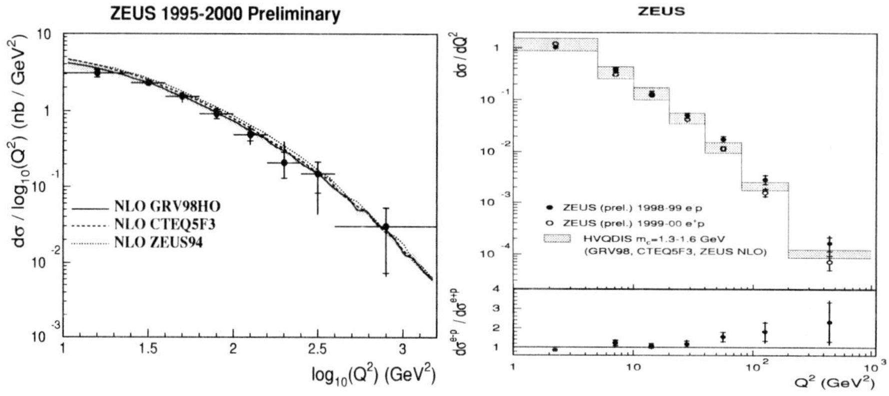

FIGURE 4. Charm production cross section measured by ZEUS; (left) comparison with NLO precictions for different structure functions; (right) comparison of results from electron-p with positron-p collisions.

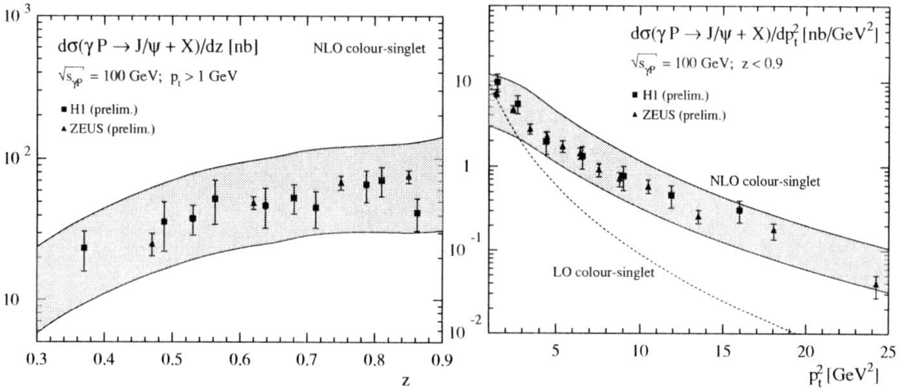

FIGURE 5. Inelastic J/ψ photoproduction cross-sections as functions of (left) z and (right) $p_t^2(J/\Psi\ )$. The data from H1 and ZEUS are shown together with NLO CSM calculations. The band reflects the uncertainty due to the charm mass ($1.3 < m_c < 1.5\,\mathrm{GeV}$) and $0.1175 < \alpha_s(M_Z) < 0.1225$.

The results on inelastic J/Ψ photoproduction from the H1 [29] and ZEUS [30] collaborations are based on integrated luminosities of $22\,\mathrm{pb}^{-1}$ and $38\,\mathrm{pb}^{-1}$, respectively, collected in the years 1996/97. The kinematic regions are characterized by $Q^2 \leq 1\ \mathrm{GeV}^2$, $50 < W_{\gamma p} < 180$ GeV, $0.4 < z < 0.9$ for the ZEUS data and $60 < W_{\gamma p} < 180\,\mathrm{GeV}^2$, $0.3 < z < 0.9$ for H1.

In Fig. 5, the z and the p_t^2 distributions are shown for data from H1 [29] and ZEUS [30], which agree well with each other. The prediction of the CSM in next-to-leading order [28] is overlaid, with a band reflecting the major theoretical uncertainties. Data are pretty well described. That NLO contributions must be included is apparent from the distribution of $p_{t,\psi}^2$ in Fig. 5(right). The theoretical band including NLO contri-

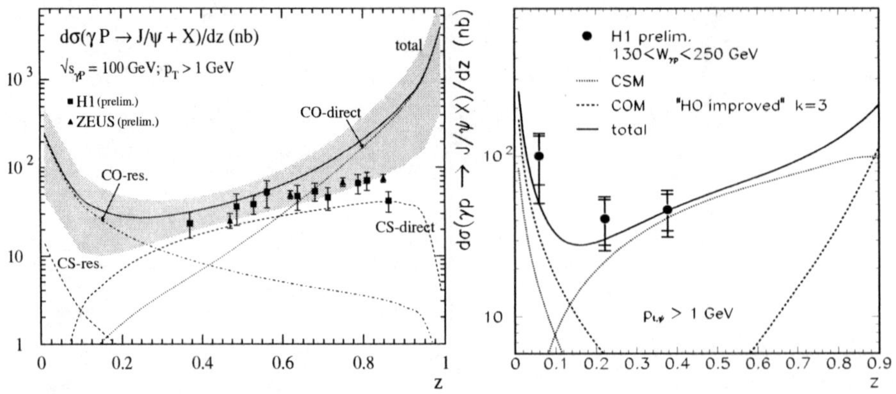

FIGURE 6. Differential cross section for photoproduction of J/Ψ mesons as a function of z in two different z domains. (Left) predictions in LO from Kraemer [28] are overlaid, including direct and resolved contributions. The band depicts the uncertainties, which are mainly due to the long range matrix elements. (Right) H1 data, in a different $W_{\gamma p}$ range compared to medium z. The LO theoretical predictions are from Kniehl [32].

butions describes the data well. It is in this distribution that the discrepancy was found in $p\overline{p} \rightarrow J/\psi + X$ [27], leading to the introduction of CO contributions.

Figure 6 (left) depicts again the differential cross section $d\sigma/dz$, however in a comparison with the theoretical predictions within the NRQCD/factorisation approach calculated in *leading order* [28]. Direct photon contributions as well as resolved processes are included and shown separately. At z above 0.3 direct photon contributions dominate and the CO contribution is dominant at $z \geq 0.6$. The uncertainties of the total prediction shown as a band are mainly due to the uncertainties of the long range matrix elements [28]. The strong rise predicted at large z values is not seen in the data which has led to several theoretical attempts for an explanation. The large z region is dominated by diffraction and no experimental results on inelastic processes are available.

The NRQCD calculations shown in Fig. 6(left) neglect the energy smearing of the transition of the $c\overline{c}$ state to the J/ψ via emission of soft gluons. This effect may possibly change the shape at large z [31], but no definite conclusions can be drawn yet.

An analysis of H1 addressed the low z region with $z < 0.45$, $130 < W_{\gamma p} < 250\,\mathrm{GeV}$ and $p_t > 1\,\mathrm{GeV}$. In Fig. 6(right) a comparison of the data with predictions from [32] is shown. Resolved photon contributions are anticipated to be substantial at such low z values, and CO contributions are expected to exceed the CS contributions. The number of parameters entering this comparison is large, and thus the presently available measurements do not allow one to distinguish between different models yet.

STUDIES OF BEAUTY PRODUCTION

Production of open beauty was measured at HERA through the detection of semileptonic decays and compared with theory based on FO NLO pQCD. Previous data suggest that

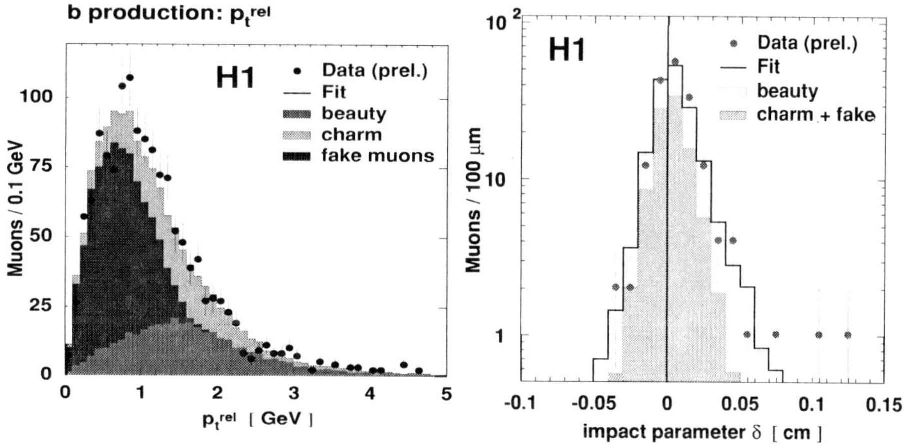

FIGURE 7. (left) H1: fit to the $p^{\mu}_{T,rel}$ distribution using three components, shown separately: beauty, charm and background. in photoproduction; (right) H1: fit to the impact parameter distribution in DIS.

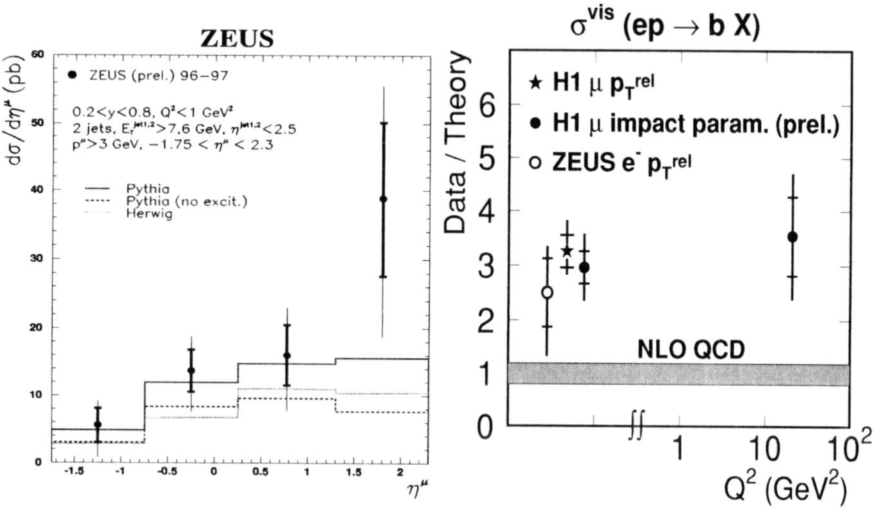

FIGURE 8. (left) ZEUS : differential b-production cross section as function of rapidity, obtained from $p^{\mu}_{T,rel}$ fits in each η-bin. (right) summary of the b-production measurements compared to NLO pQCD predictions.

pQCD calculations are quite accurate in describing charm production, both in shape and normalisation. Because the beauty quark mass, and hence the scale involved, is considerably larger, pQCD is anticipated to be more accurate in describing the beauty case.

The main strategy for detecting beauty is to look for semileptonic decays of the b-quark, $b \to l\nu_l X$, with its particular decay kinematics of a hard lepton (μ or e). In the H1

TABLE 2. ZEUS: measured b-production cross sections $\sigma^{vis}_{b\bar{b}}(e^+p \rightarrow dijet + l^{\pm} + X)$ in nb, compared with LO MC predictions.

Region	$\eta(lepton)$	σ (pb)	HERWIG	PYTHIA	CASCADE
Central μ	$-1.75 < \eta < 1.3$	$36.4 \pm 5.2^{+10.4}_{-9.1}$	12	30	24.5
Forward μ	$1.4 < \eta < 2.4$	$20.5 \pm 6.5^{+6.3}_{-5.7}$	4.1	9.6	5.9
Central e	$-1.1 < \eta < 1.1$	$24.9 \pm 6.4^{+4.2}_{-7.3}$	8	18	18

analysis [33], events with at least two jets with $E_T^{jet} > 6\,\text{GeV}$ and one high momentum muon with $p_t > 2$ GeV are selected. Two different strategies served to distinguish beauty production from other processes such as charm and light quark decays. The first discriminating variable is the so-called $p^{\mu}_{T,rel}$, describing the relative transverse momentum of the muon with respect to the jet axis. The $p^{\mu}_{T,rel}$ distribution is fitted to a superposition of three components: beauty, charm and background (BG). The shapes for the b and c distributions are modeled by simulation. The BG shape is determined from data, and its normalisation is kept fixed for the fit. Such a fit, shown for the H1 analysis with $E_T()jet) > 5$ GeV in Fig. 7(left), yields a considerable b-fraction of 26%.

The second method exploits the long lifetime of the B-mesons. In the case of a semileptonic decay, the muon particle trajectory usually does NOT pass through the production vertex, but has a sizeable impact parameter δ_{μ} [34]. The electroproduction cross section obtained from a combined likelihood fit in the $(p^{\mu}_{T,rel}, \delta_{\mu})$-plane in the visible range, defined by $35° < \theta^{\mu} < 130°$, $p^{\mu}_{T,lab} > 2.0$ GeV, $Q^2 < 1\,\text{GeV}^2$, $95 \leq W_{\gamma p} \leq 270\,\text{GeV}$, is measured to be $\sigma^{vis}(ep \rightarrow b\bar{b}X \rightarrow \mu X') = (160 \pm 16 \pm 29)$ pb, which confirms the published [33] value of $(176 \pm 16^{+26}_{-17})$ pb.

The corresponding predictions by NLO QCD [3] are (54 ± 9) pb, assuming $m_b = 4.75\,\text{GeV}$ and the MRSG and GRV-HO parton density functions for the proton and the photon, respectively. The LO prediction is 38 pb by AROMA and 67 pb by CASCADE programs. The charm production cross section obtained in parallel is in agreement with previous measurements.

A new measurement using this $p^{\mu}_{T,rel}$ technique was recently also performed in the high Q^2 DIS-domain of $2 < Q^2 < 100\,\text{GeV}^2$ and $0.05 < y < 0.7$. The impact parameter fit to the H1 data [35], shown in Fig. 7(right), yielded a cross section of $\sigma^{vis}_{(ep \rightarrow b\bar{b}X \rightarrow \mu X')} = (39 \pm 8 \pm 10)$ pb. This is also much larger than the QCD predictions of (11 ± 2) pb (NLO), 9 pb (LO AROMA) and 15 pb (LO CASCADE).

The ZEUS collaboration [36] has measured b-production in a similar way, via the $p^{\mu}_{T,rel}$ variable in semileptonic decays in different rapidity regions. One analysis of the ZEUS collaboration [37] is based on the electron semileptonic decay. In a total of 38.5 pb^{-1} luminosity, the final data sample of dijet events with at least one electron is obtained by statistical subtraction of a hadron-enriched and an electron-enriched sample. The enrichment criteria rely on the energy loss dE/dx and the ratio E_{el}/E_{tot} of electromagnetic to total deposited energy. Again, the p_T^{rel} distribution is fitted to extract the production cross section for the visible range. The various b-production cross sections for the visible ranges measured by ZEUS and the corresponding LO MC predictions are given in Table 2.

In a recent contribution, ZEUS [38] presented preliminary measurements of the differential b-production cross sections as a function of p_t and rapidity η, using the semi-muonic decays and fitting the p_T^{rel}-spectra for different bins. The results for the photoproduction regime are shown for the η_μ distribution in Fig. 8(left). The overlaid LO predictions based on PYTHIA and HERWIG appear to be lower, in particular in the forward region.

The overall situation is summarized in Fig. 8(right) where the ratio of the measured cross section to that predicted by NLO QCD is displayed. Clearly, the predictions are significantly below the data even at large Q^2.

SUMMARY AND CONCLUSIONS

The production of charm and beauty quarks in ep-collisions at HERA has been studied by identifying charm in the form of D- and J/Ψ -mesons and b through semileptonic decays.

In the overall picture, the hard nature of the charm production process appears to be firmly established, and thus yields reasonable predictions for the total production rates. The shape of differential distributions is in general reasonably well predicted by the pQCD calculations, with possible exceptions at large η-, small Q^2-, and small x-values. The charm contribution to F_2 is found to be very large, driving a substantial part of the scaling violation.

In the case of inelastic J/Ψ meson production, the NLO NRQCD calculations with colour-singlet contributions alone describe the data well within errors, thus exhibiting no indications of colour-octet contributions.

The reasonable agreement in the charm sector is in strong contrast to the case of beauty production, where discrepancies of factors 2-3 have been observed, in contradiction to the naive expectation of perturbation theory.

This raises the need for calculations at the next higher order (NNLO), and for a better understanding of non-perturbative effects, and more detailed measurements on the experimental side.

ACKNOWLEDGMENTS

It is my pleasure to thank the organisers for letting me participate in this very stimulating and fruitful symposium, despite some of the unfortunate circumstances.

REFERENCES

1. H1 collab., I. Abt, *et al*, DESY 96-001 (1996).
2. ZEUS collab., J. Breitweg, *et al*, DESY 95-193 (1995).
3. S. Frixione, *et al*, *Phys. Lett. B* **348**, 633 (1995).

4. B.W. Harris and J. Smith, FSU-HEP-970527, 1997.
5. J. Binnewies, B.A. Kniehl and G. Kramer, DESY-97-012, 1997; B.A. Kniehl, G. Kramer and M. Spira, DESY-96-210, 1996.
6. M. Cacciari, *et al*, *Phys. Rev. D* **55**, 2734 (1997); *Z. Phys. C* **69**, 459 (1995).
7. M. Cacciari and M. Greco, *Phys. Rev. D* **55**, 7134 (1997).
8. E. Laenen, *et al*, *Nucl. Phys. B* **392**, 162 (1993) ; *Nucl. Phys. B* **392**, 229 (1993).
9. S. Frixione, M. Cacciari and P. Nason, EPS-conference 2001, Budapest, 2001.
10. M.A.G. Aivazis, *et al*, SMU-HEP-92-04; *Phys. Rev. D* **50**, 3102 (1994).
11. H. Jung and G.P. Salam, *Eur. Phys. J. C* **19**, 351 (2001).
12. M. Ciafaloni, *Nucl. Phys. B* **296**, 49 (1988); S. Catani, F. Fiorani and G. Marchesini, *Phys. Lett. B* **234**, 339 (1990).
13. C.F. Weizsäcker, *Z. Phys.* **88**, 612 (1934); E.J. Williams, *Phys. Rev.* **45**, 729 (1934).
14. ZEUS collab., contribution 497 to EPS-HEP conference, Budapest 2001.
15. ZEUS collab., contribution 854 to ICHEP conference, Osaka 2000.
16. ZEUS collab., J. Breitweg, *et al*, *Phys. Lett. B* **481**, 213 (2000).
17. ZEUS collab., contribution 501 to EPS-HEP conference, Budapest 2001.
18. H1 collab., S. Aid *et al*, *Nucl. Phys. B* **545**, 21 (1999).
19. ZEUS collab., J. Breitweg *et al*, *Eur. Phys. J. C* **6**, 67 (1999); contribution 528 to EPS-HEP conference, Tampere 1999.
20. H1 collab., C. Adloff, *et al*, DESY-01/100, 2000.
21. ZEUS collab. J. Breitweg, *et al*, *Eur. Phys. J. C* **12**, 35 (2000); contribution to ICHEP, Osaka 2000.
22. ZEUS collab., contribution 493 to EPS-HEP conference, Budapest 2001.
23. G.T. Bodwin *et al*, *Phys. Rev. D* **51**, 1125 (1995), erratum *Phys. Rev. D* **55** 5853 (1997).
24. M. Krämer, *et al*, *Phys. Lett. B* **348**, 657 (1995).
25. M. Krämer, *Nucl. Phys. B* **459**, 3 (1996).
26. M. Cacciari, M. Krämer, DESY 96-005 (1995).
27. F. Abe *et al*, CDF collab., *Phys. Rev. Lett.* **69**, 3704 (1992).
28. M. Krämer, hep-ph/0106120, to be published in Progress in Particle and Nuclear Physics, Vol. 47, issue 1; *Phys. Lett. B* **348**, 657 (1995); *Nucl. Phys. B* **459**, 3 (1996).
29. H1 collab., contributions to EPS-HEP conference, Tampere, Finland, 1999; and to ICHEP conference, Osaka, Japan, 2000.
30. ZEUS collab., contribution 814 to ICHEP98, Vancouver, 1998; and contribution 851 to ICHEP2000, Osaka, Japan.
31. M. Beneke, G. A. Schuler and S. Wolf, *Phys. Rev. D* **62** (2000) 034004, [hep-ph/0001062].
32. B.A. Kniehl and G. Kramer, *Eur. Phys. J.* C **6**, 1999 (493)
33. H1 collab., C. Adloff, *et al*, *Phys. Lett. B* **467**, 156 (1999); erratum submitted to *Phys. Lett. B*.
34. H1 collab., contribution 979 to ICHEP 2000, Osaka 2000.
35. H1 collab., contribution to Moriond QCD, 2001.
36. ZEUS collab., contribution 498 to EPS-HEP conference, Tampere, Finland, 1999.
37. ZEUS collab., J.Breitweg, *et al*, *Eur. Phys. Journal C* **18**, 625 (2001).
38. ZEUS collab., contribution 496 to EPS-HEP conference, Budapest 2001.

Supersymmetry Explanation for the Puzzling Bottom Quark Production Cross Section

Edmond L. Berger

High Energy Physics Division, Argonne National Laboratory, Argonne, IL 60439

Abstract. It has been known for a long time that the cross section for bottom-quark b production at hadron collider energies exceeds theoretical expectations. An additional contribution from pair-production of light gluinos $\tilde{g}$, of mass 12 to 16 GeV, with two-body decays into bottom quarks and light bottom squarks $\tilde{b}$, helps to obtain a b production rate in better agreement with data. The masses of the $\tilde{g}$ and $\tilde{b}$ are restricted further by the ratio of like-sign to opposite-sign leptons at hadron colliders. Constraints on this scenario from other data are examined, and predictions are made for various processes such as Upsilon decay into $\tilde{b}'s$.

INTRODUCTION

The cross section for bottom-quark production at hadron collider energies exceeds the central value of predictions of next-to-leading order (NLO) perturbative quantum chromodynamics (QCD) by about a factor of two [1, 2]. This longstanding discrepancy has resisted fully satisfactory resolution within the standard model [3]. The NLO contributions are large, and it is not excluded that a combination of further higher-order effects in production and/or fragmentation may resolve the discrepancy. However, the disagreement is surprising because the relatively large mass of the bottom quark sets a hard scattering scale at which fixed-order perturbative QCD computations of other processes are generally successful. The photoproduction cross section at DESY's HERA [4] and and the cross section in photon-photon reactions at CERN's LEP [5] also exceed NLO expectations. The data invite the possibility of a contribution from "new physics".

SUPERSYMMETRY INTERPRETATION

The properties of new particles that can contribute significantly to the bottom quark cross section are fairly well circumscribed. To be produced with enough cross section the particles must interact strongly and have relatively low mass. They must either decay into b quarks or be close imitators of b's in a variety of channels of observation. They must evade constraints based on precise data from measurements of Z^o decays at CERN's LEP and SLAC's SLC, and from many lower-energy e^+e^- collider experiments. The minimal supersymmetric standard model (MSSM) is a favorite candidate in many quarters for physics beyond the standard model. It offers a well-motivated theoretical framework and is reasonably well-explored phenomenologically. An explanation within the context of the MSSM [6] can satisfy all of the stated criteria.

CP618, *Heavy Flavor Physics: Ninth International Symposium*, edited by A. Ryd and F. C. Porter

In Ref. [6], the existence is assumed of a relatively light color-octet gluino $\tilde{g}$ (mass $\simeq 12$ to 16 GeV) that decays with 100% branching fraction into a bottom quark b and a light color-triplet bottom squark $\tilde{b}$ (mass $\simeq 2$ to 5.5 GeV). The $\tilde{g}$ and the $\tilde{b}$ are the spin-1/2 and spin-0 supersymmetric partners of the gluon (g) and bottom quark. In this scenario the $\tilde{b}$ is the lightest SUSY particle, and the masses of all other SUSY particles are arbitrarily heavy, i.e., of order the electroweak scale or greater. The $\tilde{b}$ is either long-lived or decays via R-parity violation into a pair of hadronic jets. Improved agreement is obtained with hadron collider rates of bottom-quark production, and several predictions are made that can be tested readily with forthcoming data.

Differential Cross Section

The light gluinos are produced in pairs via standard QCD subprocesses, dominantly $g+g \to \tilde{g}+\tilde{g}$ at Tevatron and Large Hadron Collider (LHC) energies. The $\tilde{g}$ has a strong color coupling to b's and $\tilde{b}$'s and, as long as its mass satisfies $m_{\tilde{g}} > m_b + m_{\tilde{b}}$, the $\tilde{g}$ decays promptly to $b+\tilde{b}$. The magnitude of the b cross section, the shape of the b's transverse momentum p_{Tb} distribution, and the CDF measurement [7] of $B^0-\bar{B}^0$ mixing are three features of the data that help to establish the preferred masses of the $\tilde{g}$ and $\tilde{b}$. In Ref. [6], contributions are included from both $q+\bar{q} \to \tilde{g}+\tilde{g}$ and $g+g \to \tilde{g}+\tilde{g}$. The subprocess $g+b \to \tilde{g}+\tilde{b}$ contributes insignificantly. Shown in Fig. 1 is the integrated p_{Tb} distribution of the b quarks that results from $\tilde{g} \to b+\tilde{b}$, for $m_{\tilde{g}}$ =14 GeV and $m_{\tilde{b}}$ = 3.5 GeV. The results are compared with the cross section obtained from next-to-leading order (NLO) perturbative QCD and CTEQ4M parton distribution functions (PDF's) [8], with m_b = 4.75 GeV, and a renormalization and factorization scale $\mu = \sqrt{m_b^2 + p_{Tb}^2}$. SUSY-QCD corrections to $b\bar{b}$ production are not included as they are not available and are generally expected to be somewhat smaller than the standard QCD corrections. A fully differential NLO calculation of $\tilde{g}$-pair production and decay does not exist either. Therefore, the $\tilde{g}$-pair cross section is computed from the leading order (LO) matrix element with NLO PDF's [8], $\mu = \sqrt{m_{\tilde{g}}^2 + p_{T_{\tilde{g}}}^2}$, and a two-loop expression for the strong coupling α_s. To account for NLO effects, this $\tilde{g}$-pair cross section is multiplied by 1.9, the ratio of inclusive NLO to LO cross sections [9].

A relatively light gluino is necessary in order to obtain a bottom-quark cross section comparable in magnitude to the pure QCD component. Values of $m_{\tilde{g}} \simeq 12$ to 16 GeV are chosen because the resulting $\tilde{g}$ decays produce p_{Tb} spectra that are enhanced primarily in the neighborhood of $p_{Tb}^{\min} \simeq m_{\tilde{g}}$ where the data show the most prominent enhancement above the QCD expectation. Larger values of $m_{\tilde{g}}$ yield too little cross section to be of interest, and smaller values produce more cross section than seems tolerated by the ratio of like-sign to opposite-sign leptons from b decay, as discussed below. The choice of $m_{\tilde{b}}$ has an impact on the kinematics of the b. After selections on $p_{Tb}^{\min}$, large values of $m_{\tilde{b}}$ reduce the cross section and, in addition, lead to shapes of the p_{Tb} distribution that agree less well with the data.

After the contributions of the NLO QCD and SUSY components are added, the magnitude of the bottom-quark cross section and the shape of the integrated $p_{Tb}^{\min}$ distribution

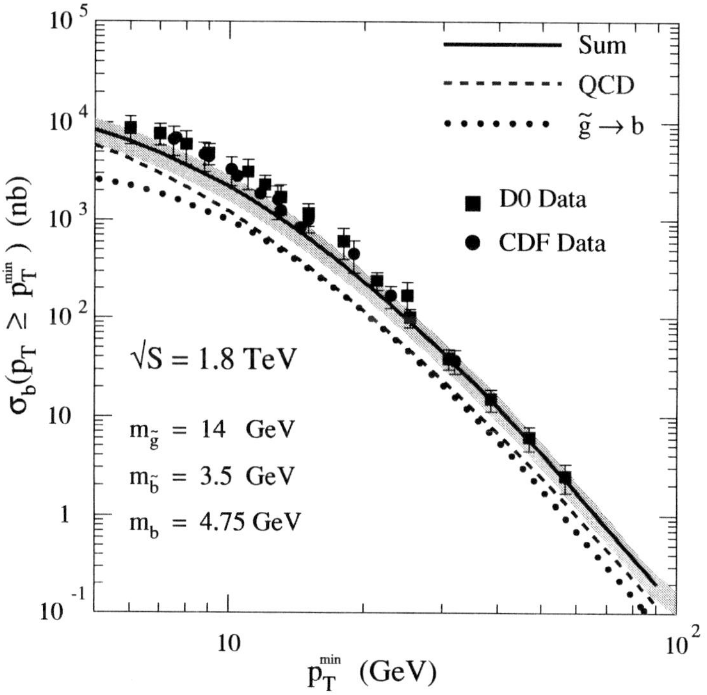

FIGURE 1. Bottom-quark cross section in $p\bar{p}$ collisions at $\sqrt{S} = 1.8$ TeV for $p_{Tb} > p_{Tb}^{\rm min}$ with a gluino of mass $m_{\tilde{g}} = 14$ GeV and a bottom squark of mass $m_{\tilde{b}} = 3.5$ GeV. The dashed curve is the central value of the NLO QCD prediction. The dotted curve shows the p_T spectrum of the b from the SUSY processes. The solid curve is the sum of the QCD and SUSY components. The shaded band represents an uncertainty of roughly $\pm 30\%$ associated with variations of the renormalization and factorization scales, the b mass, and the parton densities. The rapidity cut on the b's is $|y_b| \leq 1$. Data are from Ref. 1.

are described well. Very good agreement is obtained also with data from the UA1 experiment (not shown) [2]. The SUSY process produces bottom quarks in a four-body final state and thus their momentum correlations are different from those of QCD. Angular correlations between muons that arise from decays of b's have been measured [7, 10]. The angular correlations between b's in the SUSY case are nearly indistinguishable from those of QCD once experimental cuts are applied.

The energy dependence of the bottom cross section is a potentially important constraint on models in which new physics is invoked to interpret the observed excess bottom quark yield. Since the assumed $\tilde{g}$ mass is larger than the mass of the b, the $\tilde{g}$ pair process will turn on more slowly with energy than pure QCD production of $b\bar{b}$ pairs. The new physics contribution will depress the ratio of cross sections at 630 GeV and 1.8 TeV from the pure QCD expectation. An explicit calculation with CTEQ4M parton densities and the b rapidity selection $|y| < 1$, yields a pure QCD prediction at NLO of 0.17 +/- 0.02 for $p_{Tb}^{\rm min} = 10.5$ GeV, and 0.16 +/- 0.02 after inclusion of the gluino pair contribution. Either of these numbers is consistent with forthcoming data from CDF on this ratio [11].

Same-sign to Opposite-sign Leptons

Since the $\tilde{g}$ is a Majorana particle, its decay yields both quarks and antiquarks. Gluino pair production and subsequent decay to b's will generate bb and $\bar{b}\bar{b}$ pairs, as well as the $b\bar{b}$ final states that appear in QCD production. When a gluino is highly relativistic, its helicity is nearly the same as its chirality. Therefore, selection of $\tilde{g}$'s whose transverse momentum is greater than their mass will reduce the number of like-sign b's. In the intermediate p_T region, however, the like-sign suppression is reduced. The cuts chosen in current hadron collider experiments for measurement of the ratio of like-sign to opposite-sign muons result in primarily unpolarized $\tilde{g}$'s, and, independent of the $\tilde{b}$ mixing angle, an equal number of like-sign and opposite-sign b's is expected at production. The SUSY mechanism leads therefore to an increase of like-sign leptons in the final state after semi-leptonic decays of the b and $\bar{b}$ quarks. This increase could be confused with an enhanced rate of $B^0-\bar{B}^0$ mixing.

Time-integrated mixing analyses of lepton pairs observed at hadron colliders are interpreted in terms of the quantity $\bar{\chi} = f_d\chi_d + f_s\chi_s$, where f_d and f_s are the fractions of B_d^0 and B_s^0 hadrons in the sample of semi-leptonic B decays, and χ_f is the time-integrated mixing probability for B_f^0. Conventional $b\bar{b}$ pair production determines the quantity $LS_c = 2\bar{\chi}(1-\bar{\chi})$, the fraction of $b\bar{b}$ pairs that decay into like-sign leptons. The SUSY mechanism leads to a new expression

$$LS = \frac{1}{2}\frac{\sigma_{\tilde{g}\tilde{g}}}{\sigma_{\tilde{g}\tilde{g}}+\sigma_{\mathrm{qcd}}} + LS_c\frac{\sigma_{\mathrm{qcd}}}{\sigma_{\tilde{g}\tilde{g}}+\sigma_{\mathrm{qcd}}} = 2\bar{\chi}_{\mathrm{eff}}(1-\bar{\chi}_{\mathrm{eff}}). \quad (1)$$

The factor $1/2$ arises because $N(bb+\bar{b}\bar{b}) \simeq N(b\bar{b})$ in the SUSY mechanism for the selections on p_{Tb} made in the CDF run I analysis. Defining $G = \sigma_{\tilde{g}\tilde{g}}/\sigma_{\mathrm{qcd}}$, the ratio of SUSY and QCD bottom-quark cross sections after cuts, and solving for the effective mixing parameter, one obtains

$$\bar{\chi}_{\mathrm{eff}} = \frac{\bar{\chi}}{\sqrt{1+G}} + \frac{1}{2}\left[1-\frac{1}{\sqrt{1+G}}\right]. \quad (2)$$

The CDF measurement is interpreted in Ref. [6] as a determination of $\bar{\chi}_{\mathrm{eff}} = 0.131 \pm 0.02 \pm 0.016$ [7]. This number is marginally larger than the world average value $\bar{\chi} = 0.118 \pm 0.005$ [12], assumed to be the contribution from the pure QCD component only.

The ratio G is determined in the region of phase space where the measurement is made [7], with both final b's having $p_{Tb} \geq 6.5$ GeV and rapidity $|y_b| \leq 1$. With $m_{\tilde{b}} = 3.5$ GeV, $G = 0.37$ and 0.28 for gluino masses $m_{\tilde{g}} = 14$ and 16 GeV, respectively. The predictions are $\bar{\chi}_{\mathrm{eff}} = 0.17 \pm 0.02$ for $m_{\tilde{g}} = 14$ GeV, and $\bar{\chi}_{\mathrm{eff}} = 0.16 \pm 0.02$ with $m_{\tilde{g}} = 16$ GeV. Additional theoretical uncertainties arise because there is no fully differential NLO calculation of gluino production and subsequent decay to b's. The choice $m_{\tilde{g}} > 12$ GeV leads to a calculated $\bar{\chi}_{\mathrm{eff}}$ that is consistent with the data within experimental and theoretical uncertainties. With $\sigma_{\tilde{g}\tilde{g}}/\sigma_{\mathrm{qcd}} \sim 1/3$, the mixing data and the magnitude and p_T dependence of the b production cross section can be satisfied.

CONSTRAINTS FROM OTHER DATA

An early study by the UA1 Collaboration [13] excludes $\tilde{g}$'s in the mass range $4 < m_{\tilde{g}} < 53$ GeV, but it starts from the assumption that there is a light neutralino $\tilde{\chi}_1^0$ whose mass is less than the mass of the gluino. The conclusion is based on the absence of the expected decay $\tilde{g} \to q + \bar{q} + \not{E}_T$, where $\not{E}_T$ represents the missing energy associated with the $\tilde{\chi}_1^0$. In the scenario discussed above, this decay process does not occur since the bottom squark is the LSP, the SUSY particle with lowest mass, and the $\tilde{\chi}_1^0$ mass is presumed to be large (*i.e.*, > 50 GeV). An analysis of 2- and 4-jet events by the ALEPH collaboration [14] disfavors $\tilde{g}$'s with mass $m_{\tilde{g}} < 6.3$ GeV but not $\tilde{g}$'s in the mass range relevant for the SUSY interpretation of the bottom quark production cross section. A similar analysis is reported by the OPAL collaboration [15]. A light $\tilde{b}$ is not excluded by the ALEPH analysis. The exclusion by the CLEO collaboration [16] of a $\tilde{b}$ with mass 3.5 to 4.5 GeV does not apply since their analysis focuses only on the decays $\tilde{b} \to c\ell\tilde{\nu}$ and $\tilde{b} \to c\ell$. The $\tilde{b}$ need not decay leptonically nor into charm. On the other hand, these data might be reinterpreted in terms of a bound on the R-parity violating lepton-number violating decay of $\tilde{b}$ into $c\ell$. It would be interesting to study the hadronic decays $\tilde{b} \to cq$, with $q = d$ or s, and $\tilde{b} \to us$ with the CLEO data. The DELPHI collaboration's [17] search for long-lived squarks in their $\gamma\gamma$ event sample is not sensitive to $m_{\tilde{b}} < 15$ GeV.

There are important constraints on couplings of the bottom squarks from precise measurements of Z^0 decays. A light $\tilde{b}$ would be ruled out unless its coupling to the Z^0 is very small. The squark couplings to the Z^0 depend on the mixing angle θ_b. As described in Ref. [18], if the light bottom squark $(\tilde{b}_1)$ is an appropriate mixture of left-handed and right-handed bottom squarks, its lowest-order (tree-level) coupling to the Z^0 can be arranged to be small if $\sin^2\theta_b \sim 1/6$. The couplings $Z_{\tilde{b}_1\tilde{b}_2}$ and $Z_{\tilde{b}_2\tilde{b}_2}$ survive, where $\tilde{b}_2$ is the heavier bottom squark. However, as long as the combination of the masses $m_{\tilde{b}_1} + m_{\tilde{b}_2}$ is less than the maximum center-of-mass energy explored at LEP, these couplings present no difficulty. This condition roughly implies $m_{\tilde{b}_2} > 200$ GeV. However, much lower masses of $\tilde{b}_2$ might be tolerated. A careful phenomenological analysis is needed of expected $\tilde{b}_2$ decay signatures, along with an understanding of detection efficiencies and expected event rates, before one knows the admissible range of masses consistent with LEP data. At higher-order, unless the $\tilde{b}_2$ mass is of order 100 GeV, contributions from loop processes in which light gluinos are exchanged may produce significant deviations from measurements of the ratios A_b, the forward-backward b asymmetry at the Z^0, and R_b, the hadronic branching ratio of the Z^0 into b quarks [19, 20].

Bottom squarks make a small contribution to the inclusive cross section for $e^+e^- \to$ hadrons, in comparison to the contributions from quark production, and $\tilde{b}\bar{\tilde{b}}$ resonances are likely to be impossible to extract from backgrounds [21]. The angular distribution of hadronic jets produced in e^+e^- annihilation can be examined in order to bound the contribution of scalar-quark production. Spin-1/2 quarks and spin-0 squarks emerge with different distributions, $(1 \pm \cos^2\theta)$, respectively. The angular distribution measured by the CELLO collaboration [22] is consistent with the production of a single pair of charge-1/3 squarks along with five flavors of quark-antiquark pairs. A new examination of the angular distribution with greater statistics would be valuable.

PREDICTIONS AND IMPLICATIONS

Υ Decay into Bottom Squarks

If the bottom squark mass is less than half the mass of one of the Upsilon states, then Upsilon decay to a pair of bottom squarks might proceed with sufficient rate for experimental observation or exclusion of a light bottom squark. In Ref. [23], the expected rate for $\Upsilon \to \tilde{b}\tilde{b}^*$ is computed as a function of the masses of the bottom squark and the gluino. The mass of the gluino enters because the $\tilde{g}$ is exchanged in the decay subprocesses. The electronic width of the Υ is used as the source of absolute normalization.

The data sample is largest at the $\Upsilon(4S)$. For a fixed gluino mass of 14 GeV, the branching fraction into a pair of bottom squarks is about 10^{-3}, for $m_{\tilde{b}} = 2.5$ GeV, and about 10^{-4} for $m_{\tilde{b}} = 4.85$ GeV. A sample as large as 10,000 may be available in current data from runs of the CLEO detector.

The predicted decay rates for the $\Upsilon(nS)$, $n = 1, 3$ are shown in Fig. 2 These curves can be read as predictions of the width for the corresponding values of $m_{\tilde{b}}$ and $m_{\tilde{g}}$, or as lower limits on the sparticle masses given known bounds on the branching fractions. The current experimental uncertainties on the hadronic widths of the Υ's are compatible with the range of values of $m_{\tilde{b}}$ and $m_{\tilde{g}}$ favored in the work on the bottom quark production cross section in hadron reactions described above. The analysis of $\Upsilon(nS)$ decays shows nevertheless that tighter experimental bounds on the bottom squark fraction are potentially powerful for the establishment of lower bounds on $m_{\tilde{b}}$ and $m_{\tilde{g}}$.

The hadronic width of the Υ is calculated in conventional QCD perturbation theory from the three-gluon decay subprocess, $\Upsilon \to 3g$, and $\Gamma_{3g} \propto \alpha_s^3$. The SUSY subprocess adds a new term to the hadronic width from $\Upsilon \to \tilde{b} + \bar{\tilde{b}}$. If this new subprocess is present but ignored in the analysis of the hadronic width, the true value of $\alpha_s(\mu = m_b)$ will be smaller than that extracted from a standard QCD fit by the factor $(1 - \Gamma_{\mathrm{SUSY}}/\Gamma_{3g})^{\frac{1}{3}}$. For a contribution from the $\tilde{b}\bar{\tilde{b}}$ final state that is 25% of $\Gamma^{\Upsilon}(1S)$ (*i.e.*, a ratio $R^{\Upsilon}_{\tilde{b}\bar{\tilde{b}}} = 10$ in Fig. 2), the value of α_s extracted will be reduced by a factor of 0.9, at the lower edge of the approximately 10% uncertainty band on the commonly quoted value of $\alpha_s(m_b)$ [24]. A thorough analysis would require the computation of next-to-leading order contributions in SUSY-QCD to both the $3g$ and $\tilde{b}\bar{\tilde{b}}$ amplitudes and the appropriate evolution of $\alpha_s(\mu)$ with inclusion of a light gluino and a light bottom squark.

Direct observation of Upsilon decay into bottom squarks requires an understanding of the ways that bottom squarks may manifest themselves, discussed in more detail below. Possible baryon-number-violating R-parity-violating decays of the bottom squark lead to $u + s$; $c + d$; and $c + s$ final states. These final states of four light quarks should be distinguishable from conventional hadronic final states mediated by the three-gluon intermediate state. If the $\tilde{b}$ lives long enough, it will pick up a light quark and turn into a B-mesino, $\widetilde{B}$. Charged B-mesino signatures in Υ decay include single back-to-back equal momentum tracks in the center-of-mass; measurably lower momentum than lepton pairs (< 4 GeV/c *vs.* $\simeq 5$ GeV/c for muons and electrons); $1 + \cos^2\theta$ angular distribution; and ionization, time-of-flight, and Cherenkov signatures consistent with a particle whose

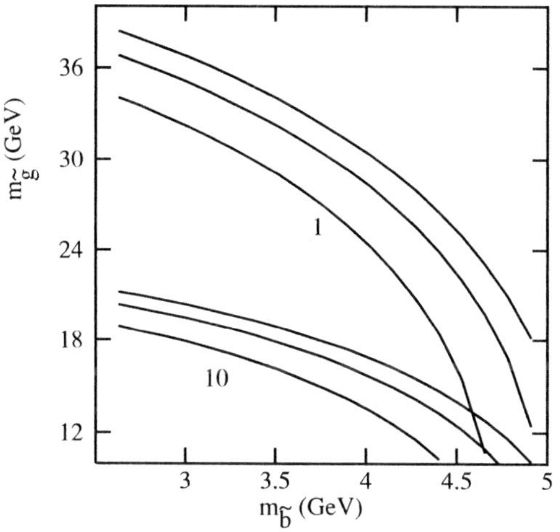

FIGURE 2. Loci in the $m_{\tilde{b}} - m_{\tilde{g}}$ plane for which the rate for Upsilon decay into a pair of bottom squarks is either 10 times $\Gamma_{\ell\bar{\ell}}$ for each $\Upsilon(nS)$, with $n = 1-3$ (lower set of curves) or equal to $\Gamma_{\ell\bar{\ell}}$ (upper set). Above the curves, the rate would be less. Within each set of curves, the order is (bottom to top) $1S, 2S, 3S$.

mass is heavier than that of a proton. At stake is discovery, or new limits on the mass, of the $\tilde{b}$ as well as measurement of or new limits on the R-parity violation couplings of the $\tilde{b}$.

Pseudoscalar η_b decay into a pair of bottom squarks is forbidden, but the higher-order process of η_b decay into a pair of B-mesinos can proceed. Decays of the χ_{b0} and χ_{b2} into a pair of bottom squarks are allowed.

Hadron Reactions

Among the predictions of this SUSY scenario, the most clearcut is pair production of like-sign charged B mesons at hadron colliders, B^+B^+ and B^-B^-. To verify the underlying premise, that the cross section exceeds expectations of conventional perturbative QCD, a new measurement of the absolute rate for b production in run II of the Tevatron is important. A very precise measurement of $\bar{\chi}$ in run II is desirable. Since the fraction of b's from gluinos changes with p_{Tb}, a change of $\bar{\chi}$ is expected when the cut on p_{Tb} is changed. The b jet from $\tilde{g}$ decay into $b\tilde{b}$ will contain the $\tilde{b}$, implying unusual material associated with the $\tilde{b}$ in some fraction of the $b\bar{b}$ data sample. The existence of light $\tilde{b}$'s means that they will be pair-produced in partonic processes, leading to a slight increase ($\sim 1\%$) in the hadronic dijet rate.

The SUSY approach increases the b production rate at HERA and in $\gamma\gamma$ collisions at LEP by a small amount, not enough perhaps if early experimental indications in these cases are confirmed [4, 5]. Full NLO SUSY-QCD studies should be undertaken. In these two cases, the apparent discrepancy may find at least part of its resolution in the fact that $b\bar{b}$ production occurs very near threshold where fixed-order QCD calculations are not obviously reliable. Uncertain parton densities of photons may play a significant role.

Running of α_s

The presence of a light gluino and a light bottom squark slow the running of the strong coupling strength $\alpha_s(\mu)$. Above gluino threshold, the β function of (SUSY) QCD is

$$\beta(\alpha_S) = \frac{\alpha_S^2}{2\pi}\left(-11+\frac{2}{3}n_f+\frac{1}{6}n_s+2\right). \tag{3}$$

The $\widetilde{b}$ (color triplet scalar) contributes little to the running, equivalent to that of 1/4'th of a new flavor, but the $\widetilde{g}$ (color octet fermion) is much more significant, equivalent to that of 3 new flavors of quarks. A precise determination of $\beta(\alpha_S)$ appears to be the best way to exclude the presence of a light gluino or to establish its possible presence.

In the standard model, a global fit to all observables provides an indirect measurement of α_s at the scale of the Z boson mass M_Z. The value $\alpha_s(M_Z) \simeq 0.1184 \pm 0.006$ describes most observables properly [24]. Extrapolation from M_Z to a lower scale μ with inclusion of a light gluino reduces $\alpha_s(\mu)$ from its pure QCD value. The presence of a light gluino, with or without a light bottom squark, also requires reanalysis of the phenomenological determinations of $\alpha_s(\mu)$ at all scales to take into account SUSY processes and SUSY-QCD corrections to the amplitudes that describe the relevant processes. To date, a systematic study of this type has not been undertaken, but, as mentioned above, consistency is achieved for Υ decays. A lesser value of $\alpha_s(m_b)$ leads, under slower evolution, to the same $\alpha_s(M_Z)$.

Slower running of $\alpha_s(Q)$ also means a slower evolution of parton densities at small x, an effect that might be seen in HERA data for $Q > m_{\tilde{g}}$. Presence of a scalar $\tilde{b}$ in the proton breaks the Callan-Gross relation and yields a non-zero leading-twist longitudinal structure function $F_L(x,Q)$ at leading-order.

$\widetilde{b}$-onia

Bound states of bottom squark pairs could be seen as $J^P = 0^+$, 1^-, 2^+, ... mesonic resonances in $\gamma\gamma$ reactions and in $p\bar{p}$ formation, with masses in the 4 to 10 GeV range. They could show up as narrow states in the $\mu^+\mu^-$ invariant mass spectra at hadron colliders, between the J/Ψ and Υ. At an e^+e^- collider, the intermediate photon requires production of a $J^{PC} = 1^{--}$ state. Bound states of low mass squarks with charge $2/3$ were studied with a potential model [21]. The small leptonic widths were found to preclude bounds for $m_{\tilde{q}} > 3$ GeV. For bottom squarks with charge $-1/3$, the situation is more difficult.

$\tilde{b}$ lifetime and observability

Strict R-parity conservation in the MSSM forbids $\tilde{b}$ decay unless there is a lighter supersymmetric particle. R-parity-violating and lepton-number-violating decay of the $\tilde{b}$ into at least one lepton is disfavored by the CLEO data [16] and would imply the presence of an extra lepton, albeit soft, in some fraction of b jets observed at hadron colliders. The baryon-number-violating R-parity-violating ($\not{R}_p$) term in the MSSM superpotential is $\mathcal{W}_{\not{R}_p} = \lambda''_{ijk} U^c_i D^c_j D^c_k$; U^c_i and D^c_i are right-handed-quark singlet chiral superfields; and i, j, k are generation indices. The limits on individual $\not{R}_p$ and baryon-number violating couplings λ'' are relatively weak for third-generation squarks [25, 26], $\lambda''_{ijk} < 0.5$ to 1.

The possible $\not{R}_p$ decay channels for the $\tilde{b}$ are $123 : \bar{\tilde{b}} \to u + s$; $213 : \bar{\tilde{b}} \to c + d$; and $223 : \bar{\tilde{b}} \to c + s$. The hadronic width is [26]

$$\Gamma(\tilde{b} \to \text{jet} + \text{jet}) = \frac{m_{\tilde{b}}}{2\pi} \sin^2\theta_{\tilde{b}} \sum_{j<k} |\lambda''_{ij3}|^2. \tag{4}$$

If $m_{\tilde{b}} = 3.5$ GeV, $\Gamma(\tilde{b} \to ij) = 0.08|\lambda''_{ij3}|^2$ GeV. Unless all λ''_{ij3} are extremely small, the $\tilde{b}$ will decay quickly and leave soft jets in the cone around the b. b-jets with an extra c are possibly disfavored by CDF, but a detailed simulation is needed.

If the $\tilde{b}$ is relatively stable, the $\tilde{b}$ could pick up a light $\bar{u}$ or $\bar{d}$ and become a $\tilde{B}^-$ or $\tilde{B}^0$ "mesino" with $J = 1/2$, the superpartner of the B meson. The mass of the mesino would fall roughly in the range 3 to 7 GeV for the interval of $\tilde{b}$ masses we consider. The charged mesino could fake a heavy muon if its hadronic cross section is small and if it survives passage through the hadron calorimeter and exits the muon chambers. Extra muon-like tracks would then appear in a fraction of the $b\bar{b}$ event sample, but tracks that leave some activity in the hadron calorimeter. The mesino has baryon number zero but acts like a heavy $\bar{p}$ – perhaps detectable with a time-of-flight apparatus. A long-lived $\tilde{b}$ is not excluded by conventional searches at hadron and lepton colliders, but an analysis [27] similar to that for $\tilde{g}$'s should be done to verify that there are no additional constraints on the allowed range of $\tilde{b}$ masses and lifetimes.

ACKNOWLEDGMENTS

I am indebted to Brian Harris, David E. Kaplan, Zack Sullivan, Tim Tait, Carlos Wagner and Lou Clavelli for their collaboration and suggestions. I have benefitted from valuable discussions with Barry Wicklund, Tom LeCompte, Bruce Berger, and Harry Lipkin. The research reported here was supported by the U.S. Department of Energy under Contract W-31-109-ENG-38. This paper was prepared at the invitation of the organizers of the 9th

International Symposium on Heavy Flavor Physics, Caltech, Pasadena, CA, September 10 - 13, 2001. I thank David Hitlin, Frank Porter, and the other members of the Local Organizing Committee for an excellent, timely, and most enjoyable meeting.

REFERENCES

1. CDF Collaboration, F. Abe *et al.*, Phys. Rev. Lett. **71**, 500 (1993); *ibid* **79**, 572 (1997); *ibid* **79**, 572 (1997); *ibid* **75**, 1451 (1995); CDF Collaboration, D. Acosta *et al.*, http://arXiv.org/abs/hep-ph/0111359; D0 Collaboration, B. Abbott *et al.*, Phys. Lett. B **487**, 264 (2000) and Phys. Rev. Lett. **85**, 5068 (2000).
2. CERN UA1 Collaboration, C. Albajar *et al.*, Phys. Lett. B **213**, 405 (1988); **256**, 121 (1991). Early comparisons of theoretical expectations show good agreement with these data but are based on DFLM and other sets of parton densities now considered outmoded.
3. P. Nason *et al.*, in *Proceedings of the 1999 CERN Workshop on Standard Model Physics (and more) at the LHC*, edited by G. Altarelli and M. L. Mangano, CERN Yellow Report No. CERN 2000-004, p. 231.
4. H1 Collaboration, C. Adloff *et al.*, Phys. Lett. B **467**, 156 (1999); ZEUS Collaboration, J. Breitweg *et al.*, Eur. Phys. J. C **18**, 625 (2001).
5. L3 Collaboration, M. Acciarri *et al.*, Phys. Lett. B **503**, 10 (2001); OPAL Collaboration, A. Csilling *et al.*, Photon 2000, Lancaster University, UK, 2000.
6. E. L. Berger, B. W. Harris, D. E. Kaplan, Z. Sullivan, T. Tait, and C. E. M. Wagner, arXiv:hep-ph/0012001, Phys. Rev. Lett. **86**, 4231 (2001).
7. CDF Collaboration, F. Abe *et al.*, Phys. Rev. D **55**, 2546 (1997).
8. CTEQ Collaboration, H. L. Lai *et al.*, Phys. Rev. D **55**, 1280 (1997).
9. W. Beenakker, R. Hopker, M. Spira, P. M. Zerwas, Nucl. Phys. **B492**, 51 (1997); W. Beenakker, R. Hopker, and M. Spira, arXiv:hep-ph/9611232.
10. D0 Collaboration, B. Abbott *et al.*, Phys. Lett. B **487**, 264 (2000) and references therein.
11. CDF Collaboration, paper to appear.
12. Particle Data Group, Eur. Phys. J. C **15** 1, (2000).
13. CERN UA1 Collaboration, C. Albajar *et al.*, Phys. Lett. B **198**, 261 (1987).
14. ALEPH Collaboration, R. Barate *et al.*, Z. Phys. **C76**, 1 (1997); ALEPH contribution to the 2001 Summer conferences ALEPH 2001-042 CONF 2001-026.
15. OPAL Collaboration, G. Abbiendi *et al.*, Eur. Phys. J. C **20**, 601 (2001).
16. CLEO Collaboration, V. Savinov *et al.*, Phys. Rev. D **63**, 051101 (2001).
17. DELPHI Collaboration, P. Abreu *et al.*, Phys. Lett. B **444**, 491 (1998).
18. M. Carena, S. Heinemeyer, C. E. M. Wagner, and G. Weiglein, arXiv:hep-ph/0008023, Phys. Rev. Lett. **86**, 4463 (2001).
19. J. Cao, Z. Xiong, and J. M. Yang, arXiv:hep-ph/0111144.
20. T. M. P. Tait and C. E. M. Wagner, work in progress.
21. C. R. Nappi, Phys. Rev. D **25**, 84 (1982); S. Pacetti and Y. Srivastava, arXiv:hep-ph/0007318.
22. CELLO Collaboration, H.-J. Behrend *et al.*, Phys. Lett. B **183**, 400 (1987).
23. E. L. Berger and L. Clavelli, Phys. Lett. B **512**, 115 (2001), arXiv:hep-ph/0105147.
24. S. Bethke, J. Phys. **G26**, R27 (2000).
25. B. C. Allanach, A. Dedes, and H. K. Dreiner, Phys. Rev. D **60**, 075014 (1999).
26. E. L. Berger, B. W. Harris, and Z. Sullivan, Phys. Rev. Lett. **83**, 4472 (1999) and arXiv:hep-ph/0012184, Phys. Rev. D **63**, 115001 (2001).
27. H. Baer, K. Cheung, and J. F. Gunion, Phys. Rev. D **59**, 075002 (1999).

Nonrelativistic Bound States in Quantum Field Theory

Aneesh V. Manohar

University of California, San Diego, 9500 Gilman Drive, La Jolla, CA 92093-0319

Abstract. Nonrelativistic bound states are studied using an effective field theory with an expansion in the velocity v of the bound state constituents. Logarithms of v can be summed using the the velocity renormalization group. This calculational method provides new insights into the nature of the perturbation series for QED bound states such as Hydrogen and Positronium, and allows one to compute the energy series up to order $\alpha^8 \ln^3 \alpha$, which is necessary for the present comparison between theory and experiment. The formalism can also be applied to nonrelativistic QCD problems such as the Υ, and $\bar{t}t$ production near threshhold.

INTRODUCTION

Nonrelativistic bound states in QCD and QED provide an interesting and highly non-trivial problem to which effective field theory methods can be applied [1, 2]. The QCD bound states we will consider are heavy $\bar{Q}Q$ states such as $\bar{t}t$ bound states or the Υ system. In QED, the classic examples are Hydrogen, muonium ($\mu^+ e^-$), and positronium. Each of these systems has three important scales, m the fermion mass, mv the fermion momentum, and mv^2, the fermion energy. (For Hydrogen and muonium, m is the electron mass or the reduced mass of the two particles.) The velocity v is of order the coupling constant (α_s or α), and we will only consider the case $v \ll 1$, $mv^2 \gg \Lambda_{\rm QCD}$ so that nonperturbative effects are small.

The goal is to correctly separate the scale m, mv and mv^2 for nonrelativistic bound state problems using an effective field theory, and to sum large logarithms using the renormalization group. The large logarithms in this case are $\ln p/m$, $\ln E/m$ and $\ln p/E$ which are proportional to $\ln v$, and lead to $\ln \alpha$ contributions to bound state energies. Furthermore, for QCD, the effective theory also determines the scale of the strong coupling constant, i.e. whether one should use $\alpha_s(m)$, $\alpha_s(mv)$ or $\alpha_s(mv^2)$. The nonrelativistic effective theory, NRQCD/NRQED, has been studied extensively in the past [1–17]. What is new is the precise formulation of the effective theory, and the way in which the renormalization group is scaling is implemented.

The results presented here can be applied to the study of $\bar{t}t$ production in the threshold region. There is a large ratio of scales, $m_t \sim 175$ GeV, $m_t v \sim 26$ GeV and $m_t v^2 \sim 4$ GeV, where $v \sim 0.15$ is the typical velocity in the nonrelativistic bound state. Clearly $\alpha_s \ln v$ is not small, and summing logarithms is important in this case. This calculation is described in I. Stewart's talk. The results are also useful in QED. While $\alpha \ln \alpha$ is small, it is important to compute to high orders because the experiments have high precision. The Hydrogen Lamb shift of 1057.845 MHz is known to an accuracy of 9 KHz [18], the

CP618, *Heavy Flavor Physics: Ninth International Symposium*, edited by A. Ryd and F. C. Porter

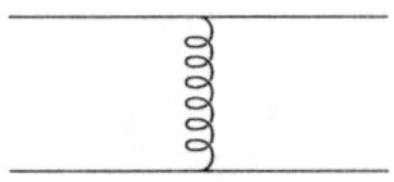

FIGURE 1. A potential gauge boson exchange. The typical momentum and energy transfered are mv and mv^2.

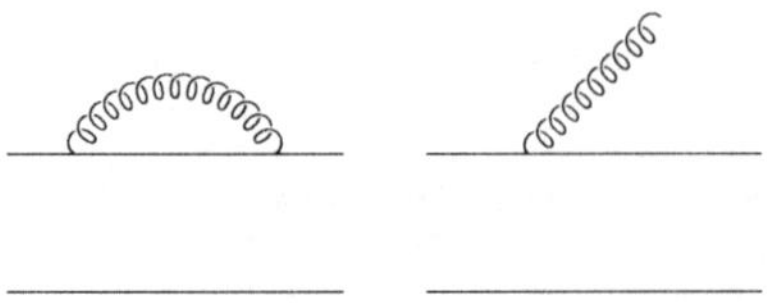

FIGURE 2. Graphs containing ultrasoft photons, with energy and momentum of order mv^2.

Hydrogen hyperfine splitting is measured to be 1420.405 751 766 7(9) MHz [19], and the muonium hyperfine splitting is 4463.302 776(55) MHz [20]. The binding energy of Hydrogen, $m_e\alpha^2/2$ is 2×10^{10} MHz, so the experimental error in the Lamb shift of 10 ppm is a part in 10^{12} of the binding energy. We will be able to compute corrections of order $m_e\alpha^8 \ln^3 \alpha/(4\pi)^2 \sim 5$ KHz to the Lamb shift, which are relevant for the present comparison between theory and experiment. [The counting of 4π factors for bound states is a little different than the conventional counting [21]. Potential loops give powers of α whereas soft and ultrasoft loops give powers of $\alpha/(4\pi)$.] A detailed comparison of theory and experiment for QED can be found in Refs. [22, 23, 24].

MOMENTUM REGIONS AND DEGREES OF FREEDOM

The basic problem can be seen by drawing a few Feynman diagrams. A typical gauge boson exchange in the t channel such as Fig. 1 has momentum transfer of order $p \sim mv$. A wavefunction graph or radiated gauge boson graph such as Figs. 2 have gauge boson momenta of order $E \sim mv^2$. More interesting diagrams such as those in Figs. 3 involve gauge bosons with momenta of order p and order E. In a graph such as Fig. 4, the vacuum polarization insertions make the effective coupling of the two gluons $\alpha_s(mv)$ and $\alpha_s(mv^2)$ respectively. One result which should be clear from Fig. 4 is that graphs can involve $\alpha_s(mv)$ and $\alpha_s(mv^2)$ *simultaneously*. We will return to this important point later on.

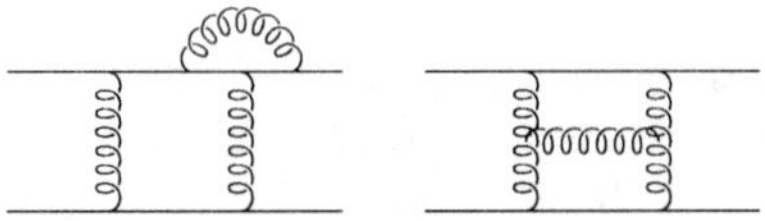

FIGURE 3. Graphs containing gauge bosons carrying momentum of order mv and mv^2.

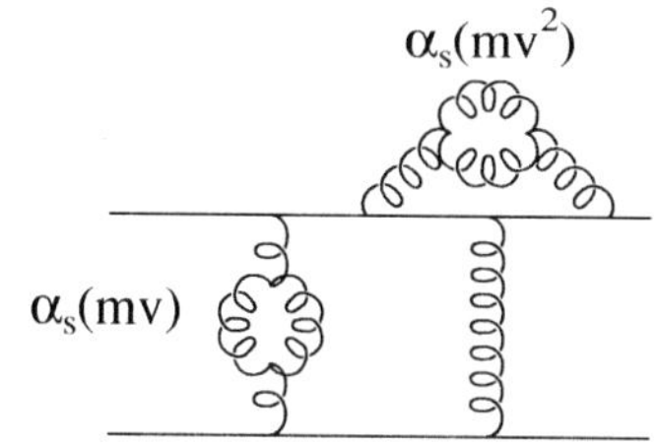

FIGURE 4. An example of a graph involving both $\alpha(mv)$ and $\alpha(mv^2)$.

The Feynman integrals in the full theory can be evaluated using the threshold expansion [29]. The important momentum regions (in Feynman gauge) are referred to in the literature as hard ($E \sim m$, $p \sim m$), potential ($E \sim mv^2$, $p \sim mv$), ultrasoft ($E \sim mv^2$, $p \sim mv^2$) and soft ($E \sim mv$, $p \sim mv$). The threshold expansion momentum regions are often used to describe bound state computations; however it is important to note that *the threshold expansion is not an effective field theory*. To construct an effective field theory, one needs to include only modes that can be on-shell. The effective theory therefore has nonrelativistic fermions (which are potential modes), and soft and ultrasoft gauge boson modes. The hard fermion and gauge boson momentum regions, the soft fermion momentum region, and the potential gauge boson momentum region do not require modes in the effective theory.

The desired effective theory is valid for energies and momenta much smaller than the fermion mass m. One can try expanding in powers of E/m and p/m as in heavy quark effective theory, so that the expansion parameter is $1/m$. For example, the dispersion relation $E = \sqrt{\mathbf{p}^2 + m^2}$ gives terms in the Lagrangian of the form

$$L = \psi^\dagger \left(E - \frac{\mathbf{p}^2}{2m} + \frac{\mathbf{p}^4}{8m^3} + \ldots \right) \psi. \tag{1}$$

The lowest order propagator is $1/(E + i\varepsilon)$, which gives $\theta(t)$ in position space. This is the static propagator of HQET: fermions propagate forward in time, but do not move in space. This propagator is acceptable for some calculations involving heavy quarks. For example, one can compute the static potential between fixed sources using this propagator. However, for $\bar{t}t$ production, the quarks are produced at the same point, and they remain at the same point for all time if the static propagator is used. This is too singular, and the HQET expansion breaks down. In general, it is essential for treating nonrelativistic bound states that the heavy fermions move. For this to occur, the lowest order propagator should be $1/(E - \mathbf{p}^2/2m + i\varepsilon)$, so that E and $\mathbf{p}^2/2m$ are of the same order in the effective theory power counting. This implies that the $1/m$ expansion cannot be used; instead one must use an expansion in powers of v, where E and $\mathbf{p}^2/2m$ are both of order v^2 [1, 2].

The effective theory expansion parameter is the velocity v, and formally, α must also be treated as order v. Thus order α^2 radiative corrections to the leading term are just as

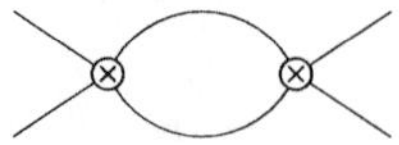

FIGURE 5. An iteration of two potentials in the effective theory.

important as order v^2 relativistic corrections. The effective theory below the scale m has:

- Nonrelativistic fermions with propagator

$$\frac{1}{E - \mathbf{p}^2/2m + i\varepsilon}$$

- Ultrasoft gauge bosons coupled via interactions that are multipole expanded [6].
- Potentials $V(\mathbf{p},\mathbf{p}')$ for the scattering of an incoming Q and $\bar{Q}$ with momenta $\mathbf{p}$ and $-\mathbf{p}$ to outgoing Q and $\bar{Q}$ with momenta $\mathbf{p}'$ and $-\mathbf{p}'$.
- Soft gauge bosons. The importance of introducing soft fields in the effective theory was first pointed out by Griesshammer [30].

The effective theory has two different gauge boson fields, soft bosons and ultrasoft bosons. This does not lead to any double counting if graphs are evaluated in dimensional regularization.

The static theory is not the $m \to \infty$ limit or the $v \to 0$ limit of the effective theory. For this reason, the static potential and the effective theory potential are not equal.

POWER COUNTING

The power counting parameter of the effective theory is the velocity v. If one expands the dispersion relation as in Eq. (1), then E and $\mathbf{p}^2/2m$ are both of order v^2, and $\mathbf{p}^4/8m^3$ is of order v^4, i.e. of order v^2 relative to the leading term.

The potential $V(\mathbf{p},\mathbf{p}')$ also has an expansion in powers of v. The leading term is the Coulomb potential, $V(\mathbf{p},\mathbf{p}') \propto \alpha/|\mathbf{k}|^2$, where $k = \mathbf{p}' - \mathbf{p}$ is the momentum transfer. Since momentum is of order mv, the Coulomb potential is naively of order α/v^2. However, the potential is a four-fermion operator, whereas the kinetic energy is a two-fermion operator. This leads to an additional factor of v from the power counting factors for the fields, so that the Coulomb potential is of order α/v in the effective theory. One can then determine the power counting for all the other potentials by comparing with the Coulomb potential. The hyperfine interaction $\propto \alpha \mathbf{S}_1 \cdot \mathbf{S}_2/m^2$ is generated by one-photon exchange, and is of order v^2 relative to the Coulomb interaction, so it is of order αv in the power counting, as are the spin-orbit, tensor and contact (Darwin) interactions. At one-loop, there are also potentials that are proportional to odd-powers of $\mathbf{k}$. The first such potential is proportional to $\alpha^2/|\mathbf{k}|$, and is of order $\alpha^2 v^0$ in the power counting.

A loop graph such as Fig. 5 of the time-ordered product of two potentials of order $\alpha^{a_1} v^{b_1}$ and $\alpha^{a_2} v^{b_2}$ is of order $\alpha^{a_1+a_2} v^{b_1+b_2}$. One can now see that the static potential differs from the $m \to \infty$ or $v \to 0$ limit of the effective theory potentials. For example, the loop graph of Fig. 5 with one $1/(m|\mathbf{k}|)$ and one Coulomb potential is of order

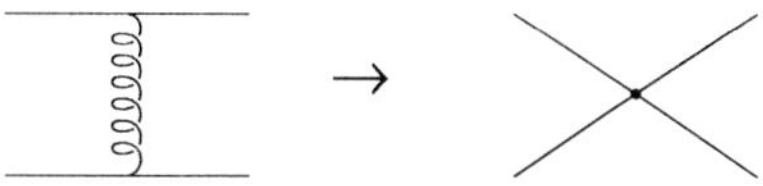

FIGURE 6. Tree-level matching for the potential.

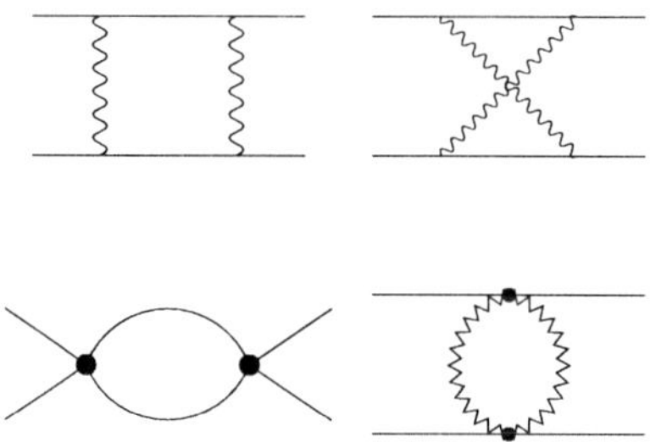

FIGURE 7. One-loop matching for the potential in QED. The first line gives examples of the full theory graphs. The second line gives examples of effective theory graphs: an iteration of two potentials, and a soft photon graph. The difference of the two sets of graphs gives the one-loop correction to the potential.

$\alpha^2 v^0 \times \alpha/v = \alpha^3/v$, and is of the same order in v as the Coulomb potential. The two particle intermediate state propagator $1/(E - \mathbf{p}^2/2m) = 2m/(2mE - \mathbf{p}^2)$ produces a factor of m in the numerator, that cancels the $1/m$ at the vertex. In the static theory, the $1/(m|\mathbf{k}|)$ potential is set to zero before the loop integration, so that the graph of Fig. 5 is not present in the static theory. As a result, the NRQCD potential [26] differs from the static potential.

MATCHING CONDITIONS

The method of calculating matching conditions is the same as in any effective theory. One computes the graphs in the full theory at the scale $\mu = m$, and subtracts the corresponding graphs in the effective theory. The graph in Fig. 6 gives the matching condition for the fermion potential. The full theory amplitude,

$$\frac{[\bar{u}(\mathbf{p}')\gamma^\mu u(\mathbf{p})]\,[\bar{u}(-\mathbf{p}')\gamma_\mu u(-\mathbf{p})]}{(\mathbf{p}-\mathbf{p}')^2} \tag{2}$$

is expanded in powers of $\mathbf{p}, \mathbf{p}'$, to give the potential in the effective theory. At one-loop, the difference of the full theory and effective theory graphs in Fig. 7 give the one-loop corrections to the matching potential. The only difference at this stage between Hydrogen and positronium is that there are annihilation contributions to the positronium potential from graphs such as Fig. 8. The graphs can have an imaginary part, that give the positronium decay width.

FIGURE 8. Annihilation contributions to the positronium potentials. The second graph has an imaginary part. [There is also a one-loop crossed box in the annihilation channel.]

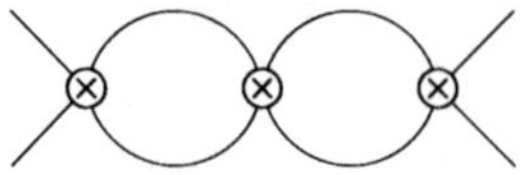

FIGURE 9. An ultraviolet divergent two-loop graph involving the iteration of three potentials.

RENORMALIZATION GROUP EVOLUTION

The nonrelativistic bound state system has three important mass scales, m, mv and mv^2.

Two-stage running

The conventional method of implementing the renormalization group is as follows

- Start at $\mu = m$
- Scale μ from m to mv
- Integrate out the soft modes at mv
- Scale μ from mv to mv^2

This is referred to as the two-stage method, because there are two-stages of renormalization group evolution. Consider a loop graph involving time-ordered products of potentials, such as Fig. 9. This graph contains a logarithm of the form $\ln \sqrt{mE}/\mu$. When μ is set to mv, this logarithm has the from $\ln \sqrt{E/mv^2}$, and is small. Thus the logarithms in the graph are summed by renormalization group evolution of μ from m to mv.

The graph in Fig. 10 involving an ultrasoft photon exchange contains a logarithm of the form $\ln E/\mu$. The μ in this ultrasoft graph is scaled all the way down (in two stages) to mv^2, at which point the logarithm is $\ln E/mv^2$, and also small.

However, this two-stage method of implementing the renormalization group turns out to be incorrect for nonrelativistic bound states. The reason is that the scales mv and mv^2

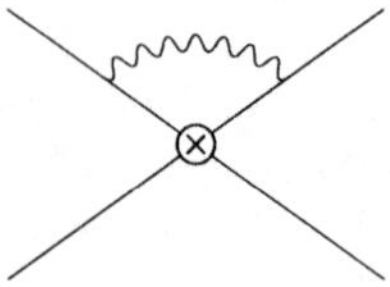

FIGURE 10. One-loop ultrasoft photon renormalization of the potential.

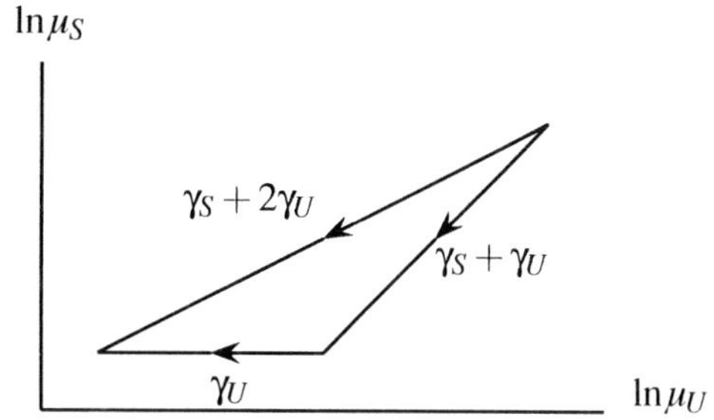

FIGURE 11. Paths in the (μ_U, μ_S) plane for one-stage and two-stage running.

are correlated—one cannot be varied independently of the other. Instead one needs to use an alternative one-stage scaling procedure.

One-stage running

In one stage running, one introduces two different μ parameters, μ_S and μ_U [11]. In dimensional regularization in $4-2\varepsilon$ dimensions, the soft photon coupling is multiplied by μ_S^{ε}, the ultrasoft photon coupling by μ_U^{ε}, and the potentials by $\mu_S^{2\varepsilon}$. Note that this is only possible because we have two different photon fields to represent the soft and ultrasoft photons in the effective theory. Then

- Set $\mu_S = m\nu$, $\mu_U = m\nu^2$
- Start at $\nu = 1$ and scale to $\nu = v$.

This procedure is referred to as the velocity renormalization group, because one runs in velocity ν rather than momentum [11]. The logarithms in Figs. 9 and 10 are now $\ln \sqrt{mE}/m\nu$ and $\ln E/m\nu^2$, which are minimized when $\nu = v$. Thus this method also minimizes logarithms in the diagrams, and sums them by renormalization group evolution.

The difference between the two renormalization group methods can be seen in Fig. 11 [25]. In two-stage running, there is only a single μ, so that $\mu_S = \mu_U = \mu$, and they are lowered together from m to mv. At this point, the soft modes are integrated out, and μ_U for the ultrasoft modes is lowered to mv^2. The integration path in Fig. 11 is along the lower edges of the triangle. In one-stage running, the integration path is along the diagonal. It is convenient to define two anomalous dimensions, γ_S and γ_U by taking the derivatives of Green's functions with respect to $\ln \mu_S$ and $\ln \mu_U$, respectively. One can show by explicit calculation that

- The two paths give different answers. The integration is path dependent because $\nabla \times \gamma \neq 0$.
- One-stage running using the velocity renormalization group agrees with explicit QED calculations at order $\alpha^3 \ln^2 \alpha$, $\alpha^7 \ln^2 \alpha$ and $\alpha^8 \ln^3 \alpha$.

The moral is that for nonrelativistic bound states, one should run in velocity rather than momentum.

The difference between the two integration methods can be made more precise. In the two-stage method, one first integrates $\gamma_S + \gamma_U$ from $\mu = m$ to $\mu = mv$, and then integrates γ_U from $\mu = mv$ to $\mu = mv^2$. In the one-stage method, one integrates $\gamma_S + 2\gamma_U$ (since $\ln \mu_U$ runs twice as fast as $\ln \mu_S$) from $\nu = 1$ to $\nu = v$. If the anomalous dimensions are constant, the two methods give

$$(\gamma_S + \gamma_U)\ln \tfrac{mv}{m} + \gamma_U \ln \tfrac{mv^2}{mv} \qquad \text{two-stage}$$

$$(\gamma_S + 2\gamma_U)\ln v \qquad \text{one-stage}$$

and agree with each other. However, in general anomalous dimensions are not constant, but can depend on coupling constants V_i, that themselves run. As a result, one finds that the $\ln v$ terms agree, but the higher order terms differ. For example, consider a $\ln^2 v$ term that depends on the product of γ_S and γ_U. For two-stage running, the contribution is proportional to $\gamma_S\gamma_U + 0\gamma_U = \gamma_S\gamma_U$ from the two pieces of the path. For one-stage running, the contribution is $\gamma_S(2\gamma_U)$, which differs by a factor of two. Similarly, a $\gamma_S\gamma_U^2 \ln^3 v$ contribution differs by a factor of four, and so on.

RUNNING POTENTIALS

The running potential $V(\mathbf{p}, \mathbf{p}')$ has an expansion

$$V(\mathbf{p}, \mathbf{p}') = V^{(-1)} + V^{(0)} + V^{(1)} + V^{(2)} + \ldots \tag{3}$$

where $V^{(n)}$ is of order v^n in the velocity power counting. The first three terms in the expansion have the form

$$V^{(-1)} = \frac{U_c}{\mathbf{k}^2} \qquad V^{(0)} = \frac{U_k}{|\mathbf{k}|}, \tag{4}$$

$$V^{(1)} = U_2 + U_s \mathbf{S}^2 + \frac{U_r(\mathbf{p}^2 + \mathbf{p}'^2)}{2\mathbf{k}^2} - \frac{i\mathbf{U}_\Lambda \cdot (\mathbf{p}' \times \mathbf{p})}{\mathbf{k}^2} + U_t\left(\sigma_1 \cdot \sigma_2 - \frac{3\mathbf{k}\cdot\sigma_1\,\mathbf{k}\cdot\sigma_2}{\mathbf{k}^2}\right),$$

where $V^{(0)} \sim 1/m$, and $V^{(1)} \sim 1/m^2$. In QCD, each of the coefficients can be written as $U \to U^{(1)} 1 \otimes 1 + U^{(T)} T^A \otimes \bar{T}^A$, where 1 and $T^A/\bar{T}^A$ are color matrices acting on the quark/antiquark lines. The anomalous dimensions for the coefficients U_c–U_t have been computed, and the details are given in Refs. [12, 13, 14]. The renormalization group improved static potential was computed in Ref. [31]. An important point to note is that graphs can involve both soft and ultrasoft gluons, so that the anomalous dimensions involve *both* $\alpha_s(mv)$ and $\alpha_s(mv^2)$. As an example, the running of $U_2^{(1)}$ is given by

$$m^2 U_2^{(1)}(\nu) = \frac{14C_1}{3}\alpha_s(m\nu)\alpha_s(m)\ln\left(\frac{m\nu}{m}\right) - \frac{32\pi C_1}{3\beta_0}\alpha_s(m)\ln\left[\frac{\alpha_s(m\nu)}{\alpha_s(m\nu^2)}\right] \tag{5}$$

TABLE 1. Table of potentials for QED. The first column is the potential, the second gives typical terms in the potential, the third gives the power counting in α and v, the fourth gives the order in the v counting scheme when $v \sim \alpha$, and the fifth gives the contribution of the potential to the bound state energy.

		Power Counting	Order	E
$V^{(-1)}$	$\frac{\alpha}{\mathbf{k}^2}$	$\frac{\alpha}{v}$	1	α^2
$V^{(0)}$	$\frac{\alpha}{m\lvert\mathbf{k}\rvert}$	α^2	α^2	α^4
$V^{(1)}$	$\frac{\alpha}{m^2}, \frac{\alpha \mathbf{S}^2}{m^2}$	αv	α^2	α^4
$V^{(2)}$	$\frac{\alpha\lvert\mathbf{k}\rvert}{m^3}$	$\alpha^2 v^2$	α^4	α^6
$V^{(3)}$	$\frac{\alpha \mathbf{k}^2}{m^4}$	αv^3	α^4	α^6
$\vdots$	$\vdots$	$\vdots$	$\vdots$	$\vdots$

where $C_1 = 2/9$ for QCD. Note that Eq. (5) depends on $\alpha_s(m)$, $\alpha_s(m\nu)$, and $\alpha_s(m\nu^2)$.

The renormalization group improved potentials can be used to calculate the renormalization group improved cross-section for $\bar{t}t$ production in the threshold region. It leads to a reduction in the theoretical uncertainty by a factor of ten. $\bar{t}t$ production is discussed in detail in I. Stewart's talk.

QED

The velocity renormalization group method gives very interesting and important results when applied to QED [15]. The basic potentials we will need for QED are summarized in Table 1. The last column gives the contribution to the bound state energy levels due to the given potential. The fourth column gives the order of a given potential, treating v as order α. Since the Coulomb potential is of order unity, one finds the obvious result that the Coulomb potential must be summed to all orders, and cannot be treated as a perturbation. The potentials $V^{(-1)}$, $V^{(1)}$, $V^{(3)}$, are first generated at tree-level, and are of order α, whereas the potentials $V^{(0)}$, $V^{(2)}$, $V^{(4)}$, are first generated at one-loop, and are of order α^2.

The bound state energy levels can be determined to order α^4 by computing the matrix elements of $V^{(0)}$ and $V^{(1)}$ between Coulomb wavefunctions. Time-ordered products of two potentials, such as $T\left[V^{(0)}V^{(0)}\right]$, $T\left[V^{(0)}V^{(1)}\right]$ and $T\left[V^{(1)}V^{(1)}\right]$ first contribute at order α^6. In principle, to obtain the energy levels to order α^4, one also needs the one- and two-loop matching corrections to the Coulomb potential. However, such corrections vanish in QED. As a result, the first correction to the order α^2 binding energy is of order α^4, and is given by the matrix element of $V^{(0)} + V^{(1)}$. There are no order α^3 corrections to the energy levels in QED.

Define the leading and next-to-leading order anomalous dimensions of a potential to

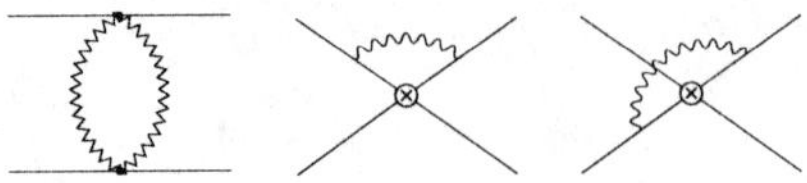

FIGURE 12. One-loop running of $V^{(1)}$ in QED. The first graph has a soft photon, and the other graphs have ultrasoft photons and a potential.

be the anomalous dimension from graphs at one and two higher orders in α than the potential itself. For $V^{(-1)}$ and $V^{(1)}$ which are of order α, the leading order anomalous dimension is of order α^2, and the next-to-leading order anomalous dimension is of order α^3. For $V^{(0)}$ which is of order α^2, the leading order anomalous dimension is of order α^3, and the next-to-leading order anomalous dimension is of order α^4. Since different terms in the potential are of different orders in α, the terms leading and next-to-leading order are not related to the number of loops.

Integrating the renormalization group equations for $V^{(0)}$ and $V^{(1)}$ using the leading order anomalous dimension gives a series of the form

$$\alpha\left(1+\alpha\ln\alpha+\alpha^2\ln^2\alpha+\alpha^3\ln^3\alpha+\ldots\right),$$

which contributes

$$\alpha^4\left(1+\alpha\ln\alpha+\alpha^2\ln^2\alpha+\alpha^3\ln^3\alpha+\ldots\right)$$

to the energy. Integrating the next-to-leading order anomalous dimensions gives

$$\alpha^4\alpha\left(1+\alpha\ln\alpha+\alpha^2\ln^2\alpha+\alpha^3\ln^3\alpha+\ldots\right)$$

terms in the energy. The next-to-next-to-leading anomalous dimension gives

$$\alpha^4\alpha^2\left(1+\alpha\ln\alpha+\alpha^2\ln^2\alpha+\alpha^3\ln^3\alpha+\ldots\right),$$

terms in the energy, which are the same order as those obtained by using the leading order anomalous dimension for the $V^{(2)}$ and $V^{(3)}$ potentials which first contribute at order α^6. Thus one can compute the

$$\begin{array}{llll} \alpha^5\ln\alpha & \alpha^6\ln^2\alpha & \alpha^7\ln^3\alpha & \ldots \\ \alpha^6\ln\alpha & \alpha^7\ln^2\alpha & \alpha^8\ln^3\alpha & \ldots \end{array}$$

series in the energy using $\gamma_{\rm LO}$, $\gamma_{\rm NLO}$ for $V^{(0,1)}$.

LEADING ORDER

The Coulomb potential and $V^{(0)}$ do not run in QED at leading and next-to-leading order, so one is left with the running of $V^{(1)}$. The anomalous dimensions are evaluated for a particle of mass m_1 and charge $-e$ interacting with a second particle of mass m_2 and charge Ze. Evaluating the graphs in Fig. 12 gives

$$\nu\frac{dU_2}{d\nu} = \frac{14Z^2\alpha^2}{3m_1m_2} + \frac{2\alpha}{3\pi}\left(\frac{1}{m_1}+\frac{Z}{m_2}\right)^2 U_c \tag{6}$$

where the first term is the soft contribution from Fig. 12a and the second is the ultrasoft contribution from Fig. 12b,c. Note that the ultrasoft contribution has been multiplied by two, since the anomalous dimension for the velocity renormalization group is $\gamma_S + 2\gamma_U$. The other coefficients in $V^{(1)}$ (U_r, U_s, U_Λ, U_t) have zero anomalous dimension at this order.

Since the Coulomb potential and α do not run in QED, one can combine the two terms,

$$\nu\frac{dU_2}{d\nu} = \gamma_0 U_c, \qquad \gamma_0 = \frac{2\alpha}{3\pi}\left(\frac{1}{m_1^2}+\frac{Z}{4m_1m_2}+\frac{Z^2}{m_2^2}\right). \tag{7}$$

γ_0 is a constant in QED since α does not run. Integrating Eq. (7) gives

$$U_2(\nu) = U_2(1) + \gamma_0 U_c \ln\nu, \tag{8}$$

where U_2 is evaluated at $\nu = v = \alpha$. Since γ_0 is a constant, $U_2(\nu)$ only has a $\ln\nu$ term, and terms of the form $\ln^n\nu$, with $n > 1$ vanish. As a result, the leading order energy series Eq. () terminates after a single term, so one has an $\alpha^5\ln\alpha$ contribution to the energy, but the $\alpha^6\ln^2\alpha$, etc. terms vanish. At low orders, the absence of terms other than $\alpha^5\ln\alpha$ in the leading order series has been noticed before by an explicit examination of Feynman graphs. This is the first general proof that all the terms beyond $\alpha^5\ln\alpha$ in the leading order series vanish

The matrix element of U_2 gives the energy shift

$$\Delta E = \langle U_2(\nu)\rangle = \gamma_0 U_c \ln\nu\,|\psi(\mathbf{0})|^2 = -\frac{8Z^4\alpha^5 m_R^3}{3\pi n^3}\left(\frac{1}{m_1^2}+\frac{Z}{4m_1m_2}+\frac{Z^2}{m_2^2}\right)\ln Z\alpha,$$

where we have used

$$|\psi(\mathbf{0})|^2 = \frac{(m_R Z\alpha)^3}{\pi n^3} \tag{9}$$

for the nS state, and m_R is the reduced mass. This is the famous $\alpha^5\ln\alpha$ correction to the Lamb shift first computed by Bethe, including all recoil corrections.

NEXT-TO-LEADING ORDER

At next-to-leading order, the anomalous dimension for $V^{(1)}$ is

$$\nu\frac{dU_{2+s}}{d\nu}\bigg|_{\rm NLO} = \rho_{ccc}U_c^3 + \rho_{cc2}U_c^2\left(U_{2+s}+U_r\right) + \rho_{c22}U_c\left(U_{2+s}^2 + 2U_{2+s}U_r + \frac{3}{4}U_r^2 - 5U_t^2\mathbf{S}^2\right)$$
$$+\rho_{ck}U_cU_k + \rho_{k2}U_k\left(U_{2+s}+U_r/2\right) + \rho_{c3}U_c\left(U_3 + U_{3s}S^2 + \frac{1}{2}U_{rk}\right), \tag{10}$$

where $U_{2+s} = U_2 + U_s\mathbf{S}^2$, and the coefficients are

$$\begin{aligned}
\rho_{ccc} &= -\frac{m_R^4}{64\pi^2}\left(\frac{1}{m_1^3}+\frac{1}{m_2^3}\right)^2, & \rho_{c22} &= -\frac{m_R^2}{4\pi^2}, \\
\rho_{cc2} &= -\frac{m_R^3}{8\pi^2}\left(\frac{1}{m_1^3}+\frac{1}{m_2^3}\right), & \rho_{c3} &= \frac{2m_R}{\pi^2}, \\
\rho_{ck} &= \frac{m_R^2}{2\pi^2}\left(\frac{1}{m_1^3}+\frac{1}{m_2^3}\right), & \rho_{k2} &= \frac{2m_R}{\pi^2}.
\end{aligned} \tag{11}$$

The anomalous dimension Eq. (10) can be integrated by substituting the leading order running, Eq. (8) for the coefficients on the right-hand side. Since only U_2 runs at leading order, the right hand side has at most a $\ln^2\nu$, so that the integral has at most a $\ln^3\nu$ term. This implies that the next-to-leading order series Eq. () terminates after the first three terms, $\alpha^6\ln\alpha$, $\alpha^7\ln^2\alpha$, and $\alpha^8\ln^3\alpha$.

$\ln^3\alpha$

The only term that contributes to the $\ln^3\alpha$ correction is the U_2^2 term of Eq. (10). Integrating gives a contribution to $U_2(\nu)$ of the form

$$\frac{1}{3}\gamma_0^2\rho_{c22}U_c^3(1)\ln^3\nu, \tag{12}$$

which is spin-independent, and has no imaginary part. There is no contribution to the decay width or hyperfine splitting at this order. The Lamb shift at this order is obtained by multiplying Eq. (12) by the matrix element of the unit operator, $|\psi(\mathbf{0})|^2$, to give

$$\Delta E = \frac{64m_R^5\alpha^8Z^6}{27\pi^2n^3}\ln^3(Z\alpha)\left(\frac{1}{m_1^2}+\frac{Z}{4m_1m_2}+\frac{Z^2}{m_2^2}\right)^2 \tag{13}$$

which is approximately 8 KHz for the $2P$–$2S$ Lamb shift in Hydrogen. Substituting $Z=1$ and $m_1 = m_2 = m_e$ gives the $\alpha^8\ln^3\alpha$ Lamb shift for positronium

$$\Delta E = \frac{3m_e\alpha^8\ln^3\alpha}{8\pi^2n^3}. \tag{14}$$

The positronium Lamb shift is a new result, as are the recoil terms in the Hydrogen Lamb shift. In the limit $m_1/m_2 \to 0$, the Hydrogen Lamb shift has been computed previously by several groups. There is an analytic computation by Karshenboim [32] and a numerical computation by Goidenko et al. [33] that agree with our result. There are also numerical computations by Malampalli and Sapirstein [34], and by Yerokhin [35] which agree with each other, but disagree with the other results. Recently, there has been a computation by Pachucki [36] that agrees with our result. Yerokhin [37] has emphasized that the complete $\alpha^8\ln^3\alpha$ Lamb shift might not be contained in the loop-after-loop calculations of Refs. [34, 35].

The other calculations rely on extracting the logarithm from four-loop diagrams such as Fig. 13. The velocity renormalization group factors the graph into the product of a two-loop anomalous dimension ρ_{c22}, and the square of a one-loop anomalous dimension γ_0^2.

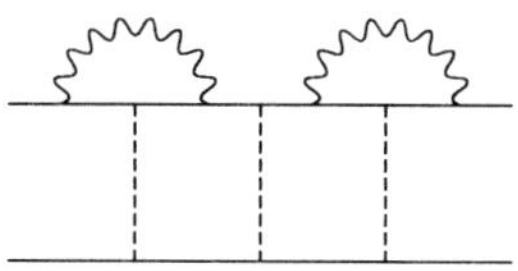

FIGURE 13. Four-loop diagram that contributes to the $\alpha^8 \ln^2 \alpha$ Lamb shift

$\ln^2 \alpha$ and $\ln \alpha$

One can also compute the $\alpha^7 \ln^2 \alpha$ and $\alpha^6 \ln \alpha$ corrections from the renormalization group equations, and the results agree with existing calculations for the Lamb shift, hyperfine splitting, and decay widths for Hydrogen and Positronium [38–41].

CONCLUSIONS

The methods presented here give a systematic way of separating scales in nonrelativistic bound state problems. All large logarithms are summed using the velocity renormalization group. The method provides a universal description of QED logarithms. The agreement with known results at order $\alpha^5 \ln \alpha$, $\alpha^6 \ln \alpha$, $\alpha^7 \ln^2 \alpha$, and $\alpha^8 \ln^3 \alpha$ is a highly non-trivial check of the formalism. In QED, one finds that the leading order series terminates after one term, and the next-to-leading order series terminates after three terms. In addition, the method resolves a controversy about the $\alpha^8 \ln^3 \alpha$ Lamb shift for Hydrogen, and gives the first calculation of the $\alpha^8 \ln^3 \alpha$ energy shift for positronium.

In QCD, one can distinguish $\alpha_s(mv)$ and $\alpha_s(mv^2)$, and both can appear simultaneously in the same anomalous dimension. The renormalization group improved potentials can be used to compute $\bar{t}t$ production, and reduce the scale uncertainties by a factor of ten.

The velocity renormalization group should also be applicable to other problems with correlated scales. In the bound state problem, one can generate the scale mv in loop graphs from the scale m and mv^2, $mv = \sqrt{m \times mv^2}$. Similar effects can occur at finite temperature, where one has the scales T, gT and g^2T, and some of the ideas described here might be applicable to that problem as well.

ACKNOWLEDGMENTS

This work was supported in part by the Department of Energy under grant DOE-FG03-97ER40546 and by NSERC of Canada.

REFERENCES

1. Caswell, W. E., and Lepage, G. P., *Phys. Lett.*, **B167**, 437 (1986).
2. Bodwin, G. T., Braaten, E., and Lepage, G. P., *Phys. Rev.*, **D51**, 1125–1171 (1995).
3. Labelle, P., *Phys. Rev.*, **D58**, 093013 (1998).

4. Luke, M. E., and Manohar, A. V., *Phys. Rev.*, **D55**, 4129–4140 (1997).
5. Manohar, A. V., *Phys. Rev.*, **D56**, 230–237 (1997).
6. Grinstein, B., and Rothstein, I. Z., *Phys. Rev.*, **D57**, 78–82 (1998).
7. Luke, M. E., and Savage, M. J., *Phys. Rev.*, **D57**, 413–423 (1998).
8. Pineda, A., and Soto, J., *Nucl. Phys. Proc. Suppl.*, **64**, 428–432 (1998).
9. Pineda, A., and Soto, J., *Phys. Rev.*, **D58**, 114011 (1998).
10. Pineda, A., and Soto, J., *Phys. Rev.*, **D59**, 016005 (1999).
11. Luke, M. E., Manohar, A. V., and Rothstein, I. Z., *Phys. Rev.*, **D61**, 074025 (2000).
12. Manohar, A. V., and Stewart, I. W., *Phys. Rev.*, **D62**, 014033 (2000).
13. Manohar, A. V., and Stewart, I. W., *Phys. Rev.*, **D62**, 074015 (2000).
14. Manohar, A. V., and Stewart, I. W., *Phys. Rev.*, **D63**, 054004 (2001).
15. Manohar, A. V., and Stewart, I. W., *Phys. Rev. Lett.*, **85**, 2248–2251 (2000).
16. Brambilla, N., Pineda, A., Soto, J., and Vairo, A., *Nucl. Phys.*, **B566**, 275 (2000).
17. Kniehl, B. A., and Penin, A. A., *Nucl. Phys.*, **B563**, 200–210 (1999).
18. Lundeen, S., and Pipkin, F., *Phys. Rev. Lett.*, **46**, 232 (1981).
19. Hellwig, H., *I.E.E.E Trans.*, **IM-19**, 200 (1970).
20. Liu, W., et al., *Phys. Rev. Lett.*, **82**, 711–714 (1999).
21. Manohar, A., and Georgi, H., *Nucl. Phys.*, **B234**, 189 (1984).
22. Kinoshita, T., editor, *Quantum electrodynamics*, Advanced series on directions in high energy physics 7, World Scientific, 1990.
23. Pachucki, K., *Hyp. Int.*, **114**, 55 (1998).
24. Eides, M. I., Grotch, H., and Shelyuto, V. A., *Phys. Rept.*, **342**, 63–261 (2001).
25. Manohar, A. V., Soto, J., and Stewart, I. W., *Phys. Lett.*, **B486**, 400–405 (2000).
26. Hoang, A. H., Manohar, A. V., and Stewart, I. W., *Phys. Rev.*, **D64**, 014033 (2001).
27. Hoang, A. H., Manohar, A. V., Stewart, I. W., and Teubner, T., *Phys. Rev. Lett.*, **86**, 1951–1954 (2001).
28. Hoang, A. H., Manohar, A. V., Stewart, I. W., and Teubner, T., *hep-ph/0107144* (2001).
29. Beneke, M., and Smirnov, V. A., *Nucl. Phys.*, **B522**, 321–344 (1998).
30. Griesshammer, H. W., *Nucl. Phys.*, **B579**, 313–351 (2000).
31. Pineda, A., and Soto, J., *Phys. Lett.*, **B495**, 323–328 (2000).
32. Karshenboim, S. G., *Sov. Phys. JETP*, **76**, 541 (1993).
33. Goidenko, I., Labzowsky, L., Nefiodov, A., Plunien, G., and Soff, G., *Phys. Rev. Lett.*, **83**, 2312 (1999).
34. Mallampalli, S., and Sapirstein, J. R., *Phys. Rev. Lett.*, **80**, 5297 (1998).
35. Yerokhin, V. A., *Phys. Rev.*, **A62**, 012508 (2000).
36. Pachucki, K. (2000), unpublished.
37. Yerokhin, V. A., *Phys. Rev. Lett.*, **86**, 1990–1993 (2001).
38. Caswell, W. E., and Lepage, G. P., *Phys. Rev.*, **A20**, 36 (1979).
39. Khriplovich, I. B., and Yelkhovsky, A. S., *Phys. Lett.*, **B246**, 520–522 (1990).
40. Labelle, P., Ph.D. thesis, Cornell University (1994).
41. Melnikov, K., and Yelkhovsky, A., *Phys. Lett.*, **B458**, 143–151 (1999).

Threshold Top Quark Production

Iain W. Stewart

*Department of Physics, University of California at San Diego,
9500 Gilman Drive, La Jolla, CA 92093-0319, USA*

Abstract. Predictions for the total cross section for $e^+e^- \to t\bar{t}$ near threshold are reviewed. The renormalization group improved results at NNLL order have improved convergence and reduced scale dependence relative to fixed order results at NNLO. Prospects for measurements of the top-quark mass, width, and Yukawa coupling are discussed.

INTRODUCTION

The production of top quark pairs is among the important projects of a future linear collider. The large top quark width $\Gamma_t \sim 1.5$ GeV makes the threshold cross section look quite different from that of charm or bottom pairs. The large value of the width prohibits the production of toponium states, and at the same time serves as an infrared cutoff from sensitivity to non-perturbative effects. Thus, perturbative methods can be used to describe the top-antitop dynamics to a very high degree of precision. This makes the threshold region an ideal place for extracting fundamental top quark parameters such as the top mass, width, and Yukawa coupling (for a light higgs).

In the threshold region $\sqrt{s} \simeq 2m_t \pm 10\,\mathrm{GeV}$, the t and $\bar{t}$ move with non-relativistic velocities. Defining the energy $m_t \mathrm{v}^2 = \sqrt{s} - 2m_t$ we see that this region of s corresponds to velocities $|\mathrm{v}| \lesssim \alpha_s$. In this region an exact treatment of QCD Coulomb singularities $(\alpha_s/\mathrm{v})^k$ is required, ruling out a pure α_s expansion. However, a combined expansion in powers of v and α_s can be performed. Schematically, for the $e^+e^- \to t\bar{t}$ cross section, $\sigma_{t\bar{t}}(s)$, we want an expansion of the form

$$R = \frac{\sigma_{t\bar{t}}}{\sigma_{\mu^+\mu^-}} = \mathrm{v} \sum_k \left(\frac{\alpha_s}{\mathrm{v}}\right)^k \times \Big[\underset{\text{LO}}{1} + \underset{\text{NLO}}{\{\alpha_s, \mathrm{v}\}} + \underset{\text{NNLO}}{\{\alpha_s^2, \alpha_s \mathrm{v}, \mathrm{v}^2\}} + \ldots \Big]. \tag{1}$$

The power counting for these corrections can be implemented in a simple and systematic way using the effective theory framework of Non-Relativistic QCD (NRQCD).

The leading order (LO) prediction for R is shown in Fig. 1. The large top width gives the threshold cross section a smooth line-shape, with a single bump from the remnant of the 1S toponium state. The characteristics of the cross section are sensitive to the top parameters. In particular, the top mass determines the location of the rise/peak, the top width determines the slope of the rise and shape of the peak, and the overall normalization provides information on $\alpha_s(m_t)$ and the top-Yukawa coupling for a light

CP618, *Heavy Flavor Physics: Ninth International Symposium,* edited by A. Ryd and F. C. Porter

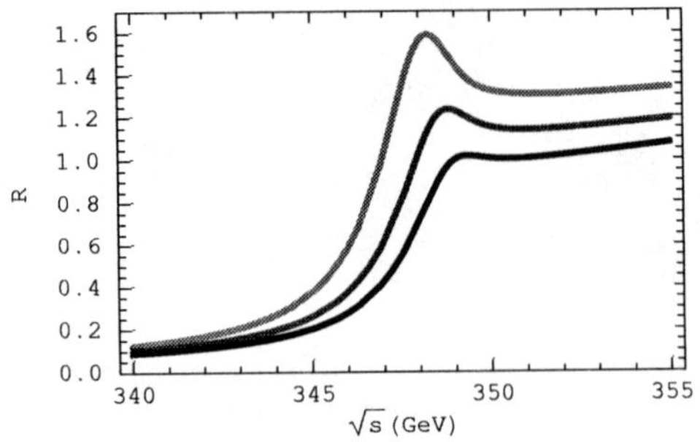

FIGURE 1. Leading order predictions for R. Results are shown for three renormalization scales of order the momentum transfer in the Coulomb potential, $\mu = 15\,\text{GeV}$ (upper), $30\,\text{GeV}$ (middle), and $60\,\text{GeV}$ (lower).

higgs. However, at LO the prediction for R suffers from considerable scale uncertainty as shown by the three curves in Fig. 1. This illustrates the importance of including higher order terms. Several groups have computed the next-to-next-to-leading order (NNLO) QCD corrections to the total cross section (see Ref. [1] for a summary and comparison of these results). Using "threshold" top-quark mass parameters, an accurate prediction for the location of the peak in the cross section was obtained. It was concluded that infrared safe top masses could be determined with a precision of < 200 MeV. However, surprisingly the NNLO corrections to the cross section normalization were big, with large scale uncertainty. The residual scale dependence was estimated to make the normalization of the cross section uncertain to $\approx 20\%$ [1]. These large corrections seemed to jeopardize measurements of the top width, strong coupling, and Yukawa coupling.

To understand the source of this large scale dependence it is useful to recall that the dynamics of the top-antitop system are governed by vastly different energy scales. These quarks have mass $m_t \sim 175\,\text{GeV}$, typical momenta $\mathbf{p} \simeq m_t \mathrm{v} \sim 25\,\text{GeV}$, and energies $E \simeq m_t \mathrm{v}^2 \sim 4\,\text{GeV}$. The scattering amplitudes therefore involve logarithms

$$\ln\left(\frac{\mu^2}{m_t^2}\right), \qquad \ln\left(\frac{\mu^2}{\mathbf{p}^2}\right), \qquad \ln\left(\frac{\mu^2}{E^2}\right), \tag{2}$$

which are not all small for a single choice of μ. In the expansion for R these large logarithms appear as logs of the velocity. This suggests that a better expansion might involve a summation of these logarithms

$$R = \mathrm{v}\sum_k \left(\frac{\alpha_s}{\mathrm{v}}\right)^k \sum_j (\alpha_s \ln \mathrm{v})^j \times \Big[\underset{\text{LL}}{1} + \underset{\text{NLL}}{\{\alpha_s, \mathrm{v}\}} + \underset{\text{NNLL}}{\{\alpha_s^2, \alpha_s \mathrm{v}, \mathrm{v}^2\}} + \ldots\Big]. \tag{3}$$

The additional summation of logarithms can be performed using renormalization group equations (RGE). This is similar to how logarithms of m_W/m_b are summed for the electroweak Hamiltonian. In our case the summation is complicated by the presence of two low energy scales that are coupled by the equations of motion, $E = \mathbf{p}^2/(2m_t)$. This complication can be dealt with by using a renormalization group with a subtraction velocity ν [2, 3, 4, 5, 6]. In this framework there are two renormalization group parameters in the effective Lagrangian: μ_S for soft gluons and μ_U for ultrasoft gluons, where $\mu_S = m_t\nu$ and $\mu_U = m_t\nu^2$. Running from $\nu = 1$ to $\nu \sim v$ sums logarithms of v and minimizes both $\ln(\mu_S^2/\mathbf{p}^2)$ and $\ln(\mu_U^2/E^2)$ terms in the amplitudes. (The $\ln(\mu^2/m_t^2)$ terms are minimized by matching QCD onto the effective theory near $\mu = m_t$.) In Refs. [7, 8] it was shown that the expansion in Eq. (3) is better behaved than that in Eq. (1). Furthermore, the normalization of the cross section at NNLL has significantly smaller scale uncertainty ($\approx 3\%$) than the NNLO result. This review focuses on these results; for more details and further references see Ref. [8].

Below, the steps in the renormalization group improved calculation are briefly described. This is followed by predictions for the cross section as well as a description of the prospects for measurements of the top parameters in light of the reduced theoretical uncertainties at NNLL order.

RENORMALIZATION GROUP IMPROVED CALCULATION

The computation of the renormalization group improved cross section can be divided into three parts:

1. Matching of QCD onto an effective theory for non-relativistic top quarks. Determine the Wilson coefficients $C(\nu)$ at $\nu = 1$ ($\mu = m_t$) to the desired order in $\alpha_s(m_t)$.
2. Scaling $C(\nu)$ from $\nu = 1$ to $\nu = \nu_0 \simeq C_F\alpha_s$ by calculating anomalous dimensions and using the renormalization group. This scaling sums the terms of the form $[\alpha_s \ln(v)]^j$ in Eq. (3).
3. Computing the cross section using the Lagrangian and currents renormalized at the low scale $\nu = \nu_0$.

Each of these parts will be discussed in turn.

1. Matching onto the Effective Theory

For non-relativistic scattering the relevant momentum regions can be classified by their typical energy and momenta (k^0,$\mathbf{k}$). They include hard modes with momenta $\sim (m,m)$, potential modes with momenta $\sim (mv^2,mv)$, soft modes with momenta $\sim (mv,mv)$, and ultrasoft modes with momenta $\sim (mv^2,mv^2)$. Fluctuations involving hard or offshell momenta are integrated out, while effective theory fields are introduced for modes with nearly on-shell momenta. The degrees of freedom therefore include potential top and anti-top quarks ($\psi_\mathbf{p}$ and $\chi_\mathbf{p}$), soft gluons and light quarks (A_q^μ and φ_q), and ultrasoft gluons and light quarks (A^μ and ϕ_{us}). Soft energies and momenta appear as labels on the fields while ultrasoft momenta are represented by explicit coordinate dependence [2]. This enables us to distinguish the size of momenta, for instance a derivative $\partial^\mu\psi_\mathbf{p}(x) \sim mv^2\psi_\mathbf{p}(x)$.

In this framework the action for non-relativistic top quarks has terms

$$\begin{aligned}
\mathcal{L} &= \sum_{\mathbf{p}} \psi_{\mathbf{p}}^{\dagger}(x)\Big\{ iD^0 - \frac{(\mathbf{p}-i\mathbf{D})^2}{2m} + \ldots \Big\} \psi_{\mathbf{p}}(x) + (\psi \to \chi) \\
&\quad - \sum_{\mathbf{p},\mathbf{p}'} F(\mathbf{p},\mathbf{p}') \Big[\psi_{\mathbf{p}'}^{\dagger}(x)\, \psi_{\mathbf{p}}(x)\, \chi_{-\mathbf{p}'}^{\dagger}(x)\, \chi_{-\mathbf{p}}(x) \Big] \\
&\quad -2\pi\alpha_s(m\nu) \sum_{\mathbf{p},\mathbf{p}',q,q'} \psi_{\mathbf{p}'}^{\dagger} [A_{q'}^{\alpha}, A_q^{\beta}] U_{\alpha\beta}\, \psi_{\mathbf{p}} + (\psi \to \chi)\,,
\end{aligned} \tag{4}$$

where color and spin indices are suppressed. The covariant derivatives in the first line involve only ultrasoft gluons. The function $F(\mathbf{p},\mathbf{p}')$ in the second line contributes to the potential between quarks and anti-quarks. For our purposes

$$F(\mathbf{p},\mathbf{p}') = \frac{1}{\mathbf{k}^2}\mathcal{V}_c(\nu) + \frac{\pi^2}{m|\mathbf{k}|}\mathcal{V}_k(\nu) + \frac{1}{m^2}\mathcal{V}_2(\nu) + \frac{\mathbf{S}^2}{m^2}\mathcal{V}_s(\nu) + \frac{(\mathbf{p}^2+\mathbf{p}'^2)}{2m^2\mathbf{k}^2}\mathcal{V}_r(\nu)\,, \tag{5}$$

where $\mathbf{k} = \mathbf{p}' - \mathbf{p}$, $\mathbf{S}$ is the total spin operator, and all coefficients are in the color singlet channel. The matching for these Wilson coefficients is needed at two loops for $\mathcal{V}_c$, one loop for $\mathcal{V}_k$, and tree level for $\mathcal{V}_{2,s,r}$ [5, 6]. Finally, the third line in Eq. (4) is an example of the type of interaction that occurs between potential quarks and soft gluons, with $U_{\alpha\beta}(\mathbf{p},\mathbf{p}',q,q')$ a matching function of the label momenta. At NNLO (and NNLL order) time ordered products of two of these soft interactions also contribute terms to the potential giving

$$\begin{aligned}
V_{\text{soft}}(\mathbf{p},\mathbf{p}') &= -\frac{C_F\,\alpha_s^2(m\nu)}{\mathbf{k}^2}\Big[-\beta_0 \ln\Big(\frac{\mathbf{k}^2}{m^2\nu^2}\Big) + a_1 \Big] \\
&\quad - \frac{C_F\,\alpha_s^3(m\nu)}{4\pi\mathbf{k}^2}\Big[\beta_0^2 \ln^2\Big(\frac{\mathbf{k}^2}{m^2\nu^2}\Big) - \Big(2\beta_0 a_1 + \beta_1\Big) \ln\Big(\frac{\mathbf{k}^2}{m^2\nu^2}\Big) + a_2 \Big]\,,
\end{aligned} \tag{6}$$

where β_i are coefficients of the QCD β-function and the constants a_i can be found in Ref. [9]. The complete potential is then $V(\mathbf{p},\mathbf{p}') = F(\mathbf{p},\mathbf{p}') + V_{\text{soft}}(\mathbf{p},\mathbf{p}')$.

We also need to take into account that the top quarks decay. We will assume that the decay products are hard and can be integrated out. At lowest order this induces the operators

$$\mathcal{L} = \sum_{\mathbf{p}} \psi_{\mathbf{p}}^{\dagger} \frac{i}{2}\Gamma_t\, \psi_{\mathbf{p}} + \sum_{\mathbf{p}} \chi_{\mathbf{p}}^{\dagger} \frac{i}{2}\Gamma_t\, \chi_{\mathbf{p}}\,. \tag{7}$$

In the Standard Model the dominant decay channel is $t \to bW^+$ and gives a width of $\Gamma_t = 1.43\,\text{GeV}$ which we will use as our central value. Counting $\Gamma_t \sim m_t\nu^2$, Eq. (7) gives a consistent next-to-leading order treatment of electroweak effects for the total cross section [10, 11]. Thus, we will not include electroweak decay related effects to the same order as the QCD corrections. From partial knowledge of these corrections [12], the missing terms are expected to be at the few percent level.

Besides the effective Lagrangian we also need external currents to produce the top-antitop pair. Since we wish to describe $e^+e^- \to \{\gamma^*, Z^*\} \to t\bar{t}$ these currents are induced by both electromagnetic and weak interactions. The relevant vector current is $\mathbf{J}_{\mathbf{p}}^{\mathrm{v}} = c_1(\nu)\, O_{\mathbf{p},1} + c_2(\nu)\, O_{\mathbf{p},2}$, where

$$O_{\mathbf{p},1} = \psi_{\mathbf{p}}^{\dagger}\, \boldsymbol{\sigma}(i\sigma_2)\chi_{-\mathbf{p}}^{*}, \qquad O_{\mathbf{p},2} = \frac{1}{m^2}\psi_{\mathbf{p}}^{\dagger}\, \mathbf{p}^2\boldsymbol{\sigma}(i\sigma_2)\chi_{-\mathbf{p}}^{*},$$

and the relevant axial-vector current is $\mathbf{J}_{\mathbf{p}}^{a} = c_3(\nu)\, O_{\mathbf{p},3}$, where

$$O_{\mathbf{p},3} = \frac{-i}{2m}\psi_{\mathbf{p}}^{\dagger}\, [\boldsymbol{\sigma}, \boldsymbol{\sigma}\cdot\mathbf{p}]\, (i\sigma_2)\chi_{-\mathbf{p}}^{*}. \tag{8}$$

The matching for these Wilson coefficients is needed at two loops for c_1 and tree level for $c_{2,3}$. The two-loop matching for c_1 is scheme dependent [13, 14, 15], and in the $\overline{\mathrm{MS}}$ scheme with our definition of the operators, can be found in Ref. [8].

2. Renormalization group scaling

To sum the $(\alpha_s \ln \nu)^j$ terms in R we must determine the anomalous dimensions for the Wilson coefficients $\mathcal{V}_{c,k,2,s,r}$ in Eq. (5) and the current coefficients $c_{1,2,3}$. The anomalous dimensions for $\mathcal{V}_c$, $\mathcal{V}_k$, and $\mathcal{V}_{2,s,r}$ are required at three, two, and one loop respectively. These have been computed in Refs. [3, 5, 6, 16], and due to mixing depend on the one-loop running of the HQET terms up to $1/m^2$ [17]. For example we have

$$\begin{aligned}
\nu\frac{\partial}{\partial\nu}\mathcal{V}_r &= \frac{8}{3}(\beta_0+8)\,\alpha_s^2(m_t\nu) - \frac{128}{3}\alpha_s(m_t\nu)\alpha_s(m_t\nu^2) - \frac{64}{3}\alpha_s^2(m_t\nu)\ln\left[\frac{\alpha_s(m\nu^2)}{\alpha_s(m\nu)}\right], \\
\nu\frac{\partial}{\partial\nu}\mathcal{V}_s &= \left(\frac{28}{3} - \frac{8}{9}\beta_0\right) c_F^2(\nu)\,\alpha_s^2(m_t\nu).
\end{aligned} \tag{9}$$

For $\mathcal{V}_{c,k,2,r}$ both soft and usoft loops contribute, which explains why the anomalous dimension for $\mathcal{V}_r$ depends on both $\alpha_s(m_t\nu)$ and $\alpha_s(m_t\nu^2)$. The anomalous dimension for $\mathcal{V}_s$ comes only from soft loops but also involves the mixing of the $\bar{\psi}\boldsymbol{\sigma}\cdot\mathbf{B}\psi$ operator whose Wilson coefficient is $c_F(\nu)$. Of the current coefficients, c_3 has no anomalous dimension, while $c_2(\nu)$ has contributions only from ultrasoft loops [7]. More interesting is the anomalous dimension for $c_1(\nu)$ which starts at two-loop order from purely potential loops. At this order [2]

$$\nu\frac{\partial}{\partial\nu}\ln[c_1(\nu)] = -\frac{\mathcal{V}_c(\nu)}{16\pi^2}\left(\frac{\mathcal{V}_c(\nu)}{4} + \mathcal{V}_2(\nu) + \mathcal{V}_r(\nu) + 2\mathcal{V}_s(\nu)\right) + \frac{\mathcal{V}_k(\nu)}{2}, \tag{10}$$

so the solution for c_1 depends on the solutions for the potential coefficients. At three loops there are new contributions to the anomalous dimension for $c_1(\nu)$ coming from mixed potential-ultrasoft and potential-soft loops. These contribute at NNLL order but are currently unknown. In Ref. [8] these unknown terms were estimated to affect the cross section at the 2% level (this rough estimate was based on the size of known terms and using dimensional analysis and parameter dependence to estimate the size of the contributions which are unknown). Of all the Wilson coefficients the one which is the

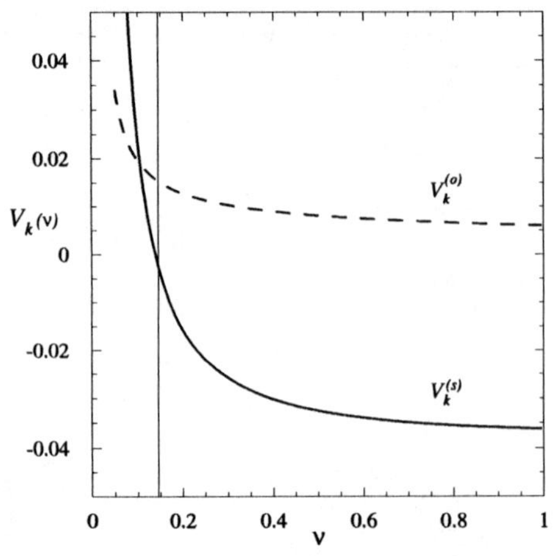

FIGURE 2. Running of the color singlet (s) and octet (o) Wilson coefficients for the $\pi^2/(m|\mathbf{k}|)$ potential from Ref. [5]. The solid vertical line marks the Coulombic region where $\nu \simeq \nu_0$.

most responsible for the difference in the NNLO and NNLL cross sections is $\mathcal{V}_k(\nu)$ which changes by an order of magnitude between $\nu = 1$ to $\nu = 0.15$. The two-loop running of this coefficient is shown in Fig. 2.

3. Cross section computation

The total cross section for $e^+e^- \to t\bar{t}$ is given by

$$\sigma_{\rm tot}^{\gamma,Z}(s) = \frac{4\pi\alpha^2}{3s}\Big[F^v(s)R^v(s) + F^a(s)R^a(s)\Big], \tag{11}$$

where F^v and F^a are trivial functions depending on the charge and weak isospin of the fermions, and $\sin\theta_W$. R^v and R^a are determined by

$$R^v(s) = \frac{4\pi}{s}\,\mathrm{Im}\Big[c_1^2(\nu)\,\mathcal{A}_1(v,m,\nu) + 2c_1(\nu)c_2(\nu)\,\mathcal{A}_2(v,m,\nu)\Big], \tag{12}$$

$$R^a(s) = \frac{4\pi}{s}\,\mathrm{Im}\Big[c_3^2(\nu)\,\mathcal{A}_3(v,m,\nu)\Big],$$

where $\mathcal{A}_i$ are time-ordered products of effective theory currents $[\hat{q} = (\sqrt{s} - 2m_t, 0)]$

$$\mathcal{A}_1(v,m,\nu) = i\sum_{\mathbf{p},\mathbf{p}'}\int d^4x\, e^{i\hat{q}\cdot x}\Big\langle 0\Big|T\ \mathcal{O}_{\mathbf{p},1}(x)\ \mathcal{O}^\dagger_{\mathbf{p}',1}(0)\Big|0\Big\rangle,$$

$$\mathcal{A}_2(v,m,\nu) = \frac{i}{2}\sum_{\mathbf{p},\mathbf{p}'}\int d^4x\, e^{i\hat{q}\cdot x}\Big\langle 0\Big|T\Big[\ \mathcal{O}_{\mathbf{p},1}(x)\ \mathcal{O}^\dagger_{\mathbf{p}',2}(0) + \ \mathcal{O}_{\mathbf{p},2}(x)\ \mathcal{O}^\dagger_{\mathbf{p}',1}(0)\Big]\Big|0\Big\rangle,$$

$$\mathcal{A}_3(\nu,m,\nu) \;=\; i\sum_{\mathbf{p},\mathbf{p}'}\int d^4x\, e^{i\hat{q}\cdot x}\Big\langle 0\Big| T\; O_{\mathbf{p},3}(x)\; O^{\dagger}_{\mathbf{p}',3}(0)\Big|0\Big\rangle. \tag{13}$$

These time-ordered products can be evaluated in terms of non-relativistic Greens functions to give

$$\begin{aligned}\mathcal{A}_1(\nu,m,\nu) &= 18\Big[G^c(\nu,m,\nu)+(\mathcal{V}_2(\nu)+2\mathcal{V}_s(\nu))\,\delta G^{\delta}(\nu,m,\nu)\\ &\qquad +\mathcal{V}_r(\nu)\,\delta G^{r}(\nu,m,\nu)+\mathcal{V}_k(\nu)\,\delta G^{k}(\nu,m,\nu)+\delta G^{\rm kin}(\nu,m,\nu)\Big],\\ \mathcal{A}_2(\nu,m,\nu) &= \mathrm{v}^2\mathcal{A}_1(\nu,m,\nu)\,,\qquad \mathcal{A}_3(\nu,m,\nu)=12G^1(\nu,m,\nu)/m^2\,,\end{aligned} \tag{14}$$

Here G^c are evaluated numerically with the $1/\mathbf{k}^2$ term in $F(\mathbf{p},\mathbf{p}')$ and $V_{\rm soft}(\mathbf{p},\mathbf{p}')$ [18]. In Ref. [8] we analytically evaluated $\delta G^{\delta,r,k,kin}$ with a single insertion of the corresponding potentials or $\mathbf{p}^4$ kinetic energy correction. The P-wave Greens function G^1 was also evaluated in closed form. The analytic calculations enabled all ultraviolet subdivergences to be subtracted in $\overline{\rm MS}$ which is necessary to be consistent with the scheme dependence of the Wilson coefficients. In Eq. (14) the velocity $\mathrm{v}=[\sqrt{s}-2m_t+i\Gamma_t]^{1/2}/m_t^{1/2}$ and $m=m_t$ is the pole mass. The Greens functions depend on the subtraction velocity through $\ln(\mathrm{v}^2/\nu^2)$ and these logarithms are not large when the Greens functions are evaluated at the low scale $\nu\simeq\nu_0$. At this scale all large logarithms have been resummed in the potential and current Wilson coefficients. Typically, $\nu_0\simeq 0.15-0.2$, but to numerically test the remaining scale dependence we use the larger range $\nu_0=0.1-0.4$.

RESULTS

Soon after the NNLO results were derived it was realized that the inherent uncertainty in the top-quark pole mass due to infrared renormalons causes problems for predictions for the peak in the cross section. Therefore, for precision predictions the pole mass is not a suitable mass parameter. The $\overline{\rm MS}$ top-mass provides an infrared safe alternative, however it complicates the non-relativistic power counting. Essentially, it shifts E by an amount $\sim m_t\alpha_s$ which is much larger than the original size of the energy $E\sim m_t\alpha_s^2$. Both of these problems can be addressed by switching to a "threshold mass", defined as an infrared safe mass parameter which differs from $m_t^{\rm pole}$ by $\sim m_t\alpha_s^2$. Several possible threshold masses were suggested, including the PS mass [19], kinetic mass [20], and 1S mass [21, 12]. In Ref. [1] it was concluded that threshold masses could be determined with a precision of < 200 MeV from the total cross section. Converting the result to an $\overline{\rm MS}$ top-mass would then lead to a similar precision for this parameter. In this section predictions will be given using the 1S mass parameter. For a detailed description of how the NNLL pole mass expressions are converted to the 1S mass in a manner consistent with the power counting see Ref. [8].

We begin by comparing results in the fixed order and renormalization group improved expansions. We concentrate on R^v since R^a gives only a small contribution to the cross section, and is essentially identical in the two approaches. The R^v results are shown in Fig. 3 and use $M_t^{1S}=175$ GeV, $\alpha_s(m_Z)=0.118$ and $\Gamma_t=1.43$ GeV. At each order

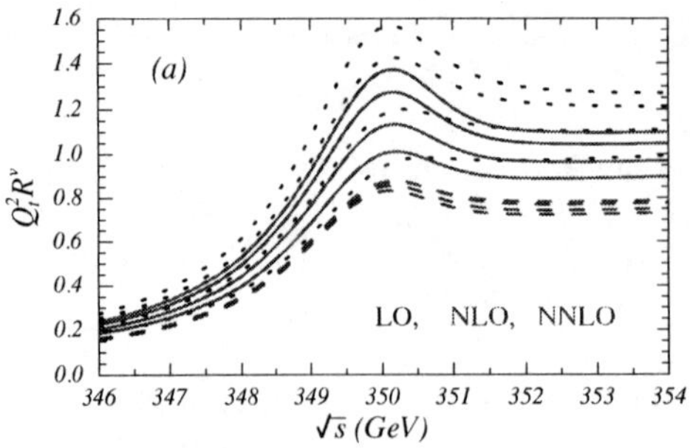

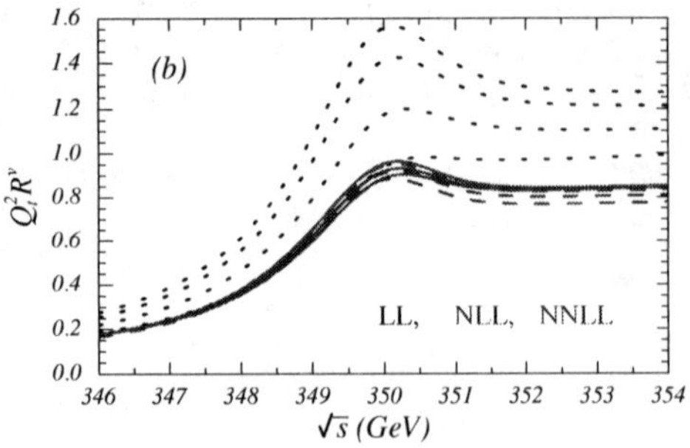

FIGURE 3. Comparison of $Q_t^2 R^v$ with fixed M_t^{1S} mass for the fixed order and resummed expansions. The dotted, dashed, and solid curves in a) are LO, NLO, and NNLO, and in b) are LL, NLL, and NNLL order. For each order four curves are plotted for $\nu = 0.1, 0.125, 0.2$, and 0.4.

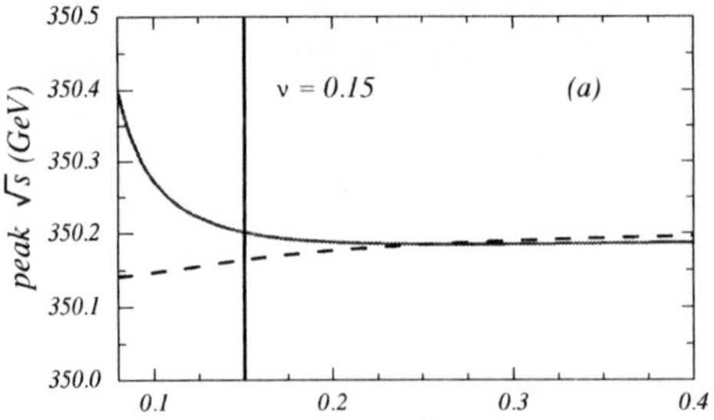

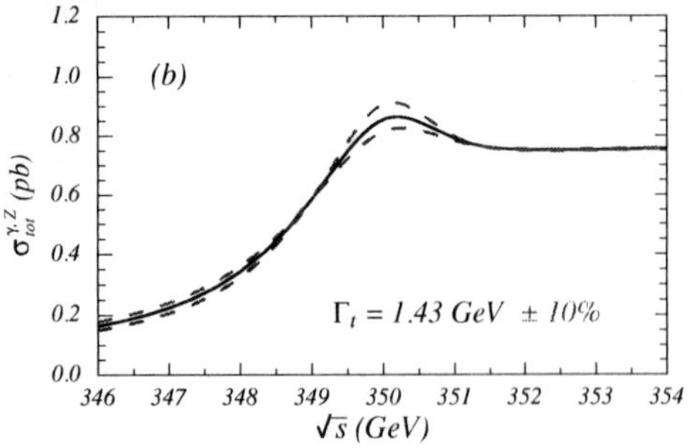

FIGURE 4. (a) Position of the peak in the cross-section versus ν at NNLO (dashed) and NNLL (solid). The vertical line at $\nu = 0.15$ is a physically motivated endpoint for the running. (b) Variation of the NNLL cross section for a $\pm 10\%$ change in the value of the top quark width.

in the expansions four curves are shown which correspond to $\nu = 0.1, 0.125, 0.2$, and 0.4. It is clearly visible that the NNLL results in Fig. 3(b) have much smaller scale dependence than the NNLO results in Fig. 3(a). It should be noted that our NNLO results shown in Fig. 3(a) agree quantitatively with those presented in Ref. [1]. The uncertainty in these results stems to a large extent from the uncertainty in the choice of the renormalization scales in the NNLO contributions. Essentially what the anomalous dimensions and renormalization group do is remove this uncertainty. Also, more than half of the improved convergence of the NNLL result is due to the reduced size of $\mathcal{V}_k(\nu = \nu_0)$ compared to $\mathcal{V}_k(1)$.

From the remaining scale uncertainty and the size of some higher order QCD corrections, the uncertainty in the NNLL cross section was conservatively estimated to be $\pm 3\%$ [8]. This level of precision should enable extractions of various top parameters from the cross section with fairly good precision. In Fig. 4 we show the scale dependence of the peak position for the NNLO (dashed) and NNLL (solid) predictions. The NNLL prediction is slightly less scale dependent than the NNLO prediction until we get to small ν. For $\nu < 0.15$ the larger scale dependence at NNLL is explained by the fact

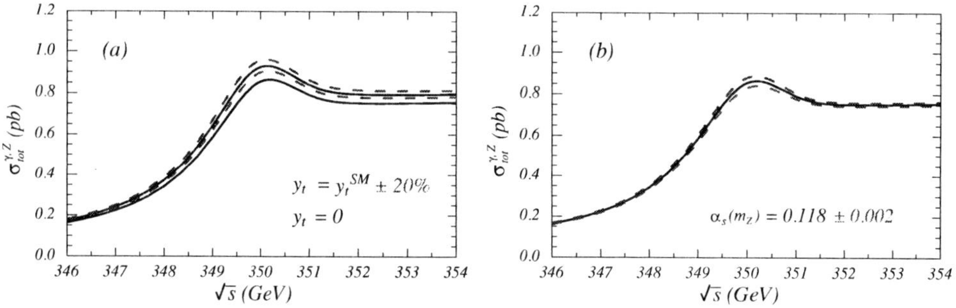

FIGURE 5. Variation of the NNLL cross section for (a) the inclusion of a Standard Model (SM) Higgs boson and (b) the value of the strong coupling. The relative changes are shown by red dashed lines. For (a) the lower black solid line is the decoupling limit for the Higgs boson, and the upper blue solid line is for a SM Higgs with mass $m_H = 115\,\text{GeV}$.

that these predictions depend on the coupling $\alpha_s(m_t \nu^2)$, while the NNLO predictions do not. Also shown in Fig. 4 are the NNLL predictions for the total cross section varying the width by $\pm 10\%$. The size of the variations indicate that a measurement with better than 10% precision is definitely feasible. In Fig. 5 the dependence of the cross section on the top Yukawa y_t (for a Higgs mass $m_H = 115\,\text{GeV}$) and on $\alpha_s(m_Z)$ are shown. It looks quite promising that a $\pm 20\%$ variation in y_t gives a larger change in the cross section than our estimate for the remaining theoretical uncertainty. It should be kept in mind that since both y_t and $\alpha_s(m_Z)$ mainly effect the normalization, at some level these parameters cannot be fixed independently using only the total cross section.

CONCLUSION

In this talk I have discussed predictions for the threshold $e^+e^- \to t\bar{t}$ cross section at NNLO and NNLL order as defined by the expansions in Eq. (1) and Eq. (3). The NNLL predictions made in Refs. [7, 8] sum large logarithms of the velocity using results for the renormalization group improved Wilson coefficients from Refs. [2, 3, 5, 6]. One missing ingredient is the three loop anomalous dimension for c_1, for which only partial results are known. However, rough estimates indicate that this missing anomalous dimension is unlikely to affect the cross section at more than the 2% level [8]. The stability of predictions for the peak in the cross section are very similar at NNLO and NNLL, so that measurements with $\delta m_t < 200\,\text{MeV}$ for short distance masses are feasible with either expansion. The size of the NNLL normalization corrections and variation of the NNLL cross section for various choices of the renormalization parameter are an order of magnitude smaller than the results of earlier NNLO calculations. A conservative estimate of the remaining theoretical uncertainty in the total cross section is $\pm 3\%$ [8]. With such small uncertainty, measurements of top parameters with uncertainties $\delta\alpha_s(m_Z) \sim 0.002$, $\delta\Gamma_t/\Gamma_t \sim 5\%$, and $\delta y_t/y_t \sim 20\%$ appear feasible. However, realistic simulation studies should be done to see how these numbers hold up once effects such as initial state radiation, beamstrahlung, and the beam energy spread are taken into account.

ACKNOWLEDGMENTS

I would like to thank A. Hoang, A. Manohar, and T. Teubner for their collaboration on the results presented here.

REFERENCES

1. A. H. Hoang, *et al.*, *Eur. Phys. J. direct*, **C3**, 1–22 (2000).
2. M.E. Luke, A.V. Manohar, and I.Z. Rothstein., *Phys. Rev.*, **D61**, 074025 (2000).
3. A.V. Manohar and I.W. Stewart, *Phys. Rev.*, **D62**, 014033 (2000).
4. A.V. Manohar and I.W. Stewart, *Phys. Rev.*, **D62**, 074015 (2000).
5. A.V. Manohar and I.W. Stewart, *Phys. Rev.*, **D63**, 054004 (2001).
6. A.H. Hoang, A.V. Manohar, I.W. Stewart , *Phys. Rev.*, **D64**, 014033 (2001).
7. A. H. Hoang, A. V. Manohar, I. W. Stewart, and T. Teubner, *Phys. Rev. Lett.*, **86**, 1951 (2001).
8. A. H. Hoang, A.V. Manohar, I.W. Stewart, and T. Teubner, *Phys. Rev.*, **D65**, 014014 (2001).
9. Y. Schroder, *Phys. Lett.*, **B447**, 321–326 (1999).
10. K. Melnikov and O.I. Yakovlev, *Phys. Lett.*, **B324**, 217–223 (1994).
11. V.S. Fadin, V.A. Khoze, and A.D. Martin, *Phys. Rev.*, **D49**, 2247–2256 (1994).
12. A.H. Hoang and T. Teubner, *Phys. Rev.*, **D60**, 114027 (1999).
13. A.H. Hoang, *Phys. Rev.*, **D56**, 7276 (1997).
14. A. Czarnecki and K. Melnikov, *Phys. Rev. Lett.*, **80**, 2531 (1998).
15. M. Beneke, A. Signer and V. A. Smirnov, *Phys. Rev. Lett.*, **80**, 2535 (1998).
16. A. Pineda, hep-ph/0109117 (2001).
17. C.W. Bauer and A.V. Manohar, *Phys. Rev.*, **D57**, 337–343 (1998).
18. M. Jezabek, J.H. Kuhn, and T. Teubner, *Z. Phys.*, **C56**, 653–660 (1992).
19. M. Beneke, *Phys. Lett.*, **B434**, 115–125 (1998).
20. I. Bigi, M.A. Shifman, N. Uraltsev, and A.I. Vainshtein, *Phys. Rev.*, **D56**, 4017–4030 (1997).
21. A.H. Hoang, Z. Ligeti, and A.V. Manohar, *Phys. Rev. Lett.*, **82**, 277–280 (1999).

Production & Decay of Quarkonium

Sean Fleming

Department of Physics, Carnegie Mellon University, Pittsburgh PA 15213, USA

Abstract. In this talk I review NRQCD predictions for the production of charmonium at the Tevatron. After a quick presentation of the NRQCD factorization formalism for production and decay I review some old results and discuss how they compare to recent data. Following this I discuss some recent work done with Adam Leibovich and Ira Rothstein.

INTRODUCTION

Heavy quarkonia have proven fruitful in helping us gain a better understanding of QCD. Early theoretical analyses of quarkonia decay were based on the color-singlet model (CSM) [1]. The underlying assumption of this model is that the heavy-quark–antiquark pair has the same quantum numbers as the quarkonium meson. (For example the $b\bar{b}$ that forms an Υ must be in a color-singlet $^3S_1$ configuration.) One consequence of such a restrictive assumption is that theoretical predictions based on the CSM are simple, depending on only one nonperturbative parameter. However, the CSM does not provide a systematic approach to studying quarkonium. This is clear in P-wave decays where infrared divergences signal the breakdown of the CSM [2].

In order to systematically study nonrelativistic systems long distance physics needs to be separated from short distance physics. This can be accomplished with a proper effective field theory which provides a power counting that determines relevant operators. In most effective theories the power counting is based upon dimensional analysis, however, for non-relativistic QCD (NRQCD) [3, 4] this is not the case. Instead, it is an expansion in the parameter v, the relative velocity of the heavy quarks. This leads to the result that operators of the same dimension may be of different orders in the power counting. Inclusive decay rates and production cross sections are now understood within the framework of NRQCD factorization [3], where decay rates and production cross sections are predicted in a systematic double expansion in α_s and v. These predictions have met with varying degrees of success.

In the first half of my talk I give a quick review of NRQCD, and the NRQCD factorization formalism for production and decay. This gives the background needed to understand theoretical predictions for production and decay of J/ψ and ψ'. I do not attempt to review all of the predictions, instead I focus on the transverse momentum distribution of J/ψ and ψ' at the Tevatron. For the reason that here lies the earliest success and the greatest challenge to NRQCD factorization. In Ref. [5] the ψ' 'anomaly' (a factor of 30 discrepancy between the CSM prediction and date) was resolved using NRQCD. However, the initial data on the polarization of these states at large transverse momentum [6] seems to be at odds with the NRQCD prediction.

CP618, *Heavy Flavor Physics: Ninth International Symposium,* edited by A. Ryd and F. C. Porter

In the second half of the talk I will discuss a possible resolution to the charmonium polarization puzzle at the Tevatron. This is based on recent work [7] with Ira Rothstein and Adam Leibovich wherein we propose an alternative power counting for charmonium. I do not present all the pieces of evidence which seems to tell us that the effective field theory which best describes the J/ψ system may not be the same theory which best describes the Υ. For that the reader is directed to the literature. I merely give a quick review of the new power counting, and then proceed to discuss how this changes the predictions for J/ψ and ψ' polarization at the Tevatron.

NRQCD

The power counting depends upon the relative size of the four scales $(m, mv, mv^2, \Lambda_{\rm QCD})$. If we take $m > mv > mv^2 \simeq \Lambda_{\rm QCD}$ the bound state dynamics will be dominated by exchange of Coulombic gluons with $(E \simeq mv^2, \vec{p} = m\vec{v})$. This hierarchy has been assumed in the NRQCD calculation of production and decay rates and is probably a reasonable choice for the Υ system, where $mv \sim 1.5$ GeV. However, whether or not it is correct for the J/ψ, where $mv \sim 700$ MeV remains to be seen.

The power counting can be established in a myriad of different ways. Here I will follow the construction of [4], which I now briefly review. There are three relevant gluonic modes [8]: the Coulombic (mv^2, mv), soft (mv, mv) and ultrasoft (mv^2, mv^2). The soft and Coulombic modes can be integrated out leaving only ultrasoft propagating gluons. In the process of integrating out these modes large momenta must be removed from the quark field. This is accomplished by rescaling the heavy quark fields by a factor of $\exp(i\mathbf{p}\cdot\mathbf{x})$ and labeling them by their three momentum $\mathbf{p}$. The ultrasoft gluon can only change residual momenta and not labels on fields. This is analogous to HQET, where the four-velocity labels the fields and the nonperturbative gluons only change the residual momenta [9]. This rescaling must also be done for soft gluon fields [10] which, while they cannot show up in external states, do show up in the Lagrangian. After this rescaling a matching calculation leads to the following tree level Lagrangian [4]

$$
\begin{aligned}
\mathcal{L} \;=\; & \sum_{\mathbf{p}} \psi^\dagger_{\mathbf{p}} \left\{ iD^0 - \frac{\mathbf{p}^2}{2m} \right\} \psi_{\mathbf{p}} - 4\pi\alpha_s \sum_{q,q'\mathbf{p},\mathbf{p}'} \left\{ \frac{1}{q^0} \psi^\dagger_{\mathbf{p}'} \left[A^0_{q'}, A^0_q\right] \psi_{\mathbf{p}} \right. \\
& \left. + \frac{g^{\nu 0}(q'-p+p')^\mu - g^{\mu 0}(q-p+p')^\nu + g^{\mu\nu}(q-q')^0}{(\mathbf{p}'-\mathbf{p})^2} \psi^\dagger_{\mathbf{p}'} \left[A^\nu_{q'}, A^\mu_q\right] \psi_{\mathbf{p}} \right\} \\
& + \psi \leftrightarrow \chi,\; T \leftrightarrow \bar{T} + \sum_{\mathbf{p},\mathbf{q}} \frac{4\pi\alpha_s}{(\mathbf{p}-\mathbf{q})^2} \psi^\dagger_{\mathbf{q}} T^A \psi_{\mathbf{p}} \chi^\dagger_{-\mathbf{q}} \bar{T}^A \chi_{-\mathbf{p}} + \ldots
\end{aligned}
\tag{1}
$$

where we have retained the lowest order terms in each sector of the theory. The matrices T^A and $\bar{T}^A$ are the color matrices for the $\mathbf{3}$ and $\bar{\mathbf{3}}$ representations, respectively. Note the last term is the Coulomb potential, which is leading order and must be resummed in the four-quark sector, while the other non-local interactions arise from soft gluon scattering.

All the operators in the Lagrangian have a definite scaling in v, and the spin symmetry, which will play such a crucial role in the polarization predictions, is manifest. The two subleading interactions which will dominate my discussion are the electric dipole ($E1$)

$$\mathcal{L}_{E1} = \psi_{\mathbf{p}}^{\dagger} \frac{\mathbf{p}}{m} \cdot \mathbf{A} \psi_{\mathbf{p}}, \tag{2}$$

and the magnetic dipole ($M1$)

$$\mathcal{L}_{M1} = c_F\, g \psi_{\mathbf{p}}^{\dagger} \frac{\sigma \cdot \mathbf{B}}{2m} \psi_{\mathbf{p}}. \tag{3}$$

The $E1$ interaction is down by a factor of v while the $M1$ is down by a factor of v^2.

NRQCD FACTORIZATION FORMALISM

In the NRQCD factorization formalism developed in Ref. [3] decay rates and production cross sections are written as a sum of products of Wilson coefficients encoding short distance physics and NRQCD matrix elements describing long distance physics. In this formalism a general decay process is written as

$$\Gamma_{J/\psi} = \sum C_{2S+1_{L_J}}(m, \alpha_s) \langle \psi \mid O^{(1,8)}({}^{2S+1}L_J) \mid \psi \rangle. \tag{4}$$

The matrix element represents the long distance part of the rate and may be thought of as the probability of finding the heavy quarks in the relative state n, while the coefficient $C_{2S+1_{L_J}}(m, \alpha_s)$ is a short distance quantity calculable in perturbation theory. The sum over operators may be truncated as an expansion in the relative velocity v. Similarly, production cross sections may be written as

$$d\sigma = \sum_n d\sigma_{i+j \to Q\bar{Q}[n]+X} \langle 0 \mid O_n^H \mid 0 \rangle. \tag{5}$$

Here $d\sigma_{i+j \to Q\bar{Q}[n]+X}$ is the short distance cross section for a reaction involving two partons, i and j, in the initial state, and two heavy quarks in a final state, labeled by n, plus X. This part of the process is calculable in perturbation theory, up to possible structure functions in the initial state. The production matrix elements, which differ from those used in the decay processes, describe the probability of the short distance pair in the state n to hadronize, inclusively, into the state of interest. The relative size of the matrix elements in the sum are again fixed by the power counting which we will discuss in more detail below.

The formalism for decays is on the same footing as the operator product expansion (OPE) for non-leptonic decays of heavy quarks, while the production formalism assumes factorization, which is only proven, and in some applications of production this is not even the case, in perturbation theory [1] [11]. The trustworthiness of factorization depends upon the particular application. I have reviewed these results here to emphasize the point that when the theory is tested one is really testing both the factorization hypothesis as well the validity of the effective theory as applied to the J/ψ system. Thus, care must be taken in assigning blame when theoretical predictions do not agree with data.

---

[1] For a discussion of factorization in NRQCD see Refs. [3, 12].

TABLE 1. Scaling of matrix elements relevant for ψ production in NRQCD$_b$ and NRQCD$_c$.

	$\langle O_1^\psi(^3S_1)\rangle$	$\langle O_8^\psi(^3S_1)\rangle$	$\langle O_8^\psi(^1S_0)\rangle$	$\langle O_8^\psi(^3P_0)\rangle$
NRQCD$_b$	v^0	v^4	v^4	v^4
NRQCD$_c$	$(\Lambda_{QCD}/m_c)^0$	$(\Lambda_{QCD}/m_c)^4$	$(\Lambda_{QCD}/m_c)^2$	$(\Lambda_{QCD}/m_c)^4$

CHARMONIUM PRODUCTION AT THE TEVATRON

Having introduced NRQCD and the factorization formalism I will now turn to theoretical predictions of J/ψ and ψ' production at the Tevatron. The leading order in v contribution to J/ψ production is through the color-singlet matrix element $\langle O_1^\psi(^3S_1)\rangle$, since the quantum numbers of the short distance quark pair matches those of the final state. All other matrix elements need insertions of operators into time ordered products to give a non-zero result, and are therefore suppressed compared to the color-singlet matrix element above. For instance, the matrix element $\langle O_8^\psi(^1S_0)\rangle$ vanishes at leading order. The first non-vanishing contribution comes from the insertion of two $M1$ operators into time ordered products, thus giving a v^4 suppression. The scaling of the relevant matrix elements for ψ production are shown in Table 1 under NRQCD$_b$ (for reasons which I will explain later). It appears from just the v counting that only the color-singlet contribution is important. However, other contributions can be enhanced by kinematic factors. At large transverse momentum, fragmentation type production dominates [13], and only the $\langle O_8^\psi(^3S_1)\rangle$ contribution is important. Without the color-octet contributions (*i.e.*, the Color-Singlet Model), the theory is below experiment by about a factor of 30. By adding the color-octet contribution the fit to the data is very good [5].

Once the color-octet matrix elements are fit to the unpolarized date it is possible to make a parameter free prediction for the polarization of J/ψ and ψ' at the Tevatron: they are predicted to be transversely polarized at large p_T. This is because at large transverse momentum, the dominant production mechanism is through fragmentation from a nearly on shell gluon to the octet $^3S_1$ state. The quark pair inherits the polarization of the fragmenting gluon, and is thus transversely polarized [14]. The leading order transition to the final state goes via two $E1$, spin preserving, gluon emissions. Higher order perturbative fragmentation contributions [15], fusion diagrams [16, 17], and feed-down for the J/ψ [18] dilute the polarization some, but the prediction still holds that as p_T increases so should the transverse polarization. Indeed, for the ψ', at large $p_T \gg m_c$, we expect nearly pure transverse polarization. This prediction seems to be at odds with the initial data which seems to suggest that the J/ψ and the ψ' are unpolarized or slightly longitudinally polarized as p_T increases [6].[2] If, after the statistics improve, this trend continues one is left with two obvious possibilities: 1) The power counting of NRQCD does not apply to the J/ψ system. 2) Factorization is violated "badly", meaning that there are large power corrections.

NRQCD$_C$

In this section I discuss the work done with Adam Leibovich and Ira Rothstein in Ref.[7]. In that work we marshaled evidence that the NRQCD power counting might not apply to the J/ψ system. We did not consider the second possibility mentioned at the end of the previous section: that factorization is violated. I will not be considering this possibility either.

The standard NRQCD methodology, which is based upon the hierarchy $m > mv > mv^2 \simeq \Lambda_{\rm QCD}$, has been applied to the J/ψ as well as the Υ systems. While it seems quite reasonable to apply this power counting to the Υ system, it is not clear that it should apply to the J/ψ system. Indeed, I believe that the data is hinting toward the possibility that a new power counting is called for in the charmed system.

If NRQCD does not apply to the J/ψ system, then one must ask: is there another effective theory which does correctly describe the J/ψ? One good reason to believe that such a theory does exist is that NRQCD, as formulated, does correctly predict the ratios of decay amplitudes for exclusive radiative decays. Using spin symmetry the authors of [14] made the following predictions:

$$\begin{aligned}
&\Gamma(\chi_{c0} \to J/\psi+\gamma) : \Gamma(\chi_{c1} \to J/\psi+\gamma) : \Gamma(\chi_{c2} \to J/\psi+\gamma) : \Gamma(h_c \to \eta_c+\gamma) \\
&= 0.095 : 0.20 : 0.27 : 0.44 \quad \text{(theory)} \\
&= 0.092 \pm 0.041 : 0.24 \pm 0.04 : 0.27 \pm 0.03 : \text{unmeasured} \quad \text{(experiment)}. \qquad (6)
\end{aligned}$$

Thus, an alternative formulation of NRQCD must preserve these predictions yet yield different predictions in other relevant processes.

Let me now consider the alternate hierarchy $m > mv \sim \Lambda_{\rm QCD}$. One might be tempted to believe that in this case the power counting should be along the lines of HQET, where the typical energy and momentum exchanged between the heavy quarks is of order $\Lambda_{\rm QCD}$. However, this leads to an effective theory which does not correctly reproduce the infrared physics. With this power counting, the leading order Lagrangian would simply be

$$\mathcal{L}_{\rm HQET} = \psi_v^\dagger D_0 \psi_v, \qquad (7)$$

where the fields are now labeled by their four velocity. This is a just a theory of time-like Wilson lines (static quarks) which does not produce any bound state dynamics. Thus I am forced to conclude that the typical momentum is of order $\Lambda_{\rm QCD}$, whereas the typical energy is $\Lambda_{\rm QCD}^2/m$, so that $D^2/(2m)$ is still relevant. I will call this theory NRQCD$_c$, and will refer to the traditional power counting as NRQCD$_b$ as I assume that it describes the bottom system.

---

[2] The data still has rather large error bars, so we should withhold judgment until the statistics improves.

The power counting of this theory is now along the lines of HQET where the expansion parameter is $\Lambda_{\rm QCD}/m_Q$. However the residual energy of the quarks is order $\Lambda^2_{\rm QCD}/m_Q$, while the residual three momentum is $\Lambda_{\rm QCD}$. Thus one must be careful in the power counting to differentiate between time and spatial derivatives acting on the quark fields. As far as the phenomenology is concerned, perhaps the most important distinction between the power counting in NRQCD$_c$ and NRQCD$_b$ is that the magnetic and electric gluon transitions are now of the same order in NRQCD$_c$. This difference in scaling does not disturb the successes of the standard NRQCD$_b$ formulation but does seem help in some of its shortcomings.

NRQCD$_C$ PREDICTIONS

The relative size of the different matrix elements change in NRQCD$_c$. In particular, the $M1$ transition is now the same order as the $E1$ transition. The new scaling is shown in Table 1. [3] Due to the dominance of fragmentation at large transverse momentum, we need to include effects up to order $(\Lambda_{\rm QCD}/m_c)^4$, since the $\langle O_8^\psi(^3S_1)\rangle$ matrix element will still dominate at large p_T.

Is this consistent? The size of the matrix elements is a clue. Extraction of the matrix elements uses power counting to limit the number of channels to include in the fits. Calculating J/ψ and ψ' production up to order $(\Lambda_{\rm QCD}/m_c)^4$ in NRQCD$_c$ requires keeping the same matrix elements as in NRQCD$_b$. Previous extractions of the matrix elements only involve the linear combination

$$M_r^\psi = \langle O_8^\psi(^1S_0)\rangle + \frac{r}{m_c^2}\langle O_8^\psi(^3P_J)\rangle, \tag{8}$$

with $r \approx 3-3.5$, since the short-distance rates have similar size and shape. In the new power-counting, I can just drop the contribution from $\langle O_8^\psi(^3P_J)\rangle$, since it is down by $(\Lambda_{\rm QCD}/m_c)^2 \sim 1/10$ compared to $\langle O_8^\psi(^1S_0)\rangle$. It is the same order as $\langle O_8^\psi(^3S_1)\rangle$, but is not kinematically enhanced by fragmentation effects. The extraction from [18] would then give for the J/ψ and ψ' matrix elements

$$\begin{aligned} \langle O_8^{J/\psi}(^1S_0)\rangle : \langle O_8^{J/\psi}(^3S_1)\rangle &= (6.6\pm0.7)\times10^{-2} : (3.9\pm0.7)\times10^{-3} \approx 17:1\,, \\ \langle O_8^{\psi'}(^1S_0)\rangle : \langle O_8^{\psi'}(^3S_1)\rangle &= (7.8\pm3.6)\times10^{-3} : (3.7\pm0.9)\times10^{-3} \approx 2:1. \end{aligned} \tag{9}$$

Other extractions have various values of the hierarchy, ranging from 3 : 1 to 20 : 1 [21]. While the relation of the color-octet matrix elements in the J/ψ system is indeed in agreement with the NRQCD$_c$ power counting, the ψ' does not look to be hierarchical. However, it should be noted that the statistical errors in the ψ' extraction, quoted above, are quite large. Furthermore, there are also large uncertainties introduced in the parton distribution function. The above ratios used the CTEQ5L parton distribution functions. If we take the central values from [18] for the MRST98LO distribution functions, we

[3] These results reproduce those given in [20] when λ is taken to be 1 in this reference.

find the ratio 3 : 1. On the other hand, the J/ψ extraction is much less sensitive to the choice of distribution function. Given the statistical and theoretical errors, it clear that the ψ' ratio is not terribly illuminating.

Let me now consider the extraction of these color-octet matrix elements in the Υ sector [22], where according to NRQCD$_b$ power counting there is should be no hierarchy:

$$\begin{aligned}
\langle O_8^{\Upsilon(3S)}(^1S_0)\rangle : \langle O_8^{\Upsilon(3S)}(^3S_1)\rangle &= (5.4 \pm 4.3^{+3.1}_{-2.2}) \times 10^{-2} : (3.6 \pm 1.9^{+1.8}_{-1.3}) \times 10^{-2} \\
&\approx 1:1, \\
\langle O_8^{\Upsilon(2S)}(^1S_0)\rangle : \langle O_8^{\Upsilon(2S)}(^3S_1)\rangle &= (-10.8 \pm 9.7^{-3.4}_{+2.0}) \times 10^{-2} : (16.4 \pm 5.7^{+7.1}_{-5.1}) \times 10^{-2} \\
&\approx 1:1, \\
\langle O_8^{\Upsilon(1S)}(^1S_0)\rangle : \langle O_8^{\Upsilon(1S)}(^3S_1)\rangle &= (13.6 \pm 6.8^{+10.8}_{-7.5}) \times 10^{-2} : (2.0 \pm 4.1^{-0.6}_{+0.5}) \times 10^{-2} \\
&\approx 6:1. \qquad (10)
\end{aligned}$$

For the $\Upsilon(3S)$ and $\Upsilon(2S)$ we observe that there is indeed no hierarchy, while for the $\Upsilon(1S)$ it appears there may be a hierarchy. However, it is not possible to draw any strong conclusions from these data because the errors on the extractions are large. In fact the ratio for the $\Upsilon(1S)$ color-octet matrix elements is 1 : 1 within the one sigma errors. Furthermore, these matrix elements are those extracted subtracting out the feed down from the higher states. While phenomenologically it is perfectly reasonable to define the subtracted matrix elements, I believe that, since the matrix elements are inclusive, one should not subtract out the feed down from hadronic decays when checking the power counting. In principle this subtraction should not change things by orders of magnitude, but nonetheless it can have a significant effect. Indeed, if one compares the ratios for inclusive matrix elements, which do not have the accumulated error, then the ratios come out to be 1 : 1, even for the $\Upsilon(1S)$ [22].

With NRQCD$_c$, the intermediate color-octet $^3S_1$ states hadronize through the emission of either two $E1$ or $M1$ dipole gluons, at the same order in $1/m_c$. Since the magnetic gluons do not preserve spin, the polarization of ψ produced through the $\langle O_8^{\psi}(^3S_1)\rangle$ can be greatly diluted. The net polarization will depend on the ratio of matrix elements

$$R_{M/E} = \frac{\int \prod_\ell d^4x_\ell \langle 0 \mid T(M_1(x_1)M_1(x_2)\psi^\dagger T^a \sigma_i \chi)\, a_H^\dagger a_H\, T(M_1(x_3)M_1(x_4)\chi^\dagger T^a \sigma_i \psi) \mid 0\rangle}{\int \prod_\ell d^4x_\ell \langle 0 \mid T(E_1(x_1)E_1(x_2)\psi^\dagger T^a \sigma_i \chi)\, a_H^\dagger a_H\, T(E_1(x_3)E_1(x_4)\chi^\dagger T^a \sigma_i \psi) \mid 0\rangle} \qquad (11)$$

where

$$a_H^\dagger a_H = \sum_X \mid H + X\rangle\langle H + X \mid . \qquad (12)$$

The leads to the polarization leveling off at large p_T at some value which is fixed by $R_{M/E}$. In Fig. 1, we show the prediction for J/ψ and ψ' polarization at the Tevatron. The data is from [6]. The three lines correspond to different values for $R_{M/E}$=(0 (dashed), 1 (dotted), ∞ (solid)). The dashed line is also the prediction for NRQCD$_b$. The residual transverse polarization for J/ψ at asymptotically large p_T is due to feed down from χ states. The non-perturbative corrections to our predictions are suppressed by $\Lambda_{\rm QCD}^4/m^4$.

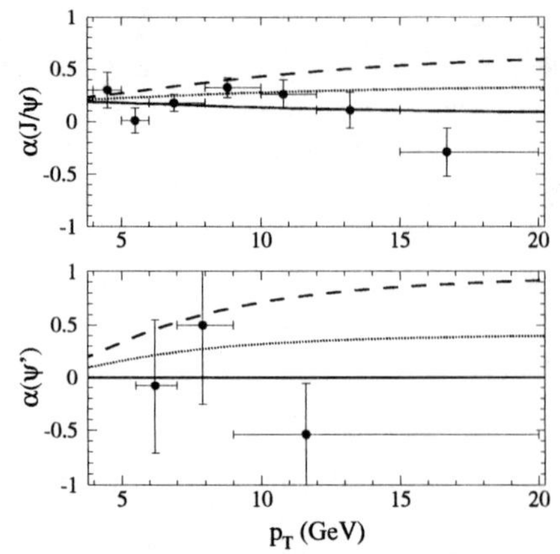

FIGURE 1. Predicted polarization in NRQCD$_c$ for J/ψ and ψ' at the Tevatron as a function of p_T. The three lines correspond to $R_{M/E}$=(0 (dashed), 1 (dotted), ∞ (solid)). The dashed line is also the prediction for NRQCD$_b$.

CONCLUSION

In this talk I have reviewed NRQCD, and the NRQCD factorization formalism which is used to make predictions for the production and decay of charmonium and bottomonium. I did not discuss all of these predictions. Instead I focused on what I believe to be the most important prediction of NRQCD factorization: the transverse momentum distribution of unpolarized and polarized J/ψ and ψ' produced at the Tevatron. Because the unpolarized data can be used to determine the unknown color-octet matrix elements, it is possible to make a parameter free prediction for polarized production. This provides a clean test of the NRQCD factorization formalism. Moreover the quality of the data for unpolarized production is good, and while the data for polarized production has large error bars it is expected to get better.

The NRQCD factorization formalism predicts the J/ψ and ψ' to be transversely polarized at large p_T. This is because at large transverse momentum, the dominant production mechanism is through fragmentation from a nearly on shell gluon to the octet $^3S_1$ state. The quark pair inherits the polarization of the fragmenting gluon, and is thus transversely polarized [14]. The leading order transition to the final state goes via two $E1$, spin preserving, gluon emissions. Various corrections dilute the polarization some, but the prediction still holds that as p_T increases so should the transverse polarization. Indeed, for the ψ', at large $p_T \gg m_c$, we expect nearly pure transverse polarization. The current experimental results [6] show no or a slight longitudinal polarization, as p_T increases. If, after the statistics improve, this trend continues, then it will be the smoking gun that leads us to conclude that either NRQCD is not the correct effective field theory for charmonia, or that factorization fails in these processes.

The possibility that NRQCD is not the correct effective theory for charmonium leads me to ask: is there any reason to believe that there is any effective theory to correctly describe the J/ψ? I believe that the spin symmetry predictions for the ratio of χ decays clearly answers this question in the affirmative. Assuming that such an effective theory exists, then is it NRQCD_c or NRQCD_b? As I have shown the two theories do indeed make quite disparate predictions, which in principle should be easy to test.

However, these tests can be clouded by the issues of factorization and the convergence of the perturbative expansion. One would be justified to worry about the breakdown of factorization in hadro-production at small transverse momentum. However, for large transverse momentum one would expect factorization to hold, with non-factorizable corrections suppressed by powers of m_c/p_T.As far as the perturbative expansion is concerned, it seems that for most calculations the next-to-leading order results are indeed smaller than the leading order result [23, 24, 25], though, the NNLO calculation performed, in the leptonic decay width [26], is not well behaved at this order.

In the end I believe the data will be the final arbiter. The polarization measurement may fall in line with the NRQCD_b prediction. Or the data may result in longitudinal polarization for J/ψ and ψ', in which case it may be that neither NRQCD_c nor NRQCD_b are the correct theory.

ACKNOWLEDGMENTS

I would like that thank my collaborators Adam Leibovich and Ira Rothstein. Thanks also go to the organizers of this conference. This work was supported in part by the Department of Energy under grant number DOE-ER-40682-143 and DE-AC02-76CH03000.

REFERENCES

1. W. . Buchmuller, *Amsterdam, Netherlands: North-Holland (1992) 316 p. (Current physics-sources and comments, 9).*
2. G. T. Bodwin, E. Braaten and G. P. Lepage, *Phys. Rev. D* **46**, 1914 (1992) [arXiv:hep-lat/9205006].
3. G. T. Bodwin, E. Braaten and G. P. Lepage, *Phys. Rev. D* **51**, 1125 (1995) [Erratum-ibid. *D* **55**, 5853 (1995)] [hep-ph/9407339].
4. M. E. Luke, A. V. Manohar and I. Z. Rothstein, *Phys. Rev. D* **61**, 074025 (2000) [hep-ph/9910209].
5. E. Braaten and S. Fleming, *Phys. Rev. Lett.* **74**, 3327 (1995) [arXiv:hep-ph/9411365].
6. T. Affolder *et al.* [CDF Collaboration], *Phys. Rev. Lett.* **85**, 2886 (2000) [arXiv:hep-ex/0004027].
7. S. Fleming, I. Z. Rothstein and A. K. Leibovich, *Phys. Rev. D* **64**, 036002 (2001) [arXiv:hep-ph/0012062].
8. M. Beneke and V. A. Smirnov, *Nucl. Phys. B* **522**, 321 (1998) [arXiv:hep-ph/9711391].
9. A. V. Manohar and M. B. Wise, *Cambridge Monogr. Part. Phys. Nucl. Phys. Cosmol.* **10**, 1 (2000).
10. H. W. Griesshammer, *Phys. Rev. D* **58**, 094027 (1998) [arXiv:hep-ph/9712467].
11. See ref. [3] and related proofs in, J. C. Collins, D. E. Soper and G. Sterman, *Adv. Ser. Direct. High Energy Phys.* **5**, 1 (1988).
12. M. Beneke and I. Z. Rothstein, *Phys. Rev. D* **54**, 2005 (1996) [Erratum-ibid. D **54**, 7082 (1996)] [arXiv:hep-ph/9603400].
13. E. Braaten and T. C. Yuan, *Phys. Rev. Lett.* **71**, 1673 (1993) [arXiv:hep-ph/9303205].
14. P. L. Cho and M. B. Wise, *Phys. Lett. B* **346**, 129 (1995) [arXiv:hep-ph/9411303].
15. M. Beneke and I. Z. Rothstein, *Phys. Lett. B* **372**, 157 (1996) [Erratum-ibid. *B* **389**, 769 (1996)] [arXiv:hep-ph/9509375].
16. M. Beneke and M. Kramer, *Phys. Rev. D* **55**, 5269 (1997) [arXiv:hep-ph/9611218].
17. A. K. Leibovich, *Phys. Rev. D* **56**, 4412 (1997) [arXiv:hep-ph/9610381].

18. E. Braaten, B. A. Kniehl and J. Lee, *Phys. Rev. D* **62**, 094005 (2000) [arXiv:hep-ph/9911436]. B. A. Kniehl and J. Lee, *Phys. Rev. D* **62**, 114027 (2000) [arXiv:hep-ph/0007292].
19. E. Braaten and J. Lee, *Nucl. Phys. B* **586**, 427 (2000) [arXiv:hep-ph/0004228].
20. M. Beneke, arXiv:hep-ph/9703429.
21. P. L. Cho and A. K. Leibovich, *Phys. Rev. D* **53**, 150 (1996) [arXiv:hep-ph/9505329]. *ibid. D* **53**, 6203 (1996) [arXiv:hep-ph/9511315].
B. Cano-Coloma and M. A. Sanchis-Lozano, *Nucl. Phys. B* **508**, 753 (1997) [arXiv:hep-ph/9706270].
C. G. Boyd, A. K. Leibovich and I. Z. Rothstein, *Phys. Rev. D* **59**, 054016 (1999) [arXiv:hep-ph/9810364].
B. A. Kniehl and G. Kramer, *Eur. Phys. J. C* **6**, 493 (1999) [arXiv:hep-ph/9803256].
22. E. Braaten, S. Fleming and A. K. Leibovich, *Phys. Rev. D* **63**, 094006 (2001) [arXiv:hep-ph/0008091].
23. M. Beneke, I. Z. Rothstein and M. B. Wise, *Phys. Lett. B* **408**, 373 (1997) [arXiv:hep-ph/9705286].
24. F. Maltoni and A. Petrelli, *Phys. Rev. D* **59**, 074006 (1999) [arXiv:hep-ph/9806455].
25. A. Petrelli, M. Cacciari, M. Greco, F. Maltoni and M. L. Mangano, *Nucl. Phys. B* **514**, 245 (1998) [arXiv:hep-ph/9707223].
E. Braaten and Y. Q. Chen, *Phys. Rev. D* **55**, 7152 (1997) [arXiv:hep-ph/9701242].
26. M. Beneke, A. Signer and V. A. Smirnov, *Phys. Rev. Lett.* **80**, 2535 (1998) [arXiv:hep-ph/9712302].
27. E. Braaten and Y. Q. Chen, *Phys. Rev. D* **57**, 4236 (1998) [Erratum-ibid. D **59**, 079901 (1998)] [arXiv:hep-ph/9710357].
G. T. Bodwin and Y. Q. Chen, *Phys. Rev. D* **60**, 054008 (1999) [arXiv:hep-ph/9807492].

BTeV

Sheldon Stone representing the BTeV collaboration

Physics Dept., Syracuse Univ., Syracuse, NY 13244

Abstract. The BTeV program at the Fermilab collider will make precision measurements of CP violation, mixing and rare decays in the charm and bottom systems. The aim is observe and investigate "New Physics" phenomena, beyond the Standard Model.

PHYSICS RATIONALE

Our ultimate goal is to use b and c decays to elucidate or find new physics phenomena. To do this we need to make a complete set of measurements of CP violation, mixing and rare decays. Recent Babar and Belle determinations of $\sin(2\beta)$ have unambiguously shown large CP violating effects in the b system [1] [2]. However, there are great many measurements that need to be done with high accuracies, requiring large samples.

I now describe some of these objectives. The unitarity of the CKM matrix [3] allows us to construct six relationships. These equations may be thought of triangles in the complex plane. They are shown in Fig. 1.

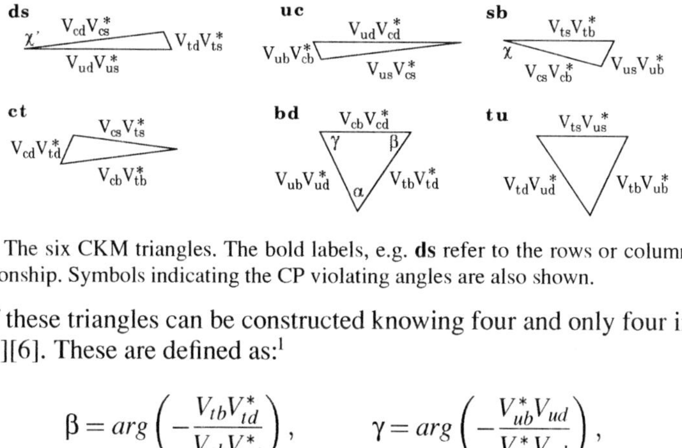

FIGURE 1. The six CKM triangles. The bold labels, e.g. **ds** refer to the rows or columns used in the unitarity relationship. Symbols indicating the CP violating angles are also shown.

All six of these triangles can be constructed knowing four and only four independent angles [4][5][6]. These are defined as:[1]

$$\beta = arg\left(-\frac{V_{tb}V_{td}^*}{V_{cb}V_{cd}^*}\right), \qquad \gamma = arg\left(-\frac{V_{ub}^*V_{ud}}{V_{cb}^*V_{cd}}\right),$$
$$\chi = arg\left(-\frac{V_{cs}^*V_{cb}}{V_{ts}^*V_{tb}}\right), \qquad \chi' = arg\left(-\frac{V_{ud}^*V_{us}}{V_{cd}^*V_{cs}}\right). \tag{1}$$

[1] Since $\alpha+\beta+\gamma=\pi$ any two of the these angles can be used.

CP618, *Heavy Flavor Physics: Ninth International Symposium,* edited by A. Ryd and F. C. Porter

Two of the phases β and γ are probably large while χ is estimated to be small $\approx$0.02, but measurable, while χ' is likely to be much smaller.

It has been pointed out by Silva and Wolfenstein [4] that measuring β and α may not be sufficient to detect new physics. For example, suppose there is new physics that arises in $B^o - \overline{B}^o$ mixing. Let us assign a phase θ to this new contribution. If we then measure CP violation in $B^o \to J/\psi K_S$ and eliminate any Penguin pollution problems in using $B^o \to \pi^+\pi^-$, then we actually measure $2\beta' = 2\beta + \theta$ and $2\alpha' = 2\alpha - \theta$. So while there is new physics, we miss it, because $2\beta' + 2\alpha' = 2\alpha + 2\beta$ and $\alpha' + \beta' + \gamma = 180°$.

The angle χ, defined in equation 1, can be extracted by measuring the time dependent CP violating asymmetry in the reaction $B_s \to J/\psi\eta^{(\prime)}$, or if one's detector is incapable of quality photon detection, the $J/\psi\phi$ final state can be used. However, here there are two vector particles in the final state, making this a state of mixed CP requiring a time-dependent angular analysis to find χ that requires large statistics.

Measurements of the magnitudes of CKM matrix elements all come with theoretical errors. Some of these are hard to estimate; we now try and view realistically how to combine CP violating phase measurements with the magnitude measurements to best test the Standard Model.

The most accurately measured magnitude is $\lambda = |V_{us}/V_{ud}| = 0.2205 \pm 0.0018$. Silva and Wolfenstein [4] [5] show that the Standard Model can be checked in a profound manner by seeing if:

$$\sin\chi = \left|\frac{V_{us}}{V_{ud}}\right|^2 \frac{\sin\beta\,\sin\gamma}{\sin(\beta+\gamma)} \quad . \tag{2}$$

Here the precision of the check will be limited initially by the measurement of $\sin\chi$, not of λ. This check can reveal new physics, even if other measurements have not shown any anomalies. There are other checks using $|V_{ub}/V_{cb}|$ or $|V_{td}/V_{ts}|$ [7].

New physics can also appear in inconsistencies in determinations of CKM parameters (for example $\rho - \eta$ [8]) using different physical processes. These include (1) using magnitudes of CKM elements using $|V_{ub}/V_{cb}|$ and B_d mixing; (2) using B_d mixing in $J/\psi K_S$ and $\rho\pi$ to determine $\sin(2\beta)$ and $\sin(2\alpha)$, respectively; (3) using B_s mixing in $D_s^{\pm}K^{\mp}$ and $J/\psi\eta^{(\prime)}$ to determine γ and $\sin(2\chi)$, respectively.

There are a few key measurements that can directly identify new physics. Comparing the CP asymmetry in $B^o \to \phi K_S$ with $J/\psi K_S$ whould show new physics that would arise from a CP violating phase in the decay amplitude. Hinchliff and Kersting [9] have shown in a specific model of supersymmetry that this new physics would also appear as an asymmetry between $B^- \to \phi K^-$ and $B^+ \to \phi K^+$ that would be equal to $\left(\frac{M_W}{m_{squark}}\right)^2 \sin(\phi)$, where m_{squark} is the mass of the lightest squark and ϕ is the weak decay phase. In the Standard Model this asymmetry is zero. Their model also predicts an asymmetry in $B_s \to J/\psi\eta$ that is about ten times the Standard Model value.

Other interesting places to look for new physics are the CP asymmetry in $D^o \to K^-\pi^+$ that is zero in the Standard Model but finite in other models [10], and the polarization in $B \to K^*\ell^+\ell^-$ [11].

B PRODUCTION AT HADRON COLLIDERS

To make precision measurements, large samples of b's and c's are necessary. Fortunately, these are available. With the Fermilab Main Injector, the Tevatron collider will produce $\approx 4\times10^{11}$ b hadrons/10^7 s at a luminosity of $2\times10^{32}\mathrm{cm}^{-2}\mathrm{s}^{-1}$. These compare very favorably to e^+e^- machines operating at the $\Upsilon(4S)$. At a luminosity of 10^{34} they would produce 2×10^8 B's/10^7 s. Furthermore B_s, Λ_b and other b-flavored hadrons are accessible for study at hadron colliders. Also important are the large charm rates, $\sim$10 times larger than the b rate.

In order to understand the detector design it is useful to examine the characteristics of b quark production at at $p\bar{p}$ collider. It is often customary to characterize heavy quark production in hadron collisions with the two variables p_t and η, where $\eta = -ln(\tan(\theta/2))$, and θ is the angle of the particle with respect to the beam direction. According to QCD based calculations of b quark production, the B's are produced "uniformly" in η and have a truncated transverse momentum, p_t, spectrum characterized by a mean value approximately equal to the B mass [12]. The distribution in η is shown in Fig. 2(a). Note that at larger values of $|\eta|$, the B boost, $\beta\gamma$, increases rapidly (b).

The "flat" η distribution hides an important correlation of $b\bar{b}$ production at hadronic colliders. In Fig. 2(c) the production angles of the hadron containing the b quark is plotted versus the production angle of the hadron containing the $\bar{b}$ quark according to the Pythia generator. Many important measurements require the reconstruction of a b decay and the determination of the flavor of the other $\bar{b}$, thus requiring both b's to be observed in the detector. There is a very strong correlation in the forward (and backward) direction: when the B is forward the $\overline{B}$ is also forward. This correlation is not present in the central region (near zero degrees). By instrumenting a relative small region of angular phase space, a large number of $b\bar{b}$ pairs can be detected. Furthermore the B's populating the forward and backward regions have large values of $\beta\gamma$.

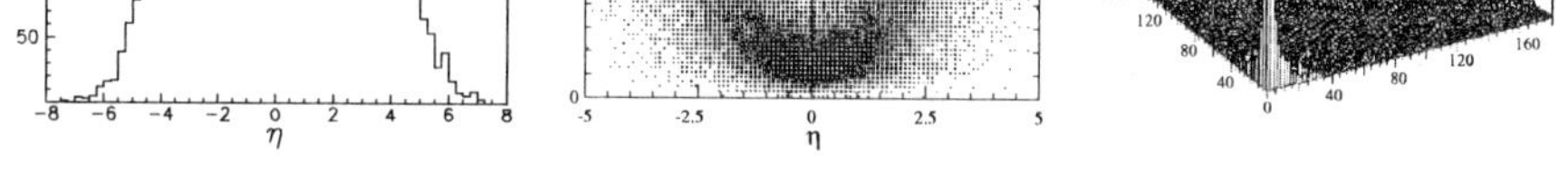

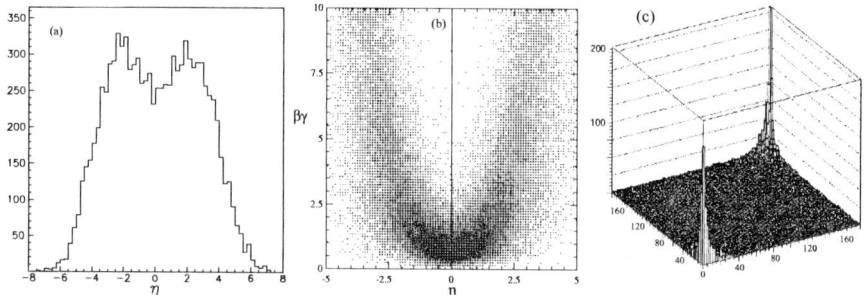

FIGURE 2. (a) The B yield versus η. (b) $\beta\gamma$ of the B versus η. (c) The production angle (in degrees) for the hadron containing a b quark plotted versus the production angle for a hadron containing $\bar{b}$ quark.

BTeV uses two forward spectrometers (along both the p and $\bar{p}$ directions) that utilize the boost of the B's at large rapidities. This is of crucial importance because the main way to distinguish b decays is by the separation of decay vertices from the main interaction.

DETECTOR DESCRIPTION

There are difficulties that heavy quark experiments at hadron colliders must overcome. First of all, the hugh b rate is accompanied by an even larger rate of uninteresting interactions. At the Tevatron the b-fraction is only 1/500. In searching for rare processes, at the level of parts per million, the background from b events is dominant. (Of course all b experiments have this problem.) The large data rate of b's must be handled. For example, BTeV, has 1 kHz into the detector, and these events must be selected and written out. The electromagnetic calorimeter must robust enough to deal with the particles from the underlying event and still have useful efficiency. Furthermore, radiation damage can destroy detector elements.

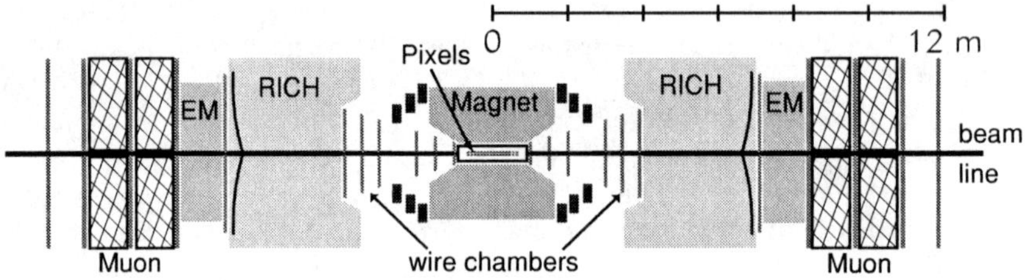

FIGURE 3. Schematic of the BTeV detector.

The BTeV Detector is shown in Fig. 3. The central part of the detector has a pixel detector inside a 1.5 T dipole magnet. The pixel detector provides precision space points for use in both the offline analysis and the trigger. The pixel geometry is sketched in Fig. 4(a). Pulse heights are measured on each pixel. Prototype detectors where tested in a beam at Fermilab; excellent resolutions were obtained, expecially when reading out pulse heights [13] (see Fig. 4(b). The final design uses a 3-bit ADC.

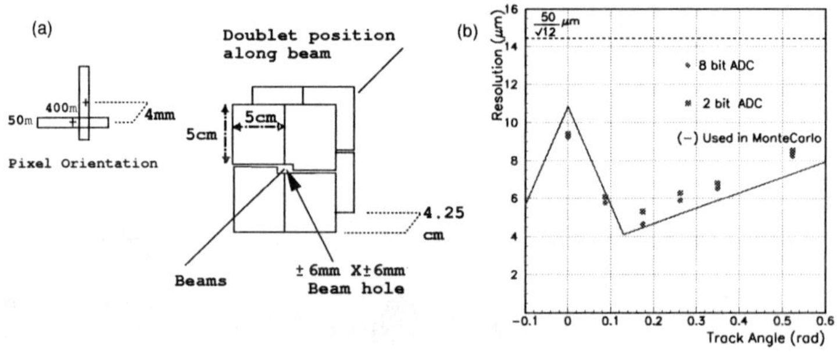

FIGURE 4. (a) Pixel detector geometry. (b) The spatial resolution as a function of the incident track angle for both 2-bit and 8-bit ADC's as measured in an 800 GeV/c pion beam. The straight lines are piecewise fits to the data used in the Monte Carlo simulation. The dotted line near the top indicates the resolution obtainable without using pulse height information.

The pixels provide excellent vertex resolution, good enough to trigger on events with detached vertices characteristic of b or c decays. We obtain a rejection of 100:1 for minimum bias events in the first trigger level while accepting about 50% of the useable b decays. A "movie" describing the trigger can be found at http://www-

TABLE 1. A sample calculation: the error in the CP asymmetry for $B^o \to \pi^+\pi^-$

Cross-section	Luminosity	# of $B^o/10^7$s	$\mathcal{B}(B^o \to \pi^+\pi^-)$
100 μb	$2 \times 10^{32}\text{cm}^{-2}\text{s}^{-1}$	1.5×10^{11}	4.3×10^{-5}
Reconstruction eff.	**Particle I. D. eff.**	**Triggering eff.**†	$\#(\pi^+\pi^-)$
0.08	0.82	0.57	23,700
$\varepsilon \cdot D^2$ **(flavor tags)**	**# of tagged** $(\pi^+\pi^-)$	**Signal/Background**	**Error in asymmetry**
0.1	2,370	3:1	± 0.024

† After all other cuts for all trigger levels.

btev.fnal.gov/public_documents/animations/Animated_Trigger/ . Further trigger levels reduce the background by about a factor of twenty while decreasing the b sample by only 10%. The trigger system stores data in a pipeline that is long enough to ensure no deadtime. The data acquisition system has sufficient throughput to accommodate an output of 1 kHz of b's, 1 kHz of c's and 2 kHz of junk.

Charged particle identification is done using a Ring Imaging CHerenkov detector. A gaseous C_4F_{10} radiator is used with a large mirror that focuses light on plane of photon detectors; these current are Hybrid Photo-Diodes. They have a photocathode and a 20 KV potential difference between the photocathode and a silicon diode. The photoelectron is accelerated and focused onto the segmented photodiode yielding position information for the initial photon. The system will provide four standard-deviation kaon/pion separation between 3-70 GeV/c, electron/pion separation up to 22 GeV/c and pion/muon separation up to 15 GeV/c. Because protons don't radiate until 9 GeV/c they can't be distinguished from kaons below this momenta. We are considering an additional liquid C_6F_{14} radiator, 1 cm thick, in the front of the gas along with a proximity focused phototube array along the sides of the gas volume. to resolve this ambiguity.

Furthermore BTeV has an excellent Electromagnetic calorimeter made from $PbWO_4$ crystals, based on the design of CMS. Finally, the Muon system rounds out the detector. It used to both identify muons and also provides an independent trigger on dimuons [14].

To show how the detector performs we list in Table 1, the relevant quantities necessary to determine the precision in measuring the CP asymmetry in $B^o \to \pi^+\pi^-$.

The signal efficiencies and background levels have been simulated. The trigger code includes full pattern recognition. The effective flavor tagging efficiency, $\varepsilon \cdot D^2$, is taken as 10%. We see that in one "Snowmass" year of running we could measure the CP violating asymmetry with an error of $\pm 2.4\%$. This would not be possible without the using the K/π separation in the RICH as can be seen in Fig. 5.

PHYSICS REACH

The errors on measured quantities related to CKM tests are given in Table 2 for a "Snowmass year," defined as 10^7 seconds, with a luminosity of $2 \times 10^{32}\text{cm}^{-2}\text{s}^{-1}$. In every case the accuracies are excellent and will vastly increase our knowledge.

We also list the some of the events samples relevant for new physics studies in

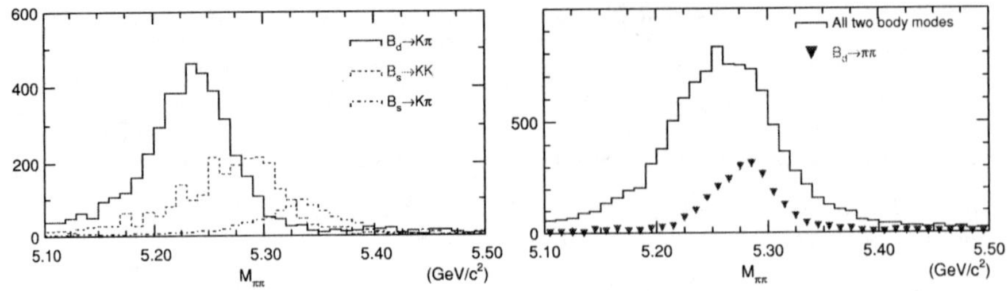

FIGURE 5. Two-body B meson decays reconstructed as $\pi^+\pi^-$ in the absence of RICH particle identification. (left) The background contributions only. (right) The signal and the sum of signal plus background.

TABLE 2. Physics Reach (CKM) in 10^7s

Reaction	# of Events	S/B	Parameter	Error or (Value)
$B^o \to \pi^+\pi^-$	24,000	3	Asymmetry	0.024
$B_s \to D_s^+ K^-$	13,100	7	γ	7°
$B^o \to J/\psi K_S$	485,000	10	$\sin(2\beta)$	0.013
$B_s \to D_s^+ \pi^-$	103,000	3	x_s	(75)
$B^- \to \overline{D}^o(K^+\pi^-)K^-$	300	1		
$B^- \to D^o(K^+K^-)K^-$	1,800	>10	γ	10°
$B^- \to K_S\pi^-$	8,000	1		
$B^o \to K^+\pi^-$	108,000	20	γ	$< 5°$
$B^o \to \rho^+\pi^-$	9,400	4.1		
$B^o \to \rho^o\pi^o$	1,350	0.3	α	$\approx 10°$
$B_s \to J/\psi\eta$	4,600	15		
$B_s \to J/\psi\eta'$	17,500	30	$\sin(2\chi)$	0.023

Table 3. Here we compare with what can be expected from an e^+e^- B factory that has accumulated a 500 fb^{-1} data sample, approximately what might be expected before BTeV starts. We see that BTeV will easily surpass this sample in one year in all of these final states, and the BTeV samples accumulated over several years will allow for very precise tests for new physics.

It is also interesting to compare projections from BTeV with those of the competing

TABLE 3. Reconstructed Events in New Physics Modes for BTeV and e^+e^- B-factories

Mode	BTeV (10^7 s)			B-factories (500 fb^{-1})		
	Yield	Tagged	S/B	Yield	Tagged	S/B
$B_s \to J/\psi\eta^{(\prime)}$	22000	2200	> 15	-	-	
$B^- \to \phi K^-$	11000	11000	> 10	700	700	4
$B^o \to \phi K_S$	2000	200	5.2	250	75	4
$B^o \to K^{*o}\mu^+\mu^-$	4400	4400	11	~50	~50	?
$D^{*+} \to \pi^+ D^o$, $D^o \to K^-\pi^+$	$\sim 10^8$	$\sim 10^8$	large	8×10^5	8×10^5	large

TABLE 4. Two comparisons with LHCb

Mode	$\mathcal{B}$	BTeV		LHCb	
		Yield	S/B	Yield	S/B
$B_s \to D_s^{\pm} K^{\mp}$	3.0×10^{-4}	13100	7	7660	7
$B^o \to \rho^{\pm}\pi^{\mp}$	2.8×10^{-5}	9400	4.1	2140	0.8

experiment, LHCb. In Table 4 two modes are compared, one with a π^o in the final and one without. In both case BTeV is superior, more so in the $\rho\pi$ mode. This is to be expected because BTeV has a much better trigger and electromagnetic calorimeter.

CONCLUSIONS

BTeV can effectively use the large b and c rates at the Tevatron. A complete set of measurements on CP violation, mixing and rare decays in the b and c systems will be performed. It is hoped that these studies will reveal or elucidating physics beyond the Standard Model. BTeV was approved by Fermilab in June, 2000. We are currently completing the necessary R&D and plan for a baseline review in the spring of 2002. We are looking for new collaborators.

ACKNOWLEDGMENTS

This meeting was held during the tragic events of Sept. 11, 2001. I thank my fellow attendees and the organizers for providing a suitably supportive enviornment. I thank my BTeV collaborators for providing the BTeV design and simulation results used here. This work was supported by the National Science Foundation.

REFERENCES

1. A. Abashian et al. (BELLE), *Phys. Rev. Lett.*, **86**, 2509 (2001).
2. B. Aubert et al. (BABAR), *Phys. Rev. Lett.*, **87**, 091801 (2001).
3. M. Kobayashi and T. Maskawa, *Prog. Theor. Phys.*, **49**, 652 (1973).
4. J. P. Silva and L. Wolfenstein, *Phys. Rev. D*, **55**, 5331 (1997).
5. R. Aleksan, B. Kayser and D. London, *Phys. Rev. Lett.*, **73**, 18 (1994).
6. I. I. Bigi and A. I. Sanda, *UND-HEP-99*, **hep-ph/9909479**, 1 (1999).
7. S. Stone, *Plenary Talk, Heavy Flavours 8*, **hep-ph/9904350**, 1 (1999).
8. L. Wolfenstein, *Phys. Rev. D*, **31**, 2381 (1985).
9. I. Hinchliffe and N. Kersting, *Phys. Rev. D*, **63**, 015003 (2001).
10. Y. Nir, *IASSNS-HEP-99-96*, **hep-ph/9911321**, 1 (1999).
11. G. Buchalla, G. Hiller, G. Isidori, *Phys. Rev. D*, **63**, 014015 (2001).
12. M. Artuso, *Experimental Facilities for b-Quark Physics*, **in *B Decays* revised 2nd Edition, Ed. S. Stone, World Scientific, Sinagapore**, 80 (1994).
13. J. A. Appel et al., *sub. to Nucl. Instr. & Meth.*, **Fermilab-Pub-01/229-e**, hep–ex/0108014 (2001).
14. More information on the BTeV Detector can be found in the proposal at, **http://www-btev.fnal.gov/public_documents/btev_proposal/index.html** (2000).

Status of the LHCb Experiment

Ulrich D. Straumann

Physik-Institut, Universität Zürich, Winterthurerstrasse 190, CH-8057 Zürich

Abstract. LHCb is a second generation experiment on b physics which will run from the beginning of the LHC operation. The goals of the experiment are systematic measurements of CP violation and rare decays in the B-meson system with unprecedented precisions. By measuring CP violation in many different decay modes of B_d and B_s mesons and comparing the results with the predictions from the Standard Model, the experiment will search for new physics. This talk summarises the present status of the preparation and construction of the LHCb experiment. Details can be found in the technical design reports of the respective subsystems.

A second generation b physics experiment

The study of CP violation is at present clearly the most active area in experimental particle physics, and it will certainly remain an important research field for many years. b quark systems provide the easiest experimental access and the cleanest theoretical interpretation for studying all basic questions concerning the fundamental nature of CP violation. Present experimental results on the direct measurement of the angle β [1, 2] are consistent (albeit with large uncertainties) with a complete standard model analysis to determine the numerical values of the V_{CKM} Matrix from all other available experimental information [3].

In the center of any strategy for planning future particle physics experiments stands the search for new physics beyond the standard model. Using systems containing b quarks to look for virtual contributions of new particles is very promising, since there are many b decay processes available that are very rare in the standard model due to the numerical configuration of the V_{CKM} matrix. These are mainly processes that are described by standard model box or penguin diagrams. However these type of studies need obviously very large data samples.

Experiments at e^+e^- colliders running on a $b\bar{b}$ resonance are presently investigating CP violation in the decay of the B_d meson. They will eventually determine the angle β rather accurately, the expected final systematical error will be about 0.04. Much more statistics is needed to measure the angle γ, due to smaller branching ratios of the decay channels involved. Among others, Lunghi [4] has recently pointed out the importance of an accurate measurement of γ for the determination of the parameters of supersymmetric models.

Only hadron colliders with high enough energy produce large number of b quarks through gluon gluon fusion and gluon splitting processes. Since these processes depend strongly on the collider energy, we need to look for the largest possible energy. In near future, this will be provided by the LHC. The expected b production cross section in the

CP618, *Heavy Flavor Physics: Ninth International Symposium,* edited by A. Ryd and F. C. Porter

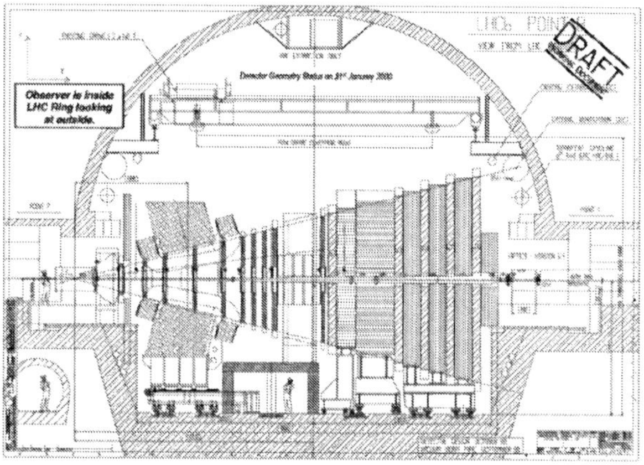

FIGURE 1. A draft technical drawing, showing how the LHCb experiment fits into one of the existing experimental halls (the former DELPHI hall) at CERN. The pp interaction region is within the Vertex Locator on the left side of the drawing. The spectrometer magnet and the tracking chambers are located in the left half of the hall. The right half contains the electromagnetic and hadronic calorimeters and the muon filter (compare with figure 2).

LHC collider is about 1/160 of the total inelastic cross section, leading to a production of more than three orders of magnitude more b quarks per year than at the present day e^+e^- colliders. Furthermore the study of the B_s system is not possible at e^+e^- colliders, since B_s mesons can only be produced efficiently at high energy hadron colliders.

Since at LHC the b production is concentrated to small angles with respect to the collider beam axis, the LHCb experiment is planned as an axial spectrometer. It needs good momentum and vertex position resolution to fully reconstruct the decay channels. A RICH system, hadronic and electromagnetic calorimeters and a muon system allow for precise particle identifications and for triggering. The experiment is designed to fit in the existing experimental hall at the interaction point at the former DELPHI experiment (see figure 1).

A similar experiment, BTeV, is being discussed in future at the FERMILAB collider.

Expected performance

As examples for the expected performance of LHCb, two typical decay channels allowing a measurement of the angle γ are worth mentioning here:

In $B_d^0 \to D^{*-}\pi^+$ two interfering diagrams produce a time dependant asymmetry of the decay amplitude, which is proportional to $\sin(2\beta+\gamma)$. Since the oscillation amplitude is, however, very small (only about 1%), large statistics are necessary. LHCb will produce about 260'000 reconstructed and identified events per year, which allows the determination of the angle $2\beta+\gamma$ with an accuracy of about 12° per year [5].

If we replace both d quarks by s quarks, we arrive at the reaction $B_s \to D_s^- K^+$. The strength of the two interfering diagrams becomes now of similar magnitude, therefore the interference and the amplitude of the asymmetry become larger. The absolute rate is, however, smaller due to the smaller production rate of B_s and due to the smaller vertex factor (V_{us} instead of V_{ud}). The mixing diagram contains now V_{ts} instead of V_{td} and is thus sensitive to $\delta\gamma = \arg V_{ts}$ instead of β. LHCb will produce about 2400 reconstructed and identified events per year, allowing to determine $\gamma - 2\delta\gamma$ with an accuracy of about 8° per year [5].

These measurements of γ will complete the study of the unitarity triangle with high precision. In addition the direct comparison to various other methods to measure γ, will allow to observe even very small possible inconsistencies due to new physics.

Status of the construction of LHCb

After the technical proposal [6] was accepted by the CERN program committees in 1998, the construction of all larger detector components has been studied in detail, and separate technical design reports (TDR) have been produced:

Magnet [7]	approved
Calorimeters [8]	approved
RICH [9]	approved
Muon System [10]	approved
Vertex Locator [11]	approved
Outer Tracking [12]	submitted Sep. 2001
Inner Tracking	to be submitted in Sep. 2002

There will also be TDR's for trigger, for data acquisition and for data handling. All project planning is done such that the experiment should be fully operational at the startup of LHC in 2006.

In general, the experiment is still close to the description in the technical proposal. However detailed simulation studies of the tracking system have shown, that the number of tracking stations can be reduced from 11 to 9 (see figure 2) without significant losses in the pattern recognition efficiency, while the savings in radiation length and nuclear interaction length improve the overall spectrometer performance. Alternative track reconstruction algorithms are presently being studied, that may allow to reduce the number of tracking stations even further.

Starting from the backside of the experiment, we discuss shortly the progress of the detector developments for the various subdetectors (compare with figure 2):

The **muon system** is used as a trigger and particle identification device and consists of five stations (M1 ... M5) with a total detector surface of more than 400 m^2. For the regions of higher particle fluxes four-gap multi wire proportional chambers both with cathode pad and wire readout are foreseen. For the low flux region two-gap resistive plate chambers have been chosen. The total of 120'000 physical channels are combined to 26'000 logical channels to be used in the triggering system. The size of the logical channels is determined by occupancy and momentum resolution and varies from $1 \times 2.5\ cm^2$ to $16 \times 20\ cm^2$.

The **calorimeter** is to be used for identifying electrons and photons and should deliver trigger signals for electrons, photons and hadrons. It is thus split into three parts.

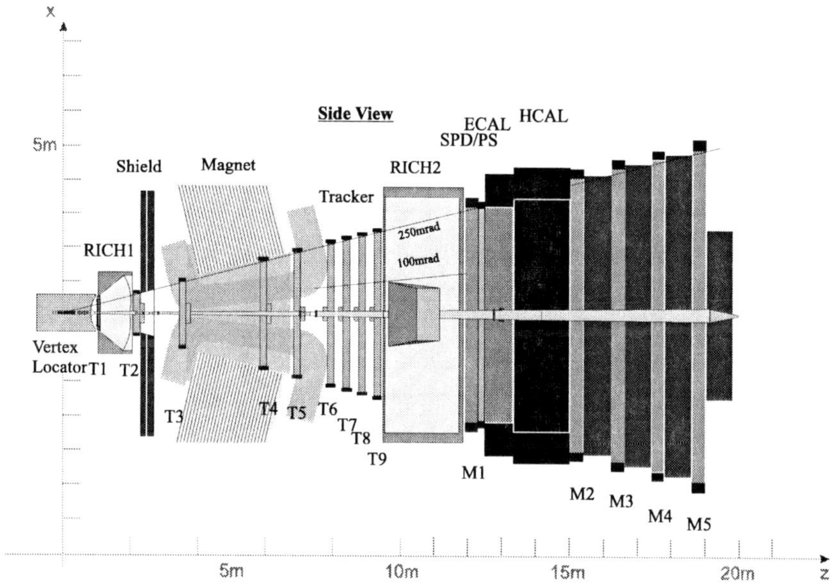

FIGURE 2. Schematic drawing, of the major elements of the LHCb experiment. (compare figure 1)

A scintillator lead preshower system (SPD/PS) of two radiation lengths allows to separate photons from charged particles. It is followed by a shashlik type electromagnetic (ECAL) and an iron scintillator tile hadronic section (HCAL). The expected resolutions are $0.10/\sqrt{E}+0.015$ and $0.80/\sqrt{E}+0.10$ for the electromagnetic and hadronic section respectively. These numbers were verified in prototype beam tests. The start of the mass production of the calorimeters is imminent.

Many interesting rare B decay channels need an accurate particle identification for charged Kaons and Pions. The required momentum range (from about 3 up to 200 GeV/c) and angular coverage can be reasonably satisfied only with an arrangement of two **RICH** systems containing three different radiators. Hybrid photo tubes (HPD) consisting of a classical photocathode and a silicon pixel array including frontend electronics in the vacuum of the tube to detect the photoelectrons have been chosen as the baseline option for the photon detector. Multianode photomultipliers (MAPMT) are kept as a backup solution.

The spectrometer dipole **magnet** is powered by warm aluminum conductors. Its gap is adjusted to the spectrometer acceptance, in order to keep the total field volume reasonably small. The integrated field along the spectrometer axis amounts to 4 Tm. All contracts with industry for the construction have been signed.

In the present version the **tracking** detectors are arranged in nine stations (T1 ... T9). Optimisation of their locations along the beam required high pattern recognition efficiency, small number of ghost tracks and good momentum resolution. Not unexpectedly most of the stations sit downstream of the magnet to get a precise track definition, while multiple scattering would affect the resolution most within the magnet volume.

Each tracking station will be constructed from two different technologies. In the innermost area, where particle density is high, silicon strip detectors with a pitch of 200 μm are arranged in ladders of about 20 cm length. In the outer part straw tubes with 5 mm diameter will be used. For both technologies beam test results are available using full length prototypes, albeit not yet with the final electronics. The border line between the inner part and the outer part has been reoptimised for acceptable detector channel occupancies. Also the beam pipe construction has evolved: The presently foreseen flanges and bellow positions and the Al-Be alloy used as beam pipe material reduces the background occupancy of the tracking stations, mainly electrons from photon conversions, to an acceptable level.

300 μm n-on-n double metal layer silicon microstrip detectors with two different geometrical arrangements are used as a **vertex locator** (VELO). R measuring sensors have circular strips, while ϕ measuring sensors have slightly tilted radial strips. The sensors have a semi circular shape, such that each of them covers half of a measuring plane. A total number of 100 sensors are arranged in such a way, that every track within the spectrometer acceptance has at least three measuring points.

The LHCb **trigger system** is based on a four level system, where the first level (L0) runs synchronously with the 40 MHz bunch crossing rate and selects high p_T electrons, hadrons, photons and muons. A pileup detector, using two of the VELO detector planes, allows to veto on multiple pp event vertices in the same bunch crossing. The second level (L1) does a complete analysis of the VELO data and allows to select secondary vertices, while running at an input rate of 1 MHz. This system requires a two stage hardware pipeline to store the data of all subdetectors during the latency of the first two trigger levels. The third and fourth level of event selection is done after the data has been read out to a processor farm.

In conclusion the LHCb experiment is well under way. TDR's are being produced according to schedule and we can expect, that LHCb will be operational at the startup of LHC. As a second generation B experiment it will provide high precision and large data samples for studying many different rare B decays, looking for new physics beyond the standard model.

REFERENCES

1. The BaBar Collaboration (2001), hep-ex/0107013.
2. The BELLE Collaboration (2001), hep-ex/0107061.
3. A. Stocchi *et al.*, 2000 CKM-Triangle Analysis: A Critical Review with Updated Experimental Inputs and Theoretical Parameters (2001), hep-ph/0012308, JHEP 0107, 013.
4. E. Lunghi, Extended Minimal Flavour Violating MSSM (2001), hep-ph/0108066.
5. P. Ball *et al.*, B Decays at the LHC (2000), hep-ph/0003238, CERN-TH/2000-101.
6. The LHCb technical proposal (1998), CERN/LHCC/1998-4.
7. LHCb Magnet: Technical Design Report (1999), CERN/LHCC/2000-007.
8. LHCb Calorimeter, Technical Design Report (2000), CERN/LHCC/2000-036.
9. LHCb RICH, Technical Design Report (2000), CERN/LHCC/2000-037.
10. LHCb Muon System: Technical Design Report (2001), CERN/LHCC/2001-010.
11. LHCb Vertex Locator: Technical Design Report (2001), CERN/LHCC/2001-011.
12. LHCb Outer Tracking: Technical Design Report (2001), CERN/LHCC/2001-024.

CLEO-c and CESR-c: A New Frontier in Weak and Strong Interactions

Ian Shipsey

1396 Physics Building, West Lafayette, IN 47907, U.S.A.

Abstract. We report on the physics potential of a proposed conversion of the CESR machine and the CLEO detector to a charm and QCD factory: "CLEO-c and CESR-c" that will make crucial contributions to flavor physics in this decade and offers our best hope for mastering non-perturbative QCD which is essential if we are to understand strongly coupled sectors in the new physics that lies beyond the Standard Model.

EXECUTIVE SUMMARY

A focused three year program of charm and QCD physics with the CLEO detector operating in the 3-5 GeV energy range at the Cornell Electron Storage Ring in Ithaca, NY, is proposed. The CLEO-c physics program includes a set of measurements that will substantially advance our understanding of important processes within the Standard Model of particle physics and set the stage for understanding the larger theory in which we imagine the Standard Model to be embedded. The program will be preceded by one year of bottomium running with CLEOIII in 2002.

The program revolves around the strong interaction and the pressing need to develop sufficiently powerful tools to deal with an intrinsically non-perturbative theory. At the present time, and for the last twenty years, progress in weak interaction physics and the study of heavy flavor physics has been achieved primarily by seeking the few probes of weak-scale physics that succesfully evade or minimise the role of strong interaction physics. The pre-eminence of the mode $B \to J/\psi K_S^o$ in measuring $sin2\beta$ arises from the absence of complications due to the strong interaction. Similarly, Heavy Quark Symmetry and the development of HQET enabled the identification of the zero-recoil limit as the best place to measure $b \to c\ell\nu$ exclusive decays as the strong intercation effects at this kinematic point are small and calculable. This method has become one of the two preferred ways to extract V_{cb}. If we had similar strategies that would allow us to extract V_{ub} from $b \to u\ell\nu$ without form factor uncertainties, and V_{td} and V_{ts} from B_d and B_s mixing respectively, without decay constant and bag parameter uncertainties, we would be well on the way to understanding the CKM matrix at the few per cent level.

The goal of flavor physics is to test the Standard Model description of CP violation by probing the unitarity of the CKM matrix. However this goal is in jeopardy. Across the spectrum of heavy quark physics the study of weak-scale phenomena and the extractiion of quark mixing paramters remain fundamentally limited by our restricted capacity to

CP618, *Heavy Flavor Physics: Ninth International Symposium*, edited by A. Ryd and F. C. Porter

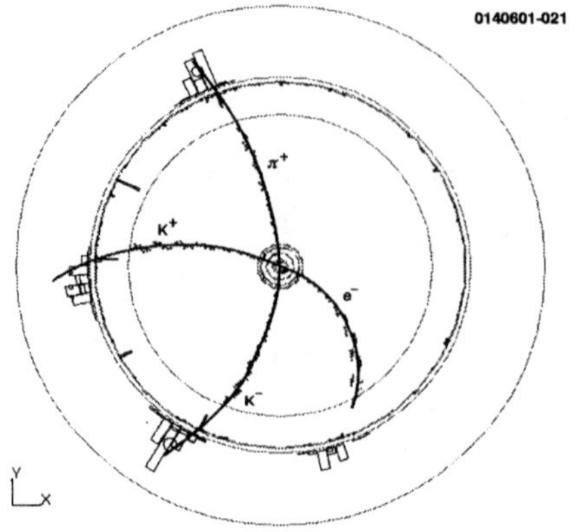

FIGURE 1. A doubly tagged event at the $\psi(3770)$

deal with the non-perturbative strong interaction dynamics.

Moreover, as we look to the future beyond the Standard Model, we anticipate that the larger theory in which the Standard Model is embedded will likely be strongly coupled, or have strongly coupled sectors. Both Technicolor, which is modeled on QCD and is ab initio strongly coupled, and Supersymmetry, which employs strongly coupled sectors to break the supersymmetry, are prime examples of candidates for physics beyond the Standard Model. Strong coupling is a phenomenon to be expected, weak coupling is the exception in field theory not the norm. However our ability to compute reliably in a strongly coupled theory is not well developed. Techniques such as lattice gauge theory that deal squarely with strongly coupled theories will eventually determine our progress on all fronts in particle physics. At the present time the absence of adequate theoretical tools significantly limits the physics we can obtain from the study heavy quarks.

Recent advances in Lattce QCD (LQCD) may offer hope. Algorithmic advances and improved computer hardware have produced a wide variety of non-perturbative results with accuracies at the 10-20% level for systems involving one or two heavy quark such as B and D mesons, and Ψ and Υ quarkonia. First generation unquenched calculations have been completed for decay constants and semileptonic form factors, for mixing and spectra. There is very strong interest in the LQCD community in pursuing much higher precision and the techniques needed to reduce uncertainties to 1-2% precision. But the push towards higher precision is hampered by the absence of accurate charm data against which to test and calibrate the new theoretical techniques.

CLEO-c will address this challenge in the charm system at threshold where the experimental conditions are optimal. We will obtain data samples several hundred times larger than any previous experiment and with a detector which is an order of magnitude better than any previous detector to operate at charm threshold. We will supplement the charm and charmonium data with bottomium data taken starting Fall 2001 with CLEO III prior to the conversion to CLEO-c. Decay constants, form factors, spectroscopy of

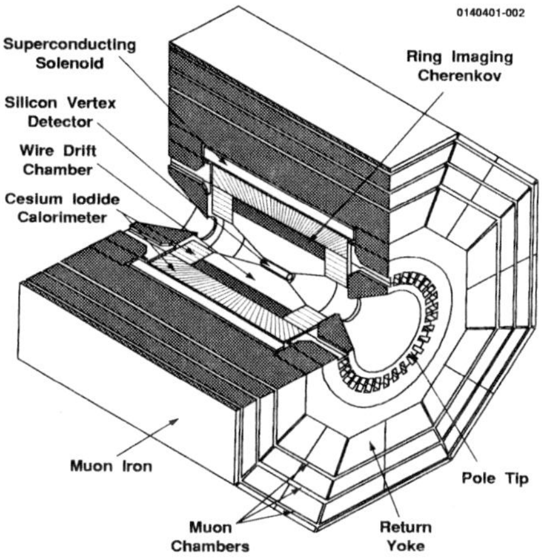

FIGURE 2. The CLEO III detector.

open and hidden charm and hidden bottom, and an immense variety of absolute branching ratio measurements will be obtained with accuracies at the 1-2% level. Precision measurements will demand precision theory.

The measurements described here are an essential and integral part of the global program in heavy flavor physics of this decade, and the larger program of the as yet unknown physics of the next decade.

INTRODUCTION

For many years, the CLEO experiment at the Cornell Electron Storage Ring, CESR, operating on the ϒ(4S) resonance, has provided most of the world's information about the B_d and B_u mesons. At the same time, CLEO, using the copious continuum pair production at the ϒ(4S) resonance has been a leader in the study of charm and τ physics. Now that the asymmetric B factories have achieved high luminosity, CLEO is uniquely positioned to advance the knowledge of heavy flavor physics by carrying out several measurements near charm threshold, at center of mass energies in the 3.5-5.0 GeV region. These measurements address crucial topics which benefit from the high luminosity and experimental constraints which exist near threshold but have not been carried out at existing charm factories because the luminosity has been too low, or have been carried out previously with meager statistics. They include:

1. Charm Decay constants f_D, f_{D_s}
2. Charm Absolute Branching Fractions
3. Semileptonic decay form factors
4. Direct determination of V_{cd} & V_{cs}

TABLE 1. Summary of CLEO-c charm decay measurements.

Topic	Reaction	Energy (MeV)	L (fb^{-1})	current sensitivity	CLEO-c sensitivity
Decay constant					
f_D	$D^+ \to \mu^+ \nu$	3770	3	UL	2.3%
f_{D_s}	$D_s^+ \to \mu^+ \nu$	4140	3	14%	1.9%
f_{D_s}	$D_s^+ \to \mu^+ \nu$	4140	3	33%	1.6%
Absolute Branching Fractions					
$Br(D^0 \to K\pi)$		3770	3	2.4%	0.6%
$Br(D^+ \to K\pi\pi)$		3770	3	7.2%	0.7%
$Br(D_s^+ \to \phi\pi)$		4140	3	25%	1.9%
$Br(\Lambda_c \to pK\pi)$		4600	1	26%	4%

5. QCD studies including:
   Charmonium and bottomonium spectroscopy
   Glueball and exotic searches
   Measurement of R between 3 and 5 GeV, via scans
   Measurement of R between 1 and 3 GeV, via ISR
6. Search for new physics via charm mixing, CP violation and rare decays
7. τ decay physics

The CLEO detector can carry out this program with only minimal modifications. The CLEO-c project is described at length in[1 - 11]. A very modest upgrade to the storage ring is required to achieve the required luminosity. Below, we summarize the advantages of running at charm threshold, the minor modifications required to optimize the detector, examples of key analyses, a description of the proposed run plan, and a summary of the physics impact of the program.

Advantages of running at charm threshold

The B factories, running at the $\Upsilon(4S)$ will have produced 500 million charm pairs by 2005. However, there are significant advantages of running at charm threshold:

1. Charm events produced at threshold are extremely clean;
2. Double tag events, which are key to making absolute branching fraction measurements, are pristine;
3. Signal/Background is optimum at threshold;
4. Neutrino reconstruction is clean;
5. Quantum coherence aids D mixing and CP violation studies

These advantages are dramatically illustrated in Figure 1, which shows a picture of a simulated and fully reconstructed $\psi(3770) \to D\bar{D}$ event.

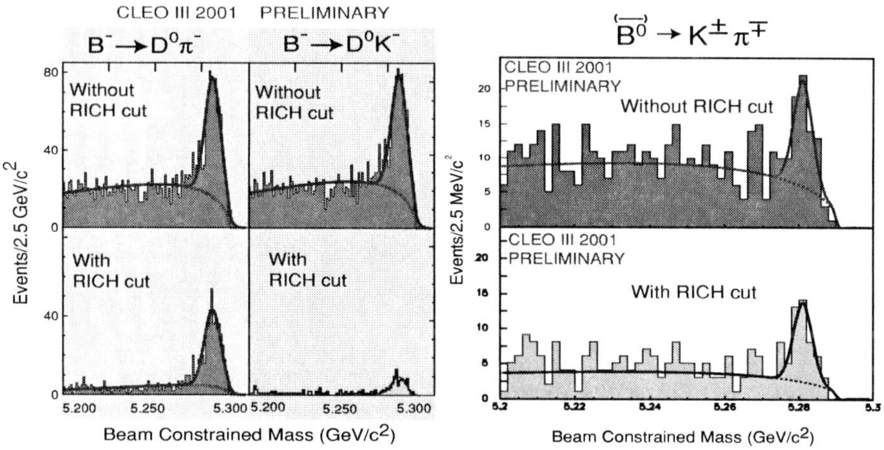

FIGURE 3. (Left) Beam constrained mass for the Cabibbo allowed decay $B \to D\pi$ and the Cabibbo suppressed decay $B \to DK$ with and without RICH information. The latter decay was extremely difficult to observe in CLEO II/II.V which did not have a RICH detector. (Right) The penguin dominated decay $B \to K\pi$. Both of these modes are observed in CLEO III with branching ratios consistent with those found in CLEO II/II.V.

The CLEO-III Detector : Performance, Modifications and issues

The CLEO III detector, shown in Figure 2, consists of a new silicon tracker, a new drift chamber, and a Ring Imaging Cherenkov Counter (RICH), together with the CLEO II/II.V magnet, electromagnetic calorimeter and muon chambers. The upgraded detector was installed and commissioned during the Fall of 1999 and Spring of 2000. Subsequently operation has been very reliable (see below for a caveat) and a very high quality data set has been obtained. To give an idea of the power of the CLEO III detector in Figure 3 (left plot) the beam constrained mass for the Cabibbo allowed decay $B \to D\pi$ and the Cabibbo suppressed decay $B \to DK$ with and without RICH information is shown. The latter decay was extremely difficult to observe in CLEO II/II.V which did not have a RICH detector. In the right plot of Figure 3 the penguin dominated decay $B \to K\pi$ and the tree dominated decay is shown. Both of these modes are observed in CLEO III with branching ratios consistent with those found in CLEO II/II.V. and are also in agreement with recent Belle and BABAR results. Figure 3 is a demonstration that CLEO III performs very well indeed.

Unfortunately, there is one detector subsystem that is not performing well. The CLEO III silicon has experienced an unexpected and unexplained loss of efficiency The situation is under constant evaluation. It is likely that the silicon detector will be replaced with a wire vertex chamber for CLEO-c. We note that if one was to design a charm factory detector from scratch the tracking would be entirely gas based to ensure that the detector material was kept to a minimum. CLEO-c simulations indicate that a simple six layer stereo tracker inserted into the CLEO III drift chamber as a silicon replacement would provide a system with superior momentum resolution to the current CLEO III tracking system. We propose to build such a device for CLEO-c.

Due to machine issues we plan to lower the solenoid field strength to 1 T from 1.5 T. All other parts of the detector do not require modification. The dE/dx and Ring Imaging Cerenkov counters are expected to work well over the CLEO-c momentum range. The electromagnetic calorimeter works well and has fewer photons to deal with at 3-5 GeV than at 10 GeV. Triggers will work as before. Minor upgrades may be required of the Data Acquisition system to handle peak data transfer rates. CESR conversion to CESR-c requires 18 m of wiggler magnets, to increase transverse cooling, at a cost of $\sim$ \$4M. The conclusion is that, with the addition of the replacement wire chamber, CLEO is expected to work well in the 3-5 GeV energy range at the expected rates.

Examples of analyses with CLEO-c

The main targets for the CKM physics program at CLEO-c are absolute branching ratio measurements of hadronic, leptonic and semileptonic decays. The first of these provides an absolute scale for all charm and hence all beauty decays. The second measures decay constants and the third measures form factors and in combination with theory allows the determination of V_{cd} and V_{cs}.

Absolute branching ratios

The key idea is to reconstruct a D meson in any hadronic mode. This, then, constitutes the tag. Figure 4 shows tags in the mode $D \to K\pi$. Note the y axis is a log scale. Tag modes are very clean. The signal to background ratio is $\sim$ 5000/1 for the example shown. Since $\psi(3770) \to D\bar{D}$, reconstruction of a second D meson in a tagged event to a final state X, corrected by the efficiency which is very well known, and divided by the number of D tags, also very well known, is a measure of the absolute branching ratio $Br(D \to X)$. Figure 5 shows the $K^-\pi^+\pi^+$ signal from doubly tagged events. It is essentially background free. The simplicity of $\psi(3770) \to D\bar{D}$ events combined with the absence of background allows the determination of absolute branching ratios with extremely small systematic errors. This is a key advantage of running at threshold.

Leptonic decay $D_s \to \mu\nu$

This is a crucial measurement because it provides information which can be used to extract the weak decay constant, f_{D_s}. The constraints provided by running at threshold are critical to extracting the signal.

The analysis procedure is as follows:

1. Fully reconstruct one D;
2. Require one additional charged track and no additional photons;

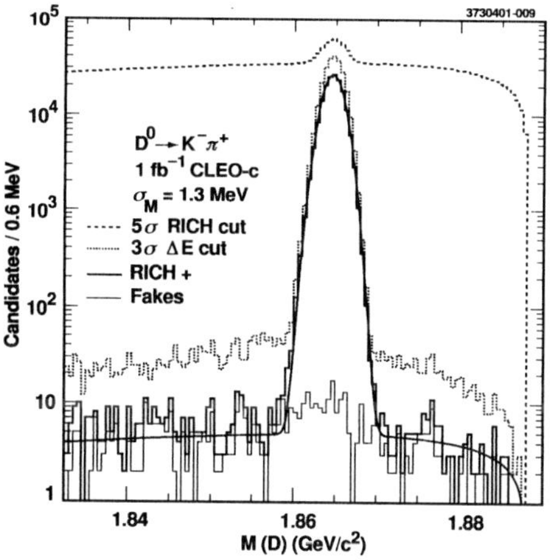

FIGURE 4. $K\pi$ invariant mass in $\psi(3770) \to D\bar{D}$ events showing a strikingly clean signal for $D \to K\pi$. The y axis is logarithmic. The S/N ~ 5000/1.

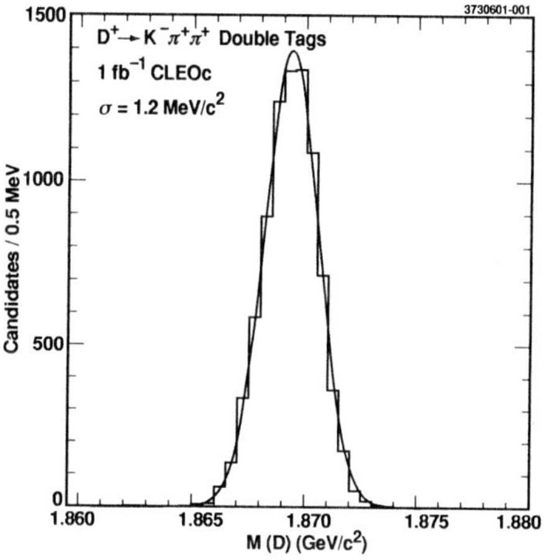

FIGURE 5. $K\pi\pi$ invariant mass in $\psi(3770) \to D\bar{D}$ events where the other D in the event has already been reconstructed. A clean signal for $D \to K\pi\pi$ is observed and the absolute branching ratio $Br(D \to K\pi\pi)$ is measured by counting events in the peak.

3. Compute the missing mass squared (MM2) which peaks at zero for a decay where only a neutrino is unobserved.

The missing mass resolution, which is of order $\sim M_{\pi^0}$, is sufficient to reject the backgrounds to this process as shown in Fig. 6. There is no need to identify muons, which helps reduce the systematic error. One can inspect the single prong to make sure it is not an electron. This provides a check of the background level since the leptonic decay to an electron is severely helicity-suppressed and no signal is expected in this

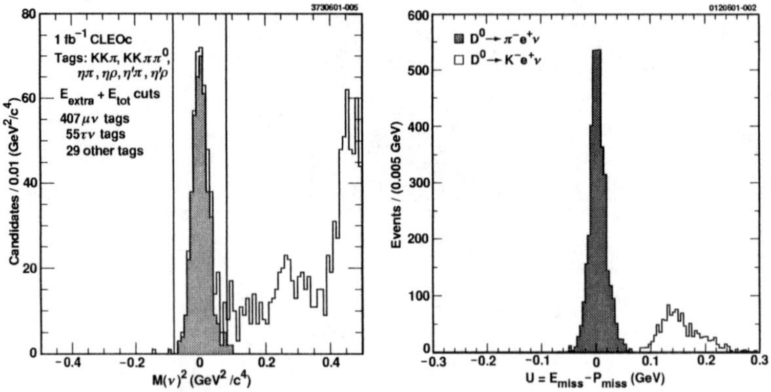

FIGURE 6. (Left) Missing mass for D_sD_s tagged pairs produced at $\sqrt{s} = 4100$ MeV. Events due to the decay $D_s \to \mu\nu$ are shaded. (Right) The difference between the missing energy and missing momentum in $\psi(3770)$ tagged events for the Cabibbo suppressed decay $D \to \pi\ell\nu$ (shaded). The unshaded histogram arises from the ten times more copiously produced Cabibbo allowed transition $D \to K\ell\nu$ where the K is outside the fiducial volume of the RICH.

mode.

Semileptonic decay $D \to \pi\ell\nu$

The analysis procedure is as follows:

1. Fully reconstruct one D
2. Identify one electron and one hadronic track.
3. Calculate the variable, $U = E_{miss} - P_{miss}$, which peaks at zero for semileptonic decays.

Using the above procedure results in the right plot of Figure 6. With CLEO-c for the first time it will become possible to make absolute branching ratio and absolute form factor measurements of every charm meson semileptonic pseudoscalar to pseudoscalar and pseudoscalar to vector transition. This will be a lattice calibration data set without equal. Figure 7 graphically shows the improvement in absolute semileptonic branching ratios that CLEO-c will make.

Run Plan

CLEO-c must run at various center of mass energies to achieve its physics goals. The "run plan" currently used to calculate the physics reach is given below. Note that item 1 is prior to machine conversion and the remaining items are post machine conversion.

1. 2002 : Υ's – 1-2 fb^{-1} each at $\Upsilon(1S), \Upsilon(2S), \Upsilon(3S)$

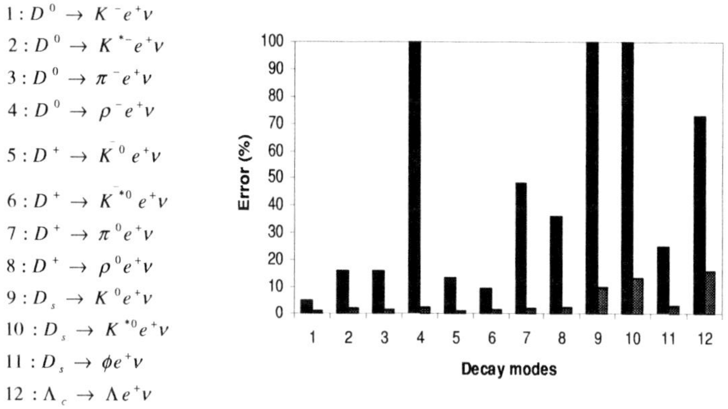

FIGURE 7. Absolute branching ratio current precision from the PDG (left entry) and precision attainable at CLEO-c (right entry) for twelve semileptonic charm decays.

Spectroscopy, electromagnetic transition matrix elements, the leptonic width. Γ_{ee}, and searches for the yet to be discovered h_b, η_b with 10-20 times the existing world's data sample.

2. 2003 : $\psi(3770) - 3\ fb^{-1}$
   30 million events, 6 million tagged D decays (310 times MARK III)
3. 2004 : 4100 MeV – 3 fb^{-1}
   1.5 million $D_s D_s$ events, 0.3 million tagged D_s decays (480 times MARK III, 130 times BES)
4. 2005 : $J/\psi - 1\ fb^{-1}$
   1 Billion J/ψ decays (170 times MARK III, 20 times BES II)

Physics Reach of CLEO-c

Tables 1, 2 , and 3, and Figures 7 and 8 summarize the CLEO-c measurements of charm weak decays, and compare the precision obtainable with CLEO-c to the expected precision at BABAR which expects to have recorded 500 million charm pairs by 2005. CLEO-c clearly achieves far greater precision for many measurements. The reason for this is the ability to measure absolute branching ratios by tagging and the absence of background at threshold. In those analyses where CLEO-c is not dominant it remains comparable or complementary to the B factories.

Also shown in Table 3 is a summary of the data set size for CLEO-c and BES II at the J/ψ and ψ', and the precision with which R, the ratio of the e^+e^- annihilation cross section into hadrons to mu pairs, can be measured. The CLEO-c data sets are over an order of magnitude larger, the precision with which R is measured is a factor of three higher, in addition the CLEO detector is vastly superior to the BES II detector. Taken together the CLEO-c datasets at the J/ψ and ψ' will be qualitatively and quantitatively superior to any previous dataset in the charmonium sector thereby providing discovery potential for glueballs and exotics without equal.

TABLE 2. Summary of direct CKM reach with CLEO-c

Topic	Reaction	Energy (MeV)	L (fb^{-1})	current sensitivity	CLEO-c sensitivity
V_{cs}	$D^0 \to K\ell^+\nu$	3770	3	16%	1.6%
V_{cd}	$D^0 \to \pi\ell^+\nu$	3770	3	7%	1.7%

TABLE 3. Comparision of CLEO-c reach to BaBar and BES

Quantity	CLEO-c	BaBar	Quantity	CLEO-c	BES-II
f_D	2.3%	10-20%	$\#J/\psi$	10^9	5×10^7
f_{D_s}	1.7%	5-10%	ψ'	10^8	3.9×10^6
$Br(D^0 \to K\pi)$	0.7%	2-3%	4.14 GeV	$1 fb^{-1}$	$23 pb^{-1}$
$Br(D^+ \to K\pi\pi)$	1.9%	3-5%	3-5 R Scan	2%	6.6%
$Br(D_s^+ \to \phi\pi)$	1.3%	5-10%			

CLEO-c Physics Impact

CLEO-c will provide crucial validation of Lattice QCD, which will be able to calculate with accuracies of 1-2%. The CLEO-c decay constant and semileptonic data will provide a "golden", and timely test while CLEO-c QCD and charmonium data provide additional benchmarks.

CLEO-c will provide dramatically improved knowledge of absolute charm branching fractions which are now contributing significant errors to measurements involving b's in a timely fashion. CLEO-c will significantly improve knowledge of CKM matrix elements which are now not very well known. V_{cd} and V_{cs} will be determined directly by CLEO-c data and LQCD, or other theoretical techniques. V_{cb}, V_{ub}, V_{td} and V_{ts} will be

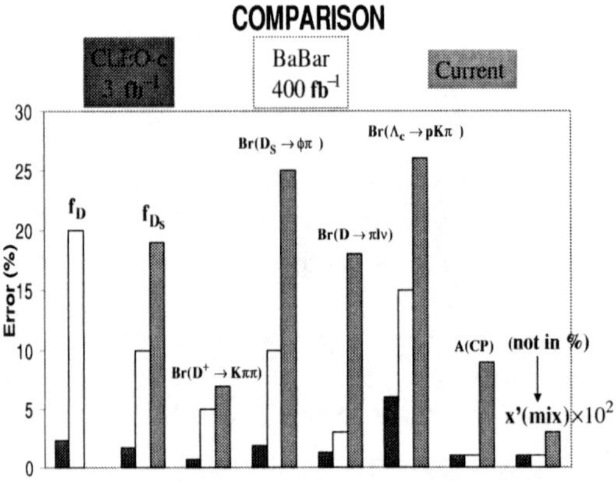

FIGURE 8. Comparison of CLEO-c (left) BaBar (center) and PDG2001 (right) for eight physics quantities indiocated in the key.

TABLE 4. Current knowledge of CKM matrix elements (row one). Knowledge of CKM matrix elements after CLEO-c (row two). See the text for further details.

V_{cd}	V_{cs}	V_{cb}	V_{ub}	V_{td}	V_{ts}
7%	16%	5%	25%	36%	39%
1.7%	1.6%	3%	5%	5%	5%

determined with enormously improved precision using B factory data once the CLEO-c program of lattice validation is complete. Table 4 give a summary of the situation.

CLEO-c data alone will also allow new tests of the unitarity of the CKM matrix. The unitarity of the second row of the CKM matrix will be probed at the 3% level which is comparable to our current knowledge of the first row. CLEO-c data will also test unitarity by measuring the ratio of the long sides of the squashed *cu* triangle to 1.3%.

Finally the potential to observe new forms of matter; glueballs, hybrids, etc in J/ψ decays and new physics through sensitivity to charm mixing, CP violation, and rare decays provides a discovery component to the program.

ACKNOWLEDGMENTS

I would like to thank Dave Hitlin and Frank Porter for the superb organization of this conference.

REFERENCES

1. "CLEO-c and CESR-c : A New Frontier of Weak and Strong Interactions", CLNS 01/1742.
2. "E2 Snowmass Report", I. Shipsey on behalf of the E2 convenors, G. Burdman, J. Butler, I. Shipsey, and H. Yamamoto. Talk at the final plenary session of Snowmass, 2001, "The Future of Particle Physics", Snowmass, CO, July 2001. All E2 working group talks (refs. 2-11) may be found at http://www.physics.purdue.edu/Snowmass2001_E2/
3. "Another look at Charm: the CLEO-c physics program", M. Artuso, talk to the E2 Working Group.
4. "CLEO-c and CESR-c : A New Frontier of Weak and Strong Interactions", I. Shipsey, talk to a joint E2/P2/P5 Working Group session.
5. "Projected Non-perturbative QCD Studies with CLEO-c", S. Dytman, talk to the E4 Working Group.
6. "CLEO-c and R measurements", L. Gibbons, talk to the E2 Working Group.
7. "An Introduction to CLEO-c ", L. Gibbons, talk to the E2 Working Group.
8. "CLEO-C reach in D meson Decays : Measuring absolute D meson branching fractions, D decay constants,and CKM matrix elements", D. Cassel, talk to the E2 Working Group.
9. "Beyond the Standard Model: the clue from charm", M. Artuso, talk to the E2 Working Group.
10. "A case for running CLEO-C at the ψ' ($\sqrt(s) = 3686$ MeV)", S. Pordes, talk to the E2 Working Group.
11. "Experimental Aspects of Tau Physics at CLEO-c", Y. Maravin, talk to the E2 Working Group.

An Asymmetric B Factory at 10^{36} Luminosity

David G. Hitlin

California Institute of Technology, 356-48 Caltech, Pasadena, CA 91125 USA

Abstract. The physics opportunities at an asymmetric B Factory operating at the unprecedented luminosity of 10^{36} cm^{-2}s^{-1} are unique and attractive. The accelerator appears to be practical and the challenges of performing a sensitive experiment in this environment can be met.

PHYSICS MOTIVATION

The physics of flavor is central to our understanding of the structure of matter. Recent developments in the quark and neutrino sectors have deepened our understanding of the Standard Model and pointed the way to further experimental studies in the search for new physics. This presentation addresses the future of the study of quark physics in e^+e^- annihilation, at the $\Upsilon(4S)$ and $\Upsilon(5S)$, over the next decade and beyond. A more detailed treatment can be found in Ref. 1.

The search for new physics in the quark sector involves direct searches for new particles (*e.g.* squarks or sleptons), precise tests of Standard Model predictions for rare decay branching fractions and decay distributions to search for new amplitudes in loop processes, and overconstrained tests of the CKM matrix. A very high luminosity asymmetric B Factory can make unique contributions to these studies, as well as providing capabilities complementary to those of experiments at hadronic machines.

Precision tests of CKM unitarity require the percent level precision on the measurement of sin2β obtainable at a 10^{36} asymmetric B Factory as well as the several percent precision obtainable on sin2α and γ with very large samples of rare hadronic B decays. In particular, measurements of the separate branching ratios of $B^0 \to \pi^0\pi^0$ and $\overline{B}^0 \to \pi^0\pi^0$ decays (Fig. 1), possible only at an e^+e^- B Factory, are vital to obtain a precise value of α with minimal theoretical assumptions [2]. Taken together with concomitant improvements in our understanding of the magnitudes of CKM matrix elements, which require new techniques involving tagging and exclusive reconstruction of B semileptonic decays, as well as anticipated improvements in lattice gauge calculations, this program is capable of tests of CKM unitarity of exquisite precision. Measurements of the third angle γ are difficult, but can be done to excellent precision at a 10^{36} machine, both by the comparison of $b \to c\bar{u}s$ and $b \to u\bar{c}s$ decays using $B \to DK$ transitions at the $\Upsilon(4S)$ [3]-[5], and measurements of $B_s \to D^{(*)\pm}K^{(*)\mp}$ decays at the $\Upsilon(5S)$ [6] (see Fig. 2).

CP618, *Heavy Flavor Physics: Ninth International Symposium*, edited by A. Ryd and F. C. Porter

Table 1 summarizes the precision obtainable on the angles of the unitarity triangle at an e^+e^- experiment with 500 fb^{-1} and 10 ab^{-1}, corresponding to 1 year of running at 10^{36} luminosity, with that obtainable with planned and proposed experiments at hadronic colliders. The determination of γ using the $\Upsilon(5S)$ requires a separate data set of 1 ab^{-1}. In general, multiple complementary measurements of the CKM angles are possible. The 10^{36} collider has the capability to measure all three *CP*-violating angles of the unitarity triangle with superb precision.

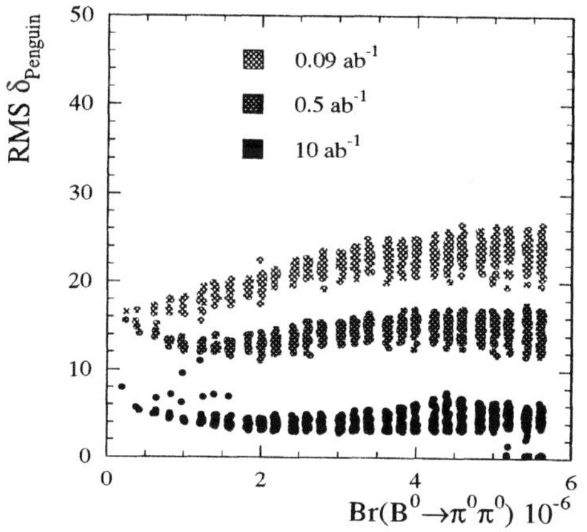

FIGURE 1. Uncertainty in the Penguin angle δPenguin = α_{eff}-α in degrees *vs.* the branching ratio for 90 fb^{-1}, 500 fb^{-1} and 10 ab^{-1}. (A. Roodman)

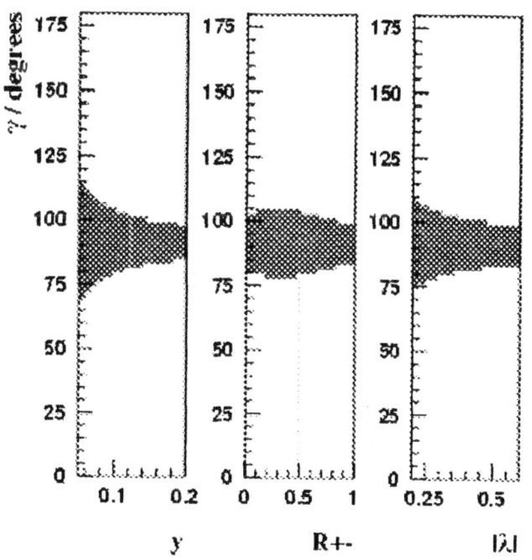

FIGURE 2. Statistical resolution on γ as a function of y_s, $R_\pm$ and $|\lambda|$ for 10,000 $D_s^{(*)\pm}K(*)^\mp$ events, corresponding to 1 ab^{-1} at the $\Upsilon(5S)$. A central value of γ=90° is assumed. (S. Petrak)

TABLE 1. Summary of estimated precision of CKM angle measurements for *BABAR* and Super*BABAR*, compared to planned experiments at hadronic colliders.

CKM Angle	*BABAR* (0.5ab^{-1})	Super*BABAR* (10 ab^{-1})	*B*TeV	LHC*b*	ATLAS/CMS
$\sin 2\beta(B^0 \to J/\psi K_s^0)$	0.037	0.008	0.025	0.014	0.021/0.025
$\sin 2\beta(B^0 \to \phi K_s^0)$	0.25	0.056			
$\sin 2\alpha(B^0 \to \pi^+\pi^-)$	0.14	0.032	0.024	0.056	0.10/0.17
$\alpha_{eff} - \alpha(B \to \pi^0\pi^0)$	< 18°	< 7°	-	-	-
$\sin(2\beta+\gamma)(B^0 \to D^*\pi)$	0.15	0.03			
$\gamma(B \to DK)$		<2.5°	<10°	<19°	
$\gamma(B_s \to D_sK)$		<15°	<7°	<13°	

Measurements to this precision would not be interesting if theoretical uncertainties affecting the determination of the precision of the sides of the unitarity triangle could not be commensurately reduced. Fortunately, improvements in lattice gauge theory techniques, through the replacement of quenched with unquenched calculations, improved actions and larger lattices, promise to improve on a time scale which keeps pace with the next generation of experiments [7]. Table 2 shows the expected improvement in both experimental and theoretical determinations of CKM matrix elements over the coming decade. Improvements in the determination of experimental inclusive and exclusive semileptonic branching ratios depend on larger data sets to reduce statistical errors, but also on new techniques, such as complete event reconstruction, made possible by the large data sets, that have smaller systematic and theoretical uncertainties.

TABLE 2. Projections for improvement in the experimental and theoretical contributions to the precision of CKM matrix elements.

V_{ij}	Experimental Measurement	σ 2001 stat/sys (%)	σ 2006 stat/sys (%)	σ stat/sys 2011 (%)	Theoretical Quantity	σ 2001 quenched (%)	σ 2-5 years unquenched (%)	σ 4-10 years unquenched (%)
V_{ub}	$B(B \to \rho\ell\nu)$	4.3/8	8.6/2.4	1.4/2.4	$f_+(E_\pi)$	18	15	5
	$B(B \to u\ell\nu)$	3.4/16	4.0/2.4	2.8/2.4	$f_B{}^\dagger$	10-15	10	2
	$B(B \to \tau\nu)$		24	5	$\bar{\Lambda},\lambda_1,\lambda_2{}^*$	see note	see note	see note
V_{cb}	$B(B \to D^*\ell\nu)$	3.1/4	0.4/2	0.1/1	$\mathcal{F}(1)$ ‡	2-4	2-4	1
	$B(B \to c\ell\nu)$	2.5/2	0.3/1	0.07/0.5	$\bar{\Lambda},\lambda_1,\lambda_2{}^*$	25	15	5
V_{us}	$B(K \to \pi\ell\nu)$	0.8	0.8	0.8	$f_+(q^2)$	15	15	2-5
V_{cd}	$B(D \to \pi\ell\nu)$	7.1	1		$f_+(E_\pi)$	15	15	2-5
	$B(D \to \ell\nu)$		2		$f_D{}^\dagger$	10-15	10	2
V_{cs}	$B(D \to K\ell\nu)$		0.4		$f_+(E_K)$	15	15	2-5
	$B(D_s \to \ell\nu)$		1		$f_{Ds}{}^\dagger$	10-15	10	2
V_{td}	Δm_d	1/1	0.2/0.5	0.05/0.2	$f_{B_d}\sqrt{B_{B_d}}$ #	~20	15	5
V_{ts}	Δm_s				$f_{B_s}\sqrt{B_{B_s}}$ #	~20	15	5

† 50% of the error on f_B / f_{D_s}

* from experiment: λ_2 from $m_{B^*} - m_B$; $\bar{\Lambda}$ and λ_1 from moments of $B \to c l\nu$ and $B \to s\gamma$ spectra

‡ lattice measures $\mathcal{F}(1)$ -1

$\xi = \dfrac{f_{B_s}/\sqrt{B_s}}{f_{B_d}/\sqrt{B_d}}$ error divided by 1.5-2

The last major area of experimental interest is rare decays, which provide sensitivity to new physics through loop diagrams. Table 3, also developed at Snowmass, compares the sensitivity of e^+e^- experiments with 500 fb^{-1}, the target total sample for *BABAR*, and with 10 ab^{-1}, corresponding to 1 year of running at 10^{36} luminosity, with that obtainable with planned and proposed experiments at hadronic colliders. The 10^{36} collider compares quite favorably with hadronic experiments, in

TABLE 3. Comparison of the number of reconstructed rare B decays in hadronic and e^+e^- experiments.

		Hadron Collider Experiments			e^+e^- B Factories	
Decay Mode	**Branching Fraction**	**CDF/D0 (2 fb^{-1})**	***B*TeV/LHC*b* (10^7 s)**	**ATLAS/CMS (1 y)**	***BaBar*/Belle (0.5ab^{-1})**	**Super*BaBar* (10 ab^{-1})**
$B \to X_s \gamma$	$(3.3 \pm 0.3)\times 10^{-4}$				11K	220K
					1.7K (*B* tagged)	34K (*B* tagged)
$B \to K^* \gamma$	5×10^{-5}	170	25K		6K	120K
$B \to \rho(\omega)\gamma$	2×10^{-6}				300	6K
$B \to X_s\mu^+\mu^-$	$(6.0 \pm 1.5)\times 10^{-6}$		3.6K		300	6K
$B \to X_s e^+e^-$					350	7K
$B \to K^*\mu^+\mu^-$	$(2 \pm 1)\times 10^{-6}$	60-150	2.2K/4.5K	665/4.2K	120	2.4K
$B \to K^* e^+e^-$					150	3K
$B \to X_s\nu\bar{\nu}$	$(4.1 \pm 0.9)\times 10^{-5}$				8	160
$B \to K^*\nu\bar{\nu}$	5×10^{-6}				1.5	30
$B_d^0 \to \tau^+\tau^-$	10^{-7}					
$B_s^0 \to \mu^+\mu^-$	10^{-9}	5/1.5-6	5/11	9/7		
$B_d^0 \to \mu^+\mu^-$	8×10^{-11}	0/0	1/2	0.7/20		
$B \to \tau\nu$	5×10^{-5}				17	350
$B \to \mu\nu$	1.6×10^{-7}				8	150
$B^0 \to \gamma\gamma$	10^{-8}				0.4	8

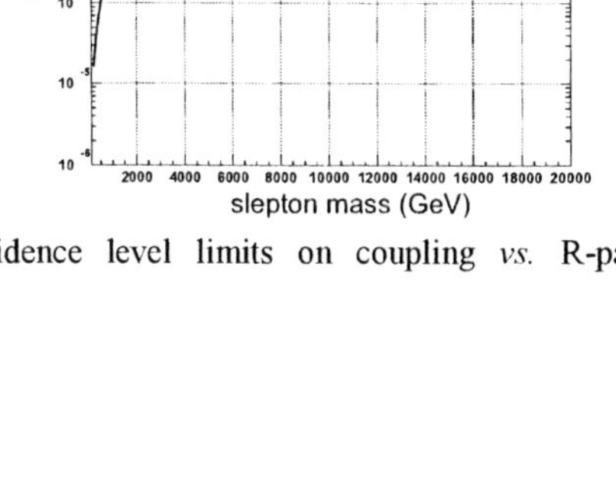

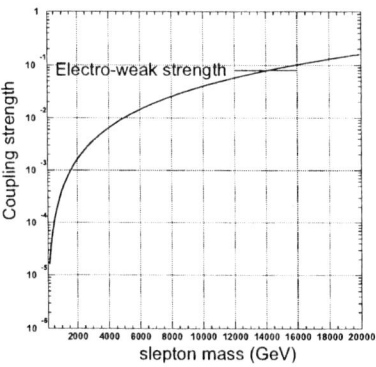

FIGURE 3. 90% confidence level limits on coupling *vs.* R-parity-violating *slepton* mass. (S. Yang).

rare inclusive and exclusive modes, and particularly in radiative modes. Note also that only an e^+e^- experiment can produce a sample of tagged $B \rightarrow X_s \gamma$ decays. Fig. 3 shows that the *slepton* mass sensitivity exceeds 10 TeV at standard electro-weak coupling for a $B_d^0 \rightarrow e^+e^-$ branching fraction limit of 10^{-9}, corresponding to 10 ab^{-1}.

SuperPEP-II

The next generation *B* Factory requires a significant increase in luminosity, approaching $10^{36} cm^{-2}s^{-1}$, well beyond the already record-setting performance of PEP-II and KEKB. It appears that such a luminosity is feasible; initial parameters of SuperPEP-II, a very high luminosity e^+e^- *B* Factory are being developed, incorporating several new ideas from the successful operation of the present generation accelerators [8][9]. In this regime, the luminosity lifetime is primarily determined by the collisions themselves requiring continuous injection. This has a positive consequence: the ratio of average to peak luminosity in SuperPEP-II can be increased by 30% due to continuous injection, thereby directly improving the ability to integrate luminosity. With continuous injection, the operation of this accelerator will be qualitatively different from present e^+e^- colliders.

The next generation e^+e^- Factory will operate mainly at the $\Upsilon(4S)$ with a center-of-mass energy of 10.58 GeV with an energy asymmetry similar to those currently used, but a period of operation at the $\Upsilon(5S)$ may also be desirable. For the present study the PEP-II tunnel geometry was used as were the PEP-II beam energies of 9.0 and 3.1 GeV: a reduced energy asymmetry would reduce RF costs. To increase the luminosity about two orders of magnitude the beam currents must be raised an order of magnitude and the beam cross sectional area reduced an order of magnitude while keeping the beam-beam tune shifts under control. The parameters shown in Table 4 are self-consistent but further optimization is certainly possible.

The observed beam-beam tune shifts in PEP-II now approach 0.07 [10]. The expected tune shifts in this new accelerator should be larger for two reasons: the use of round beams at the collision point will increase the tune shifts by about a factor of two (there may be increased backgrounds from round beam operation, but significantly more backgrounds are expected from other sources as well), and it has been observed in PEP-II that by adjusting the tunes the luminosity can be increased significantly 10% at the expense of the beam lifetime [11]. (This beam lifetime will be called the beam-beam lifetime.) Higher luminosity for the same current means higher tune shifts. The new accelerator can take advantage of continuous injection to push the tune shifts to significantly higher values and consequently the beam-beam lifetimes to significantly lower values. The beam-beam lifetime in present colliders is about 100 minutes. The design assumption is that the tune shifts can be increased from 0.07 to 0.14 by reducing the beam-beam lifetime from 100 minutes to 10 minutes and by adopting round beams at the collision point.

The interaction region will likely have a geometry similar to that of PEP-II [12]. The cone angle separating the accelerator and detector components can be about 300 mrad, as at present. The LER quadrupoles for this accelerator can be moved significantly closer to the IP than in PEP-II using superconducting Q1 and Q2 magnets with stronger gradients, such as those used in the HERA upgrade [13]. The HER quadrupoles can also be moved closer because the LER quadrupoles have been moved. A crossing angle of about ±1.5 mrad is used to help separate the beams at the first parasitic beam-beam crossing. The beams are horizontally separated by about 12 σ_x at the first parasitic crossing.

TABLE 4. Parameters for a 10^{36} asymmetric *B* Factory in the PEP-II tunnel

Parameter	High Energy Ring (HER)		Low Energy Ring (LER)
Beam Energy (GeV)	9.0		3.1
Beam Particle	e^+		e^-
Center of mass energy (GeV)		10.58	
Circumference		2200	
RF frequency (MHz)		476	
RF voltage (MV)	50		30
Synchrotron radiation power (MW)	23		12
Number of bunches		3492	
Total beam current (A)	6.6		19.2
Number of beam particles	3.0×10^{14}		8.8×10^{14}
$\beta^*_{y/x}$ (cm)		0.32/0.32	
Emittance (y/x) (nm)		22/22	
Momentum compaction	0.001		0.0013
Bunch length (mm)	3.5		3.5
Approx. AC power (MW)	50		27
Beam lifetime (min)	5		5
Injected particles per pulse	7.3×10^{10}		5.3×10^{10}
Continuous injection rate (Hz)	20		80
Ring particles lost per second	1.1×10^{12}		3.2×10^{12}
Beam-beam tune shift	0.14		0.14
Transverse beam size (μm)	8.4		8.4
Luminosity ($cm^{-2}s^{-1}$)		10^{36}	

The HER vacuum system must dissipate over 16 kW/m of synchrotron radiation power. The chambers will likely be made with an antechamber with a continuous built-in photon stop. The design of bellows (expansion) modules would be very difficult for these high currents and short bunch lengths. Instead, the plan is to use a concept investigated for the PEP-II rings but not implemented. The vacuum system would be a continuous extrusion welded together with no bellows but with rigid supports to constrain thermal stresses [14]. A similar technique is used to build very long welded railroad tracks. The beam impedance will improve without bellows. The stainless steel chambers in the straight sections will have to be changed to a lower resistance material to reduce the resistive wall effect for the LER.

The beam lifetime from the beam-beam interaction will be reduced to about five minutes to maximize the beam-beam tune shifts.

Injection must be a continuous process because the beam lifetimes are short. Taking the SLAC site, the beams would come from the damping ring and linac complex. The SLAC system was built to provide about 1×10^{11} electrons per pulse at 120 Hz and about half that rate for positrons. The damping ring cavity RF frequency will be changed from 714 MHz to 476 MHz. In the damping rings, the particle bunches will be distributed uniformly over about half the circumference (35 m) in about 30 bunches. The other half of the ring circumference is used by the injection and extraction kicker rise times. The linac can operate at 120 Hz. The electron injection rate would likely be 80 Hz, the positron injection rate 20 Hz and the remaining 20 Hz used for positron production. Injection losses can cause detector problems. However, the damped injected beam will have transverse emittances smaller than the stored beam emittances. The linac bunch length and energy spread are well matched to the stored bunches, promising a relatively clean injection process. As some injection collimation will likely be needed, however, the injection efficiencies were assumed to be 75%.

SUPER*BABAR*

Doing a precision experiment at a 10^{36} asymmetric *B* Factory requires a new detector to cope with backgrounds and radiation levels[15]. Initial studies at Snowmass indicate that this is a tractable problem. A detector based on an all silicon tracking detector, and using short radiation length calorimeter crystals, would be more compact than *BABAR*. The higher physics and background rates are dealt with by employing detector systems with high segmentation and short integration times, such as pixel devices and fast scintillating crystals.

The overall scale of the detector studied at Snowmass, called Super*BABAR*, and shown in Fig. 4, is determined by the crystal calorimeter. We require scintillating crystals with greater radiation hardness and faster decay times than the CsI(Tl) deployed in current generation devices. Crystals such as LSO and GSO have these desirable characteristics, as well as shorter radiation length and smaller Molière radius. Thus a calorimeter having energy and angular resolution comparable to existing devices can be quite a bit smaller in volume.

This, of course, poses a challenge to the resulting smaller radius tracking system. The tracking detector would combine two pixel layers with seven layers of arched double-sided silicon strips, providing track and vertex reconstruction in a single system. A 3 Tesla superconducting solenoid provides momentum resolution comparable to that of larger current combined silicon/drift chamber systems in a weaker field, *albeit* with different contributions from measurement error and multiple coulomb scattering.

A DIRC Cherenkov device, based on that used in *BABAR*, but with a new compact readout that is much less sensitive to backgrounds, provides $\pi/K/p$ identification over the full kinematic range.

A straightforward extension of the open trigger approach traditionally employed in e^+e^- experiments appears to be quite practical. Techniques for detector readout pioneered for the new generation of experiments at the Tevatron and LHC appear to be generally applicable to a 10^{36} e^+e^- storage ring, allowing unbiased triggering on essentially all events of interest.

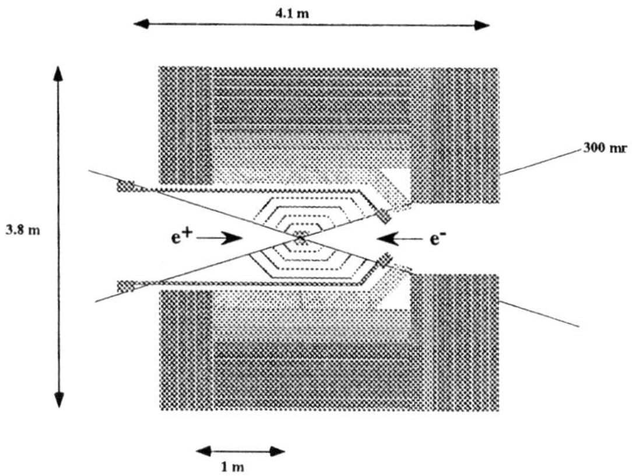

FIGURE 4. Elevation view of a concept for Super*BABAR*, a detector designed for a 10^{36} collider.

CONCLUSION

In summary, the physics case for a 10^{36} asymmetric *B* Factory is quite strong. The program has many unique aspects and is complementary to the programs at hadronic machines. The details of machine and detector design are far from mature, but both machine and detector appear to present reasonable challenges. Undoubtedly, developments in both theory and experiment over the next several years will sharpen our vision and allow a clearer determination of the importance of pushing flavor physics investigations to this new level in rare decays and precision measurements.

REFERENCES

1. "Physics at a 10^{36} Asymmetric *B* Factory", SLAC-PUB-8970, Aug. 2001. To be published in *Proceedings of the 2001 DPF Snowmass Summer Study on the Future of Particle Physics*.
2. Gronau, M. and London, D., *Phys. Rev. Lett.* **65**, 3381 (1990).
3. Gronau, M. and Wyler, D., *Phys. Lett.* **B265**, 172, (1991).
4. Atwood, D., Dunietz, I., and Soni, A., *Phys. Rev. Lett.* **78**, 3257 (1997).
5. Soffer, A., *Phys. Rev.*, **D60**, 54032 (1999).
6. Petrak, S., *Proceedings of the 4th International Conference on B Physics and CP Violation,* Ise, Japan, edited by Oshima, T. and Sanda, A.I., Singapore: World Scientific, 2001, pp 124-129.
7. The projections of the improvement of lattice gauge calculations with time in Table 2 are those of Kronfeld, A. and Mackenzie, P. in Ref. 1.
8. Seeman, J., "Initial Parameters for a 10^{36} cm^{-2}s^{-1} Luminosity e^+e^- *B* Factory", SLAC-PUB-8787, March 2001.
9. Seeman, J., "Higher Luminosity *B* Factories", *Proceedings of PAC 2001.*

10. Seeman, J., "Status Report on PEP-II Performance", *EPAC 2000*, 38, and *Proceedings of PAC 2001*.
11. Sullivan, M., Wienands, U., private communication.
12. *PEP-II Conceptual Design Report*, SLAC Report 418, June 1993.
13. Parker, B. *et al.*, "Superconducting Magnets for use inside the HERA *ep* Interaction Regions", *PAC 1999*, 308.
14. Klaisner, L., private communication.
15. *BABAR* Upgrades Required to Address PEP-II Operations Through 3×10^{34} $cm^{-2}s^{-1}$, Report of the Ad-hoc Committee to Investigate Detector Upgrades, April 2000. See http://www.slac.stanford.edu/BFROOT/www/Detector/Upgrades/FINALREPORT.pdf

Status of Tau Physics

Alan J. Weinstein

California Institute of Technology, Pasadena, California 91125, USA

Abstract. We review, very briefly, many of the important results arising in the study of tau lepton production and decay, with emphasis on recent results. We discuss the studies of topological and exclusive branching fractions, inclusive semi-leptonic tau decay spectral functions, extraction of QCD parameters including α_S and the strange quark mass, exclusive semileptonic decays and low energy meson dynamics, and rare and forbidden (neutrinoless) decays. Special attention is paid to recent searches for CP non-conservation in tau pair production and tau decay.

INTRODUCTION

The tau is considered to be a heavy flavor, hence its token appearance at this conference. That it is "heavy" is borne out by the fact that the tau has many decay modes, and thus a rich phenomenology. Its mass is large compared to the scale of hadronization: $m_\tau \gg \Lambda_{QCD}$. Thus, decays of the tau can be used to probe QCD in the perturbative regime, as discussed in the section on inclusive semileptonic decays.

Most importantly, the tau appears to be a pointlike fundamental fermion, and its electroweak couplings can be measured with precision that, in some cases, can exceed that of the lighter leptons. It is thus a useful tool for probing some of the least well understood aspects of particle physics: the dynamics of light mesons at low energies, and physics beyond the Standard Model (SM).

Tau physics is a large field. Since this is a status report, I endeavor to cover many of the important results, but necessarily, in a very brief manner. Special attention is paid to recent searches for CP violation in tau production and decay. Explicitly omitted from this report are discussions of tau neutrino properties, leptonic decays, tests of lepton universality, Michel parameters, and neutral current couplings.

Progress since Heavy Flavors 8

Progress in tau physics has tapered off in the last couple of years. The LEP program is over, and final results in for $Z^0 \to \tau^+\tau^-$ from LEP I and $W^\pm \to \tau^\pm\nu_\tau$ from LEP II (all results are nicely consistent with the SM). There were many beautiful results presented at the TAU2000 workshop in Victoria, Canada [1].

There have been recent results from LEP on Michel parameters from DELPHI [2] and ALEPH [3], on precision topological branching fractions from L3 [4] and DELPHI [5], as well as final results from all LEP collaborations on tau polarization at the Z^0 peak.

CP618, *Heavy Flavor Physics: Ninth International Symposium*, edited by A. Ryd and F. C. Porter

The results on hadronic contributions to the muon anomalous magnetic moment, based on ALEPH data from tau decays [6], have fueled the debate over whether the recent precision measurement of a_μ from BNL are in conflict with the SM.

CLEO has published results on structure in hadronic decays containing charged kaons [7], the branching fraction for the rare decay to $6\pi\nu$ [8], and two new papers on searches for CP violation in tau decays [9, 10].

Belle has presented first results on tau physics at this summer's conferences, including searches for neutrinoless tau decays [11, 12], for CP violation in tau decays [13], and for a CP-violating EDM in tau pair production [14].

TOPOLOGICAL AND EXCLUSIVE BRANCHING FRACTIONS

There are new precision measurements on tau decay topological (charged particle multiplicity) branching fractions from L3 [4] and DELPHI [5], with errors of the same order as the PDG2000 world averages [15]. In these measurements, charged tracks from K_S^0 decays, Dalitz decays, and photon conversions do not contribute to the charged particle multiplicity. The results from L3 are:

$$\mathcal{B}(1-prong) = (85.27 \pm 0.11 \pm 0.07)\% \tag{1}$$
$$\mathcal{B}(3-prong) = (14.56 \pm 0.11 \pm 0.08)\% \tag{2}$$
$$\mathcal{B}(5-prong) = (0.17 \pm 0.02 \pm 0.03)\% \tag{3}$$

All conventional exclusive final states are reasonably well measured. The final states used in the PDG fit include:

$$e^-\nu\nu_\tau,\ \mu^-\nu\nu_\tau,\ \pi^-\nu_\tau, \pi^-\pi^0\nu_\tau,\ \pi^-2\pi^0\nu_\tau,\ \pi^-3\pi^0\nu_\tau,\ \pi^-4\pi^0\nu_\tau,$$
$$\pi^-\pi^+\pi^-\nu_\tau,\ \pi^-\pi^+\pi^-\pi^0\nu_\tau,\ (5\pi)^-\nu_\tau,\ (6\pi)^-\nu_\tau,$$
$$K^-\nu_\tau,\ K^-\pi^0\nu_\tau,\ \pi^-\bar{K}^0\nu_\tau,\ (K\pi\pi)^-\nu_\tau,\ (K3\pi)^-\nu_\tau,\ K^-\bar{K}^0\nu_\tau,$$
$$(KK\pi)^-\nu_\tau,\ (KK2\pi)^-\nu_\tau,\ \eta\pi^-\pi^0\nu_\tau,\ \eta K^-\nu_\tau.$$

Exclusive final states saturate these topological branching fractions nicely, and sum to one. The constrained PDG 2000 fit to the branching fractions agree well with the averages of the direct measurements (in contrast to previous years).

INCLUSIVE SEMI-LEPTONIC DECAYS

The semi-leptonic decays of the tau to ν_τ plus hadrons provides a clean source of hadrons with invariant mass below m_τ for the study of low-energy meson dynamics. The hadronic system in $\tau \to \nu_\tau X$ can be studied in terms of its exclusive final state (*e.g.*, π, 2π, 3π, $K\pi$, *etc.*), or inclusively, broken down by invariant mass $s = m_X^2$, strangeness, and whether the final state proceeds via the vector V or axial-vector A part of the charged weak current.

Inclusive studies can be cast in terms of the production-independent vector and axial-vector spectral functions $v^{(s)}$ and $a^{(s)}$, where the superscript denotes whether the final state has strangeness or not. One can extract these spectral functions directly from the data; *e.g.*, the $\pi\pi$ contribution to $v(s)$ is obtained from

$$v^{\pi\pi}(s) \sim \underbrace{\frac{B_{\pi\pi^0}}{B_e}}_{\text{Branching ratios}} \underbrace{\frac{1}{N_{\pi\pi^0}} \frac{dN_{\pi\pi^0}}{ds}}_{\text{Mass spectrum}} \underbrace{\frac{m_\tau^2}{(1-s/m_\tau^2)^2(1+2s/m_\tau^2)}}_{\text{Kinematic factor}}. \tag{4}$$

The Conserved Vector Current (CVC) hypothesis relates the vector part of the hadronic spectral function measured in weak decays to the same quantity measured in electromagnetic processes. An isospin rotation connects the charged weak hadronic spectral function to the isospin-1 part of the neutral hadronic spectral function. Thus, the vector spectral function measured in tau decays, $\tau^- \to \nu_\tau W^-$, $W^- \to$ hadrons, can be compared with that measured in, *e.g.*, $e^+e^- \to \gamma^*$, $\gamma^* \to$ hadrons:

$$\sigma(e^+e^- \to \pi^+\pi^-) = \frac{4\pi\alpha^2}{s} v_{I=1}^{\pi\pi} \tag{5}$$

as a test of CVC and isospin symmetry. In this example, the $I = 0$ part of the e^+e^- cross section ($e^+e^- \to \omega \to \pi^+\pi^-$) must be subtracted out. A comparison of the $v^{\pi\pi}(s)$ measured in these two ways [16] is shown in Fig. 1, as a test of CVC.

Assuming CVC and isospin symmetry, the vector spectral functions measured in tau decays can be used to improve the estimate of the contribution of hadronic vacuum polarization to the running QED coupling constant $\alpha_{QED}(m_Z^2)$ and the muon anomalous magnetic moment a_μ^{had}. The values of these quantities so obtained [6] constitute some of the most accurate estimates of these quantities.

The total ratio of hadrons to leptons in tau decays can be extracted from the world average [15] leptonic branching fractions alone:

$$R_\tau \equiv \frac{\Gamma(\tau^- \to \nu_\tau \text{hadrons})}{\Gamma(\tau^- \to \nu_\tau e^- \bar{\nu}_e)} = \frac{1 - \mathcal{B}_e - \mathcal{B}_\mu}{\mathcal{B}_e} = 3.642 \pm 0.019. \tag{6}$$

The dependence of $R_\tau(s_0)$ on the maximum hadronic invariant mass can be extracted from the inclusive vector and axial-vector spectral functions. These have been measured/estimated in tau decays by ALEPH [17] and OPAL [18], with good agreement. The results from ALEPH are shown in Fig. 2.

The fact that these spectral functions, and their integrals, flatten out as s_0 approaches m_τ^2 is qualitative evidence that m_τ^2 is "QCD asymptopia", so that QCD perturbative methods and the Operator Product Expansion can be applied to understand the tau data at the tau mass scale. In this sense, we can regard the tau as a "heavy flavor".

The inclusive spectral functions can be used to extract QCD perturbative parameters ($\alpha_S(m_\tau^2)$), quark masses ($m_q(m_\tau^2)$), and non-perturbative condensates [19]:

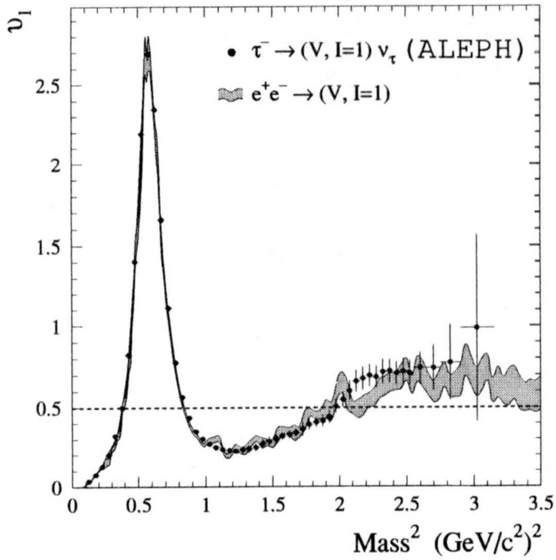

FIGURE 1. Total hadronic vector current spectral function from tau decays measured by ALEPH (data points), and the corresponding distribution calculated from e^+e^- isovector states using isospin symmetry (shaded band). The dashed line corresponds to the naive isovector quark-parton prediction. Taken from [16] with permission, © Springer-Verlag 1997.

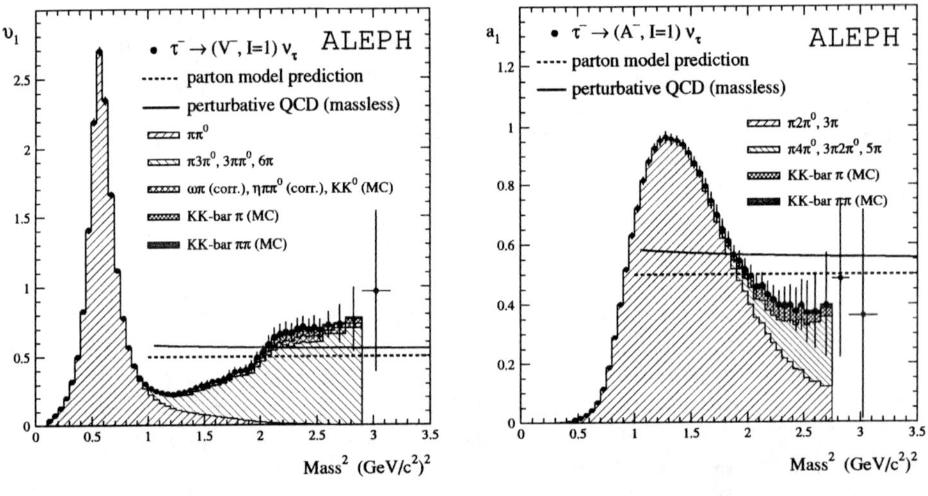

FIGURE 2. The total vector (left) and axial-vector (right) spectral functions from tau decays measured by ALEPH (data points), broken down by exclusive final state contributions (some of which are estimated from Monte Carlo simulations). The lines show the predictions from the naive parton model and from massless QCD using $\alpha_S(M_Z^2) = 0.120$. Taken from [17] with permission, © Springer-Verlag 1998.

$$R_{\tau V,A} = \frac{3}{2}|V_{ud}|^2 \underbrace{S_{EW}}_{known} \left(1 + \underbrace{\delta^{(0)}(\alpha_S(m_\tau^2))}_{perturb} + \underbrace{\delta'_{EW}}_{known} + \underbrace{\delta^{(2,m)}_{V,A}(m^2_{ud})}_{mass-dependence} + \underbrace{\sum_{D=4,6,\cdots} \delta^{(D)}_{ud,V,A}}_{non-perturb} \right). \tag{7}$$

The dependence of the perturbative correction $\delta^{(0)}$ on $\alpha_S(m_\tau^2)$ is calculated up to fourth order in fixed-order perturbation theory [19]. The quark mass dependent non-perturbative correction $\delta^{(2,m)}_{V,A}(m^2_{ud})$ has been calculated [20] up to next-to-leading order in α_S.

From fits to R_τ and the moments of $R_\tau(s_0)$, ALEPH extracts [17] $\alpha_S(m_\tau^2)$, and runs the value up to the Z^0 mass scale:

$$\alpha_S(m_\tau^2) = 0.334 \pm 0.022 \quad \Longrightarrow \quad \alpha_S(m_Z^2) = 0.1202 \pm 0.0027. \tag{8}$$

This agrees well with other measures of $\alpha_S(m_Z^2)$ [21], and constitutes one of the most precise measurements of this quantity.

Strange quark mass

Considering the decays of the tau to final states containing net strangeness (K^-, $(K\pi)^-$, $(K2\pi)^-$, *etc.*), the contribution to $R_{\tau,S}$ depends on $\delta^{(2,m)}_{V,A}(m_s^2)$, which in turn depends on the strange quark mass evaluated at the tau mass scale, $m_s(m_\tau^2)$. The perturbative expansion exhibits bad behavior, so care must be taken to extract a meaningful estimate of the quark mass [22]. Using data from ALEPH, CLEO, and world-average branching fractions for $\tau^- \to \nu_\tau(Kn\pi)^-$ decays, Ref. [22] obtains

$$R_{\tau,S} = 0.1630 \pm 0.0057 \quad \text{and} \quad m_s(m_\tau^2) = 120^{+21}_{-26}\,\text{MeV}. \tag{9}$$

EXCLUSIVE SEMILEPTONIC DECAYS

Semi-leptonic tau decays occur at relatively low momentum transfer. Although the inclusive sum of all the final state modes approaches a limit where QCD perturbative expansions can be applied, the study of exclusive final states relies instead on models of low-energy meson dynamics, resonance dominance, and to some extent, chiral perturbation theory. ARGUS, CLEO, and the LEP experiments have performed many detailed studies of the resonant structure of exclusive final states, and have extracted resonance parameters, sub-resonances, and couplings between vector, axial-vector, scalar and tensor resonances. These measurements have been used to test CVC, test the spin structure in tau decay, look for second-class (isospin-violating) currents, *etc.*

There are many resonances that contribute to the structure of of hadronic final states in tau decays. The $\pi^-\pi^0$ final state proceeds through the $\rho(770)$ and its radial excitations. The 3π final state proceeds through the $a_1(1260)$, which decays to $\rho\pi$ and $\rho'\pi$ via both s-wave and d-wave amplitudes, and to isoscalars $\sigma(400-1200)\pi$, $f_2(1270)\pi$ and $f_0(1370)\pi$ via p-wave amplitudes. There have been (unsuccessful) searches for a contribution from $\pi'(1300) \to \rho\pi$ or $\sigma\pi$. The 4π final state proceeds through the ρ', decaying to $\omega\pi$, $a_1\pi$, and $\rho\pi\pi$. There have been (unsuccessful) searches for a contribution from $b_1(1235) \to \omega\pi$, which would be an isospin-violating second-class current [23]. The $\eta\pi^-$ final state [24, 25] is forbidden by isospin (it is a second-class current), and the $\eta\pi^-\pi^0$ final state proceeds via the Wess-Zumino chiral anomaly [25]. The 5π and 6π final state resonant substructure is so complex, and the data samples so limited, that they have yet to be explored with any precision [26].

Final states containing kaons have also been studied [27]. The $K\pi$ final state is dominated by the $K^*(892)$, although redial excitations (the $K^*(1410)$) may also contribute. The $K\pi\pi$ final state proceeds through the $K_1(1270)$ and, to a lesser extent, the $K_1(1400)$, both of which decay to $K^*\pi$ and $K\rho$. The $K_1(1270)$ and $K_1(1400)$ are mixtures of the K_{1a} (in the $J^{PC} = 1^{++}$ octet) and the K_{1b} (in the $J^{PC} = 1^{+-}$ octet); the latter couples to the weak charged current only through $SU(3)_f$ violation. The $KK\pi$ final states can proceed through the axial-vector a_1. The ηK^- final state has been observed [24]; it is an $SU(3)_f$ violating transition.

This brief status report cannot do justice to this rich subject; refer to Ref. [28] for more details.

RARE AND FORBIDDEN DECAYS

There are a variety of rare but allowed decays that have been seen or searched for in tau decays. Some interesting recent measurements include:

$$
\begin{aligned}
\mathcal{B}(\tau \to \eta 3\pi\nu_\tau) &= (4.8 \pm 1.1) \times 10^{-4} && \text{(CLEO 98 [29])} \\
\mathcal{B}(\tau \to \eta K\nu_\tau) &= (2.7 \pm 0.6) \times 10^{-4} && \text{(CLEO\&ALEPH [24])} \\
\mathcal{B}(\tau \to \eta K\pi\nu_\tau) &= (5.0 \pm 1.2) \times 10^{-4} && \text{(CLEO 98 [24])} \\
\mathcal{B}(\tau \to e^-e^+e^-\nu\nu_\tau) &= (2.8 \pm 1.5) \times 10^{-5} && \text{(CLEO 98 [30])} \\
\mathcal{B}(\tau \to \pi^-K^0\bar{K}^0\pi^0\nu_\tau) &= (3.1 \pm 2.3) \times 10^{-4} && \text{(ALEPH 99 [22])} \\
\mathcal{B}(\tau \to \pi^-K^+K^-\pi^0\nu_\tau) &= (4.0 \pm 1.6) \times 10^{-4} && \text{(CLEO\&ALEPH [27])} \\
\mathcal{B}(\tau \to K^-K^+K^-\nu_\tau) &< 1.9 \times 10^{-4} && \text{(ALEPH 98 [31])} \\
\mathcal{B}(\tau \to 7\text{-prong } \nu_\tau) &< 2.4 \times 10^{-6} && \text{(CLEO 97 [32])} \\
\mathcal{B}(\tau \to \eta\pi\nu_\tau) &< 1.4 \times 10^{-4} && \text{(CLEO 96 [24])} \\
\mathcal{B}(\tau \to (\omega\pi)_{1^{++}}\nu_\tau) &< 1.3 \times 10^{-3} && \text{(CLEO 99 [23])}
\end{aligned}
\tag{10}
$$

The last two upper limits correspond to isospin-violating second-class currents.

Limits on neutrinoless decays

There are several classes of decay modes that are forbidden in the SM, usually because they produce no ν_τ in the final state. These include:

- Lepton Flavor Violating (LFV) decays like $\tau^- \to \mu^-\gamma$, $\tau^- \to e^-e^+e^-$
- Lepton number violating (LV) decays like $\tau^- \to \mu^+\pi^-\pi^-$, $\mu^+e^-e^-$
- B-L conserving modes like $\tau^- \to p^-\gamma$
- Exotic modes like $\tau^- \to \pi^-\nu_{heavy}$, e^-G^0, where ν_{heavy} is a neutral heavy fermion and G^0 is a neutral heavy boson
- Angular momentum violating modes like $\tau^- \to \pi^-\pi^0$, $\pi^-\gamma$

CLEO [33] has set limits on a long list of such decays, typically resulting in branching fraction upper limits of a few $\times 10^{-6}$ at 95% C.L. The angular momentum violating modes, and the e^-G^0, μ^-G^0 modes were studied by ARGUS [34]. Belle has recently searched and found no evidence for the decays to e^-K^0 and μ^-K^0, with branching fraction upper limits of 1.8×10^{-6} at 95% C.L. for both modes [12].

Many extensions of the SM predict that the decay $\tau^- \to \mu^-\gamma$ should proceed, and some SUSY and LR-symmetric models predict an enhancement of this decay rate relative to $\mu^- \to e^-\gamma$ of $10^5 - 10^6$. Recent searches for these modes give:

$$\begin{array}{lcll} \mathcal{B}(\mu^- \to e^-\gamma) & < & 1.2 \times 10^{-11} & \text{(MEGA '99 [35])} \\ \mathcal{B}(\tau^- \to \mu^-\gamma) & < & 1.1 \times 10^{-6} & \text{(CLEO '00 [36])} \\ \mathcal{B}(\tau^- \to \mu^-\gamma) & < & 1.0 \times 10^{-6} & \text{(Belle '01 [11])} \end{array} \qquad (11)$$

CP NON-CONSERVATION IN TAU PHYSICS

In the SM, there is no mechanism for CP violation in the lepton sector. If neutrinos mix, then the analog of the CKM matrix for neutrinos can produce non-zero but immeasurably small CP-violating effects. The search for CP violation in tau physics is a search for physics beyond the SM.

CP non-conservation in $e^+e^- \to \gamma^*/Z^{0*} \to \tau^+\tau^-$

A simple extension to the SM description of tau pair production in e^+e^- collisions is to add tensor couplings, producing anomalous dipole couplings. For example, well below the Z^0 pole, the QED Lagrangian can be written as [37]:

$$\mathcal{L} = \mathcal{L}_{SM} + \mathcal{L}_E + \mathcal{L}_M; \quad \mathcal{L}_{SM} = eF_1(q^2)\bar{\psi}\gamma^\mu\psi A_\mu; \qquad (12)$$

$$\mathcal{L}_E = \frac{1}{2}\frac{eF_2(q^2)}{2m_\tau}\bar{\psi}\sigma^{\mu\nu}\psi F_{\mu\nu}; \quad \mathcal{L}_M = \frac{-i}{2}\frac{eF_3(q^2)}{2m_\tau}\bar{\psi}\sigma^{\mu\nu}\gamma_5\psi F_{\mu\nu}, \qquad (13)$$

where the anomalous magnetic and electric dipole moments are given by

$$a_\tau \equiv F_2(q^2=0); \quad d_\tau \equiv \frac{eF_3(q^2=0)}{2m_\tau}. \tag{14}$$

Analogous forms which couple the leptons to the Z^0 permit the definition of the weak anomalous magnetic and electric dipole moments; there is no general connection between the electromagnetic and weak dipole moments. The electric and weak electric dipole moments (EDMs) are CP-violating. The small CP violation due to the CKM phase in the quark sector produces negligibly small values for these dipole moments in the SM.

Anomalous dipole moments due to beyond-SM physics can have a small(2nd order) effect on the total cross section σ and the differential cross section $d\sigma/d\cos\theta$. The effect on the tau spins is first order in these dipole moments; thus, the most sensitive searches for measurable dipole moments come from studies in which the tau spins are inferred from their decays.

The $\tau\tau Z^0$ coupling is studied at LEP-I and SLD, using a set of optimal CP-odd tensor observables [38], and limits have been placed on anomalous weak dipole moments $Re(a_\tau^W)$, $Im(a_\tau^W)$, $Re(d_\tau^W)$, $Im(d_\tau^W)$. The LEP limits on the CP-violating weak electric dipole moment are [39] (at 95% C.L.):

$$\begin{aligned} |Re(d_\tau^W)| &< 3.0\times10^{-18} \quad e\cdot\text{cm} \\ |Im(d_\tau^W)| &< 9.2\times10^{-18} \quad e\cdot\text{cm} \\ |(d_\tau^W)| &< 9.4\times10^{-18} \quad e\cdot\text{cm} \end{aligned} \tag{15}$$

The optimal observables at the Z^0 exploit the parity violating net longitudinal spin asymmetry of the produced taus. The optimal observables are different well below the Z^0, where there is no net longitudinal spin asymmetry, but there exist non-zero transverse spin correlations between the two taus (at the high LEP/SLD energies, these transverse spin correlations are extremely small). These spin correlations are illustrated in Fig. 3. The optimal observables at, *e.g.*, $E_{cm} \approx 10$ GeV exploit the transverse spin asymmetry due to a CP-violating electric dipole moment d_τ^γ.

ARGUS has searched for an anomalous EDM using 277K tau pairs produced near $E_{cm} \approx 10.6$ GeV [40]. They estimate the tau spin polarization on both sides of the event using "polarimeter vectors" constructed from $\tau \to e, \mu, \rho$ decays. They construct optimal observables, and find expectation values for these that are consistent with zero (no CP violation). From this, they extract limits on the tau EDM of $|Re(d_\tau)| < 4.6\times10^{-16}$ e cm, $|Im(d_\tau)| < 1.8\times10^{-16}$ e cm, both at 95% C.L. Belle has searched for an anomalous EDM using 9.4 million tau pairs [14], and establishes limits of $|Re(d_\tau)| < 6.7\times10^{-17}$ e cm, $|Im(d_\tau)| < 2.2\times10^{-17}$ e cm, both at 95% C.L. These results can be compared with analogous limits for the electron and muon, $|d_e| < 4\times10^{-27}$ e cm, and $|d_\mu| < 9\times10^{-19}$ e cm.

An anomalous magnetic moment for the tau would lead to enhancement of radiated photons in the reaction $e^+e^- \to \gamma^*/Z^0 \to \tau^+\tau^-\gamma$. L3 has searched for and found no evidence of anomalous photon production at LEP, and sets the limits [41] $-0.052 < Re(a_\tau) < 0.058$ and $|Re(d_\tau)| < 3.1\times10^{-16}$ e cm, both at 95% C.L. In Ref. [42], data on

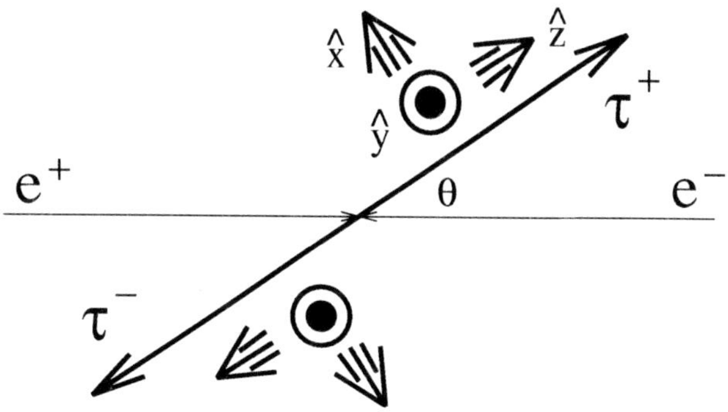

FIGURE 3. Illustration of axes along which tau spins produced in e^+e^- collisions are correlated. The longitudinal ($\hat{z}$) spin polarization has a net non-zero value at the peak of the Z^0, but the transverse ($\hat{x}$, $\hat{y}$) spin polarizations are small. There is no net spin polarization well below the Z^0, but there are strong spin correlations along all three axes.

the $e^+e^- \to \tau^+\tau^-$ cross sections measured at and above the Z^0 peak, the tau polarization and polarization asymmetry measured at the Z^0 peak, and the $W \to \tau\nu$ branching fraction measured at LEPII and the Tevatron, are combined to yield limits on the anomalous magnetic moments of $-0.007 < Re(a_\tau) < 0.005$ and $-0.0024 < Re(a_\tau^W) < 0.0025$, both at the 2σ level.

CP non-conservation in tau decay

One can observe CP violation in tau decay if it proceeds via two interfering amplitudes (with different angular distributions) and complex relative phases. For example, a charged Higgs could couple to the tau and then to a scalar hadronic system such as a scalar $a_0^-(980) \to \pi^-\pi^0$, as illustrated in Fig. 4. If the A_W and A_H amplitudes have a relative complex phase, the interference term is CP-odd.

The CP-violating effect is proportional to isospin violation proportional to $(m_u - m_d)$ in the decay $\tau \to \pi^-\pi^0\nu_\tau$, but the high rate could make such small CP-violating effects observable. In the decay $\tau \to (K\pi)^-\nu_\tau$, the effect requires SU(3)-violation proportional to $(m_u - m_s)$, but the rate is much smaller due to Cabibbo suppression.

Both Belle and CLEO have searched for CP violation in tau decays. Belle, using 2.1 million produced tau pairs, looks at single $\tau \to \pi^-\pi^0\nu_\tau$ decays. They search for an asymmetry in a CP-odd angular observable, different for τ^+ and τ^-. They see no effect, and set a limit [13] on the asymmetry.

CLEO, using 12.2 million produced tau pairs, searches for CP violation in tau decays to both $\pi^-\pi^0\nu_\tau$ [9] and $\pi^-K_S^0\nu_\tau$ [10]. In the first analysis, both taus are required to decay to $\pi^\pm\pi^0$. All dynamical information in the event is used to construct an optimal observable sensitive to CP-violating asymmetries in the spin correlations between the two decays. No asymmetry in the optimal observable is observed. This is used to set a limit on a model-dependent dimensionless coupling constant $Im(\Lambda)$ which measures the strength of a Higgs-like coupling relative to the W coupling. A limit of $-0.046 <$

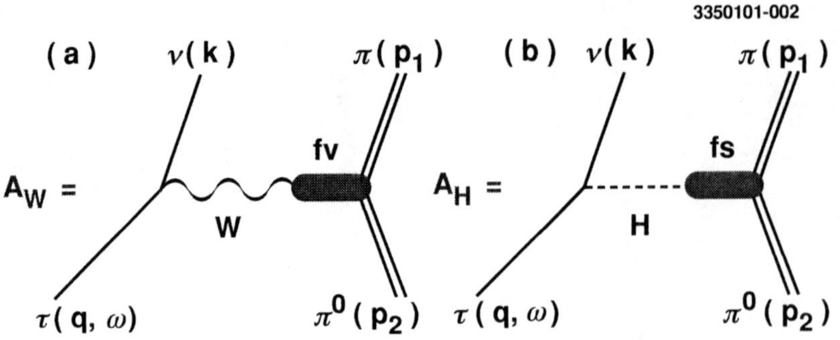

FIGURE 4. Amplitudes for (a) standard W exchange, and (b) scalar (*e.g.*, charged Higgs) exchange, mediating the decay $\tau \to \pi^-\pi^0\nu_\tau$. The vector form factor f_V is dominated by the rho meson propagator, while the scalar form factor f_S might be dominated by the scalar $a_0^-(980)$. The second diagram could have a weak phase which flips sign under CP.

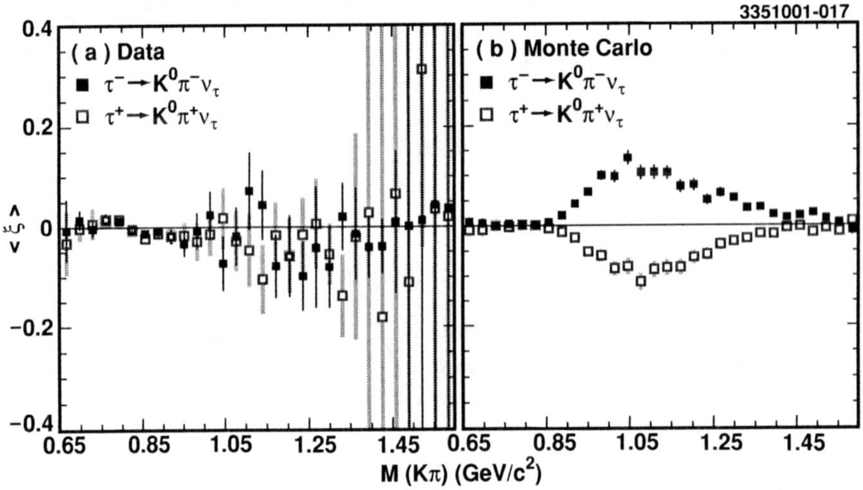

FIGURE 5. Average value of the CP-odd observable $\langle\xi\rangle$ in $\tau \to \pi^- K_S^0 \nu_\tau$ decays as a function of the $(\pi^- K_S^0)$ invariant mass, for (a) CLEO data and (b) Monte Carlo with maximum CP violation, $Im(\Lambda) = 1$. From [10].

$Im(\Lambda) < 0.022$ at 90% C.L. is obtained.

In the second analysis, one tau decay to $\pi^\pm K_S^0 \nu_\tau$ is reconstructed; the other tau is used only to tag the event. A CP-odd optimal observable sensitive to the (model-dependent) interference between a vector K^* and a scalar $K^{*0}(1430)$ is constructed. No asymmetry in this observable is seen for either τ^+ or τ^- decays, at any value of $m(K\pi)$ (Fig. 5), and the result is again expressed as a limit on a dimensionless coupling constant $Im(\Lambda)$ which measures the strength of a Higgs-like coupling relative to the W coupling: $-0.174 < Im(\Lambda) < 0.049$ at 90% C.L.

FUTURE OF TAU PHYSICS, AND CONCLUSIONS

The B-Factories (BaBar, Belle) will have produced tau pair samples approaching 10^8 events. This will enable them to push harder on rare decays, make high statistics measurements using modes with kaons, and pursue higher precision on many different topics, such as the Michel parameters.

CLEO-c [43] will run near tau threshold, which will allow it to measure the mass of the tau, and the tau neutrino, with greater precision. It will also be able to measure the Michel parameters using the unique spin correlations at tau threshold.

High statistics will allow rare decays to be seen and their Lorentz structure studied; *e.g.*, the isospin violating mode $\tau \to \eta\pi\nu$, and the Wess-Zumino coupling in $\eta\pi\pi$.

High statistics will yield more stringent limits on (and perhaps an observation of) neutrinoless decays, anomalous (eg, CP violating) couplings in weak decay or in QED production, or exotica such as $\tau^- \to \pi^-\nu_{heavy}$, e^-G^0.

Taus will be used increasingly as a tool to study electroweak physics at the Tevatron and LHC, *e.g.*, in Higgs decays $H^0 \to \tau^+\tau^-, H^- \to \tau^-\bar{\nu}_\tau$.

Taus will continue to be a powerful probe of electroweak physics, tests of lepton universality, low energy meson dynamics, beyond-SM effects, and more. The B-Factories and CLEO-c will probe tau physics with ever greater precision and sensitivity. The study of tau lepton production and decay is still a very rich and active field.

ACKNOWLEDGMENTS

The author would like to thank all the physicists who have contributed to the explorations of tau physics that are surveyed here. Thanks also go to the organizers of this conference, who provided a pleasant environment and very stimulating program.

REFERENCES

1. TAU 2000, Proceedings of the Sixth International Workshop of Tau Lepton Physics, Victoria, British Columbia, Canada, September 2000. Published in *Nucl. Phys.* **B (Proc. Suppl.) 98**, (2001).
2. DELPHI Collaboration (P. Abreu *et al.*), *Eur. Phys. J.* **C16**, 229 (2000).
3. ALEPH Collaboration (A. Heister *et al.*), CERN EP/2001-035 (Submitted to *Eur. Phys. J.* C, 2001).
4. L3 Collaboration (P. Achard *et al.*), *Phys. Lett.* **B519**, 189 (2001).
5. DELPHI Collaboration (P. Abreu *et al.*), *Eur. Phys. J.* **C20**, 617 (2001).
6. M. Davier and A. Höcker, *Phys. Lett.* **B435**, 427 (1998); R. Alemany, M. Davier and A. Höcker, *Eur. Phys. J.* **C12**, 123 (1998); M. Davier and A. Höcker, *Phys. Lett.* **B419**, 419 (1998).
7. CLEO Collaboration (D. Asner *et al.*), *Phys. Rev.* **D62**, 072006 (2000).
8. CLEO Collaboration (A. Anastassov *et al.*), *Phys. Rev. Lett.* **86**, 4467 (2001).
9. CLEO Collaboration (P. Avery *et al.*), *Phys. Rev.* **D64**, 092005 (2001).
10. CLEO Collaboration (G. Bonvicini *et al.*), CLNS 01/1766 (2001), (submitted to *Phys. Rev. Lett.*, 2001).
11. Belle Collaboration (K. Abe *et al.*), BELLE-CONF-0118 (2001), obtainable at http://belle.kek.jp/conferences/LP01-EPS/.
12. Belle Collaboration (K. Abe *et al.*), BELLE-CONF-0120 (2001).
13. Belle Collaboration (K. Abe *et al.*), BELLE-CONF-0019 (2000).
14. Belle Collaboration (K. Abe *et al.*), BELLE-CONF-0119 (2001).

15. Particle Data Group (D.E. Groom *et al.*), *Eur. Phys. J.* **C15**, 1 (2000), and 2001 off-year partial update for the 2002 edition available on the PDG WWW pages (URL: http://pdg.lbl.gov/).
16. ALEPH Collaboration (B. Barate *et al.*), *Z. Phys.* **C76**, 15 (1997).
17. ALEPH Collaboration (B. Barate *et al.*), *Eur. Phys. J.* **C4**, 409 (1998).
18. OPAL Collaboration (K. Ackerstaff *et al.*), *Eur. Phys. J.* **C7**, 571 (1999).
19. E. Braaten, S. Narison, and A. Pich, *Nucl. Phys.* **B373**, 581 (1992);
F. Le Diberdier and A. Pich, *Phys. Lett.* **B289**, 165 (1992) and **B286**, 147 (1992).
20. C. Becchi, S. Narison, E. de Rafael, F.J. Yndurain, *Z. Phys.* **C8**, 335 (1981); D.J. Broadhurst, *Phys. Lett.* **B101**, 423 (1981); S.C. Generalis, *J. Phys.* **G15**, L225 (1989).
21. Particle Data Group (D.E. Groom *et al.*), *Eur. Phys. J.* **C15**, 1 (2000), pages 85-94, and references therein.
22. ALEPH Collaboration (R. Barate *et al.*), *Eur. Phys. J.* **C11**, 599 (1999);
S. Chen, M. Davier, E. Gamiz, A. Höcker, A. Pich, J. Prades, *Eur. Phys. J.* **C22**, 31 (2001).
23. CLEO Collaboration (K. W. Edwards *et al.*), *Phys. Rev.* **D61**, 072003 (2000).
24. CLEO Collaboration (J. Bartelt *et al.*), *Phys. Rev. Lett.* **76**, 4119 (1996);
ALEPH Collaboration (D. Buskulik *et al.*), *Z. Phys.* **C74**, 263 (1997);
CLEO Collaboration (M. Bishai *et al.*), *Phys. Rev. Lett.* **82**, 281 (1999).
25. ALEPH Collaboration (D. Buskilic *et al.*), *Z. Phys.* **C74**, 263 (1997);
CLEO Collaboration (M. Artuso *et al.*), *Phys. Rev. Lett.* **69**, 3278 (1992).
26. CLEO Collaboration (A. Anastassov *et al.*), *Phys. Rev. Lett.* **86**, 4467 (2001).
27. OPAL Collaboration (G. Abbiendi *et al.*), *Eur. Phys. J.* **C13**, 197 (2000);
CLEO Collaboration (S. Richichi *et al.*), *Phys. Rev.* **D60**, 112002 (1999);
ALEPH Collaboration (R. Barate *et al.*), *Eur. Phys. J.* **C1**, 65 (1998).
28. See, for example, submissions by A. Weinstein, G. Chen, and J. van Eldik, all in Proceedings of the Sixth International Workshop of Tau Lepton Physics, Victoria, British Columbia, Canada, September 2000, Published in *Nucl. Phys.* **B (Proc. Suppl.) 98**, (2001).
29. CLEO Collaboration (T. Bergfeld *et al.*), *Phys. Rev. Lett.* **79**, 2406 (1997).
30. CLEO Collaboration (M.S. Alam *et al.*), *Phys. Rev. Lett.* **76**, 2637 (1996).
31. ALEPH Collaboration (R. Barate *et al.*), *Eur. Phys. J.* **C1**, 65 (1998).
32. CLEO Collaboration (K. Edwards *et al.*), *Phys. Rev.* **D56**, 5297 (1997).
33. CLEO Collaboration (R. Godang *et al.*), *Phys. Rev.* **D59**, 091303 (1999);
CLEO Collaboration (D. Bliss *et al.*), *Phys. Rev.* **D57**, 5903 (1998);
CLEO Collaboration (G. Bonvicini *et al.*), *Phys. Rev. Lett.* **79**, 1221 (1997);
CLEO Collaboration (K. Edwards *et al.*), *Phys. Rev.* **D55**, 3919 (1997).
34. ARGUS Collaboration (H. Albrecht *et al.*), *Z. Phys.* **C68**, 25 (1995);
ARGUS Collaboration (H. Albrecht *et al.*), *Z. Phys.* **C55**, 179 (1992).
35. MEGA Collaboration (M.L. Brooks *et al.*), *Phys. Rev. Lett.* **83**, 1521 (1999).
36. CLEO Collaboration (S. Ahmed *et al.*), *Phys. Rev.* **D61**, 071101 (2000).
37. W. Bernreuther, U. Löw, J.P. Ma, O. Nachtmann, *Z. Phys.* **C43**, 117 (1989).
38. W. Bernreuther, O. Nachtmann, *Phys. Rev. Lett.* **63**, 2787 (1989);
W. Bernreuther, G.W. Botz, O. Nachtmann, P. Overmann, *Z. Phys.* **C52**, 567 (1991).
39. A. Zalite, in TAU '98, Proceedings of the Fifth International Workshop of Tau Lepton Physics, Santander, Spain, September 1998. Published in *Nucl. Phys.* **B (Proc. Suppl.) 76**, (1999).
40. ARGUS Collaboration (H. Albrecht *et al.*), *Phys. Lett.* **B485**, 37 (2000).
41. L3 Collaboration (M. Acciari *et al.*), *Phys. Lett.* **B434**, 169 (1998).
42. G.A. Gonzalez-Sprinberg *et al.*, *Nucl. Phys.* **B582**, 3 (2000).
43. CLEO-c Collaboration (R. Briere *et al.*), CLNS-01-1742 (2001),
obtainable at http://www.lns.cornell.edu/public/CLEO/spoke/CLEOc/.

Lattice QCD and the Unitarity Triangle

Andreas S. Kronfeld

Fermi National Accelerator Laboratory, P.O. Box 500, Batavia, Illinois, USA

Abstract. Theoretical and computational advances in lattice calculations are reviewed, with focus on examples relevant to the unitarity triangle of the CKM matrix. Recent progress in semi-leptonic form factors for $B \to \pi l \nu$ and $B \to D^* l \nu$, as well as the parameter ξ in B^0-$\bar{B}^0$ mixing, are highlighted.

INTRODUCTION

To test the CKM picture of CP and flavor violation, a combination of theory and experiment is needed. A vivid way to summarize the need for redundant information is the unitarity triangle (UT), sketched in Fig. 1. It depicts two triangles, $B\gamma A$, which can (in principle) be determined from tree processes (e.g., semi-leptonic decays), and $\alpha C \beta$, in which the amplitude for B^0-$\bar{B}^0$ mixing is involved [1]. The "tree" triangle can, for the sake of argument, be taken as a measurement of the CKM matrix. Then, the "mixing" triangle tests whether new physics must be invoked to explain B^0-$\bar{B}^0$ mixing.

To obtain the sides A, B, and C of these triangles one must calculate hadronic properties from first principles of QCD. Here "calculate" implies that a reliable error estimate is given: one that includes all sources of uncertainty. Lattice gauge theory is well suited to the task: semi-leptonic form factors and the mixing matrix elements are conceptually straightforward. Indeed, according to Martin Beneke [2], "[the] standard UT fit is now entirely in the hands of lattice QCD (up to, perhaps, $|V_{ub}|$)."

Until recently, lattice QCD has been burdened by something called the quenched approximation (explained below). A bit of good news is that the available computer power is now sufficient to get rid of this approximation [3, 4, 5]. Another bit of good news is that several lattice groups have used the quenched approximation in the spirit of a blind analysis: although quenching could change the central value, one can analyze all other uncertainties as if the underlying numerical data were real QCD. This exercise has left us with several quantities, including those needed in the UT fits, with full error analyses, apart from quenching. In the next few years, we should have full QCD calculations

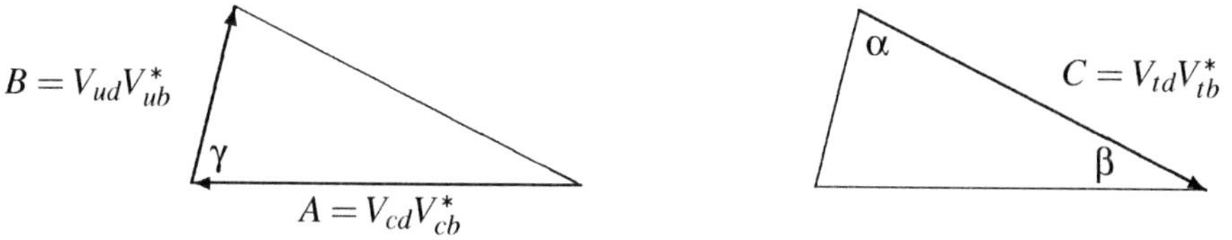

FIGURE 1. Unitarity triangles: on the left is the "tree" triangle; on the right, the "mixing" triangle.

CP618, *Heavy Flavor Physics: Ninth International Symposium*, edited by A. Ryd and F. C. Porter

with a complete assessment of the uncertainties, and thereafter the uncertainties can be incrementally reduced.

Because it is important to understand the theoretical uncertainties, this talk will start by sketching where they arise. Most reviews focus too much on central values, so, even when turning to numerical results, the focus here is on the error bars.

The paper ends with a short swim through the "Octopus's Garden", that is, some recent work by Nathan Isgur, who passed a way a few weeks before Heavy Flavors 9 convened. The symposium, and this paper, are dedicated to his memory.

UNCERTAINTIES IN LATTICE CALCULATIONS

Lattice QCD calculates matrix elements by integrating the functional integral, using a Monte Carlo with importance sampling. The Monte Carlo leads to (correlated) statistical error bars. This part of the method is well understood for quenched QCD and, these days, rarely leads to controversy. When conflicting results arise, they originate in different treatments of the systematics. The consumer probably does not need to know how the Monte Carlo works, but should develop an intuition of how the systematics work.

The main tool for controlling systematics is effective field theory. In this talk, we are concerned mostly with three classes of effects: those connected with the lattice spacing a, the heavy quark mass(es) m_Q, and the light quark mass(es) m_q. Inside the computer there is a hierarchy of scales

$$m_q \ll \Lambda \ll m_Q,\ \pi/a, \tag{1}$$

although in practice π/a is a only several GeV, and the "light" quarks are never as light as the up and down quarks. The QCD scale Λ gauges the size of power corrections; experience suggests it is 700 MeV, give or take a factor of $\sqrt{2}$. With familiar techniques of effective field theory, lattice theorists can control the extrapolation of artificial, numerical data to the real world, provided the data start "near" enough. Similarly, non-experts usually have an intuition of how effective field theories work, so they can check, on the back of the envelope, whether systematic errors have been treated sensibly.

The notable exception to the rule of effective field theory is the valence, or quenched, approximation. Consider the pictures in Fig. 2. The one on the left depicts a meson made of a valence quark and anti-quark, bound by a shmear of gluons. The gluons can also create virtual quark-antiquark pairs, leading to the picture on the right. These are computationally very costly. The quenched approximation omits them but compensates the omission with a shift in the bare couplings. This is analogous to a dielectric, where

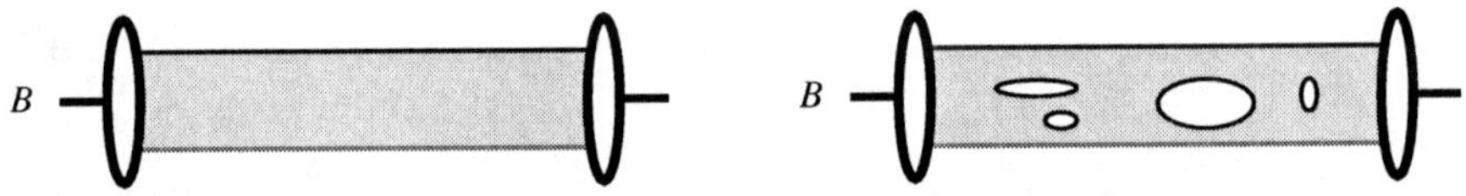

FIGURE 2. Quenched approximation: processes on the left are incorporated "exactly", whereas processes on the right are not computed, but modeled with a shift in the bare couplings.

$g_0^2 \to g_0^2\varepsilon$. Quenching retains many effects, such as retarded gauge potentials, that are omitted in, say, the quark model. It is also the first term in a systematic expansion [6].

The quenched approximation can fall short of reproducing chiral logarithms of the form $\ln(\Lambda^2/m_q^2)$ [7, 8, 9, 10]. Figure 3 shows some quark-line configurations that generate, at the hadronic level, meson loops. The quenched approximation includes (b), but not (a) or (c). As a consequence, some pion loops are omitted, and η' loops are mistreated. A better situation is a "partially quenched" theory, with dynamical quark loops, but, possibly, $m_{\text{valence}} \neq m_{\text{sea}}$. Then features like the η'-π splitting emerge, and it is possible to relate the partially quenched theory to full QCD using chiral perturbation theory. We shall return to chiral logs in the next section, when discussing B^0-$\bar{B}^0$ mixing.

To understand lattice spacing effects Symanzik suggested matching lattice gauge theory to continuum QCD [12]:

$$\mathcal{L}_{\text{lat}} \doteq \mathcal{L}_{\text{QCD}} + \sum_i a^{s_i} K_i(g^2, m_q a; \mu) O_i(\mu). \tag{2}$$

The right-hand side is a local effective Lagrangian (LE$\mathcal{L}$) renormalized in some continuum scheme; details of the scheme are not important. Discretization effects are, of course, short-distance effects, so as usual in an effective field theory, they are lumped into the coefficients K_i.

There are two key points to the Symanzik effective field theory. First, if Λa is small enough, the $\sum_i$ can be treated as a perturbation. For example, for the proton mass

$$m_p(a) = m_p - aK_{\sigma F}(c_{\text{SW}})\langle p|\bar{\psi}\sigma \cdot F\psi|p\rangle \tag{3}$$

using the leading term for Wilson fermions as an example. Here c_{SW} is the so-called clover coupling, and $K_{\sigma F} = (1 - c_{\text{SW}})/4 + O(\alpha_s)$. Second, to reduce lattice spacing effects, one can tune c_{SW} so that $K_{\sigma F}$ vanishes or, in practice, is $O(\alpha_s^\ell)$ or $O(a)$. Thus, the Symanzik effective field theory shows that lattice artifacts can be reduced through short-distance process-independent methods. Indeed, for light hadrons a combination of this Symanzik improvement and extrapolation in a^2 gives continuum QCD results with very well understood uncertainties (modulo quenching).

With the bottom and charmed quarks, the mass is large in lattice units: $m_b a \sim$1–2 and $m_{ch} a$ about a third of that. The split in the LE$\mathcal{L}$ between QCD and small correction breaks down, because there are terms in the $\sum_i$ in Eq. (3) that go like $(m_Q a)^n$. It will not be possible to reduce a enough to make $m_b a \ll 1$ for a long time: Moore's Law suggests 15–25 years. There are, nevertheless, several ways to treat heavy-light hadrons in lattice calculations, all of which appeal to HQET in some way. Table 1 lists the most

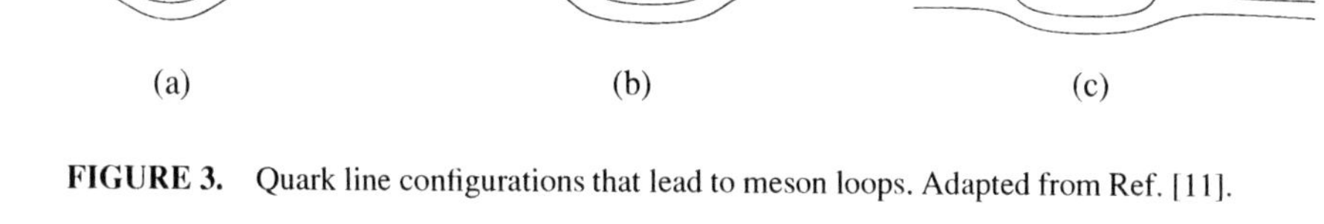

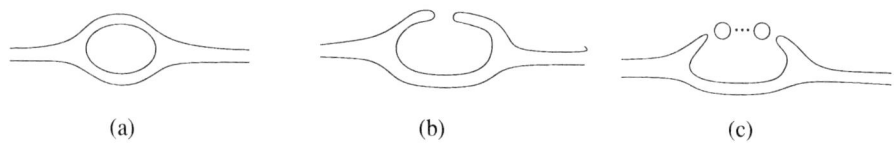

FIGURE 3. Quark line configurations that lead to meson loops. Adapted from Ref. [11].

TABLE 1. Widely used methods for heavy quarks in lattice gauge theory.

#	method	Ref.	how HQET enters
1.	static approximation	[13, 14]	$\mathcal{L}_{\text{static}} = -\bar{h}D_4h$
2.	lattice NRQCD	[15, 16]	discretization of first few terms of HQET
3.	extrapolation from below charm	*ad hoc*	results with small m_Q fit to Taylor series in $1/m_Q$
3′.	3 + 1	*ad hoc*	as in method 3
4.	"Fermilab"	[17, 18]	match lattice to QCD, term-by-term in HQET

widely used methods. In lattice NRQCD and in the Fermilab method, it is possible to set $m_Q = m_b$, even when $a^{-1} \sim m_b$. There are, of course, uncertainties involved, but a grasp of the basics of heavy-quark theory is enough understand them.

A convenient way to contrast the uncertainties in the various methods is to match lattice gauge theory to (continuum) HQET [18]. Instead of Eq. (3) one writes

$$\mathcal{L}_{\text{lat}} \doteq \mathcal{L}_{\text{HQET}} = \sum_n C_n^{\text{lat}} O_n = -m_1\bar{h}h - \bar{h}D_4h + \frac{\bar{h}\boldsymbol{D}^2h}{2m_2} + \frac{\bar{h}i\boldsymbol{\Sigma}\cdot\boldsymbol{B}h}{2m_\mathcal{B}} + \cdots. \quad (4)$$

The operators O_n on the right hand side are the same as in the HQET description of continuum QCD. The short-distance coefficients C_n^{lat}, on the other hand, are not the same, because there are two short distances, a and $1/m_Q$. Heavy-quark lattice artifacts are, thus, isolated into the mismatch $C_n^{\text{lat}} - C_n^{\text{cont}}$. In lattice NRQCD and in the Fermilab method, these mismatches can be reduced, along the same lines as reducing $K_{\sigma F}$ in the Symanzik program. In recent work, they are controlled to several percent or less.

Most work with method 3 has chosen normalization conditions for which the mismatch in the kinetic and chromomagnetic terms, although formally $O(m_Qa)^2$, is quite large in practice. It is also not well understood how this mismatch is amplified when extrapolating linearly or quadratically in $1/m_Q$ from $m_Q \sim 1$ GeV up to m_b.

LATTICE CALCULATIONS

We now turn our attention to some of the most interesting recent lattice calculations in B physics. The discussion focuses on the error bars, and central values are deferred to the next section. We consider the ratio ξ for B-$\bar{B}$ mixing in unquenched QCD, and the semi-leptonic decays $B \to \pi l \nu$ and $B \to D^* l \nu$ for $|V_{ub}|$ and $|V_{cb}|$ in quenched QCD.

Neutral Meson Mixing

The mass difference of neutral B meson CP eigenstates is

$$\Delta m_{B_q^0} = \frac{G_F^2 m_W^2 S_0}{16\pi^2} |V_{tq}^* V_{tb}|^2 \eta_B \langle \bar{B}_q^0 | Q_q^{\Delta B=2} | B_q^0 \rangle + \text{new physics?} \quad (5)$$

where $q \in \{d, s\}$, S_0 is an Inami-Lim function, η_B is a short-distance QCD correction, and $Q_q^{\Delta B=2}$ is a four-quark operator in the $\Delta B = 2$ electroweak Hamiltonian. Note that new physics could compete with the Standard Model. The matrix element of $Q_q^{\Delta B=2}$ is usually written

$$\langle \bar{B}_q^0 | Q_q^{\Delta B=2} | B_q^0 \rangle = \frac{8}{3} m_{B_q}^2 f_{B_q}^2 B_{B_q}, \tag{6}$$

but lattice QCD gives $\langle \bar{B}_q^0 | Q_q^{\Delta B=2} | B_q^0 \rangle$ and f_{B_q} (from $\langle 0 | A_{bd}^{\mu} | B_q^0 \rangle$) individually. This traditional separation turns out to be useful when looking at chiral logs.

Conventional wisdom says that the uncertainties in B_{B_q} and in $\xi^2 = f_{B_s}^2 B_{B_s} / f_{B_d}^2 B_{B_d}$ are "easy" to control, because they are ratios. At *Lattice '01*, Norikazu Yamada of JLQCD reported on new results [19] that suggest otherwise. The JLQCD collaboration is mounting a large project to carry out calculation with the loops of $n_f = 2$ flavors of quarks. Figure 4 compares their previous quenched work for f_{B_q} with their new (and still preliminary) work with $n_f = 2$. In the spirit of a blind analysis, the definition of $\Phi_{f_{Bq}}$ is not so important. The important matter is whether the curve is a straight line, or whether there is any curvature as a function of the light quark mass. (Because the pseudoscalar $m_\pi^2 \propto m_q$, the horizontal axis is, essentially, the light quark mass.)

The quenched approximation (diamonds in Fig. 4) shows no evidence of curvature. With $n_f = 2$ (triangles in Fig. 4), however, allowing for curvature obtains a better fit. In fact, curvature is expected from chiral logarithms. Meson loops with fixed spectator mass m_s and varying sea quark mass m_f contribute to f_{B_s} [20]

$$\Delta f_{B_s} = c_0 + c_1 m_{\eta_{ff}}^2 - \frac{1+3g^2}{(4\pi f_\pi)^2} m_{\eta_{sf}}^2 \ln\left(\frac{m_{\eta_{sf}}^2}{m_{\eta_{ss}}^2}\right), \tag{7}$$

where $\eta_{qq'}$ denotes a pseudoscalar meson with constituents q and q'. From the D^* width, $3g^2 \approx 1.04$ [21]. The second plot in Fig. 4 shows that Eq. (7) describes the points well.

It is worth making a few remarks about this kind of analysis. The formula for f_{B_d} (where the spectator and sea quarks have the same mass) is slightly different, so it is difficult to display all chiral log effects on one plot. Nevertheless, one can see from Fig. 4 that the effect is to increase f_{B_s}/f_{B_d}. The chiral logs for B_{B_q} are multiplied by

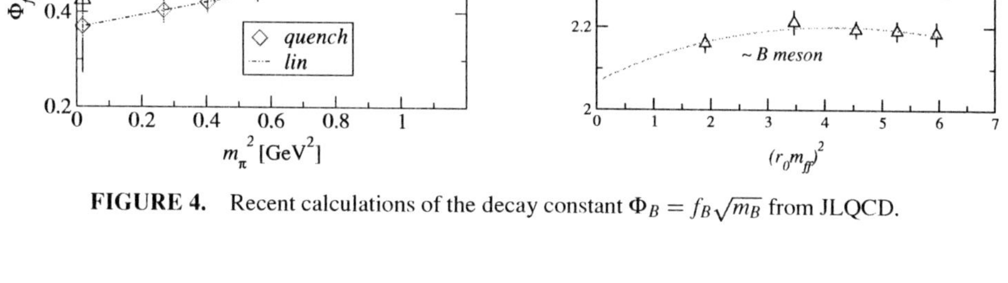

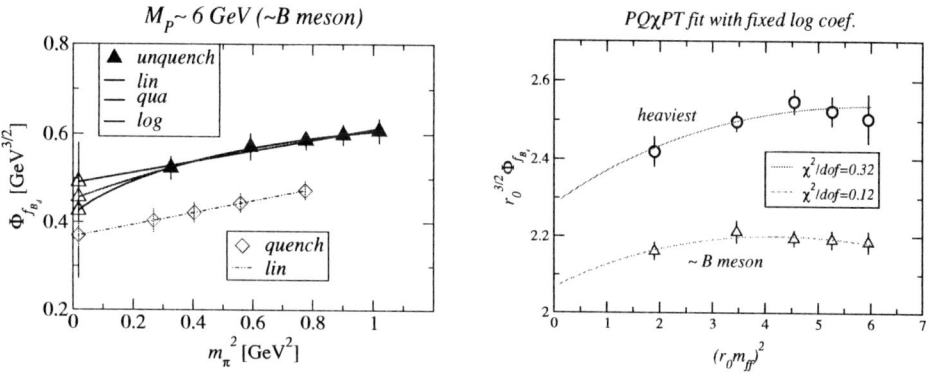

FIGURE 4. Recent calculations of the decay constant $\Phi_B = f_B\sqrt{m_B}$ from JLQCD.

$1 - 3g^2 \approx 0.04$, so the B_{B_s}/B_{B_d} is almost insensitive to them. Therefore, the chiral logs in the ratio ξ come almost completely from f_{B_s}/f_{B_d}. One concludes that ξ may have been underestimated in the quenched approximation. A new estimate, based on JLQCD and previous work, is given below.

$B \to \pi l \nu$ and V_{ub}

The semi-leptonic decay $B \to \pi l \nu$ is mediated by a $b \to u$ transition. The decay rate is

$$\frac{d\Gamma}{dE} = \frac{G_F^2 m_B p^3}{12\pi^3} |V_{ub}|^2 |f_+(E)|^2, \tag{8}$$

where $v = p_B/m_B$ is the B meson's velocity, $E = v \cdot p_\pi$, and $p^2 = E^2 - m_\pi^2$. The form factor $f_+(E)$ is a linear combination of the form factors $f_\perp(E)$ and $f_\parallel(E)$, defined through the matrix element of the vector current

$$\langle \pi | V^\mu | B \rangle = \sqrt{2m_B}\left[v^\mu f_\parallel(E) + p_\perp^\mu f_\perp(E)\right]. \tag{9}$$

The pion energy E is related to $q^2 = m_B^2 + m_\pi^2 - 2m_B E$. Chiral symmetry and heavy-quark symmetry are simpler to follow with $f_\parallel$ and $f_\perp$ considered as functions of E, rather than f_+ considered as a function of q^2. For example, heavy-quark symmetry suggests relations between B and D decays with the same E. In principle, one would like to compare the E dependence of experimental measurements with lattice calculations. Discretization uncertainties are smallest for low E, where phase space suppresses the event rate. Therefore, lattice calculations and experimental measurements will have to be combined in the way that minimizes the error on $|V_{ub}|$.

In the past year or so, four new quenched calculations of $f_+(E)$ for $B \to \pi l \nu$ appeared, using several different methods [22, 23, 24, 25]. It is, thus, timely to compare and contrast. In addition to discretization effects at large pa, there is evidence for considerable dependence on the light spectator quark mass [24].

The calculations of UKQCD [22] and APE [23] use method 3, so they have discretization errors of order $(m_Q a)^2$. They keep $m_Q < 1.2 m_c$ and extrapolate up to m_b linearly or quadratically in $1/m_Q$. The quark masses used in Ref. [22] are shown in Table 2. Those in Ref. [23] are a bit larger, which is better for HQET but worse for the discretization errors. In these papers models are used to extrapolate to the full kinematic range of E. It is better to think of these calculations as computing the model parameters in D decays, and then invoking HQET to make predictions for B decays. The error associated with the assumptions are, at least to me, not transparent.

TABLE 2. Heavy quark masses used in Ref. [22]. Note that $m_{\text{pole}} < m^{\text{RI}}$.

κ	am_Q	m_Q^{RI} (GeV)	M_P (GeV)
0.1200	0.485	1.52	2.035(5)
0.1233	0.374	1.23	1.771(5)
0.1266	0.268	1.02	1.483(5)
0.1299	0.168	0.69	1.157(5)

The calculation of El-Khadra *et al.* [24] uses the Fermilab method, and the calculation of JLQCD [25] uses lattice NRQCD. Since the matching procedures build in the heavy-quark expansion, both calculate directly at $m_Q = m_b$. To avoid introducing models of the

E dependence, these two papers advocate comparing lattice and experiment in a region where the two overlap. For example, one can look at

$$T_B(E_{\rm max}) = \int_{m_\pi}^{E_{\rm max}} dE\, p^3|f_+(E)|^2 = \frac{1}{|V_{ub}|^2} \frac{12\pi^3}{G_F^2 m_B} \int_{m_\pi}^{E_{\rm max}} dE\, \frac{d\Gamma}{dE}, \tag{10}$$

where $E_{\rm max}$ is an upper kinematic cut. At present $E_{\rm max} \sim 1$ GeV, but, with future increases in computer power, it may be possible to raise the cut. This method is sketched in Fig. 5 using the form factor from Ref. [24]. Uncertainties are still about 15–20% on $|V_{ub}|$. A strategy for reducing them are given in Ref. [24].

Ref. [24] finds strong dependence on the light quark mass, as illustrated for $f_\perp(0.7\text{ GeV})$ in Fig. 5. This is the only calculation to reduce the light quark down to $m_q = 0.4m_s$. The extrapolation to physical quark mass is the largest source of uncertainty quoted by Ref. [24], and a similar uncertainty is presumably present in all four calculations. One would like even lighter quarks, which take more CPU time.

Because of the discretization errors at higher E, the large light quark mass dependence, etc., a sanity check on $f_+(E)$ for $B \to \pi l\nu$ would be welcome. The calculation of similar form factors in D decays encounters many of the same issues. A compelling test would be to compare lattice and experiment for $f_+^{D\to K}/f_{D_s}$ and $f_+^{D\to\pi}/f_D$, as a function of E. New physics is unlikely to alter the underlying processes, and the CKM matrix drops out. The CLEO-c program [26] promises to measure these ratios with an uncertainty of a few percent.

$B \to D^* l\nu$ and V_{cb}

The exclusive semi-leptonic decay $B \to D^* l\nu$ is a good way to determine $|V_{cb}|$, but an estimate of the hadronic transition is needed. The differential decay rate is

$$\frac{d\Gamma}{dw} = \frac{G_F^2}{4\pi^3}\sqrt{w^2-1}m_{D^*}^3(m_B - m_{D^*})^2 \mathcal{G}(w)|V_{cb}|^2|\mathcal{F}_{B\to D^*}(w)|^2, \tag{11}$$

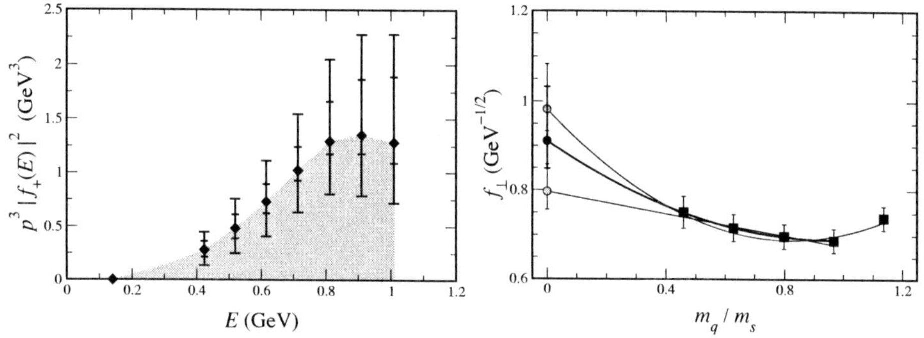

FIGURE 5. $B \to \pi l\nu$: (a) Integrand of Eq. (10). (b) Dependence of the form factor $f_\perp$ on the light quark mass, in the range $0.4m_s < m_q < 1.2m_s$.

where $w = v \cdot v'$ and $\mathcal{G}(1) = 1$. At zero recoil ($w = 1$) heavy-quark flavor symmetry forbids terms of order $1/m_Q$. The exclusive determination of $|V_{cb}|$ therefore proceeds by extracting $|V_{cb}\mathcal{F}_{B\to D^*}(1)|$ from a fit to the measurement of $d\Gamma/dw$, and then taking a theoretical estimate of $\mathcal{F}_{B\to D^*}(1)$. One would prefer the latter not to depend on models.

To see what is needed from the theoretical calculation, let us review the anatomy of $\mathcal{F}_{B\to D^*}(1)$. From HQET one can show that, at zero recoil,

$$\mathcal{F}_{B\to D^*}(1) = \eta_A \left[1_{\text{Isgur-Wise}} + 0_{\text{Luke}}/m + \delta_{1/m^2} + \delta_{1/m^3}\right] \tag{12}$$

where η_A is a short-distance radiative correction, and the δ_{1/m^n} contain the long-distance properties of the bound states. The numerical values of η_A and the δ_{1/m^n} depend on how HQET is renormalized. It is less important *which* renormalization scheme is chosen than it is to use the *same* scheme for both. The $1/m_Q^n$ corrections take the form

$$\delta_{1/m^2} = -\frac{\ell_V}{(2m_c)^2} + \frac{2\ell_A}{(2m_c)(2m_b)} - \frac{\ell_P}{(2m_b)^2}, \tag{13}$$

$$\delta_{1/m^3} = -\frac{\ell_V^{(3)}}{(2m_c)^3} + \frac{\ell_A^{(3)}\Sigma + \ell_D^{(3)}\Delta}{(2m_c)(2m_b)} - \frac{\ell_P^{(3)}}{(2m_b)^3}, \tag{14}$$

where $\Sigma = 1/(2m_c) + 1/(2m_b)$ and $\Delta = 1/(2m_c) - 1/(2m_b)$.

With brute force alone, a sufficiently precise calculation of $\mathcal{F}_{B\to D^*}(1)$ lies beyond reach. (See Ref. [27] for details.) Hashimoto *et al.* [27] have devised a method to extract all ℓs in Eq. (13) and all but $\ell_D^{(3)}$ in Eq. (14). The key is the heavy-quark mass dependence of three double ratios of matrix elements. As in Eqs. (12)–(14), HQET implies

$$\frac{\langle D|\bar{c}\gamma^4 b|B\rangle\langle B|\bar{b}\gamma^4 c|D\rangle}{\langle D|\bar{c}\gamma^4 c|D\rangle\langle B|\bar{b}\gamma^4 b|B\rangle} = \eta_V^2 \left[1 - \Delta^2\left(\ell_P + \ell_P^{(3)}\Sigma\right)\right]^2, \tag{15}$$

$$\frac{\langle D^*|\bar{c}\gamma^4 b|B^*\rangle\langle B^*|\bar{b}\gamma^4 c|D^*\rangle}{\langle D^*|\bar{c}\gamma^4 c|D^*\rangle\langle B^*|\bar{b}\gamma^4 b|B^*\rangle} = \eta_V^2 \left[1 - \Delta^2\left(\ell_V + \ell_V^{(3)}\Sigma\right)\right]^2, \tag{16}$$

$$\frac{\langle D^*|\bar{c}\gamma^j\gamma_5 b|B\rangle\langle B^*|\bar{b}\gamma^j\gamma_5 c|D\rangle}{\langle D^*|\bar{c}\gamma^j\gamma_5 c|D\rangle\langle B^*|\bar{b}\gamma^j\gamma_5 b|B\rangle} = \frac{\eta_{A^{cb}}\eta_{A^{bc}}}{\eta_{A^{cc}}\eta_{A^{bb}}} \left[1 - \Delta^2\left(\ell_A + \ell_A^{(3)}\Sigma\right)\right]^2. \tag{17}$$

In particular, heavy-quark spin symmetry requires the same ℓs to appear in Eqs. (15)–(17) as in Eqs. (13) and (14).

In a lattice calculation of these double ratios most of the statistical and sytematic uncertainties cancel. The main difference from continuum QCD is that the short-distance coefficients are different [18, 27]. But, just like in continuum QCD, the short-distance behavior can be computed in perturbation theory. Thus, the analysis proceeds by removing the lattice short-distance contribution, fitting to the prediction of HQET, and then reconstituting $\mathcal{F}_{B\to D^*}(1)$ from the ℓs and η_A. The result is

$$\mathcal{F}_{B\to D^*}(1) = 0.9??^{+0.024}_{-0.017} \pm 0.016^{+0.003+0.000+0.006}_{-0.014-0.016-0.014}. \tag{18}$$

To encourage the reader to think about the uncertainties, the central value is not revealed until the end. The error bars come from, in order, statistics and fitting, matching lattice gauge theory with HQET to QCD, lattice spacing dependence, light quark mass effects, *and* the quenched approximation. The secret to the small error on $\mathcal{F}$ is that they all scale as $\mathcal{F} - 1$, by design. As a fraction of $\mathcal{F} - 1$ the uncertainties are still sizable: 5–25%.

The most novel aspect of this analysis is how seriously it takes the idea of matching lattice gauge theory to QCD through HQET. A central part of the analysis is to calculate short-distance properties, which Ref. [27] does partly in perturbation theory. The second error bar reflects the associated uncertainty. It is reducible, but through traditional theoretical physics, rather than intensive computation.

AN OCTOPUS'S GARDEN

Although it does not bear directly the unitarity triangle, I would like to discuss some recent work by Nathan Isgur. I did not know Nathan well, but his enthusiasm for physics, especially the strong interactions, made a big impression on me. He struck me as the kind of person who had a hand all sorts of things. He must have needed eight arms to keep all his projects moving.

In the last year of his life, one of his projects was to understand whether instanton-like gauge fields or, on the other hand, disordered gauge fields play a larger role in QCD [28, 29, 11]. This work reflected Nathan's broad knowledge of the phenomenological and theoretical sides of the strong interactions, touching on the OZI rule, chiral zero modes, the η' mass, instantons, the quark model, the AdS/CFT correspondence, lattice QCD, and the large N_c limit. A particularly striking passage notes first how recent work by Witten on the AdS/CFT correspondance favors disordered, confining gauge fields (although those sentences were probably written by Nathan's collaborator Hank Thacker), and then how details of how to treat strong decays in the quark model favor the disorder also. That part was *certainly* written by Nathan.

These issues surround the quantum ground state, or vacuum, of QCD. Gauge theories have many classical ground states, one for each integer. Instantons are the semi-classical configurations that tunnel from one ground state to another. In a quantum mechanical situation, fluctuation-dominated configurations could also mediate tunneling. Nathan proceeded by devising tests that could distinguish whether the quantum vacuum is obtained via such disordered gauge fields or via a gas or liquid of instantons. He then used (quenched) lattice QCD to see which way the gauge fields behaved.

One test stemmed from the observation, from empirically successful quark models of strong decays, that quark pairs pop out of the vacuum with scalar (*i.e.*, $^3P_0$) quantum numbers. That implies that the OZI rule should fail for $J^{PC} = 0^{++}$ quantum numbers. (The OZI rule says that decays, like $\phi \to \pi\pi$, in which quarks annihilate, are weaker than those, like $\phi \to KK$, in which they do not.) Empirically, the OZI rule fails in the 0^{-+} sector; that is how the η' mass is split away from the other pseudoscalar mesons. For the 0^{++} sector, however, there is a dearth of experimental data. Isgur and Thacker's idea to to compute the quark annihilation process with lattice QCD, for various J^{PC}, which is possible at leading order in the OZI approximation even in quenched QCD [28]. They found a small amplitude where the OZI rule holds empirically, *e.g.*, for $J^{PC} = 1^{-+}$. But for scalar and pseudoscalar channels the amplitude is large. This is evidence against an instanton-dominated vacuum, because instantons couple to quarks through the axial anomaly, that is, preferentially to the pseudoscalar channel.

UNBLINDING

Instead of a paragraph of bland conclusions, the reward for readers who have made it this far consists of the most interesting numerical results. The preliminary results of JLQCD on B physics with $n_f = 2$ are [19]

$$f_{B_d} = 190(14)(7)(19)\,\text{MeV}, \tag{19}$$
$$f_{B_s}/f_{B_d} = 1.184(26)(20)(15), \tag{20}$$
$$\xi = 1.183(27)(20)(15), \tag{21}$$

where ξ is a combination of their results for f_{B_s}/f_{B_d} and B_{B_s}/B_{B_d}. In her review at *Lattice '01* [30], Sinéad Ryan took stock of the ratio ξ, which enters into UT fits. Although JLQCD's result is still preliminary, it cannot be denied that the chiral logarithms could affect f_{B_s}/f_{B_d} and, hence, ξ. Ryan's average, which strikes me as reasonable, is

$$\xi = 1.15 \pm 0.06^{+0.07}_{-0.00} \tag{22}$$

which retains the central value in common usage, but allows for a future upward revision, once the chiral behavior is fully understood and controlled.

For the zero-recoil form factor needed to determine $|V_{cb}|$ from $B \to D^* l\nu$, Ref. [27] finds

$$\mathcal{F}_{B\to D^*}(1) = 0.913^{+0.024+0.017}_{-0.017-0.030}, \tag{23}$$

with systematics added in quadrature. Note that this result does *not* include the QED correction of +0.007, which is included, for example, in the *BaBar* Physics Book. This result still relies on the quenched approximation. Nevertheless, it is probably still under better control than estimates based on the quark model.

A hallmark of these two results, shared by some work on $B \to \pi l\nu$ for $|V_{ub}|$ [24, 25], is that they attempt a full analysis of the uncertainties, including those of heavy quarks, the chiral behavior, and quenching. Thus, they are suitable templates for the next round of lattice calculations, from which one can expect serious *unquenched* calculations with a direct impact on our knowledge of the unitarity triangle.

ACKNOWLEDGMENTS

It is a pleasure to thank Shoji Hashimoto, Hank Thacker, and Norikazu Yamada for helpful correspondance while preparing this talk. The organizers and the staff at CalTech managed the symposium superbly, at a time when we were all distracted and appalled by world events. Fermilab is operated by Universities Research Association Inc., under contract with the U.S. Department of Energy.

REFERENCES

1. Kronfeld, A. S., "B and D mesons in lattice QCD", in *30th International Conference on High-Energy Physics*, edited by C. S. Lim and T. Yamanaka, World Scientific, 2001, hep-ph/0010074.
2. Beneke, M., talk at XIX International Symposium on Lattice Field Theory, URL `http://www.ifh.de/~latt2001/plenary-transp.html`.
3. Ali Khan, A., et al. [CP-PACS Collaboration], *Phys. Rev.* **D64**, 034505 (2001) [hep-lat/0010009].
4. Ali Khan, A., et al. [CP-PACS Collaboration], *Phys. Rev.* **D64**, 054504 (2001) [hep-lat/0103020].
5. Bernard, C., et al., *Nucl. Phys. Proc. Suppl.* **94**, 346 (2001) [hep-lat/0011029]; hep-lat/0110072.
6. Sexton, J., and Weingarten, D., *Phys. Rev.* **D55**, 4025 (1997).
7. Morel, A., *J. Phys. (France)* **48**, 1111 (1987).
8. Sharpe, S. R., *Phys. Rev.* **D41**, 3233 (1990).
9. Bernard, C. W., and Golterman, M. F. L., *Phys. Rev.* **D46**, 853 (1992) [hep-lat/9204007].
10. Bernard, C. W., and Golterman, M. F. L., *Phys. Rev.* **D49**, 486 (1994) [hep-lat/9306005].
11. Bardeen, W., Duncan, A., Eichten, E., Isgur, N., and Thacker, H., hep-lat/0106008.
12. Symanzik, K., *Nucl. Phys.* **B226**, 187 (1983); 205 (1983).
13. Eichten, E., "Heavy Quarks on the Lattice", in *Field Theory on the Lattice*, edited by A. Billoire et al., North-Holland, 1988.
14. Eichten, E., and Hill, B., *Phys. Lett.* **B234**, 511 (1990).
15. Caswell, W. E., and Lepage, G. P., *Phys. Lett.* **B167**, 437 (1986);
    Lepage, G. P., and Thacker, B. A., "Effective Lagrangians for Simulating Heavy Quark Systems", in *Field Theory on the Lattice*, edited by A. Billoire et al., North-Holland, 1988.
16. Thacker, B. A., and Lepage, G. P., *Phys. Rev.* **D43**, 196 (1991).
17. El-Khadra, A. X., Kronfeld, A. S., and Mackenzie, P. B., *Phys. Rev.* **D55**, 3933 (1997) [hep-lat/9604004].
18. Kronfeld, A. S., *Phys. Rev.* **D62**, 014505 (2000) [hep-lat/0002008].
19. Yamada, N., et al. [JLQCD Collaboration], hep-lat/0110087.
20. Sharpe, S. R., and Zhang, Y., *Phys. Rev.* **D53**, 5125 (1996) [hep-lat/9510037].
21. Anastassov, A., et al. [CLEO Collaboration], hep-ex/0108043.
22. Bowler, K. C., et al. [UKQCD Collaboration], *Phys. Lett.* **B486**, 111 (2000) [hep-lat/9911011].
23. Abada, A., et al., hep-lat/0011065.
24. El-Khadra, A. X., Kronfeld, A. S., Mackenzie, P. B., Ryan, S. M., and Simone, J. N., *Phys. Rev.* **D64**, 014502 (2001) [hep-ph/0101023].
25. Aoki, S., et al. [JLQCD Collaboration], *Phys. Rev.* **D64**, 114505 (2001) [hep-lat/0106024].
26. Briere, R. A., et al., "CLEO-c and CESR-c: A New Frontier of Weak and Strong Interactions", CLNS-01-1742;
    Shipsey, I. P. J., these proceedings.
27. Hashimoto, S., Kronfeld, A. S., Mackenzie, P. B., Ryan, S. M., and Simone, J. N., hep-ph/0110253.
28. Isgur, N., and Thacker, H. B., *Phys. Rev.* **D64**, 094507 (2001) [hep-lat/0005006].
29. Horvath, I., Isgur, N., McCune, J., and Thacker, H. B., hep-lat/0102003.
30. Ryan, S., hep-lat/0111010.

Summary and Outlook for 9th International Symposium on Heavy Flavor Physics

Helen R. Quinn

Stanford Linear Accelerator Center, Menlo Park, CA 94025

Abstract. This is the summary talk of a meeting held at the California Institute of Technology Sept 10–13, 2001. I do not attempt to summarize all the beautiful experimental results we have seen this week, nor to repeat the lively theoretical discussions that have occurred. Rather I will present my own biased perspective on what we have learned, and on the important tasks that need our attention as we work to make the most of the rapidly accumulating data in this field.

I want to begin by thanking our hosts for presenting a well-organized and very effective meeting in a week where such tasks were not easy. Given the dreadful events of last Tuesday we were all a little distracted. None of us will forget where we were in this week, nor that we were well taken care of here. Despite the disaster (which, as one small side-effect, shut down Caltech for the day) a new location for our meeting on Tuesday was found, new arrangements to feed us made quickly and smoothly, and our meeting carried on. The technology continued to function, the talks were all given. We all appreciate the excellent hospitality we were given throughout the week, and the support we are now offered as we all try to figure out how we will get home. We owe a vote of thanks to David Hitlin, Frank Porter, Gregory Dubois-Felsmann, and Anders Ryd and their support staff for their efforts. They made it look easy, but I am sure it was not.

To turn to the physics, I start with a personal comment. I have greatly enjoyed being part of the process of development of the B factory and the BaBar experiment almost from the beginning. This was a fascinating and humbling experience for a theorist. I have seen how much the naive first estimates of what such a facility can achieve are transformed as the reality of building an actual experiment and analyzing real data take over from the back-of-the-envelope guesses with which we begin the process. On both the theory side and the experimental side we (and from here on we means not just BaBar collaborators but all working in this field) have learned where our first approximations are insufficient. The process could be disheartening, except that there are enough clever and determined people involved that somehow we continue to make progress. Despite the difficulties, we manage to maintain optimism and work hard enough that real progress is made. The results on both the theory side and the experimental side that we have seen this week are a proof of that. The job is far from done, but it is well begun. We have begun to unravel the physics of B meson decays.

I mention BaBar first only because it is there that I sit and so I have seen the process there first hand. The same hard work and persistence in other laboratories has also produced new and interesting results reported this summer. We have heard many results

CP618, *Heavy Flavor Physics: Ninth International Symposium,* edited by A. Ryd and F. C. Porter

this week (even a few actually new ones) and over this last summer. In B physics CLEO, Belle and BaBar are reporting results on some branching ratios as small as 10^{-6} [1], and both BaBar and Belle have reported evidence that the CP violating parameter sin (2β) (or sin $(2\phi_1)$ which is the same thing) is non-zero [2]. With current accuracy, the result is consistent with the range predicted by the Standard Model. Results on charm, strange and tau physics have also been reported here. I do not intend to summarize all these results, they speak for themselves in the many excellent presentations we have heard [3]. Ongoing experiments will bring more new results and we can expect improved precision on interesting measurements for some time to come.

So what have we learned from all these results? What more do we expect to learn by pursuing the physics of flavor over the next few years? We certainly have learned that persistence pays, and indeed is required to reach our goals. Both in theory and in experiment we must continue to work hard, for several more years at least, to truly pin down all the details of this sector. We have seen that in B-physics as in previous flavor physics there is no free lunch. The golden channel of ψK_s and its cousins ($b \to c\bar{c}s$ decays of B_d) is interesting. We were not lucky enough to find any challenge to the Standard Model in this first observation of CP violation in the neutral B meson decays. Since one of our major motivations for pursuing this physics is to test the Standard Model story on CP violation that simply means that we must continue with the rest of the program, implementing physics analyses of many more channels.

In our struggle to understand the physics of flavor and with it the physics of CP violation, for at least within the Standard Model the two are intimately linked, we have, once again, run into the fact that we cannot isolate quarks. This means that to study their decays we must inevitably also study some aspects of strong interaction hadronic physics. There are few "golden channels" in which the physics of interest to us for testing the Standard Model can be cleanly separated from the hadronic physics. Our calculational tools are then in need of help. The theory work advances steadily, but the net result is that the predictions we need to be able to make depend on quantities that we cannot directly calculate. We must therefore take a multichannel approach, where theory and experiment feed information to one another.

Theoretical predictions, once hadronic physics enters the picture, depend on some inputs that can only be obtained from measurement or models. Interpretation of measurements depend on input from theory. The process of learning is murky and iterative. In the rest of this talk I will illustrate this with some examples from the theory talks we have heard this week. I will draw a few morals about the way this game must be played if it is to succeed. Being an equal opportunity moralist I have advice for theorists and for experimentalists alike. In the end it comes down to the same advice—a plea for logical clarity and honest revelations about what is an input and what is an output in presenting any theory work or any experimental result. I will make myself clearer on this after the examples. I am going to say a lot of obvious things in this talk, at the risk of sounding preachy and naive. The reason I do this is because it seems sometimes that our tools become so sophisticated that we forget some obvious things. I guess I'm old enough to get away with being preachy on occasion.

My first example is the measurement of charmless semileptonic B decays and the extraction of V_{ub}. Here we heard of new theoretical work from Christain Bauer [4] and ideas on using the spectrum of $B \to s\gamma$ data as a way to measure some input parameters from Ira Rothstein [5]. The CLEO experiment has begun to implement some of these ideas [6]. You all know the problem. To extract V_{ub} we must measure charmless decays. The cuts on the data that must be introduced to remove the background from the much more common decays to charm bring with them a price for the theorists. The fraction of the data kept after any cuts must be calculated. This fraction is sensitive to details of the spectrum that may not be reliably predicted by a quark level calculation.

Christian Bauer showed us how a combination of cuts can be used to limit the sensitivity to this problem and suggested how one can also gain some tests of that sensitivity by varying the cut prescription. He advised minimizing sensitivity to theory. That is impossible; we are trying to extract a theory parameter. But what he really meant was minimizing sensitivity to the uncertainties in the theory that arise from soft physics. That can be done! I would further suggest that the data should be presented in two ways. The cuts may be tuned based on theory input, but the result should first be stated in as theory independent a fashion as possible. Only after that should the analysis that gives the theoretical parameter be introduced. In this case this separation is readily made. One has simply to quote a rate of events in a given kinematic region, or perhaps a table of such numbers for different choices of kinematic cuts. The second step, of turning that table into a best estimate for V_{ub}, is also needed. It should be kept separate. In this step theory and experiment become inextricably mixed together. The reason I plead for the first step, the presentation of data with no theory in the numbers, is that that is what will allow us to come back at a later date (even possibly after the collaboration presenting the results is long disbanded) and re-analyze the data with new theory inputs. Experiments should not become so theory-driven that they only present results for theoretically interesting quantities. I know there are cases where my advice is simply too naive, where there is essentially no way to present a theory-free set of numbers. It is just this fact that has led us into the bad habit of confounding the two steps, that of measurement and that of interpretation. The pattern is perhaps reinforced when experiments publish only as letters; there is not room in a letter for the two steps, and the second number is regarded as the real result of the measurement. In the short term this is true, in the long term the theory independent result has more staying power! My advice to experiments is to pay attention to whether or not they are forced into this procrustean bed, or whether they are allowing themselves to fall there simply because they have not tried to hard enough avoid it!

I want to return to the idea that Ira Rothstein [5] talked about, that of using one set of channels to measure some soft physics parameters and then using those parameters in interpreting another set of channels. This may sound obvious, it is in practice neither obvious nor simple to implement. The issue is that these parameters are not physical quantities in the technical sense—they are definition and convention dependent. This means that one must in fact do some higher order QCD calculation to understand how the spectrum in say radiative B decays is related to that in semileptonic B decays. This is ongoing work, and the "best" or even standard conventions for defining the relevant B physics parameters may not yet have emerged. This is a technical point, not a problem. However when experiments quote numbers for a theoretically-defined quantity, such as the parameter $\bar{\Lambda}$ of the heavy quark effective theory, they must at the same time tell us what definition of this parameter (or equivalently, in this case, of quark mass) they are using. Without such a definition the quantity is meaningless.

A second topic where we saw some interesting theoretical analysis was on the subject of $D\bar{D}$ mixing, long advertized by theorists as a good place to look for new physics effects. In the talk by Zoltan Ligeti [7] we heard how the naively predicted pattern of the operator product expansion contributions to x and y is possibly disrupted because of SU(3) breaking effects. Since the otherwise-leading operator is suppressed by SU(3) symmetry, the symmetry-breaking effects can be significant. Indeed they may alter the expected relative sizes of x and y. The size of expected Standard Model effects is also enhanced once SU(3) breaking is considered. So the lesson here is not to trust first rough estimates of theoretical uncertainties, these things always some study. However the effects here are still at most at the few percent level, so this remains a place to look for new physics effects.

No method can completely remove the issues of theoretical uncertainties. No matter what technical improvements are made, in the end we must are rely on some quark-level calculations and some version of quark-hadron duality. Now it is the job of the theorists to estimate how big the remaining uncertainties could be. Unfortunately there is often no rigorous approach that gives a definite answer. I believe that theorists must give skeptical answers here, but that it is important to try to be quantitative. It is just as bad to give an off-the-cuff overestimate of the uncertainties as it is to claim that everything is under control with an optimistic estimate of uncertainties.

So let us spend a little time considering how big can the violations of quark hadron duality be? What sets the scale of these effects? What level of averaging is needed to remove the detailed dependence on resonance masses in a spectrum and hence to get results that depend only on the underlying quark diagrams? To make this point I will digress a little and talk about a case where I know something about the answers, that is the rate of hadron production in e^+e^- annihilation. This is typically expressed as a ratio $R_{e^+e^-}$ of hadron to non-resonant muon-pair production [8]. One can prove quite formally that an integral over this ratio along the entire real physical cut can be given by the relevant quark-loop graph calculated in the deep Euclidean region. Indeed one can use the analytic properties of the quark graph to calculate an integral that weights a small segment of the cut. In the Euclidean region there are no small denominators from near-mass shell propagators and so the power series expansion is well behaved.

What limits the reliability of the calculation is that as one approaches a threshold on the physical cut certain propagators in the diagram give small denominator factors. We know which diagrams have to be summed to all orders to produce the onium resonances just below a new quark threshold. Right on resonance the violations of duality are huge if we do not make this resummation. We can avoid the resummation only if we calculate an integral that can be determined without the knowledge of the function close to the threshold. That is the essence of the "global" duality –it gives us a correct averaged cross-section, but not the detailed threshold and sub-threshold structure. On the other hand, once we are well above threshold we can approach quite close to the physical cut without any small denominators appearing in any diagram. Indeed we find that the averaged rate and the naive "local" quark calculation agree well in such a region. There are two main points that this example teaches us. The first is that the violations of local duality are very large if we are foolish enough to try to use it where some quark propagator goes on-shell, but quite small when we are far from any threshold. In a properly averaged quantity, the "duality violations" are small. The second point is that a careful enough examination of the quark diagrams revealed what went wrong with the perturbation series at the threshold, and hence indicated the range over which averaging was needed to avoid these problems. The size of duality violations is then controlled by the averaging scale compared to Λ_{QCD} (and also to light quark masses which are smaller).

These features then generalize to two questions. The first is how much do we need to average over hadronic mass spectra to ensure that corrections to a quark-hadron duality estimate are under control? The second is to find what effects can give large violations for the process in question, and what sets the scale of these effects in a suitably averaged quantity. In an ideal world we could give rigorous answers to these questions, in the real world different theorists reach different conclusions. For example for the case of the inclusive semileptonic decay rate to particle with charm, where no significant kinematic cuts are required to select events, these corrections have been discussed in some detail and are generally agreed to be be small. The range of lepton momenta achieves some averaging over the hadron mass spectrum. The argument among theorists is then over whether small means a few percent or of order 10^{-3}. Bigi and Uraltsev have argued that the latter number is appropriate [9]. Their argument is plausible; no one has shown specific effects which they have ignored. However this same type of argumentation leads to other results that are discrepant with experiment, for example the differences in the lifetimes of different b-containing mesons and baryons are larger than duality-based arguments would predict. The conundrum is then whether we are at the stage where we should say these are serious violations of Standard Model predictions, or simply that violations of quark-hadron duality are bigger than expected. Any skeptical observer takes the latter position. But the results certainly show how important it is to try to gain a better understanding of how to quantify such duality violations reliably. We may lose the ability to recognize many signals of new physics if we do not do better on this front. Theorists generally divide into those who have made the calculations and are willing to make estimates of the size of the breaking, and skeptics who say we cannot reliably estimate these effects. The skeptics typically cannot point to any specific error in the calculations that have been made. I admit that I have generally been among the latter group—but at this stage of the game it is an inadequate position to take. We theorists need to keep attempting to do better.

However, at the same time as I preach that skeptical theorists should take on this challenge, I warn the experimentalists that the most aggressive (optimistic) estimates of theoretical uncertainties are often not reliable. The problem is that the diseases of the quark-level perturbation theory calculation can be very subtle and do not always manifest themselves in an obvious way. For the case of B decays, we will only learn how to limit the size of duality violating effects by comparing predictions with data in many, many channels. In particular it is important to test the sensitivity of results to variation of any experimental cuts or changes in procedure. When there is significant sensitivity in the parameter extraction to cut-variation or to input models or assumptions then we know that we cannot trust the duality-based calculation. However, even when the sensitivity to variation in input is small we still cannot be certain that the effects of duality violations are small. This all leaves a very murky path to finding a distinction between new physics effects and unexpectedly large corrections to our calculations. I hope and expect that we will be able to find ways to separate these two things, but there are no guarantees.

One class of processes for which this process is now at a well-developed stage is the calculation of two-pseudo-scalar decay channels using qcd-improved factorization or the alternative formulation called perturbative qcd. We have all seen the papers and heard talks by two groups working on this topic. It has taken me some time to understand some of the details of this work. The results of the two groups are quite different, but it appears at first glance that they are pursuing the same methods. The methods both keep leading order terms in Λ_{QCD}/m_b and calculate the leading $\alpha_s(m_b)$ corrections. The groups get quite different results on some points. The problem is that even the conclusions on the power counting for the relative sizes of certain contributions are dependent on assumptions about the quark distribution function end-point behavior. In my opinion the fact that two independent groups have been working on this is important. We have seen that the estimates of theoretical uncertainties of one group have evolved due to the work of the other group. This is what we need, honest efforts to give good error estimates, and, to find what parts of those estimates are sensitive to assumptions, more than one group doing independent work on the same problem. This then leads to an iterative process that can eventually give us some confidence that we have a good estimate of theoretical uncertainties.

The group of Beneke *et al.* [10], represented in this meeting in the talk by Matthias Neubert, assume that the end point behavior is a power of x, and do not include any Sudakov suppression factor. They argue this effect does not play a significant role at the actual B mass scale. They then use data and models to give the needed input for matrix elements; for example, the transition matrix element for B to pseudoscalar meson is taken from semileptonic decay measurements. They make an honest effort to test a range of assumptions, for example for quark distribution functions, and see how their results vary as those input assumption are varied. This is the model we all must follow. But no matter how honest a group might be, it is important to also have another group or groups think independently about what input assumptions are reasonable. Only in this way can there be a serious discussion about whether the estimates of theory uncertainties are conservative or aggressive.

The group of Keum *et al.* [11], represented here by the talk by Y-Y. Keum, make a different set of assumptions. They include transverse-momentum dependence via Sudakov factor in the quark distribution functions. This suppresses the contribution which is the leading term in the Beneke *et al.* approach, so that in this so-called perturbative QCD approach the dominant term is an order $\alpha_s(m_b)$ hard-gluon exchange to the spectator quark. Then the inputs are light-cone quark distribution functions for the mesons, which they argue are calculable. I find that last statement quite doubtful (see the arguments on this subject in a paper that appeared shortly after this talk [12]). However one can simply regard these inputs as an alternate (and stronger) set of assumptions. One then must ask whether these are reasonable assumptions. They choose the parameters of the Sudakov term so that the scale of average transverse momenta in the quark distributions is $k_\perp^2 = m_b\Lambda_{QCD}$. This is an incorrect choice! Even for the B meson, and certainly for the pion, the transverse momenta should be scaled only by Λ_{QCD}. This choice has significant implications for their numerical results, which I therefore find suspect.

However even if their results are not yet reliable, there are some points that have been raised by these authors that have led to some revisions of the error estimates of the former group. This reinforces my statement that it is always useful to have two groups approaching the problem independently, to test assumptions about theoretical uncertainties. The major issue raised by the second set of calculations is the impact of, and control over, the (Λ_{QCD}/m_b)-suppressed contributions. The contribution of annihilation graphs is one such term. In the Beneke *et al.* formalism it turns out that this contribution is infrared singular and hence dependent on the cut-offs introduced to control this singularity. Recent papers of this group include an larger estimate of the uncertainties due to these terms than earlier papers (as far as I can tell). In the Keum *et al.* approach the infrared behavior of this contribution is softened by the $k_\perp$ dependence of quark propagators, and the term is found to be numerically significant. In particular it contributes to a large imaginary part, the size of which is important for any estimate of direct CP-violation. These numerical results depend on the incorrect scale for the average $k_\perp$ mentioned above, and so I do not trust them. However the fact that this contribution needs to be estimated carefully is an important point that was raised by this work. The public debate between the two groups has also helped those not directly involved to learn what are the critical issues. That is an important impact of having two sets of calculations.

This example provides a model for our path into the future. It is not possible to avoid all hadronic physics questions. We need theorists to tackle them and to make some assumptions in order to do so. We then must do a serious job of thinking about the range of input assumptions. Human nature being what it is, we must have more than one group tackle that job. That helps ensure a public discussion of all relevant issues. However much this is done, there will always be some remaining tension between those who have made the estimates and skeptics who doubt their methods. The history of the $\Delta I = 1/2$ enhancement and $\frac{\varepsilon'}{\varepsilon}$ provide a warning that favors the skeptics, but B-decays offer a new regime where more systematic expansions are available. There is ongoing progress on the theory of this regime.

So this effort must continue, both on the theory front in estimating uncertainties, and on the experimental front in presenting measurements in as theory-independent a fashion as possible. Then the clever application of theory to experimental data is needed, to provide the cleanest possible tests of the Standard Model. That requires close cooperation of theorists and experimentalists. Even after all the hard work is done the question remains whether we can be confident enough of our uncertainty estimates that we can be sure a discrepant measurement tells us there is new physics. The pessimistic view is that we will always be able to adjust the theory uncertainties to cover any measured results. The optimistic view is that eventually we will gain enough confidence in our methods to recognize true discrepancies if such exist. I suspect that any one result will not be convincing. It will take a pattern of discrepancies in fitting many channels to tell us that the Standard Model is failing. Both theorists and experiments still have much work to do. I have been impressed up till now by the persistence on both fronts in this work. This persistence has yielded steady progress, as we have seen in this meeting. I am hopeful that this progress will continue.

REFERENCES

1. w4.lns.cornell.edu/public/CLEO/, bsunsrv1.kek.jp/, www.slac.stanford.edu/BFROOT/www/Public/index.html
2. Aubert, B. *et al.* [BaBar Collaboration], *Phys. Rev. Lett.* **87**, 091801 (2001) [hep-ex/0107013]. Abe, K. *et al.* [Belle Collaboration], *Phys. Rev. Lett.* **87**, 091802 (2001) [hep-ex/0107061].
3. Proceedings of the 9th International Symposium on Heavy Flavor Physics California Institute of Technology, Sept 10-13 2001. (hereafter referred to as "these proceedings")
4. Bauer, C. W., Ligeti, Z. and Luke, M., hep-ph/0107074. See also talk by C. Bauer in these proceedings.
5. Leibovich, A. K., Low, I. and Rothstein, I. Z., *Phys. Lett. B* **513**, 83 (2001) [hep-ph/0105066]. See also talk by I. Rothstein in these proceedings.
6. Cronin-Hennessy, D. *et al.* [CLEO Collaboration], arXiv:hep-ex/0108033. See the talk by R. Briere in these proceedings.
7. Falk, A. F., Grossman, Y., Ligeti, Z. and Petrov, A. A., arXiv:hep-ph/0110317. See also the talk by Z. Ligeti in these proceedings.
8. Poggio, E. C., Quinn, H. R., and Weinberg, S., *Phys. Rev. D* **13**, 1958 (1976).
9. Bigi, I. I. and Uraltsev, N., hep-ph/0106346.
10. Beneke, M., Buchalla, G., Neubert, M. and Sachrajda, C. T., *Phys. Rev. Lett.* **83**, 1914 (1999) [hep-ph/9905312]. Beneke, M., Buchalla, G., Neubert, M. and Sachrajda, C. T., *Nucl. Phys. B* **606**, 245 (2001) [hep-ph/0104110].
11. Keum, Y. Y., Li, H., and Sanda, A. I., *Phys. Rev. D* **63**, 054008 (2001) [hep-ph/0004173].
12. Descotes-Genon, S. and Sachrajda, C. T., arXiv:hep-ph/0109260.

Participants

Brad Abbott
University of Oklahoma
440 W. Brooks St
Norman OK 73019
USA
babbott@fnal.gov

Jeffrey Appel
Fermilab
Mail Stop 122
P. O. Box 500
Batavia IL 60510
USA
appel@fnal.gov

Christian Bauer
UC San Diego
9500 Gilman Drive
La Jolla CA 92093
USA
bauer@einstein.ucsd.edu

Edmond Berger
Argonne National Laboratory
High Energy Physics Division
Argonne IL 60439
USA
berger@anl.gov

Stefano Bertolini
INFN and SISSA
Via Beirut 4
I-34014 Trieste
ITALY
bertolin@he.sissa.it

Gerard Bonneaud
Ecole Polytechnique Lab. de Physique
Nucleaire et de Hautes Engeries
Route de Saclay
F-91128 PalaiseauCEDEX
FRANCE
gerard.bonneaud@in2p3.fr

Adolf Bornheim
California Institute of Technology
M/C 256-48
Pasadena, CA 91125
USA
bornheim@hep.caltech.edu

Stefan Bosch
Max-Planck-Inst. fur Physik
Fohringer Ring 6
D-80805 Munich
GERMANY
bosch@mppmu.mpg.de

Thorsten Brandt
Techn. Univ. Dresden
Zellescher Weg 19
D-01189 Dresden
GERMANY
tbrandt@slac.stanford.edu

Roy Briere
Carnegie Mellon Phyiscs Dept.
5000 Forbes Ave.
Pittsburgh PA 15213
USA
briere@mail.lns.cornell.edu

Thomas Browder
University of Hawaii-Physics Dept.
2505 Correa Road
Honolulu HI 96822
USA
teb@phys.hawaii.edu

David Brown
LBNL
Mail Stop 50A-2160
Berkeley CA 94720
USA
Dave_Brown@LBL.gov

Gustavo Burdman
LBNL
1 Cyclotron Rd.
Berkeley CA 94720
USA
gaburdman@lbl.gov

Edward Chen
California Institute of Technology
Caltech M/C 356-48
Pasadena CA 91125
USA
edward@hep.caltech.edu

Chih-hsiang Cheng
Stanford University
Physics Department,
382 Via Pueblo Mall
Stanford CA 94305-4060
USA
chcheng@slac.stanford.edu

Harry Cheung
Fermilab
Wilson Road,
P.O. Box 500, MS 122
Batavia IL 60510-0500
USA
cheung@fnal.gov

Cheng-Wei Chiang
Argonne National Laboratory
and U. of Chicago
9700 South Cass Avenue,
Bldg. 362
Argonne IL 60439
USA
chengwei@hep.uchicago.edu

David Cinabro
Wayne State University Department of Physics
and Astronomy
666 W. Hancock
Detroit MI 48202
USA
cinabro@physics.wayne.edu

Carlo Dallapiccola
University of Massachusetts, Amherst
LGRT, Dept. of Physics,
Amherst MA 01003
USA
carlod@physics.umass.edu

Gregory Dubois-Felsmann
Caltech High Energy Physics,
M/C 356-48
Pasadena CA 91125-4800
USA
gpdf@hep.caltech.edu

Alexei Dvoretskii
California Institute of Technology
Caltech M/C 356-48
Pasadena CA 91125
USA
dvoretsk@hep.caltech.edu

Gerald Eigen
University of Bergen-Dept. of Physics
Allegaten 55
N-5007 Bergen
NORWAY
eigen@asfys2.fi.uib.no

Greg Feild
Yale University
1050 Terace Lake Drive
Aurora IL 60504
USA
feild@fnal.gov

Robert Fleischer
Deutsches Elektronen-Synchrotron
DESY Theorie-Gruppe
Notkestrasse 85
D-22607 Hamburg
GERMANY
Robert.Fleischer@desy.de

Sean Fleming
Carnegie Mellon University
Physics Department
Pittsburgh PA 15213
USA
fleming@kayenta.phys.cmu.edu

Christoph Grab
Institute for Particle Physics of ETH
Zurich/HPK F26
CH-8093 Zurich
SWITZERLAND
grab@particle.phys.ethz.ch

Andrei Gritsan
LBNL
1 Cyclotron Rd,
Mail Stop 50A-2160
Berkeley CA 94720
USA
gritsan@mh1.lbl.gov

Michael Gronau
Physics Department
Technion
Haifa 32000
ISRAEL
gronau@physics.technion.ac.il

Monika Grothe
UC Santa Cruz
SLAC, Mail Stop 59
2575 Sand Hill Road
Menlo Park CA 94025
USA
monika@slac.stanford.edu

Robert Hetrick
25629 Wilson Drive
Dearborn Heights MI 48127
USA
rhetrick@mediaone.net

Gudrun Hiller
SLAC, Mail Stop 81,
2575 Sand Hill Road
Menlo Park CA 94025
USA
ghiller@slac.stanford.edu

David Hitlin
California Institute of Technology
Caltech M/C 356-48
Pasadena CA 91125
USA
hitlin@hep.caltech.edu

Klaus Honscheid
The Ohio State University
174 W 18th Avenue
Columbus OH 43210
USA
kh@mps.ohio-state.edu

George W.S. Hou
National Taiwan U.
No. 1, Sec. 4 Roosevelt Road
Taipei, Taiwan 106
Republic of China
wshou@phys.ntu.edu.tw

Yee Bob Hsiung
Fermilab
Mail Stop 12, P.O. Box 500
Batavia IL 60510
USA
hsiung@fnal.gov

Alex Kagan
Physics Dept. ML-11
University of Cincinnati
Cincinnati, OH 45221
USA
kagan@physics.uc.edu

Michael Kelsey
SLAC
P.O. Box 4349
Stanford CA 94309
USA
kelsey@slac.stanford.edu

Yong-Yeon Keum
Institute of Physics, Academia Sinica
Nankang, Taipei
Taipei, Taiwan 11529
Republic of China
keum@phys.sinica.edu.tw

Heejong Kim
Dept. of Physics,
Yonsei Univ., Shin-chon dong
134, Seo-dae-mun gu
Seoul Korea 120-749
Korea
heejong@bsunsrv1.kek.jp

Andreas Kronfeld
Fermilab P.O. Box 500
Batavia IL 60510-0500
USA
ask@fnal.gov

Frank Kruger
Physik-Department T31
TU Munich
James-Franck-Str.
D-85748 Garching
GERMANY
fkrueger@ph.tum.de

Heiko Lacker
Laboratoire de l'Accelerateur Lineaire,
IN2P3-CNRS et Universite de Paris
F-91405 Orsay
FRANCE
lacker@lal.in2p3.fr

David Lange
LLNL
Mail Stop L-050
7000 East Ave.
Livermore CA 94550
USA
Lange6@llnl.gov

Adam Leibovich
Fermilab
Mail Stop 106, P.O. Box 500
Batavia IL 50610
USA
adam@fnal.gov

Zoltan Ligeti
Lawrence Berkeley Lab
Mail Stop 50A-5101
Berkeley CA 94720
USA
zligeti@lbl.gov

Elliot Lipeles
California Institute of Technology
Caltech M/C 256-48
Pasadena CA 91125
USA
lipeles@hep.caltech.edu

Harry J. Lipkin
Weizmann Institute of Science
Rehovot 76100
ISRAEL
lipkin@theory.hep.anl.gov

Lawrence Littenberg
Physics Department
Brookhaven National Laboratory
Upton NY 11973
USA
littenberg@bnl.gov

Ian Low
Harvard University
17 Oxford St.
Cambridge MA 02138
USA
ian@feynman.harvard.edu

David MacFarlane
UC San Diego
Department of Physics
9500 Gilman Dr.
La Jolla CA 92093
USA
dbmacf@slac.stanford.edu

Aneesh Manohar
UC San Diego
Department of Physics
9500 Gilman Drive
La Jolla CA 92093
USA
amanohar@ucsd.edu

Flavio Marchetto
Instituto di Fisica
via Pietro Giuria, 1
I-10125 Torino
ITALY
flavio.marchetto@cern.ch

Thomas Mehen
Ohio State
174 W. 18th Ave.
Columbus OH 43210
USA
mehen@mps.ohio-state.edu

Timothy Meyer
Stanford University
1318 Woodside Road
Redwood City CA 94061
USA
meyertim@stanford.edu

Hans-Günther Moser
Max-Planck-Inst. fur Physik
Fohringer Ring 6
D-80805 Munich
GERMANY
moser@mppmu.mpg.de

Roberto Mussa
INFN Torino
via P. Giuria 1
I-10126 Torino
ITALY
mussa@to.infn.it

Mikihiko Nakao
KEK-Physics Department
1--1 Oho, Tsukuba-shi
Ibaraki-ken 305-0801
JAPAN
mikihiko.nakao@kek.jp

Ilya Narsky
California Institute of Technology
M/C 356-48
Pasadena CA 91125
USA
narsky@hep.caltech.edu

Homer Neal
Yale University
P.O. Box 208120
New Haven CT 06520
USA
homer.neal@yale.edu

Matthias Neubert
Cornell University
318 Newman Lab
Ithaca NY 14853-5001
USA
neubert@mail.lns.cornell.edu

Hitoshi Ozaki
KEK, Physics Department
1-1 Oho, Tsukuba-shi
Ibaraki-ken 305-0801
JAPAN
hitoshi.ozaki@kek.jp

Stephen Pappas
California Institute of Technology
Caltech M/C 256-48
Pasadena CA 91125
USA
pappas@hep.caltech.edu

Fabrizio Parodi
Genoa University/I.N.F.N.
via Dodecaneso 33
I-16146 Genova
ITALY
Fabrizio.Parodi@ge.infn.it

Popat Patel
McGill University
Rutherford Physics Bldg.
3600 University
Montreal, OC H9X-2X4
CANADA
patel@hep.physics.mcgill.ca

Kevin Pitts
University of Illinois
1110 W. Green St.
Urbana IL 61801
USA
kpitts@uiuc.edu

Frank Porter
California Institute of Technology
Caltech M/C 356-48
Pasadena CA 91125
USA
fcp@hep.caltech.edu

Soeren Prell
UC San Diego
SLAC, Mail Stop 17,
P.O. Box 4349
Stanford CA 94309
USA
prell@slac.stanford.edu

Helen Quinn
SLAC, Mail Stop 81
Stanford CA 94028
USA
quinn@slac.stanford.edu

Jeffrey Richman
UC Santa Barbara
Dept of Physics
Santa Barbara CA 93106
USA
richman@charm.physics.ucsb.edu

Chris Roat
Stanford University
796 Escondido Road
Rains Apt 16K
Stanford CA 94305
USA
croat@stanford.edu

Enrico Robutti
University & INFN
Genova Universita' degli Studi di Genova
Dipartimento di Fisica
via Dodecaneso 33
I-16146 Genova
ITALY
robutti@ge.infn.it

Ira Rothstein
Carnegie Mellon University
Physics Dept
Pittsburgh PA 15213
USA
ira@cmuhep2.phys.cmu.edu

James Russ
Carnegie Mellon University
Physics Dept
Pittsburgh PA 15213
USA
russ@cmphys.phys.cmu.edu

Anders Ryd
California Institute of Technology
M/C 356-48
Pasadena CA 91125
USA
ryd@hep.caltech.edu

Alex Samuel
California Institute of Technology
M/C 356-48
Pasadena CA 91125
USA
samuel@hep.caltech.edu

Klaus R. Schubert
Physics Department
Techn. Univ. Dresden
Inst. fur Kern-u. Teilchenphysik
Zellescher Weg 19
D-01062 Dresden
GERMANY
schubert@physik.tu-dresden.de

Ullrich Schwanke
UC San Diego-Physics Department
9500 Gilman Drive
La Jolla CA 92093
USA
schwanke@ucsd.edu

Anna Shapiro
California Institute of Technology
Caltech M/C 256-48
Pasadena CA 91125
USA
anna@hep.caltech.edu

Vivek Sharma
UC San Diego-Physics Department
9500 Gilman Drive
La Jolla CA 92093
USA
vsharma@ucsd.edu

Vasilii Shelkov
Lawrence Berkeley National Lab
1 Cyclotron Road, Mail Stop 50A-2160
Berkeley CA 94720
USA
vshelk@yahoo.com

Ian Shipsey
Purdue University
1396 Physics Building
West Lafayette IN 47907-1396
USA
shipsey@physics.purdue.edu

Kevin Stenson
Vanderbilt University
VU Station B 351807,
6301 Stevenson Center
Nashville TN 37235
USA
stenson@fnal.gov

Sheldon Stone
Syracuse University
Physics Dept.
201 Physics Bldg.
Syracuse NY 13244-1130
USA
stone@phy.syr.edu

Ulrich Straumann
University of Zurich
Physik-Institut
Winterthurerstr. 190
CH-8057 Zurich
SWITZERLAND
strauman@physik.unizh.ch

Denis Suprun
University of Chicago
1649 E. 50th St., #18 C
Chicago IL 60615
USA
d-suprun@uchicago.edu

Shiro Suzuki
Yokkaichi University
1200 Kayo-cho
Yokkaichi 512-8512
JAPAN
shiro@yokkaichi-u.ac.jp

Jun-ichi Tanaka
University of Tokyo
Dept. of Phys., Fac. of Sci.
7-3-1 Hongo, Bunkyo-ku
Tokyo 113-0033
JAPAN
jtanaka@hep.phys.s.u-tokyo.ac.jp

Jian-Chun Wang
Department of Physics
Syracuse University
Syracuse NY 13244
USA
jwang@phy.syr.edu

Matthew Weaver
SLAC, Mail Stop 59
2575 Sand Hill Rd.
Menlo Park CA 94025
USA
weaver@hep.caltech.edu

Alan Weinstein
California Institute of Technology
Caltech M/C 256-48
Pasadena CA 91125
USA
ajw@caltech.edu

Mark Wise
California Institute of Technology
M/C 452-48
Pasadena CA 91125
USA
wise@theory.caltech.edu

Douglas Wright
LLNL
L-50 P.O. Box 808
Livermore CA 94551
USA
doug.wright@llnl.gov

Frank Wuerthwein
MIT 24-514,
Cambridge MA 02139
USA
fkw@mit.edu

Songhoon Yang
California Institute of Technology
M/C 356-48
Pasadena CA 91125
USA
songhoon@slac.stanford.edu

Author Index